FOR ALL PRACTICAL PURPOSES

PROJECT DIRECTOR

Solomon Garfunkel, *Consortium for Mathematics and Its Applications*

CONTRIBUTING AUTHORS

PART I   MANAGEMENT SCIENCE
Joseph Malkevitch, *York College, CUNY*
Rochelle Meyer, *Nassau Community College*
Walter Meyer, *Adelphi University*

PART II   STATISTICS: THE SCIENCE OF DATA
David S. Moore, *Purdue University*

PART III   CODING INFORMATION
Joseph Gallian, *University of Minnesota–Duluth*

PART IV   SOCIAL CHOICE AND DECISION MAKING
Steven J. Brams, *New York University*
Bruce P. Conrad, *Temple University*
Alan D. Taylor, *Union College*

PART V   ON SIZE AND SHAPE
Paul J. Campbell, *Beloit College*

FOR ALL PRACTICAL PURPOSES

INTRODUCTION TO CONTEMPORARY MATHEMATICS *fourth edition*

W. H. FREEMAN AND COMPANY
NEW YORK

Acquisitions Editor: Holly Hodder

Development Editor: Randi Rossignol

Project Editor: Mary Louise Byrd

Text Designer: Circa 86, Inc.

Cover Designer: Circa 86, Inc.

Photo Editor: Travis Amos

Photo Researchers: Larry Marcus

Mary Teresa Giancoli

Cover/Part Opener Illustrations: Amy Wasserman

Illustration Coordinator: Bill Page

Illustration: Academy Artworks, Inc.

Production Coordinator: Paul W. Rohloff

Composition: Progressive Information Technologies

Manufacturing: RR Donnelley & Sons Company

Library of Congress Cataloging-in-Publication Data

For all practical purposes: introduction to contemporary mathematics
/ by COMAP — 4th ed.
p.   cm.
Includes index.
ISBN 0-7167-2841-9
1. Mathematics.   I. Consortium for Mathematics and Its Applications (U.S.)
QA7.F68 1996
510 — dc20   96-4941
                                                                CIP

Printed in the United States of America

Second printing 1997

CONTENTS

PART II

STATISTICS: THE SCIENCE OF DATA

CHAPTER 5   PRODUCING DATA   179

CHAPTER 6   DESCRIBING DATA   217

CHAPTER 7   PROBABILITY: THE MATHEMATICS OF CHANCE   **267**

CHAPTER 8   STATISTICAL INTERFERENCE   **309**

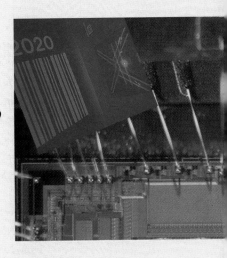

PART

V
ON SIZE AND SHAPE

PREFACE

A PRACTICAL APPROACH

It is difficult to believe that this is the fourth edition of *For All Practical Purposes*. *For All Practical Purposes* still seems like a new idea—an upstart textbook trying to change what we teach and how we teach it. But *FAPP* is an established textbook and is the leading mathematics for liberal arts text, having been used at over 500 colleges.

The Consortium for Mathematics and Its Applications continues to see its role as that of bringing a wide range of new contemporary applications of mathematics into the classroom. A new edition of *FAPP* is an opportunity to demonstrate to students and faculty alike the dynamic and ever-changing face of our subject and the new problems we can model and solve.

A new edition is also our opportunity to evolve with students' and teachers' needs. To that end, our mission for this revision was to make *FAPP* a textbook that is easier to use—both for instructors and students. Each sentence and paragraph has been scrutinized for clarity and overall consistency of style. Each new term and concept is carefully defined the first time it appears in the text. New format features, such as additional subheads, help readers follow the progress of the material. We have made these and other pedagogic improvements without sacrificing the integrity of the *FAPP* philosophy. A basic statement of this philosophy appeared in the preface to the first edition:

> Every mathematician at some time has been called upon to answer the innocent question, "Just what is mathematics used for?" With understandable frequency, usually at social gatherings, the question is raised in similar ways: "What do mathematicians do, practice, or believe in?" At a time when success in our society depends heavily on satisfying the need for developing quantitative skills and reasoning ability, the mystique surrounding mathematics persists. *For All Practical Purposes: Introduction to Contemporary Mathematics* is our response to these questions and our attempt to fill this need.

For All Practical Purposes represents our effort to bring the excitement of contemporary mathematical thinking to the nonspecialist, as well as help him or her develop the capacity to engage in logical thinking and to read critically the technical information with which our contemporary society abounds. For the study of mathematics we attempt to implement Thomas Jefferson's notion of an "enlightened citizenry," in which individuals having acquired a broad knowledge of topics exercise sound judgment in making personal and political decisions. Environmental and economic issues dominate modern life, and behind these issues are complex matters of science, technology, and mathematics that call for an awareness of fundamental principles.

To achieve these goals, *For All Practical Purposes* stresses the connections between contemporary mathematics and modern society. Since the technological explosion that followed World War II, a cluster of mathematical sciences has emerged. This encompasses statistics, computer science, operations research, and decision science. At the same time, the more traditional areas of mathematics have enjoyed spectacular growth both in theoretical power and in their applications. In science and industry, mathematical models are the tools par excellence for solving complex problems. In this book our goal is to convey the power of mathematics, as illustrated by the great variety of problems that can be modeled and solved by analytic and quantitative means.

NEW CONTENT FEATURES OF THE FOURTH EDITION

Every chapter in this new edition has been carefully revised, with attention to greater clarity of exposition and to contemporary issues and examples. Major changes and additions to this edition have come in Part IV, Social Choice and Decision Making. We gratefully acknowledge the efforts of Steven Brams, Bruce Conrad, and Alan Taylor for the inclusion of exciting new material that has never appeared before in any undergraduate text. For example, this section offers a new procedure for divorce settlements called the adjusted winner procedure, a whole chapter on the theory of moves in game theory, and the first publication of the Banzhaf power index for the electoral college for the 1990s.

Content changes make a point. Mathematics is dynamic; new mathematics is being invented and applied in new ways everyday. Our goal and our text must be flexible enough to vividly demonstrate the contemporary nature of our subject and its applications to our daily lives. Some of the more important content changes are as follows:

▶ **Part I, Management Science,** features a greatly simplified discussion of linear programming (Chapter 4). We responded to the advice of students and instructors and now begin the chapter with a manufacturing problem involving only one resource. Minimum quantities are introduced for the first time in this edition, making the problems more realistic while retaining algebraic simplicity. All the algebraic steps are now shown and described. After the student has confidence in solving one-resource problems, two-resource problems are introduced.

▶ **Part II, Statistics: The Science of Data,** is better integrated. In Chapter 5, Producing Data, a new section on statistical estimation carefully introduces the concepts of sampling variability and margin of error, which are fortified with examples of Gallup poll surveys on crime. This section prepares the reader for the more formal treatment of probability in Chapter 7, Probability: The Mathematics of Chance, and inference in Chapter 8. Chapter 6, Describing Data, more fully outlines the strategy of data analysis. The discussion of variance and standard deviation is clearer and supported by examples comparing metabolic rates for men and women.

▶ **Part III, Coding Information,** is also more focused and accessible. For instance, we now discuss only the important bar codes: ZIP codes and UPC codes.

▶ **Part IV, Social Choice and Decision Making,** has evolved considerably to reflect changes in contemporary society *and* to make the material more accessible to students. Chapter 11, Social Choice: The Impossible Dream, is reorganized so that the chapter begins with elections involving only two alternatives and then moves to elections involving three or more alternatives. This latter treatment is more focused than in previous editions, with specific voting systems, such as the Hare system currently used in Ireland and Australia, tied to specific defects, such as a failure of the Condorcet winner criterion. The chapter's concluding section, Insurmountable Difficulties: From Paradox to Impossibility, now includes a comprehensive discussion of Arrow's impossibility theorem.

Chapter 12, Weighted Voting Systems, breaks ground with the inclusion of the Banzhaf power index for the electoral college.

Chapter 13, Fair Division, opens with a new section on a recently developed scheme for handling divorces (called the adjusted winner procedure), which uses the Donald and Ivana Trump divorce to illustrate the procedure. This is followed by a discussion of the Knaster inheritance procedure for dividing an inheritance if there are more than two heirs. An engaging new section, Divide-and-Choose, sets the stage for the remainder of the chapter, which is devoted to cake cutting.

A refocused Chapter 14, Apportionment, more clearly compares and contrasts the U.S. House of Representatives apportionment methods proposed by Hamilton, Jefferson, and Webster by showing specifically the apportionment that each method gives and the paradoxes they create. The chapter ends with a new section on the method currently used to apportion the House (called the Hill–Huntington method) and a summary section discussing which apportionment method is best. In response to questions about the relevance of the Alabama paradox, which occurs only if the House size is allowed to change, a discussion of the population paradox has been included.

The ideas presented in Chapter 15, Game Theory: The Mathematics of Competition, are brought up to date in a new Chapter 16, called Theory of Moves. Theory of moves shows how competitors from Samson and Delilah to Leno and Letterman can resolve conflicts using a farsighted strategy and, perhaps, deception.

▶ **For Part V, On Size and Shape,** we have honed our coverage to include only those topics that are most popular and appropriate to the book. For instance, to illustrate hyperbolic geometry (Chapter 19), Beltrami's pseudosphere has been replaced with an example on the geometry of political alliances. Chapter 18, Geometric Growth, features many new financial growth models, including certificates of deposit, money market accounts, savings plans, sinking funds, inflation, and depreciation. Part V concludes with an invitation to readers to attempt their own Escher-like tilings (Chapter 21).

IMPROVED PEDAGOGY

All chapters have been carefully reviewed and revised to make learning the material easier for students. Some new pedagogic features are as follows:

▶ Key concepts and algorithms are now highlighted in color to reinforce their importance. This new feature will reassure students about the chapters' "take home messages."

▶ Chapter-ending exercises are now categorized by subject matter, which reflects the subject-matter divisions in the chapter. This new feature will help instructors with their assignments and help students succeed at solving problems.

▶ More than 150 new exercises and writing projects have been included in this edition. For instance, new statistics exercises present attractive data on such issues as whether drinking wine helps prevent heart attacks (6-48) and whether recent years have improved the plight of the endangered Florida manatee (6-21 and 6-25). New Exercise 14-3 highlights one of the weirdest paradoxes of apportionment: the population paradox. Also particularly interesting are those exercises that ask students to use free software programs to create models for hyperbolic geometry (see Chapter 19).

SUPPLEMENTS

The supplements package for *FAPP* has been extensively revamped for the fourth edition. Existing supplements have been updated and improved and new supplements have been added for both instructors and students. The components of this package are described below. For more information about any of the supplements, please contact your local W. H. Freeman sales representative.

▶ An exciting new CD-ROM/study guide learning package, called *FAPP Interactive,* for students combines printed and electronic materials that provide innovative approaches to the topics treated in the textbook. Each chapter of *FAPP Interactive* is divided into three parts. One part focuses on summaries of the topics within each chapter. These summaries are paired with questions and answers designed to determine whether or not the student sufficiently understands the material being presented. The second part uses an impressive array of tools—easy-to-use software, video clips, animations, and original examples—that are applied to more questions. Exploring these tools and questions will give students a deeper understanding of the topics in the textbook. Each chapter also contains an electronic multiple-choice quiz that is designed to help students learn the material and at the same time become more comfortable taking multiple-choice tests. Each option of every question provides instant feedback for the student: choosing an incorrect option reveals a hint; choosing a correct option reveals positive reinforcement. Students at all levels will benefit a great deal from this exciting and functional student supplement.

▶ The *Instructor's Guide* has long been a useful tool for instructors switching from a more traditional teaching approach to the *FAPP* approach. It has been thoroughly revised to reflect the changes in the new edition of the text. Popular features retained from earlier editions include extensive classroom-tested teaching hints, solutions to even-numbered text exercises, detailed summaries, and suggested skill objectives. New to this edition will be strategies for using the collaborative learning approach in teaching most chapters.

▶ For the first time, a full *Test Bank* has been developed to accompany *FAPP.* Each chapter will contain approximately 50 multiple-choice and 25 short-answer questions, for a total of approximately 1500 questions. The *Test Bank* will be available in printed, Macintosh, and IBM versions.

▶ A set of *Transparency Masters* that contain approximately 110 key figures from the textbook is also being offered.

There is also an Annenberg video series of 26 half-hour programs called *For All Practical Purposes Telecourse.* For more information on telecourse preview, purchase, or rental, please call 1-800-LEARNER, or write The Annenberg CPB Project, P.O. Box 2345, South Burlington, VT 05407-2345.

We thank the individuals who have contributed to the supplements package, including the authors of supplements to previous editions. Following is a list of all the instructors who have been involved, some of whom have worked on more than one title in this package:

John Emert, Ball State University
Chris Leary, St. Bonaventure College
Eli Passow, Temple University
Dan Reich, Temple University
Kay Meeks Roebuck, Ball State University
Ken Ross, University of Oregon
Sandra H. Savage, Orange Coast College
Jerad Zimmerman, Tacoma Community College

The staff of COMAP and Lew Hollerbach of Infon, Inc., have also contributed a great deal to the supplements package.

ACKNOWLEDGMENTS

Since the inception of *For All Practical Purposes,* we have benefited from the interest and contributions of many people, and this edition was no exception. We wish to thank our friends and colleagues who offered suggestions, comments, and corrections:

Kathy Bavelas, Manchester Community College, Connecticut
J. Patrick Brewer, University of Oregon
John Bruder, University of Alaska—Bristol Bay
Michelle Clement, Louisiana State University
John Emert, Ball State University, Indiana
Rich France, Millersville University, Pennsylvania
Ira Gessel, Brandeis University, Massachusetts
Henry Gore, Morehouse College, Georgia
Edwin Herman, University of Oregon
Frederick Hoffman, Florida Atlantic University
Alec Ingraham, New Hampshire College
Karla Karstens, University of Vermont
Darrell Kent, Washington State University
Antonio Lopez, Loyola University, Louisiana
John Montgomery, University of Rhode Island
John L. Orr, University of Nebraska—Lincoln
Diane Radin, University of Texas—Austin
Edward Thome, Murray State University, Kentucky
Cynthia Wyels, Weber State University, Utah

We are also grateful to the following people, who evaluated previous editions of *For All Practical Purposes:*

Mark S. Anderson, Rollins College
Jerry W. Bradford, Wright State University
Helen Burrier, Kirkwood Community College
Lothar A. Dohse, University of North Carolina—Asheville
Sandra Fillebrown, Saint Joseph's University
William Gratzer, Iona College
Rodger Hammons, Morehead State University
Sherman Hunt, Community College of Finger Lakes
Phillip E. Johnson, University of North Carolina—Charlotte
Carmelita R. Keyes, Broome Community College
Christopher McCord, University of Cincinnati
Bennett Manvel, Colorado State University
John G. Michaels, SUNY—Brockport
John Oprea, Cleveland State University
James Osterburg, University of Cincinnati
Margaret A. Owens, California State University, Chico
Sandra H. Savage, Orange Coast College
Richard Schwartz, College of Staten Island
Joanne R. Snow, Saint Mary's College

We owe our appreciation to the people at W. H. Freeman and Company who participated in the preparation of this book. We wish especially to thank the editorial staff for their tireless efforts and support. Among them are Holly Hodder, Acquisitions Editor; Randi Rossignol, Development Editor; Mary Louise Byrd, Project Editor; Larry Marcus, Photo Researcher; Patrick Shriner, Supplements Editor; Trumbull Rogers, Copy Editor; Walter Hadler, Proofreader; Melina LaMorticella, Senior Editorial Assistant; and Diana Gill, Editorial Assistant.

We also thank the production staff at Freeman: Paul Rohloff, Production Coordinator; Maria Epes, Art Director; Bill Page, Illustration Coordinator; and Carmen DiBartolomeo at Circa 86, Inc., Designer.

The efforts of the COMAP staff must be recognized. To the production and administrative staff—Roger Slade and Roland Cheyney—go all our thanks. And finally, we recognize the contribution of Laurie Aragon, the manager, who kept this project, as she does all of COMAP, running smoothly and efficiently. To everyone who helped make our purposes practical, we offer our appreciation for an exciting, exhausting, and exhilarating time.

Solomon Garfunkel, COMAP
September 1996

TO THE STUDENT

Have fun. Enjoy. That is what this book is all about. You may at one time or another have wondered why people go into mathematics. Do they have different genes, that they like this stuff? This text is one attempt to show some of the payoffs for working with mathematics. The areas we discuss and the problems we tackle are real. Most of the people you will meet in these pages are alive and much of their work has been done in the last twenty or thirty years. We have tried to show you mathematics as we see it—solving problems we care about and have fun doing.

We don't expect you to become a mathematician. We understand that you won't remember much about the techniques presented here in a year or two. That's not the point. What we hope is that you will do some mathematics, solve some problems, and gain an appreciation for what we do and why. To a large extent mathematics is not about formulas and equations; it is about CDs and CAT scans, and parking meters, and presidential polls, and computer graphics. Mathematics is about looking at our world and creating representations we can work with to solve problems that matter—and we hope it is in this spirit that you use this book. Welcome to the world of contemporary mathematics.

FOR ALL
PRACTICAL
PURPOSES

Neil Armstrong's first step onto the moon's surface in July 1969 was a triumph for American science and technology and the culmination of a national quest that had begun in the office of President John F. Kennedy. It was Kennedy's goal to put a man on the moon before the decade was out, a goal realized in the Nixon administration.

Before the 1960s, no one really knew if rockets would ever be able to carry humans into space. When NASA commissioned the Apollo module, it was asking several hundred companies to design, build, test, and deliver components and systems that had never been built before. But alongside the physical science challenges of building all these components, there were management science challenges: how to make it all happen economically and on schedule.

MANAGEMENT SCIENCE

The *Apollo 11* launch is certainly not the only project in which efficiency and timeliness are important. In a variety of modern projects, ranging from the making of a Hollywood movie to the building of the "Chunnel" (a tunnel under the English Channel), the difference between success and failure depends on whether the project is on time and on budget. Chapters 2 and 3 present some project planning and scheduling techniques that were pioneered in the Apollo program.

But projects are not the only area where organization and efficiency are valuable. Day-to-day operations, such as the delivery of city services discussed in Chapter 1, and the production problems faced by manufacturing firms discussed in Chapter 4, offer big opportunities for cost savings using management science.

Spotlight

Apollo 11 *Launch Owes Success to Management Science*

Captain Robert F. Freitag, who later became director of Policy and Plans for NASA's space station, in 1969 headed the team responsible for landing the *Apollo 11* safely on the moon. The success of the lunar mission can be traced to management science techniques that ensured that thousands of small tasks would come together to meet a single giant objective. Freitag shares his observations about that historic event:

I think the feeling most of us in NASA shared was, "My gosh, now we really have to do it." When you think that the enterprise we were about to undertake was ten times larger than any that had ever been undertaken, including the Manhattan Project, it was a pretty awesome event. But we knew it was the kind of thing that could be broken down into manageable pieces and that if we could get the right people and the right arrangement of these people, it would be possible.

To begin with, there are five major pieces: the launch site, the launch vehicles, the spacecraft, the lunar module, and worldwide tracking networks. Then, once these pieces are broken down, you assign them to one organization or another. They, in turn, take those small pieces, like the rocket, and break it down into engines or structures or guidance equipment. And this breakdown, or "tree," is the really tough part about managing.

Robert F. Freitag

You need to be sure that the pieces come together at the right time, and that they work when put together. Management science helps with that. The total number of people who worked on the *Apollo* was about 400,000 to 500,000, all working toward a single objective. But that objective was clear when President Kennedy said, "I want to land a man on the moon and have him safely returned to the earth, and to do so within the decade." Of course, Congress set aside $20 billion. So you had cost, performance, and schedule, and you knew what the job was in one simple sentence. It took a lot of effort to make that happen.

Naturally, efficiency is not a modern invention. The Romans were masters of it (for their time), and it is doubtful that their empire would have been as successful as it was without this remarkable flair for organization. Until management science was founded in the modern era, however, efficiency was pursued by trial-and-error, "seat-of-the-pants," approaches. The distinctive contribution of management science is to add mathematical analysis to the tools available for organizing and decision making.

The value of having scientific principles for operations management was first perceived in World War II. The founders of management science were mathematicians and industrial technicians associated with the armed services who worked together to improve military operations. In applying quantitative techniques to project planning, these pioneers founded a new science whose impact reaches beyond the military to many corners of our lives. We will explore some of this new science in the chapters ahead.

STREET NETWORKS

T he underlying theme of **management science,** also called **operations research,** is finding the best method for solving some problem — what mathematicians call the **optimal solution.** In some cases, it may be to finish a job as quickly as possible. In other situations, the goal might be to maximize profit or minimize cost. In this chapter our goal is to save time in traversing a street network while checking parking meters, delivering mail, or carrying out some similar task.

Let's begin by concentrating on the parking department of a city government. Most cities and many small towns have parking meters that must be regularly checked for parking violations or emptied of coins. We will use an imaginary town to show how management science techniques can help to make parking control more efficient.

EULER CIRCUITS

The street map in Figure 1.1 is typical of many towns across the United States, with streets, residential blocks, and a village green. Our job, or that of the commissioner of parking, is to find the most efficient route for the parking-control officer, who travels on foot, to check the meters in an area. Our map shows only a small area, allowing us to start with an easy problem. But the problem occurs on a larger scale in all cities and towns and, for larger areas, there are almost unlimited possibilities for parking-control routes.

FIGURE 1.1 A street map for part of a town.

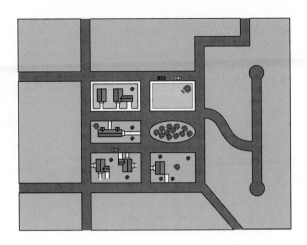

The commissioner has two goals in mind: (1) the parking-control officer must cover all the sidewalks that have parking meters without retracing any more steps than are necessary; and (2) the route should end at the same point from which it began, perhaps where the officer's patrol car is parked. To be specific, suppose there are only two blocks that have parking meters, the two blue-shaded blocks that are side by side toward the top of Figure 1.1. Suppose further that the parking-control officer must start and end at the upper left corner of the left-hand block. You might enjoy working out some routes by trial and error and evaluating their good and bad points. We are going to leave this problem for the moment and establish some concepts that will give us a better method to deal with this problem than trial and error.

A **graph** is a finite set of dots and connecting links. The dots are called **vertices** (a single dot is called a *vertex*), and the links are called **edges.** Each edge must connect two different vertices. A **path** is a connected sequence of edges showing a route on the graph that starts at a vertex and ends at a vertex; a path is usually described by naming in turn the vertices visited in traversing it. A path that starts and ends at the same vertex is called a **circuit.** A graph can represent our city map, a communications network, or even a system of air routes.

EXAMPLE ▶ *Parts of a Graph*

We can see examples of these technical terms in Figure 1.2. The vertices represent cities, and the edges represent nonstop airline routes between them. We see that there is a nonstop flight between New Orleans and Phoenix, but no such flight between Seattle and New York. There are several paths that describe

FIGURE 1.2 The edges of this graph show nonstop routes that an airline might offer.

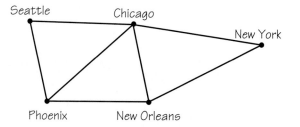

how a person might travel with this airline from Seattle to New York. The path that seems most direct is Seattle, Chicago, New York, but Seattle, Phoenix, New Orleans, New York is also such a path. An example of a circuit is Phoenix, New Orleans, Chicago, Phoenix. It is a circuit because the path starts and ends at the same vertex. In this chapter we are especially interested in circuits, just as we are in real life; most of us end our day in the same location where we start it—at home! ◆

Returning to the case of parking control in Figure 1.1, we can use a graph to represent the whole territory to be patrolled: think of each street intersection as a vertex and each sidewalk that contains a meter as an edge, as in Figure 1.3. Notice in Figure 1.3b that the street separating the blocks is not explicitly represented; it has been shrunk to nothing. In effect, we are simplifying our problem by ignoring any distance traveled in crossing streets.

The numbered sequence of edges in Figure 1.4a shows one circuit that covers all the meters (note that it is a circuit because its path returns to its starting point). But Figure 1.4b shows another solution that is better because its circuit covers every edge (sidewalk) exactly once. In Figure 1.4b no edge is

FIGURE 1.3 (a) A graph superimposed upon a street map. The edges show which sidewalks have parking meters. (b) The same graph enlarged.

FIGURE 1.4 (a) A circuit and (b) an Euler circuit.

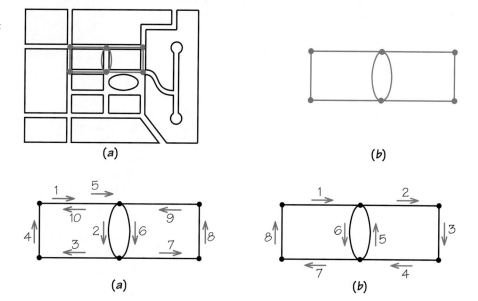

covered more than once, or *deadheaded* (a term borrowed from shipping, which means making a return trip without a load).

Circuits that cover every edge only once are called **Euler circuits.**

Figure 1.4b shows an Euler circuit. These circuits get their name from the great eighteenth-century mathematician Leonhard Euler (pronounced oy′ lur), who first studied them (see Spotlight 1.1). Euler was the founder of the theory of graphs. One of his first discoveries was that some graphs have no Euler circuits at all. For example, in the graph in Figure 1.5b, it would be impossible to start at one point and cover all the edges without retracing some steps: if we try to start a circuit at the leftmost vertex, we discover that once we have left the vertex, we have "used up" the only edge meeting it. We have no way to return to our starting point except to reuse that edge. But this is not allowed in an Euler circuit. If we try to start a circuit at one of the other two vertices, we

Spotlight   *Leonhard Euler*

Leonhard Euler

Leonhard   Euler (1707–1783)   was one of those rare individuals who was remarkable in many ways. He was extremely prolific, publishing over 500 works in his lifetime. But he wasn't devoted just to mathematics; he was a people person too. He was extremely fond of children and had thirteen of his own, of whom only five survived childhood. It is said that he often wrote difficult mathematical works with a child or two in his lap.

Human interest stories about Euler have been handed down through two centuries. He was a prodigy at doing complex mathematical calculations under less than ideal conditions, and continued to do them even after he became totally blind later in life. His blindness diminished neither the quantity nor the quality of his output. Throughout his life, he was able to mentally calculate in a short time what would have taken ordinary mathematicians hours of pencil-and-paper work. A contemporary claimed that Euler could calculate effortlessly, "just as men breathe, as eagles sustain themselves in the air."

Euler's mathematical mind found new mathematics in everyday life. In the old German town of Königsberg, people frequently tried to take a Sunday stroll whose route crossed each of the seven bridges in the town exactly once. Euler analyzed this local pastime using what are now known as Euler circuits.

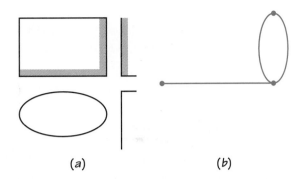

FIGURE 1.5 (a) The three shaded sidewalks cannot be covered by an Euler circuit. (b) The graph of the shaded sidewalks in part (a).

(a)   (b)

likewise can't complete it to form an Euler circuit.

Since we are interested in finding circuits, and Euler circuits are the most efficient ones, we will want to know how to find them. If a graph has no Euler circuit, we will want to develop the next best circuits, those having minimum deadheading. These topics make up the rest of this chapter.

FINDING EULER CIRCUITS

Now that we know what an Euler circuit is, we are faced with two obvious questions:

1.   Is there a way to tell by calculation, not by trial and error, if a graph has an Euler circuit?
2.   Is there a method, other than trial and error, for finding an Euler circuit when one exists?

Euler answered these questions in 1735 by using the concepts of valence and connectedness.

> The **valence** of a vertex in a graph is the number of edges meeting at the vertex.

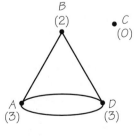

FIGURE 1.6 Valences of vertices.

Figure 1.6 illustrates the concept of valence, with vertices A and D having valence 3, vertex B having valence 2, and vertex C having valence 0. (Isolated vertices such as vertex C are an annoyance in Euler circuit theory. Because they don't occur in typical applications, we henceforth assume that our graphs have no vertices of valence 0.)

Figure 1.3b has four vertices of valence 2, namely, the outer corners of the graph. This graph also has two vertices of valence 4. Notice that each vertex has a valence that is an even number. We'll soon see that this is very significant.

A graph is said to be **connected** if for every pair of its vertices there is at least one path of one or more edges connecting the two vertices.

Given a graph, if we can find even one pair of vertices not connected by a path, then we say that the graph is not connected. For example, the graph in Figure 1.7 is not connected because we are unable to join A to D with a path of edges. However, the graph does consist of two "pieces" or connected components, one containing the vertices A, B, F, and G, the other containing C, D, and E. A connected graph will contain a single connected component. Notice that the parking-control graph of Figure 1.3b is connected.

We can now state Euler's theorem, his simple answer to the problem of detecting when a graph G has an Euler circuit:

$Spotlight$ **1.2**

The Human Aspect of Problem Solving

Thomas Magnanti, professor of operations research and management, heads the Department of Management Science at MIT's Sloan School of Management. Here are some of his observations:

Typically, a management science approach has several different ingredients. One is just structuring the problem—understanding that the problem is an Euler circuit problem or a related management science problem. After that, one has to develop the solution methods.

But one should also recognize that you don't just push a button and get the answer. In using these underlying mathematical tools, we never want to lose sight of our common sense, of understanding, intuition, and judgment. The computer provides certain kinds of insights. It deals with some of the combinatorial complexities of these problems very nicely. But a model such as an Euler circuit can never capture the full essence of a decision-making problem.

Thomas Magnanti

Typically, when we solve the mathematical problem, we see that it doesn't quite correspond to the real problem we want to solve. So we make modifications in the underlying model. It is an interactive approach, using the best of what computers and mathematics have to offer and the best of what we, as human beings, with our own decision-making capabilities, have to offer.

Figure 1.7 A nonconnected graph.

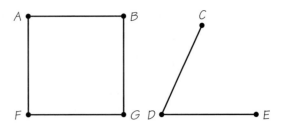

1. If G is connected and has all valences even, then G has an Euler circuit.
2. Conversely, if G has an Euler circuit, then G must be connected and all its valences must be even numbers.

In the optional section "Proving Euler's Theorem," you will find an outline of a proof of this theorem.

Since the parking-control graph of Figure 1.3b conforms to the connectedness and even-valence conditions, Euler's theorem tells us that it has an Euler circuit. We already have found an Euler circuit for Figure 1.4b by trial and error. For a very large graph, however, trial and error may take a long time. It is usually quicker to check whether the graph is connected and even-valent than to find out if it has an Euler circuit.

Once we know there is an Euler circuit in a certain graph, how do we find it? Many people find that, after a little practice, they can find Euler circuits by trial and error, and they don't need detailed instructions on how to proceed. At this point you should see if you can develop this skill by trying to find Euler circuits in Figure 1.8a, Figure 1.9a, and Figure 1.10. In doing your experiments, draw your graph in ink and the circuit in pencil so you can erase. Make your graph big and clear so you won't get confused.

If you would like more guidance on how to find an Euler circuit without trial and error, here is a method that works: never use an edge that is the only

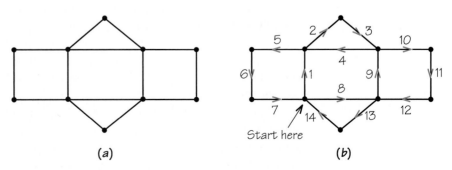

Figure 1.8 (a) A graph having (b) an Euler circuit.

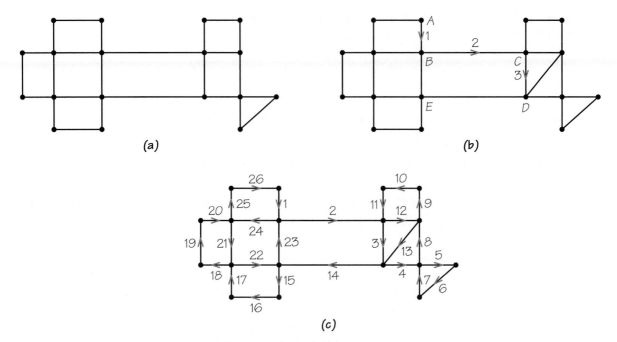

FIGURE 1.9 A crucial decision point in finding an Euler circuit.

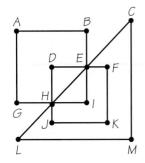

FIGURE 1.10 A graph with an Euler circuit.

link between two parts of the graph that still need to be covered. Figure 1.9b illustrates this. Here we have started the circuit at *A* and gotten to *D* via *B* and *C*, and we want to know what to do next. Going to *E* would be a bad idea because the uncovered part of the graph would then be disconnected into left and right portions. You will never be able to get from the left part back to the right part because you have just used the last remaining link between these parts. Therefore you should stay on the right side and finish that before using the edge from *D* to *E*. This kind of thinking needs to be applied every time you need to choose a new edge.

Let's see how this works, starting at the beginning at *A*. From vertex *A* there are two possible edges, and neither of them disconnects the unused portion of the graph. Thus, we could have gone either to the left or down. Having gone down to *B*, we now have three choices, none of which disconnects the unused part of the graph. After choosing to go from *B* to *C*, we find that any of the three choices at *C* is acceptable. Can you complete the Euler circuit? Figure 1.9c shows one of many ways to do this.

The method just described leaves many edge choices up to you. When there are many acceptable edges for your next step, you can pick one at random. You might even flip a coin. When computers carry out algorithms of this sort, they use random number generators, which mimic the flipping of a coin.

EXAMPLE ▶ *Finding an Euler Circuit*

Check the valences of the vertices and the connectivity of the graph in Figure 1.8a to verify that the graph does have an Euler circuit. Now try to find an Euler circuit for that graph. You can start at any vertex. When you are done, compare your solution with the Euler circuit given in Figure 1.8b. If your path covers each edge exactly once and returns to its original vertex (is a circuit), then it is an Euler circuit, even if it is not the same as the one we give. ◆

Optional ▶ **Proving Euler's Theorem**

We'll start by proving that if a graph has an Euler circuit, then it must have only even valences and it must be connected. Let X be any vertex of the graph. We will show that the edges at X can be paired up, and this will prove that the valence is even. Every edge at X is used by the Euler circuit as an outgoing edge (leaving from X) or an incoming edge (arriving at X). As we follow the Euler circuit, each time we arrive at X, the next step involves an outgoing edge that we pair up with the previous incoming one. Since all edges are used by the Euler circuit, none more than once, this pairs up the edges. For example, in Figure 1.11 at vertex B we would pair up edges 2 and 3 and edges 9 and 10. At vertex C we would pair up edges 4 and 5 and edges 8 and 9. Can you see how the pairings would work at D? How about vertex A? (In studying this example, you might think it would be simpler to count the edges at a vertex to see that the valence is even. True, but our pairing method works for a graph about which we know nothing except that it has an Euler circuit.)

To see that a graph with an Euler circuit is connected, note that by following the Euler circuit around we can get from one edge to any other edge (it covers them all) using a portion of the Euler circuit. Since every vertex is on an edge (there are no vertices of 0 valence), we can get from any vertex to any other using a portion of the Euler circuit.

FIGURE 1.11 An Euler circuit starting and ending at A.

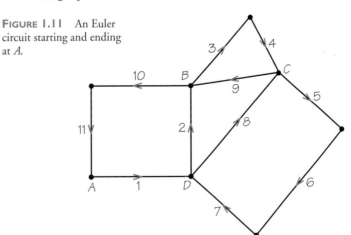

So far, this is not a complete proof of Euler's theorem. It is also necessary to prove that if a graph has all vertices even-valent and is connected, then an Euler circuit can be found for it. The book by Malkevitch and Meyer in the Suggested Readings section contains an elementary proof of this. ◆

CIRCUITS WITH REUSED EDGES

Now let's see what Euler's theorem tells us about the three-block neighborhood with parking meters, represented by dots in Figure 1.12a. Figure 1.12b shows the corresponding graph. (Since we only use edges to represent sidewalks along which the officer must walk, the sidewalk with no meters is not represented by any edge in the graph.) This graph has two odd valences, so Euler's theorem tells us that there is no Euler circuit for this graph.

Since we must reuse some edges in this graph in order to cover all edges in a circuit, for efficiency we need to keep the total length of reused edges to a minimum. This type of problem, in which we want to minimize the length of a circuit by carefully choosing which edges to retrace, is often called the **Chinese postman problem** (like parking-control routes, mail routes need to be efficient). The problem was first studied by the Chinese mathematician Meigu Guan in 1962; hence the name. Although the Euler circuit theory doesn't deal directly with reused edges or edges of different lengths, we can extend the theory to help solve the Chinese postman problem. The remainder of this chapter is dedicated to solving the Chinese postman problem and discussing applications besides parking control.

In a realistic Chinese postman problem, we need to consider the lengths of the sidewalks, streets, or whatever the edges represent, since we want to minimize the total length of the reused edges. However, to simplify things at the start, we can suppose that all edges represent the same length. (This is often called the *simplified* Chinese postman problem.) In this case, we need only count reused edges and need not add up their lengths. To solve the problem, we want to find a circuit that covers each edge and that has the minimal number of reuses of edges already covered.

FIGURE 1.12 (a) A street network and (b) its graphic representation.

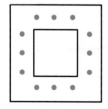 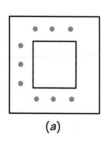 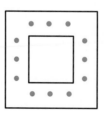

(a)

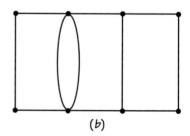

(b)

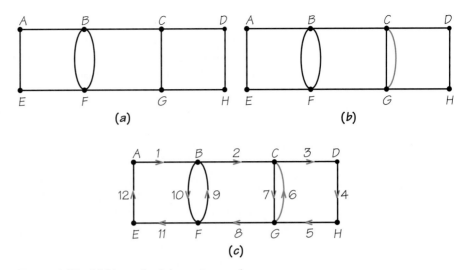

FIGURE 1.13 Making a circuit by reusing an edge.

To follow the procedure we are going to develop, look at the graph in Figure 1.13a, which is the same graph as in Figure 1.12b, but with labeled vertices. This graph has no Euler circuit, but there is a circuit that has only one reuse of an edge (*CG*), namely, *ABCDHGCGFBFEA*. Let's draw this circuit so that when edge *CG* is about to be reused, we install a new, extra, blue edge in the graph for the circuit to use. By duplicating edge *CG*, we can avoid reusing the edge. To duplicate an edge, we must add an edge that joins the two vertices that are already joined by the edge we want to duplicate. (It is not a good idea to join vertices that are not already connected by an edge; see Figure 1.15.) We have now created the graph of Figure 1.13b. In this graph the original circuit can be traced as an Euler circuit, using the new edge when needed. The circuit is shown in Figure 1.13c. Our theory will be based on using this idea in reverse, as follows:

1. Take the given graph and add edges by duplicating existing ones, until you arrive at a graph that is connected and even-valent.

> Adding edges to a graph so as to make all valences even is called **eulerizing** the graph.

We call this process eulerizing a graph, because the graph we produce will have an Euler circuit. (In our graphs, the edges we add are in color, and thus can be distinguished from the original edges, which are black. You may want to create a system to help you remember which edges are original and which are duplicates.)

2. Find an Euler circuit on the eulerized graph.
3. "Squeeze" this Euler circuit from the eulerized graph onto the original graph by reusing an edge of the original graph each time the circuit on the eulerized graph uses an added edge.

EXAMPLE ▶ *Eulerizing a Graph*

Suppose we want to eulerize the graph of Figure 1.14a. When we eulerize a graph, we first locate the vertices with odd valence. The graph in Figure 1.14a has two, *B* and *C*. Next, we add one end of an edge at each such vertex, matching the new edge up with an existing edge in the original graph. Figure 1.14b shows one way to eulerize the graph. Note that *B* and *C* have even valence in the second graph. After eulerization, each vertex has even valence. To see an Euler circuit on the eulerized graph in Figure 1.14c, simply follow the edges in numerical order and in the direction of the arrows, beginning and ending at vertex *A*. The final step, shown in Figure 1.14d, is to "squeeze" our Euler circuit into the original graph. There are two reuses of previously covered edges. Notice that each reuse of an edge corresponds to an added edge. ◆

In the previous example we noticed that we could count how many reuses we needed by counting added edges. This is generally true in this type of problem: *if you add the new edges correctly, the number of reuses of edges equals the number of edges added during eulerization.*

Adding new edges correctly means adding only edges that are duplicates of existing edges. Doing this makes the rule, just stated in italics, always true, and so it is easy to count the needed reuses.

To see why we add only duplicate edges, examine Figure 1.15a. We need to give *X* and *Y* even valences. Adding one long edge from *X* to *Y* (Figure 1.15b) might seem like an attractive idea, but such an edge is not a duplicate of an existing edge—it runs along a series of existing edges. (Remember, to duplicate

FIGURE 1.14 Eulerizing a graph.

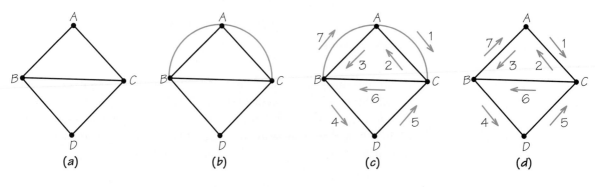

(a)　　　(b)　　　(c)　　　(d)

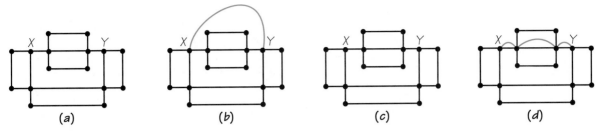

FIGURE 1.15   Eulerizing when the vertices are more than one edge apart.

an edge means to add an edge that joins two vertices that are already joined.) Suppose we added this long edge anyhow and applied the rule in italics. We would conclude that we only needed to reuse one edge. You can see that this is wrong by imagining an Euler circuit that uses this long edge. When you squeeze the Euler circuit back into the original graph, the alternative to using the long edge will be the whole series of three edges stretching from X to Y (shown by the heavy line in Figure 1.15c).

If you don't add the long edge, but instead follow the rule that an added edge must duplicate an existing edge, you'll add the three edges in Figure 1.15d. Counting them will tell you the number of reuses you need (three, in this case).

Now that we have learned to eulerize, the next step is to try to get a best eulerization we can—one with the fewest added edges. It turns out that there are many ways to eulerize a graph. It is even possible that the smallest number of added edges can be achieved with two different eulerizations. This is the reason we use the phrase "a best eulerization" rather than "the best eulerization." Remember, we want a best eulerization because this enables us to find the circuit for the original graph that has the minimum number of reuses of edges.

EXAMPLE ▶ *A Better Eulerization*

In Figure 1.16a, we begin with the same graph as in Figure 1.14, but we eulerize it in a different way—by adding only one edge (see Figure 1.16b). Figure 1.16c shows an Euler circuit on the eulerized graph, and in Figure 1.14d we see how it is squeezed onto the original graph. There is only one reuse of an edge, because we added one edge during eulerization. ◆

The solution in Figure 1.16 is better than the solution in Figure 1.14 because one reuse is better than two. These examples suggest the following addition to our solution procedure: try to find the eulerization with the smallest number of added edges. This extra requirement makes the problem both more interesting and more difficult. For large graphs, a best eulerization may not be obvious. We can try out a few and pick the best among the ones we find, but there may be an even better one that our haphazard search does not turn up.

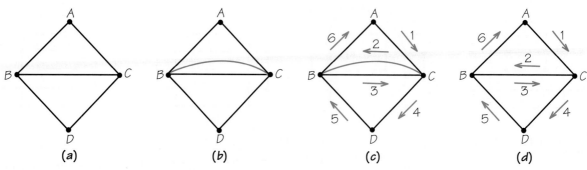

FIGURE 1.16 A better eulerization of Figure 1.14.

A systematic procedure for finding a best eulerization does exist, but the process is complicated. There is an especially easy technique for eulerizing the following special category of networks often found in our neighborhoods.

> If a street network is composed of a series of rectangular blocks that form a large rectangle a certain number of blocks high by a certain number of blocks wide, the network is called *rectangular.*

Examples of rectangular street networks (a 3-by-3, a 3-by-4, and a 4-by-4) are shown in Figure 1.17. The graph on the right in each pair shows a best eulerization for the rectangular street network on the left. There appear to be three different eulerization patterns, depending upon whether the rectangle height and width in the original graph are odd or even numbers. In Figure 1.17a, both lengths are 3, both odd; in Figure 1.17b, one length is odd (3) and one is even (4); in Figure 1.17c, both lengths are 4, an even number.

Although the patterns appear different, one technique can be used to create all of them. This technique can be thought of as involving an "edge walker" who walks around the outer boundary of the large rectangle in some direction, say, clockwise. He starts at any corner, say, the upper left corner. As he goes around, he adds edges by the following rules. When he comes to an odd-valent vertex, he links it to the next vertex with an added edge. This next vertex now becomes either even or odd. If it became even, he skips it and continues around, looking for an odd vertex. If it became odd (this could only happen at a corner of the big rectangles), the edge walker links it to the next vertex and then checks this vertex to see whether it is even or odd. Each of the three parts of Figure 1.17 has been eulerized by this method.

In a street network that is not rectangular, the eulerization process is started by locating all the vertices with odd valence and then pairing these vertices with each other and finding the length of the shortest path between each

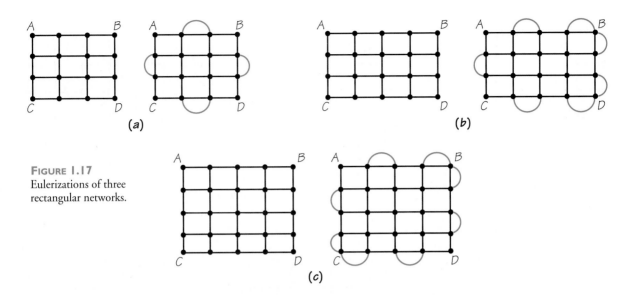

FIGURE 1.17
Eulerizations of three
rectangular networks.

pair. We look for the shortest paths, since each edge on the connecting paths will be duplicated. The idea is to choose the pairings cleverly so that the sum of the lengths of those paths is the smallest it can be. With a little practice, most people can find a best or nearly best eulerization using only this idea, trial and error, and a little ingenuity. Those interested in a further discussion can read the following optional section "Finding Good Eulerizations."

O p t i o n a l ▶ **Finding Good Eulerizations**

Suppose we want a perfect procedure for eulerizing a graph. What theoretical ideas and methods could we use to build such a tool?

One building block we could use is a method for finding the shortest path between two given vertices of a graph. For example, let us focus on vertices X and Y in Figure 1.18a; both have odd valence. We can connect them with a pattern of duplicate edges, as in Figure 1.18b. The cost of this is the length of the path we duplicated from X to Y. A shorter path from X to Y, such as the one shown in Figure 1.18c, would be better. Fortunately, the *shortest-path problem* has been well studied, and we have many good procedures for solving it exactly, even in large, complex graphs. These procedures are discussed in some of the suggested readings given at the end of this chapter, but are beyond the scope of this text.

But there is more to eulerizing the graph in Figure 1.18a than dealing with X and Y. Notice that we have odd valences at Z and W. Should we connect X and Y with a path, and then connect Z and W, as in Figure 1.18d? Or should we connect X to Z and Y to W, as in Figure 1.18e? Another alternative is to use connections X to W and Y to Z, as in Figure 1.18f. It turns out that the

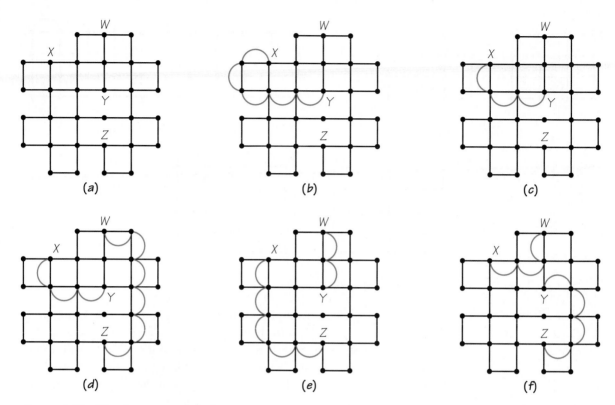

Figure 1.18 Choosing among eulerizations.

alternatives in both Figures 1.18e and 1.18f are preferable to the one in Figure 1.18d, since they involve seven added edges, whereas Figure 1.18d uses nine.

At the start, it is often not clear which alternatives are best. The problem is how to pair up vertices for connection to get a set of paths whose total length is minimal. This problem, called the *matching problem,* has also been studied and solved. As with the shortest-path problem, refer to Suggested Readings for further details. ◆

CIRCUITS WITH MORE COMPLICATIONS

Euler circuits and eulerizing have many more practical applications than just checking parking meters. Almost any time services must be delivered along streets or roads, our theory can make the job more efficient. Examples include collecting garbage, salting icy roads, plowing snow, inspecting railroad tracks, and reading electric meters (see Spotlight 1.3).

Spotlight 1.3

Israel Electric Company Reduces Meter-Reading Task

The Beersheba branch of Israel's major electric company wanted to make the job of meter reading more efficient. When the branch managers decided to minimize the number of people required to read the electric meters in the houses of one particular neighborhood, they set a precedent by applying management science. Formerly, each person's route had been worked out by trial and error and intuition, with no help from mathematics. The whole job required 24 people, each doing a part of the neighborhood in a five-hour shift.

At first, it looks as though one would find a more efficient way of doing the work the same way as in the Chinese postman problem, but there are two important differences. First, the neighborhood was big enough to negate any possibility of having only one route assigned to one person. Instead, it was necessary to find a number of routes that, taken together, covered all the edges (sidewalks). Second, a meter reader who was done with a route was allowed to return home directly. Thus, there was no reason for the individual routes to return to their starting points; therefore, routes could be paths instead of circuits.

The Beersheba researchers found solutions to these problems by modifying the basic ideas we have described in this chapter. They managed to cover the neighborhood with 15 five-hour routes, a 40% reduction of the original 24 five-hour routes. Altogether, these routes involve a total of 4338 minutes of walking time, of which 41 minutes (less than 1%) is deadheading.

Each of these problems has its own special requirements that may call for modifications in the theory. For example, in the case of garbage collection, the edges of our graph will represent streets, not sidewalks. If some of the streets are one-way, we need to put arrows on the corresponding edges, resulting in a directed graph, or **digraph.** The circuits we seek will have to obey these arrows. In the case of salt spreaders and snowplows, each lane of a street needs to be modeled as a directed edge, as shown in Figure 1.19. Note that the arrows on the map and digraph are not in color because these arrows denote restrictions in traversal possibilities, not parts of circuits.

Like salt spreaders, street-sweeping trucks can travel in only one lane at a time and need to obey the direction of traffic. Street sweepers, however, have an additional complication: parked cars. It is very difficult to clean the street if cars are parked along the curb. Yet for overall efficiency, those who are responsible for routing street sweepers want to interfere with parking as little as possible. The common solution is to post signs specifying times when parking is prohibited, such as Thursday between 8 A.M. and 2 P.M. Because the parking-time factor is a constraint on street sweeping, it is important not only to find an Euler circuit, or a circuit with very few duplications, but a circuit that can be completed in the time available. Once again, the theory can be modified to handle this constraint.

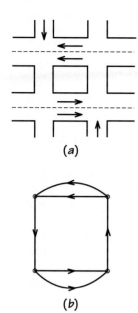

(a)

(b)

FIGURE 1.19   (a) Salt-spreading route, where each east–west street has two traffic lanes in the same direction, and (b) an appropriate digraph model.

Finally, because towns and cities of any size will have more than one street sweeper, parking officer, or garbage truck, a single best route will not suffice. Instead, they will have to divide the territory into multiple routes. The general goal is to find optimal solutions while taking into account traffic direction, number of lanes, time restrictions, and divided routes (see Figure 1.20).

Management science makes all this possible. For example, a pilot study done in the 1970s in New York City showed that applying these techniques to street sweepers in just one district could save about $30,000 per year. With 57 sanitation districts in New York, this would amount to a savings of more than $1.5 million in a single year. In addition, the same principles could be extended to garbage collection, parking control, and other services carried out on street networks.

This plan was not adopted when first proposed. Because city services take place in a political context, several other factors come into play. For example, union leaders try to protect the jobs of city workers, bureaucrats might try to keep their departmental budgets high, and elected politicians rarely want to be accused of cutting the jobs of their constituents. Thus political obstacles can overrule management science. As mentioned in Spotlight 1.2, such human factors often arise when applying management science. Perhaps a more acceptable street-sweeping plan would have been devised for New York if more attention had been paid to the human factors earlier.

Despite the complications of real-world problems, management science principles provide ways to understand these problems by using graphs as models. We can reason about the graphs and then return to the real-world problem with a workable solution. The results we get can have a lasting effect on the efficiency and economic well-being of any organization or community.

FIGURE 1.20   (a) Fairfield, California, USA. Today, finding optimal routes within complex street networks is often done with sophisticated computer-based color graphics systems. (b) A computer-generated street network.

(a)

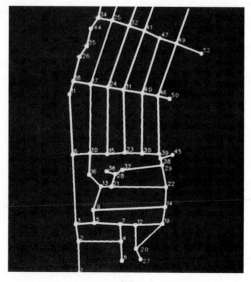

(b)

REVIEW VOCABULARY

Chinese postman problem The problem of finding a circuit on a graph that covers every edge of the graph at least once and that has the shortest possible length.

Circuit A path that starts and ends at the same vertex.

Connected graph A graph is connected if it is possible to reach any vertex from any specified starting vertex by traversing edges.

Digraph A graph in which each edge has an arrow indicating the direction of the edge. Such directed edges are appropriate when the relationship is "one-sided" rather than symmetric (e.g., one-way streets as opposed to regular streets).

Edge A link joining two vertices in a graph.

Euler circuit A circuit that traverses each edge of a graph exactly once.

Eulerizing Adding new edges to a graph so as to make a graph that possesses an Euler circuit.

Graph A mathematical structure in which points (called vertices) are used to represent things of interest, and in which links (called edges) are used to connect vertices, denoting that the connected vertices have a certain relationship.

Management science A discipline in which mathematical methods are applied to management problems in pursuit of optimal solutions that cannot readily be obtained by common sense.

Operations research Another name for management science.

Optimal solution When a problem has various solutions that can be ranked in preference order (perhaps according to some numerical measure of "goodness"), the optimal solution is the best-ranking solution.

Path A connected sequence of edges in a graph.

Simple circuit A circuit in which every vertex has a valence equal to two.

Valence (of a vertex) The number of edges touching that vertex.

Vertex A point in a graph where one or more edges end.

SUGGESTED READINGS

BELTRAMI, EDWARDS J. *Models for Public Systems Analysis,* Academic Press, New York, 1977. Section 5.4 deals with material similar to that in this chapter. The rest of the book gives a nice selection of applications of mathematics to plant location, manpower scheduling, providing emergency services, and other public service areas. The mathematics is somewhat more advanced than in this chapter.

COZZENS, MARGARET B., AND RICHARD P. PORTER. *Mathematics and Its Applications,* Heath, Lexington, Mass., 1987. Includes a nice discussion of Euler circuit ideas applied to DNA fragments.

MALKEVITCH, JOSEPH, AND WALTER MEYER. *Graphs, Models, and Finite Mathematics,* Prentice Hall, Englewood Cliffs, N.J., 1974. An introductory text, which includes much the same material as in this chapter, but with a little more detail. A different algorithm for finding Euler circuits is given.

The following two references discuss the shortest-path and matching problems in depth; they are suitable for advanced students and for faculty.

ROBERTS, FRED S. *Applied Combinatorics,* Prentice Hall, Englewood Cliffs, N.J., 1984.

TUCKER, ALAN. *Applied Combinatorics,* 2nd ed., Wiley, New York, 1984.

EXERCISES

▲ *Optional.* ▦ *Advanced.* ◆ *Discussion.*

Basic Concepts

1. In the graph below, the vertices represent cities and the edges represent roads connecting them. What are the valences of the vertices in this graph? (Keep in mind that *E* is part of the graph.) What might the valence of city *E* be showing about the geography?

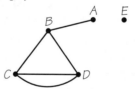

2. In the two graphs below, the vertices represent cities and the edges represent roads connecting them. In which graphs could a person located in city *A* choose any other city and then find a sequence of roads to get from *A* to that other city?

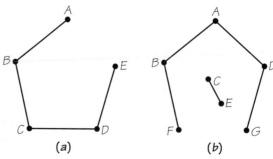

(a)   (b)

3.   Draw a graph to represent five cities A, B, C, D, and E having one road connecting each pair of cities in the following list (when a pair of cities appears twice in the list, there are two roads): (A, B), (A, C), (B, C), (B, C), (C, E), (D, E). What is the valence of each vertex in your graph? What is a real-world consequence of whether or not the graph is connected?

4.   Draw a graph to represent six cities A, B, C, D, E, and F having one road connecting each pair of cities in the following list: (A, B), (B, C), (D, E), (D, F). What is the valence of each vertex in your graph? What is a real-world consequence of whether or not the graph is connected?

5.   In the graph in Figure 1.8, find the smallest possible number of edges you could remove that would disconnect the graph.

6.   In the graph in Figure 1.17, find the smallest possible number of edges you could remove that would disconnect the graph.

7.   Is it possible that a street network gives rise to a disconnected graph? If so, draw such a network of blocks and streets and parking meters (in the style of Figure 1.12a). Then draw the disconnected graph it gives rise to.

Modeling

▲   8.   For the street network in Exercise 7, draw the graph that would be useful for routing a garbage truck. Assume that all streets are two-way and that passing once down a street suffices to collect from both sides.

◆   9.   A postal worker is supposed to deliver mail on all streets represented by edges in the graph below by traversing each edge exactly once. The first day the worker traverses the numbered edges in the order shown in (a), but the supervisor is not satisfied—why? The second day the worker follows the path indicated in (b), and the worker is unhappy—why? Is the original job description realistic? Why?

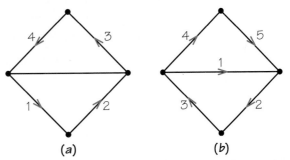

(a)   (b)

▲   10.   For the street network in Exercise 9, draw the graph that would be useful for routing a snowplow. Assume that all streets are two-way, one lane in each direction and that you need to pass down each lane separately.

▲ 11. For the street network below, draw the graph that would be useful for finding an efficient route for checking parking meters. (*Hint:* Notice that not every sidewalk has a meter; see Figure 1.12 in the text.)

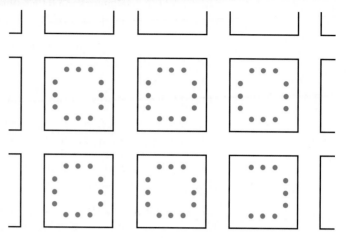

▲ 12. For the street network in Exercise 11, draw a graph that would be useful for routing a garbage truck. Assume that all streets are two-way and that passing once down a street suffices to collect from both sides.

13. For the street network below, draw the graph that would be useful for finding an efficient route for checking parking meters. (*Hint:* Notice that not every sidewalk has a meter; see Figure 1.12 in the text.)

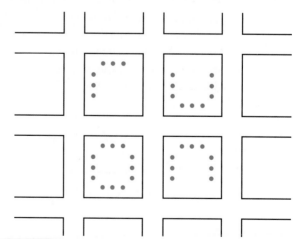

Euler Circuits

14. Examine the paths represented by the numbered sequences of edges in both parts of the figure at the top of the next page. Determine whether each path is a circuit. If it is a circuit, determine if it is an Euler circuit.

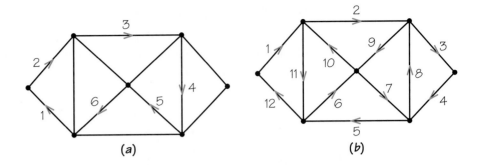

(a)   (b)

15. In Figure 1.13c, suppose we started an Euler circuit using this sequence of edges: 9, 10, 8, 7 (ignore existing arrows on the edges). What does our guideline for finding Euler circuits tell you not to do next?

16. In Figure 1.8b, suppose we started an Euler circuit using this sequence of edges: 14, 13, 8, 1, 4 (ignore existing arrows on the edges). What does our guideline for finding Euler circuits tell you not to do next?

17. Find an Euler circuit on the graph of Figure 1.15d (including the blue edges).

18. Find an Euler circuit on the right-hand graph in Figure 1.17a.

19. In the graph below, we see a territory for a parking-control officer that has no Euler circuit. Which sidewalk (edge) could be dropped in order to enable us to find an Euler circuit?

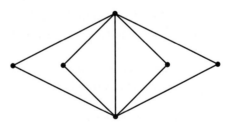

20. In the graph below, we see a territory for a parking-control officer that has no Euler circuit. Which sidewalk (edge) could be dropped in order to enable us to find an Euler circuit?

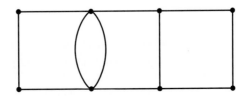

Eulerization and Squeezing

21. Find an Euler circuit on the eulerized graph (b) of the following figure. Use it to find a circuit on the original graph (a) that covers all edges and only reuses edges five times.

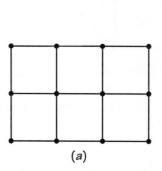
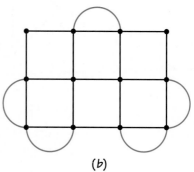

(a)   (b)

22. In the graph of the figure below, add one or more edges to produce a graph that has an Euler circuit.

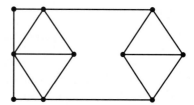

23. Can you find an eulerization with seven added edges for a 2-by-5-block rectangular street network? Can you do better than seven?

24. Squeeze the circuit shown in graph (a) below onto graph (b). Show your answers by numbered arrows on the edges.

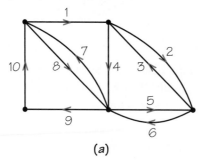
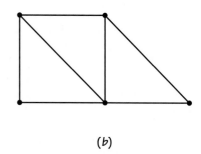

(a)   (b)

Then squeeze the circuit shown in graph (a) (at the top of the next page) onto graph (b). Show your answers by numbered arrows on the edges.

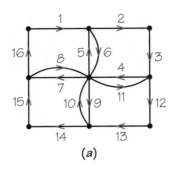

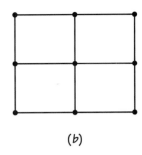

(a)   (b)

▲ 25. Eulerize these rectangular street networks using the same patterns that would be used by the "edge walker" described in the text.

    (a)   A 5 × 5 rectangle
    (b)   A 5 × 4 rectangle
    (c)   A 6 × 6 rectangle

▲ 26. Find good eulerizations for these graphs, using as few duplicated edges as you can. See the optional section "Finding Good Eulerizations" for hints.

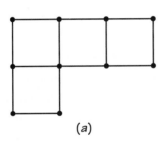

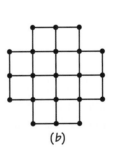

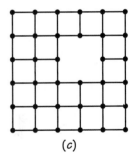

(a)   (b)   (c)

Minimum Duplication Circuits

27. A college campus has a central square with sidewalks arranged like the edges in the graph at the left. Show how all the sidewalks can be traversed efficiently in one circuit; your circuit will have to repeat some edges.

28. The figure at the top of page 30 shows a river, some islands, and bridges connecting the islands and riverbanks. A charity fund-raiser is sponsoring a puzzle race in which entrants have to start and end at *A*, go over every bridge at least once, and end at *A*. The first one back to *A* gets a prize. Draw a graph that would be useful for finding a route that requires the least recrossing of bridges. Show what that route would be.

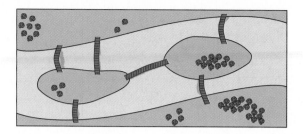

▲ 29. The year after the puzzle race described in Exercise 28, the sponsors want to do it again, but they need a new puzzle so that last year's participants won't have an unfair advantage over newcomers because of their previous experience. They can't construct new bridges or tear old ones down, so they add this rule: each of the bridges touching the right-hand island needs to be crossed at least twice. Draw a graph that would be useful for finding a route that requires the least recrossing of bridges. Show what that route would be.

▲ 30. Find a circuit in the graph below that covers every edge and has as few reuses as possible. See the optional section "Finding Good Eulerizations" for hints.

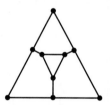

▲ 31. In the figure below, all blocks are 1000-by-1000 feet, except for the middle column of blocks, which are 1000-by-4000 feet. Find a circuit of minimum total length that covers all edges.

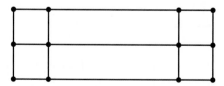

▲ 32. In the figure below, all blocks are 1000-by-1000 feet, except for the middle column of blocks, which are 1000-by-4000 feet. Find a circuit of minimum total length that covers all edges.

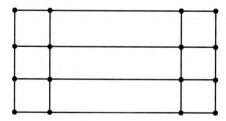

Additional Exercises

33. Which graphs in the figure below have Euler circuits? In the ones that do, find the Euler circuits by numbering the edges in the order the Euler circuit uses them. For the ones that don't, explain why no Euler circuit is possible.

(a)

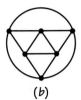

(b)

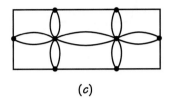

(c)

34. Find a minimum duplication circuit in a 3-by-5 block rectangular street network.

▲ 35. Draw a graph with four vertices and all of the valences odd.

36. Find an Euler circuit on the eulerized graph (b) of the following figure. Use it to find a circuit on the original graph (a) that covers all edges and only reuses edges four times.

(a)

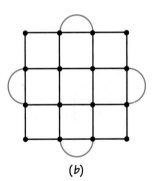

(b)

37. Which of these graphs are connected?

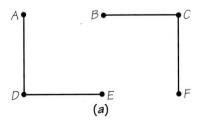

(a)

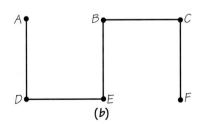

(b)

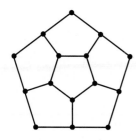

▲ 38. In the graph at the left, find a circuit that covers every edge and has as few reuses as possible. See the optional section "Finding Good Eulerizations" for hints.

▪ 39. A graph *G* represents a street network to be traveled by a postal worker who must traverse every street twice, once for each side of the street. In graph *G*, the edges represent sidewalks. Does such a graph always have an Euler circuit? Explain your answer.

▪ 40. Suppose for a certain graph that it is possible to disconnect it by removing one edge. Explain why such a graph (before the edge is removed) must have at least one vertex of odd valence. (*Hint:* Show that it cannot have an Euler circuit.)

▲ 41. Find the best eulerizations you can for the two graphs below.

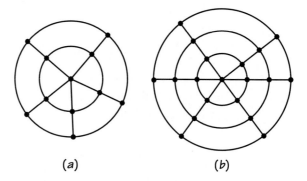

(*a*)   (*b*)

42. Each graph below represents the sidewalks to be cleaned in a fancy garden (one pass over a sidewalk will clean it). Can the cleaning be done using an Euler circuit? If so, show the circuit in each graph by numbering the edges in the order the Euler circuit uses them. If not, explain why no Euler circuit is possible.

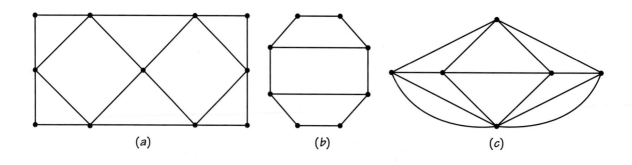

(*a*)   (*b*)   (*c*)

43. If an edge is added to an already existing graph, connecting two vertices already in the graph, explain why the number of vertices with odd valence has the same parity before and after. This means if it was even before, it is even after, while if it was odd before, it remains odd.

44. Any graph can be built in the following fashion: put down dots for the vertices, then add edges connecting the dots as needed. When you have put down the dots, and before any edges have been added, is the number of vertices with odd valence an even number or an odd number? What can you say about the number of vertices with odd valence when all the edges have been added (see Exercise 43)?

45. Draw the graph for the parking-control territory shown in the figure below. Label each vertex with its valence and determine if the graph is connected.

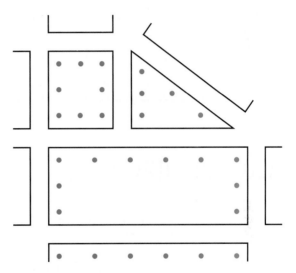

▲ 46. Eulerize these rectangular street networks using the same patterns that would be used by the "edge walker" described in the text.

    (a)   A 6 × 5 rectangle
    (b)   A 6 × 6 rectangle
    (c)   A 5 × 3 rectangle

◆ 47. The word *valence* is also used in chemistry. Find out what it means in chemistry and explain how this usage is similar to the use we make of it here.

48. For the street network below, draw a graph that represents the sidewalks with meters. Then find the minimum-length circuit that covers all sidewalks with

meters. If you drew the graph as we recommended in this section, you would find that the shortest circuit has length 18 (it reuses every edge). But the meter checker comes to you and says: "I don't know anything about your theories, but I have found a way to cover the sidewalks with meters using a circuit of length 10. My trick is that I don't rule out walking on sidewalks with no meters." Explain what he means and discuss whether his strategy can be used in other problems.

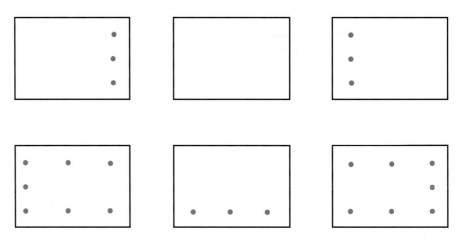

WRITING PROJECTS

1 ▶   Write a memo of three double-spaced typewritten pages to your local department of parking control (or police department) in which you suggest that management science techniques like the ones in this chapter be used to plan routes. Assume that the person to whom you are writing is not extensively trained in mathematics, but is willing to read through some technical material, provided you make it seem worth the trouble.

2 ▶   Do the same as in Writing Project 1, but to the department in charge of spreading salt on roads after snowstorms.

3 ▶   If you were making a recommendation to the mayor of New York City concerning proposed new street-sweeping routes, designed using the theory of this chapter, would you recommend that the changes be adopted or not? Write a memo (three double-spaced typewritten pages) that outlines the pros and cons as fairly as you can, and then conclude with your recommendation.

CHAPTER 2

VISITING VERTICES

I n the last chapter, we saw that it is relatively easy to determine if there is a circuit traversing the edges of a graph exactly once—for example, a route for street sweepers that covers the streets in a section of a city exactly once. However, the situation changes radically if we make an apparently innocuous change in the problem: When is it possible to find a route along distinct edges of a graph that visits each *vertex* once and only once in a simple circuit? For example, the wiggly line in Figure 2.1a shows a circuit we can take to tour that graph, visiting each vertex once and only once. This tour can be written *ABDGIHFECA.* Note that another way of writing the same circuit would be *EFHIGDBACE.* A different circuit visiting each vertex once and only once would be *CDBIGFEHAC* (Figure 2.1b). Do not be confused because *C* is written twice when we write down this list of vertices. We can think of the circuit as starting at any of its vertices, but we do start and end at the same vertex.

HAMILTONIAN CIRCUITS

A tour, like the ones marked by wiggly edges in Figure 2.1, that starts at a vertex of a graph and visits each vertex once and only once, returning to where it started, is called a **Hamiltonian circuit.**

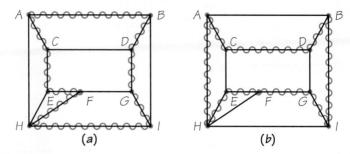

FIGURE 2.1   Wiggled edges illustrate Hamiltonian circuits.

The concept is named for the Irish mathematician William Rowan Hamilton (1805–1865), who was one of the first to study it. (We now know that the concept was discovered somewhat earlier by Thomas Kirkman [1806–1895], a British minister with a penchant for mathematics.) This problem is typical of the many new problems mathematicians create as a consequence of their professional training. In the situation here, motivated by our success in solving the problem of traversing all the *edges* of a graph, we will investigate visiting all the *vertices* of a graph.

The concepts of Euler and Hamiltonian circuits are similar in that both forbid reuse: the Euler circuit of edges, the Hamiltonian circuit of vertices. However, it is far more difficult to determine which connected graphs admit a Hamiltonian circuit than to determine which connected graphs have Euler circuits. As we saw in Chapter 1, looking at the valences of vertices tell us if a connected graph has an Euler circuit, but we have no such simple method for telling whether or not a graph has a Hamiltonian circuit.

Some special classes of graphs are known to have Hamiltonian circuits, and some special classes of graphs are known to lack them. For example, here is a method to construct an infinite family of graphs where each graph in the family cannot have a Hamiltonian circuit. Construct a vertical column of m vertices and a parallel column of n vertices, where m is bigger than n, as shown in Figure 2.2 The figure illustrates the typical case where $m = 4$ and $n = 2$. Now join each vertex on the left in the figure to every vertex on the right. As m and n vary, we get a family of different graphs.

Any graph obtained in this manner cannot have a Hamiltonian circuit. If a Hamiltonian circuit existed, it would have to include alternately vertices on the left and right of the figure. This is not possible, since the number of vertices on the left and right, m and n, respectively, are not the same. Unfortunately, it is unlikely that a method will ever be found to easily determine whether or not an arbitrarily chosen graph has a Hamiltonian circuit. As we shall see, if Hamiltonian circuits were easy to find in any graph at all, many applied problems could be solved in a less costly way.

FIGURE 2.2 An example
of one graph from a family
of graphs that has no
Hamiltonian circuit. The
number of vertices *m* on
the left is chosen to be
greater than the number of
vertices *n* on the right. The
case *m* = 4 and *n* = 2 is
shown.

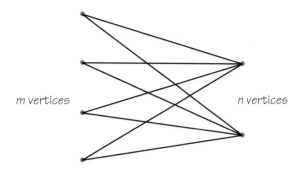

m vertices   n vertices

The Hamiltonian circuit problem and the Euler circuit problem are both
examples of graph theory problems. Although we posed the Hamiltonian circuit
problem merely as a variant of another graph theory problem with many
applications (i.e., the Euler circuit problem), the Hamiltonian circuit problem
itself has many applications. This is not unusual in mathematics. Often mathematics
used to solve a particular real-world problem leads to new mathematics
that suggests applications to other real-world situations.

Suppose inspections or deliveries need to be made at each vertex (rather
than along each edge) of a graph. An "efficient" tour of the graph would be a
route that started and ended at the same vertex and passed through all the vertices
without reuse, or repetition; that is, the route would be a Hamiltonian
circuit. Such routes would be useful for inspecting traffic signals or for delivering
mail to drop-off boxes, which hold heavy loads of mail so urban postal

Mailman.

carriers do not have to carry it long distances. There are many similar examples, but rather than pursue problems involving Hamiltonian circuits in general graphs, we will study instead of a more important class of related problems.

EXAMPLE ▶ *Vacation Planning*

Let's imagine that you are a college student studying in Chicago. During spring break you and a group of friends have decided to take a car trip to visit other friends in Minneapolis, Cleveland, and St. Louis. There are many choices as to the order of visiting the cities and returning to Chicago, but you want to design a route that minimizes the distance you have to travel. Presumably, you also want a route that cuts costs, and you know that minimizing distance will minimize the cost of gasoline for the trip. (Similar problems with different complications would arise for railroad or airplane trips.)

Imagine now that the local automobile club has provided you with the intercity driving distances between Chicago, Minneapolis, Cleveland, and St. Louis. We can construct a graph model with this information, representing each city by a vertex and the legs of the journey between the cities by edges joining the vertices. To complete the model, we add a number called a **weight** to each graph edge, as in Figure 2.3. In this example, the weights represent the distances between the cities, each of which corresponds to one of the endpoints

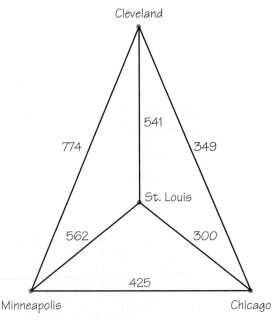

FIGURE 2.3   Road mileages between four cities.

of the edges in the graph. (In other examples the weight might represent a cost, time, or profit.) We want to find a minimal-cost tour that starts and ends in Chicago and visits each other city once. Using our earlier terminology, what we wish to find is a **minimum-cost Hamiltonian circuit**—a Hamiltonian circuit with the lowest possible sum of the weights of its edges.

How can we determine which Hamiltonian circuit has minimum cost? There is a conceptually easy **algorithm,** or mechanical step-by-step process, for solving this problem:

1. Generate all possible Hamiltonian tours (starting from Chicago).
2. Add up the distances on the edges of each tour.
3. Choose the tour of minimum distance.

Steps 2 and 3 of the algorithm are straightforward. Thus, we need worry only about step 1, generating all the possible Hamiltonian circuits in a systematic way. To find the Hamiltonian tours, we will use the *method of trees,* as follows. Starting from Chicago, we can choose any of the three cities to visit after leaving Chicago. The first stage of the enumeration tree is shown in Figure 2.4. If Minneapolis is chosen as the first city to visit, then there are two possible cities to visit next, Cleveland and St. Louis. The possible branchings of the tree at this stage are shown in Figure 2.5. In this second stage, however, for each choice of first city we visited, there are two choices from this city to the second city visited. This would lead to the diagram in Figure 2.6.

Having chosen the order of the first two cities to visit, and knowing that no revisits (reuses) can occur in a Hamiltonian circuit, there is only one choice left for the next city. From this city we return to Chicago. The complete tree diagram showing the third and fourth stages for these routes is given in Figure 2.7. Notice, however, that because we can traverse a circular tour in either of

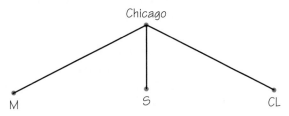

FIGURE 2.4 First stage in finding vacation-planning routes.

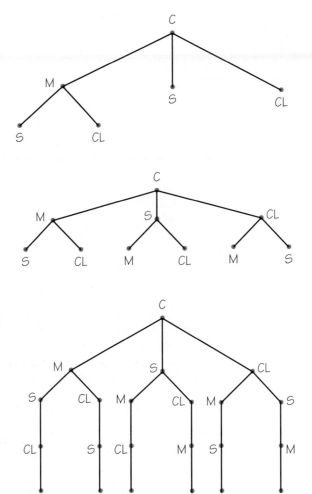

FIGURE 2.5 Part of the second stage in finding vacation-planning routes.

FIGURE 2.6 Complete second stage in finding vacation-planning routes.

FIGURE 2.7 Completed tree enumeration of routes for vacation-planning problem.

two directions, the paths enumerated in the tree diagram of Figure 2.7 do *not* correspond to different Hamiltonian circuits. For example, the leftmost path (C−M−S−CL−C) and the rightmost path (C−CL−S−M−C) represent the same Hamiltonian circuit. Thus, among what appear to be six different paths in the tree diagram, in fact, only three correspond to different Hamiltonian circuits. These three distinct Hamiltonian circuits are shown in Figure 2.8.

Note that in generating the Hamiltonian circuits we disregard the distances involved. We are concerned only with the different patterns of carrying out the visits. To find the optimal route, however, we must add up the distances on the edges to get each tour's length. Figure 2.8 shows that the optimal tour is Chicago, Minneapolis, St. Louis, Cleveland, Chicago. The length of this tour is 1877 miles. ◆

FIGURE 2.8 The three Hamiltonian circuits for the vacation-planning problem of Figure 2.3.

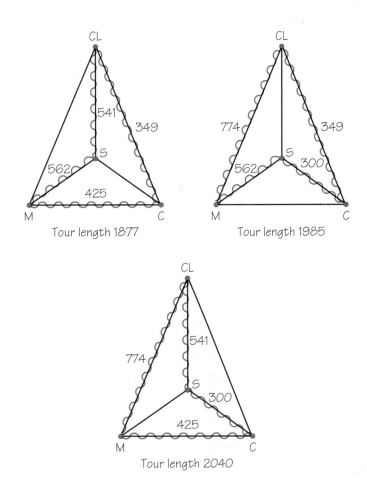

Tour length 1877

Tour length 1985

Tour length 2040

The method of trees is not always as easy to use as our example suggests. Instead of doing our analysis for four cities, consider the general case of *n* cities. The graph model similar to that in Figure 2.3 would consist of a weighted graph with *n* vertices, with every pair of vertices joined by an edge. Such a graph is called **complete** because the edge between any pair of vertices is present in the graph. A complete graph with five vertices is illustrated in Figure 2.9.

Fundamental Principle of Counting

How many Hamiltonian circuits are in a complete graph of *n* vertices? We can solve this problem by using the same type of analysis that emerged in the enumeration tree. The **method of trees** is a visual application of the *fundamental principle of counting*, a procedure for counting outcomes in multistage processes. Using this procedure we can count how many patterns occur in a situation by looking at the number of ways the component parts can occur. For

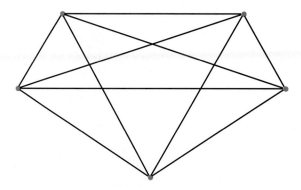

FIGURE 2.9 A complete graph with five vertices. Every pair of vertices is joined by an edge.

example, if Jack has 9 shirts and 4 pair of trousers, he can wear $9 \times 4 = 36$ shirt–pants outfits. Each shirt can be worn with any of the pants. (This can be verified by drawing a tree diagram, but such a diagram is cumbersome for big numbers.)

> In general, the **fundamental principle of counting** can be stated this way: if there are a ways of choosing one thing, b ways of choosing a second after the first is chosen, . . . , and z ways of choosing the last item after the earlier choices, then the total number of choice patterns is $a \times b \times c \times \cdots \times z$.

E X A M P L E ▶ *Counting*

Here are some other examples of how to use the fundamental principle of counting:

1. In a restaurant there are 4 kinds of soup, 12 entrees, 6 desserts, and 3 drinks. How many different four-course meals can a patron choose from? The four choices can be made in 4, 12, 6, and 3 ways, respectively. Hence, applying the fundamental principle of counting, there are $4 \times 12 \times 6 \times 3 = 864$ possible meals.
2. In a state lottery a contestant gets to pick a four-digit number that contains no zero followed by an uppercase or lowercase letter. How many such sequences of digits and a letter are there? Each of the four digits can be chosen in 9 ways (that is, 1, 2, . . . , 9), and the letter can be chosen in 52 ways (that is, *A, B,* . . . , *Z, a, b,* . . . , *z*). Hence there are $9 \times 9 \times 9 \times 9 \times 52 = 341{,}172$ possible patterns.
3. A corporation is planning a musical logo consisting of four different notes from the scale C, D, E, F, G, A, and B. How many logos are there to chose from? The first note can be chosen in 7 ways, but since

reuse is not allowed, the next note can be chosen in only 6 ways. The remaining two notes can be chosen in 5 and 4 ways, respectively. Using the fundamental principle of counting, $7 \times 6 \times 5 \times 4 = 840$ musical logos are possible. ◆

Again, in the problem of enumerating Hamiltonian circuits for the complete graph with n vertices, the city visited first after the home city can be chosen in $n - 1$ ways, the next city in $n - 2$ ways, and so on, until only one choice remains. Using the fundamental principle of counting, there are $(n - 1)! = (n - 1)(n - 2) \cdots \times 3 \times 2 \times 1$ routes. The exclamation mark in "$(n - 1)!$" is read "factorial" and is shorthand notation for the product $(n - 1)(n - 2) \times \cdots \times 3 \times 2 \times 1$. For example, $5! = 5 \times 4 \times 3 \times 2 \times 1 = 120$.

As we saw in Figure 2.7, pairs of routes correspond to the same Hamiltonian circuit because one route can be obtained from the other by traversing the cities in reverse order. Thus, although there are $(n - 1)!$ possible routes, there are only half as many, or $(n - 1)!/2$, different Hamiltonian circuits. Now, if we have only a few cities to visit, $(n - 1)!/2$ Hamiltonian circuits can be listed and examined in a reasonable amount of time. Analysis of a 6-city problem would require generation of $(6 - 1)!/2 = 5!/2 = 120/2 = 60$ tours. But for, say, 25 cities, $24!/2$ is approximately 3×10^{23}. Even if these tours could be generated at the rate of 1 million a second, it would take 10 billion years to generate them all. Since large vacation-planning problems would take so long to solve using this method, despite its conceptual ease, it is sometimes referred to as the **brute force method** (that is, trying all the possibilities).

TRAVELING SALESMAN PROBLEM

If the only benefit were saving money and time in vacation planning, the difficulty of finding a minimum-cost Hamiltonian circuit in a complete graph with n vertices for large values of n would not be of great concern. However, the problem we are discussing is one of the most common in operations research, the branch of mathematics concerned with getting governments and businesses to operate more efficiently. It is usually called the **traveling salesman problem (TSP)** because of its early formulation: determine the trip of minimum cost that a salesperson can make to visit the cities in a sales territory, starting and ending the trip in the same city.

Many situations require solving a TSP:

1. A lobster fisherman has set out traps at various locations and wishes to pick up his catch.

2. The telephone company wishes to pick up the coins from its pay telephone booths. (To avoid the high cost of picking up these coins, phone companies in many countries have adopted a system that uses prepurchased phone cards to operate phones. This means that there are no coins to collect!)

3. The electric (or gas) company needs to design a route for its meter readers.

4. A minibus must pick up six day campers and deliver them to camp, and later in the day return them home.

5. In drilling holes in a series of plates, the drill press operator (perhaps a robot) must drill the holes in a predetermined order.

The meaning of cost can vary from one formulation of TSP to another. We may measure cost as distance, airplane ticket prices, time, or any other factor that is to be optimized.

In many situations, the TSP arises as a subproblem of a more complicated problem. For example, a supermarket chain may have a very large number of stores to be served from a single large warehouse. If there are fewer trucks than stores, the stores must be grouped into clusters so that one truck serves each cluster. If we then solve the TSP for every truck, we can minimize total costs for the supermarket chain. Similar vehicle-routing problems for dial-a-ride services and for delivering children to their schools or camps often involve solving the TSP as a subproblem.

STRATEGIES FOR SOLVING THE TRAVELING SALESMAN PROBLEM

Because the traveling salesman problem arises so often in situations where the associated complete graphs would be very large, we must find a faster method than the brute force method we have described. We need to look at our original problem in Figure 2.3, and try to find an alternative algorithm for solving it. Recall that our goal is to find the minimum-cost Hamiltonian circuit.

Nearest-Neighbor Algorithm

Let's try a new approach: starting from Chicago, first visit the nearest city, then visit the nearest city that has not already been visited. We return to the start city when no other choice is available. This approach is called the **nearest-neighbor algorithm.**

Applying this algorithm to the TSP in Figure 2.3 quickly leads to the tour of Chicago, St. Louis, Cleveland, Minneapolis, and Chicago, with a length of 2040 miles. Here is how this tour is determined. Since we are starting in Chicago, there is a choice of going to a city that is 425, 300, or 349 miles away. Since the smallest of these numbers is 300, we next visit St. Louis, which is the nearest neighbor of Chicago not already visited. At St. Louis, we have a choice of visiting next cities that are 541 or 562 miles away. Hence, Cleveland, which is nearer (541), is visited. To complete the tour, we visit Minneapolis and return to Chicago, thereby adding 774 and 425 miles to the length of the tour.

The nearest-neighbor algorithm is an example of a **greedy algorithm,** because at each stage a best (greedy) choice, based on an appropriate criterion, is made. Unfortunately, this is not the optimal tour which we saw C–M–S–Cl–C was, for a total of 1877 miles. Making the best choice at each stage may not yield the best "global" solution. Even for a large TSP, one can always find a nearest-neighbor route quickly.

Figure 2.10 again illustrates the ease of applying the nearest-neighbor algorithm, this time to a weighted complete graph with five vertices. Starting at vertex A, we get the tour $ADECBA$ (cost 2800) (Figure 2.10a). Note that the nearest-neighbor algorithm starting at vertex B yields the tour $BCADEB$ (cost 3050) (Figure 2.10b).

This example illustrates that a nearest-neighbor tour can be computed for each vertex of the complete graph being considered, and that different nearest-neighbor tours can be gotten starting at different vertices. Note that even though we may seek a tour starting at a particular vertex, say, A in Figure 2.10, since a Hamiltonian circuit can be thought of as starting at any of its vertices, we can apply the nearest-neighbor procedure, if we wish, starting at vertex B (rather than at A). The Hamiltonian circuit we get can still be thought of as beginning at vertex A rather than B. Even for complete graphs with a large number of vertices, it would still be faster to apply nearest neighbor for each vertex and pick the cheapest of the tours generated (though such a tour might not be optimal) than to apply the brute force method.

FIGURE 2.10   (a) A weighted complete graph with five vertices that illustrates the use of the nearest-neighbor algorithm (starting at A). (b) TSP tour generated by the nearest-neighbor algorithm (starting at B).

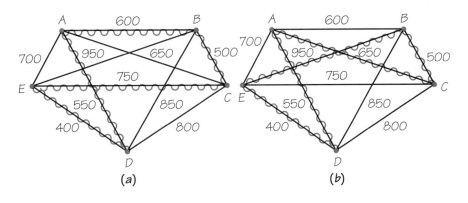

Sorted-Edges Algorithm

Perhaps some other easy method would yield optimal solutions.

> We might start by sorting or arranging the edges of the complete graph in order of increasing cost (or, equivalently, arrange the intercity distances in order of increasing distance). Then we can select at each stage that edge of least cost that (1) never requires that three used edges meet at a vertex (since a Hamiltonian circuit uses up exactly two edges at each vertex), and that (2) never closes up a circular tour that doesn't include all the vertices. This algorithm is called the **sorted-edges algorithm.**

Applying the sorted-edges algorithm to the TSP in Figure 2.3 yields the tour Chicago, St. Louis, Minneapolis, Cleveland, and Chicago, since the edges chosen would be 300, 349, 562, 774. Here are the details of how the sorted-edges method works in this example.

First, the six weights on the edges listed in increasing order would be 300, 349, 425, 541, 562, and 774. Since the cheapest edge in this sorted list is 300, this is the first edge that we place into the tour we are building. Next we add the edge with weight 349 to the tour. The next cheapest edge would be 425, but using this edge together with those already selected would result in having three edges at a vertex (see Figure 2.11a), which is not consistent with having a Hamiltonian circuit. Hence, we do not use this edge. The next cheapest edge, 541, used together with the edges already selected, would create a circuit (see Figure 2.11b) that does not include all the vertices. Thus, this edge, too, would be skipped over. However, we are able to add the edges 562 and 774 without either creating a circuit shorter than one including all the vertices or having

FIGURE 2.11 (a) When three shortest edges are added in order of increasing distance, three edges at a vertex are selected, which is not allowed as part of a Hamiltonian circuit. (b) When the edges of distances 300, 349, and 541 are selected, a circuit that does not include all vertices results.

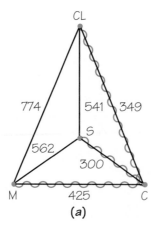
(a)

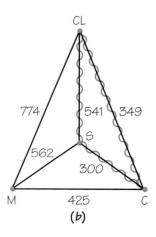
(b)

three edges at a vertex. Hence, the tour we arrive at is Chicago, St. Louis, Minneapolis, Cleveland, and Chicago. Again, this solution is not optimal because its length is 1985. Note that this algorithm, like the nearest neighbor, is greedy.

Although the edges selected by applying the sorted-edges method to the example in Figure 2.3 are connected to each other at every stage, this does not always happen. For example, if we apply the sorted-edges algorithm to the graph in Figure 2.10a, we build up the tour first with edge *ED* (400) and then edge *BC* (500), which do not touch. The edges that are then selected are *AD*, *AB*, and *EC*, giving the circuit *EDABCE*, which is the same as the nearest-neighbor circuit starting at vertex *A*.

Although many "quick and dirty" methods for solving the TSP have been suggested and some methods give an optimal solution in some cases, none of these methods guarantees an optimal solution. Surprisingly, most experts believe that no efficient method that *guarantees* an optimal solution will ever be found (see Spotlight 2.1).

Recently, mathematical researchers have adopted a somewhat different strategy for dealing with TSP problems. If finding a fast algorithm to generate optimal solutions for large problems is unlikely, perhaps one can show that the "quick and dirty" methods, usually called *heuristic algorithms,* come close enough to giving optimal solutions to be important for practical use. For example,

Spotlight NP-Complete Problems

2.1

Steven Cook, a computer scientist at the University of Toronto, showed in 1971 that certain hard, frustrating problems are equivalently difficult. This class of problems, now referred to as **NP-complete problems,** has the following characteristic: if a "fast" algorithm for solving one of these problems could be found, then a fast method would exist for all these problems.

In this context, "fast" means that as the size n of the problem grows (the number of cities gives the problem size in the traveling salesman problem), the amount of time needed to solve the problem grows no more rapidly than a polynomial function in n. (A polynomial function has the form $a_k n^k + a_{k-1} n^{k-1} + \cdots + a_1 n + a_0$.) On the other hand, if it could be shown that any problem in the class of NP-complete problems required an amount of time that grows faster than any polynomial (an exponential function, like 3^n, is an example of a function that grows faster than any polynomial) as the problem size increased, then all problems in the NP-complete class would share this characteristic. If some NP-complete problems had fast solutions, it seems likely that at least one such fast solution would have been found by now. It has been known for some time that the traveling salesman problem is an NP-complete problem; for this reason it is generally thought that no "fast" algorithm for an optimal solution for the TSP will ever be found.

Spotlight 2.2

Solving the Elusive Traveling Salesman Problem

More than 20 years ago, Shen Lin, executive consultant and retired head of AT&T Laboratories Network Configuration and Planning Department, Murray Hill, New Jersey, set out to solve the traveling salesman problem. Although the TSP defies any quick mathematical solution, Shen Lin has discovered a method that works fairly quickly on many practical problems. Today, he applies this knowledge to designing private telephone networks for AT&T's corporate customers. Here, Shen Lin recalls his earlier work on the TSP:

I started trying to solve the traveling salesman problem back in 1965 and ended in 1972, collaborating with a colleague, Brian Kernahan, in an algorithm that up to this day is considered to be the most efficient, practical method of solving this problem.

In 1965, we knew very little about which kinds of problems were hard and which were easy. So naively, I went in and thought I could solve this problem. Of course, I couldn't. I published my first paper on computer approximations by using the so-called iterative techniques [repeating some operation over and over] to solve them.

Today, I essentially use a heuristic method. The difference between this and the so-called classical method is that exact-solution methods usually take too long when you are trying to solve a hard problem like the traveling salesman problem. Also, they are formulated for problems that are very well defined. But in real life, we frequently find that no problem is so well defined, so we need some method of solving problems that is more flexible. We want something that can get at approximate solutions, which may turn out to be optimum solutions. The method must be powerful enough to solve a variety of problems. We cannot guarantee that the solution is the absolute and best one, because to find the best solution would take billions of years of computation, even by the world's fastest computer.

Within a few seconds, however, I could give you the solution to the traveling salesman problem, say, for a few hundred [geographical] points. And I'm sure no human being could look at the map, connect those points, and achieve the same result. I can't prove that it's the best solution that could ever be found, but it's usually quite close—within 1 to 2% of the optimum.

suppose one could prove that the nearest-neighbor heuristic was never off by more than 25% in the worst case or by more than 15% in the average case. For a medium-size TSP, one would then have to choose whether to spend a lot of time (or money) to find an optimal solution or to use instead a heuristic algorithm to obtain a fairly good solution. Researchers at AT&T Bell Laboratories have developed many remarkably good heuristic algorithms (see Spotlight 2.2). The best-known guarantee for a heuristic algorithm for a TSP is that it yields a cost that is no worse than one and a half times the optimal cost. Interestingly, this heuristic algorithm involves solving a Chinese postman problem (see Chapter 1), for which a "fast" algorithm is known to exist.

Throughout our discussion of the TSP we have concentrated on the goal of minimizing the cost (or time) of a tour that visited each of a variety of sites once and only once. One of the things that makes mathematical modeling exciting, however, is the subtle issues that arise in specific real-world situations (or that provide contrast between seemingly similar situations). For example, suppose the TSP situation is that of picking up day campers and taking them to and from the camp. From the point of view of the camp, it may wish to minimize the total length of time that the bus needs to pick up the campers. From the point of view of the parents of the campers, however, they may like the time their children spend on the bus to be as little as possible. For some problems, the tour that minimizes the mean (average) time that a child spends on the bus may not be the same tour that minimizes the total time of the tour. (Specifically, if the bus goes first to pick up the child the farthest from the camp, and then picks up the other children, this may yield a relatively short time on the bus for the kids, but a relatively long time for the tour itself.) It is these subtleties between problems that mathematicians go back to at a later time to examine, after the basic structure of the main problem itself is well understood. It is in this way that mathematics continues to grow, explore new ideas, and find new applications.

MINIMUM-COST SPANNING TREES

The traveling salesman problem is but one of many graph theory optimization problems that have grown out of real-world problems in both government and industry. Here is another.

EXAMPLE ▶ *Pictaphone Service*

Imagine that Pictaphone service (e.g., telephone service where a video image of the callers is provided) is to be set up on an experimental basis between five cities. The graph in Figure 2.12 shows the possible links that might be included in the Pictaphone network, with each edge showing the cost in millions of dollars to create that particular link. To send a Pictaphone message between two cities, a direct communication link is not necessary because it is possible to send a message indirectly via another city. Thus, in Figure 2.12, sending a message from *A* to *C* could be achieved by sending the message from *A* to *B*, from *B* to *E*, and from *E* to *C*, provided the links *AB*, *BE*, and *EC* are part of the network. We assume that the cost of relaying a message, compared with the direct communication link cost, is so small that we can neglect this amount. The problem that concerns us, therefore, is to provide service between any pair of cities in a way that minimizes the total cost of the links.

FIGURE 2.12 Costs (in millions of dollars) of installing Pictaphone service between five cities.

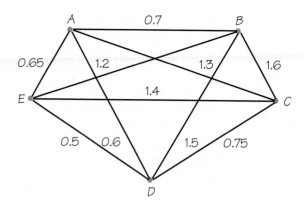

Our first guess at a solution is to put in the cheapest possible links between cities first, until all cities could send messages to any other city. Such an approach would be analogous to the sorted-edges method that was used to study the traveling salesman problem. In our example, if the cheapest links are added until all cities are joined, we obtain the connections shown in Figure 2.13a.

The links were added in the order *ED, AD, AE, AB, DC.* However, because this graph contains the circuit *ADEA* (wiggly edges in Figure 2.13b), it has redundant edges: we can still send messages between any pair of cities using relays after omitting the most expensive edge in the circuit—*AE.* After deleting an edge of a circuit, a message can still be relayed among the cities of the circuit by sending signals the long way around. After *AE* is deleted, messages from *A* to *E* can be sent via *D* (Figure 2.13c).

This procedure suggests a modified algorithm for our problem, **Kruskal's algorithm:** add the links in order of cheapest cost so that no circuits form and so that every vertex belongs to some link added (Figure 2.13d).

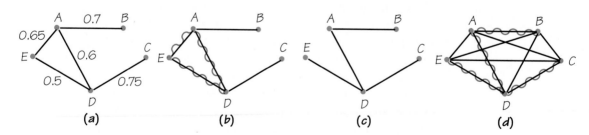

FIGURE 2.13 (a) Cities are linked in order of increasing cost until all cities are connected. (b) Circuit in part (a) highlighted. (c) Most expensive link in circuit in part (a) deleted. (d) Highlighted edges show, as a subgraph of the original graph, those links connecting the cities with minimum cost obtained using Kruskal's algorithm.

Picturephone (or Pictaphone).

As in the sorted-edges method for the TSP, the edges that are added need not be connected to each other until the end.

A subgraph formed in this way will be a tree; that is, it will consist of one piece and contain no circuits. It will also include all the vertices of the original graph. A subgraph that is a tree and that contains all the vertices of the original graph is called a **spanning tree** of the original graph.

To understand these concepts better, consider the graph G in Figure 2.14a. The wiggled edges in Figure 2.14b would constitute a subgraph of G that is a tree (since it is connected and has no circuit), but this tree would not be a spanning tree of G since the vertices D and E would not be included. On the other hand, the wiggled edges in Figure 2.14c and d show subgraphs of G that include all the vertices of G but are not trees because the first is not connected and the second contains a circuit. Figure 2.14e shows a spanning tree of G; the wiggled edges are connected and contain no circuit, and every vertex of the original graph is an endpoint of some wiggled edge. ◆

Finding a **minimum-cost spanning tree,** that is, a spanning tree whose edge weights sum to a minimum value, solves the Pictaphone problem. Note that having a different goal in the Pictaphone problem led to a different mathematical question from that of finding a Chinese Postman tour or TSP tour. In

Spotlight 2.3

Some Reminiscences on Shortest Spanning Trees

Joseph Kruskal has written the following thoughts on the minimum-cost spanning algorithm, which he refers to as shortest spanning trees.

In 1951, I entered the graduate mathematics program at Princeton. Among those active in combinatorial mathematics at Princeton then were Al Tucker, who soon became chairman of the department, Harold Kuhn, and Roger Lyndon. For a while, there was a joint Princeton–Bell Laboratories combinatorics seminar, which met alternately at two locations. My first introduction to Bell Laboratories, where I subsequently have spent several decades, came in this connection.

One day, someone in the combinatorics community handed me a blurry, carbon-copy typescript on flimsy lightweight paper about shortest spanning trees. It was in German, and appeared to be the foreign-language summary of a paper written in some other language. I had the typescript only for a limited period, and then passed it on to someone else. The typescript contained no date, and at the time I had no idea of whether it had been published or where. To this day, I have no idea where the typescript had come from, and why it was circulating. While writing my own paper stimulated by this material, I must have been told that the typescript was (part of?) a translation of a 1926 paper by Otakar Boruvka, for I give an exact citation in my paper. However, I have no memory of who told me this, nor do I remember ever seeing the full paper.

By the way, I regret that the phrase "minimum-spanning tree" has become dominant. What is actually meant is the minimum length spanning tree (MST), which is more concisely rendered by shortest spanning tree (SST). That is the phrase I used in 1954, and continue to use. If you wish to stick with MST rather than SST, however, at least do not make the common error of misconstruing MST as minimal, rather than minimum, spanning tree. The thing being minimized, length, has a unique minimum under very general conditions. [In this book, the term "minimum-cost spanning tree" is used rather than MST, as is common in the technical mathematics literature.]

The chief point of Boruvka's paper, I believe, was the uniqueness of the SST of a graph if the edges of the graph all have distinct lengths. His proof was based on a construction of the SST that I had difficulty in fully grasping. Although the construction has a certain spare elegance, I decided that it was difficult to follow because it was unnecessarily complicated. That led me to devise some much simpler methods, which are described in my 1956 paper. In 1985 Graham and Hell wrote a full history of the SST and closely related ideas. Well over 100 papers dating back to 1909 are cited, though the earliest paper specifically on the SST is Boruvka's work in 1926.

It is interesting to note that I was very uncertain about whether the material in my simple paper, which occupies only two and a half printed pages and requires few equations, was worth writing up for publication. I was a graduate student in my second year, and did not have a good sense of what was publishable. I asked a friend for advice (I don't remember whom), and fortunately received encouragement. It was my second published paper.

Although we have mentioned many routing problems in graphs, we have not discussed one of the most obvious: finding the path between two specified, distinct vertices with the sum of the weights of the edges in the path as small as possible. (Here there is no need to cover all vertices or to cover all edges.) We have seen that the weights on the edges have many possible interpretations, including time, distance, and cost. Following are some of the many possible applications:

1. Design routes to be used by a fire engine to get to a fire as quickly as possible.
2. Design delivery routes that minimize gasoline use.
3. Design routes to bring soldiers to the front as quickly as possible.
4. Design a route for a truck carrying nuclear waste.

The need to find shortest paths seems natural. Next we investigate a situation where finding a longest path is the right tool.

CRITICAL-PATH ANALYSIS

One of the delights of mathematics is its ability to confirm the obvious in certain situations while showing that our intuition is wrong in other circumstances. Our next group of mathematical applications will illustrate this point.

One of the characteristics of recent American life seems to be its fast pace. People are interested in getting things done quickly and efficiently. This means that when you take your car in to be repaired before work, you want to know for sure that the repairs will be done when you go to pick the car up. You want the trains and the bus that take you to your doctor's appointment to run on time; and when you arrive at the doctor's office, you want a nurse to be free to take a blood sample and a throat culture. You want your outpatient appointment for an X ray at the local hospital to occur on schedule. You want the X ray to be interpreted quickly and the results reported back to your internist.

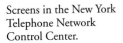

Screens in the New York Telephone Network Control Center.

Spotlight 2.4

AT&T Manager Explains How Long-Distance Calls Run Smoothly

Although long-distance calls are now routine, it takes great expertise and careful planning for a company like AT&T to handle its vast amounts of telephone traffic. Rich Wetmore was district manager of AT&T's Communications Network Operations Center in Bedminster, New Jersey. Here are his responses to questions about how AT&T handles its huge volumes of long-distance traffic and how it tracts its operations to keep things running smoothly.

How do you make sure that a customer doesn't run into a delayed signal when attempting a long-distance call?

We monitor the performance of our AT&T network by displaying data collected from all over the country on a special wallboard. The wallboard is configured to tell us if a customer's call is not going through because the network doesn't have enough capacity to handle it.

That's when we step in and take control to correct the problem. The typical control we use is to reroute the call. Instead of sending the customer's call directly to its destination, we'll route it via a third city—to someplace else in the country that has the capacity to complete the call.

It would seem that routing via another city would take longer. Is the customer aware of this process?

Routing a call via a third city is entirely transparent [imperceptible] to the customer. I'm an expert about the network, and even when I make a phone call, I have no idea how that individual call was routed. It's transparent both in terms of how far away the other person sounds and in how quickly the telephone call gets set up. With the signaling network we use, it takes milliseconds for switching systems to "talk" to each other to set up a call. So the fact that you are involved in a third switch in some distant city is something you would never know.

You want to be sure to keep costs down while supplying enough service to customers. So how do you balance company benefits with customer benefit?

In terms of making the network efficient, we want to do two things. First, we want our customers to be happy with our service and for all their calls to go through, which means we must build enough capacity in the network to allow that to happen. Second, we want to be efficient for our stockholders and not spend more money than we need to for the network to be at the optimum size.

There are basically two costs in terms of building the network. There is the cost of switching systems and the cost of the circuits that connect the switching systems. Basically, you can use operations-research techniques and mathematics to determine cost trade-offs. It may make sense to build direct routing between two switching systems and use a lot of circuits, or maybe to involve three switching systems, with fewer circuits between the main two, and so on.

Scheduling machines and people is a big part of modern life. It is important in your own personal daily activities as well as to businesses and governments. Scheduling is involved in running a school, a hospital, an airline, or in landing a person on Mars. Perhaps surprisingly, modern mathematics is a big part of what is involved in solving scheduling problems.

Part of what makes scheduling complicated is that when one performs the tasks that make up a job, the tasks usually cannot be done in a random order. For example, to make Thanksgiving dinner one must buy and prepare the turkey before putting it in the oven, and one must set the table before serving the food.

If the tasks cannot be performed in arbitrary sequence or order, we can specify the order in an **order-requirement digraph.** Digraph is short for directed graph. A digraph is a geometrical tool similar to a graph except that each edge has an arrow on it to indicate a direction for that edge. Digraphs can be used to model the fact that traffic on a street must go in one direction, or that certain tasks in a job must be completed prior to other tasks. A typical example of an order-requirement digraph is shown in Figure 2.16. There is a vertex in this digraph for each task. If one task must be done immediately before another, we draw a directed edge, or arrow, from the prerequisite task to the subsequent task. The numbers within the circles representing vertices are the times it takes to complete the tasks. In Figure 2.16 there is no arrow from T_1 to T_5 because task T_2 intervenes. Also, T_1, T_7, and T_8 have no tasks that must precede them. Hence, if there are at least three processors (i.e., people or machines) available, tasks T_1, T_7, and T_8 can be worked on simultaneously at the start of the job.

Let us investigate a typical scheduling problem faced by a business.

E X A M P L E ▶ *Turning a Plane Around*

Consider an airplane that carries both freight and passengers. The plane must have its passengers and freight unloaded and new passengers and cargo loaded before it can take off again. Also, the cabin must be cleaned before departure

FIGURE 2.16 Typical order-requirement digraph.

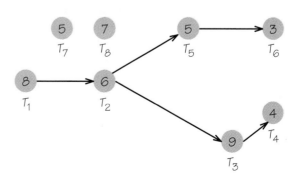

can occur. Thus, the job of "turning the plane around" requires completion of five tasks:

TASK *A*	Unload passengers	13 minutes
TASK *B*	Unload cargo	25 minutes
TASK *C*	Clean cabin	15 minutes
TASK *D*	Load new cargo	22 minutes
TASK *E*	Load new passengers	27 minutes

The order-requirement digraph for the problem of turning an airplane around is shown in Figure 2.17. The presence or absence of an edge in the order-requirement digraph depends on the analysis made as part of the modeling process for the problem. It seems natural that one needs to have an arrow between task *A* and task *C*, since before one can clean the cabin, the passengers on the plane should have been unloaded. However, the presence of some arrows may not seem natural, say, perhaps the arrow from task *B* (unload the cargo) to task *E* (load new passengers). This arrow may be due to government rules or requirements. What matters is that the mathematics of solving the problem does not depend on the reason that the order-requirement digraph looks the way that it does. The person solving the problem constructs the order-requirement digraph and then the mathematical techniques we will develop can be applied, regardless of whether or not some business faced with a similar problem might model the problem in a different way.

Because we want to find the earliest completion time, it might seem that finding the shortest path through the digraph (i.e., path *BD* with time length $25 + 22 = 47$) would solve the problem. But this approach shows the danger of ignoring the relationship between the mathematical model (the digraph) and the original problem.

The time required to complete all the tasks, *A* through *E*, must be at least as long as the time necessary to do the tasks on any particular path. Consider the path *BD*, which has length $25 + 22 = 47$. Recall that here *length* of a path refers to the sum of the times of the tasks that lie along the path. Since task *B* must be done before task *D* can begin, the two tasks *B* and *D* cannot be completed before time 47. Hence, even if work on other tasks (such as *A*, *C*, and *E*) is proceeding during this period, all the tasks cannot be finished before the tasks on path *BD* are finished. The same statement is true for every other path in the order-requirement digraph. Thus, the earliest completion time actually corresponds to the length of the longest path. In the airplane example, this ear-

FIGURE 2.17 Order-requirement digraph for turning an airplane around after landing.

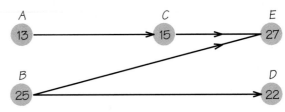

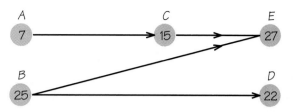

FIGURE 2.18 Order-requirement digraph for turning an airplane around with reduced times due to construction of new jetway.

liest completion time is 55 (= 13 + 15 + 27) minutes, corresponding to the path *ACE*. We call *ACE* the *critical path* because the times of the tasks on this path determine the earliest completion time.

> A **critical path** in an order-requirement digraph is a longest path. The length is measured in terms of summing the task times of the tasks making up the path.

Note that if none of these tasks could go on simultaneously, the time to complete all the tasks would be 13 + 25 + 15 + 22 + 27 = 102 minutes. However, even though tasks may go on simultaneously, the length of the critical path being 55 shows that completion of the tasks in less than 55 minutes is not possible. Only by speeding up the times to complete the critical-path tasks themselves can a completion time earlier than 55 time units be achieved.

Suppose it were desirable to speed the turnaround of the plane to below 55 minutes. One way to do this might be to build a second jetway to help unload passengers. For example, we could unload passengers (task *A*) in 7 minutes instead of 13. However, reducing task *A* to 7 minutes does not reduce the completion time by 6 minutes because in the new digraph (Figure 2.18) *ACE* is no longer the critical (i.e., longest) path. The longest path is now *BE*, which has a length of 52 minutes. Thus, shortening task *A* by 6 minutes results in only a 3-minute saving in completion time. This may mean that building a new jetway is uneconomical. Note also that shortening the time to complete tasks that are not on the original critical path *ACE* will not shorten the completion time at all. Speeding tasks on the critical path will shorten completion time of the job only up to the point where a new critical path is created. Also note that a digraph may have more than one longest path. ◆

Not all order-requirement digraphs are as simple as the one shown in Figure 2.17. The order-requirement digraph in Figure 2.19 has 12 paths, which can be found by exhaustive search. Examples of such paths are $T_1 T_2 T_3$, $T_1 T_5 T_9$, $T_4 T_5 T_9$, and $T_7 T_5 T_3$. (Although we have not discussed them here, fast algorithms for finding longest and shortest paths in graphs are known.) The critical path is $T_7 T_8 T_6$ (length 21), and the earliest completion time for all nine tasks is time 21.

These examples are typical of many scheduling problems that occur in practice (see Spotlight 2.5). Perhaps the most dramatic use of critical-path

Spotlight Every Moment Counts in Rigorous Airline Scheduling

2.5

When people think of airline scheduling, the first thing that comes to mind is how quickly a particular plane can safely reach its destination. But using ground time efficiently is just as important to an airline's timetable as the time spent in flight. Bill Rodenhizer was the manager of control operations for an airline that provided shuttle service between Boston and New York. He is considered to be an expert on airplane turnaround time, the process by which an airplane is prepared for almost immediate takeoff once it has landed. He tells us how this well-orchestrated effort works:

Scheduling, to the airline, is just about the whole ball game. Everything is scheduled right to the minute. The whole fleet operates on a strict schedule. Each of the departments responsible for turning around an aircraft has an allotted period of time in which to perform its function. Manpower is geared to the amount of ground time scheduled for that aircraft. This would be adjusted during off-weather or bad-weather days or during heavy air-traffic delays.

Most of our aircraft in Boston are scheduled for a 42- to 65-minute ground time. Boston is the end of the line, so it is a "terminating and originating station." In plain talk, that means almost every aircraft that comes in must be fully unloaded, refueled, serviced, and dispatched within roughly an hour's time.

This is how the process works: in the larger aircraft, it takes passengers roughly 20 minutes to load and 20 minutes to unload. During this period, we will have completely cleaned the aircraft and unloaded the cargo, and the caterers will have taken care of the food. The ramp service may take 20 to 30 minutes to unload the baggage, mail, and cargo from underneath the plane, and it will take the same amount of time to load it up again. We double-crew those aircraft with heavier weights so that the work load will fit the time it takes passengers to load and unload upstairs.

While this has been going on, the fueler has fueled the aircraft. As to repairs, most major maintenance is done during the midnight shift, when all but 20 of [our] several hundred aircraft are inactive.

We all work under a very strict time frame. There are four functional departments. If any of the four cannot fit its work into its time frame, then it advises us at the control center, and we adjust the departure time or whatever, so that the other departments can coordinate their activities accordingly.

FIGURE 2.19 An order-requirement digraph with 12 paths, to examine how to find the length of the longest path.

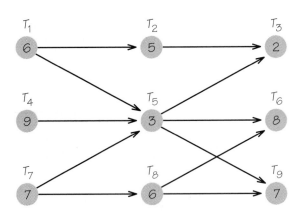

analysis is in the construction trades. No major new building project is now carried out without first performing a critical-path analysis to ensure that the proper personnel and materials are available at the right times in order to have the project finished as quickly as possible. Many such problems are too large and complicated to be solved without the aid of computers.

The critical-path method was popularized and came into wider use as a consequence of the Apollo project. As we saw in the part introduction, this project, which aimed at landing a man on the moon within 10 years of 1960, was one of the most sophisticated projects in planning and scheduling ever attempted. The dramatic success of the project can be attributed partly to the use of critical-path ideas and the related program evaluation and review technique (PERT), which helped keep the project on schedule.

REVIEW VOCABULARY

Algorithm   A step-by-step description of how to solve a problem.

Brute force method   The method that solves the traveling salesman problem (TSP) by enumerating all the Hamiltonian circuits and then selecting the one with minimum cost.

Complete graph   A graph in which every pair of vertices is joined by an edge.

Critical path   The longest path in an order-requirement digraph. The length of this path gives the earliest completion time for all the tasks making up the job consisting of the tasks in the digraph.

Fundamental principle of counting   A method for counting outcomes of multistage processes.

Greedy algorithm   An approach for solving an optimization problem, where at each stage of the algorithm the best (or cheapest) action is taken. Unfortunately, greedy algorithms do not always lead to optimal solutions.

Hamiltonian circuit   A circuit using distinct edges of a graph that starts and ends at a particular vertex of the graph and visits each vertex once and only once. A Hamiltonian circuit can be thought of as starting at any one of its vertices.

Heuristic algorithm   A method of solving an optimization problem that is "fast," but that does not guarantee an optimal answer to the problem.

Kruskal's algorithm   An algorithm developed by Joseph Kruskal (AT&T Bell Laboratories) that solves the minimum-cost spanning-tree problem by selecting edges in order of increasing cost, but so that no edge forms a circuit with edges chosen earlier. It can be proved that this algorithm always produces an optimal solution.

Method of trees   A visual method of carrying out the fundamental principle of counting.

Minimum-cost Hamiltonian circuit A Hamiltonian circuit in a graph with weights on the edges, for which the sum of the weights of the edges of the Hamiltonian circuit is as small as possible.

Minimum-cost spanning tree A spanning tree of a weighted connected graph having minimum cost. The cost of a tree is the sum of the weights on the edges of the tree.

Nearest-neighbor algorithm An algorithm for attempting to solve the TSP that begins at a "home" vertex and visits next that vertex not already visited that can be reached most cheaply. When all other vertices have been visited, the tour returns to home. This method may not give an optimal answer.

NP-complete problems A collection of problems, which includes the TSP, that appear to be very hard to solve quickly for an optimal solution.

Order-requirement digraph A directed graph that shows which tasks precede other tasks among the collection of tasks making up a job.

Sorted-edges algorithm An algorithm for attempting to solve the TSP where the edges added to the circuit being built up are selected in order of increasing cost, but no edge is added that would prevent a Hamiltonian circuit's being formed. These edges must all be connected at the end, but not necessarily at earlier stages. The tour obtained may not have lowest possible cost.

Spanning tree A subgraph of a connected graph that is a tree and includes all the vertices of the original graph.

Traveling salesman problem (TSP) The problem of finding a minimum-cost Hamiltonian circuit in a complete graph where each edge has been assigned a cost (or weight).

Tree A connected graph with no circuits.

Weight A number assigned to an edge of a graph that can be thought of as a cost, distance, or time associated with that edge.

SUGGESTED READINGS

BURR, STEFAN. *The Mathematics of Networks,* American Mathematical Society, Providence, R.I., 1982. A collection of articles dealing with applications of networks.

DOLAN, ALAN, AND JOAN ALDUS. *Networks and Algorithms: An Introductory Approach,* Wiley, Chichester, 1993. An excellent introduction to graph theory algorithms.

LAWLER, EUGENE, J. LENSTRA, RINNOY KAN, AND D. SHMOYS, EDS. *The Traveling Salesman Problem,* Prentice Hall, Englewood Cliffs, N.J., 1985. Includes survey and technical articles on all aspects of the TSP.

LUCAS, WILLIAM, FRED ROBERTS, AND ROBERT THRALL, EDS. *Discrete and Systems Models,* vol. 3: *Modules in Applied Mathematics,* Springer-Verlag, New York, 1983. Chapter 6, "A Model for Municipal Street Sweeping Operations," by A. Tucker and L. Bodin, describes street sweeping and related models in detail. Other models detail many recent applications of mathematics.

ROBERTS, FRED. *Applied Combinatorics,* Prentice Hall, Englewood Cliffs, N.J., 1984. The chapters on graphs and related network optimization problems are excellent.

ROBERTS, FRED. *Graph Theory and Its Applications to Problems of Society,* Society for Industrial and Applied Mathematics, Philadelphia, 1978. A very readable account of how graph theory is finding a wide variety of applications.

EXERCISES   ▲ *Optional.*   ▦ *Advanced.*   ◆ *Discussion.*

Hamiltonian Circuits

1. For each graph below, write a Hamiltonian circuit starting at X_1.

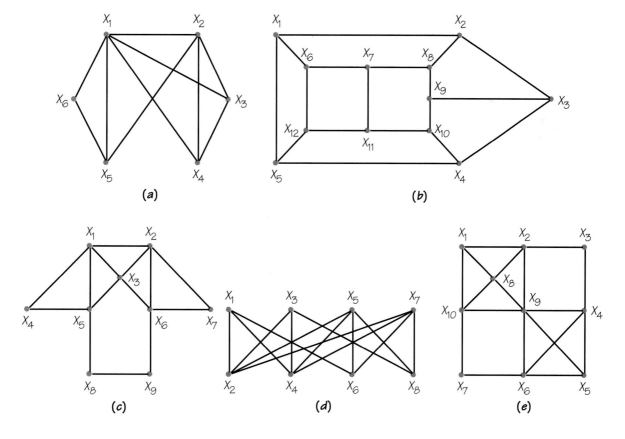

(a)   (b)

(c)   (d)   (e)

2.   For each of the graphs below, add wiggly edges to indicate a Hamiltonian circuit.

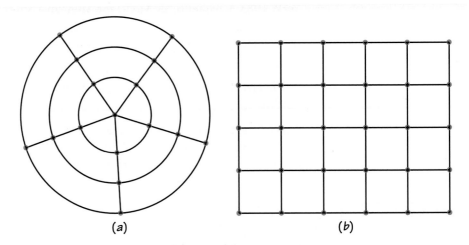

(a)   (b)

3.   Suppose two Hamiltonian circuits are considered different if the collections of edges that they use are different. How many other Hamiltonian circuits can you find in the graph in Figure 2.1 different from the two discussed?

4.   If the edge $X_2 X_3$ is erased from each of the graphs in Exercise 1, does the resulting graph still have a Hamiltonian circuit?

5.   Explain why the tour *ACEDCBA* is not a Hamiltonian circuit for the graph at the left. Does this graph have a Hamiltonian circuit?

6.   Do the following graphs have Hamiltonian circuits? If not, can you demonstrate why not?

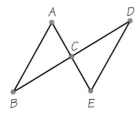

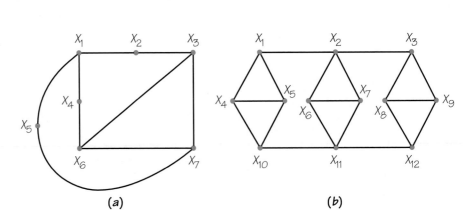

(a)   (b)

7. For each of the graphs below, determine if there is a Hamiltonian circuit.

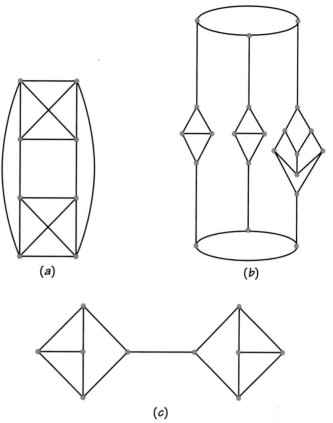

(a)   (b)

(c)

8. (a) The graph below is known as a four spokes and three concentric circles graph. What conditions on m and n guarantee that an m spokes and n concentric circles graph has a Hamiltonian circuit? (Assume $m \geq 2$, $n \geq 1$.)

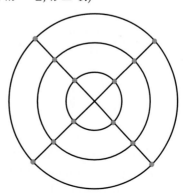

(b) The graph below is known as a 3 × 4 grid graph. What conditions on *m* and *n* guarantee that an *m* × *n* grid graph has a Hamiltonian circuit?

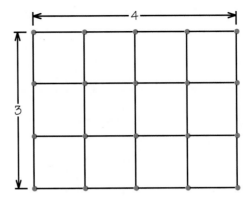

Can you think of a real-world situation in which finding a Hamiltonian circuit in an *m* × *n* grid graph would represent a solution to the problem? If an *m* × *n* grid graph has no Hamiltonian circuit, can you find a tour that repeats a minimum number of vertices and starts and ends at the same vertex?

■ 9. The *n*-dimensional cube is obtained from two copies of an $(n − 1)$-dimensional cube by joining corresponding vertices. (The process is illustrated for the 3-cube and the 4-cube in the following figure.)

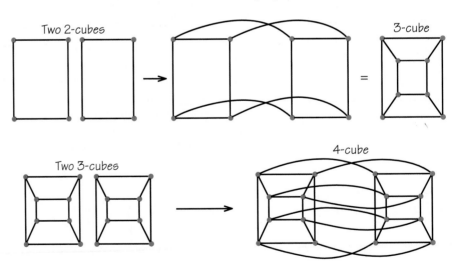

Find formulas for the number of vertices and the number of edges of an *n*-cube. Can you show that every *n*-cube has a Hamiltonian circuit? (*Hint:* Show that if you know how to find a Hamiltonian circuit on an $(n − 1)$-cube, then you can use two copies of this to build a Hamiltonian circuit on an *n*-cube.)

10. To practice your understanding of the concepts of Euler circuits and Hamiltonian circuits, determine for each graph below if there is an Euler circuit and/or a Hamiltonian circuit. If so, write it down.

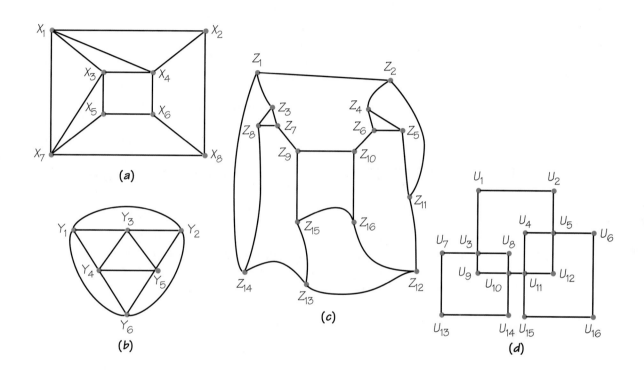

(a)

(b)

(c)

(d)

◆ 11. Give examples of real-world situations that can be modeled using a graph and for which finding a Hamiltonian circuit in the graph would be of interest.

Counting Problems

12. A lottery game ticket requires that a person select an upper- or lowercase letter followed by five different two-digit numbers (where the digits cannot both be zero). How many different ways are there to fill out a ticket?

13. (a) In designing a security system for its accounts, a bank asks each customer to choose a five-digit decimal number, all the digits to be distinct and nonzero. How many choices can the customer make?

(b) A suitcase with a liquid crystal display allows one to select a combination to lock it by choosing three capital letters that are not necessarily different. How many different choices would a thief have to go through to be sure that all the possibilities had been tried? How does this compare to a "standard" combination lock?

14. (a) For going outside on a cold winter day, Jill can choose from three winter coats, four wool scarfs, four pairs of boots, and three ski hats. How many outfits might her friends see her in?

(b) If Jill insists on always wearing her green wool scarf, how many outfits might her friends see her in?

◆ 15. (a) In New York State one type of license plate has three letters followed by a three-digit number. Suppose the digits can be chosen from 0, 1, . . . , 9, except that all three digits being zero is not allowed and any letter from A to Z (repeats allowed) can be chosen. How many plates are possible?

(b) Investigate what schemes for license plates are used in your state and determine how many different plates are possible.

16. A restaurant offers 6 appetizers, 10 entrees, and 8 desserts. How many different choices for a meal can a customer make if one selection is made from each category? If three of the desserts are pies and the customer will never order pie, how many meals can the customer choose?

Traveling Salesman Problems

17. Draw complete graphs with four, five, and six vertices. How many edges do these graphs have? Can you generalize to n vertices? How many TSP tours would these graphs have? (Tours yielding the same Hamiltonian circuit are considered the same.)

18. Calculate the values of 5!, 6!, 7!, 8!, 9!, and 10!. Then find the number of TSP tours in the complete graph with 10 vertices.

19. The table below shows the mileage between four cities: Springfield, Ill. (S), Urbana, Ill. (U), Effingham, Ill. (E), and Indianapolis, Ind. (I).

	E	I	S	U
E	—	147	92	79
I	147	—	190	119
S	92	190	—	88
U	79	119	88	—

(a)   Represent this information by drawing a weighted complete graph on four vertices.

(b)   Use the weighted graph in part (a) to find the cost of the three distinct Hamiltonian circuits in the graph. (List them starting at U.)

(c)   Which circuit gives the minimum cost?

(d)   Would there be any difference in parts (b) and (c) if the start vertex were at I?

(e)   If one applies the nearest-neighbor method starting at U, what circuit would be obtained? Does the answer change if one applies the nearest-neighbor algorithm starting at S? At E? At I?

(f)   If one applies the sorted-edges method, what circuit would be obtained? Does one get the optimal answer?

20.   After a party at her house, Francine (F) has agreed to drive Mary (M), Rachel (R), and Constance (C) home. If the times (in minutes) to drive between her friends' homes are shown below, what route gets Francine back home the quickest?

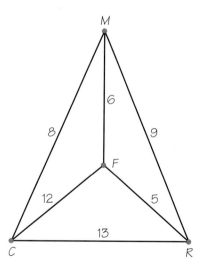

21.   Starting from the location where she moors her boat (M), a fisherwoman wishes to visit three areas A, B, and C where she has set fishing nets. If

the times (in minutes) between the locales are given in the figure below, what route to visit the three sites and return to the mooring place would be optimal?

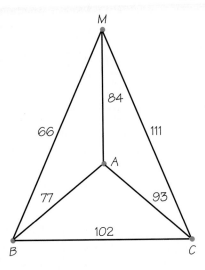

22. (a) For each of the complete graphs that follow, find the costs of the nearest-neighbor tour starting at A and of the tour generated using the sorted-edges algorithm.

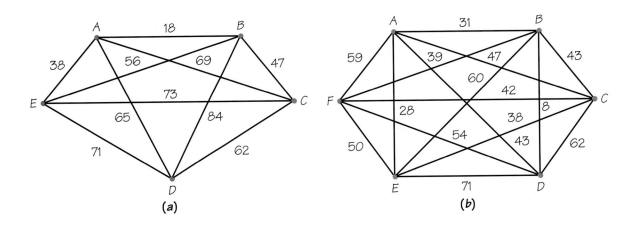

(b) How many Hamiltonian circuits would have to be examined to find a shortest route for part (a) by the brute force method?

(c) Can you invent an algorithm different from the sorted-edges and nearest-neighbor algorithms that is easy to apply for finding TSP solutions? (See Lawler et al., in Suggested Readings.)

23. An airport limo must take its six passengers to different downtown hotels from the airport. Is this a traveling salesman problem, Chinese postman problem, or an Euler circuit problem?

24. (a) Solve the six-city TSP shown in the diagram using the nearest-neighbor algorithm starting at vertex A; starting at vertex B.
     (b) Apply the sorted-edges method.

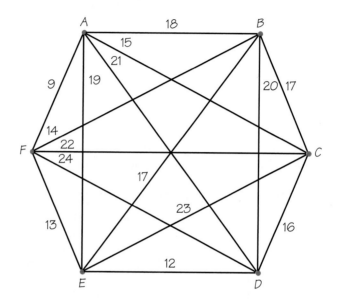

25. Construct an example of a complete graph on five vertices, with distinct weights on the edges for which the nearest-neighbor algorithm starting at a particular vertex and the sorted-edges algorithm yield different solutions for the traveling salesman problem. Can you find a five-vertex complete graph with weights on the edges in which the optimal solution, the nearest-neighbor solution, and the sorted-edges algorithm solution are all different?

26. If the brute force method of solving a 20-city TSP is employed, use a calculator to determine how many Hamiltonian circuits must be examined. How long would it take to determine the minimum-cost tour if the cost of tours could be computed at the rate of 1 billion per second? (Convert your answer to years by seeing how many years are equivalent to a billion seconds!)

27. Suppose one has found an optimal tour for a given 15-city TSP problem to have weight 3460. Now suppose the weights on the edges of the complete graph are increased by 30. What can you say about the optimal tour and its weight?

Trees and Spanning Trees

28. Which of the graphs below are trees?

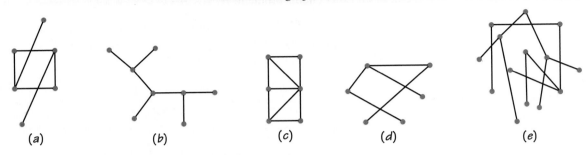

(a)   (b)   (c)   (d)   (e)

29. For each of the diagrams below explain why the wiggled edges are not

    (a)   a spanning tree.
    (b)   a Hamiltonian circuit.

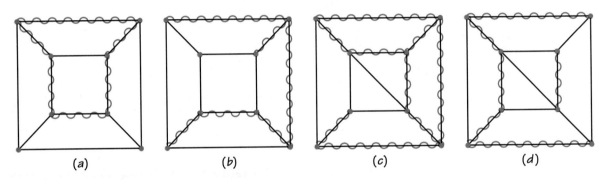

(a)   (b)   (c)   (d)

30. Find all the spanning trees in the graphs below.

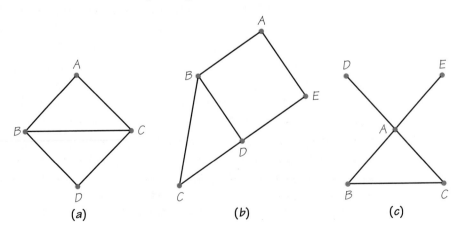

(a)   (b)   (c)

31.   Use Kruskal's algorithm to find a minimum-cost spanning tree for graphs (a), (b), (c), and (d).

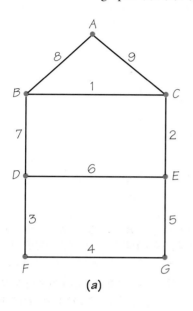

(a)

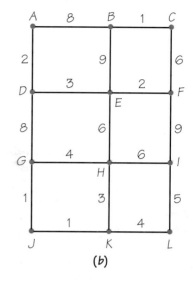

(b)

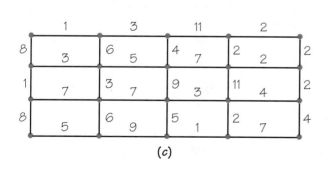

(c)

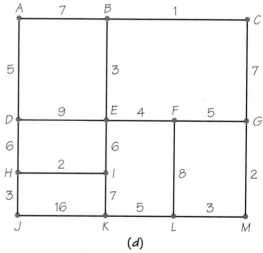

(d)

32.   A large company wishes to install a pneumatic tube system that would enable small items to be sent between any of 10 locales, possibly by relay. If the nonprohibitive costs (in $100) are shown in the graph model below, between which sites should the tube be installed to minimize the total cost?

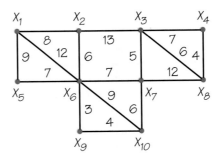

33.   If the weight of each edge in Exercise 32 is increased by 1, will the tree that achieves minimum cost for the new collection of weights be the same as the one that achieves minimum cost for the original set of weights?

◆ 34.   Give examples of real-world situations that can be modeled using a weighted graph and for which finding a minimum-cost spanning tree for the graph would be of interest.

▨ 35.   Can Kruskal's algorithm be modified to find a maximum-weight spanning tree? Can you think of an application for finding a maximum-weight spanning tree?

◆ 36.   Find the cost of providing a relay network between the six cities with the largest population in your home state, using the road distances between the cities as costs. Does it follow that the same solution would be obtained if air distances are used instead?

▨ 37.   Would there ever be a reason to find a minimum-cost spanning tree for a weighted graph in which the weights on some of the edges were negative? Would Kruskal's algorithm still apply?

38.   Let G be a graph with weights assigned to each edge. Consider the following algorithm:

(a)   Pick any vertex V of G.
(b)   Select an edge E with a vertex at V that has a minimum weight. Let the other endpoints of E be W.
(c)   Contract the edge VW so that edge VW disappears and vertices V and W coincide (see the following figures).

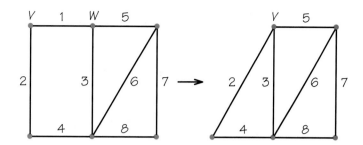

If in the new graph two or more edges join a pair of vertices, delete all but the cheapest. Continue to call the new vertex V.

(d)   Repeat steps (b) and (c) until a single point is obtained. The edges selected in the course of this algorithm (called Prim's algorithm) form a minimum-cost spanning tree. Apply this algorithm to the following graphs.

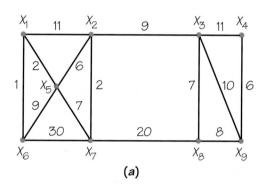

(a)

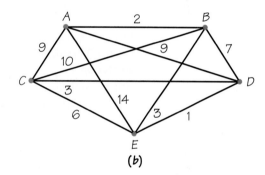

(b)

39.   Determine whether each of the following statements is true or false for a minimum-cost spanning tree T for a weighted connected graph G:

(a)   T contains a cheapest edge in the graph.
(b)   T cannot contain a most expensive edge in the graph.
(c)   T contains one fewer edge than there are vertices in G.
(d)   There is some vertex in T to which all others are joined by edges.
(e)   There is some vertex in T that has valence 3.

40. In the following graphs, the number in the circle for each vertex is the cost of installing equipment at the vertex if relaying must be done at the vertex, while the number on an edge indicates the cost of providing service between the endpoints of the edge.

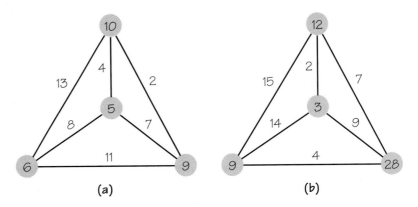

(a)　　　　　　　　(b)

In each case, find the minimum cost (allowing relays) for sending messages between any pair of vertices, taking vertex relay costs into account. Would your answer be different if vertex relay costs are neglected? (*Warning:* Kruskal's algorithm cannot be used to answer the first question. This problem illustrates the value of having an algorithm overrelying on "brute force.")

41. Two spanning trees of a (weighted) graph are considered different if they use different edges. Show that the graph below has different minimum-cost spanning trees, though all these different trees have the same cost.

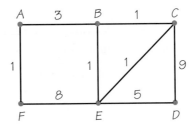

42. Suppose G is a graph such that all the weights on its edges are different numbers. Show that there is a unique minimum-cost spanning tree.

43. Find a minimum-cost spanning tree for the complete graphs in Exercise 22.

Scheduling

44.   Find the earliest completion time and critical paths for the three order-requirement digraphs below.

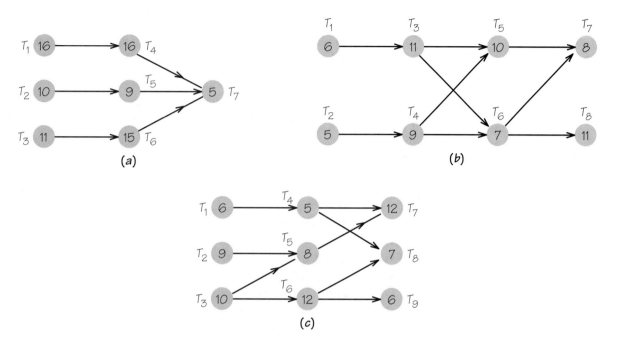

(a)

(b)

(c)

45.   Construct an example of an order-requirement digraph with three different critical paths.

46.   In the following order-requirement digraph, determine which tasks, if shortened, would reduce the earliest completion time and which would not. Then find the earliest completion time if task T_5 is reduced to time length 7. What is the new critical path?

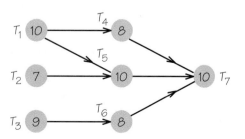

47.   To build a new addition on a house, the following tasks must be completed.

(a)   Lay foundation.
(b)   Erect sidewalls.
(c)   Erect roof.
(d)   Install plumbing.
(e)   Install electric wiring.
(f)   Lay tile flooring.
(g)   Obtain building permits.
(h)   Put in door that adjoins new room to existing house.
(i)   Install track lighting on ceiling.
(j)   Install wall air-conditioner.

Construct reasonable time estimates for these tasks and a reasonable order-requirement digraph. What is the fastest time in which these tasks can be completed?

48.   At a large toy store, scooters arrive unassembled in boxes. To assemble a scooter the following tasks must be performed:

TASK 1.   Remove parts from the box.
TASK 2.   Attach wheels to the footboard.
TASK 3.   Attach vertical housing.
TASK 4.   Attach handlebars to vertical housing.
TASK 5.   Put on reflector tape.
TASK 6.   Attach bell to handlebars.
TASK 7.   Attach decals.
TASK 8.   Attach kickstand.
TASK 9.   Attach safety instructions to handlebars.

Give reasonable time estimates for these tasks and construct a reasonable order-requirement digraph. What is the earliest time by which these tasks can be completed?

49.   For the order-requirement digraph below, find the critical path and the task(s) in the critical path whose time, when reduced the least, creates a new critical path.

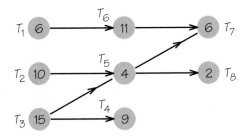

50.   Construct an order-requirement digraph with six tasks that has three critical paths of length 24.

Additional Exercises

51. (a) Each of the graphs below has no Hamiltonian circuit. Is it possible to add a single new edge to these graphs and obtain a new graph that has a Hamiltonian circuit?

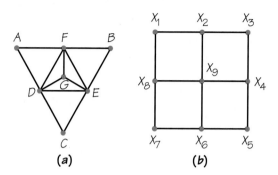

(a)   (b)

(b) Find an example of a graph that has no Hamiltonian circuit and yet no matter what single edge is added to the graph the result is a new graph that still has no Hamiltonian circuit.

(c) For a graph to have a Hamiltonian path, one relaxes the condition for a Hamiltonian circuit of having to start and end at the same vertex but still requires that the path include all the vertices of the graph. Do the graphs here have Hamiltonian paths?

(d) Can you think of applications situations where the concept of a Hamiltonian path would be useful?

52. The following figure represents a town where there is a sewer located at each corner (where two or more streets meet). After every thunderstorm, the department of public works wishes to have a truck start at its headquarters (at vertex H) and make an inspection of sewer drains to be sure that leaves are not clogging them. Can a route start and end at H that visits each corner exactly once? (Assume that all the streets are two-way streets.) Does this problem involve finding an Euler circuit or a Hamiltonian circuit?

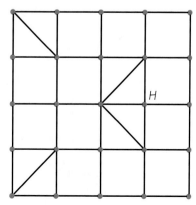

Assume that at equally spaced intervals along the blocks in this graph there are storm sewers that must be inspected after each thunderstorm to see if they are clogged. Is this a Hamiltonian circuit problem, an Euler circuit problem, or a Chinese Postman problem? Can you find an optimal tour to do this inspection?

53.   In the last several years regions that contain large cities that have had telephone service provided via only one area code have had to be divided into service areas with more than one area code. What is the largest number of different phone numbers that can be served using one area code? If an area code cannot begin with a zero, how many different area codes are possible?

54.   For each of the graphs with weights below, apply the nearest-neighbor method (starting at vertex *A*) and the sorted-edges method to find (it is hoped) a cheap tour.

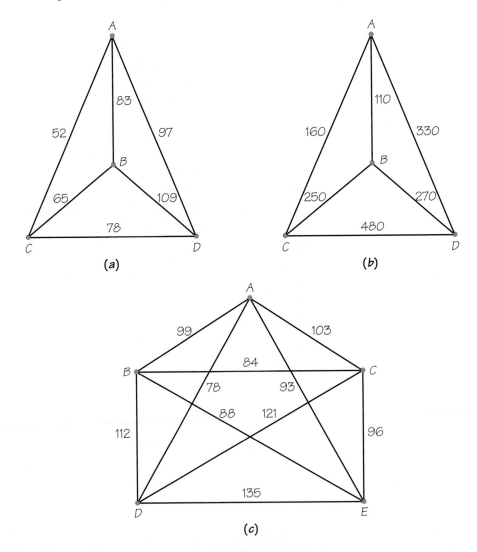

(a)

(b)

(c)

55. Suppose one has found an optimal tour for a given 10-city TSP problem to have weight 4520. Now suppose the weights on the edges of the complete graph are doubled. What can you say about the optimal tour and its weight?

56. Suppose that Pictaphone service is not possible between two towns where there is a hill of more than 800 feet along a straight line that runs between them. In constructing a model for solving a relay network problem, how would you handle the question of putting a link between two cities with a 1200-foot hill between them?

57. Suppose that a letter requires postage of p (positive integer) and that stamps of various denominations are available, say, $d_1, \ldots, d_v$ (positive integers). We are interested in finding the minimum number of stamps to choose with the available denominations to obtain the postage p exactly.

(a) Give an example to show that unless other conditions are put on p and $d_1, \ldots, d_v$, it might happen that no selection of stamps will achieve the desired postage.

(b) Show that even if the desired postage can be obtained for some choice of stamps, it does not follow that a minimum number of stamps can be achieved by using a greedy algorithm, that is, by selecting at each stage the largest denomination possible.

(c) Show that for at least one choice of denominations, a greedy algorithm will produce the fewest stamps for any given postage!

58. Draw an order-requirement digraph for the following set of tasks, giving a reasonable estimate for the times to do the tasks involved. Find a critical path for the order-requirement digraph that you obtain.

Kitchen remodeling project.

Tasks: Clean the kitchen; scrape walls to remove old paint; prime walls; install wallpaper on walls; scrape paint on ceiling; paint ceiling; replace old floor with new floor tiles; install new stove; install new sink.

WRITING PROJECTS

1 ▶ Write an essay about a variety of situations in which you are personally involved for which a solution of the TSP is (perhaps implicitly) required. Explain under what circumstances it might be valuable to carry out a formal mathematical solution to such TSPs rather than use an ad hoc solution.

2 ▶ Pick a situation that involves the traveling salesman model and discuss how closely the mathematics describes this situation and what features of the problem are likely to be important but have been neglected in the TSP model.

3 ▶ Construct an example, of the kind suggested on pages 48–49, that shows that in a situation where three day campers must be picked up and brought to camp, it may make a difference if the optimization criterion is minimizing distance traveled by the camp bus versus minimizing average time that the children spend on the bus.

4 ▶ Determine the six largest cities in the state in which you live. By consulting a road atlas (or by some other means) construct the graph that represents the road distances between your hometown and these six other cities. Now apply (a) the nearest-neighbor method, (b) the sorted-edges method, and (c) the nearest neighbor from each city, and pick the minimum tour method to solve the associated TSP. Do you have reason to believe that the answers you get might include an optimum solution among them?

CHAPTER 3

PLANNING AND SCHEDULING

In a society as complex as ours, everyday problems such as providing services efficiently and on time require accurate planning of both people and machines. Take the example of a hospital in a major city. Around-the-clock scheduling of nurses and doctors must be provided to guarantee that people with particular expertise are available during each shift. The operating rooms must be scheduled in a manner flexible enough to deal with emergencies. Equipment used for X-ray, CAT, or MRI scans must be scheduled for maximal efficiency.

Although many scheduling problems are often solved on an ad hoc basis, we can also use mathematical ideas to gain insight into the complications that arise in scheduling. The ideas we develop in this chapter have practical value in a relatively narrow range of applications, but they throw light on many characteristics of more realistic and hence more complex scheduling problems.

SCHEDULING TASKS

Assume that a certain number of identical **processors** (machines, humans, or robots) work on a series of tasks that make up a job. Associated with each task is a specified amount of time required to complete the task. For simplicity, we assume that any of the processors can work on any of the tasks. Our problem, known as the **machine-scheduling problem,** is to decide how the tasks should be scheduled so that the completion time for the tasks collectively is as early as possible.

Nurses scheduling patient care.

Even with these simplifying assumptions, complications in scheduling will arise. Some tasks may be more important than others and perhaps should be scheduled first. When "ties" occur, they must be resolved by special rules. As an example, suppose we are scheduling patients to be seen in a hospital emergency room staffed by one doctor. If two patients arrive simultaneously, one with a bleeding foot, the other with a bleeding arm, which patient should be processed first? Suppose the doctor treats the arm patient first, and while treatment is going on, a person in cardiac arrest arrives. Scheduling rules must establish appropriate priorities for cases such as these.

Another common complication arises with jobs consisting of several tasks that cannot be done in an arbitrary order. For example, if the job of putting up a new house is treated as a scheduling problem, the task of laying the foundation must precede the task of putting up the walls, which in turn must be completed before work on the roof can begin.

Assumptions and Goals

To simplify our analysis, we need to make clear and explicit assumptions:

1. If a processor starts work on a task, the work on that task will continue without interruption until the task is completed.

2. No processor stays voluntarily idle. In other words, if there is a processor free and a task available to be worked on, then that processor will immediately begin work on that task.

3. The requirements for ordering the tasks are given by an order-requirement digraph. (A typical example is shown in Figure 3.1, with task times highlighted within each vertex. The ordering of the tasks imposed by the order-requirement digraph represents constraints of physical reality. For example, you cannot fly a plane until it has taken fuel on board.)

4. The tasks are arranged in a priority list that is independent of the order requirements. (The **priority list** is an ordering of the tasks according to some criterion of "importance," which may in no way reflect physical reality. For example, imagine a construction job with several tasks. Task *A* may have to be done before task *B*, but when task *B* is done, a monetary payment will be made. Thus, *B* may be given a higher priority than *A*. The priority list represents, from some point of view, an ordering of the tasks. Another such point of view is to order the tasks in a manner that will help the algorithm being used construct schedules with early completion times. Usually, different points of view for giving priority to tasks are not consistent. Mathematical analysis may sometimes assist in clarifying trade-offs implicit in these different points of view.)

When considering a scheduling problem, there are various goals one might wish to achieve. Among these are

1. Minimizing the completion time of the job
2. Minimizing the total time that processors are idle
3. Finding the minimum number of processors necessary to finish the job by a specified time

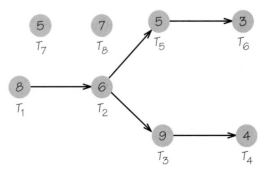

FIGURE 3.1 A typical order-requirement digraph.

Air traffic controllers must
be experts at scheduling.

For the moment we will concentrate on goal 1, finishing all the tasks at the earliest possible time. Note, however, that optimizing with respect to one criterion or goal may not optimize with respect to another. Our discussion here goes beyond what was discussed in Chapter 2 (see the section Critical-Path Analysis, pages 55–61) by dealing with how to assign tasks in a job to the processors that do the work.

List-Processing Algorithm

The scheduling problem we have described sounds more complicated than the traveling salesman problem (TSP). Indeed, like the TSP, it is known to be NP-complete. This means that it is unlikely that anyone will ever find a computationally fast algorithm that can find an optimal solution. Thus, we will be content to seek a solution method that is computationally fast and gives only approximately optimal answers.

The algorithm we use to schedule tasks is the **list-processing algorithm.** In describing it, we will call a task **ready** at a particular time if all its predecessors as indicated in the order-requirement digraph have been completed at that time. In Figure 3.1 at time 0 the ready tasks are T_1, T_7, and T_8, while task T_2 cannot be ready until 8 time units after T_1 is started. The algorithm works as follows: at a given time, assign to the lowest-numbered free processor the first task on the priority list that is *ready* at that time and that hasn't already been assigned to a processor.

In applying this algorithm, we will need to develop skill at coordinating the use of the information in the order-requirement digraph and the priority list. It will be helpful to cross out the tasks in the priority list as they are assigned to a processor to keep track of which tasks remain to be scheduled. Let's apply this algorithm to one possible priority list, T_8, T_7, T_6, . . . , T_1, using two processors and the order-requirement digraph in Figure 3.1. The result is the schedule shown in Figure 3.2, where idle processor time is indicated by white. How does the list-processing algorithm generate this schedule?

Because T_8 (task 8) is first on the priority list and ready at time 0, it is assigned to the lowest-numbered free processor, processor 1. Task 7, next on the priority list, is also ready at time 0 and thus is assigned to processor 2. The first processor to become free is processor 2 at time 5. Recall that by assumption 1, once a processor starts work on a task, its work cannot be interrupted until the task is complete. Task 6, the next unassigned task on the list, is not ready at time 5, as can be seen by consulting Figure 3.1. The reason task 6 is not ready at time 5 is that task 5 has not been completed by time 5. In fact, at time 5, the only ready task on the list is T_1, so that task is assigned to processor 2. At time 7, processor 1 becomes free, but no task becomes ready until time 13. Thus, processor 1 stays idle from time 7 to time 13. At this time, because T_2 is the first ready task on the list not already scheduled, it is assigned to processor 1. Processor 2, however, stays idle because no other ready task is available at this time. The remainder of the scheduling shown in Figure 3.2 is completed in this manner.

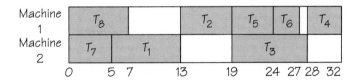

FIGURE 3.2   The schedule produced by applying the list-processing algorithm to the order-requirement digraph in Figure 3.1 using the list T_8, T_7, . . . , T_1.

As the priority list is scanned from left to right to assign a processor at a particular time, we pass over tasks that are not ready to find ones that are ready. If no task can be assigned in this manner, we keep one or more processors idle until such time that, reading the priority list from the left, there is a ready task not already assigned. After a task is assigned to a processor, we resume scanning the priority list, starting over at the far left, for unassigned tasks.

When Is a Schedule Optimal?

The schedule in Figure 3.2 has a lot of idle time, so it may not be optimal. Indeed, if we apply the list-processing algorithm for two processors to another possible priority list $T_1, \ldots, T_8$, using the digraph in Figure 3.1, the resulting schedule is that shown in Figure 3.3.

Here are the details of how this schedule was arrived at. Remember we must coordinate the list $T_1, T_2, \ldots, T_8$ with the information in the order-requirement digraph shown in Figure 3.3a. At time 0, task T_1 is ready, so this task is assigned to processor 1. However, at time 0, tasks $T_2, T_3, \ldots, T_6$ are not ready since their predecessors are not done. For example, T_2 is not ready at time 0 since T_1, which precedes it, is not done at time 0. The first ready task on the list, reading from left to right, that is not already assigned is T_7, so task T_7 gets assigned to processor 2. Both processors are now busy until time 5, at which point processor 2 becomes idle (Figure 3.3b).

Tasks T_1 and T_7 have been assigned; reading from left to right along the list, the first task not already assigned whose predecessors are done by time 5 is T_8, so this task is started at time 5 on processor 2; processor 2 will continue to work on this task until time 12, since the task time for this task is 7 time units. At time 8, processor 1 becomes free, and reading the list from left to right we find that T_2 is ready (since T_1 has just been completed). Thus, T_2 is assigned processor 1, which will stay busy on this task until time 14. At time 12,

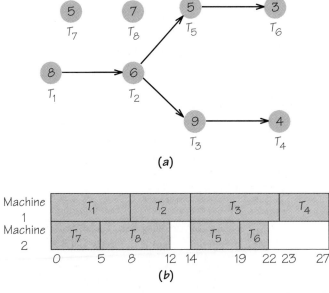

(a)

(b)

FIGURE 3.3   (a) A typical order-requirement digraph (repeat of Figure 3.1). (b) The schedule produced by applying the list-processing algorithm to the order-requirement digraph in Figure 3.3a using the list $T_1, T_2, \ldots, T_8$.

processor 2 becomes free, but the tasks that have not already been assigned from the list, T_3, T_4, T_5, T_6, are not ready, since they depend on T_2 being completed before these tasks can start. Thus, processor 2 stays idle involuntarily until time 14. At this time, T_3 and T_5 become ready. Since both processors 1 and 2 are idle at time 14, the lower numbered of the two, processor 1, gets to start on T_3 because it is the first ready task left to be assigned on the list scanned from left to right. Task T_5 get assigned to processor 2 at time 14. The remaining tasks are assigned in a similar manner.

The schedule shown in Figure 3.3b is optimal because the path T_1, T_2, T_3, T_4, with length 27, is the critical path in the order-requirement digraph. As we saw in Chapter 2, the earliest completion time for the job made up of all the tasks is the length of the longest path in the order-requirement digraph.

There is another way of relating optimal completion time for a scheduling problem to the completion time that is yielded by the list-processing algorithm. Suppose that we add all the task times given in the order-requirement digraph and divide by the number of processors. The completion time using the list-processing algorithm must be at least as large as this number. For example, the task times for the order-requirement digraph in Figure 3.3a sum to 47. Thus, if these tasks are scheduled on two processors, the completion time is at least $47/2 = 23.5$ (in fact, 24, since the list-processing algorithm applied to integer task times yields an integer answer), while for three processors the completion time is at least $47/3$ (in fact, 16).

Why is it helpful to take the total time to do all the tasks in a job and divide this number by the number of processors? Think of each task that must be scheduled as a rectangle that is 1 unit high and t units wide, where t is the time allotted for the task. Think of the scheduling diagram with m processors as a rectangle that is m units high and whose width W is the completion time for the tasks. The scheduling diagram is to be filled up by the rectangles that represent the tasks. How small can W be? The area of the rectangle that represents the scheduling diagram must be at least as large as the sum of all the rectangles representing tasks that are "packed" into it. The area of the scheduling diagram rectangle is mW. The combined areas of all the tasks, plus the area of rectangles corresponding to idle time, will equal mW. Width W is smallest when the idle time is zero. Thus, W must be at least as big as the sum of all the task times divided by m. Sometimes the estimate for completion time given by the list-processing algorithm from the length of the critical path gives a more useful value than the approach based on adding task times, and sometimes the opposite is true. For the order-requirement digraph in Figure 3.1, except for a schedule involving one processor, the critical-path estimate is superior. For some scheduling problems, both these estimates may be poor.

The number of priority lists that can be constructed if there are n tasks is $n!$, as can be computed using the fundamental principle of counting.

For example, for eight tasks, $T_1, \ldots, T_8$, there are $8 \times 7 \times 6 \cdots \times 1 = 40{,}320$ possible priority lists. For different choices of the priority list, the list-processing algorithm we are using will schedule the tasks, subject to the constraints of the order-requirement digraph, in different ways. Different lists may yield schedules with different completion times or different schedules with the same completion time. Different priority lists may yield identical scheduling of tasks (and hence identical completion times). A little later we will see a method that can be used to select a list that, if we are lucky, will give a schedule with a relatively good completion time. In fact, no method is known, except for very specialized cases, of how to choose a list that can be guaranteed to give rise to an optimal schedule when the list algorithm is applied to it.

Strange Happenings

The list-processing algorithm involves four factors that affect the final schedule. The answer we get depends on the

1. Times of the tasks
2. Number of processors
3. Order-requirement digraph
4. Ordering of the tasks on the list

To see the interplay of these four factors, consider another scheduling problem, this one associated with the order-requirement digraph shown in Figure 3.4 (the highlighted numbers are task time lengths).

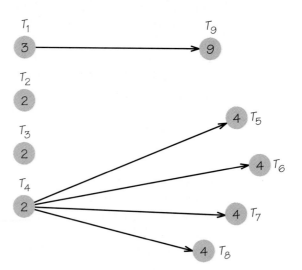

FIGURE 3.4 An order-requirement digraph designed to help illustrate some paradoxical behavior produced by the list-processing algorithm.

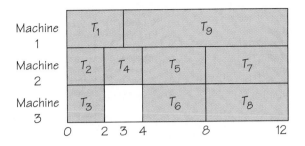

FIGURE 3.5   The schedule produced by applying the list-processing algorithm to the order-requirement digraph in Figure 3.4 using the list $T_1, T_2, \ldots, T_9$ with three processors.

The schedule generated by the list-processing algorithm applied to the list $T_1, T_2, \ldots, T_9$, using three processors, is given in Figure 3.5.

Treating the list $T_1, \ldots, T_9$ as fixed, how might we make the completion time earlier? Our alternatives are to pursue one or more of these strategies:

1. Reduce task times.
2. Use more processors.
3. "Loosen" the constraints of the order-requirement digraph.

Let's consider each alternative in turn, changing one feature of the original problem at a time, and see what happens to the resulting schedule. If we adopt strategy 1, reducing the time of each task by one unit, it seems intuitively clear that the completion time would go down. Figure 3.6 shows the new order-requirement digraph, and Figure 3.7 shows the schedule produced for this problem, using the list-processing algorithm with three processors applied to the list $T_1, \ldots, T_9$. The completion time is now 13, longer than the completion time of 12 for the case (Figure 3.5) with longer task times. Here is something unexpected! Let's explore further and see what happens.

FIGURE 3.6   The order-requirement digraph obtained from the one in Figure 3.4 by reducing by one unit each of the task times shown there.

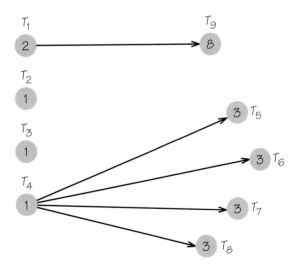

Figure 3.7 The schedule produced by applying the list-processing algorithm to the order-requirement digraph in Figure 3.6 using the list $T_1, T_2, \ldots, T_9$ with three processors.

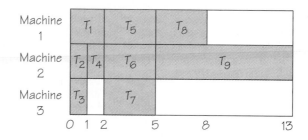

Next we consider strategy 2, increasing the number of machines. Surely this should speed matters up. When we apply the list-processing algorithm to the original graph in Figure 3.4, using the list $T_1, \ldots, T_9$ and four machines, we get the schedule shown in Figure 3.8. The completion time is now 15, *an even later completion time than for the previous alteration!*

Finally, we consider strategy 3, trying to shorten completion time by erasing all constraints (edges with arrows) in the order-requirement digraph shown in Figure 3.4. By increasing flexibility of the ordering of the tasks, we might guess we could finish our tasks more quickly. Figure 3.9 shows the schedule using the list $T_1, \ldots, T_9$; now it takes 16 units! This is the worst of our three strategies to reduce completion time.

The failures we have seen here appear paradoxical at first glance, but they are typical of what can happen when a situation is too complex to analyze with naive intuition. Sometimes our common sense leads us astray. The value of using mathematics rather than intuition or trial and error to study scheduling and other problems is that it points up flaws that can occur in unguarded intuitive reasoning.

The paradoxical behavior we see here is a consequence of the rules we set up for generating schedules. Such paradoxical behavior for the list-processing algorithm will not occur for every example you try. In fact, one has to be quite clever to design such examples. The list-processing algorithm has many nice features, including the fact that it is easy to understand and fast to implement. However, the results of the model in some cases can appear strange. Since we have been explicit about our assumptions, we could go back and make changes

Figure 3.8 The schedule produced by applying the list-processing algorithm to the order-requirement digraph in Figure 3.4 using the list $T_1, T_2, \ldots, T_9$ with four processors.

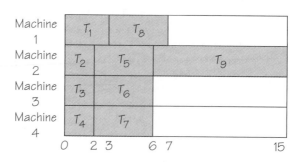

FIGURE 3.9 The schedule produced by applying the list-processing algorithm to the order-requirement digraph in Figure 3.4, modified by erasing all its directed edges, using the list $T_1, T_2, \ldots, T_9$ with three processors.

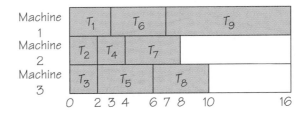

in these assumptions in hopes of eliminating the strange behavior. But the price we may pay is more time spent in constructing schedules and perhaps even new types of strange behavior. Unfortunately for modern society with its increasing concern with economical and efficient scheduling, recent mathematical research suggests that scheduling is an intrinsically hard problem.

CRITICAL-PATH SCHEDULES

In our discussion so far, we have acted as though the priority list used in applying the list-processing algorithm was given to us in advance based on external considerations. We might, however, consider the question of whether there is a systematic method of *choosing* a priority list that yields optimal or nearly optimal schedules. We will show how to construct a specific priority list based on this principle, to which the list-processing algorithm can then be applied.

Recall from our discussion of critical-path analysis in Chapter 2 that no matter how a schedule is constructed, the finish time cannot be earlier than the length of the longest path in the order-requirement digraph. This suggests that we should try to schedule first those tasks that occur early in long paths, because they might be a bottleneck for the other tasks.

EXAMPLE ▶ *Scheduling Two Processors*

To illustrate this method, consider the order-requirement digraph in Figure 3.10a. Suppose we wish to schedule these tasks on two processors. Initially, there are two critical paths of length 64: T_1, T_2, T_3 and T_1, T_4, T_3. Thus, we place T_1 first on the priority list. With T_1 "gone," there is a new critical path of length 60 (i.e., T_5, T_6, T_4, T_3) that starts with T_5, so T_5 is placed second on the priority list. At this stage, with T_1 and T_5 removed, we have the residual order-requirement digraph shown in Figure 3.10b. In this diagram there are paths of length 50 (T_2, T_3), 56 (T_6, T_4, T_3), 36 (T_6, T_4, T_7), and 24 (T_8, T_4, T_{10}). Since T_6 heads the path that is currently longest in length, it gets placed third in the priority list. Once T_6 is removed from Figure 3.10b, there is a tie for which is the longest path remaining, since both T_2, T_3 and T_4, T_3 are paths of length 50.

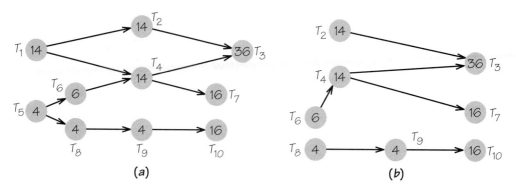

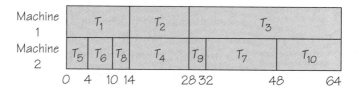

FIGURE 3.10 (a) An order-requirement digraph used to illustrate the critical-path scheduling method. (b) Residual order-requirement digraph after tasks T_1 and T_5 have been removed.

When there is a tie between two longest paths, we place next on the priority list the lowest-numbered task heading a longest path. In the example shown here, this means that T_2 is placed next into the priority list, to be followed by T_4. Continuing in this fashion, we obtain the priority list T_1, T_5, T_6, T_2, T_4, T_3, T_8, T_9, T_7, T_{10}. Note that the order of T_7 and T_{10} was decided using the rule for breaking ties. The list-processing algorithm is now applied using this priority list and the order-requirement digraph in Figure 3.10a. We obtain the schedule in Figure 3.11. ◆

FIGURE 3.11 The optimal schedule produced by applying the critical-path scheduling method to the order-requirement digraph in Figure 3.10. The list used was T_1, T_5, T_6, T_2, T_4, T_3, T_8, T_9, T_7, T_{10}.

This example illustrates what is called critical-path scheduling. The algorithm for critical-path scheduling can be described for use with the order-requirement digraph defining any particular scheduling problem. The algorithm applies the list-processing algorithm using the priority list L obtained as follows:

1. Find a task that heads a critical (longest) path in the order-requirement digraph. If there is a tie, choose the task with the lower number.
2. Place the task found in step 1 next on the list L. (The first time through the process this task will head the list.)
3. Remove the task found in step 1 and the edges attached to it from the current order-requirement digraph, obtaining a new (modified) order-requirement digraph.
4. If there are no vertices left in the new order-requirement digraph, the procedure is complete; if there are vertices left, go to step 1.

This procedure will terminate when all of the tasks in the original order-requirement digraph have been placed on the list L.

The example above shows that critical-path scheduling can sometimes yield optimal solutions. Unfortunately, this algorithm does not always perform well. For example, the critical-path method employing four processors applied to the order-requirement digraph shown in Figure 3.12 yields the list T_1, T_8, T_9, T_{10}, T_{11}, T_5, T_6, T_7, T_{12}, T_2, T_3, T_4 and then the schedule in Figure 3.13. (Note that T_5, T_6, T_7 are thought of as heading paths of length 10.) In fact, there can be no worse schedule than this one. An optimal schedule is shown in Figure 3.14.

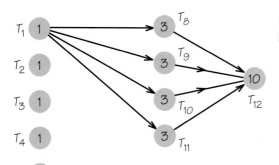

FIGURE 3.12 An order-requirement digraph used to illustrate how poorly the critical-path scheduling method can sometimes behave.

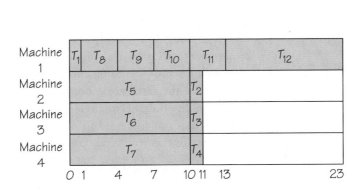

FIGURE 3.13 The schedule produced by applying the critical-path scheduling method to the order-requirement digraph in Figure 3.12 using four processors. The list used was T_1, T_8, T_9, T_{10}, T_{11}, T_5, T_6, T_7, T_{12}, T_2, T_3, T_4.

FIGURE 3.14 An optimal schedule for the order-requirement digraph in Figure 3.12 using four processors.

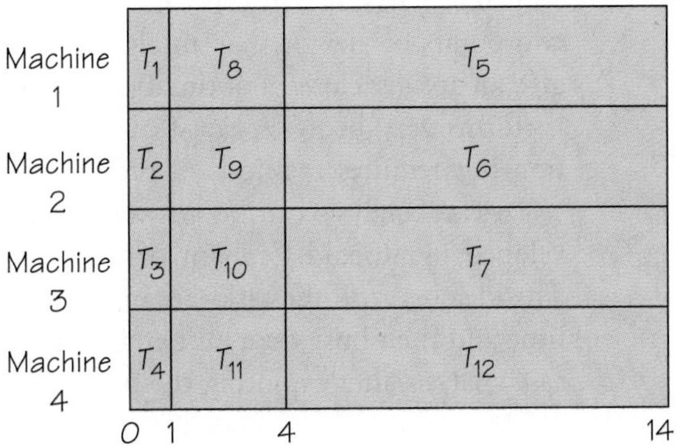

Many of the results we have examined so far are negative because we are dealing with a general class of problems that defy our using computationally efficient algorithms to find an optimal schedule. But we can close on a more positive note. Consider an arbitrary order-requirement digraph, but assume all the tasks take equal time. It turns out that we can always construct an optimal schedule using two processors in this situation. Ironically, we can choose among many algorithms to produce these optimal schedules. The algorithms are easy to understand (though not easy to prove optimal) and have all been discovered since 1969! Many people think that mathematics is a subject that is no longer alive, and that all its ideas and methods were discovered hundreds of years ago. As we have just seen this is not true. In fact, more new mathematics has been discovered and published in the last 30 years than during any previous 30-year period.

INDEPENDENT TASKS

Mathematicians suspect that no computationally efficient algorithm for solving general scheduling problems optimally will ever be found. Due to our limited success in designing algorithms for finding optimal schedules for general order-requirement digraphs, we will consider a special class of scheduling problems for which the order-requirement digraph has no edges with arrows. In this case we say that the tasks are independent of each other, since they can be performed in any order. (No edges with arrows in the order-requirement digraph indicates that no tasks need to precede others; that is, the tasks can be done in any order.) In this section we consider the problem of scheduling **independent tasks.**

There are two approaches we can consider. To study **average-case analysis** we might ask if the average (mean) of the completion times arrived at by using the list-processing algorithm with all the possible different lists is close to the optimal possible completion time. To study **worst-case analysis** we can ask how far from optimal a schedule obtained using the list-processing algorithm with one particular priority list can be. What is being contrasted with these two points of view is that an algorithm may work well most of the time (i.e., give an answer close to optimal) even though there may be a few cases where it performs very badly. Average-case analysis is amenable to mathematical solution but requires methods of great sophistication. For independent tasks, the worst-case analysis can be answered using a surprisingly simple argument developed by Ronald Graham of AT&T Bell Laboratories (see Spotlight 3.1). The idea is that if the tasks are independent, no processor can be idle at a given time and then busy on a task at a later time. We will return to Graham's worst-case analysis after exploring the problem of independent tasks in more detail.

FIGURE 3.15 (a) A nonoptimal way to schedule independent tasks of time lengths 10, 4, 5, 9, 7, 7 using two processors. (b) An optimal way to schedule independent tasks of time lengths 10, 4, 5, 9, 7, 7 using two processors.

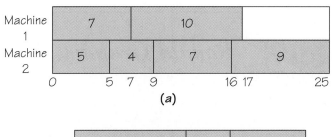

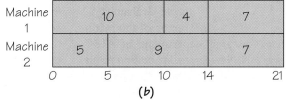

Geometrically, we can think of the independent tasks as rectangles of height 1 whose lengths are equal to the time length of the task. Finding an optimal schedule amounts to packing the task rectangles into a longer rectangle whose height equals the number of machines. For example, Figure 3.15 shows two different ways to schedule tasks of length 10, 4, 5, 9, 7, 7 on two machines. (For convenience, the rectangles in the case of independent tasks are labeled with their task times rather than their task numbers.) Scheduling basically means efficiently packing the task rectangles into the machine rectangle. Finding the optimal answer among all possible ways to pack these rectangles is like looking for a needle in a haystack. The list-processing algorithm produces a packing, but it may not be a good one.

What Graham's worst-case analysis showed for independent tasks is that no matter which list L one uses, if the optimal schedule requires time T, then the completion time for the schedule produced by the list-processing algorithm applied to list L with m processors is less than or equal to $(2 - 1/m)T$. For example, for two machines ($m = 2$), if an optimal schedule yields completion at time 30, then no list would ever yield a completion later than

$$(2 - \tfrac{1}{2})(30) = 45$$

Although it is of great theoretical interest, Graham's result does not provide much comfort to those who are trying to find good schedules for independent tasks.

Decreasing-Time Lists

Is there some way of choosing a priority list that consistently yields relatively good schedules? The surprising answer is yes! The idea is that when long tasks appear toward the end of the list, they often seem to "stick out" on the right end, as in Figure 3.15a.

Spotlight
3.1

Ronald Graham on Mathematics and Mathematicians

Ronald Graham is director of the mathematics sciences research department of AT&T Bell Labs, Murray Hill, New Jersey. In addition to his own research, he supervises some of the country's most distinguished mathematicians as they investigate the puzzles of management science. Here are some of his comments:

On Scheduling

Scheduling is a very interesting area. You have some number of processors that you can think of as computers or as people, and you have a number of tasks or jobs that you would like to get done. One might think that adding more processors to the system would guarantee that you would be done sooner. In fact, just the opposite can happen. It's an example of the old adage, "Too many cooks can spoil the broth." You can add more machines or decrease the amount of time it takes for each job, or relax the constraints between the various jobs, but all of these can actually end up taking longer. To try to understand why that is and how to avoid it is the arena of mathematics.

One of the earliest applications of scheduling came in looking at the Apollo moon shot. In that case, there were three different processors, namely, the three astronauts, who had a large collection of tasks that they were supposed to do—various experiments to run on the way to and from the moon. And of course, there were various constraints. You want the astronauts to sleep a little every day. Other constraints you might not think of. For example, you have to rotate the capsule so that the same side isn't facing the sun all the time, because of heating problems. Then, within these constraints, you would next like to know: If you sequence the experiments in different orders, how much time could you save over sequencing them in orders that aren't so good?

It turns out that in fact NASA had developed very good sequencing already. All I could show them was that they couldn't improve on it very much, no matter what they might try. And of course this is one of the problems in management science where it's just impossible to enumerate the possibilities. So that's where the mathematics comes in: it enables you to say that no matter what you do, you'll never be able to "do any better than this" or "do it any faster."

You might well ask, how can you know about every one of the possibilities even though you didn't try them all? That's an interesting question, and it really goes to the root of how mathematics is used to analyze the real world. There are several well-known steps in going from a real-world problem to a model of the problem—a model in which you try to capture the essential aspects of the problem, but in a way that can be dealt with mathematically. Once you convert the problem into the world of mathematics, you can say something about all the mathematical possibilities. Then you translate it back to the real-world situation. How well it works depends crucially on how accurate the translation of the model was.

Now it's always good to check a few of the things that you've predicted are going to happen. If the translation is good and you've captured the essence of the problem, then there is a good chance that you can say something sensible about the thing you've studied. If it isn't, you can get some pretty bizarre conclusions. There are famous examples of this. One that comes to mind is the case where the people who first analyzed how bees fly were able to prove mathematically that bees couldn't fly. That didn't bother the bees, of course, and the model was eventually modified to show that bees could fly after all.

On the Spirit of Investigation

I think in order to do the best work in any area, you have to enjoy it and be intrigued by it. You have to wonder why this is happening. There are many examples where something slightly out of the ordinary occurs and 99 people notice it yet go on to something else. But one in a hundred—or fewer—is fascinated by that slightly anomalous or even bizarre behavior in one mathematical area or another. They start probing it more deeply and soon find that it's really the tip of a giant iceberg. Once you start to melt it, you have a much deeper understanding about what is really going on. There are those who feel mathematics is a giant game, albeit one that is amazingly relevant to everything that is going on in the scientific world.

Often, half the battle in trying to solve a difficult problem is knowing the right question to ask. If you know what it is you are trying to look for, you can often be very far along in finding the solution. It's useful for me, and for many others who are working on a difficult problem, to look at a special case first. You can work your way up to the full problem by trying easier special cases that you still can't do. You may be bouncing the idea around a bit and not forcing it, and it's amazing how often it happens that the next day or next week something will seem obvious that was very hard to imagine even a week earlier.

On Managing the Mathematicians

One of the rules or axioms that we have here [at AT&T Labs] is that each person is best equipped to know how he or she functions optimally. Some people are night people, for example, and they just don't function before 12 o'clock.

In our mathematics group, there are roughly 65 professionals, and quite a few are leading-edge, world-class researchers. People of this caliber you don't so much manage as try to keep up with what they are doing. In many cases, you act as a sounding board, because one of the most useful things you can do with a colleague who has some mathematical ideas is to act as a sympathetic listener. As the person explains the ideas or where he is stuck, he will more often than not gain a much clearer insight into what he's doing and where he's going.

Many times there is much more interactive collaboration. There is quite a lot of joint work that goes on here, not just from within the mathematics area, but among mathematicians, computer scientists, physicists, and chemists. It is a very interactive environment which, I think, works to everyone's benefit.

This suggests that before one tries to schedule a collection of tasks, the tasks should be placed in a list so that longest tasks are listed first. The list-processing algorithm applied to a list arranged in this fashion is called the **decreasing-time-list algorithm.**

If we apply it to the set of tasks listed previously (10, 4, 5, 9, 7, 7), we obtain the times 10, 9, 7, 7, 5, 4 and the schedule (packing) shown in Figure 3.16. This packing is again optimal, but it is different from the optimal scheduling in Figure 3.15b. It is worth noting that the decreasing-time list and the list obtained by the critical-path method discussed earlier will coincide in the case of independent tasks. The decreasing-time list can also be constructed for the case where the tasks are not independent, but for general order-requirement digraphs, the decreasing-time list does not produce particularly good schedules.

It is important to remember that the decreasing-time-list algorithm *does not guarantee* optimal solutions. This can be seen by scheduling the tasks with times 11, 10, 9, 6, 4 (Figure 3.17). The schedule has completion time 21. However, the rearranged list 9, 4, 6, 11, 10 yields the schedule in Figure 3.18, which finishes at time 20. This solution is obviously optimal, since the machines finish at the same time and there is no idle time. Note that when tasks are independent, if there are *m* machines available, the completion time cannot be less than the sum of the task times divided by *m*.

The problems we have encountered in scheduling independent tasks seem to have taken us a bit far from our goal of applying the mathematics we have developed. Sometimes mathematicians will pursue their mathematical ideas even though they have reached a point where there appear to be no applications. Fortunately, it is very common to be able to find applications for the "abstract" extensions. This is the case in the current instance.

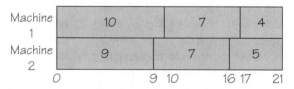

FIGURE 3.16 The optimal schedule resulting from applying the decreasing-time-list algorithm to a collection of independent tasks. The list used, written down in terms of task times only, is 10, 9, 7, 7, 5, 4.

EXAMPLE ▶ *Photocopy Shop and Typing Pool Problems*

Imagine a photocopy shop with three photocopiers. Photocopying tasks that must be completed overnight are accepted until 5 P.M. The tasks are to be done in any manner that minimizes the finish time for all the work. Because

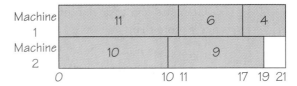

FIGURE 3.17   The nonoptimal schedule resulting from applying the decreasing-time-list algorithm to a collection of independent tasks. The list used, written down in terms of task times only, is 11, 10, 9, 6, 4.

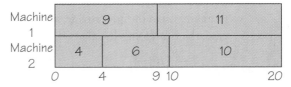

FIGURE 3.18   The optimal schedule resulting from applying the list-processing algorithm to a collection of independent tasks. The list used, written down in terms of task times only, is 9, 4, 6, 11, 10.

this problem involves scheduling machines for independent tasks, the decreasing-time-list algorithm would be a good heuristic to apply.

For another example, consider a typing pool at a large corporation or college, where individual typing tasks can be assigned to any typist. In this setting, however, the assumption that the processors (typists) are identical in skill is less likely to be true. Hence, the tasks might have different times with different processors. This phenomenon, which occurs in real-world scheduling problems, violates one of the assumptions of our mathematical model.

Typing pool at a Western Union office.

Graham's result for the list-processing algorithm—that the finishing time is never more than

$$(2 - 1/m) T$$

(where T represents optimal completion time and m the number of processors)—offers us the small comfort of knowing that even the worst choice of priority list will not yield a completion time worse than twice the optimal time. Compared with the list-processing algorithm, the decreasing-time-list algorithm seems to improve completion times. Thus, it is not surprising that Graham was able to provide an improved bound, or time estimate, for this case: the decreasing-list algorithm gives a completion time of no more than

$$[\tfrac{4}{3} - 1/(3m)] T$$

where m is the number of processors and T is the optimal time in which the tasks can be completed. In particular, when the number of processors is 2, the schedule produced by the decreasing-time-list algorithm is never off by more than 17%! Usually, the error is much less. This result is a remarkable instance of the value of mathematical research into applied problems. Note that the optimal completion time T depends on m and that Graham's theoretical analysis is necessary precisely because there is no known algorithm to compute T easily. ◆

BIN PACKING

Suppose you plan to build a wall system for your books, records, and stereo set. It requires 24 wooden shelves of various lengths: 6, 6, 5, 5, 5, 4, 4, 4, 4, 2, 2, 2, 2, 3, 3, 7, 7, 5, 5, 8, 8, 4, 4, and 5 feet. The lumberyard, however, sells wood only in boards of length 9 feet. If each board costs $8, what is the minimum cost to buy sufficient wood for this wall system?

Because all shelves required for the wall system are shorter than the boards sold at the lumberyard, the largest number of boards needed is 24, the precise number of shelves needed for the wall system. Buying 24 boards would, of course, be a waste of wood and money because several of the shelves you need could be cut from one board. For example, pieces of length 2, 2, 2, and 3 feet can be cut from one 9-foot board.

To be more efficient, we think of the boards as bins of capacity W (9 feet in this case) into which we will pack (without overlap) n weights (in this case, lengths) whose values are $w_1, \ldots, w_n$, where each $w_i \leq W$. We wish to find the minimum number of bins into which the weights can be packed. In this formulation, the problem is known as the **bin-packing problem.** Thus *bin packing* refers to finding the minimum number of bins of weight capacity W

into which weights $w_1, \ldots, w_n$ (each less than or equal to W) can be packed.

At first glance, bin-packing problems may appear unrelated to the machine-scheduling problems we have been studying; however, there is a connection.

Let's suppose we want to schedule independent tasks so that each machine working on the tasks finishes its work by time W. Instead of fixing the number of machines and trying to find the earliest completion time, we must find the minimum number of machines that will guarantee completion by the fixed completion time (W). Despite this similarity between the machine-scheduling problem and the bin-packing problem, the discussion that follows will use the traditional terminology of bin packing.

By now, it should come as no surprise to learn that no one knows a fast algorithm that always picks the optimal (smallest) number of bins (boards). In fact, the bin-packing problem belongs to the class of NP-complete problems (see Spotlight 2.1, page 47), which means that most experts think it unlikely that any fast optimal algorithm will ever be found.

Bin-Packing Heuristics

We will think of the items to be packed, in any particular order, as constituting a list. In what follows we will use the list of 24 shelf lengths given for the wall system. We will consider various **heuristic** algorithms, namely, methods that can be carried out quickly but cannot be guaranteed to produce optimal results. Probably the easiest approach is simply to put the weights into the first bin until the next weight won't fit, and then start a new bin. (Once you open a new bin, don't use leftover space in an earlier, partially filled bin.) Continue in the same way until as many bins as necessary are used. The resulting solution is shown in Figure 3.19. This algorithm, called **next fit (NF),** has the advantage of not requiring knowledge of all the weights in advance; only the remaining space in the bin currently being packed must be remembered. The disadvantage of this heuristic is that a bin packed early on may have had room for small items that come later in the list.

FIGURE 3.19 The list 6, 6, 5, 5, 5, 4, 4, 4, 4, 2, 2, 2, 2, 3, 3, 7, 7, 5, 5, 8, 8, 4, 4, 5 packed in bins using next fit.

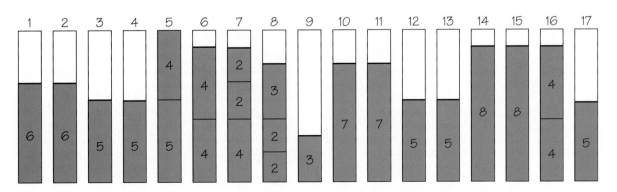

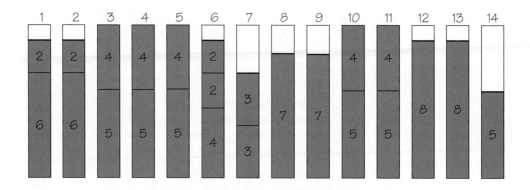

FIGURE 3.20   The list 6, 6, 5, 5, 5, 4, 4, 4, 4, 2, 2, 2, 2, 3, 3, 7, 7, 5, 5, 8, 8, 4, 4, 5 packed in bins using first fit. Worst fit would yield a packing that would look identical.

Our wish to avoid permanently closing a bin too early suggests a different heuristic—**first fit (FF):** put the next weight into the first bin already opened that has room for this weight; if no such bin exists, start a new bin. Note that a computer program to carry out first fit would have to keep track of how much room was left in all the previously opened bins. For the 24 wall-system shelves the first-fit algorithm would generate a solution that uses only 14 bins (see Figure 3.20) instead of the 17 bins generated by the next-fit algorithm.

If we are keeping track of how much room remains in each unfilled bin, we can put the next item to be packed into the bin that currently has the most room available. This heuristic will be called **worst fit (WF).** The name worst fit refers to the fact that an item is packed into a bin with the most room available, that is, into which it fits "worst," rather than into a bin that will leave little room left over after it is placed in that bin (i.e., "best fit"). The solution generated by this approach looks the same as that shown in Figure 3.20. Although this heuristic also leads to 14 bins, the items are packed in a different order. For example, the first item of size 2, the tenth item in the list, is put into bin 6 in worst fit, but into bin 1 in first fit.

Decreasing-Time Heuristics

One difficulty with all three of these heuristics is that large weights that appear late in the list can't be packed efficiently. Therefore, we should first sort the items to be packed in order of decreasing size, assuming that all items are known in advance. We can then pack large items first and then the smaller items into leftover spaces. This approach yields three new heuristics: **next-fit decreasing (NFD), first-fit decreasing (FFD),** and **worst-fit decreasing (WFD).** Here is the original list sorted by decreasing size: 8, 8, 7, 7, 6, 6, 5, 5, 5, 5, 5, 5, 4, 4, 4, 4, 4, 4, 3, 3, 2, 2, 2, 2. Packing using first-fit-decreasing order yields the solution in Figure 3.21. This solution uses only 13 bins.

Is there any packing that uses only 12 bins? No. In Figure 3.21, there are only 2 free units (1 unit each in bins 1 and 2) of space in the first 12 bins, but 4 occupied units (two 2s) in bin 13. We could have predicted this by dividing

FIGURE 3.21 The bin packing resulting from applying first-fit decreasing to the wall-system numbers. The list involved, which uses the original list sorted in decreasing order, is 8, 8, 7, 7, 6, 6, 5, 5, 5, 5, 5, 5, 4, 4, 4, 4, 4, 3, 3, 2, 2, 2, 2.

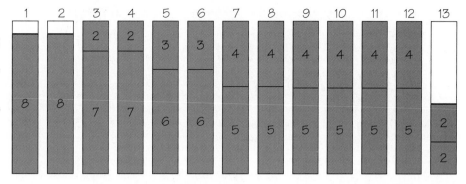

the total length of the shelves (110) by the capacity of each bin (board): $\frac{110}{9} = 12\frac{2}{9}$. Thus, no packing could squeeze these shelves into 12 bins—there would always be at least 2 units left over for the 13th bin. (In Figure 3.21, there are 4 units in bin 13 because of the 2 wasted empty spaces in bins 1 and 2.) Even if this division had created a zero remainder, there would still be no guarantee that the items could be packed to fill each bin without wasted space. For example, if the bin capacity is 10 and there are weights of 6, 6, 6, 6, and 6, the total weight is 30; dividing by 10, we get 3 bins as the minimum requirement. Clearly, however, 5 bins are needed to pack the five 6s.

None of the six heuristic methods shown will necessarily find the optimal number of bins for an arbitrary problem. How can we decide which heuristic to use? One approach is to see how far from the optimal solution each method might stray. Various formulas have been discovered to calculate the maximum discrepancy between what a bin-packing algorithm actually produces and the best possible result. For example, in situations where a large number of bins are to be packed, first fit (FF) can be off as much as 70%, but first-fit decreasing (FFD) is never off by more than 22%. Of course, FFD doesn't give an answer as quickly as FF, because extra time for sorting a large collection of weights may be considerable. Also, FFD requires knowing the whole list of weights in advance, whereas FF does not. It is important to emphasize that 22% is a worst-case figure. In many cases, FFD will perform much better. Recent results obtained by computer simulation indicate excellent average-case performance for this algorithm.

When modeling real-world problems, we always have to look at the relationship between mathematics and the real world. Thus, first-fit decreasing usually results in fewer bins than next fit, but next fit can be used even when all the weights are not known in advance. Next fit also requires much less computer storage than first fit, because once a bin is packed, it need never be looked at again. Fine-tuning of the conditions of the actual problem often results in better practical solutions and in interesting new mathematics as well. (See Spotlight 3.2 for a discussion of some of the tools mathematicians use to verify and even extend mathematical truths.)

Spotlight 3.2 — Using Mathematical Tools

No Euler circuit
Nonconnected
Even-valent

No Euler circuit
Connected
Not even-valent

The tools of a carpenter include the saw, T square, level, and hammer. A mathematician also requires tools of the trade. Some of these tools are the proof techniques that enable verification of mathematical truths. Another set of tools consists of strategies to sharpen or extend the mathematical truths already known. For example, suppose that if A and B hold, then C is true. What happens if only A holds? Will C still be true? Similarly, if only B holds, will C still be true?

This type of thinking is of value because such questions will result either in more general cases where C holds or in examples showing that B alone and/or A alone can't imply C. For example, we saw that if a graph G is connected (hypothesis A) and even-valent (hypothesis B), then G has a tour of its edges using each edge only once (conclusion C). If either hypothesis is omitted, the conclusion fails to hold. The figures illustrate this point. On the left is an even-valent but nonconnected graph, and on the right, a connected graph with two odd-valent vertices; neither graph has an Euler circuit.

Here is another way that a mathematician might approach extending mathematical knowl-

edge. If A and B imply C, will A and B imply both C and D, where D extends the conclusion of C? For example, not only can we prove that a connected, even-valent (hypotheses A and B) graph has an Euler circuit, but we can also show that the first edge of the Euler circuit can be chosen arbitrarily (conclusions C and D). It turns out that being able to specify the first two edges of the Euler circuit may not always be possible. Mathematicians are trained to vary the hypotheses and conclusions of results they prove, in an attempt to clarify and sharpen the range of applicability of the results.

We have seen that machine scheduling and bin packing are probably computationally difficult to solve because they are NP-complete. A mathematician could then try to find the simplest version of a bin-packing problem that would still be NP-complete: What if the items to be packed can have only eight weights? What if the weights are only one and two? Asking questions like these is part of the mathematician's craft. Such questions help to extend the domain of mathematics and hence the applications of mathematics.

CRYPTOGRAPHY

Not so many years ago secret codes existed primarily for the purpose of allowing diplomats and military personnel to exchange information without their messages being read by people from other countries. Due to changes in telecommunications, banking, and lifestyles, secret codes are now widely used to protect computer files, electronic transfers of funds, and electronic mail.

We conclude with a variant of the bin-packing problem that is of interest to those who make and break secret codes. Given numbers (weights) $w_1, \ldots, w_n$ and a number W, find an algorithm that selects a *subcollection* of the weights whose sum is exactly W. This problem, called the **subset sum problem,** is known to be NP-complete. It is related to the **knapsack problem** where we find the largest collection of weights $w_1, \cdots, w_n$ that will fit in a knapsack of capacity W. As tame as these problems sound, they are related to revolutionary new ideas for ensuring the security of data stored in computers, such as bank records and classified military information, and for credit card transactions on the Internet.

The security of data has become an important new branch of *cryptography,* the study of codes and ciphers and how to break them. Traditional cryptography is based on a single key that is used to transform or encrypt a message (plaintext) into garbled form (ciphertext). The same key used to encrypt the message is used to decipher the message. However, in the new scheme, called *public-key cryptography,* there are two keys. One key, made public, is used by anyone wishing to send a secure message to X. The other key is known only to X, the receiver of the message. The keys are designed so that knowledge of the public key does not compromise the private key.

The idea behind public-key cryptography is that certain processes are very easy to carry out in one direction, but very hard to reverse. For example, one can easily add $708 + 259 + 871 + 1836 + 82$ to get 3756, but it is not so easy to find a subcollection of the numbers 1086, 708, 82, 259, 589, 871, and 1836 that add to $W = 3756$ (i.e., subset sum problem). As another example, multiply 2993 by 1362 using pencil and paper or a calculator. Now try to find two numbers whose product is 2,414,203. In case you ran out of patience, the answer is 1111×2173. If you carried out these two problems, you might guess that factoring is "harder" than multiplying.

Because knapsack problems are thought to be computationally hard to solve, schemes have been suggested for applying these ideas to public-key systems. Although recent work has shown that knapsack systems are faulty as a base for public-key systems, other public-key systems based on factoring large numbers appear to be viable. Furthermore, the idea of NP-completeness (see Spotlight 2.1, page 47), which arose as a concept in the abstract analysis of the complexity of algorithms, turned out to be related to the mundane issue of how to protect money being transferred between banks and how to guarantee that a photograph entered into evidence in a court hearing has not been tampered with! We can see once more how the use of clever (though not necessarily impenetrable) mathematics can affect and enrich our lives. If you are interested in learning more about cryptography and codes used for purposes other than maintaining secrecy, you will find more information about this important subject in Chapter 10.

REVIEW VOCABULARY

Average-case analysis   The study of the list-processing algorithm (more generally, any algorithm) from the point of view of how well it performs on all the types of problems it may be used on and seeing on average how well it does. See also worst-case analysis.

Bin-packing problem   The problem of determining the minimum number of containers of capacity W into which objects of size $w_1, \ldots, w_n (w_i \leq W)$ can be packed.

Critical-path scheduling   A heuristic algorithm for solving scheduling problems where the list-processing algorithm is applied to the priority list obtained by listing next in the priority list a task that heads a longest path in the order-requirement digraph. This task is then deleted from the order-requirement digraph, and the next task placed in the priority list is obtained by repeating the process.

Cryptography   The study of how to make and break codes. These codes are now used primarily for data and computer security rather than for national security.

Decreasing-time-list algorithm   The heuristic algorithm that applies the list-processing algorithm to the priority list obtained by listing the tasks in decreasing order of their time length.

First fit (FF)   A heuristic algorithm for bin packing in which the next weight to be packed is placed in the lowest-numbered bin already opened into which it will fit. If it fits in no open bin, a new bin is opened.

First-fit decreasing (FFD)   A heuristic algorithm for bin packing where the first-fit algorithm is applied to the list of weights sorted so that they appear in decreasing order.

Heuristic algorithm   An algorithm that is fast to carry out but that doesn't necessarily give an optimal solution to an optimization problem.

Independent tasks   Tasks are independent when there are no edges in the order-requirement digraph.

Knapsack problem   Given a knapsack of size W and a collection of weights $w_1, \ldots, w_n$, find a largest collection of weights that will fit in the knapsack.

List-processing algorithm   A heuristic algorithm for assigning tasks to processors: assign the first ready task on the priority list that has not already been assigned to the lowest-numbered processor that is not working on a task.

Machine scheduling   The problem of assigning tasks to processors so as to complete the tasks by the earliest time possible.

Next fit (NF)   A heuristic algorithm for bin packing in which a new bin is opened if the weight to be packed next will not fit in the bin that is currently being filled; the current bin is then closed.

Next-fit decreasing (NFD)   A heuristic algorithm for bin packing where the next-fit algorithm is applied to the list of weights sorted so that they appear in decreasing order.

Priority list   An ordering of the collection of tasks to be scheduled for the purpose of attaining a particular scheduling goal. One such goal is minimizing completion time when the list algorithm is applied.

Processor   A person, machine, robot, operating room, or runway whose time must be scheduled.

Ready task   A task is called ready at a particular time if its predecessors as given by the order-requirement digraph have been completed by that time.

Subset sum problem   Given an integer W and integer numbers $w_1, \ldots, w_n$, find a subcollection of the numbers whose sum is W.

Worst-case analysis   The study of the list-processing algorithm (more generally, any algorithm) from the point of view of how well it performs on the hardest problems it may be used on. See also average-case analysis.

Worst fit (WF)   A heuristic algorithm for bin packing in which the next weight to be packed in placed into the open bin with the largest amount of room remaining. If the weight fits in no open bin, a new bin is opened.

Worst-fit decreasing (WFD)   A heuristic algorithm for bin packing where the worst-fit algorithm is applied to the list of weights sorted so that they appear in decreasing order.

SUGGESTED READINGS

BRUCKER, P. *Scheduling Algorithms,* Springer-Verlag, Heidelberg, Germany, 1995. A detailed mathematical look at scheduling.

DEMILLO, R., ET AL., EDS. *Applied Cryptology,* American Mathematical Society, Providence, R.I., 1983. A relatively recent survey of research work.

FRENCH, SIMON. *Sequencing and Scheduling,* Wiley, New York, 1982. A detailed account of a wide variety of scheduling models, most of them different from the ones treated in this chapter.

GRAHAM, RONALD. Combinatorial scheduling theory. In Lynn Steen (ed.), *Mathematics Today,* Springer-Verlag, New York, 1978. This essay on scheduling is one of many excellent accounts of recent developments in mathematics in this book.

GRAHAM, RONALD. The combinatorial mathematics of scheduling. *Scientific American,* March 1978, pp. 124–132. A very readable introduction to scheduling and bin packing.

LAWLER, E., ET AL. Sequencing and scheduling algorithms and complexity. In S. C. Graves et al. (eds.), *Handbooks in OR and MS,* vol. 4, pp. 445–522, Elsevier, New York, 1993. A recent survey of results about scheduling.

EXERCISES ▲ *Optional.* ▆ *Advanced.* ◆ *Discussion.*

Scheduling

◆   1.   In order to get to a ski resort for a weekend vacation Jocelyn must accomplish a variety of things. She will leave work early at 1 P.M. and must get to the airport to be on a 5 o'clock shuttle to Boston. She then hopes to take a bus to get to the resort. Discuss some of the tasks that must be accomplished to get Jocelyn to the resort by 10 P.M. What are the different types of processors that are involved in getting these tasks done? Can any of these tasks be done simultaneously?

2.   For the situation where a family with three children is preparing a Thanksgiving meal for 10 guests, list tasks that must be completed and the types of processors that are involved. Can any of these tasks be done simultaneously?

3.   List as many scheduling situations as you can for these environments:

(a)	Hospital	(e)	Machine shop
(b)	Railroad station	(f)	Your home
(c)	Airport	(g)	Your school
(d)	Automobile repair garage	(h)	Police station
		(i)	Firehouse

Compare and contrast the scheduling issues involved in these situations.

◆   4.   Discuss scheduling problems for which it is not reasonable to assume that once a processor starts a task, it would always complete that task before it works on any other task. Give examples for which this approach would be reasonable.

Using the List-Processing Algorithm

5.   (a)   Use the list-processing algorithm to schedule the tasks in the following order-requirement digraph on three processors, using the list $T_1, \ldots, T_{11}$. From the schedule so constructed, for each task list the start and finish time for that task.

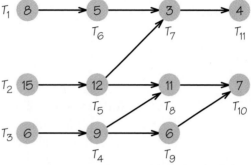

(b)   Compare your answers in part (a) with scheduling these tasks on two processors with the same list.

6.   For the accompanying order-requirement digraph, apply the list-processing algorithm, using three processors for lists (a) and (b). How do the completion times obtained compare with the length of the critical path?

(a) $T_1, T_2, T_3, T_4, T_5, T_6, T_7, T_8$
(b) $T_1, T_3, T_5, T_7, T_2, T_4, T_6, T_8$

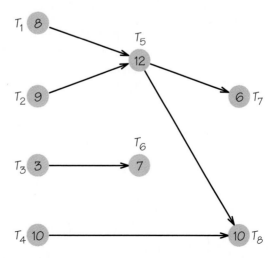

7. Consider the following order-requirement digraph:

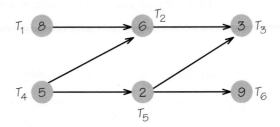

(a) Find the critical path(s).

(b) Schedule these tasks on one processor using the critical-path scheduling method.

(c) Schedule these tasks on one processor using the priority list obtained by listing the tasks in order of decreasing time.

(d) Do either of these schedules have idle time? How do their completion times compare?

(e) If two different schedules have the same completion time, what criteria can be used to say one schedule is superior to the other?

(f) Schedule these tasks on two processors using the order-requirement digraph shown and the priority list from part (b).

(g) Does the schedule produced in part (f) finish in half the time that the schedule in part (b) did, which might be expected, since the number of processors has doubled?

(h) Schedule the tasks on (i) one processor and (ii) two processors (using the decreasing-time list), assuming that each task time has been reduced by one. Do the changes in completion time agree with your expectations?

8. To prepare a meal quickly involves carrying out the tasks shown (time lengths in minutes) in the following order-requirement digraph:

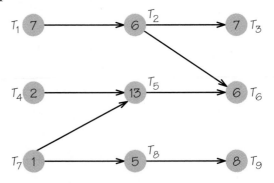

(a) If Mike prepares the meal alone, how long will it take?

(b) If Mike can talk Mary into helping him prepare the meal, how long will it take if the tasks are scheduled using the list T_5, T_9, T_1, T_3, T_2, T_6, T_8, T_4, T_7 and the list-processing algorithm?

(c) If Mike can talk Mary and Jack into helping him prepare the meal, how long will it take if the tasks are scheduled using the same list as in part (b)?

(d) What would be a reasonable set of criteria for choosing a priority list in this situation?

◆   9. Discuss different criteria that might be used to construct a priority list for a scheduling problem.

10. If two schedules have the same completion time, can one schedule have more idle time than the other?

Independent Tasks and Other Issues

11. For the following schedules, can you produce a list so that the list-processing algorithm produces the schedule shown when the tasks are independent? What are the task times for each task?

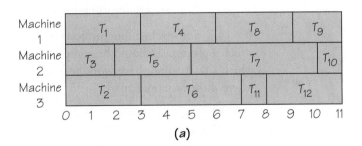

(a)

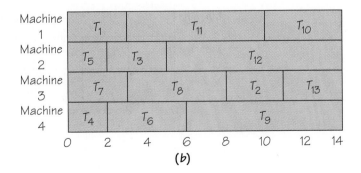

(b)

◆   12.   Once an optimal schedule has been found for independent tasks (e.g., see diagrams in Exercise 11), usually the scheduling of the tasks can be rearranged and the same optimal time achieved (i.e., one can, among other things, reorder the tasks done by a particular processor). Discuss criteria that might be used in implementing the rearrangement process.

13.   At a large toy store, scooters arrive unassembled in boxes. To assemble a scooter the following tasks must be performed:

TASK 1.   Remove parts from the box.
TASK 2.   Attach wheels to the footboard.
TASK 3.   Attach vertical housing.
TASK 4.   Attach handlebars to vertical housing.
TASK 5.   Put on reflector tape.
TASK 6.   Attach bell to handlebars.
TASK 7.   Attach decals.
TASK 8.   Attach kickstand.
TASK 9.   Attach safety instructions to handlebars.

(a)   Give reasonable time estimates for these tasks and construct a reasonable order-requirement digraph. What is the earliest time by which these tasks can be completed?
(b)   Schedule this job on two processors (humans) using the decreasing-time-list algorithm.

◆   14.   Some scheduling projects have due dates for tasks (i.e., times by which a given task should be completed and release dates (i.e., times before which a task cannot have work begun on it). Give examples of circumstances where these situations might arise.

15.   Can you find a schedule (use the order-requirement digraph in Figure 3.4) with a completion time earlier than 12 as shown in the following figure, using a list other than $T_1, \ldots, T_9$? If not, why not?

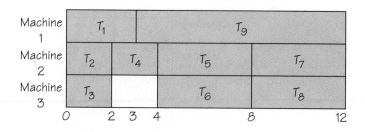

16. Given the accompanying order-requirement digraph:

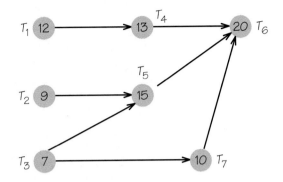

   (a) Use the list-processing algorithm to schedule these seven tasks on two processors using these lists:
      (i) T_1, T_3, T_7, T_2, T_4, T_5, T_6
      (ii) T_1, T_3, T_2, T_4, T_5, T_6, T_7
      (iii) The list obtained by listing the tasks in order of decreasing time
   (b) Try to determine if any of the resulting schedules are optimal.
   (c) Schedule the tasks using the critical-path scheduling method. Try to determine if this schedule is optimal.

17. Describe possible modifications for the list-processing algorithm that would allow for a

   (a) machine to be "voluntarily" idle.
   (b) machine to interrupt work on a task once it has begun work on the task.

18. (a) Find the completion time for independent tasks of length 8, 11, 17, 14, 16, 9, 2, 1, 18, 5, 3, 7, 6, 2, 1 on three processors, using the list-processing algorithm.
    (b) Find the completion time for the tasks in part (a) on three processors, using the decreasing-time-list algorithm.
    (c) Does either algorithm give rise to an optimal schedule?
    (d) Repeat for tasks of lengths 19, 19, 20, 20, 1, 1, 2, 2, 3, 3, 5, 5, 11, 11, 17, 18, 18, 17, 2, 16, 16, 2.

19. A photocopy shop must schedule independent batches of documents to be copied. The times for the different sets of documents are (in minutes): 12, 23, 32, 13, 24, 45, 23, 23, 14, 21, 34, 53, 18, 63, 47, 25, 74, 23, 43, 43, 16, 16, 76.

    (a) Construct a schedule using the list-processing algorithm on three machines.

    (b) Construct a schedule using the list-processing algorithm on four machines.

    (c) Repeat parts (a) and (b), but use the decreasing-time-list algorithm.

    (d) Suppose union regulations require that an 8-minute rest period be allowed for any photocopy task over 45 minutes. Use the decreasing-time-list algorithm, with the preceding times modified to take into account the union requirement, to schedule the tasks on three human-operated machines.

20. Find a list that produces the following optimal schedule when the list-processing algorithm is applied to this list. (Assume the tasks are independent.)

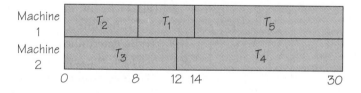

What completion time and schedule are obtained when the decreasing-time-list algorithm is applied to this list?

21. Can you think of situations other than those mentioned in the text where scheduling independent tasks on processors occurs?

22. Can you think of real-world scheduling situations in which all the tasks have the same time and are independent? Can you find an algorithm for solving this problem optimally? (If there are n independent tasks of time length k, when will all the tasks be finished?)

23. (a) Show that when tasks to be scheduled are independent, the critical-path method and the decreasing-time-list method are identical.

    (b) The (usually unknown) optimal time to complete a specific collection of independent tasks on three machines turns out to be 450 minutes.

(i) Estimate the worst possible completion time when the list-processing algorithm is used with the worst choice of priority list.

(ii) Estimate the longest possible completion time using the list-processing algorithm and the decreasing-time list.

Bin Packing

24. A radio station's policy allows advertising breaks of no longer than 2 minutes, 15 seconds. Using algorithms (a) and (b) below, determine the minimum number of breaks into which the following ads will fit (lengths given in seconds): 80, 90, 130, 50, 60, 20, 90, 30, 30, 40. Can you find the optimum solution? Do the same for these ads: 60, 50, 40, 40, 60, 90, 90, 50, 20, 30, 30, 50.

(a) First fit

(b) First-fit decreasing

25. It takes 4 seconds to photocopy one page. Manuscripts of lengths 10, 8, 15, 24, 22, 24, 20, 14, 19, 12, 16, 30, 15, and 16 pages are to be photocopied. How many photocopy machines would be required, using the first-fit-decreasing algorithm, to guarantee that all manuscripts are photocopied in 2 minutes or less? Would the solution differ if worst-fit decreasing were used?

26. Two wooden wall systems are to be made with pieces of wood with lengths shown in the accompanying diagram. If wood is sold in 10-foot planks and can be cut with no waste, what number of boards would be purchased if one uses the first-fit-decreasing, next-fit-decreasing, and worst-fit-decreasing heuristics, respectively?

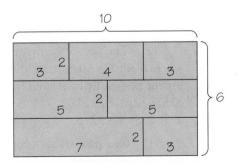

In solving this problem, does it make a difference if the 10-foot horizontal shelves and 6-foot vertical boards employ single-length pieces as compared with using pieces of boards that add up to 10- and 6-foot lengths?

27.   Fiberglass insulation comes in 36-inch precut sections. A plumber must install insulation in a basement on piping that is interrupted often by joints. The distances between the joints on the stretches of pipe that must be insulated are 12, 15, 16, 12, 9, 11, 15, 17, 12, 14, 17, 18, 19, 21, 31, 7, 21, 9, 23, 24, 15, 16, 12, 9, 8, 27, 22, 18 inches. How many precut sections would he have to use to provide the insulation if he bases his decision on

   (a)   next fit?
   (b)   next-fit decreasing?
   (c)   worst fit?
   (d)   worst-fit decreasing?

28.   The files that a company has for its employees dealing with utilities occupy 100, 120, 60, 90, 110, 45, 30, 70, 60, 50, 40, 25, 65, 25, 55, 35, 45, 60, 75, 30, 120, 100, 60, 90, 85 sectors. If, after operating systems are installed, a disk can store up to 480 sectors, determine the number of disks to store the utilities if each of these heuristics is used to pack the disk with files:

   (a)   NF
   (b)   NFD
   (c)   FF
   (d)   FFD

◆   29.   We have described two algorithms for bin packing called "worst fit" and "best fit" (see page 104). The words "best" and "worst" have connotations in English. However, the performance of algorithms depends on their merits as algorithms, not on the names we given them.

   (a)   On the basis of experiments you perform with the best-fit and worst-fit algorithms, which one do you think is the "better" of the two?
   (b)   Can you construct an example where worst fit uses fewer bins than best fit?

30.   The best-fit heuristic (see page 104) also has a "decreasing" version, where the list is first sorted in decreasing order. Using bins of capacity 10, apply the best-fit heuristic and its decreasing version to the following list: 6, 9, 5, 8, 3, 2, 1, 9, 2, 7, 2, 5, 4, 3, 7, 6, 2, 8, 3, 7, 1, 6, 4, 2, 5, 3, 7, 2, 5, 2, 3, 6, 2, 7, 1, 3, 5, 4, 2, 6.

▪   31.   One pianist's recording of the complete Mozart piano sonatas takes the following times (given in minutes and seconds): 13:46, 6:15, 3:29, 5:37, 7:52, 2:55, 5:00, 4:28, 4:21, 7:39, 7:55, 6:42, 4:23, 3:52, 4:21, 4:20, 5:46, 6:29, 5:34, 6:23, 6:39, 7:19, 5:54, 6:54, 2:58, 5:22, 1:42, 5:00, 1:29, 5:47, 7:30, 8:19, 4:44, 4:57, 4:09, 14:31, 3:55, 4:04, 4:01, 6:06, 6:50, 5:27, 4:28,

5:40, 2:52, 5:16, 5:34, 3:10, 7:22, 4:40, 3:08, 6:32, 4:47, 6:59, 5:38, 7:57, 3:38. If the maximal time that can be recorded on a compact disk is 70:30, can all the music be performed on 4 compact disks? Can all the music be performed on 5 compact disks?

■ 32. In the wall-system example in the text (see Figure 3.20), first fit and worst fit required equal numbers of bins. Can you find an example where first fit and worst fit yield different numbers of bins? Can you find an example where first fit, worst fit, and next fit yield answers with different numbers of bins?

◆ 33. A common suggestion for heuristics for the bin-packing problem with bins of capacity W involves finding weights that sum to exactly W. Discuss the pros and cons of a heuristic of this type.

■ 34. A record company wishes to record all the Beethoven string quartets (16 quartets, each consisting of several consecutive parts called movements) on LPs. It wishes to complete the project on as few records as possible. Recording can be done on two sides as long as the movements are consecutive. Is this an example of a bin-packing problem? (Defend your answer.) If the project were to record the quartets on (standard) tape cassettes or compact disks, would your answer be different?

■ 35. Two-dimensional bin packing refers to the problem of packing rectangles of various sizes into a minimum number of $m \times n$ rectangles, with the sides of the packed rectangles parallel to those of the containing rectangle.

    (a) Suggest some possible real-world applications of this problem.
    (b) Devise a heuristic algorithm for this problem.
    (c) Give an argument to show that the problem is at least as hard to solve as the usual bin-packing problem.
    (d) If you have $1 \times m$ rectangles with total area W to be packed into a single rectangle of area $p \times q = W$, can the packing always be accomplished?

◆ 36. In what situations would packing bins of different capacities be the appropriate model for real-world situations? Suggest some possible algorithms for this type of problem.

■ 37. Can you find an example of weights that, when packed into bins using first fit, use fewer bins than the number of bins used when the first-fit algorithm is applied with the first weight on the list removed?

◆ 38. Can you formulate "paradoxical" situations for bin packing that are analogous to those we found for scheduling processors?

Additional Exercises

39. Give examples of scheduling problems in which

   (a) processors available can be treated as if they are identical.
   (b) processors available cannot be treated as if they are identical.

40. Consider the accompanying order-requirement digraph:

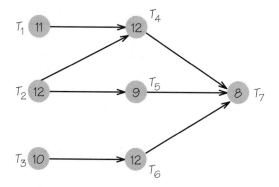

   (a) Find the length of the critical path.
   (b) Schedule these seven tasks on two processors using the list algorithm and the lists
      (1) $T_1, T_2, T_3, T_4, T_5, T_6, T_7$
      (2) $T_2, T_1, T_3, T_6, T_5, T_4, T_7$
   (c) Does either list lead to a completion time that equals the length of the critical path?
   (d) Show that no list can ever lead to a completion time equal to the length of the critical path (providing the schedule uses two processors).

41. Could the following schedule have arisen from the list-processing algorithm? Could it have arisen from the application of the list-processing algorithm to a collection of independent tasks?

Machine 1	T_1		T_5		T_8	
Machine 2	T_2				T_9	
Machine 3	T_3		T_6		T_{10}	
Machine 4	T_4		T_7		T_{11}	

42. Can you give examples of scheduling problems for which it seems reasonable to assume that all the task times are the same?

43. A typing pool gets in 30 (independent) tasks that will take the following amounts of time (in minutes) to type: 25, 18, 13, 19, 30, 32, 12, 36, 25, 17, 18, 26, 12, 15, 31, 18, 15, 18, 16, 19, 30, 12, 16, 15, 24, 16, 27, 18, 9, 14.

    (a) Using these times as a priority list:
        (i) Use the list-processing algorithm to find the completion time for scheduling these tasks with four secretaries; with five secretaries.
        (ii) Repeat the scheduling using the decreasing-time-list algorithm.
        (iii) Can you show that any of the schedules that you get are optimal?

    (b) If one needs to finish the typing in one hour:
        (i) Use the FFD heuristic to find how many typists would be needed.
        (ii) Repeat for the NFD and WFD heuristics.
        (iii) Can you show that any of the solutions you get are optimal?

44. Find the minimum number of bins necessary to pack items of size 8, 5, 3, 4, 3, 7, 8, 8, 6, 5, 3, 2, 1, 2, 1, 2, 1, 3, 5, 2, 4, 2, 6, 5, 3, 4, 2, 6, 7, 7, 8, 6, 5, 4, 6, 1, 4, 7, 5, 1, 2, 4 in bins of capacity (a) through (d) using the first-fit and first-fit-decreasing algorithms. Can you determine if any of the packings you get are optimal?

    (a) 9
    (b) 10
    (c) 11
    (d) 12

45. Advertisements for the TV show Q are permitted to last up to a total of 8 minutes, and each group of ads can last up to 2 minutes. If the ads slated for Q last 63, 32, 11, 19, 24, 87, 64, 36, 27, 42, 63 seconds, determine if FF and FFD yield acceptable configurations for the ads.

46. Consider the following heuristic for packing bins known as *best fit*. Keep track of how much room remains in each unfilled bin and put the next item to be packed into that bin that would leave the least room left over after the item is put into the bin. (For example, suppose that bin 4 had 6 units left, bin 7 had 5 units, and bin 9 had 8 units left. If the next item in the list had size 5, then first fit would place this item in bin 4, worst fit would place the

item in bin 9, while best fit would place the item in bin 7.) If there is a tie, place the item into the bin with the lowest number. Apply this heuristic to the list: 8, 7, 1, 9, 2, 5, 7, 3, 6, 4, where the bins have capacity 10.

47. Can you find a list that gives rise to the optimal schedule shown in Figure 3.14 for the order-requirement digraph in Figure 3.12?

48. Give an example to show that scheduling to minimize completion time may not minimize total idle time. (*Hint:* Assume independent tasks and use two machines for one schedule and three machines for the other schedule.)

WRITING PROJECTS

1 ▶ Scheduling is important for hospitals, schools, transportation systems, police services, and fire services. Pick one of these areas and write an essay about the different scheduling situations that come up, types of processors, and extent to which the assumptions of the list-processing model hold for the area you pick.

2 ▶ Write an essay that compares and contrasts the basic scheduling problem we investigated with the scheduling version of the bin-packing problem.

3 ▶ One of the oversimplifications made in our discussion of scheduling was that there were no "due dates" involved for the tasks making up a job. Develop an algorithm for solving a scheduling problem under the assumption that each task has a due date as well as a time length. You will probably want to decide on a penalty amount that will occur when a due date is exceeded.

4 ▶ Consider the problem of scheduling tasks on a single machine. Design different algorithms for achieving different goals. You will probably wish to assume that each task has a due date, such that if the task is not finished by this date, some penalty payment must be made.

5 ▶ Suppose that one has found that the optimal solution to a bin packing problem with bins of size W requires p bins. What can one say about the number of bins needed when the bin size is $2W$? (*Hint:* Be careful!)

CHAPTER 4

LINEAR PROGRAMMING

A manager's job often calls for making very complicated decisions. One set of decisions involves planning what products the business is to make and determining what resources are needed. In the modern business world, diversification of products provides a company with stability in a climate of changing tastes and needs. So it is not surprising that companies would produce many products, some of which share resource needs. For example, any bakery uses many resources, among which are butter, sugar, eggs, and flour, to make its products: cookies, cakes, pies, and breads.

Resources can include more than just raw materials. A labor force with appropriate skills, farmland, time, and machinery are also resources. Typically, resources are limited: a farmer owns only so much land; there are only so many hours in a day; in a year of drought the wheat crop is very small. Resource availability is also limited by location and competition.

Because resources are limited, management faces important questions: How should the available resources be shared among the possible products? One goal of management is to maximize profit. How can that determine how much of each product should be produced? There are usually so many alternative product mixes that it is impossible to evaluate them all individually. Despite this complexity, millions of dollars may ride on management's decision.

In this chapter we learn about **linear programming,** a management science technique that helps a business allocate the resources it has on hand to make a particular mix of products that will maximize profit. The technique is so powerful that linear programming is said to account for over 50% and perhaps as much as 90% of all computing time used for management decisions in business.

Automated car assembly.

Linear programming is an example of "new" mathematics. It did not originate with the ancient Egyptians or Greeks, nor was it developed in Europe during the time the Western Hemisphere was being explored and settled. Linear programming came into being, along with many other management science techniques, during and shortly after World War II, in the 1940s; it is quite young as intellectual ideas go. Yet, during its short history, linear programming has changed the way businesses make decisions, from "seat-of-the-pants" methods based on guesswork and intuition to using an algorithm based on available data and guaranteed to produce an optimal decision.

Linear programming has saved businesses millions of dollars. Of all the management science techniques presented in this book, linear programming is far and away the most frequently used. It can be applied in a variety of situations, in addition to the one we study in this chapter. Some of the problems studied in Chapters 1, 2, and 3 can be viewed as linear programming problems, and examples of other uses are in Spotlight 4.1. Linear programming is an excellent example of a mathematical technique useful for solving many different kinds of problems that at first do not seem to be similar problems at all. It has been suggested that without linear programming, management science would not exist.

Spotlight

Case Studies in Linear Programming

4.1

Linear programming is not limited to mixture problems. Here are two case studies that do not involve mixture problems, yet where applying linear programming techniques produced impressive savings:

- The Exxon Corporation spends several million dollars per day running refineries in the United States. Because running a refinery takes a lot of energy, energy-saving measures can have a large effect. Managers at Exxon's Baton Rouge plant had over 600 energy-saving projects under consideration. They couldn't implement them all because some conflicted with others, and there were so many ways of making a selection from the 600 that it was impossible to evaluate all selections individually.

  Exxon used linear programming to select an optimal configuration of about 200 projects. The savings are expected to be about $100 million over a period of years.

- American Edwards Laboratories uses heart valves from pigs to produce artificial heart valves for human beings. Pig heart valves come in different sizes. Shipments of pig heart valves often contain too many of some sizes and too few of others; however, each supplier tends to ship roughly the same imbalance of valve sizes on every order, so the company can expect consistently different imbalances from the different suppliers. Thus, if they order shipments from all the suppliers, the imbalances could cancel each other out in a fairly predictable way. The amount of cancellation will depend on the sizes of the individual shipments. Unfortunately, there are too many combinations of shipment sizes to consider all combinations individually.

  American Edwards used linear programming to figure out which combination of shipment sizes would give the best cancellation effect. This reduced their annual cost by $1.5 million.

MIXTURE PROBLEMS

In this chapter we study one application of linear programming, solving mixture problems.

> In a **mixture problem,** limited resources must be combined into products so that the profit from selling those products is a maximum.

Mixture problems are widespread because nearly every product in our economy is created by combining resources. Widely different industries can share the need to solve mixture problems as part of their planning. Analyzing a mixture problem is mostly a matter of careful reading and logic, so you might be fooled into thinking that it is not mathematics at all. But actually, analysis is an extremely important skill in the practice of applied mathematics. Without analysis we cannot solve any linear programming problem.

We will analyze small problems like those that might confront a toy or a beverage manufacturer. Both manufacturers can sell many different products on which it makes profits. There could be dozens of possible products and many resources. A manufacturer must periodically look at the quantities and prices of resources and then determine which products should be produced in which quantities in order to gain the greatest, or optimum, profit. This is an enormous task, usually requiring a computer to solve.

What does it mean to find a solution to a linear programming mixture problem? A solution to a mixture problem is a production policy that tells us how many units of each product to make.

An **optimal production policy** has two properties. First, it is possible; that is, it does not violate any of the limitations under which the manufacturer operates, such as availability of resources. Second, the optimal production policy gives the maximum profit.

Having studied the previous chapters on management science, you may sense that there must be some algorithm that will give us the optimal production policy. There are indeed such algorithms, most of which are very algebraic. Since the algebraic details of these algorithms are easily forgotten, and there are computer programs available to carry out those details, we are most interested here in the ideas behind the algorithms.

At the heart of every algorithm for linear programming are geometric ideas. This is somewhat surprising, since there seem to be no geometric ideas used to describe a mixture problem. We can see these geometric ideas clearly if we solve small linear programming problems, involving just one or two products and one or two resources, and draw some appropriate pictures, or graphs. This visualization of linear programming is readily understandable by nonmathematicians, and it has been valuable to theorists and practitioners who develop solution algorithms for linear programming. Those algorithms are used to solve problems with dozens or hundreds of products and equally large numbers of resources typical of most business applications.

We start our discussion with some very simple examples and work up to more involved ones, all of which we can solve by drawing graphs and making some relatively simple calculations. We end this chapter with a discussion of what larger problems look like and how they are typically solved.

Mixture Problems Having One Resource

One Product and One Resource: Making Skateboards

Suppose a toy manufacturer has 60 containers of plastic and wants to make and sell skateboards. The "recipe" for one skateboard requires 5 containers of plastic, plus paint and decals, which for simplicity we assume are available in essentially unlimited quantities. The profit on one skateboard is $1.00, and in order to keep things simple, we will assume that there will be customers for every skateboard produced. So the manufacturer must decide how many skateboards to make.

We see that the manufacturer can make $\frac{60}{5} = 12$ skateboards. And there seems to be no particular reason not to do exactly that, earning a profit of $1.00(12) = 12.00. We use the variable x to stand for the number of skateboards made; we see that x could be any value between 0 and 12, or algebraically, $0 \le x \le 12$. Those values are the *feasible set,* the particular values of our variable x that are feasible, or possible, given the available resources. Figure 4.1 shows a number line, or x-axis, with the feasible set indicated by a thick line. The problems we will discuss in this chapter will be simple enough that we will always be able to draw a picture of the feasible set. Sometimes the feasible set is called the *feasible region,* so we will use the terms *feasible set* and *feasible region* interchangeably.

FIGURE 4.1 Feasible region for the skateboard problem.

x-Axis

> The **feasible set,** also called the **feasible region,** is the set of all possible solutions to a linear programming problem.

There are several features to note about our feasible region and the point within it that gives the maximum profit:

1. There are no negative values of x in the feasible region. That makes physical sense: How could one make a negative number of skateboards?
2. Any point within the feasible region represents a possible **production policy**—that is, it gives the number of skateboards (product) that it is possible to produce with the limited supply of containers of plastic

But wait, the manufacturer is in this for profit. The company needs to know how much profit it will get from one doll. Suppose it is $0.55. Now, since we used x for the number of skateboards made, we will use y for the number of dolls. So the profit from the dolls will be 0.55y$. The total profit will come from the sale of x skateboards plus y dolls, so the profit formula is 1.00x$ + 0.55y$. We want to find a pair of numbers (x, y) that makes that profit formula as high as possible. But we don't even know what (x, y) pairs are even possible! We need to locate those **feasible points,** the points that make up the feasible set, before we even think about maximizing profit. In order to construct that region, it is helpful to summarize our problem in a *mixture chart.*

Mixture Charts

The most important skill required when solving a mixture problem is the ability to understand its underlying structure. This is as important as being able to do the subsequent algebra and arithmetic. In fact, since linear programming problems can be solved using readily available computer software, extracting the important data from the underlying structure may be the only part of the problem-solving process that must be done by a human being. Understanding the underlying structure means being able to answer these questions:

1. What are the resources?
2. What quantity of each resource is available?
3. What are the products?
4. What are the recipes for creating the products from the resources?
5. What are the unknown quantities?
6. What is the profit formula?

We will display the answers to these questions in a diagram called a **mixture chart.** Then we will translate information in our mixture chart into mathematical statements that we can use to solve the mixture problem. Figure 4.3 shows the mixture chart for Skateboards and Dolls, Part 1.

Some features are present in every mixture chart. There is a row of the chart for every product, for every type of item on which the business can make a profit. Here the products are skateboards and dolls. All the entries on a row give information about the one product to which the row belongs. There is a column for every resource, every input to the business that comes only in limited quantities. Each resource column has information about just one of the limited resources. This problem has only one resource, containers of plastic, so it has only one resource column. (As we progress to problems with more than one resource, our mixture charts will have more columns.) There is also a column for the profit data. We will formulate a mathematical statement corresponding to the profit column and to each of the resource columns.

FIGURE 4.3 Mixture chart for Skateboards and Dolls, Part 1.

RESOURCE(S)

Containers of plastic
60

PROFIT

PRODUCTS	**Skateboards** (x units)	5	$1.00
	Dolls (y units)	2	$0.55

Each problem has specifics that are put into the mixture chart. In filling in the mixture chart, we have labeled the number of skateboards as *x* and the number of dolls as *y*, just as we did before. The recipes for the two products in terms of the number of containers of plastic they use have been entered in the column for the containers of plastic. Since each skateboard uses 5 containers of plastic, there is a 5 where the row for product skateboards meets the column for the container of plastic resource. Similarly, there is a 2 where the row for the product dolls meets the column for the container of plastic resource.

We have also entered the profit numbers in the chart. For example, in the row for dolls and the column for profits, we put $0.55 to indicate that each doll brings in a $0.55 profit. We can use the mixture chart to determine a profit formula of $1.00x + $0.55y. The $1.00 and the *x* are both on the row for skateboards, and the $0.55 and the *y* are both on the row for dolls.

EXAMPLE ▶ *Making a Mixture Chart*

Make a mixture chart to display this situation. A clothing manufacturer has 60 yards of cloth available to make shirts and decorated vests. Each shirt requires 3 yards of material and provides a profit of $5. Each vest requires 2 yards of material and provides a profit of $2.

SOLUTION: See the mixture chart in Figure 4.4. ◆

FIGURE 4.4 Mixture chart for the clothing manufacturer.

RESOURCE(S)

Yards of cloth
60

PROFIT

PRODUCTS	**Shirts** (x units)	3	$5
	Vests (y units)	2	$3

Resource Constraints

Every resource in our mixture problems gives us a *resource constraint,* an algebraic statement that says what is obvious in the physical world, "You can't use more of a resource than the amount you have available." Each resource column in the mixture chart gives us one resource constraint.

We put together the information we have about the resource, containers of plastic. There are just 60 containers of plastic, so "the number of containers of plastic used must be less than or equal to 60." That statement can be rewritten as "the number of containers of plastic used" ≤ 60. To translate the words in quotation marks, we reason in much the same way we did for finding our profit formula. If the manufacturer makes x skateboards, and each skateboard requires 5 containers of plastic, $5x$ containers of plastic are used. Similarly, in making y dolls, each requiring 2 containers of plastic, $2y$ containers of plastic are used up. So making x skateboards *plus* y dolls uses up $5x$ *plus* $2y$ containers of plastic, or $5x + 2y$, which cannot exceed 60. Thus we get the resource constraint $5x + 2y \leq 60$. Note that the numbers 5, 2, and 60 are all in the "container of plastic resource column" of the mixture chart.

> A **resource constraint** is an inequality in a mixture problem that reflects the fact that no more of a resource can be used than what is available.

The resource constraint $5x + 2y \leq 60$ is really a combination of two mathematical statements: $5x + 2y < 60$ and $5x + 2y = 60$. The first statement, $5x + 2y < 60$, is an *inequality,* and tells us that the number of containers of plastic used to make the two products is *less than* the total number of containers of plastic available. The second statement, $5x + 2y = 60$, is an *equality,* or *equation,* and tells us that the number of containers of plastic used to make the two products is *equal to* the total number of containers of plastic available. So $5x + 2y \leq 60$ tells us that the number of containers of plastic used to make the two products must be *less than or equal to* the total number of containers of plastic available.

EXAMPLE ▶ *Writing a Resource Constraint and a Profit Formula*

Using the numbers in the mixture chart in Figure 4.4, write a resource constraint for the cloth resource. Also write the profit formula.

SOLUTION: The resource constraint is $3x + 2y \leq 60$. The profit formula is $\$5x + \$3y$. ◆

Graphing the Constraints to Form the Feasible Region

When we have two products in a mixture problem, we use two variables, x and y. So the feasible region for a problem having two products will be a portion of the xy, or *Cartesian*, plane.

> Every point in the feasible region is a possible solution to a linear programming problem because it satisfies every constraint of that problem. For a two-product problem, the feasible region is a part of the plane. If the problem has n products, the feasible region is a portion of n-dimensional space.

How can we use a resource constraint to help us find the feasible region? In particular, how do we graph an inequality such as $5x + 2y \leq 60$? It is not difficult to draw a graph of $5x + 2y = 60$. That equation, and all the other ones we will get from resource constraints, is the equation of a straight line. Any equation having either an x term, a y term, or both, and some numerical constant, like the 60, but no other kinds of terms, always represents a line. (The equation cannot have any squared, square root, or other kind of algebraic combination of x or y.)

Constraint inequalities are always associated with equations for lines, hence the term *linear programming*. The programming does not refer to a computer but to a well-defined sequence of steps, or program of action, that solves the kinds of problems we are exploring. Here we are using program as a synonym for algorithm. With the introduction of computers into business settings, linear programming is usually carried out by running a computer program.

In order to draw a straight line, we need to know two of the points on the line. In fact, we already know two useful points, but we may not have thought of them as points. One point says that if the manufacturer makes 0 skateboards, then there are enough containers of plastic to make 30 dolls. That point, expressed in terms of the x and y coordinates of the plane would take the general form of (x, y), and since x is the number of skateboards and y is the number of dolls, the point we want is $(0, 30)$. Another point we have already considered is the point representing 12 skateboards and 0 dolls, which we write as $(12, 0)$.

In general, when we want to draw the graph of the line portion of a resource constraint, we can substitute $x + 0$ into the equation part of the resource constraint, and find the corresponding value for y. Here's how that algebra looks:

$$5x + 2y = 60$$
$$5(0) + 2y = 60 \qquad \text{Substitute } x = 0 \text{ into the equation.}$$
$$2y = 60 \qquad \text{Multiply 5 by 0 and simplify.}$$
$$y = 30 \qquad \text{Divide both sides of the equation by 2.}$$

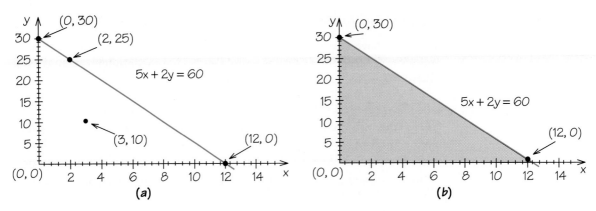

Figure 4.5 The feasible region for Skateboards and Dolls, Part 1. (a) Graph of $5x + 2y = 60$. (b) Shading the half plane $5x + 2y < 60$.

So one (x, y) point on the line is $(0, 30)$. This is the point representing our making $x = 0$ skateboards and $y = 30$ dolls.

We now follow a similar procedure for making 0 dolls; we substitute $y = 0$ into the equation $5x + 2y = 60$ and find out what x corresponds to that value of y. Starting with $5x + 2(0) = 60$ and following the same steps as in the algebra above, we get $x = 12$. So another (x, y) point on the line is $(12, 0)$. This point represents our making $x = 12$ skateboards and $y = 0$ dolls.

In Figure 4.5a we have a graph of the xy-plane showing the points $(0, 30)$ and $(12, 0)$ and a segment of the line $5x + 2y = 60$ connecting them. Every point on the line segment represents a production policy for the two products that uses up all of the containers of plastic available. Some of the points, like $(2, 25)$, give us whole products, and some of the points represent fractional products. We verify that $(2, 25)$ is on the line by substituting 2 for x and 25 for y into the equation $5x + 2y = 60$, getting $5(2) + 2(25) = 60$, which simplifies to $10 + 50 = 60$, which is a true statement. Remember, points on the xy-plane are always expressed in the form (x, y), with the x value, or coordinate, written before the y value, or coordinate.

The xy-plane has four portions, called *quadrants*. In the graph in Figure 4.5a, we only show the one in which both x and y are nonnegative because, in reality, we can never make negative quantities of our products. Reflecting that reality, we have **minimum constraints** of $x \geq 0$ and $y \geq 0$. The line segment in the graph represents all the points, or production policies, for which these properties are true: all the containers of plastic are used up, $5x + 2y = 60$, and both x and y are nonnegative. These points are part of our feasible region. We also need to identify those points corresponding to the inequality $5x + 2y < 60$; these are points corresponding to production policies that do not use up all the containers of plastic.

Any line, such as $5x + 2y = 60$, divides the xy-plane into two parts, called *half planes*. Each of those half planes corresponds to one of two inequalities, in this case $5x + 2y < 60$ and $5x + 2y > 60$. We can determine which half plane goes with which inequality by testing one point not on the line and seeing which inequality it makes true. For example, (3, 10) is on the "down" side of the line segment. When we substitute that point into the inequality $5x + 2y < 60$, which we do by replacing the x by 3 and the y by 10, we get $5(3) + 2(10) < 60$. Simplifying gives us $15 + 20 < 60$ or $35 < 60$, which is true. Substituting the same point into the other inequality, $5x + 2y > 60$, would give $35 > 60$, which is false. So we know that the "down" side of the line segment corresponds to the inequality $5x + 2y < 60$, and the "down" side plus the line segment itself corresponds to the combination inequality $5x + 2y \le 60$. The "up" side of the line corresponds to the inequality $5x + 2y > 60$, and thus is *not* part of the feasible region. In practice, the point (0, 0) is often used as a test point. (Can you see why?) In Figure 4.5b we show the feasible region for the skateboards and dolls problem as a shaded region in the quadrant where both x and y are nonnegative.

E X A M P L E ▶ *Drawing a Feasible Region*

In the earlier clothing manufacturer example, we developed a resource constraint of $3x + 2y \le 60$. Draw the feasible region corresponding to that resource constraint, using the reality minimums of $x \ge 0$ and $y \ge 0$.

SOLUTION: First we find the two points where the line, $3x + 2y = 60$, crosses the axes. When $x = 0$, we get $3(0) + 2y = 60$, giving $y = \frac{60}{2} = 30$, yielding the point (0, 30). For $y = 0$, we get $3x + 2(0) = 60$, or $x = \frac{60}{3} = 20$, so we have the point (20, 0). We draw the line connecting those points. Testing the point (0, 0) we find that the "down" side of the line we have drawn corresponds to $3x + 2y < 60$. The feasible region is shown in Figure 4.6. ◆

FIGURE 4.6 Feasible region for the clothing manufacturer.

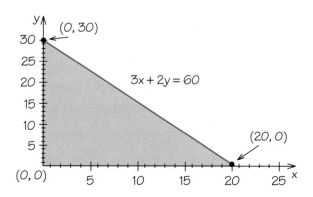

Finding the Optimal Production Policy

After all this work we may think we are done, but in fact what we have just learned how to do is to draw one type of feasible region. We still must find the *optimal production policy,* a point within that region that gives a maximum profit. There are a lot of points in that region. If you only consider points with whole numbers as values for *x* or *y,* there are many points, and in fact either *x* or *y* or both of them could be some fractional number. There are so many points in this feasible region that to consider the profit at each one of them would require us to calculate profits from now until we grow very old, and still the calculations would not be done. Here is where the genius of the linear programming technique comes in, with the *corner point principle,* which we define in terms of our mixture problems.

> The **corner point principle** states that in a linear programming problem, the maximum value for the profit formula always corresponds to a **corner point** of the feasible region. (Later on in this chapter we discuss why the principle works; for now we will accept its validity.)

The corner point principle is probably the most important insight in the theory of linear programming. The geometric nature of this principle explains the value of creating a geometric model from the data in a mixture chart.

The corner point principle gives us the following method to solve a mixture problem:

1. Determine the corner points of the feasible region.
2. Evaluate the profit at each corner point of the feasible region.
3. Choose the corner point with the highest profit as the production policy.

Let's look at the feasible region we just drew. It is a triangle having three corners, namely, (0, 0), (0, 30), and (12, 0). Now all we need to do is find out which of these three points gives us the highest value for the profit formula, which in this problem is $1.00x + $0.55y. We display our calculations in Table 4.1. The maximum profit for the toy manufacturer is $16.50, and that happens if the manufacturer makes 0 skateboards and 30 dolls. The point (0, 30) is called the *optimal production policy.*

> An **optimal production policy** is a corner point of the feasible region where the profit formula has a maximum value.

TABLE 4.1 **Calculation of the Profit Formula for Skateboards and Dolls, Part I**

Corner Point	Value of the Profit Formula: $1.00x + 0.55y
(0, 0)	$1.00(0) + $0.55(0) = $0.00 + $0.00 = $0.00
(0, 30)	$1.00(0) + $0.55(30) = $0.00 + $16.50 = $16.50
(12, 0)	$1.00(12) + $0.55(0) = $12.00 + $0.00 = $12.00

EXAMPLE ▶ *Finding the Optimal Production Policy*

Our analysis of the clothing manufacturer problem resulted in a feasible region with three corner points, (0, 0), (0, 30), and (20, 0). Which of these maximizes the profit formula, $5x + $3y, and what does that corner represent in terms of shirts and vests?

SOLUTION: The evaluation of the profit formula at the corner points is shown in Table 4.2. The maximum profit of $100 occurs at the corner point (20, 0), which represents making 20 shirts and no vests.   ◆

TABLE 4.2 **Evaluating the Profit Formula in the Clothing Example**

Corner Point	Value of the Profit Formula: $5x + $3y
(0, 0)	$5(0) + $3(0) = $0 + $0 = $0
(0, 30)	$5(0) + $3(30) = $0 + $90 = $90
(20, 0)	$5(20) + $3(0) = $100 + $0 = $100

General Shape of Feasible Regions

The shape of a feasible region for a linear programming mixture problem has some important characteristics, without which the corner point principle would not work:

1. The feasible region is a polygon in the first quadrant, where both $x \geq 0$ and $y \geq 0$. This is because the minimum constraints require that both x and y be nonnegative.
2. The region is a polygon that has neither dents (as in Figure 4.7a) nor holes (as in Figure 4.7b). Figure 4.7c is a typical example.

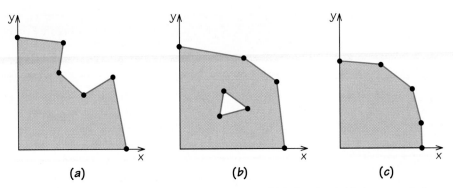

FIGURE 4.7 A feasible region may not have (a) dents or (b) holes. Graph (c) shows a typical feasible region.

The Role of the Profit Formula: Skateboards and Dolls, Part 2

In practice, there are often different amounts of resources available in different time periods. The selling price for the products can also change. For example, if competition forces us to cut our selling price, the profit per unit can decrease. In order to maximize profit, it is usually necessary for a manufacturer to redo the mixture problem calculations whenever any of the numbers change.

Suppose that business conditions change and now the profits per skateboard and doll are, respectively, $1.05 and $0.40. Let us keep everything else about the skateboards and dolls problem the same. The change in profits would give us a new profit formula of $1.05x + $0.40y$. When we evaluate the new profit formula at the corner points, we get the results shown in Table 4.3. This time the optimal production policy, the point which gives the maximum value for the profit formula, is the point (12, 0). To get the maximum profit of $12.60, the toy manufacturer should now make 12 skateboards and 0 dolls.

We see from this example that the shape of the feasible region, and thus the corner points we test, are determined by the constraint inequalities. The profit formula is used to choose an optimal point from among the corner points, so it is not surprising that different profit formulas give us different optimal production policies.

TABLE 4.3	A Different Profit Formula: Skateboards and Dolls, Part 2
Corner Point	Value of the Profit Formula: $1.05x + $0.40y$
(0, 0)	$1.05(0) + $0.40(0) = $0.00 + $0.00 = $0.00
(0, 30)	$1.05(0) + $0.40(30) = $0.00 + $12.00 = $12.00
(12, 0)	$1.05(12) + $0.40(0) = $12.60 + $0.00 = $12.60

We started the exploration of skateboard and doll production with the idea that the toy manufacturer wanted to expand the product line from one to two products. But both linear programming solutions we have found tell the manufacturer that to maximize profit, just make one product. This is probably not an acceptable result for the manufacturer, who might want to produce both products for business reasons other than profit, such as establishing brand loyalty. And it certainly would be very difficult for the manufacturer to be ready to switch back and forth between producing either skateboards or dolls every time the profit formula changed. Linear programming is a flexible enough technique that it can accommodate the desire for there to be both products in the optimal production policy. The way this is done is by specifying that there be nonzero minimum quantities for each period.

Setting Minimum Quantities for Products: Skateboards and Dolls, Part 3

Suppose the toy manufacturer has kept track of the sales of the two products, and has discovered that no matter what, every day there has been demand for at least 4 skateboards and at least 10 dolls. It seems reasonable to set the minimum number of skateboards as 4 and the minimum number of dolls as 10. We keep the same recipes and the 60 containers of plastic. We will redo the mixture chart to include these minimums, draw a new feasible region, and find its corner points. Then we will use each of the earlier profit formulas to see which corner point is the optimal production policy in each case.

Figure 4.8 gives the mixture chart for our expanded problem. A column for minimums has been added to our mixture chart. Note that there are two sets of profits. This time when we draw the feasible region, we have the same resource constraint as we did before, namely, $5x + 2y \leq 60$, so we get the same line as we did before, and the desired inequality still is on the "down" side of that line, as in Figure 4.5b. But now we have minimum constraints which are nonzero. Let us first see what that means in terms of the skateboards. The skateboards row has a 4 in the column for minimums. That says to us "Make a

		RESOURCE(S) Containers of plastic 60	MINIMUMS	PROFIT
PRODUCTS	Skateboards (x units)	5	4	(1) $1.00; (2) $1.05
	Dolls (y units)	2	10	(1) $0.55; (2) $0.40

FIGURE 4.8 Mixture chart for Skateboards and Dolls, Part 3 (with nonzero minimums).

minimum of 4 skateboards." Another way to say this is, "The number of skateboards must be equal to or greater than 4." Since x represents the number of skateboards, we get "x must be equal to or greater than 4," which becomes the mathematical statement $x \geq 4$. Instead of the "reality minimum constraint" of $x \geq 0$, which we used earlier, now we have the nonzero minimum $x \geq 4$. Similarly, the minimum for dolls translates into $y \geq 10$.

Drawing a Feasible Region When There Are Nonzero Minimum Constraints

Figure 4.9a shows the feasible region we constructed with minimum constraints reflecting the reality that $x \geq 0$ and $y \geq 0$. We now need to incorporate the nonzero minimum constraints into that feasible region. As we did with the resource constraint, we can split a minimum constraint into two parts, an equation and an inequality. We can draw the line that corresponds to the equality and then determine which side of that line matches the inequality. First we follow these steps for the skateboards minimum. The constraint $x \geq 4$ has two parts, $x = 4$ and $x > 4$. What sort of line do we draw in the xy-plane for $x = 4$? When is a point (x, y) on the line $x = 4$? Well, clearly, x must be equal to 4, so the points look like $(4, y)$.

But what about the y value? We note that any y value will "work," because there is no y in the equation $x = 4$, so no matter what y value we choose we can never substitute it into the equation and get a false statement. So not only will any y value work, every possible y value works. If we pick two y values, we get two points, and then we can draw a line. Suppose we choose $y = 0$ and $y = 30$. We can draw the two points $(4, 0)$ and $(4, 30)$ and then draw the line they determine. We do that in Figure 4.9b. In general, any line whose equation is of the form $x =$ *some number* is a vertical line passing through the point (*that number,* 0). The y-axis is a special case, with equation $x = 0$.

When we substitute a point into the inequality $x > 4$, for example, the point $(0, 0)$, we get $0 > 4$, which is false, so we know that the point $(0, 0)$ is not on the side of the line for which $x > 4$ is true. That inequality is true for points to the right of the line $x = 4$. Thus the shaded area of Figure 4.9b shows the region of the plane for which $x \geq 4$ is true. Since $y \geq 0$ is always the case in mixture problems, the figure does not show that part of the xy-plane where y has negative values.

Now we need to draw the region where both inequalities $x \geq 4$ and $5x + 2y \leq 60$ are true. We need to find the shape that is shaded in both Figures 4.9a and 4.9b, to find the shape that is shaded twice. That shape is the intersection, or overlapping, of the two shaded regions and is shown in Figure 4.9c.

We see that the new region is a triangle, and we know the coordinates of two of its three corners. However, before we calculate the coordinates of the

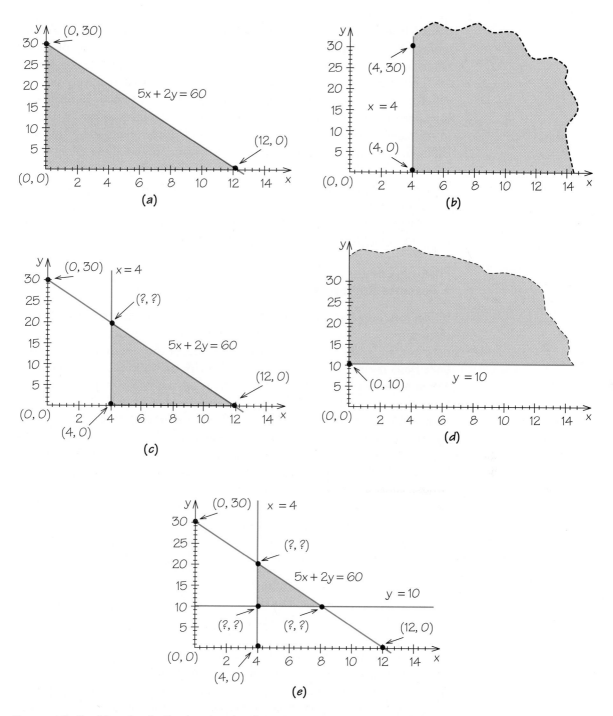

FIGURE 4.9 Feasible region for Skateboards and Dolls, Part 3 (with nonzero minimums). (a) Region where $5x + 2y \leq 60$. (b) Region where $x \geq 4$ is true (and also $y \geq 0$). (c) Region where both inequalities $x \geq 4$ and $5x + 2y \leq 60$ are true. (d) Region where $y \geq 10$ is true (and also $x \geq 0$). (e) Region where $x \geq 4$, $y \geq 10$, and $5x + 2y \leq 60$.

point labeled (?, ?) in Figure 4.9c, we should finish incorporating the minimums into the picture, so we know that our calculations are really of corner points in the final feasible region.

The other minimum constraint is $y \geq 10$. We need to graph the line $y = 10$, all of whose points are of the form $(x, 10)$. This line is horizontal; it is graphed in Figure 4.9d. The point $(0, 0)$ makes the inequality $y > 10$ false, so the region we want is the horizontal line and all the points above that line. Figure 4.9d shows the region where $y \geq 10$ is true for just those values where $x \geq 0$. In general, any line whose equation is of the form $y = $ *some number* is a horizontal line passing through the point $(0, $ *that number*$)$. A special case is the x-axis, which has the equation $y = 0$.

Now we combine the shaded regions in Figures 4.9c and 4.9d, giving us, in Figure 4.9e, the feasible region for our problem.

Finding Corner Points of a Feasible Region Having Nonzero Minimums

In Figure 4.9e, we see that the feasible region for the problem is a triangle with three corners, but as yet we do not know the coordinates of any of those corners. Not to worry. We can get lots of help from the vertical and horizontal lines from the minimum constraints. First, let us work out the coordinates of the lower left corner. Since that point is on the line $x = 4$, we know its x-coordinate must be 4. Similarly, its y-coordinate must be 10, because the point is on the line $y = 10$. So the point at the lower left is $(4, 10)$.

We proceed clockwise around the boundary of the feasible region. The next corner point of the feasible region is directly above $(4, 10)$. Its x-coordinate is also 4, but we need to do some calculation to find its y-coordinate. The point is on the intersection of the lines $x = 4$ and $5x + 2y = 60$. We need to find the y-coordinate of the point that has $x = 4$ and lies on that second line. We do this by substituting 4 for x in the equation of the second line. Here is the algebra:

$5(4) + 2y = 60$	Substitute $x = 4$ into the equation.
$20 + 2y = 60$	Multiply 5 by 4.
$2y = 40$	Subtract the 20 from both sides of the equation.
$y = 20$	Divide both sides of the equation by 2.

so the coordinates of the point are $(4, 20)$.

And, finally, we find the coordinates of the third corner point of the triangle. This point has $y = 10$, so we make that substitution into the equation of the line $5x + 2y = 60$, because the point lies on that line. We get $5x + 2(10) = 60$, and then follow the same steps as we did earlier in a similar calculation. From $5x + 20 = 60$, or $5x = 40$, we get $x = 8$. So the coordinates of the point are $(8, 10)$.

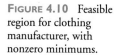

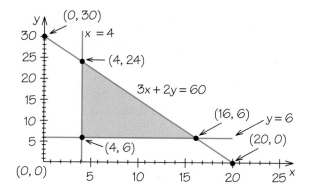

EXAMPLE ▶ *Incorporating Nonzero Minimums*

Suppose the clothing manufacturer needs to make at least 10 shirts and 6 vests. Incorporate these minimums into the feasible region shown in Figure 4.6

SOLUTION: The minimum constraint $x \geq 10$ gives us a vertical line at $x = 10$. The part of the old feasible region to the left of that line is no longer feasible. Similarly, the minimum constraint $y \geq 6$ gives us a horizontal line. The part of the old feasible region below that line is no longer feasible. The new feasible region is shown in Figure 4.10. To find the coordinates of the new corner points, we use the equations of the two lines that meet at that corner. The point (4, 6) is on the lines $x = 4$ and $y = 6$. Substituting $x = 4$ (one line) into $3x + 2y = 60$ (other line), we get $3(4) + 2y = 60$, or $2y = 60 - 12 = 48$, so $y = 24$, giving us (4, 24). We get the coordinates (16, 6) by substitution $y = 6$ (one line) into $3x + 2y = 60$ (other line).   ◆

Evaluating the Profit Formula at the Corners of a Feasible Region with Nonzero Minimums

Now that we have the corner points of the feasible region, we can find out which point gives the maximum profit. The original profit formula, $1.00x + 0.55y$, gave a production policy to make no skateboards. This time all the feasible production policies have nonzero minimums, so that kind of result cannot occur. As we see from Table 4.4, the optimum production policy in this case is to make 4 skateboards and 20 dolls, for a maximum profit of $15.00. The policy calls for the absolute minimum number of skateboards, but not zero. And the "price paid for having the minimums" in this case is $16.50 (the old profit with minimums that are zeros) − $15.00 (the profit now) = $1.50.

The second profit formula, $1.05x + 0.40y$, resulted in a production policy to make no dolls. With the nonzero minimums, that profit formula gives a production policy that says to make 8 skateboards and 10 dolls (check it by using the second formula in a table like Table 4.4), the minimum number, for a

TABLE 4.4	Evaluating One Profit Formula When There Are Nonzero Minimums
Corner Point	Value of the Profit Formula: $\$1.00x + \$0.55y$
(4, 10)	$\$1.00(4) + \$0.55(10) = \$4.00 + \$5.50 = \$9.50$
(4, 20)	$\$1.00(4) + \$0.55(20) = \$4.00 + \$11.00 = \$15.00$
(8, 10)	$\$1.00(8) + \$0.55(10) = \$8.00 + \$5.50 = \$13.50$

profit of \$12.40. And the "price paid for having the minimums" in this case is \$12.60 (the old profit with the minimums that are zeros) − \$12.40 (the profit now) = \$0.20, a very small amount.

EXAMPLE ▶ *Evaluating a Profit Formula*

Finish the clothing manufacturer problem by finding the corner point that maximizes the profit formula $\$5x + \$3y$.

SOLUTION: From Table 4.5 we see that the maximum profit of \$98 is obtained by making 16 shirts and 6 vests. Note that the optimal production policy is still slanted toward shirts, with the manufacturer making just the minimum number of vests. ◆

Summary of the Pictorial Method

Before we proceed to more involved linear programming problems, let's stop and summarize the steps we are following to find the optimal production policy in a mixture problem:

1.   Read the problem carefully. Identify the resources and the products.
2.   Make a mixture chart showing the resources (associated with limited quantities), the products (associated with profits), the recipes for creating the products from the resources, the profit from each product, and the amount of each resource on hand. If the problem has nonzero minimums, include a column for those as well.

TABLE 4.5	Evaluating the Clothing Profit Formula When There Are Nonzero Minimums
Corner Point	Value of the Profit Formula: $\$5x + \$3y$
(4, 6)	$\$5(4)\ + \$3(6)\ = \$20 + \$18 = \$38$
(4, 24)	$\$5(4)\ + \$3(24) = \$20 + \$72 = \$92$
(16, 6)	$\$5(16) + \$3(6)\ = \$80 + \$18 = \$98$

3. Assign an unknown quantity, x or y, to each product. Use the mixture chart to write down the resource constraints, minimum constraints, and the profit formula.

4. Graph the line corresponding to each resource constraint and determine which side of the line is in the feasible region. If there are nonzero minimum constraints, graph lines for them also, and determine which side of each is in the feasible region. Sketch the feasible region by putting together, that is, intersecting, the half planes from all the resource constraints plus the minimum constraints.

5. Find the coordinates of all the corner points of the feasible region. Some of these may have been calculated in order to graph the individual lines. Proceed in order around the boundary of the feasible region. Be sure that every point you consider is part of the feasible region.

6. Evaluate the profit formula for each of the corner points. The production policy that maximizes profit is the one that gives the biggest value to the profit formula.

MIXTURE PROBLEMS HAVING TWO RESOURCES

Two Products and Two Resources: Skateboards and Dolls, Part 4

We return for one last time to the toy manufacturer, now to consider two limited resources instead of one. The second limited resource will be time, the number of person-minutes available to prepare the products. Suppose that there are 360 person-minutes of labor available and that making one skateboard requires 15 person-minutes and making one doll requires 18 person-minutes. We will continue to use the original figures regarding containers of plastic, the first of our two profit formulas, and to keep the problem relatively simple, we will return to the zero minimum constraints: $x \geq 0$ and $y \geq 0$. We need a new mixture chart. In general, we will only include a column for minimums in a mixture chart if there are any nonzero minimum constraints. In Figure 4.11 we have the mixture chart for this problem.

Using the mixture chart, we can write the two resource constraints:

$$5x + 2y \leq 60 \quad \text{for containers of plastic}$$

and

$$15x + 18y \leq 360 \quad \text{for person-minutes}$$

| | RESOURCE(S) | | |
	Containers of plastic 60	Person-minutes 360	PROFIT
Skateboards (x units)	5	15	$1.00
Dolls (y units)	2	18	$0.55

(PRODUCTS)

FIGURE 4.11 Mixture chart for Skateboards and Dolls, Part 4 (two resources).

We can also write the profit formula: $1.00x + $0.55y$.

The half plane corresponding to the plastic resource is shown in Figure 4.12a. We now need to graph the half plane corresponding to the time constraint. We find where the line $15x + 18y = 360$ intersects the two axes by substituting first $x = 0$ and then $y = 0$ into that equation. Here's how that algebra looks:

$5(0) + 18y = 360$ Substitute $x = 0$ into the equation.
$18y = 360$ Simplify.
$y = 20$ Divide both sides of the equation by 18.

So one (x, y) point on the line is $(0, 20)$. This is the point representing our making $x = 0$ skateboards and $y = 20$ dolls.

FIGURE 4.12 Feasible region for Skateboards and Dolls, Part 4 (two resources). (a) Half plane for the plastic resource constraint. (b) Half plane for the time resource constraint. (c) Intersection of two half planes.

We now follow a similar procedure for making 0 dolls. We substitute $y = 0$ into the equation $15x + 18y = 360$, getting $15x + 18(0) = 360$, and carry out similar calculations to the earlier ones, getting $x = 24$. So another (x, y) point on the line is $(24, 0)$. This point represents our making $x = 24$ skateboards and $y = 0$ dolls.

The line corresponding to the time constraint contains the two points $(0, 20)$ and $(24, 0)$. When we substitute the point $(0, 0)$ into the inequality

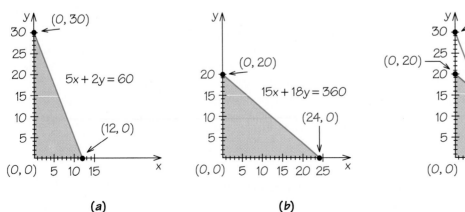

(a)   (b)   (c)

$15x + 18y < 360$, we get $15(0) + 18(0) < 360$, or $0 < 360$, which is true, so $(0, 0)$ is on the side of the line that we shade. Putting all this together, we get the half plane in Figure 4.12b as the correct half plane for the time resource constraint.

We are not permitted to exceed the supply of even a single resource; therefore, the feasible region must be made up of points that are shaded twice, both in the half plane for the plastic resource constraint, shown in Figure 4.12a, and in the half plane for the time resource constraint in Figure 4.12b. As we did before with the half planes from nonzero minimum constraints, we build our feasible region by finding the intersection, or overlap, of the individual half planes in the problem. In Figure 4.12c we show the result of intersecting the half plane from the two resource constraints. Since this problem has minimums that are zeros, the shaded region in Figure 4.12c is in fact the feasible region for the problem.

We now need to find the coordinates of the corner points of the feasible region. In general, there are points in our graph that are not corner points of the feasible region. One easy way to keep track is to shade in the feasible region and then systematically move around its boundary, working from one corner point to the next. We start at the origin, $(0, 0)$, and proceed clockwise. The next corner point is one whose coordinates we calculated in order to draw a line, so we know that it is $(0, 20)$. For the point labeled $(?, ?)$ we need to use a bit of algebra. And the last of the four corner points is again one whose coordinates are already known to us, namely, $(12, 0)$.

To find the coordinates of the point that lies on the intersection of the lines $5x + 2y = 60$ and $15x + 18y = 360$, we take these two equations with two variables, or unknowns, and eliminate one of the unknowns, leaving us with an equation having just one unknown. We have solved such an equation in earlier problems. Once we know the value of one of the unknowns, we simply substitute it into the equation of the line to get the other coordinate. Here are the worked-out details:

Write the two equations with "like terms" in columns:

$$5x + 2y = 60$$
$$15x + 18y = 360$$

Pick the variable to eliminate. Here we will eliminate y. Multiply the top equation by the coefficient of that variable in the bottom equation (the top equation gets multiplied by the 18 from the $18y$). Multiply the bottom equation by the coefficient of that variable in the top equation (the bottom equation gets multiplied by the 2 from the $2y$). Change one of the two multipliers to a negative number (the 2 became a -2):

$$18(5x + 2y = 60)$$
$$-2(15x + 18y = 360)$$

Do the actual multiplication. Note that the coefficients of the variable to be eliminated are now the same except for sign; now one is + and one is − :

$$90x + 36y = 1080$$
$$-30x + -36y = 720$$

Add the two equations by adding "like terms to like terms."

$(90 - 30)x + (36 - 36)y = (1080 - 720)$
$60x + 0y = 360$, so $60x = 360$ Simplify; the y variable "drops out."
$x = 6$ Divide both sides by 60, getting the coordinate for x.

Now that we know the value of one of the coordinates of the point, we are in the same place algebraically as we were when we needed to find the coordinates of a point where a resource line meets a minimum constraint line. We just substitute the value we have, in this case $x = 6$, into either of the two *original* equations and find the corresponding value of the other coordinate. For example, if we use the equation $5x + 2y = 60$, the substitution gives us $5(6) + 2y = 60$, which yields $y = 15$. So the point at which the two lines from resource constraints cross is $(6, 15)$. You can verify that if you picked the other original equation to find the y-coordinate, you would still get $y = 15$. In fact, substituting $(6, 15)$ into the equation $15x + 18y = 360$ and seeing that you get a true statement is a good way to check your calculations.

Some readers may notice that other pairs of numbers will work as multipliers that will cause elimination of the y variable. That is true, and if you know ways to find pairs that work, you of course may use them. Many "tricks" involving getting the smallest values for such numbers were very useful when all calculations were done by hand. Today, since calculators are available to most of us, these "tricks" are not as important.

Now we are ready to finish the problem. In Table 4.6 we have evaluated the profit formula at the four corner points of the feasible region. The optimal production policy for the toy manufacturer would be to make 6 skateboards and 15 dolls, for a maximum profit of $14.25.

TABLE 4.6   The Profit at the Four Corner Points

Corner Point	Value of the Profit Formula: $1.00x + $0.55y
(0, 0)	$1.00(0) + $0.55(0) = $0.00 + $0.00 = $0.00
(0, 20)	$1.00(0) + $0.55(20) = $0.00 + $11.00 = $11.00
(6, 15)	$1.00(6) + $0.55(15) = $6.00 + $8.25 = $14.25
(12, 0)	$1.00(12) + $0.55(0) = $12.00 + $0.0 = $12.00

EXAMPLE ▶ *Mixtures of Two Fruit Juices: Beverages, Part 1*

A juice manufacturer produces and sells two fruit beverages: 1 gallon of cranapple is made from 3 quarts of cranberry juice and 1 quart of apple juice; and 1 gallon of appleberry is made from 2 quarts of apple juice and 2 quarts of cranberry juice. The manufacturer makes a profit of 3 cents on a gallon of cranapple and 4 cents on a gallon of appleberry. Today, there are 200 quarts of cranberry juice and 100 quarts of apple juice available. How many gallons of cranapple and how many gallons of appleberry should be produced to obtain the highest profit without exceeding available supplies? We use zeros as "reality minimums." The mixture chart for this problem is shown in Figure 4.13.

For each resource, we develop a resource constraint reflecting the fact that the manufacturer cannot use more of that resource than what is available. The number of quarts of cranberry juice needed for x gallons of cranapple is $3x$. Similarly, $2y$ quarts of cranberry are needed for making y gallons of appleberry. So if the manufacturer makes x gallons of cranapple and y gallons of appleberry, then $3x + 2y$ quarts of cranberry juice will be used. Since there are only 200 quarts of cranberry available, we get the cranberry resource constraint $3x + 2y \leq 200$. Note that the numbers 3, 2, and 200 are all in the "cranberry" column. We get another resource constraint from the column for the apple juice resource: $1x + 2y \leq 100$. We also have these minimum constraints: $x \geq 0$ and $y \geq 0$.

Finally, we have the profit formula. Since $3x$ is the profit from making x units of cranapple and $4y$ is the profit from making y units of appleberry, we get the profit formula: $3x + 4y$.

We summarize our analysis of the juice mixture problem. Maximize the profit formula, $3x + 4y$, given these constraints:

cranberry: $3x + 2y \leq 200$
apple: $1x + 2y \leq 100$
minimums: $x \geq 0$ and $y \geq 0$

RESOURCE(S)

PRODUCTS	Cranberry 200 quarts	Apple 100 quarts	PROFIT
Cranapple (x gallons)	3 quarts	1 quart	3 cents/gallon
Appleberry (y gallons)	2 quarts	2 quarts	4 cents/gallon

FIGURE 4.13   A mixture chart for Beverages, Part 1.

Remember, in a mixture problem, our job is to find a production policy, (x, y), that makes all the constraints true and maximizes the profit.

Focus first on the cranberry juice constraint, $3x + 2y \leq 200$, and the associated equation $3x + 2y = 200$. We remember that when the line crosses, or intersects, the x-axis, the value of y is 0. Substituting $y = 0$ into the equation $3x + 2y = 200$ gives $3x = 200$, or $x = 200/3$, which we approximate by 66.7. So one point on the line $3x + 2y = 200$ is (66.7, 0). On the y-axis, $x = 0$. Substituting $x = 0$ into our equation $3x + 2y = 200$, we get $2y = 200$, or $y = 200/2$, giving us $y = 100$. So we have a second point on our line, namely, (0, 100). In Figure 4.14a these two points are shown and the line segment labeled $3x + 2y = 200$ was drawn by connecting them. To find out which half plane corresponds to the inequality $3x + 2y < 200$, we need only test the point (0, 0), which is not on the line. We see that $3(0) + 2(0) < 200$ is a true statement, so the inequality corresponds to the "down" side of the line, and that is the portion shaded in the Figure 4.14a.

In Figure 4.14b we show the graph of the apple constraint inequality $1x + 2y \leq 100$. Note that the two points used to draw that line are (0, 50) and (100, 0), the two intercepts. (Can you find those two intercepts by first substituting first $x = 0$ and then $y = 0$ into the inequality $1x + 2y \leq 100$?) You can check that the correct half plane, the one for which $1x + 2y < 100$, has been shaded by substituting the point (0, 0) into that inequality and seeing that the result you get is a true statement.

We know that a point in the feasible region must satisfy, or make true, every constraint inequality in the problem. Which points in the plane make both the cranberry and the apple constraint inequalities true? They are the points that are in the overlap of the shaded regions in Figures 4.14a and 4.14b. The feasible region satisfying both resource constraints is shown in Figure 4.14c. In this problem we also have two minimum constraint inequalities,

FIGURE 4.14 Feasible region for Beverages, Part 1.

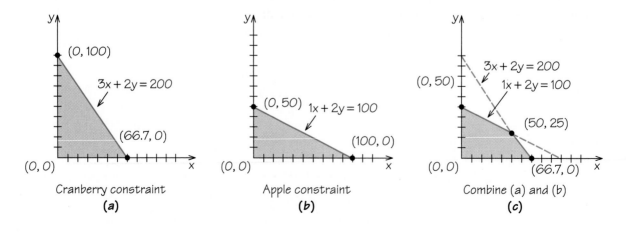

Cranberry constraint
(a)

Apple constraint
(b)

Combine (a) and (b)
(c)

$x \geq 0$ and $y \geq 0$, so our graph is drawn just in the quadrant in which both x and y are nonnegative.

We now find the corner points of the feasible region in Figure 4.14c. We start at the origin, (0, 0) and work our way clockwise around the boundary of the feasible region. Although we know the coordinates of the origin, it is useful to note that it is the intersection of two lines having equations $x = 0$, the y-axis, and $y = 0$, the x-axis, and corresponding to minimum constraints. These equations "solve" the problem of finding the coordinates. In general, we are trying to solve for values of x and y that satisfy both linear equations; then we will have the coordinates of the intersection.

The next corner is the intersection of the lines $x = 0$ and $1x + 2y = 100$. We found this point to be (0, 50) when we sketched the line $1x + 2y = 100$.

Continuing clockwise, we come to the intersection of lines $1x + 2y = 100$ and $3x + 2y = 200$. This more general type of intersection can be solved by using multiplication and addition to eliminate one unknown, solving for the remaining unknown, and then substituting that value into an original equation to get the value of the eliminated unknown.

First, we multiply one equation by a positive value and the other by a negative value so that when the two equations are added together, one unknown gets a coefficient of zero:

$$(-3)(1x + 2y = 100) = -3x - 6y = -300$$

$$(1)(3x + 2y = 200) = 3x + 2y = 200$$

Now we add the new equations together:

$$0x - 4y = -100 \quad \text{or} \quad -4y = -100$$

Dividing both sides of the equation by -4, we get $y = 25$. Substituting $y = 25$ into $1x + 2y = 100$, we get $1x + 2(25) = 100$, which gives $1x + 50 = 100$, and that simplifies to $x = 50$. The point of intersection therefore seems to be (50, 25). We can check our work in the other original equation: $3(50) + 2(25) = 200$ is a true statement.

The last corner point of this feasible region comes from the intersection of $3x + 2y = 200$ and $y = 0$. As with the other intersection of a resource constraint line and an axis, we already know the coordinates: (66.7, 0). Note that some points we used to draw the resource constraint lines are *not* corner points of the feasible region.

When we evaluate the profit formula at these four corner points (see Table 4.7), we see that the optimal production policy is to make 50 gallons of cranapple and 25 gallons of appleberry for a profit of 250 cents. ◆

TABLE 4.7	Finding the Optimal Production Policy for Beverages, Part I
Corner Point	Value of the Profit Formula: $3x = 4y$ cents
(0, 0)	$3(0) \quad + 4(0) \ = 0$ cents
(0, 50)	$3(0) \quad + 4(50) = 200$ cents
(50, 25)	$3(50) \quad + 4(25) = 250$ cents
(66.7, 0)	$3(66.7) + 4(0) \ = 200$ cents (rounded)

THE CORNER POINT PRINCIPLE

In finding solutions to our mixture problems we have been using the corner point principle, which says that the highest profit value on a polygonal feasible region is always at a corner point. A feasible region has infinitely many points, making it impossible to compute the profit for each point. The corner point principle gives us a finite set of points among which we are guaranteed to find the optimal production policy. The principle turns an impossible calculation into a possible one.

This situation is reminiscent of Alexander the Great's approach to the problem of the Gordian knot, a legendary knot so large and tight and tangled that no one had been able to untie it. Alexander's solution was to slice the knot open with his sword. Mixture problems and other linear programming problems are like Gordian knots because there are infinitely many feasible points—we can't calculate the profit for all of them. The corner point principle functions like Alexander's sword and cuts through the problem. We now look at why that principle works.

You can visualize a mathematical proof of the corner point principle by imagining that each point of the plane is a tiny light bulb that is capable of lighting up. For the juice mixture example, whose feasible region is shown in Figure 4.14e, imagine what would happen if we ask this question: Will all points with profit = 360 please light up? What geometric figure do these lit-up points form?

In algebraic terms, we can restate the profit question in this way: Will all points (x, y) with $3x + 4y = 360$ please light up? As it happens, this version of the profit question is one mathematicians learned to answer hundreds of years before linear programming was born. The points that light up make a straight line because $3x + 4y = 360$ is the equation of a straight line. Furthermore, it is a routine matter to determine the exact position of the line. We call this line

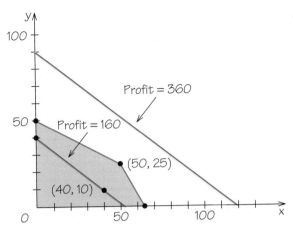

FIGURE 4.15 The profit line for 360 lies outside the feasible region, whereas the profit line for 160 passes through the region.

the **profit line** for 360; it is shown in Figure 4.15. For numbers other than 360, we would get different profit lines. Unfortunately, there are no points on the profit line for 360 that are feasible, that is, which lie in the feasible region. Therefore, the profit of 360 is impossible. *If the profit line corresponding to a certain profit doesn't touch the feasible region, then that profit isn't possible.*

Because 360 is too big, perhaps we should ask the profit line for a more modest amount, say, 160, to light up. You can see that the new profit line of 160 in Figure 4.15 is parallel to the first profit line and closer to the origin. This is no accident: All profit lines for the profit formula $3x + 4y$ have the same coefficients for x and y, 3 for x and 4 for y, and since the slope of the line is determined by those coefficients, they all have the same slope. Changing the profit value from 360 to 160 has the effect of changing where the line intersects the y-axis, but does not affect the slope.

The most important feature of the profit line for 160 is that it has points in common with the feasible region. For example, (40, 10) is on that profit line because $3(40) + 4(10) = 160$, and in addition (40, 10) is a feasible point. This means that it is possible to make 40 gallons of cranapple and 10 gallons of appleberry and that if we do so, we will have a profit of 160.

Can we do better than a 160 profit? As we slowly increase our desired profit from 160 toward 360, the location of the profit line that lights up shifts smoothly upward away from the origin. As long as the line continues to cross the feasible region, we are happy to see it move away from the origin, because the more it moves, the higher the profit represented by the line. We would like to stop the movement of the line at the last possible instant, while the line still has one or more points in common with the feasible region. It should be obvious that this will occur when the line is just touching the feasible region either

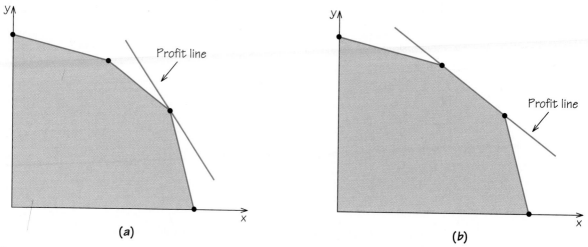

FIGURE 4.16 The highest profit will occur when the profit line is just touching the feasible region, either (a) at a corner point or (b) along a line segment.

at a corner point Figure 4.16a or along a line segment joining two corners Figure 4.16b. That point or line segment corresponds to the production policy or policies with the maximum achievable profit. This is just what the corner point principle says: the maximum profit always occurs at a corner or along an edge of the feasible region.

Here is another way to look at the situation. Suppose that each light bulb (point of the plane) has a color determined by the profit associated with that point. All points with the same profit (i.e., points on a profit line) have the same color. Furthermore, suppose the colors range continuously from violet to blue to green to yellow to red, just as they do in a rainbow. The cool colors represent low profits, the hot colors, higher ones: the higher the profit, the hotter the color. In effect, we are superimposing a straightened-out rainbow on the picture containing our feasible region. Finding the highest profit point can be thought of as finding the hottest-colored point in the feasible region Figure 4.17.

EXAMPLE ▶ *Adding Nonzero Minimums: Beverages, Part 2*

Suppose that in the beverage example the profit for cranapple changes from 3 cents per gallon to 2 cents and the profit for appleberry changes from 4 cents per gallon to 5 cents. You can verify that this change changes the optimal production policy to the point (0, 50)—no cranapple is produced. This result is not surprising: appleberry is giving a higher profit and the policy is to produce as much of it as possible. But suppose the manufacturer wants to incorporate

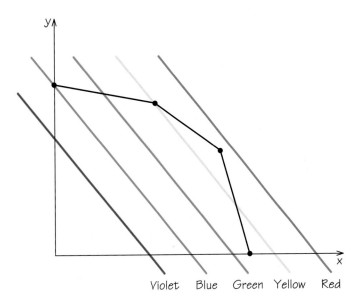

Violet   Blue   Green   Yellow   Red

FIGURE 4.17 If profit lines are colored according to the hues of the rainbow, the highest profit point will be the hottest-colored point in the feasible region.

nonzero minimums into the linear programming specifications so that there will always be both cranapple, x, and appleberry, y, produced. Specifically they decide that $x \geq 20$ and $y \geq 10$ are desirable minimums. In Figure 4.18 is the mixture chart showing the new profit formula and the nonzero minimums along with the unchanged rest of the beverage problem.

The feasible region for Beverages, Part 1, is shown in Figure 4.19a. The feasible region for Beverages, Part 2, is shown in Figure 4.19b. You can verify that, starting at the lower left corner of the new feasible region and moving clockwise around its boundary, we have the following corner points: (20, 10),

	RESOURCE(S)			
	Cranberry juice 200 quarts	Apple juice 100 quarts	**MINIMUMS**	**PROFIT**
Cranapple (x gallons)	3	1	20	2 cents
Appleberry (y gallons)	2	2	10	5 cents

PRODUCTS

FIGURE 4.18 Mixture chart for Beverages, Part 2.

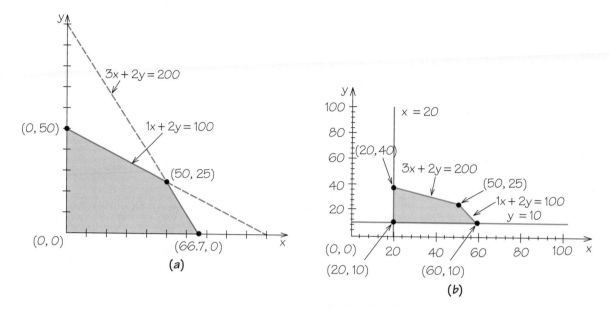

FIGURE 4.19 Feasible region for Beverages, Part 2. (a) Zero minimums. (b) Nonzero minimums.

(20, 40), (50, 25), and (60, 10). (One of those points was also a corner point of the old feasible region. Can you explain why?) Table 4.8 shows the evaluation of the profit formula at these corner points. For this modified problem the optimal production policy is to produce 20 gallons of cranapple and 40 of appleberry for a maximum profit of 240 cents.

One final note about this solution concerns the resources. The point (20, 40) is on the resource constraint line for the apple juice resource, so it represents using up all the available apple juice. We can see this by substituting into the apple juice resource constraint: $1(20) + 2(40) = 100$ is true. However, (20, 40) is *below* the line for the cranberry juice resource, indicating that there will be *slack*, or leftover, amounts of cranberry juice. Specifically, substituting (20, 40) into the cranberry juice constraint gives $3x + 2y = 3(20) + 2(40) = 60 + 80 = 140$, which is 60 quarts less than the 200 quarts available. The slack is 60 quarts of cranberry juice. Dealing with slack can be an important consideration for manufacturers. Can you see why? ◆

TABLE 4.8   Profit Evaluation for Beverages, Part 2

Corner Point	Value of the Profit Formula: $2x + 5y$
(20, 10)	$2(20) + 5(10) = 40 + 50 = 90$ cents
(20, 40)	$2(20) + 5(40) = 40 + 200 = 240$ cents
(50, 25)	$2(50) + 5(25) = 100 + 125 = 225$ cents
(60, 10)	$2(60) + 5(10) = 120 + 50 = 170$ cents

LINEAR PROGRAMMING— THE WIDER PICTURE

Characteristics of Linear Programming Algorithms

Every algorithm for solving a linear programming problem has the following three characteristics. These characteristics hold true regardless of the number of products or the number of resources in the problem.

1. The algorithm can distinguish between "good" production policies, those in the feasible set that satisfy all the constraints, and those that violate some constraint(s). There are usually many good points, each of which corresponds to some production policy; for example, "Make *x* units of product 1 and *y* units of product 2."
2. The algorithm makes use of some geometric principles—one such principle is the corner point principle—to select a special subset of the feasible set.
3. The algorithm evaluates the profit formula at points in the special subset to find which corner point actually gives the maximum profit.

The various algorithms for linear programming differ in how they find the special subset and in how quickly the algorithm finds the production policy—corner point—that gives the optimal profit. The better algorithms are faster at finding the optimal production policy.

It would be convenient if all linear programming problems yielded simple feasible regions in two-dimensional space. However, two complications arise for practical problems:

1. Sometimes, as in Figure 4.20, we have a great many corners. The more corners there are, the more calculations needed to determine the coordinates of all of them and the profit at each one.
2. It is not possible to visualize the feasible region as a part of two-dimensional space when there are more than two products. Each product is represented by an unknown, and each unknown is represented by a dimension of space. If we had 50 products, we would need 50 dimensions and couldn't visualize the feasible region.

Most practical linear programming problems present us with both complications, that is, many corners and more than two products. The number of corners literally can exceed the number of grains of sand on the earth. Even with the fastest computer, computing the profit of every corner is impossible.

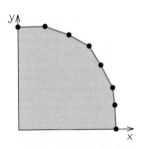

FIGURE 4.20 A feasible region with many corners.

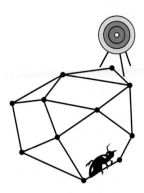

FIGURE 4.21 The simplex method can be compared to an ant crawling along the edges of a polyhedron, looking for one particular target vertex.

The Simplex Method

Several methods are used to solve the typically large linear programming problems solved in practice. The oldest method is the **simplex method,** which is still the most commonly used. Devised by the American mathematician George Dantzig (see Spotlight 4.2), this ingenious mathematical invention makes it possible to find the best corner point by evaluating only a tiny fraction of all the corners. With the use of the simplex method, a problem that might be impossible to solve if each corner point had to be checked can be solved in a few minutes or even a few seconds on a typical business computer.

The operation of the simplex method may be likened to the behavior of an ant crawling on the edges of a polyhedron (a solid with flat sides), looking for one particular target vertex (Figure 4.21). The ant cannot see where the target vertex is. As a result, if it were to wander along the edges randomly, it might take a long time to reach the target. The ant will do much better if it has a temperature clue to let it know it is getting warmer (closer to the target vertex) or colder (farther from the target vertex).

Think of the simplex method as a way of calculating these temperature hints. We begin at any randomly chosen vertex. All neighboring vertices are evaluated to see which ones are warmer and which are colder. A new vertex is chosen from among the warmer ones, and the evaluation of neighbors is repeated—this time checking neighbors of the new vertex. The process ends when we arrive at the target vertex.

Part of what the simplex method has going for it is that it works faster in practice than its worst-case behavior would lead us to believe. Although mathematicians have devised artificial cases for which the simplex method bogs down in unacceptable amounts of arithmetic, the examples arising from real applications are never like that. This may be the world's most impressive counterexample to Murphy's law, which says that if something can go wrong, it will.

Although the simplex method usually avoids visiting every vertex, it may require visiting many intermediate vertices as it moves from the starting vertex to the optimal one. The simplex method has to search along edges on the boundary of the polyhedron. If it happens that there are a great many small edges lying between the starting point and the optimal vertex, the simplex method must operate like a slow-moving bus that stops at every street corner.

In the introduction to this chapter, we noted that linear programming accounts for over 50% and possibly as much as 90% of nonroutine computer time used for management decisions. Although there are alternatives to the simplex method, as we discuss in a later section, much of that computer time is spent using the simplex method.

Spotlight 4.2

Father of Linear Programming Recalls Its Origins

George Dantzig is professor of operations research and computer science at Stanford University. He is credited with inventing the linear programming technique called the simplex method. Since its invention in the 1940s, the simplex method has provided solutions to linear programming problems that have saved both industry and the military time and money. Dantzig talks about the background of his famous technique:

Initially, all the work we did had to do with military planning. During World War II, we were planning on a very extensive scale. The civilian population and the military were all performing scheduling and planning tasks, perhaps on a larger scale than at any time in history. And this was the case up until about 1950. From 1950 on, the whole emphasis shifted from military planning to practical planning for the civilian population, and industry picked it up.

The first areas of industry to use linear programming were the petroleum refineries. They used it for blending gasoline. Nowadays, all of the refineries in the world (except for one) use linear programming methods. They are one of the biggest users of it, and it's been picked up by every other industry you can think of—the forestry industry, the steel industry—you could fill up a book with all the different places it's used.

The question of why linear programming wasn't invented before World War II is an interesting one. In the postwar period, various technologies just evolved that had never been there before. Computers were one example. These technologies were talked about before. You can go back in history and you'll find papers on them, but these were isolated cases that never went anywhere.

In the immediate postwar period, everything just fermented and began to happen, and one of the things that began to happen was linear programming. Mathematicians as well as economists and others who do practical planning and scheduling began to ask the questions: How could you formulate the process as a sort of mathematical system, and how could computers be used to make this happen?

The problems we solve nowadays have thousands of equations, sometimes a million variables. One of the things that still amazes me is to see a program run on the computer—and to see the answer come out. If we think of the number of combinations of different solutions that we're trying to choose the best of, it's akin to the stars in the heavens. Yet we solve them in a matter of moments. This, to me, is staggering. Not that we can solve them—but that we can solve them so rapidly and efficiently.

The simplex method has been used now for roughly 50 years. There has been steady work going on trying to use different versions of the simplex method, nonlinear methods, and interior methods. It has been recognized that certain classes of problems can be solved much more rapidly by special algorithms than by using the simplex method. If I were to say what my field of specialty is, it is in looking at these different methods and seeing which are more promising than others. There's a lot of promise in this—there's always something new to be looked at.

O p t i o n a l ▶ Using a Simplex Method Program

There are many computer programs available that will use the simplex method to give you an optimal production policy if you just supply the computer with the constraint inequalities and profit formula. Simplex method programs can be found in a variety of places, among which are spreadsheets, packages of mathematics programs designed for business applications or finite mathematics courses, and large "all-purpose" mathematics packages. In order to use any simplex program you will need to follow its specific instructions.

If you have access to a simplex method program and want to use it, you should know that the ways in which you, the user, give the details of your problem to the program differs greatly among the currently available packages. For example, some packages ask you to type in constraint inequalities and others ask for just the numbers in the inequality. Those that can deal with just the numbers typically begin by asking for the number of resources and the number of products in the problem.

Since a simplex method program is designed to solve all sorts of linear programming problems, not just mixture problems, it may use slightly different terminology than we have, and may offer you options that you do not need. For example, we have used the variables x and y for the quantities of the two products. Since typical applications have many products, some simplex method programs use subscripted variables such as x_1, x_2, x_3, x_4, and so on, instead of using consecutive alphabet letters. So, in Beverages, Part 1, the resource constraint for cranberry juice, which we wrote as $3x + 2y \leq 200$, would be written as $3x_1 + 2x_2 \leq 200$. Our "profit formula" is usually referred to as the "objective function" by a simplex method program. Using the subscripted variables, Beverages, Part 1, has the objective function $3x_1 + 4x_2$.

The solution from a simplex program will give the values for the variables at the optimal corner point. So, for Beverages, Part 1, the computer would present our solution of (50, 25) as $x_1 = 50$ and $x_2 = 25$, if the program uses subscripted variables, or $x = 50$ and $y = 25$, if the program uses unsubscripted variables. If a program omits a variable in the final solution, its value is zero, meaning that none of that particular product should be made. Recall that in the toy problems with zero minimums, each of our profit formulas gave a production policy with zero quantity for one of the products.

In a typical solution, other information may be given. For example, some programs also give the maximum value of the objective function, 250 cents in Beverages, Part 1. Some programs tell how much of each of the resources is left over after the optimal production policy is carried out. These leftovers are called *slacks,* and are often presented in the solution as values for subscripted variables, such as s_1, s_2, s_3, and s_4. Recall that in Beverages, Part 2, we had a slack of 60 quarts of cranberry juice. As with the variables representing products, any slack variable not mentioned in the solution has the value zero.

Earlier, we emphasized that the analysis needed to construct the mixture chart was really the heart of solving a linear programming problem. Now you see that once you have a mixture chart, you can easily use it to find the data needed for a simplex method program. We hope you have access to such a program and try to use it. For this purpose, we have included some exercises (43 through 46) in this chapter that are larger and more realistic than the ones we have been solving graphically. A graphical solution is only possible for problems limited to two products; these special exercises involve more than two products. ◆

An Alternative to the Simplex Method

In 1984, Narendra Karmarkar (see Figure 4.22), a mathematician working at Bell Laboratories, devised an alternative method for linear programming that finds the optimal corner point in fewer steps than the simplex algorithm by making use of search routes through the interior of the feasible region. The potential applications of Karmarkar's algorithm are important to a lot of industries, including telephone communications and the airlines (see Spotlight 4.3). Routing millions of long-distance calls, for example, means deciding how to use the resources of long-distance land lines, repeater amplifiers, and satellite terminals to best advantage. The problem is similar to the juice company's need to find the best use of its stocks of juice to create the most profitable mix of products.

FIGURE 4.22 Narendra Karmarkar, a researcher at AT&T Bell Laboratories, invented a powerful new linear programming algorithm that solves many complex linear programming problems faster and more efficiently than any previous method.

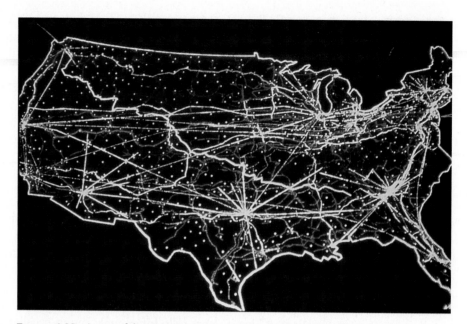

FIGURE 4.23 A map of the United States showing one conceivable network of major communication lines connecting major cities. Routing millions of calls over this immense network requires sophisticated linear programming techniques and high-speed computers.

American Airlines worked with Karmarkar to see if his algorithm could cut fuel costs. According to Thomas Cook, director of operations research for American Airlines, "It's big dollars. We're hoping we can solve harder problems faster, and we think there's definite potential."

In the 1980s, scientists at Bell Labs applied Karmarkar's algorithm to a problem of unprecedented complexity: deciding how to economically build telephone links between cities so that calls can get from any city to any other, possibly being relayed through intermediate cities. Figure 4.23 shows one such linking. The number of possible linkings is unimaginably large, so picking the most economical one is difficult. For any given linking, there is also the problem of deciding how to economically route calls through the network to reach their destinations.

Although difficult, these problems are definitely worth solving. Nat Levine, director of the transmission facilities planning center at Bell Labs, speculated that if one found the best solution, "the savings could be in the hundreds of millions of dollars." Work on these problems at Bell Labs involved a linear programming problem with about 800,000 variables, which Karmarkar's algorithm solved in 10 hours of computer time. Scientists involved believe that the problem might have taken weeks to solve if the simplex method had been used. It appears that for some kinds of linear programming problems, Karmarkar's algorithm is a big improvement over the simplex method.

Spotlight 4.3 Finding Fast Algorithms Means Better Airline Service

Linear programming techniques have a direct impact on the efficiency and profitability of major airlines. Thomas Cook, director of operations research at American Airlines, was interviewed in 1985 concerning his ideas on why optimal solutions are essential to his business:

Finding an optimal solution means finding the best solution. Let's say you are trying to minimize a cost function of some kind. For example, we may want to minimize the excess costs related to scheduling crews, hotels, and other costs that are not associated with flight time. So we try to minimize that excess cost, subject to a lot of constraints, such as the amount of time a pilot can fly, how much rest time is needed, and so forth.

An optimal solution, then, is either a minimum-cost solution or a maximizing solution. For example, we might want to maximize the profit associated with assigning aircrafts to the schedule; so we assign large aircraft to high-need segments and small aircraft to low-load segments. Whether it's a minimum or maximum solution depends on what function we are trying to optimize.

The simplex method, which was developed some 50 years ago by George Dantzig, has been very useful at American Airlines and, indeed, at a lot of large businesses. The difference between his method and Narendra Karmarkar's is speed. Finding fast solutions to linear programming problems is also essential. With an algorithm like Karmarkar's, which is 50 to 100 times faster than the simplex method, we could do a lot of things that we couldn't do otherwise. For example, some applications could be real-time applications, as opposed to batch applications. So instead of running a job overnight and getting an answer the next morning, we could actually key in the data or access the data base, generate the matrix, and come up with a solution that could be implemented a few minutes after keying in the data.

A good example of this kind of application is what we call a major weather disruption. If we get a major weather disruption at one of the hubs, such as Dallas or Chicago, then a lot of flights may get canceled, which means we have a lot of crews and airplanes in the wrong places. What we need is a way to put that whole operation back together again so that the crews and airplanes are in the right places. That way, we minimize the cost of the disruption and minimize passenger inconvenience.

In order to solve that problem in an optimal fashion we need something as fast as Karmarkar's algorithm. In the absence of that, we'll have to come up with some heuristic ways of solving it that won't be optimal.

We can also apply fast solution methods to new problems, and even to problems that we wouldn't have tried using the simplex method. I think that's the primary reason for the excitement.

REVIEW VOCABULARY

Corner point principle   The principle that states that there is a corner point of the feasible region that yields the optimal solution.

Feasible point   A possible solution (but not necessarily the best) to a linear programming problem. With just two products, we can think of a feasible point as a point on the plane.

Feasible region   The set of all feasible points, that is, possible solutions to a linear programming problem. For problems with just two products, the feasible region is a part of the plane.

Feasible set   Another term for **feasible region.**

Linear programming   A set of organized methods of management science used to solve problems of finding optimal solutions, while at the same time respecting certain important constraints. The mathematical formulation of the constraints in linear programming problems are linear equations and inequalities. Mixture problems are usually solved by some type of linear programming.

Minimum constraint   An inequality in a mixture problem that gives a minimum quantity of a product. Negative quantities can never be produced.

Mixture chart   A table displaying the relevant data in a linear programming mixture problem. The table has a row for each product and a column for each resource, for any nonzero minimums, and for the profit.

Mixture problem   A problem in which a variety of resources available in limited quantities can be combined in different ways to make different products. It is usually desired to find the way of combining the resources that produces the most profit.

Optimal production policy   A corner point of the feasible region where the profit formula has a maximum value.

Production policy   A point in the feasible set, interpreted as specifying how many units of each product are to be made.

Profit formula   The expression, involving the unknown quantities such as x and y, that tells how much profit results from a particular production policy.

Profit line   In a two-dimensional, two-product, linear programming problem, the set of all feasible points that yield the same profit.

Resource constraint   An inequality in a mixture problem that reflects the fact that no more of a resource can be used than what is available.

Simplex method   One of a number of algorithms for solving linear programming problems.

SUGGESTED READINGS

ANDERSON, DAVID R., DENNIS J. SWEENEY, AND THOMAS A. WILLIAMS. *An Introduction to Management Science: Quantitative Approaches to Decision Making,* West, St. Paul, Minn., 1985. A business management text with seven chapters covering many applications and aspects of linear programming.

BARNETT, RAYMOND A., AND MICHAEL R. ZIEGLER. *Finite Mathematics for Business, Economics, Life Sciences and Social Sciences,* 7th ed., Prentice Hall, Englewood Cliffs, N.J., 1996. Chapter 5 presents both the geometric approach done here and the simplex method, with attention given to the geometric aspects of the simplex method.

GASS, SAUL I. *An Illustrated Guide to Linear Programming,* McGraw-Hill, New York, 1970. An engagingly written beginner's approach, which introduces the algebra gently and emphasizes the formulation of problems more than algebraic technique.

GASS, SAUL I. *Decision Making, Models, and Algorithms,* Krieger, Melbourne, Fla., 1991. This unique book combines serious applied mathematics, including 12 chapters on linear programming, with a wonderful chatty style.

MEYER, WALTER. *Concepts of Mathematical Modeling,* McGraw-Hill, New York, 1984. Chapter 4 discusses several other types of linear programming problems, including minimization problems, and the "transportation problem," for which there is a special algorithm, and linear programming problems in which the corner points must have coordinates that are whole numbers, not fractions.

Note: Simplex software can be found in *Maple* (keyword is *simplex*), *Mathematica* (keyword is *LinearProgramming*), in both *Lotus 123* and *MSExcel* via *Solver,* and in other software packages, especially those intended for quantitative mathematics courses focusing on business applications.

EXERCISES ▲ *Optional.* ■ *Advanced.* ◆ *Discussion.*

Note: Restrict all graphs to the first quadrant, where both x and y are positive, since that is the only part of the xy-plane needed in our mixture problems.

Graphing

1. Using intercepts, the points where the lines cross the axes, graph each line.

    (a) $2x + 3y = 12$ (d) $7x + 4y = 42$
    (b) $3x + 5y = 30$ (e) $x = 15$
    (c) $4x + 3y = 24$ (f) $y = 4$

2. Using intercepts, the points where the lines cross axes, graph each line.

    (a) $5x + 4y = 20$ (d) $6x + 5y = 15$
    (b) $7x + 6y = 84$ (e) $y = 9$
    (c) $4x + 5y = 60$ (f) $x = 12$

3. Graph both lines on the same axes. Put a dot where the lines intersect. Use algebra to find the x- and y-coordinates of the point of intersection.

    (a) $3x + 4y = 18$ and $x = 2$
    (b) $3x + 5y = 45$ and $y = 6$

4. Graph both lines on the same axes. Put a dot where the lines intersect. Use algebra to find the x- and y-coordinates of the point of intersection.

    (a) $5x + 3y = 60$ and $x = 9$
    (b) $5x + 2y = 30$ and $y = 10$

Resource-Constraint Inequalities

5. Graph the line and half plane corresponding to the inequality, a typical constraint from a mixture problem.

    (a) $x \geq 8$ (c) $5x + 3y \leq 15$
    (b) $y \geq 5$ (d) $4x + 5y \leq 30$

6. Graph the line and half plane corresponding to the inequality, a typical constraint from a mixture problem.

    (a) $x \geq 3$ (c) $3x + 2y \leq 18$
    (b) $y \geq 11$ (d) $7x + 2y \leq 42$

7. One cake (x) requires 4 cups of flour, and one pie (y) requires 2 cups. There are 28 cups of flour available. (The unknown to use for each product is given in parentheses.)

8.   Mowing one lawn (x) takes 2 hours of gardening time, and weeding one garden (y) takes 1 hour. There are 40 hours of gardening time available.

9.   Manufacturing one package of hot dogs (x) requires 6 ounces of beef, and manufacturing one package of bologna (y) requires 4 ounces of beef. There are 240 ounces of beef available.

10.   It takes 30 feet of 12-inch board to make one bookcase (x); it takes 72 feet of 12-inch board to make one table (y). There are 420 feet of 12-inch board available.

Graphing the Feasible Region

Graph the feasible region, label each line segment bounding it with the appropriate equation, and give the coordinates of every corner point.

11. $x \geq 0; y \geq 0; 2x + y \leq 8$

12. $x \geq 0; y \geq 0; 2x + 5y \leq 50$

13. $x \geq 10; y \geq 0; 3x + 5y \leq 120$

14. $x \geq 0; y \geq 4; x + y \leq 25$

15. $x \geq 2; y \geq 6; 3x + 2y \leq 30$

16. $x \geq 8; y \geq 5; 5x + 4y \leq 80$

Finding Points in the Feasible Region

For each of the following points, determine whether it is a point of the feasible region given (a) (2, 4); (b) (10, 6).

17.   The feasible regions of Exercises 11, 13, and 15.

18.   The feasible regions of Exercises 12, 14, and 16.

19.   In the toy problem, x represents the number of skateboards and y the number of dolls. Using the version of that problem whose feasible region is presented in Figure 4.5b (page 134), and the profit formula, $\$2.30x + \$3.70y$, write a sentence giving the maximum profit and describing the production policy that gives that profit.

20.   In the toy problem, x represents the number of skateboards and y the number of dolls. Using the version of that problem whose feasible region is presented in Figure 4.5b (page 134), and the profit formula, $\$5.50x + \$1.80y$, write a sentence giving the maximum profit and describing the production policy that gives that profit.

Finding the Point of Intersection of Two Resource Constraints

21. Graph both lines on the same axes. Put a dot where the lines intersect. Use algebra to find the x- and y-coordinates of the point of intersection.

  (a) $5x + 4y = 22$ and $2x + 4y = 16$
  (b) $2x + 2y = 14$ and $3x + 4y = 24$

22. Graph both lines on the same axes. Put a dot where the lines intersect. Use algebra to find the x- and y-coordinates of the point of intersection.

  (a) $x + 2y = 10$ and $5x + y = 14$
  (b) $5x + 10y = 130$ and $7x + 4y = 112$

Graphing the Feasible Region

Graph the feasible region, label each line segment bounding it with the appropriate equation, and give the coordinates of every corner point.

23. $x \geq 0;\ y \geq 0;\ 3x + y \leq 9;\ x + 2y \leq 8$

24. $x \geq 0;\ y \geq 0;\ 2x + y \leq 4;\ 3x + 3y \leq 9$

25. $x \geq 0;\ y \geq 2;\ x + 2y \leq 10;\ 5x + y \leq 14$

26. $x \geq 4;\ y \geq 0;\ 5x + 4y \leq 60;\ x + y \leq 13$

27. $x \geq 3;\ y \geq 2;\ x + y \leq 10;\ 2x + 3y \leq 24$

28. $x \geq 2;\ y \geq 3;\ 3x + y \leq 18;\ 6x + 4y \leq 48$

Finding Points in the Feasible Region

For each of the following points, determine whether it is a point of the feasible region given (a) (4, 2); (b) (1, 3).

29. The feasible regions of Exercises 23, 25, and 27.

30. The feasible regions of Exercises 24, 26, and 28.

Mixture Problems

Exercises 31 to 42 each have several steps leading to a complete solution to a mixture problem. Practice in a specific step of the solution algorithm can be obtained by working out just that step for several problems. The steps are

  (a) Make a mixture chart for the problem.
  (b) Using the mixture chart, write the profit formula and the resource- and minimum-constraint inequalities.
  (c) Draw the feasible region for those constraints and find the coordinates of the corner points.

    (d)   Evaluate the profit information at the corner points to determine the production policy that best answers the question.

    (e)   *Requires Technology.* Compare your answer with the one you get from running the same problem on a simplex algorithm computer program.

31.   A clothing manufacturer has 600 yards of cloth available to make shirts and decorated vests. Each shirt requires 3 yards of material and provides a profit of $5. Each vest requires 2 yards of material and provides a profit of $2. The manufacturer wants to guarantee that under all circumstances there are minimums of 100 shirts and 30 vests produced. How many of each garment should be made in order to maximize profit? If there are no minimum quantities, how, if at all, does the optimal production policy change?

32.   A car maintenance shop must decide how many oil changes and how many tune-ups can be scheduled in a typical week. The oil change takes 20 minutes. The tune-up requires 100 minutes. The maintenance shop makes a profit of $15 on an oil change and $65 on a tune-up. What mix of services should the shop schedule if the typical week has available 8000 minutes for these two types of services? How, if at all, do the maximum profit and optimal production policy change if the shop is required to schedule at least 50 oil changes and 20 tune-ups?

33.   An appliance store has 90 square feet for displaying refrigerators and stoves. Suppose that each refrigerator requires 10 square feet of floor space and each stove requires 15 square feet. The store has found that it clears a daily profit of $30 for each refrigerator on display and $40 for each stove. How many of each of these appliances should the store display? How, if at all, do the maximum profit and optimal display policy change if the store is required to display at least 3 refrigerators and 2 stoves?

34.   In a certain medical office, a routine office visit requires 5 minutes of doctors' time and a comprehensive office visit requires 25 minutes of doctors' time. In a typical week, there are 1800 minutes of doctors' time available. If the medical office clears $30 from a routine visit and $50 from a comprehensive visit, how many of each should be scheduled per week? How, if at all, do the maximum profit and optimal production policy change if the office is required to schedule at least 20 routine visits and 30 comprehensive ones?

35.   A shellfish company processes (cleans, sorts, opens, and freezes) oysters and clams. In any given week, the employees process 600 bushels of shellfish, of which 100 bushels of oysters and 200 bushels of clams are required to fill standing orders from restaurants. The profit on each bushel of oysters is $8 and on each bushel of clams is $10. How many bushels of each should the company process in order to maximize profit? How, if at all, do the maximum profit and optimal production policy change if the shellfish company has no standing orders?

36.   A student has decided that passing a mathematics course will, in the long run, be twice as valuable as passing any other kind of course. The student estimates that to pass a typical math course will require 12 hours a week to study and do homework. The student estimates that any other course will require only 8 hours a week. The student has available 48 hours for study per week. How many of each kind of course should the student take? (*Hint:* The profit could be viewed as 2 "value points" for passing a math course and 1 "value point" for passing any other course.) How, if at all, do the maximum value and optimal course mix change if the student decides to take at least 2 math and 2 other courses?

Problems 37 to 42 require finding the point of intersection of two lines each corresponding to a resource constraint.

37.   A bakery produces bread and cake. To produce a loaf of bread requires 1.5 hours of oven time and 1 hour of preparation/decoration time. To produce a cake requires 1 hour of oven time and 2 hours of preparation/decoration time. In any one day there are 12 hours of oven time and 16 hours of preparation/decoration time available. The bakery makes only $0.50 per loaf of bread, but clears a profit of $2.50 per cake. What should its production policy be? How, if at all, do the maximum profit and optimal production policy change if the bakery is required to produce at least 2 breads and 3 cakes?

38.   A paper recycling company uses scrap cloth and scrap paper to make two different grades of recycled paper. A single batch of grade A recycled paper is made from 25 pounds of scrap cloth and 10 pounds of scrap paper, whereas one batch of grade B recycled paper is made from 10 pounds of scrap cloth and 20 pounds of scrap paper. The company has 100 pounds of scrap cloth and 120 pounds of scrap paper on hand. A batch of grade A paper brings a profit of $500, whereas a batch of grade B paper brings a profit of $250. What amounts of each grade should be made? How, if at all, do the maximum profit and optimal production policy change if the company is required to produce at least 1 batch of each type?

39.   Dan George has a 100-acre farm where he raises two crops, potatoes and broccoli. It costs him $20 to raise an acre of potatoes and $40 to raise an acre of broccoli, and he has $2600 to cover these costs. The profit from selling these crops is $25 per acre of potatoes and $60 per acre of broccoli. How many acres of each crop should be planted in order to maximize profit? How, if at all, do the maximum profit and optimal planting policy change if Dan George is required to plant at least 20 acres of each crop?

40. The maximum production of a soft drink bottling company is 5000 cartons per day. The company produces regular and diet drinks, and must make at least 600 cartons of regular and 1000 cartons of diet per day. Production costs are $1.00 per carton of regular and $1.20 per carton of diet. The daily operating budget is $5400. How many cartons of each type of drink should be produced if the profit is $0.10 per regular and $0.11 per diet? How, if at all, do the maximum profit and optimal bottling policy change if the company has no minimum required production?

41. Wild Things raises pheasants and partridges to restock the woodlands and has room to raise 100 birds during the season. The cost of raising one bird is $20 per pheasant and $30 per partridge. The Wildlife Foundation pays Wild Things for the birds; the latter clears a profit of $14 per pheasant and $16 per partridge. Wild Things has $2400 available to cover costs. How many of each type of bird should they raise? How, if at all, do the maximum profit and optimal planting policy change if Wild Things is required to raise at least 20 pheasants and 10 partridges?

42. Lights Afire makes desk lamps and floor lamps, on which the profits are $2.65 and $4.67, respectively. The company has 1200 hours of labor and $4200 for materials each week. A desk lamp takes 0.8 hours of labor and $4 for materials; a floor lamp takes 1.0 hours of labor and $3 for materials. What production policy maximizes profit? How, if at all, do the maximum profit and optimal production policy change if Lights Afire wants to produce at least 150 desk lamps and 200 floor lamps per week?

In Exercises 43 to 46, there are more than two products in the problem. Although you cannot solve these problems using the graphical method, you can do these steps:
  (a) Make a mixture chart for each problem.
  (b) Using the mixture chart, write the resource- and minimum-constraint inequalities. Also write the profit formula.
  (c) *Requires Technology.* If you have a simplex method program available, run the program to obtain the optimal production policy.

43. A toy company makes three types of toy, each of which must be processed by three machines: a shaper, a smoother, and a painter. Each Toy A requires 1 hour in the shaper, 2 hours in the smoother, and 1 hour in the painter, and brings in a $4 profit. Each Toy B requires 2 hours in the shaper, 1 hour in the smoother, and 3 hours in the painter, and brings in a $5 profit. Each Toy C requires 3 hours in the shaper, 2 hours in the smoother, and

1 hour in the painter, and brings in a $9 profit. The shaper can work at most 50 hours per week, the smoother 40 hours, and the painter at most 60 hours. What production policy would maximize the toy company's profit?

44.   A rustic furniture company handcrafts chairs, tables, and beds. It has three workers, Chris, Sue, and Juan. Chris can only work 80 hours per month, but Sue and Juan can each put in 200 hours. Each of these artisans has special skills. To make a chair takes 1 hour of Chris's time, 3 from Sue, and 2 from Juan. A table needs 3 hours from Chris, 5 from Sue, and 4 from Juan. A bed requires 5 hours from Chris, 4 from Sue, and 8 from Juan. Even artisans are concerned about maximizing their profit, so what product mix should they stick with if they get $100 profit per chair, $250 per table, and $350 per bed?

45.   A candy manufacturer has 1000 pounds of chocolate, 200 pounds of nuts, and 100 pounds of fruit in stock. The Special Mix requires 3 pounds of chocolate, 1 pound each of nuts and fruit, and brings in $10. The Regular Mix requires 4 pounds of chocolate, 0.5 pound of nuts, and no fruit, and brings in $6. The Purist Mix requires 5 pounds of chocolate, no nuts or fruit, and brings in $4. How many boxes of each type should be produced to maximize profit?

46.   A gourmet coffee distributor has on hand 17,600 ounces of African coffee, 21,120 ounces of Brazilian coffee, and 12,320 ounces of Colombian coffee. It sells four blends: Excellent, Southern, World, and Special on which it makes these per-pound profits, respectively: $1.80, $1.40, $1.20, and $1.00. One pound of Excellent is 16 ounces of Colombian; it is not a blend at all. One pound of Southern consists of 12 ounces of Brazilian and 4 ounces of Colombian. One pound of World requires 6 ounces of African, 8 of Brazilian, and 2 of Colombian. One pound of Special is made up of 10 ounces of African and 6 of Brazilian. What product mix should the gourmet coffee distributor prepare in order to maximize profit?

47.   Explain why finding a point of the feasible region that gives the maximum profit would be time-consuming and nearly impossible if we did not have the corner point principle.

48.   Which steps of the pictorial method are also required of a person who is solving a linear programming problem by using a simplex method computer program?

Additional Exercises

49.   A very selective store sells just tee-shirts, at a profit of $5 each, and jeans, at a profit of $4 each. It takes 4 minutes of a salesperson's time to help a customer select a tee-shirt and 6 minutes of a salesperson's time to help with the selection of jeans. If there are 240 minutes of salesperson time to be distributed, how should that time be spent so the store makes a maximum profit? How, if at all, do the maximum profit and optimal production policy change if the store decides to help at least 12 tee-shirt and 10 jeans customers?

50.   A refinery mixes high-octane and low-octane fuels to produce regular and premium gasolines. The profits per gallon on the two gasolines are $0.30 and $0.40, respectively. One gallon of premium gasoline is produced by mixing 0.5 gallon of each of the fuels. One gallon of regular gasoline is produced by mixing 0.25 gallon of high octane with 0.75 gallon of low octane. If there are 500 gallons of high octane and 600 gallons of low octane available, how many gallons of each gasoline should the refinery make? How, if at all, do the maximum profit and optimal production policy change if the refinery is required to produce at least 100 gallons of each gasoline?

51.   A toy manufacturer makes bikes, for a profit of $12, and wagons, for a profit of $10. To produce a bike requires 2 hours of machine time and 4 hours painting time. To produce a wagon requires 3 hours machine time and 2 hours painting time. There are 12 hours of machine time and 16 hours of painting time available per day. How many of each toy should be produced to maximize profit? How, if at all, do the maximum profit and optimal production policy change if the manufacturer must daily produce at least 2 bikes and 2 wagons?

52.   The planner for the office holiday party needs to get the maximum "happiness points" out of the choices made for the foods. There are two kinds of foods that could be purchased: fancy foods and junk foods. Each order of fancy food costs $50 per order and adds 60 "happiness points" to the party. Each order of junk foods costs $30 and adds 40 "happiness points" to the party. To satisfy both the formal and the casual types at the office, it is necessary to order at least 6 fancy foods and 5 junk foods. If there is $900 to spend, how many of each kind of food should the planner order? If there are no minimum quantities, how, if at all, does the food order change?

Exercises 53 to 58 are designed to highlight areas where the interface between the linear programming model and the reality of specific situations presents us with interesting discussion possibilities. These exercises do not have "right answers." Exercise 53 gives a concrete situation that could be used in the other questions.

◆ 53. You are in charge of a business that produces sandwiches for snack bars. Make a list of your products (kinds of sandwiches), the resources you would need (sandwich ingredients), and the profit you might expect to get for each product. You need not specify the recipes numerically or the amounts of the resources available.

◆ 54. Discuss the validity of the corner point principle if the solution (x, y) to mixture problem is required not only to lie in the feasible region but also to have integer coordinates. In the sandwich problem (Exercise 53), would there be a useful meaning to a fractional number of some kind of sandwich?

◆ 55. A linear programming analysis tells the sandwich company to stop making 7 of the 30 sandwiches in its product line. What are some business considerations that might suggest that this would be a bad business decision? How could linear programming still be used to help the sandwich company if it decided that it did not want to drop any of the 30?

◆ 56. In a mixture problem we view recipes as fixed, but in practice this is not always true. For example, bolognas having only slightly different percentages of beef and pork all taste the same to the customers. What might prompt the manufacturer to vary a recipe? What effect might the varying have on the profit function, the feasible region, and the optimal product mix?

◆ 57. Firms often give discounts for large-volume purchases. Does this necessarily contradict the assumption of a fixed (constant) profit on each unit sold? In examining this situation, you should note that discounts for large-volume purchases might not only apply to the products sold by a manufacturer but also to the prices paid by the manufacturer for resources.

◆ 58. We learn in economics that prices are determined by the interplay of supply and demand. For example, the price of a product may fall if a large quantity of it is available. In mixture problems, however, we assume a fixed (constant) profit regardless of how much is produced. Is there a contradiction here? Could the model be adjusted to incorporate this economic fact of life?

WRITING PROJECTS

1 ▶   Interview a local businessperson who is in charge of deciding the product mix for that business. Must this business take into consideration situations other than minimum and resource constraints? If so, what are these considerations? Find out what methods the person uses to make production policy decisions. Is linear programming used? Are other methods used? If so, what are they? Write a report of your findings, and add some of your own conclusions about the usefulness of linear programming for this business.

2 ▶   In economics, it is often useful to distinguish between a firm that has a monopoly (for example, is the only supplier of a product) and firms that supply only a small share of the market. How would the presence of a monopoly affect the relation between production and price? Would the presence of a monopoly tend to ensure the fixed-profit assumption of linear programming, or would it make it more likely that the interplay of supply and demand would have to be considered in order to have a truly realistic model?

Numerical facts, or data, make up an increasing part of the
information we need in order to understand our world. Business
executives base their decisions on data about the national
economy, financial markets, and their firms' own costs, sales, and profits.
Engineers gather data on the performance, quality, and reliability of their
products. The medical professions watch data on costs as well as data
from medical research. Advertisers fine-tune their messages in response
to market research data, and politicians use polls of public opinion to
shape their campaigns. Citizens and consumers are surrounded by data
from all these and other sources. We must be able to understand and
communicate with data just as we understand and communicate with
words.

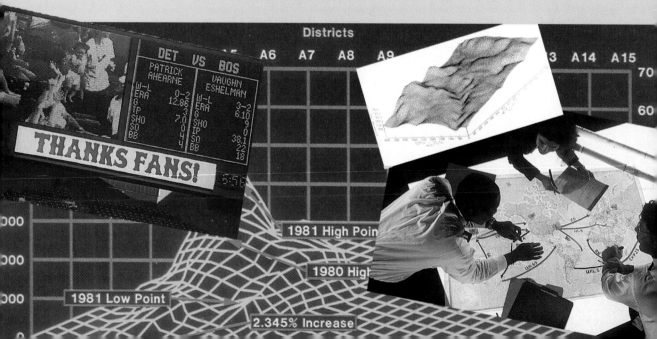

STATISTICS: THE SCIENCE OF DATA

The information conveyed by data may be as vital as the fact that 7.2% of the American labor force is out of work, or as trivial as the fact that 57% of American adults think they look younger than their true age. But whether numerical facts are vital or trivial, we must understand where they come from and whether the information they convey is trustworthy. *Statistics* is the science of data—of producing data, of putting them into clear and usable form, and of interpreting them to draw conclusions about the world around us. Just as literacy enables us to use and understand words, a basic knowledge of statistics allows us to use data honestly and skillfully.

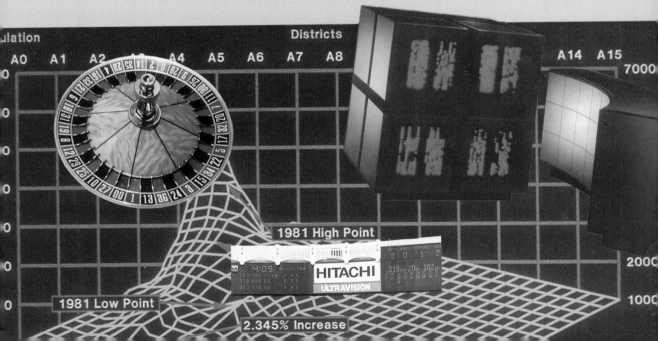

CHAPTER 5

PRODUCING DATA

The news is full of numbers. A TV newscaster informs us that the unemployment rate has dropped to 6.3%. A Gallup poll claims that 45% of Americans are afraid to go out at night because of crime. Where do these numbers come from? Most people aren't personally interviewed to determine whether they are employed. The Gallup poll asks only a few of us if fear of being mugged keeps us indoors at night.

Another day, another headline. This one says "Study Shows Aspirin Prevents Heart Attacks." We read on and learn that the study looked at 22,000 middle-aged doctors. Half took an aspirin every other day, and the other half took a dummy pill. In the aspirin group, 139 doctors died of a heart attack. The dummy pill group had 239 heart-attack deaths in the same period. Is that difference large enough to show that aspirin really does prevent heart attacks?

To escape the unpleasantness of unemployment and heart attacks, we turn to our favorite advice columnist. Ann Landers has asked her female readers whether they would be content with affectionate treatment by men with no sex ever. Over 90,000 women wrote in, with 72% answering "Yes." Can it really be true that 72% of women feel that way?

How trustworthy numbers are depends first of all on where they came from. You can trust the unemployment rate, but you ought to disbelieve Ann Landers' 72%. This chapter explains why. You will learn to recognize good and bad methods of producing numerical facts, which we call *data*. Understanding how to produce trustworthy data is the first—and the most important—step toward the ability to judge whether conclusions based on data are reliable or not. The design of trustworthy methods for producing data is our entry into the subject of statistics, the science of data.

SAMPLING

The Bureau of Labor Statistics (see Spotlight 5.1) wants to know what percent of workers are unemployed. A Gallup poll wants to know what fraction of the public stays home at night because of fear of crime. A quality engineer must estimate what percent of the bearings rolling off an assembly line are defective. In all these situations we want to gather information about a large group of people or things. It is too expensive and time-consuming to contact every worker or inspect every bearing. So we gather information about only part of the group in order to draw conclusions about the whole.

Spotlight 5.1 The Government's Statistician

Janet Norwood

The commissioner of labor statistics is one of the nation's most influential statisticians. As head of the Bureau of Labor Statistics, the commissioner supervises the collection and interpretation of data on employment, earnings, and many other economic and social trends. These data have a large impact on the U.S. economy. For example, retail price data collected by the bureau are used to adjust many federal and private payments, including social security payments and union wage scales, for the effect of inflation.

The data collected by the Bureau of Labor Statistics are often politically sensitive, as when a report released just before an election shows rising unemployment. For this reason, the bureau must remain objective and independent of political influence. To safeguard the bureau's independence, the commissioner is appointed by the president and confirmed by the Senate for a fixed term of four years. The commissioner must have statistical skill, administrative ability, and a facility for working with both Congress and the president.

Dr. Janet Norwood served three terms as commissioner, from 1979 to 1991, under three presidents. When she retired, the *New York Times* said (December 31, 1991) that she left with "a near-legendary reputation for nonpartisanship and plaudits that include one senator's designation of her as a 'national treasure.'" Norwood says, "There have been times in the past when commissioners have been in open disagreement with the Secretary of Labor or, in some cases, with the President. We have guarded our professionalism with great care."

> The entire group of individuals that we want information about is called the **population.** The individuals in a population may be people, animals, or things. A **sample** is a part of the population that we actually examine in order to gather information.

We often draw conclusions about a whole on the basis of a sample. Everyone has sipped a spoonful of soup and judged the entire bowl on the basis of that taste. But a bowl of soup is homogeneous, so that the taste of a single spoonful represents the whole. Choosing a representative sample from a large and varied population is not so easy. The first step is to say carefully just what population we want to describe. The second step is to say exactly what we want to measure. These preliminary steps can be complicated, as the following example illustrates.

EXAMPLE ▶ *The Current Population Survey*

The government's unemployment rate comes from the Current Population Survey (CPS), a sample of about 60,000 households each month. To measure unemployment, we must first specify the population we want to describe. Which age groups will we include? Will we include illegal aliens or people in prisons? What about full-time students? The CPS defines its population as all U.S. residents (whether citizens or not) 16 years of age and over who are civilians and are not in an institution like a prison. The civilian unemployment rate announced in the news refers to this specific population.

The second question is harder: What does it mean to be "unemployed"? Someone who is not looking for work—for example, a full-time student— should not be called unemployed just because she is not working for pay. If you are chosen for the CPS sample, the interviewer first asks whether you are available to work and whether you actually looked for work in the past four weeks. If not, you are neither employed nor unemployed—you are not in the labor force.

If you are in the labor force, the interviewer goes on to ask about employment. Any work for pay or in your own business the week of the survey counts you as employed. So does at least 15 hours of unpaid work in a family business. You are also employed if you have a job but didn't work because of vacation, being on strike, or other good reason. An unemployment rate of 6.3% means that 6.3% of the labor force was unemployed, using the exact CPS definitions of both "labor force" and "unemployed." ◆

Exit poll surveyor with voters.

BAD SAMPLING METHODS

How can we choose a sample that is truly representative of the population? The easiest — but not the best — way to select a sample is to choose individuals close at hand. If we are interested in finding out how many people have jobs, for example, we might go to a shopping mall and ask people passing by if they are employed. A sample selected by taking the members of the population that are easiest to reach is called a **convenience sample.** Convenience samples often produce unrepresentative data.

EXAMPLE ▶ *Convenience Samples*

A sample of mall shoppers is fast and cheap. But people at shopping malls tend to be more prosperous than typical Americans. They are also more likely to be teenagers or retired. What is more, when we decide which people to question, we will tend to choose well-dressed, respectable-looking people and we will tend to avoid poorly dressed, unfriendly, or tough-looking individuals. In short, our shopping mall interviews will not contact a sample that is representative of the entire population, and so will not accurately reflect the nation's rate of unemployment. ◆

Our shopping mall sample will almost surely overrepresent middle-class and retired people and underrepresent the poor. This will happen every time we take such a sample. That is, it is a systematic error due to a bad sampling method, not just bad luck on one sample. Such a systematic difference between the results obtained by sampling and the truth about the whole population is called *bias*.

> The design of a study is **biased** if it systematically favors certain outcomes.

EXAMPLE ▶ *Call-in Polls*

Television makes heavy use of call-in polls, which invite viewers to register their opinions by telephone. Some television stations poll the public daily, asking a question on the 6 o'clock news and reporting the responses on the 11 o'clock news. Viewers are urged to call special 900-prefix telephone numbers with their responses. Dialing one number indicates a "yes" reply; dialing the other, "no." The system works so that talking is not necessary. Completing the telephone call registers a viewer's answer. ◆

Call-in polls have several sources of bias. Households without telephones are unlikely to respond. Only about 6% of U.S. households have no phone, but the percent without phones is higher in the South, among people living alone, and among poor people. Because dialing the 900-prefix number incurs a small charge, poor people may be reluctant to pay a fee in order to telephone the station. An even more serious source of bias is that the sample consists of people who choose to call in. A sample of people who choose to respond to a general appeal is called a **voluntary response sample.** This is an almost sure source of strong bias. Voluntary response samples overrepresent people who have strong opinions, most often negative opinions, on the issue at hand. When the newscaster asks the audience if they are afraid to go out at night because of crime, people who are angry about crime are more likely to call in than those who are not. The call-in poll will almost surely overestimate the percent of all people who are afraid to go out because of crime.

SIMPLE RANDOM SAMPLES

In a voluntary response sample, people choose whether to respond. In a convenience sample, the interviewer makes the choice. In both cases, personal choice produces bias. The statistician's remedy is to allow impersonal chance to choose the sample. A sample chosen by chance allows neither favoritism by the sampler nor self-selection by respondents. Choosing a sample by chance attacks bias by giving all individuals an equal chance to be chosen. Rich and poor, young and old, black and white, all have the same chance to be in the sample.

The simplest way to use chance to select a sample is to place names in a hat (the population) and draw out a handful (the sample). This is the idea of *simple random sampling*.

> A **simple random sample (SRS)** of size *n* consists of *n* individuals from the population chosen in such a way that every set of *n* individuals has an equal chance to be the sample actually selected.

Picturing drawing names from a hat helps us understand what an SRS is. The same picture helps us see that an SRS is a better method of choosing samples than convenience sampling or voluntary response because it doesn't favor any part of the population. But writing names on slips of paper and drawing them from a hat is slow and inconvenient. That's especially true if, like the Current Population Survey, we must draw a sample of size 60,000. We can speed up the process by using a *table of random digits*. In practice, samplers use computers to do the work, but we can do it by hand for small samples.

> A **table of random digits** is a long string of the digits 0, 1, 2, 3, 4, 5, 6, 7, 8, 9 with these two properties:
>
> 1. Each entry in the table is equally likely to be any of the 10 digits 0 through 9.
> 2. The entries are independent of each other. That is, knowledge of one part of the table gives no information about any other part.

Table 5.1 is a table of random digits. The digits in the table appear in groups of five to make the table easier to read and the rows are numbered so we can refer to them, but the groups and row numbers are just for convenience. The entire table is one long string of randomly chosen digits. There are two steps in using the random digit table to choose a simple random sample.

STEP 1. **Label**   Give each member of the population a numerical label of the *same length*. Up to 100 items can be labeled with two digits, up to 1000 items can be labeled with three digits, and so on.

TABLE 5.1   Random Digits

101	03918	86495	47372	21870	28522	99445	38783	83307
102	10041	35095	66357	64569	08993	20429	28569	63809
103	43537	58268	80237	17407	89680	04655	24678	61932
104	64301	47201	31905	60410	80101	33382	95255	10353
105	43857	42186	77011	93839	28380	49296	63311	49713
106	91823	39794	47046	78563	89328	39478	04123	19287
107	34017	87878	35674	39212	98246	29735	09924	27893
108	49105	00755	39242	50472	39581	44036	54518	46865
109	72479	02741	75732	99808	02382	77201	44932	88978
110	84281	45650	28016	77753	39495	41847	19634	82681
111	61589	35486	59500	20060	89769	54870	75586	07853
112	25318	01995	87789	41212	74907	90734	31946	24921
113	40113	37395	51406	98099	43023	70195	07013	72306
114	58420	43526	15539	24845	15582	16780	95286	69021
115	18075	45894	09875	42869	20618	07699	80671	54287
116	52754	73124	93276	71521	59618	44966	37502	15570
117	05255	53579	08239	99174	75548	95776	42314	13093
118	76032	35569	28738	38092	74669	00749	17832	64855
119	97050	31553	32350	51491	53659	89336	36912	05292
120	29030	43074	84602	95131	22769	44680	68492	33987
121	28124	29686	63745	12313	15745	11570	20953	17149
122	97469	41277	90524	36459	22178	63785	20466	67130
123	91754	40784	38916	12949	76104	20556	34001	59133
124	84599	29798	57707	57392	91757	76994	43827	69089
125	06490	42228	94940	10668	62072	58983	10263	08832
126	30666	02218	89355	76117	75167	69005	42479	79865
127	87228	15736	08506	29759	74257	85594	75154	48664
128	45133	49229	32502	99698	68202	44704	39191	73740
129	55713	98670	57794	64795	27102	83420	26630	95009
130	20390	38266	30138	61250	07527	02014	43972	49370
131	13400	68249	32459	41627	56194	93075	50520	96784
132	08900	87788	73717	19287	69954	45917	80026	55598
133	86757	47905	16890	99047	78249	73739	97076	00525
134	19862	54700	18777	22218	25414	13151	54954	80615
135	96282	11576	59837	27429	60015	40338	39435	94021
136	17463	26715	71680	04853	55725	87792	99907	67156
137	44880	55285	95472	57551	24602	98311	63293	58110
138	61911	78152	96341	31473	58398	61602	38143	93833
139	07769	22819	58373	88466	71341	32772	93643	92855
140	73063	63623	29388	89507	78553	62792	89343	27401
141	24187	60720	74055	36902	22047	09091	79368	35408
142	06875	53335	91274	87824	04137	77579	54266	38762
143	23393	37710	46457	03553	58275	11138	18521	59667
144	00980	73632	88008	10060	48563	31874	90785	78923
145	46611	39359	98036	25351	88031	72020	13837	03121
146	56644	79453	49072	30594	73185	81691	29225	70495
147	98350	36891	04873	71321	29929	37145	95906	41005
148	17444	61728	86112	76261	92519	61569	65672	95772
149	45785	21301	89563	23018	60423	50801	70564	45398
150	54369	08513	36838	19805	67827	74938	66946	01206

STEP 2. **Table**   To choose a simple random sample, read from Table 5.1 successive groups of digits of the length you used as labels. Your sample contains the individuals whose labels you find in the table. This gives all individuals the same chance because all labels of the same length have the same chance to be found in the table. For example, any pair of digits in the table is equally likely to be any of the 100 possible labels 00, 01, . . . , 99. Ignore any group of digits that was not used as a label or that duplicates a label already in the sample.

Here are two examples that illustrate the technique.

EXAMPLE   ▶   *Sampling Autos*

An auto manufacturer wants to select 5 of the last 50 cars produced on an assembly line for a very detailed quality inspection. Can you see why allowing the workers to choose 5 cars is likely to cause bias? To avoid bias, we will choose a simple random sample.

STEP 1.   **Label**   Give each car a numerical label. Because two digits are needed to label 50 cars, all labels will have two digits. Let's begin with 00. The labels are 00 to 49, as shown in Figure 5.1. (It is also correct to use labels 01 to 50 if you prefer.) Be sure to say how you labeled the members of the population.

STEP 2.   **Table**   Now go to Table 5.1. Starting at line 140 (any line will do), we find

73063   63623   29388   89507   78553   62792   89343   27401

Because our labels are two digits long, we read successive two-digit groups from the table. Ignore groups not used as labels, like the initial 73. Also ignore any repeated labels, like the second 36 in this row, because we can't choose the same car twice. Our sample contains the cars labeled 06, 36, 23, 29, and 38.

◆

FIGURE 5.1   The first step in random sampling: assigning labels to 50 cars.

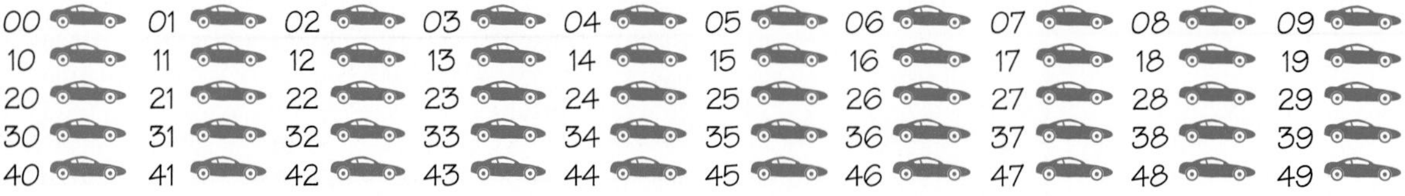

E X A M P L E ▶ *Sampling Households*

Because there is no list of every person in the United States, national samples must first choose households rather than people. A household consists of people living together at the same address. National samples usually proceed in several stages. The first stage chooses a random sample of large geographic areas like counties. Then smaller areas like towns are chosen within each county. Finally, residential addresses are chosen within each town and the households living at these addresses are interviewed.

As part of a national sample, you must choose a simple random sample of 3 of the 4756 addresses in a town. You need four digits to label 4756 addresses. Assign the labels 0000 to 4755 to the list of addresses (0001 to 4756 is also OK). Then read four-digit groups from Table 5.1. Check that if you enter at line 130, the households chosen are those labeled 2039, 0382, and 1386.   ◆

The Current Population Survey takes its monthly sample in several stages. Chance still determines the sample, but the process is more complicated than choosing a simple random sample. The Census Bureau, under contract with the Bureau of Labor Statistics, does the job. The entire country is divided into about 1900 primary sampling units, or PSUs (see Figure 5.2). Each PSU is a group of neighboring counties. The Census Bureau selects a random sample of PSUs. Within each PSU selected, smaller areas of about 500 inhabitants, called census enumeration districts, are chosen at random. Finally, the Census Bureau selects, also at random, individual households within each of the chosen districts.

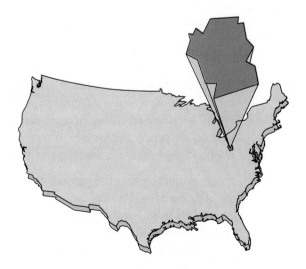

FIGURE 5.2 A primary sampling unit (PSU) for the Current Population Survey.

Such a *multistage random sample* offers several practical advantages over a simple random sample. We don't need a list of every household in the nation. A list of PSUs is used at the first stage and a list of enumeration districts at the second stage. We only need a list of the households in the relatively few enumeration districts chosen. If need be, we can make up that list by walking around these districts. Moreover, the households to be interviewed are clustered together in relatively few locations, reducing the travel costs for the interviewers. The price paid for practicality, however, is complexity in actually choosing the sample and in interpreting the results. Because simple random sampling is the essential principle behind all random sampling and because it is also the main building block for more complex samples, we will focus our study on simple random sampling.

STATISTICAL ESTIMATION

We select a sample in order to get information about the population. If the sample is chosen at random, we expect it to resemble the population. So we use a result from the sample to *estimate* a characteristic of the population.

EXAMPLE ▶ *Statistical Estimation*

A Gallup poll asked a sample of 1493 people "Are you afraid to go outside at night within a mile of your home because of crime?" Of these people, 672 said "Yes." So the percent of the sample who said "Yes" is

$$\frac{672}{1493} = 0.45 = 45\%$$

The population for the Gallup poll is all U.S. residents age 18 and over. We don't know what percent of the population would say "Yes" if we asked them about their fear of crime. Because everyone had the same chance to be in the sample, we expect the sample to represent the population. So we estimate that about 45% of all adults are afraid to go out at night because of crime. ◆

It is unlikely that the percent of the population who are afraid to go out at night is exactly 45%. All we can claim is that the sample result is probably quite close to the truth about the population. If Gallup took another sample, it would contain different people. These people would no doubt have somewhat different views on crime. If 641 of them said "Yes" to Gallup's question, we would estimate that about

$$\frac{641}{1493} = 0.43 = 43\%$$

of all adults are afraid to go outside at night because of crime. This is *sampling variability:* when we take repeated samples from the same population, the results will vary from sample to sample. Random sampling eliminates bias in choosing a sample, but it does not eliminate variability.

One sample gives 45%, another gives 43%. What if other samples give 13% or 89%? Can we trust the results of a sample when we know that we would get a different result if we took another sample? We can. To see why, we need to look more closely at sampling variability.

There are different kinds of variability. The answers obtained by sending an interviewer to a shopping mall vary in a haphazard and unpredictable way. Repeated random samples, however, vary in a regular manner because a specific chance mechanism is used to choose the sample. The long-run results are not haphazard. We see such long-run regularity in games of chance like tossing a coin many times. In fact, tossing a balanced coin 1493 times is just like choosing a simple random sample of 1493 from a large population, if the opinion in this particular population is evenly divided so that heads represents "Yes" and tails represents "No." Both tossing coins and choosing random samples produce results that vary, but we can say how much they will vary because they show a regular pattern in the long run. Let's do an experiment to look at the variation in the results of many random samples.

EXAMPLE ▶ *A Sampling Experiment*

Gallup asked a sample of 1493 adults "Are you afraid to go outside at night within a mile of your home because of crime?" Let us suppose that, unknown to Gallup, exactly 50% of all adults would answer "Yes" to this question. Can we trust a sample of 1493 to come close to this result?

To find out, we took 1000 simple random samples from a population with exactly 50% "Yes" and recorded the percent of "Yes" responses in each sample. The first sample gave 50.2%, the second 49.2%, the third 50.4%, and so on. The sample results do vary. Figure 5.3 is a **histogram** that records the results of all 1000 samples. The height of each bar in the histogram shows how often the outcomes covered by the base of that bar occurred. For example, the height of the bar covering 48% to 48.5% is 80, because 80 of our 1000 samples had between 48% and 48.5% "Yes" responses. (More details about histograms appear in the next chapter.) Study of this histogram shows why we can trust estimates from samples. ◆

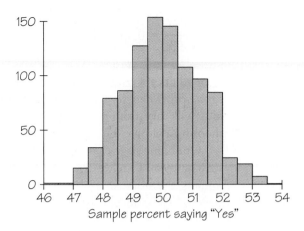

FIGURE 5.3 The results of 1000 samples of size 1493 drawn from a population in which 50% would say "Yes" to the question asked.

You can see from Figure 5.3 that all 1000 samples had between 46% and 54% "Yes" responses. That is, every sample fell within four percentage points of the truth about the population. What is more, the histogram shows a regular pattern of outcomes. The center of the pattern is at 50%. The bars are tallest in the center and get shorter as we go out from the center in either direction. That is, results that are near the truth about the population are most common, and results that are farther from the truth occur less often. The central 95% of the samples gave results between 47.6% and 52.6% "Yes." It appears that a sample of size 1493 will almost always give a result within ± 4% of the truth and will usually (95% of the time) give a result within about ± 2.5% of the truth. So we can be pretty sure that one such sample will in fact give a result close to the result for the entire population. This is wonderful—there are 195 million adults in the country, and choosing just 1493 of them at random allows us to describe their opinions quite accurately.

The regular pattern of the histogram in Figure 5.3 isn't an accident. Using chance to select samples forces the pattern of the results of a large number of samples to have the regular shape that the figure displays. We don't actually have to take thousands of samples to learn what the shape is. The mathematics of chance allows us to calculate it in advance. We'll learn more about that in Chapter 8. Here are the basic facts that explain why we can trust sample estimates:

- If we take many random samples, *the pattern of results is centered about the population truth.* That center is 50% in Figure 5.3 because the truth for this population is 50%. If we take many samples from a population in which 40% would say "Yes," the results will be centered at 40%. The center of the histogram reflects the lack of bias in random sampling. Individual samples may give results above or below the truth about the population, but there is no systematic tendency to be too high or too low.

• *The spread of the pattern is controlled by the size of the sample.* Larger samples give results that cluster closer to the population truth than smaller samples do. So the larger the sample, the more confident we can be that it estimates the population truth accurately. Opinion polls usually interview between 1000 and 4000 people. The Current Population Survey uses a sample of 60,000 households because the government wants to know the unemployment rate very accurately.

Samplers usually tell us how accurate their results are by giving a *margin of error.* We can't be *certain* that the sample results are as close to the population truth as the margin of error says. After all, chance chooses the sample, so it's possible to have terribly bad luck. An opinion poll about crime *might* have the bad luck to choose 1493 residents of high-crime urban neighborhoods rather than a sample that represents the entire population. That sample would give nearly 100% "Yes" answers to Gallup's question about fear of crime. Figure 5.3 shows that this will almost never happen if the truth about the whole population is 50%. The usual margin of error comes from looking at the central 95% of the outcomes in histograms like Figure 5.3.

The **margin of error** announced by most national samples says how close to the truth about the population the sample result would fall in 95% of all samples drawn by the method used to draw this one sample.

A news report says "A new poll shows that only 34% of all Americans approve of the way the president is handling his office. The margin of error for the poll is plus or minus three percent." That means "We got this result using a method that comes within plus or minus three percent of the truth 95% of the time." This particular sample might be one of the 5% of all samples that miss by more, but knowing that we will land within the margin of error 95% of the time gives us a good idea of the poll's accuracy.

EXAMPLE ▶ *Gallup Poll Margin of Error*

In our sampling experiment, we drew many simple random samples. The Gallup poll and the Current Population Survey use more complicated multistage samples. But because they use chance to choose their samples, the pattern of many sample results is still similar to Figure 5.3. The Gallup poll's statisticians tell us that their margin of error is

about ± 5% for samples of size about 600
about ± 4% for samples of size about 1000
about ± 3% for samples of size about 1500

Gallup interviewed 1514 adults and found that 53% of them oppose a longer school year. The margin of error is ± 3%. So we can be quite confident that between 50% (that's 53% − 3%) and 56% (that's 53% + 3%) of all adults oppose a longer school year. ◆

EXPERIMENTS

Sample surveys gather information on part of the population in order to draw conclusions about the whole. When the goal is to describe a population, statistical sampling is the right tool to use.

Suppose, however, that we want to study the response to a stimulus, to see how one variable affects another when we change existing conditions. Will a new mathematics curriculum improve the scores of sixth graders on a standard test of mathematics achievement? Will taking small amounts of aspirin daily reduce the risk of suffering a heart attack? Does smoking increase the risk of lung cancer? Observational studies, such as sample surveys, are ineffective tools for answering these questions. Instead, we prefer to carry out experiments.

> An **observational study,** such as a sample survey, observes individuals and measures variables of interest but does not attempt to influence the responses. An **experiment,** on the other hand, deliberately imposes some *treatment* on individuals in order to observe their responses.

Experiments are the preferred method for examining the effect of one variable on another. By imposing the specific treatment of interest and controlling other influences, we can pin down cause and effect. A sample survey, in contrast, may show that two variables are related, but it cannot demonstrate that one causes the other. Statistics has something to say about how to arrange experiments, just as it suggests methods for sampling.

E X A M P L E ▶ *An Uncontrolled Experiment*

The Bigfoot Mountain School District, concerned about the poor mathematics preparation of American children, adopts an ambitious new mathematics curriculum. After three years of the new curriculum, students completing sixth grade have an average achievement score 10% higher than they had before the treatment. Bigfoot Mountain pronounces the curriculum a success, and other school systems adopt it.

This experiment has a very simple design. A group of subjects (the students) were exposed to a treatment (the new curriculum), and the outcome (achievement test scores) was observed. Here is the design:

New curriculum ⟶ Observe test scores

or, in general form

Treatment ⟶ Observe response ◆

Most laboratory experiments use a design like that in the example: apply a treatment and measure the response. In the controlled environment of the laboratory, simple designs often work well. But field experiments and experiments with human subjects are exposed to more variable conditions and deal with more variable subjects. It isn't possible to control outside factors that can influence the outcome. With greater variability comes a greater need for statistical design.

In Bigfoot Mountain, a concern for education brought about a number of simultaneous changes that could influence the students' achievement test scores. Elementary school teachers were given additional training in mathematics. A parent group began to provide classroom tutors to give children individual help with mathematics. Public concern led parents to pay more attention to their children's progress and teachers to assign more homework. In these circumstances, mathematics achievement would have increased without a new curriculum. In fact, the new curriculum might even be *less* effective than the old one.

The Bigfoot Mountain experiment cannot distinguish the effects of the changes in parents and teachers from the effects of the new curriculum. We say that the new curriculum is confounded with the other changes that took place at the same time.

> Variables, whether part of a study or not, are said to be **confounded** when their effects on the outcome cannot be distinguished from each other.

RANDOMIZED COMPARATIVE EXPERIMENTS

The remedy for confounding is to do a *comparative experiment* in which some children are taught from the new curriculum and others from the old. Changes in parents' attitudes and involvement, teacher retraining, and other such variables now operate equally on both groups of students, so that direct comparison of the two curricula is possible. Most well-designed experiments compare two or more treatments.

But comparison alone isn't enough to produce results we can trust. If the treatments are given to groups that differ markedly when the experiment begins, bias will result. For example, if we allow students to volunteer for the new curriculum in Bigfoot Mountain, only adventurous children who are interested in math are likely to sign up, and these students are likely to perform well. Personal choice will bias our results in the same way that volunteers bias the results of call-in opinion polls. The solution to the problem of bias is the same for experiments and for samples: use impersonal chance to select the groups.

Let's say the Bigfoot Mountain School District decides to compare the progress of 100 students taught under the new mathematics curriculum with that of 100 students taught under the old curriculum. We select the students who will be taught the new curriculum by taking a simple random sample of size 100 from the 200 available subjects. The remaining 100 students form the **control group.** They will continue in the old curriculum.

The selection procedure is exactly the same as it is for sampling: label and table. First, tag all 200 members of the population with numerical labels, say, 000 to 199. Then go to the table of random digits and read successive three-digit groups. The first 100 labels encountered select the group that will be taught from the new curriculum. As usual, ignore repeated labels and groups of digits not used as labels. For example, if you begin at line 125 in Table 5.1, the first few students chosen are those labeled 064, 106, 102, 022, and 188.

The result is a **randomized comparative experiment** with two groups. Randomized comparative experiments are a relatively new idea. They were introduced in the 1920s by Sir R. A. Fisher (see Spotlight 5.2). Figure 5.4 outlines the design in graphical form. The experiment is comparative because it compares two treatments (the two math curricula). It is randomized because the subjects are assigned to the treatments by chance. Randomization creates groups that are similar to each other before we start the experiment. Comparison means that possible confounding variables act on both groups at once. The only difference between the groups is the different math curricula. So if we see a difference in performance, it must be due to the different curricula. That is the basic logic of randomized comparative experiments. We will see later that there are some fine points to worry about, but this basic logic shows why ex-

FIGURE 5.4 Outline of the design of a randomized comparative experiment to evaluate a new mathematics curriculum.

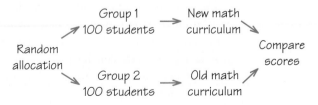

$Spotlight$

Sir Ronald A. Fisher, 1890–1962

5.2

Sir Ronald A. Fisher

While employed at the Rothamsted agricultural experiment station in the 1920s, British statistician and geneticist R. A. Fisher revolutionized the strategy of experimentation. Experimenters there were comparing the effects of several treatments, such as different fertilizers, on field crops. Because fertility and other variables can change as we move in any direction across the planted field, the experimenters used elaborate checkerboard planting arrangements to avoid bias. Fisher realized that random assignment of treatments to growing plots was simpler and better. He introduced randomization, described more complex random arrangements, such as blocks and Latin squares, and worked out the mathematics of the *analysis of variance* to analyze data from randomized comparative experiments.

Fisher contributed many other ideas, both mathematical and practical, to the new science of statistics. His influential books organized the field. Fisher was both opinionated and combative. From the 1930s until his death, he was engaged in sometimes vitriolic debates over the appropriate use of statistical reasoning in scientific inference.

periments can give good evidence that the different treatments really *caused* different outcomes. Randomized comparative experiments are used whenever environmental variables, such as changes in the behavior of Bigfoot Mountain parents and teachers, threaten to confound the results. Here is another example, this time comparing three treatments.

E X A M P L E ▶ *Raising Turkeys*

Turkeys raised commercially for food are often fed the antibiotic salinomycin to prevent infections from spreading among the birds. Salinomycin can damage the birds' internal organs, especially the pancreas. A researcher believes that adding vitamin E to the diet may prevent injury. He wants to explore the effects of three levels of vitamin E added to the diet of turkeys along with the usual dose of salinomycin. There are 30 turkeys available for the study. At the end of the study, the birds will be killed and each pancreas examined under a microscope.

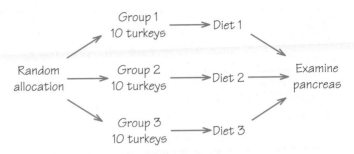

FIGURE 5.5 The design of a randomized comparative experiment to compare three diets for turkeys.

Figure 5.5 outlines a randomized comparative design that allocates 10 birds chosen at random to each of the three levels of vitamin E. The turkeys are labeled with tags marked 00 to 29 (01 to 30 is also acceptable, but be sure that each label has two digits). Read two-digit groups starting in line 115 of Table 5.1 until 10 turkeys are chosen to make up the first group. Those chosen have labels 18, 07, 09, 28, 20, 15, 24, 27, 21, and 05. Then continue in the table to choose 10 more birds for the second group. The 10 that remain form the third group. ◆

Randomized comparative experiments are common tools of industrial and academic research. They are also widely used in medical research. For example, federal regulations require that the safety and effectiveness of new drugs be demonstrated by randomized comparative experiments. Let's look at an important medical experiment.

E X A M P L E ▶ *The Physicians' Health Study*

There is some evidence that taking regular, low doses of aspirin will reduce the risk of heart attacks. Many people also suspect that regular doses of beta carotene (which the body converts into vitamin A) will help prevent some types of cancer. The Physicians' Health Study is a large experiment designed to test these claims. The subjects of this study were 22,000 male physicians over 40 years of age. Each subject took a pill every day over a period of several years. There were four treatments: aspirin alone, beta carotene alone, both, and neither. The subjects were randomly assigned to one of these treatments at the beginning of the experiment. ◆

The Physicians' Health Study example introduces several new ideas about the design of experiments. The first is the importance of the **placebo effect,** a special kind of confounding. A placebo is a fake treatment, a dummy pill that contains no active ingredient but looks and tastes like the real thing. The placebo effect is the tendency of subjects to respond favorably to any treatment, even a placebo. If subjects given aspirin, for example, are compared with subjects who receive no treatment, the first group gets the benefit of both aspirin and the placebo effect. Any beneficial effect that aspirin may have is confounded with the placebo effect. To prevent confounding, it is important that some treatment be given to all subjects in any medical experiment. In the Physicians' Health Study, all subjects took pills that looked alike. Some pills contained aspirin or beta carotene and some contained a placebo. Figure 5.6 shows the design of the experiment.

The Physicians' Health Study was a **double-blind experiment:** neither the subjects nor the experimenters who worked with them knew which treatment any subject received. Subjects might react differently if they knew they were getting "only a placebo." Knowing that a particular subject was getting "only a placebo" could also influence the researchers who interviewed and examined the subjects. So both subjects and workers were kept "blind." Only the study's statistician knew which treatment each subject received.

Finally, the Physicians' Health Study is a more elaborate experiment than our earlier examples. The Bigfoot Mountain and turkey examples compare values of a single variable (Which math curriculum? How much vitamin E?). The Physicians' Health Study looks at two distinct experimental variables: aspirin or not and beta carotene or not. A two-variable experiment, usually called a *two-factor experiment,* allows us to study the *interaction,* or joint effect, of the

FIGURE 5.6   The design of the Physicians' Health Study, an experiment with two factors.

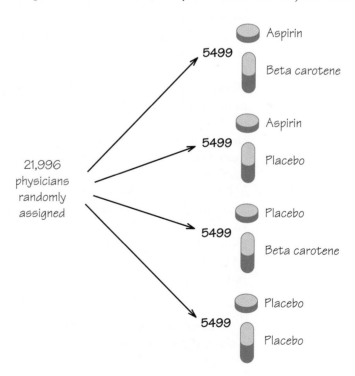

two drugs as well as the separate effects of each. For example, beta carotene may reinforce the effect of aspirin on future heart attacks. By comparing these four groups, we can study all these possible interactions. Nonetheless, the outline of the design in Figure 5.6 is similar to our earlier examples because the basic ideas of randomization and comparison of several treatments remain.

STATISTICAL EVIDENCE

A properly designed experiment, in the eyes of a statistician, is an experiment employing the principles of *comparison* and *randomization:* comparison of several treatments and randomization in assigning subjects to the treatments.

The future health of the subjects of the Physicians' Health Study, for example, may depend on age, past medical history, emotional status, smoking habits, and many other variables known and unknown. Randomization will, on the average, balance the groups simultaneously in all such variables. Because the groups are exposed to exactly the same environmental variables, except for the actual content of the pills, we can say that any differences in heart attacks or cancer among the groups are caused by the medication. That is the logic of randomized comparative experiments.

Let's be a bit more careful: any difference among the groups is due *either* to the medication *or* to the accident of chance in the random assignment of subjects. It could happen, for example, that men about to have a heart attack were, just by chance, overrepresented in one of the groups. The problem is exactly the same as in random sampling, where it could happen just by chance that an SRS chooses all Republicans. Just as in sampling, we are saved by the regular pattern of chance behavior.

If we repeat the random assignment of subjects to groups many times, differences among the groups follow a regular pattern if we don't apply different treatments. This regular pattern tells us how large the differences among the four groups are likely to be if nothing but chance is operating. If we observe differences so large that they would almost never occur just by chance, we are confident that we are seeing the effects of the treatments. So it is not *any* differences that show the results of the treatments, just differences so large that chance cannot easily account for them. Differences among the treatment groups that are so large that they would rarely occur just by chance are called *statistically significant.*

An observed effect too large to attribute plausibly to chance is called **statistically significant.**

Again as in sampling, larger numbers of subjects increase our confidence in the results. The Physicians' Health Study followed 22,000 subjects in order to be quite certain that any medically important differences among the groups would be detected and that these differences could be attributed to aspirin or to beta carotene. In fact, there were significantly fewer heart attacks among the men who took aspirin than among men who took the placebo. As a result of the Physicians' Health Study, doctors often recommend that men over age 50 take small amounts of aspirin regularly.

The logic of experimentation, the statistical design of experiments, and the laws that govern chance behavior combine to give compelling evidence of cause and effect. Only experimentation can produce fully convincing evidence of causation.

EXAMPLE ▶ *Smoking and Health*

By way of contrast, consider the statistical evidence linking cigarette smoking to lung cancer. This evidence is based on observation rather than experiment. The most careful studies have selected samples of smokers and nonsmokers, then followed them for many years, eventually recording the cause of death. These are called *prospective studies* because they follow the subjects forward in time. Prospective studies are comparative, but they are not experiments because the subjects themselves choose whether or not to smoke. Remember that an experiment must actually impose treatments on its subjects. A large prospective study of British doctors found that the death rate from lung cancer among cigarette smokers was 20 times that of nonsmokers; another study of American men aged 40 to 79 found that the lung cancer death rate was 11 times higher among smokers than among nonsmokers. These and many other observational studies show a strong connection between smoking and lung cancer. ◆

The connection between smoking and lung cancer is statistically significant. That is, it is far stronger than would occur by chance. We can be confident that something other than chance links smoking to cancer. But observation of samples cannot tell us *what* factors other than chance are at work. Perhaps there is something in the genetic makeup of some people that predisposes them both to nicotine addiction and to lung cancer. In that case, we would observe a strong link even if smoking itself had no effect on the lungs.

The statistical evidence that points to cigarette smoking as a cause of lung cancer is about as strong as nonexperimental evidence can be. First, the connection has been observed in many studies in many countries. This eliminates factors peculiar to one group of people, or to one specific study design. Second, specific ways in which smoking could cause cancer have been identified. Cigarette smoke contains tars that can be shown by experiment to cause tumors in animals. Third, no really plausible alternative explanation is available. For example, the genetic hypothesis cannot explain the increase in lung cancer among women that occurred as more and more women became smokers. Lung cancer, which has long been the leading cause of cancer deaths in men, has now passed breast cancer as the most fatal cancer for women. Moreover, genetics cannot explain why lung cancer death rates increase among nonsmokers who are exposed to cigarette smoke from other people.

This evidence is convincing to most people, and almost all physicians accept it. But it is not quite as strong as the conclusive statistical evidence we get from randomized comparative experiments.

STATISTICS IN PRACTICE

There is more to the wise use of statistics than a knowledge of such statistical techniques as simple random samples and randomized comparative experiments. These designs for data production avoid the pitfalls of voluntary response samples or uncontrolled experiments. But there are other pitfalls that can reduce the usefulness of data even when we use a sound statistical design.

EXAMPLE ▶ *Nonresponse in Sampling*

Choosing a sample at random is only the first step in carrying out a sample survey of a large human population. You must then contact the people in the sample and persuade them to cooperate. This isn't easy. Some people are rarely at home. Others don't want to talk with an interviewer. We say that *nonresponse* occurs when an individual chosen for the sample can't be contacted or refuses to cooperate. Try to learn the rate of nonresponse before putting too much trust in a sample result. Nonresponse rates of 30% or more are common. Nonresponse is higher in cities, especially in poor areas, so ignoring the people who did not respond can cause bias. Opinion polls usually substitute another person from the same neighborhood to reduce the bias.

Even the 1990 census, with the resources of the government behind it, had problems with nonresponse. The census is not a sample—it tries to count everyone in the country. The census count has some bias against cities and against minorities because of nonresponse, even though census interviewers tried six times to contact nonresponders. The Census Bureau estimates that the 1990 census missed 1.6% of the overall population, but that it failed to count 4.6% of blacks and 5.0% of Hispanics. ◆

EXAMPLE ▶ *Is the Experiment Realistic?*

The Physicians' Health Study gave pills to middle-aged men going about their everyday lives. Many experiments, however, take place in artificial environments. A psychologist studying the effects of stress on teamwork observes teams of students carrying out tasks in a psychology laboratory under different conditions. The students know it's "just an experiment" and that the stress will only last an hour. Do the conclusions of such experiments apply to real-life stress? An engineer uses a small pilot production process in a laboratory to find the choices of pressure and temperature that maximize yield from a complex chemical reaction. Do the results apply to a full-scale manufacturing plant?

These are not statistical questions. The psychologist and the engineer must use their understanding of psychology and engineering to judge how far their results apply. The statistical design enables us to trust the results for the students and the pilot production process, but not to generalize the conclusions to other settings. ◆

Spotlight **Experiments and Ethics**

5.3

Dr. Charles Hennekens, director of the Physicians' Health Study, had to concern himself with the goals, design, and implementation of his large-scale study. But other questions also arise in the course of such an experiment. Dr. Hennekens was asked about the ethics of experimenting on human health:

Charles Hennekens

Much has been made of the ethical concerns about randomized trials. There are instances where it would not be ethical to do a randomized trial. When penicillin was introduced for the treatment of pneumococcal pneumonia, which was virtually 100% fatal, the mortality rate plummeted significantly. Certainly it would have been unethical to do a randomized trial, to withhold effective treatment from people who need it.

There's a delicate balance between when to do or not to do a randomized trial. On the one hand, there must be sufficient belief in the agent's potential to justify exposing half the subjects to it. On the other hand, there must be sufficient doubt about its efficacy to justify withholding it from the other half of subjects who might be assigned to the placebos, the pills with inert ingredients. It was just these circumstances that we felt existed with regard to the aspirin and the beta carotene hypotheses.

When we are planning a statistical study, we must also face some *ethical questions*. Does the knowledge gained from an experiment or study justify the possible risk to the subjects? In the Physicians' Health Study, doctors gave their informed consent to take either aspirin, beta carotene, or a placebo, in any combination, as prescribed by the study designers. When it became clear that men taking aspirin had fewer heart attacks, the experiment was stopped so that all the subjects could take advantage of this new knowledge. In Spotlight 5.3 the director of the Physicians' Health Study explains why randomized comparative experiments are a mainstay of medical research and when such clinical trials are justified. Practical and ethical problems are never far from the surface when statistics is applied to real problems.

REVIEW VOCABULARY

Bias A systematic error that tends to cause the observations to deviate in the same direction from the truth about the population whenever a sample or experiment is repeated.

Confounding Two variables are confounded when their effects on the outcome of a study cannot be distinguished from one another.

Control group A group of experimental subjects who are given a standard treatment or no treatment (such as a placebo).

Convenience sample A sample that consists of the individuals who are most easily available, like people passing by in the street. A convenience sample is usually biased.

Double-blind experiment An experiment in which neither the experimental subjects nor the persons who interact with them know which treatment each subject received.

Experiment A study in which treatments are applied to people, animals, or things in order to observe the effect of the treatment.

Histogram A graph that displays how often various outcomes occur by means of bars. The height of each bar is the number of times an outcome or group of outcomes occurred in the data.

Margin of error As announced by most national polls, the margin of error says how close to the truth about the population the sample result would fall in 95% of all samples drawn by the method used to draw this one sample.

Observational study A study (such as a sample survey) that observes individuals and measures variables of interest but does not attempt to influence the responses.

Placebo effect The effect of a dummy treatment (such as an inert pill in a medical experiment) on the response of subjects.

Population The entire group of people or things that we want information about.

Randomized comparative experiment An experiment to compare two or more treatments in which people, animals, or things are assigned to treatments by chance.

Sample A part of the population that is actually observed and used to draw conclusions, or inferences, about the entire population.

Simple random sample A sample chosen by chance, so that every possible sample of the same size has an equal chance to be the one selected.

Statistical significance An observed effect is statistically significant if it is so large that it is unlikely to occur "just by chance" in the absence of a real effect in the population from which the data were drawn.

Table of random digits A table whose entries are the digits 0, 1, 2, 3, 4, 5, 6, 7, 8, 9 in a completely random order. That is, each entry is equally likely to be any of the 10 digits and no entry gives information about any other entry.

Voluntary response sample A sample that chooses itself by responding to a general invitation to write or call with their opinions. Such a sample is usually strongly biased.

SUGGESTED READINGS

BOX, GEORGE E. P., WILLIAM G. HUNTER, AND J. STUART HUNTER. *Statistics for Experimenters,* Wiley, New York, 1978, chapters 4, 7, and 8. This more advanced text places greater emphasis on concepts and on experimental design than most books at a similar level. It is a good source for information on designs more elaborate than those described in this chapter.

FREEDMAN, DAVID, ROBERT PISANI, ROGER PURVES, AND ANI ADHIKARI. *Statistics,* 2nd ed., Norton, New York, 1991, chapters 1, 2, 19, and 20. Excellent, but rather lengthy, conceptual discussion with good examples. Slightly higher in level than *For All Practical Purposes.*

MOORE, DAVID S. *Statistics: Concepts and Controversies,* 4th ed., Freeman, New York, 1997, chapters 1 and 2. Written for liberal arts students, this book provides more extensive discussion at about the same level as *For All Practical Purposes.*

MOORE, DAVID S. *The Basic Practice of Statistics,* Freeman, New York, 1995, chapter 3. Clear treatment of data production in a text on practical statistics at about the same level as *For All Practical Purposes.*

TANUR, JUDITH M. Samples and surveys. In David C. Hoaglin and David S. Moore (eds.), *Perspectives on Contemporary Statistics,* Mathematical Association of America, Washington, D.C., 1992, pp. 55–70. This essay describes the practice of sample surveys at a relatively nontechnical level.

EXERCISES ▲ *Optional.* ▒ *Advanced.* ◆ *Discussion.*

Sampling

1. A sociologist wants to know the opinions of employed adult women about government funding for day care. She obtains a list of the 520 members of a local business and professional women's club and mails a questionnaire to 100 of these women selected at random. Only 48 questionnaires are returned. What is the population in this study? What is the sample?

2. Home canners sometimes can vegetables in used mayonnaise jars to avoid buying special canning jars. *Organic Gardening* magazine wondered what percent of mayonnaise jars would break when used for canning. It obtained 100 mayonnaise jars and canned tomatoes in them. Only 3 of the jars broke. What is the population in this study? What is the sample?

3. A maker of electronic instruments buys 8-megabyte RAM memory chips from a supplier. The company wants to know the percent of substandard chips made by the supplier, so it tests all chips received and keeps records. Last year 32,000 out of 400,000 chips received failed to meet standards. What is the population that the company wants information about? What is the sample?

Bad Sampling Methods

◆ 4. Exercise 1 describes a sample intended to gather information about the opinions of employed adult women. The method of choosing the sample is biased. Explain why. In particular, what parts of the population will be under-represented in this sample?

◆ 5. A magazine for health foods and organic healing wants to establish that large doses of vitamins will improve health. The editors ask readers who have regularly taken vitamins in large doses to write in, describing their experiences. Of the 2754 readers who reply, 93% report some benefit from taking vitamins. Is the sample proportion of 93% probably higher than, lower than, or about the same as the percent of all adults who would perceive some benefit from large vitamin intake? Why? (In answering these questions, you have identified a source of bias in the sampling method.)

◆ 6. A newspaper advertisement for *USA Today: The Television Show* said, "Should handgun control be tougher? You call the shots in a special call-in poll tonight. If yes, call 1-900-720-6181. If no, call 1-900-720-6182. Charge is 50 cents for the first minute." Why is this opinion poll almost certainly biased?

◆ 7. Ann Landers once asked her female readers whether they would be content with affectionate treatment by men with no sex ever. Over 90,000 women wrote in, with 72% answering "Yes." Explain carefully why this sample is almost certainly biased. What is the *direction* of the bias? That is, is the percentage of all adult women who would be content with no sex ever lower or higher than the 72% in the sample?

◆ 8.   You are on the staff of a member of Congress who is considering a bill that would provide government-sponsored insurance for nursing home care. You report that 1128 letters have been received on the issue, of which 871 oppose the legislation. "I'm surprised that most of my constituents oppose the bill. I thought it would be quite popular," says the congresswoman. Are you convinced that a majority of the voters oppose the bill? How would you explain the statistical issue to the congresswoman?

Simple Random Sampling

9.   Your class in ancient Ugaritic religion is poorly taught and wants to complain to the dean. The class decides to choose three of its members at random to carry the complaint. The class list appears below. Choose a simple random sample of three using the table of random digits, beginning at line 110. Be sure to say how you labeled the class members.

Anderson	Gupta	Patnaik
Aspin	Gutierrez	Pirelli
Bennett	Harter	Rao
Bock	Henderson	Rider
Breiman	Hughes	Robertson
Castillo	Johnson	Rodriguez
Dixon	Kempthorne	Sosa
Edwards	Laskowsky	Tran
Gonzalez	Liang	Trevino
Green	Olds	Wang

10.   What kinds of programs do academic departments at a state university offer for honors students? You decide to report information from 5 randomly chosen departments in the liberal arts and sciences. Use the table of random digits starting at line 132 to select a simple random sample of 5 departments from the following list for your study. Be sure to show how you used the random digits.

Audiology	English	Physics
Biological Sciences	Foreign Languages	Political Science
Chemistry	General Sciences	Psychology
Communication	Health and Leisure Studies	Sociology
Computer Sciences	History	Statistics
Economics	Mathematics	Visual Arts
Earth Sciences	Philosophy	

11. A student wishes to study the opinions of faculty at her college on the advisability of setting up a state board of higher education to oversee all colleges in the state. The college has 380 faculty members.

   (a) What is the population in this situation?
   (b) Explain carefully how you would choose a simple random sample of 50 faculty members.
   (c) Use Table 5.1 starting at line 135 to choose *only the first 5* members of this sample.

12. The number of students majoring in political science at Ivy University has increased substantially without a corresponding increase in the number of faculty. The campus newspaper plans to interview 25 of the 450 political science majors to learn student views on class size and other issues. You suggest a simple random sample. Explain carefully how you would choose this sample. Then use Table 5.1 starting at line 120 to select *only the first 5* members of your sample.

Statistical Estimation

13. You must allocate 5 tickets to a rock concert among 25 clamoring members of your club. We will use this example to illustrate sampling variability.

   (a) Choose 5 at random to receive the tickets, using line 135 of Table 5.1 (ignore the asterisks).

Agassiz	Darwin	Herrnstein	Myrdal	Vogt*
Binet*	Epstein	Jimenez*	Perez*	Wang
Blum	Ferri	Lombrosco	Spencer*	Wilson
Chung*	Goddard*	Moll*	Thomson	Yerkes
Cuvier*	Hall	McKim*	Toulmin	Zimmer

   (b) In fact, 10 of the 25 club members are female. Their names are marked with asterisks in the list. Draw 5 at random 20 times, using a different row in Table 5.1 each time (include your sample from part a). Record the number of females in each of your samples. Make a histogram to display your results. What is the average number of females in your 20 samples?

   (c) Do you think the club members should suspect discrimination if none of the 5 tickets goes to women?

◆ 14. An opinion poll asks a sample of 1450 adults whether they jog regularly; 224 say "Yes." The margin of error for this poll is ± 3%.

(a) What percent of the sample jog?

(b) Explain to someone who knows no statistics what "margin of error" means here.

(c) What interval are you confident covers the percent of all adults who jog?

15. An opinion poll asks a sample of 1324 adults whether they believe that life exists on other planets; 609 say "Yes." What percent of the sample believes in extraterrestrial life? The polling organization announces a margin of error of ± 3%. What conclusion can you draw about the percent of all adults who believe that life exists on other planets?

◆ 16. National opinion polls such as the Gallup poll usually take weekly samples of about 1500 people.

(a) This sample size gives a margin of error of about ±3 percentage points. Explain to someone who knows no statistics what this means.

(b) Just before a presidential election, however, the polls often increase the size of their samples to about 4000 people. Is the margin of error now more than ±3%, less than ±3%, or still equal to ±3%? Why?

Experiments

The studies in Exercises 17 to 19 may produce invalid data because of confounding of outside influences with the treatment of interest. Explain in each case how confounding could influence the outcome.

◆ 17. A college student believes that rose hip tea has remarkable curative powers. To demonstrate this, she and several friends visit a local nursing home several times a week, talking with the residents and serving them rose hip tea. After a month, the head nurse reports that the residents visited are indeed more cheerful and alert.

◆ 18. A language teacher believes that study of a foreign language improves command of English. He examines the records at his high school and finds that students who elect a foreign language do indeed score higher on English achievement tests.

◆ 19. A job-training program is being reviewed. Critics claim that because the unemployment rate in the manufacturing region affected by the program was 8% when the program began and 12% four years later, the program was ineffective.

◆ 20.   It has been suggested that there is a "gender gap" in political party preference in the United States, with women more likely than men to prefer Democratic candidates. A political scientist asks each of a group of men and a group of women whether they voted for the Democratic or Republican candidate in the last congressional election. Explain carefully why this study is *not* an experiment.

◆ 21.   A study of the effect of living in public housing on family stability and other variables in poverty-level households was carried out as follows. The researchers obtained a list of all applicants for public housing during the previous year. Some applicants had been accepted, while others had been turned down by the housing authority. Both groups were interviewed and compared. Was this an observational study or an experiment? Explain your answer.

◆ 22.   Before a new variety of frozen muffin is put on the market, it is subjected to extensive taste testing. People are asked to taste the new muffin and a competing brand, and to say which they prefer. (Both muffins are unidentified in the test.) Is this an observational study or an experiment? Explain your answer.

Randomized Comparative Experiments

23.   Will reducing blood cholesterol levels prevent heart attacks? You have available a drug that will lower blood cholesterol. You also have 3000 men aged 50 to 65 who have high cholesterol and are willing to participate in a study. Outline the design of an experiment to settle the question. Your outline should follow the model of Figure 5.4. Be sure to give the sizes of the treatment groups and to indicate the outcomes you will examine.

◆ 24.   Does regular exercise reduce the risk of a heart attack? Several ways of studying this question suggest themselves.

   (a)   A researcher takes a sample of 2000 men in their 40s who recently suffered their first heart attack. He matches each with another man of the same age, occupation, and other demographic characteristics who has not had a heart attack. Both groups are questioned about their past exercise habits. Is this an experiment? Is it a prospective study? Explain your answers.

   (b)   Another researcher finds 2000 men over 40 who exercise regularly and have not had heart attacks. She matches each with a similar man who does not exercise regularly, and she follows both groups for 10 years. Is this an experiment? Is it a prospective study? Explain your answers.

   (c)   You have 4000 men over 40 who have not had a heart attack and who are willing to participate in a study. Outline the design of an experiment to investigate the effect of regular exercise on heart attacks.

(d) Explain clearly why the experiment you designed in part (c) gives better information about exercise as a preventer of heart attacks than the studies described in parts (a) and (b).

◆ 25. Should either or both the experiments in Exercises 23 and 24c be double-blind studies? Explain your answer.

26. Some investment advisers believe that charts of past trends in the prices of securities can help predict future prices. Most economists disagree. In an experiment to examine the effects of using charts, business students trade (hypothetically) a foreign currency at computer screens. There are 20 student subjects, named for convenience A, B, C, . . . , T. Their goal is to make as much as possible, and the best performances are rewarded with small prizes. The student traders have the price history of the foreign currency in dollars in their computers; they may or may not also have software that highlights trends. Describe a design for this experiment and use Table 5.1 to carry out the randomization required by your design.

◆ 27. A college allows students to choose either classroom or self-paced instruction in a basic mathematics course. The college wants to compare the effectiveness of self-paced and regular instruction. Someone proposes giving the same final exam to all students in both versions of the course and comparing the average score of those who took the self-paced option with the average score of students in regular sections.

(a) Explain why confounding makes the results of that study worthless.

(b) Given 30 students who are willing to use either regular or self-paced instruction, outline an experimental design to compare the two methods of instruction. Then use Table 5.1, starting at line 108, to carry out the randomization.

28. Below are the names of 20 patients who have consented to participate in a trial of surgical treatments for angina. Outline an experiment to compare surgical treatment with a placebo (sham surgery) and use Table 5.1, beginning at line 101, to do the required randomization. (Ignore the asterisks.)

Ashley	Cravens*	Lippmann	Strong*
Bean*	Dorfman	Mark*	Tobias
Block	Epstein	Morton*	Valenzuela*
Chen	Huang*	Popkin	Washington
Chavez*	Kidder	Spearman	Williams

29. Unknown to the researchers in Exercise 28, the eight subjects whose names are marked by asterisks will have a fatal heart attack during the study period. We can observe how sampling variability operates in a randomized experiment by keeping track of how many of these eight subjects are assigned to the group that will receive the new surgical treatment. Carry out the random assignment of 10 subjects to the treatment group 20 times, keeping track of how many asterisks are on the names you choose each time. Then make a histogram of the count of heart attack victims assigned to the treatment. What is the average number in your 20 tries?

◆ 30. Explain clearly the advantage of using several thousand subjects, rather than just 20, in the experiment of Exercise 28.

■ 31. Explain carefully how you would randomly assign the 20 subjects named in Exercise 28 to the four treatments in the Physicians' Health Study. Figure 5.6 (page 197) describes the treatments. Assign 5 of the 20 to each group. Use Table 5.1 at line 120 to carry out the randomization.

Statistical Evidence

◆ 32. A randomized comparative experiment examined whether a calcium supplement in the diet reduces the blood pressure of healthy men. The subjects received either a calcium supplement or a placebo for 12 weeks. The researchers concluded that "the blood pressure of the calcium group was significantly lower than that of the placebo group." "Significant" in this conclusion means statistically significant. Explain what statistically significant means in the context of this experiment, as if you were speaking to a doctor who knows no statistics.

◆ 33. The financial aid office of a university asks a sample of students about their employment and earnings. The report says that "for academic year earnings, a statistically significant difference was found between the sexes, with men earning more on the average. No significant difference was found between the earnings of black and white students." Explain both of these conclusions, for the effects of sex and of race on average earnings, in language understandable to someone who knows no statistics.

Additional Exercises

◆ 34. Ms. Caucus is her party's candidate in the Second Congressional District of Indiana. The party wants to know what percent of registered voters would vote for Ms. Caucus if the election were held tomorrow. A polling firm contacts 800 voters, of whom 456 say they would vote for Ms. Caucus. What is the population that the poll seeks information about? What is the sample? Explain to someone who knows no statistics the advantage of a sample of 800 voters over a sample of 200 voters.

◆ 35. Sampling from a list that contains only part of the population is a common cause of bias in sampling. In each of the following examples, explain why this source of bias may be present.

(a) To assess public opinion on a proposal to reduce welfare and unemployment payments, a polling firm selects a sample by random digit dialing. That is, they use a computer to dial residential telephone numbers at random.

(b) To assess the reaction of her constituents to the same proposal, a member of Congress uses her free-mailing privilege to send a questionnaire to every registered voter in her district.

◆ 36. The advice columnist Ann Landers regularly invites her readers to respond to questions asked in her newspaper column. On one occasion, she asked, "If you had it to do over again, would you have children?" Almost 10,000 parents wrote in, of whom 70% said "No." Shortly afterward, a national poll asked a random sample of 1400 parents the same question; 90% of this sample said "Yes." Which of these polls is more trustworthy, and why?

37. You are worried about the problem of false credentials being offered by candidates for employment at your firm. You decide to investigate some of the credentials at random from now on. Use line 123 of Table 5.1 to choose a simple random sample of four of the following group of candidates for investigation. Be sure to say how you labeled the candidates.

Adams	Edwards	Martinez	Russell
Alvarez	Frank	Michel	Sanguillen
Bartkowsky	Garcia	Miller	Toon
Bishop	Hohenstein	Ogden	Tran
Borchardt	Kim	Pierce	Ungarn
Chan	Kodaira	Pollack	Vlasov
Cleveland	LeMay	Riersol	Weinstein
Drasin	Marsden	Rubin	Wang

◆ 38. An experiment that claimed to show that meditation lowers anxiety proceeded as follows. The experimenter interviewed the subjects and rated their level of anxiety. Then the subjects were randomly assigned to two groups. The experimenter taught one group how to meditate and they meditated daily for a month. The other group was simply told to relax more. At the end of the month, the experimenter interviewed all the subjects again and rated their anxiety level. The meditation group now had less anxiety. Psychologists said that the results were suspect because the ratings were not blind. Explain what this means and how lack of blindness could bias the reported results.

39.   Ignoring all practical difficulties and moral issues, outline the design of an experiment that would settle the question of whether cigarette smoking causes lung cancer.

◆   40.   In a test of the effects of persistent pesticides, researchers will feed a diet contaminated with DDT to rats for 60 days after weaning. Then they will measure the rats' nerve responses to assess the effects of the DDT.

(a)   Explain why the experimenters should also study a control group of rats that are fed the same diet uncontaminated with DDT.

(b)   For 20 newly weaned male rats, outline the design of the experiment and use Table 5.1, starting at line 123, to carry out the randomization.

41.   Will providing child care for employees make a company more attractive to women, even those who are unmarried? You are designing an experiment to answer this question. You prepare recruiting material for two fictitious companies, both in similar businesses in the same location. Company A's brochure does not mention child care. There are two versions of Company B's material, identical except that one describes the company's on-site child-care facility. Your subjects are 40 unmarried women who are college seniors seeking employment. Each subject will read recruiting material for both companies and choose the one she would prefer to work for. You will give each version of Company B's brochure to half the women. You expect that a higher percentage of those who read the description that includes child care will choose Company B.

(a)   Outline an appropriate design for the experiment.

(b)   The names of the subjects appear below. Use Table 5.1, beginning at line 131, to do the randomization required by your design. List the subjects who will read the version that mentions child care.

Abrams	Danielson	Gutierrez	Lippman	Rosen
Adamson	Durr	Howard	Martinez	Sugiwara
Afifi	Edwards	Hwang	McNeill	Thompson
Brown	Fluharty	Iselin	Morse	Travers
Cansico	Garcia	Janle	Ng	Turing
Chen	Gerson	Kaplan	Quinones	Ullmann
Cortez	Green	Kim	Rivera	Williams
Curzakis	Gupta	Lattimore	Roberts	Wong

◆ 42.   Fizz Laboratories, a pharmaceutical company, has developed a new pain-relief medication. Sixty patients suffering from arthritis and needing pain relief are available. Each patient will be treated and asked an hour later, "About what percentage of pain relief did you experience?"

   (a)   Why should Fizz not simply administer the new drug and record the patients' responses?
   (b)   Outline the design of an experiment to compare the drug's effectiveness with that of aspirin and of a placebo.
   (c)   Should patients be told which drug they are receiving? How would this knowledge probably affect their reactions?
   (d)   If patients are not told which treatment they are receiving, the experiment is single-blind. Should this experiment be double-blind also? Explain.

▨ 43.   Is the number of days a letter takes to reach another city affected by the time of day it is mailed and whether or not the ZIP code is used as part of the address? Describe briefly the design of a two-factor experiment to investigate this question. Be sure to specify the treatments exactly and to tell how you will handle outside variables such as the day of the week on which the letter is mailed.

◆ 44.   Exercise 43 illustrates the use of a statistically designed experiment to answer questions that arise in everyday life. Select a question of interest to you that an experiment might answer and carefully discuss the design of an appropriate experiment.

▨ 45.   Corn is an important part of the feed of many farm animals. Normal corn is low in the amino acid lysine. Animals may grow faster if they eat new varieties of corn with increased amounts of lysine. Researchers conduct an experiment to compare a new variety, called floury-2, with normal corn. They mix corn-soybean meal diets using each type of corn at each of three protein levels, 12% protein, 16% protein, and 20% protein. There are thus six diets in all. Ten one-day-old male chicks are assigned to each diet, and their weight gains after 21 days are recorded.

   (a)   This experiment has two factors. What are they?
   (b)   Outline the design of the experiment. Be sure to use randomization. (You need not actually carry out the randomization required by your design.)

◆ 46.   The many connections on the bottom of electronic circuit boards are soldered by passing the board through a standing wave of molten solder. An engineer wants to study the effect the speed of the conveyor belt that carries the circuit boards has on the quality of the soldering. The speeds to be compared are 20, 25, and 30 feet per minute. The outcome variable is the number of improperly soldered connections among the 2000 connections on a circuit board.

(a) The engineer plans to process 10 boards at each conveyor speed. Why should he assign the speeds at random to the 30 boards, rather than simply process the first 10 at 20 feet per minute, the second 10 at 25 feet per minute, and so on?

(b) Outline the design of a randomized comparative experiment, beginning with boards numbered 1 through 30 in the order in which they will be soldered.

(c) Enter Table 5.1 at line 130 to carry out the randomization required. List the sequence of 30 conveyor speeds that the engineer will use when he carries out the experiment.

◆ 47. A psychologist reports that "in our sample, ethnocentrism was significantly higher among church attenders than among nonattenders." Explain what this means in language understandable to someone who knows no statistics. Do not use the word "significance" in your answer.

◆ 48. The cigarette industry has adopted a voluntary code requiring that models appearing in its advertising must appear to be at least 25 years old. Studies have shown, however, that consumers think many of the models are younger. Here is a quote from a study that asked whether different brands of cigarettes use models that appear to be of different ages:

> [Statistical analysis] revealed that the brand variable is highly significant, indicating that the average perceived age of the models is not equal across the 12 brands. As discussed previously, certain brands such as Lucky Strike Lights, Kool Milds, and Virginia Slims tended to have younger models. . . . [From Michael B. Maziz et al., Perceived age and attractiveness of models in cigarette advertisements, *Journal of Marketing*, 56 (January 1992): 22–37.]

Explain to someone who knows no statistics what "highly significant" means and why this is good evidence of differences among all advertisements of these brands even though the subjects saw only a sample of ads.

WRITING PROJECTS

1 ▶ The Current Population Survey (CPS) is the most important sample survey of the federal government. The CPS provides monthly information on employment and unemployment, and gathers information on many other economic and social issues on a less frequent basis by varying the questions asked each month.

Locate information about the design of the CPS. Here are some suggested sources. The Bureau of Labor Statistics makes information available via the Internet. The current World Wide Web location is http://stats.bls.gov (the address may change in the future). The sampling design of the CPS is described in detail in the Bureau of Labor Statistics *Handbook of Methods,* which is updated from time to time and can be found in most college libraries.

Using information from one of these sources, write a clear description of the multistage sampling design used for the CPS. Because the sampling design makes use of the new idea of *stratified sampling,* you may also wish to read the discussion of stratified sampling in Moore's *Statistics: Concepts and Controversies* (see Suggested Readings) or another text.

2 ▶   Articles in the press often describe medical findings based on an experiment. The conclusion of the Physicians' Health Study that taking aspirin regularly helps prevent heart attacks is an example. Find an article in a newspaper or magazine that deals with a recent medical study. Describe the purpose and design of the study. Was it an experiment? What were the conclusions of the study, and how well grounded do you think they are?

Add a brief critique of the news article's presentation. Does the article mention a control group? Does it mention random assignment of the subjects? If the article concerns an observational study, does it warn against causal conclusions?

3 ▶   Choose an issue of current interest to students at your school. Prepare a short (no more than five questions) questionnaire to determine opinions on this issue. Choose a sample of about 25 students, administer your questionnaire, and write a brief description of your findings. Also write a short discussion of your experiences in designing and carrying out the survey.

(Although 25 students are too few for you to be statistically confident of your results, this project centers on the practical work of a survey. You must first identify a population; if it is not possible to reach a wider student population, use students enrolled in this course. Did the subjects find your questions clear? Did you write the questions so that it was easy to tabulate the responses? At the end, did you wish you had asked different questions?)

4 ▶   Although experiments with human subjects raise special ethical questions, there are also ethical issues associated with any study that collects data from human subjects. Here are two of these issues. Address one of them in a brief essay.

- Any institution that receives federal funds must have an *Institutional Review Board (IRB)* that reviews in advance all studies that use human subjects. The IRB is charged with protecting the welfare of the subjects. What is the name of your college's IRB? Who are the members?

Are there representatives from outside the college, and if so, how are they chosen? What guidelines does your IRB follow? Do you have any suggestions for strengthening the protection of subjects offered by the IRB's review process?

• Suppose that you are conducting a sample survey that gathers opinions from subjects. Ethical standards require that you give potential subjects some information about the survey and get their *informed consent* to participate. What kinds of information should respondents be given in order to decide whether to participate? (Perhaps they expect to spend 10 minutes, but the survey takes an hour. Perhaps questions about sex and drugs appear without warning.) Should respondents always be told who is sponsoring the poll? (If so, will knowing that the Republican National Committee is the sponsor affect their answers?) Should a poll always offer to send respondents a copy of the final report so they can see how their information is being used? (That's expensive.)

CHAPTER 6

DESCRIBING DATA

A flood of data is a prominent feature of modern society. Data, or numerical facts, are essential for making decisions in almost every area of our lives. But like other great floods, the flood of numbers threatens to overwhelm us. We must control the flood by careful organization and interpretation. A corporate data base, for example, contains an immense volume of data—on employees, sales, inventories, customer accounts, equipment, and other topics. These data are useful only if we can organize them and present them so that their meaning is clear. The penalties for ignoring data can be severe—several banks have suffered billion-dollar losses from unauthorized trades in financial markets by their employees, trades that were hidden in a mass of data that the banks' management did not examine carefully.

To use data for human purposes, we must compress, summarize, and describe them. A few numbers computed from the data—averages, percents, and the like—can be very helpful. Numbers computed from the data are also the raw material of *statistical inference,* the science of drawing conclusions from data accompanied by a mathematical statement of our confidence in the conclusions. Statistical inference works best when we ask specific questions and then produce data to answer those specific questions. However, this neat process is not always possible. We may have to analyze a mass of data collected for other purposes, such as government or corporate records, before we know whether the data can help to answer our questions. In other cases, we may not even know what questions we should ask.

EXPLORING DATA

Exploratory data analysis is the art of letting the data speak, of seeing in data patterns that we may not have anticipated. Exploratory analysis is informal. Unlike inference, it does not seek answers to specific questions or use mathematics to support its conclusions. However, exploratory data analysis is not at all a diversion from the mainstream of statistics. Even in the most carefully planned experiment, exploring the data is an essential first step. It may reveal unsuspected errors or an important effect that was not anticipated.

Exploratory analysis of data combines numerical summaries with graphical display. We can grasp pictures more easily than columns of numbers, so the most powerful tools of data analysis are graphs. As long as our emphasis is on description rather than inference, pictorial display of data occupies first place.

EXAMPLE ▶ *A Graph of Napoleon's Defeat*

Displaying data pictorially is not a new idea. In 1861, the French engineer Charles Minard used an elaborate and original graph to show the impact of harsh winter conditions on Napoleon's troops during the unsuccessful invasion of Russia in 1812. Some 422,000 French soldiers entered Russia and 100,000 reached Moscow, but only 10,000 straggled back. The blue band across the map of eastern Europe in Minard's graph (Figure 6.1) shows the route of Napoleon's Grand Army. The width of the band is proportional to the number of surviving troops. As the number of survivors dwindles, the river of soldiers narrows to a trickle. The red band shows the retreat of the army, and the black line at the bottom of Minard's graph shows the temperature during the winter

FIGURE 6.1   Redrawing of Charles Minard's 1861 graph of Napoleon's Russian campaign. (Minard used the Reaumur temperature scale, in which water boils at 80° and freezes at 100°.)

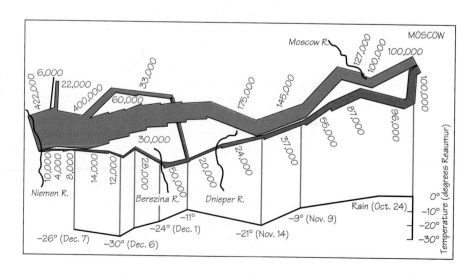

retreat from Moscow. The falling temperatures and shrinking army march together in that famous defeat. ◆

Visual representations of data abound in books and the news media, although few are as imaginative as Minard's. The progress of computer graphics has given a new emphasis to pictorial display of data in business, medicine, and technical fields. The computing power of machines allied with the unique ability of the human eye and brain to recognize visual patterns provides powerful new tools for data analysis.

In this chapter we use both numbers and pictures to explore data. Here are three principles that provide the tactics for exploratory analysis of data:

1. First examine each variable individually. Then move on to study the relationships among several variables.
2. Begin with a graph or graphs. Then add numerical summaries of specific aspects of the data.
3. When examining a graph, look first for an overall pattern in the data and then for important deviations from this pattern.

These principles also organize the material in this chapter. We start with data on a single variable, then move to relations among several variables. In each setting, we first display the data in graphs, then add numerical summaries.

DISPLAYING DISTRIBUTIONS

The **distribution** of a variable tells us what values the variable takes and how often it takes each value. Data analysis begins with graphical displays of the distribution of a single variable.

EXAMPLE ▶ *Making a Dotplot*

How many home runs must a major league baseball player hit to lead the league? Table 6.1 gives the American League leaders and their home run totals from 1972 to 1995. We can get a quick picture of the distribution of the league-leading home run totals by making a **dotplot.** First, draw a horizontal axis marked off to span the range of the home run counts. Then record each observation as a dot above the axis. Here is the result:

TABLE 6.1	American League Home Run Leaders, 1972–1995				
Year	Player	Home Runs	Year	Player	Home Runs
1972	Dick Allen	37	1984	Tony Armas	43
1973	Reggie Jackson	32	1985	Darrell Evans	40
1974	Dick Allen	32	1986	Jesse Barfield	40
1975	George Scott and Jackson	36	1987	Mark McGwire	49
1976	Graig Nettles	32	1988	Jose Canseco	42
1977	Jim Rice	39	1989	Fred McGriff	36
1978	Jim Rice	46	1990	Cecil Fielder	51
1979	Gorman Thomas	45	1991	Canseco and Fielder	44
1980	Reggie Jackson	41	1992	Julio Gonzalez	43
1981	Four players	22	1993	Julio Gonzalez	46
1982	Thomas and Jackson	39	1994	Ken Griffey, Jr.	40
1983	Jim Rice	39	1995	Albert Belle	50

We see, for example, that 32, 39, and 40 each led the league three times during these 24 seasons. Look first for the overall pattern of this distribution. All but one of the 24 observations fall between 32 and 51, but their pattern is irregular. A systematic pattern is often hard to see in small data sets such as this. What about deviations from the overall pattern? There is a clear **outlier,** an individual observation that falls far outside the range of the remaining data. In 1981, it took only 22 homers to lead the league. Outliers often point to errors in recording the data or to unusual circumstances. In fact, the 1981 baseball season was interrupted by a players' strike that reduced the number of games played from the usual 162 to 108. ◆

Dotplots are quick to draw and work well for small numbers of observations. They become awkward when there are many observations or when the observations do not have well-separated values. (Whole-number values like counts of home runs are ideal for dotplots.) When a dotplot is not satisfactory, we display a distribution by a more formal type of graph, a **histogram.** To see how histograms work, let's ask another baseball question: How well do major league batters hit?

EXAMPLE ▶ *Making a Histogram*

The simplest measure of how well a baseball player hits is his batting average, which is simply the proportion of times at bat that the player gets a hit. A player who has been at bat only a few times may have an unusually high or low batting average just because of a few lucky hits or a few unlucky outs. To eliminate these cases, we consider only players who have been at bat 200 or more times in a season. Figure 6.2 is a histogram of the distribution of batting aver-

Data displayed on the New York Yankees scoreboard.

ages for all 167 American League players who batted 200 or more times in the 1980 season. We can learn from this example how to draw a histogram of any distribution.

To make a histogram

1. *Divide the range of the data into classes of equal width.* We will use classes that cover a 10-point range of batting averages. The first two classes are

   .185 ≤ batting average < .195
   .195 ≤ batting average < .205

   Leave no space between classes and be careful about the endpoints of the classes. In this example, .194 falls in the first class and .195 in the second.

2. *Count the number of observations in each class.* These counts are called **frequencies.** For example, 3 players had batting averages between .195 and .204, so the frequency of this class is 3. The frequencies for all the batting average classes appear in Table 6.2.

3. *Draw the histogram.* The batting average scale is horizontal and the frequency scale vertical. Each bar represents a class. The base of the bar covers the class, and the bar height is the class frequency. Draw the graph with no horizontal space between the bars (unless a class is empty, so that its bar has 0 height). ◆

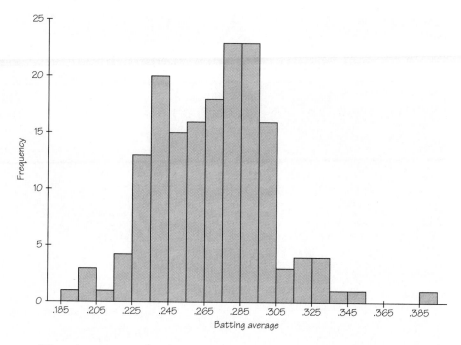

FIGURE 6.2 Histogram of 1980 American League batting averages.

There is no one right choice of the classes in a histogram. Too few classes will give a "skyscraper" graph, with all values in a few classes with tall bars. Too many will produce a "pancake" graph, with most classes having one or no observations. Neither choice will give a good picture of the shape of the distribution. You must use your judgment in choosing classes to display the shape.

Figure 6.2 is a successful histogram because the overall pattern of the distribution is immediately apparent. A typical player hit about .270, because the

TABLE 6.2	Frequencies for the Histogram in Figure 6.2		
Class	Frequency	Class	Frequency
.185 to .194	1	.295 to .304	16
.195 to .204	3	.305 to .314	3
.205 to .214	1	.315 to .324	4
.215 to .224	4	.325 to .334	4
.225 to .234	13	.335 to .344	1
.235 to .244	20	.345 to .354	1
.245 to .254	15	.355 to .364	0
.255 to .264	16	.365 to .374	0
.265 to .274	18	.375 to .384	0
.275 to .284	23	.385 to .394	1
.285 to .294	23		

Data displayed on a laptop computer.

center of the histogram lies near that value. All but a few players hit between about .225 and .305, and all but one hit between .190 and .350. What is more, the histogram has a roughly regular shape, with a few batting averages in the .220s, many between .240 and .290, then falling off to a few in the .330s and above. We just described the *center, spread,* and *shape* of the histogram. That's a good outline for describing the overall pattern.

Another aspect of the histogram is also immediately apparent. The single observation at .390 is an outlier. Outliers are often the result of errors and must be carefully investigated. It would be easy, for instance, to type .390 in place of .290 while recording the data. However, this outlier is not a mistake. George Brett of the Kansas City Royals hit .390 in 1980, the highest batting average in the major leagues since the start of World War II. Brett therefore deserves his isolated point on the histogram. Here is a summary of our tactics for looking at a distribution:

Look for the overall pattern and for deviations from the pattern. To describe the overall pattern of a distribution:

- See if the distribution has a simple *shape* that you can describe in a few words.
- Describe the *center* and the *spread* of the distribution.

One common deviation from the overall pattern in any graph of data is an *outlier,* an observation that falls outside the overall pattern of the graph.

We will soon meet some numerical ways to measure center and spread. Shape and outliers are matters of judgment. In general, don't call an observation an outlier unless it stands clearly apart from the other data. The 22 in the dotplot on page 219 is an outlier, but the points at 49, 50, and 51 don't stand clearly apart from the other observations. As for shape, keep in mind some common shapes, especially *symmetric* and *skewed*.

> A distribution is **symmetric** if the right and left sides of the histogram are approximately mirror images of each other.
>
> A distribution is **skewed to the right** if the right side of the histogram (containing the upper half of the observations) extends much farther out than the left side (containing the lower half of the observations). It is **skewed to the left** if the left side of the histogram extends much farther out than the right side.

Distributions of real data are usually only roughly symmetric. We consider Figure 6.2 (without George Brett) to be approximately symmetric. Here is an example of a skewed distribution.

EXAMPLE ▶ *How Many Operations Do Doctors Perform?*

Here are data on the number of hysterectomies performed in a year by each of a sample of 15 male doctors in Switzerland.

27   50   33   25   86   25   85   31   37   44   20   36   59   34   28

Figure 6.3 is a histogram of this distribution. The classes are 20–29, 30–39, and so on. Ten of the 15 doctors performed between 20 and 39 hysterectomies. The distribution has a long right tail not matched by corresponding observations to the left of the bulk of the data. Two doctors performed more

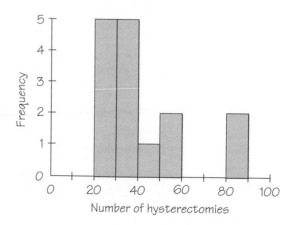

FIGURE 6.3 Histogram of the number of hysterectomies performed by a sample of male Swiss doctors.

than 80 hysterectomies. The distribution is skewed to the right. Notice that the direction of the skew is the direction of the long tail, not the direction in which most observations cluster. ◆

Deviations from a distribution's overall pattern often take the form of *outliers* or *gaps*. The baseball strike of 1981 and George Brett's exceptional performance were responsible for outliers in the distributions of home runs and of 1980 batting averages. The next example features a gap.

EXAMPLE ▶ *Quality Control*

Spotting patterns in quality control data can increase a manufacturer's productivity and profitability. Figure 6.4 presents data from a study by the quality expert W. Edwards Deming (see Spotlight 6.1). The data concern the size of steel rods used in a manufacturing process. The histogram in Figure 6.4 displays the diameters of 500 steel rods, as reported by the manufacturer's inspectors. The rod diameters are measured to the nearest thousandth of a centimeter, so that each bar in the histogram shows how often one measurement occurred.

An overall pattern is apparent in the size of the rods: the distribution is approximately symmetric, centered at 1.002 centimeters and falling off rapidly both above and below. The departure from regularity is the gap at 0.999 centimeter.

One centimeter is the lower specification limit for these rods. Rods that are any smaller than this will be loose in their bearings and should therefore be rejected. The empty 0.999 class in the histogram, together with the taller than expected bar for 1.000 centimeter, provides strong evidence that the inspectors were passing rods that measured 0.999 centimeter by recording them as 1.000 centimeter. The inspectors didn't realize that just one-thousandth of a

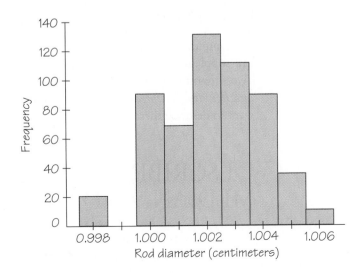

FIGURE 6.4 Deming's illustration of the effects of improper inspection: a histogram with a gap.

Spotlight *W. Edwards Deming*

6.1

From one point of view, statistics is about understanding variation. Elimination of variation in products and processes is the central theme of statistical quality control. So it is not surprising that a statistician should become the leading guru of quality management. In the final decades of his long life, W. Edwards Deming (1900–1993) was one of the world's most influential consultants to management.

Deming grew up in Wyoming and earned a doctorate in physics at Yale. Working for the U.S. Department of Agriculture in the 1930s, he became acquainted with the young field of statistics and in particular with statistical process control, newly invented by Walter Shewhart of AT&T. In 1939 he moved to the U.S. Census Bureau as an expert on sampling.

The work that made Deming famous began after he left the government in 1946. He visited

W. Edwards Deming

Japan to advise on a census, but returned to lecture on quality control. He earned a large following in Japan, which named its premier prize for industrial quality after him. As Japan's reputation for excellence in manufacturing grew, Deming's fame grew with it. Blunt-spoken and even abrasive, he told corporate leaders that most quality problems are system problems for which management is responsible. He urged breaking down barriers to worker involvement and constant search for causes of variation.

centimeter can be crucial. With better training of the inspectors, the missing 0.999 class filled in and the distribution became quite regular. ◆

Dotplots and histograms display the shape of a distribution of values. We can supplement a picture of the shape with a few carefully chosen numbers. Two important aspects of a distribution are its *center* (sometimes called *location*) and its *spread*. Center and spread are visible in a graph, but now we want to describe them using numbers.

DESCRIBING CENTER: MEDIAN AND MEAN

One simple way to describe the center of a distribution is to give the midpoint, the point with half the observations above it and half below it. This is the idea of the *median*. We will call the median M for short. You can think of it as a typical value. Although the idea of the median as midpoint or typical value is simple, we must make the idea precise by giving a specific rule for calculating the median. Here it is.

> To find the **median M** of a distribution:
>
> 1. Arrange all observations in order of size, from smallest to largest.
> 2. If the number n of observations is odd, the median M is the center observation in the ordered list. The location of the median is found by counting $(n + 1)/2$ observations up from the bottom of the list.
> 3. If the number n of observations is even, the median M is the average of the two center observations in the ordered list. The location of the median is again $(n + 1)/2$ from the bottom of the list.

Be sure to write down each individual observation in the data set, even if several observations repeat the same value. And be sure to arrange the observations in order of size before locating the median. The middle observation in the haphazard order in which the observations first come has no importance. Note that the recipe $(n + 1)/2$ gives the position of the median in the ordered list of observations, *not* the median itself. For an example that applies our rule, let's return to the study of Swiss doctors.

EXAMPLE ▶ *Calculating the Median*

Our sample data for male doctors were

27 50 33 25 86 25 85 31 37 44 20 36 59 34 28

To find the median, first arrange the observations in order:

20 25 25 27 28 31 33 34 36 37 44 50 59 85 86

There are $n = 15$ observations, so the location of the median is

$$\frac{n + 1}{2} = \frac{16}{2} = 8$$

The median is the 8th observation in the ordered list. Therefore $M = 34$.

The study also looked at a sample of 10 female Swiss doctors. The numbers of hysterectomies performed by these doctors (arranged in order) were

5 7 10 14 18 19 25 29 31 33

The location of the median is

$$\frac{n+1}{2} = \frac{11}{2} = 5.5$$

The location 5.5 means "halfway between the fifth and sixth observations in the ordered list." So the median is the average of these two observations:

$$M = \frac{18 + 19}{2} = 18.5$$

The typical female doctor performed many fewer hysterectomies than the typical male doctor. This was one of the important conclusions of the study. Notice that whenever the number of observations n is odd, the median is one of the observations in the list. When n is even, the median lies midway between two observations. ◆

Another way to measure the center of a set of observations is to find their ordinary arithmetic average, which statisticians call the *mean*. The mean is usually written as $\bar{x}$, which we read as "x-bar."

To find the **mean $\bar{x}$** of a set of observations, add their values and divide by the number of observations. If the n observations are $x_1, x_2, \ldots, x_n$, their mean is

$$\bar{x} = \frac{x_1 + x_2 + \cdots + x_n}{n}$$

EXAMPLE ▶ *Calculating the Mean*

Our sample data for 15 male doctors were

27   50   33   25   86   25   85   31   37   44   20   36   59   34   28

The mean number of operations performed by the doctors is

$$\bar{x} = \frac{27 + 50 + 33 + \cdots + 28}{15} = \frac{620}{15} = 41.3 \quad ◆$$

This example illustrates an important difference between the mean and the median. The mean is strongly influenced by a few large observations. In particular, the mean of a right-skewed distribution is larger than the median. The median number of hysterectomies performed by the male doctors was 34, but the few large values (85 and 86) in the right tail of the distribution pull the mean up to 41.3. In practice, you must ask yourself whether the "midpoint" (the median) or the "average" (the mean) is a better description of the center of the data.

DESCRIBING SPREAD: THE FIVE-NUMBER SUMMARY

The mean and median provide two different measures of the center of a distribution. But a measure of center alone can be misleading. The U.S. Bureau of Labor Statistics reports that in 1994 the median income of American households was $32,264. Half of all households had incomes below $32,264, and half had higher incomes. But these figures do not tell the whole story.

EXAMPLE ▶ *Affluentia and Spartany*

Imagine two countries with the same median household income, say, $32,264. But these countries have vastly different economic systems. In Affluentia, all households earn nearly the same amount. In Spartany, on the other hand, there are extremes of wealth and poverty. Figure 6.5 displays the two income distributions. Despite the fact that median income is identical in Affluentia and Spartany, income patterns in the two countries are dramatically different. Family incomes in Spartany have much greater spread or variation than in Affluentia. The median tells us nothing about the spread of a distribution. ◆

The simplest useful numerical description of a distribution consists of both a measure of center and a measure of spread. One way to measure spread is to calculate the *range,* the difference between the highest and lowest

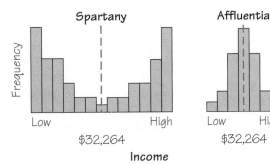

FIGURE 6.5 Two distributions with the same center but unequal spread.

observations. For example, the highest and lowest 1980 American League batting averages were .390 and .188. The range is therefore .390 − .188, or .202. This example illustrates a weakness in the range as a measure of spread: it is determined by the two most extreme observations. George Brett alone adds 40 points to the range of 1980 batting averages. In addition to noting the smallest and largest observations, we can indicate the spread of a distribution by reporting how spread out the middle 50% of the data is.

The middle 50% of the observations lie between the two *quartiles*. Here is the idea behind quartiles: 25% of the observations lie below the *first quartile*, and 75% lie below the *third quartile*. The median itself is the second quartile, because half the observations lie below it. Together the median and quartiles divide the data into quarters. Just as with the median, we need a rule to make the idea of quartiles precise. Here it is.

To calculate the **quartiles Q_1 and Q_3:**

1. Arrange the observations in increasing order and locate the median *M* in the ordered list of observations.
2. The first quartile Q_1 is the median of the observations whose position in the ordered list is to the left of the location of the overall median.
3. The third quartile Q_3 is the median of the observations whose position in the ordered list is to the right of the location of the overall median.

E X A M P L E ▶ *Calculating Quartiles*

The numbers of hysterectomies performed by our sample of 15 male doctors were (arranged in order):

20   25   25   27   28   31   33   **34**   36   37   44   50   59   85   86

There are an odd number of observations, so the median is the middle one. It is the 8th in the list, $M = 34$. The median appears in bold type in the list. The first quartile is the median of the 7 observations to the left of the median. This is the 4th of these 7 observations because for $n = 7$, $(n + 1)/2 = 8/2 = 4$. So $Q_1 = 27$. The third quartile is the median of the 7 observations to the right of the median, $Q_3 = 50$. The overall median is left out of the calculation of the quartiles when there are an odd number of observations.

Suppose that another sample gave the 10 observations

20   22   26   27   27 | 27   28   30   32   32

There are an even number of observations, so the median lies midway between the middle pair. Its location is halfway between the 5th and 6th values, marked by | in the list above. Because both these values are 27, $M = 27$. The first quartile is the median of the first 5 observations, because these are the observations below the location of the median. Be sure to note that it is the location of the median, not its numerical value, that decides which observations are included in finding the quartiles. That location is halfway between the fifth and sixth observations, even though both of these observations have the same value. Check that $Q_1 = 26$ and $Q_3 = 30$. ◆

Now we have a quick description of both the center and the spread of a set of data. The **five-number summary** consists of the median, the two quartiles, and the two extremes (the smallest and largest individual observations). Always write the five-number summary in order from the smallest number to the largest. For example, we could calculate from the list of 167 American League batting averages that the five-number summary is

.188   .244   .271   .290   .390

The five-number summary contains much useful information. We see that a typical major league regular player hit .271, and that the middle half of all such players hit between .244 and .290. A player who hit above .290 was in the top quarter of major league hitters, and will no doubt point that out when negotiating his salary. To make the five-number summary more vivid, we again turn to a graph.

A **boxplot** is a graph of the five-number summary. A central box spans the quartiles, with a line marking the median. Whiskers extend out from the box to the extremes.

The box in a boxplot shows the range of the middle half of the data. Both the box and the span of the whiskers help display the spread of the distribution, while the median marks the center. You can draw the boxes in a boxplot either horizontally or vertically, as you prefer. In either case, be sure to accompany the plot by a scale marked off in units of the variable being described. Locate the median, quartiles, and extremes on this scale in order to draw the boxplot.

Boxplots are particularly helpful for comparing several distributions. Figure 6.6, for example, displays the distributions of the number of hysterectomies performed in a year by our samples of male and female Swiss doctors. The scale in this example is the count of hysterectomies. Figure 6.6 quickly

FIGURE 6.6   Side-by-side boxplots comparing the number of hysterectomies performed by male and female Swiss doctors.

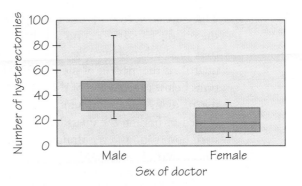

conveys a lot of information. You can tell at a glance that female doctors in general perform far fewer hysterectomies than men. In fact, the upper extreme for females falls below the male median. You can also see that the female distribution has less spread. In particular, it lacks the few very large observations that stretch out the upper whisker for the men. Figure 6.6 illustrates once again the effectiveness of visual displays. Before concluding that male doctors are more ready to perform hysterectomies, however, we need more information about the samples. Both males and females should be drawn from doctors with the same specialty and with practices of similar size. That was in fact the case for the Swiss study.

DESCRIBING SPREAD: THE STANDARD DEVIATION

Although the five-number summary is the most generally useful numerical description of a distribution, it is not the most common. That distinction belongs to the combination of the mean with the *standard deviation.* The mean, like the median, is a measure of center. The standard deviation, like the quartiles and extremes in the five-number summary, measures spread. The standard deviation and its close relative, the *variance,* measure spread by looking at how far the observations are from their mean.

EXAMPLE ▶ *Understanding the Standard Deviation*

A person's metabolic rate is the rate at which the body consumes energy. Metabolic rate is important in studies of weight gain, dieting, and exercise. Here are the metabolic rates of 7 men who took part in a study of dieting. The units are calories per 24 hours. These are the same calories used to describe the energy content of foods.

1792   1666   1362   1614   1460   1867   1439

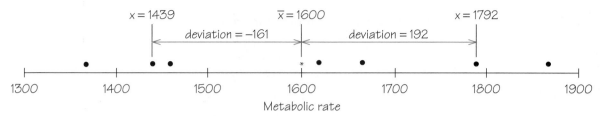

FIGURE 6.7 The variance and standard deviation measure spread by looking at the deviations of observations from their mean.

Figure 6.7 is a dotplot of these 7 metabolic rates, with their mean marked by an asterisk (*). The arrows show two of the deviations $x - \bar{x}$ of the individual observations from the mean. These deviations show how spread out the data are about their mean. Some of the deviations are positive and some are negative. Squaring the deviations makes them all positive. Observations far from the mean in either direction will have large positive squared deviations. So a reasonable measure of spread is the average of the squared deviations. This average is called the *variance*. The variance is large if the observations are widely spread about their mean; it is small if the observations are all close to the mean.

But the variance has the wrong units: if we measure metabolic rate in calories, the variance of the metabolic rates is in squared calories. Taking the square root of the variance gets us back to calories. The square root of the variance is the *standard deviation*. ◆

> The **variance s^2** of a set of observations is an average of the squares of the deviations of the observations from their mean. In symbols, the variance of n observations $x_1, x_2, \ldots, x_n$ is
>
> $$s^2 = \frac{(x_1 - \bar{x})^2 + (x_2 - \bar{x})^2 + \cdots + (x_n - \bar{x})^2}{n - 1}$$
>
> The **standard deviation s** is the square root of the variance s^2.

Notice that the "average" in the variance s^2 divides the sum by one fewer than the number of observations, that is, $n - 1$ rather than n. The reason is that the deviations $x_i - \bar{x}$ always sum to exactly 0, so that knowing $n - 1$ of them determines the last one. Some calculators offer a choice between $n - 1$ and n in calculating s. The difference is small unless there are very few observations, but $n - 1$ is the usual choice in applied statistics. You should rarely need to calculate s by hand, because most calculators allow you to obtain $\bar{x}$ and s from keyed-in data with the push of a button.

It is worthwhile doing one or two examples step by step to be certain how the recipe works. Here is one example.

E X A M P L E ▶ *Calculating the Standard Deviation*

To find the standard deviation of the 7 metabolic rates, first find the mean:

$$\bar{x} = \frac{1792 + 1666 + 1362 + 1614 + 1460 + 1867 + 1439}{7}$$

$$= \frac{11,200}{7} = 1600 \text{ calories}$$

To see clearly the nature of the variance, start with a table of the deviations of the observations from this mean.

Observations x_i	Deviations $x_i - \bar{x}$	Squared Deviations $(x_i - \bar{x})^2$
1792	$1792 - 1600 = 192$	$192^2 = 36,864$
1666	$1666 - 1600 = 66$	$66^2 = 4,356$
1362	$1362 - 1600 = -238$	$(-238)^2 = 56,644$
1614	$1614 - 1600 = 14$	$14^2 = 196$
1460	$1460 - 1600 = -140$	$(-140)^2 = 19,600$
1867	$1867 - 1600 = 267$	$267^2 = 71,289$
1439	$1439 - 1600 = -161$	$(-161)^2 = 25,921$
	sum = 0	sum = 214,870

The variance is the sum of the squared deviations divided by one less than the number of observations:

$$s^2 = \frac{214,870}{6} = 35,811.67$$

The standard deviation is the square root of the variance:

$$s = \sqrt{35,811.67} = 189.24 \text{ calories} \quad ◆$$

More important than the details of hand calculation are the properties that determine the usefulness of the standard deviation:

- s measures spread about the mean. Use s only when you choose the mean as your measure of center.
- $s = 0$ only when there is no spread, that is, when all observations have the same value. Otherwise $s > 0$, and s gets larger as the observations become more spread out.
- s has the same units of measurement as the original observations. For example, if you measure lengths in centimeters, s is also in centimeters. This is one reason to prefer s to the variance s^2, which is in squared centimeters.

- Like the mean $\bar{x}$, s is strongly influenced by a few extreme observations. For example, the standard deviation of the hysterectomy data for male doctors is 20.61. (Use your calculator to verify this.) If we omit the two extreme observations 85 and 86, the standard deviation drops to 10.97.

We now have a choice between two numerical descriptions of the center and spread of a distribution: the five-number summary, or $\bar{x}$ and s. Because of the sensitivity of $\bar{x}$ and s to extreme observations, it is best to use the five-number summary to describe skewed distributions and reserve $\bar{x}$ and s for distributions that are roughly symmetric.

Although the standard deviation is widely used, it is not a natural or convenient measure of the spread of a distribution. Why, for example, should we make our measure of spread more sensitive to a few outliers by squaring the deviations, then take the square root at the end? Why not just average the distances of the observations from $\bar{x}$? The real reason for the popularity of the standard deviation is that it is the natural measure of spread for *normal distributions,* an important class of distributions that we will meet in the next chapter.

DISPLAYING RELATIONS BETWEEN TWO VARIABLES

The examples we have looked at so far considered only a single variable, such as a hitter's batting average or the number of hysterectomies performed by a doctor. Now we will examine data for two variables, emphasizing the nature and strength of the relationship between the variables. Because it is very hard to spot a relationship from columns of numbers, graphs are essential.

EXAMPLE ▶ *Natural Gas Consumption*

Sue is concerned about the amount of energy she uses to heat her home in the Midwest. She keeps a record of the natural gas consumed over a period of nine months. The gas company measures gas consumption in hundreds of cubic feet. Because months are not all equally long, Sue divides each month's consumption by the number of days in the month. This gives the average amount of gas used per day during that month. Gas consumption should go up when the weather is colder. So from local weather records, Sue obtains the number of degree-days for each month. She divides this total by the number of days in the month, giving the average number of degree-days per day during the

TABLE 6.3	Sue's Natural Gas Consumption								
Month	Oct	Nov	Dec	Jan	Feb	Mar	Apr	May	June
Degree-days per day	15.6	26.8	37.8	36.4	35.5	18.6	15.3	7.9	0.0
Gas consumed per day (100 cubic feet)	5.2	6.1	8.7	8.5	8.8	4.9	4.5	2.5	1.1

month. (Degree-days are a measure of demand for heating. One degree-day is accumulated for each degree that a day's average temperature falls below 65°F. For example, a day with an average temperature of 30°F gives 35 degree-days.)

Table 6.3 shows Sue's data: nine measurements for each of the two variables. Looking at the numbers in the table, we can see that more degree-days (colder temperatures) go with higher gas consumption. But the shape and strength of the relationship are not fully clear. To display and interpret these data, we need a graph. Figure 6.8 is a *scatterplot* of Sue's data. ◆

A **scatterplot** shows the relationship between two variables measured on the same individuals. The values of one variable appear on the horizontal axis, and the values of the other variable appear on the vertical axis. Each individual in the data appears as the point in the plot fixed by the values of both variables for that individual.

FIGURE 6.8 Natural gas consumption versus degree-days. (a) A scatterplot. (b) A regression line and its use for prediction.

Always plot the causal or **explanatory variable** on the horizontal scale of a scatterplot and the outcome or **response variable** on the vertical scale. As a reminder, we usually call the explanatory variable x and the response variable y. Degree-days appear on the horizontal scale and gas consumption on the vertical scale in Figure 6.8a because degree-days is the explanatory variable. The weather affects gas consumption; gas consumption does not explain the weather.

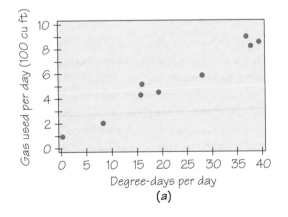

(a)

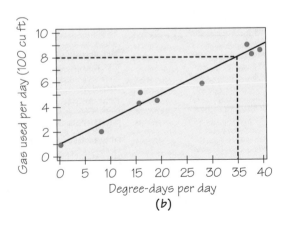

(b)

Just as when we examined distributions of a single variable, we look for an overall pattern in a scatterplot and then for any striking deviations from that pattern. Here is a suggestion for describing the overall pattern of a scatterplot: look for the *form, direction,* and *strength* of the relationship between the two variables.

The form of the relationship between degree-days and gas consumption is clear: the points have a straight-line pattern. We can represent the overall pattern of the relationship by drawing a straight line through the points of the scatterplot. Figure 6.8b shows such a line. As degree-days increase, gas consumption also increases; that is the direction of the relationship. The points in the plot lie quite close to the line, so the relationship is quite strong. That is, the number of degree-days explains most of the variation in gas consumption. A weaker straight-line relationship would show more scatter of the points about a generally straight-line pattern. The scatter reflects the effects of other factors, such as use of gas for cooking or turning down the thermostat when the family is away from home. These effects are relatively small. There are no outliers (points far outside the overall straight-line pattern) or other important deviations.

REGRESSION LINES

Sue wants to use her data to predict how much gas she will use for any outside temperature (in degree-days). She can do this by drawing a line through the straight-line pattern on the scatterplot in Figure 6.8a. A line intended to predict the value of a response variable for any given value of an explanatory variable is called a *regression line.*

The points in Figure 6.8a lie so close to a line that it is easy to draw a regression line on the graph by using a transparent straightedge. This gives us a line on the graph, but not an equation for the line. There is also no guarantee that the line we fit by eye is the best line for predicting gas consumption. There are statistical techniques for finding from the data the equation of the best line (with various meanings of "best"). The most common of these techniques, called *least squares regression,* is discussed in the next section. The line in Figure 6.8b is the least squares regression line for Sue's data. All statistical computer software packages and many calculators will calculate the least squares line for you, so that a line is often available with little work. You should therefore know how to use a fitted line even if you do not learn the details of how to get the equation from the data.

In writing the equation of a line, we use x for the explanatory variable because this is plotted on the horizontal or x-axis, and y for the response variable. Any line has an equation of the form

$$y = a + bx$$

The number b is the *slope* of the line, the amount by which y changes when x increases by one unit. The slope is usually important in a statistical setting, because it is the rate of change of the response y as x increases. The number a is the *intercept,* the value of y when $x = 0$.

EXAMPLE ▶ *Interpreting Slope and Intercept*

A computer program tells us that the least squares regression line computed from Sue's data is

$$y = 1.23 + 0.202x$$

The slope of this line is $b = 0.202$. This means that gas consumption increases by 0.202 hundred cubic feet per day when there is one more degree-day per day. The intercept is $a = 1.23$. When there are no degree-days (that is, when the average temperature is 65°F or above), gas consumption will be 1.23 hundred cubic feet per day. The slope and intercept are, of course, estimates based on fitting a line to the data in Table 6.3. We do not expect every month with no degree-days to average exactly 1.23 hundred cubic feet of gas per day. The line represents only the overall pattern of the data.   ◆

The purpose of a regression line is to predict the value of the response variable for a given value of the explanatory variable. A line drawn on a scatterplot can be used for making predictions with a straightedge and pencil. If the equation of the line is available, we can simply substitute the given value of the explanatory variable into the equation.

EXAMPLE ▶ *Predicting Gas Consumption*

How much natural gas should Sue predict that she will consume in a month with 35 degree-days per day? Figure 6.8b illustrates the use of the line drawn on the scatterplot. First locate 35 on the horizontal axis. Draw a vertical line up to the fitted line and then a horizontal line over to the gas consumption scale. As the figure shows, we predict that slightly more than 800 cubic feet per day will be consumed.

We can give a more exact prediction using the equation of the least squares regression line. This equation is

$$y = 1.23 + 0.202x$$

In this equation, x is the number of degree-days per day during a month and y is the predicted gas consumption per day, in hundreds of cubic feet. Our predicted gas consumption for a month with $x = 35$ degree-days per day is

$$y = 1.23 + (0.202)(35)$$
$$= 8.3 \text{ hundred cubic feet per day}$$

This prediction will almost certainly not be exactly correct for the next month that has 35 degree-days per day. But the past data points lie so close to the line that we can be confident that gas consumption in such a month will be quite close to 830 cubic feet per day. ◆

Sue had a practical reason for this exercise in statistics. She plans to add insulation to her house during the summer. At the end of next winter she will want to know how much she has saved in heating costs. She cannot simply compare before-and-after gas usage, because the winters before and after will not be equally severe. Nor can she do an experiment to compare the costs of insulated and uninsulated houses during the same winter, for she has only one house. However, once Sue has next winter's degree-day data, she can use the fitted line to predict how much gas she would have used before insulating. Comparing this prediction with the actual amount used after insulation will show her savings.

O p t i o n a l ▶ **Least Squares Regression**

When a scatterplot shows a straight-line relation between an explanatory variable x and a response variable y, we want to draw a line to describe the relationship. The points will rarely lie exactly on a line, so our problem is to find the line that best fits the points. To do this, we must first say what we mean by the "best-fitting" line.

Suppose that we want to use our line to predict y for given values of x, as Sue used degree-days to predict gas consumption. The error in our prediction is measured in the y, or vertical, direction. So we want to make the vertical distances of the points from the line as small as possible. A line that fits the data well does not pass entirely above or below the plotted points, so some of the errors will be positive and some negative. Their squares, however, will all be positive. The *least squares regression line* makes the sum of the squares of the errors as small as possible.

> The **least squares regression line** of y on x is the line that makes the sum of the squares of the vertical distances of the data points from the line as small as possible.

The least squares idea says what we mean by the best-fitting line. We must still learn how to find this line from the data. Given n observations on variables x and y, what is the equation of the least squares line? Here is the solution to this mathematical problem:

We have data on an explanatory variable x and a response variable y for n individuals. The **least squares regression line** of y on x is $y = a + bx$, where the slope b is

$$b = \frac{n\Sigma xy - (\Sigma x)(\Sigma y)}{n\Sigma x^2 - (\Sigma x)^2}$$

and the intercept a is

$$a = \bar{y} - b\bar{x}$$

Remember that Σ is just an abbreviation for "add them all up," so Σx means the sum of all the x values. As usual, $\bar{x}$ and $\bar{y}$ are the means of the x and y values. These algebraic formulas are shorthand for a series of operations that for clarity can be arranged as a spreadsheet-like table. Here is an example.

EXAMPLE ▶ *Gas Consumption Least Squares Line*

Sue's data for natural gas consumption y versus heating degree-days x appear in Table 6.3. Table 6.4 contains the quantities needed to obtain the least squares regression line, beginning with the values of x and y in the first two columns. There are $n = 9$ observations.

At the bottom of each column we write its sum. These sums are the building blocks for the least squares calculation. After writing down the x and y values, we follow these steps in the calculation:

TABLE 6.4   The Arithmetic of Least Squares

x	y	x^2	xy
15.6	5.2	243.36	81.12
26.8	6.1	718.24	163.48
37.8	8.7	1428.84	328.86
36.4	8.5	1324.96	309.40
35.5	8.8	1260.25	312.40
18.6	4.9	345.96	91.14
15.3	4.5	234.09	68.85
7.9	2.5	62.41	19.75
0.0	1.1	0.0	0.0
$\Sigma x = 193.9$	$\Sigma y = 50.3$	$\Sigma x^2 = 5618.11$	$\Sigma xy = 1375.0$

STEP 1. Sum the x and y columns and compute the two means. We find

$$\Sigma x = 193.9 \quad \text{and} \quad \bar{x} = \frac{193.9}{9} = 21.544$$

$$\Sigma y = 50.3 \quad \text{and} \quad \bar{y} = \frac{50.3}{9} = 5.589$$

STEP 2. Calculate x^2 and xy for each (x, y) data point, enter in the proper column, and sum.

STEP 3. Substitute $n = 9$ and the building block sums into the formulas for the slope b and then for the intercept a.

$$b = \frac{n\Sigma xy - (\Sigma x)(\Sigma y)}{n\Sigma x^2 - (\Sigma x)^2}$$

$$= \frac{(9)(1375.0) - (193.9)(50.3)}{(9)(5618.11) - (193.9)^2}$$

$$= \frac{2621.83}{12965.78} = 0.2022$$

$$a = \bar{y} - b\bar{x}$$

$$= 5.589 - (0.2022)(21.544) = 1.23$$

The equation of the least squares line is, therefore

$$y = 1.23 + 0.202x$$

It is more common in statistical practice to use a statistical calculator or computer software to obtain b and a with less work. ◆

EXPLORING DATA ON SEVERAL VARIABLES

Scatterplots are a good tool for exploring the relationship between two variables, but often the overall pattern is not as simple as the straight-line pattern in Figure 6.8. Curved relationships and clusters of points are common, for example. Sometimes we must go beyond a look at the plot. Figure 6.9 presents a scatterplot that at first glance shows no clear relationship. Let us see how using some of the numerical descriptions mentioned earlier can help us uncover the pattern in these data.

Figure 6.9 displays the results of the 1970 draft lottery. To eliminate distinctions among men eligible for the draft, Congress decided to allow chance to determine who would be selected for military service. The first draft lottery was held in 1970. Birth dates for all men born between 1943 and 1952 were to

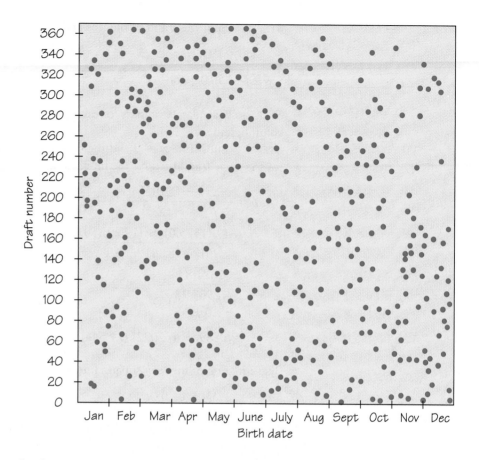

FIGURE 6.9 Scatterplot
of draft selection number
versus birth date for the
1970 draft lottery.

be drawn at random and assigned selection numbers in the order drawn. Men
whose birth date was the first drawn (selection number 1) were drafted first.
They were followed by men with numbers 2, 3, 4, and so on.

This procedure sounds fair, but the random drawing was mishandled.
Birth dates were put into identical capsules, and the capsules were placed in a
drum for the drawing. But the capsules were not mixed thoroughly enough.
December dates, which were added last, remained on top and had a greater
chance of being drawn early. Thus, men born in December were drafted and
sent to Vietnam in greater numbers than those born in January. How can we
see this sad fact in the plot?

EXAMPLE ▸ *The 1970 Draft Lottery*

Figure 6.9 plots the 366 birth dates as *x* and the selection numbers assigned to
them by the 1970 lottery as *y*. Having a low selection number, nearer the bottom of this graph, means being drafted earlier. It's hard to see the alleged association between birth dates late in the year and low selection numbers. Although the plot looks much like a random scatter of points, the mathematics

FIGURE 6.10 Side-by-side boxplots of the 1970 draft numbers for men born in each of the 12 months.

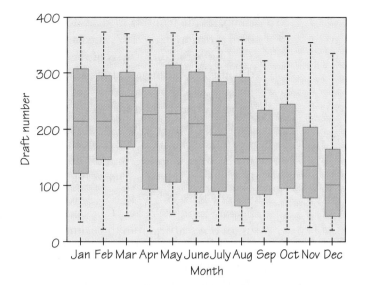

of chance behavior tells us that a result this unfair to men born late in the year would happen in less than 1 in 1000 truly random lotteries. The inequity is really there. To see it, we need to be more imaginative in looking at the data.

Figure 6.10 uses boxplots to give a more detailed picture of the draft lottery. We find the five-number summary of each month's draft numbers and present the results as 12 side-by-side boxplots. These boxplots make it easy to compare the distribution of selection numbers from month to month. Each box spans the central half of the month's draft numbers. The line within each box marks the median for the month. The downward trend of the boxes and the medians makes clear the misfortune of men born late in the year. ◆

The media noticed this inequity, and in 1971 a new and genuinely random selection process was designed by statisticians at the National Bureau of Standards. Two drums instead of one were used in the 1971 lottery: one contained birth dates; the other, selection numbers. The capsules in both drums were thoroughly mixed, and then a randomly drawn birth date was paired with a randomly drawn selection number. Was the new system fairer? Figure 6.11

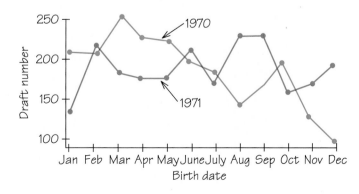

FIGURE 6.11 Median draft number by month for the 1970 and 1971 draft lotteries.

compares the monthly medians for 1970 and 1971. There is a good bit of random variation in the 1971 medians but no systematic trend as in 1970.

Our graphical analysis of the draft lottery in Figures 6.9 to 6.11 illustrates the insight we can gain using simple graphs and some basic statistical calculations. Looking at data in several ways can yield results that we wouldn't see using a single method such as a scatterplot.

So far we have looked only at scatterplots for two variables. What if we want to display a third variable in the same plot? Because we have already used the horizontal and vertical directions of the graph, there is only one direction left, moving out of and into the page. Unfortunately, three-dimensional scatterplots are very hard to see clearly unless color or motion (or both) are used to help us gain perspective. *Computer graphics* can add color and motion, allowing us to see a scatterplot in three dimensions.

Computer graphics makes it possible to see relations and detect outliers in high-dimensional data sets. Imagine a mass of points in space with a single outlier positioned far from the mass. From most viewing angles, the outlier would appear as part of the main group, blending invisibly with the mass of points. However, if we change the viewing angle by rotating the plot, the outlier would eventually be seen apart, like a moon appearing from behind a planet.

EXAMPLE ▶ *A 3-D Scatterplot*

FIGURE 6.12 Two views of a three-dimensional scatterplot. (a) From this angle, the points look like a single cloud. (b) From another angle, we see two clouds and an outlier between them.

A manufacturing process makes objects whose height, width, and depth are all important. Plot these three measurements for many objects in a three-dimensional scatterplot. Figure 6.12 shows that from most points of view, the data appear as a cloud, like gnats swarming in space. But if we rotate the image, we find that there are two clouds, one of which was hidden behind the other from our first viewpoint. What is more, an outlier lies between the two clouds. We now see that there are two groups of objects and an outlier that belongs to neither group.

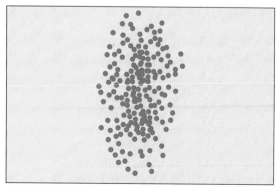

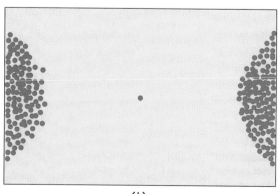

(a)   (b)

Remember that every point in this picture is positioned according to three measurements. If we tried to spot the outlier by looking at a long table with three columns of measurements for hundreds of points, we would quickly lose our bearings. This is a case where graphic display creates clarity out of the confusion of raw numbers. ◆

Computer graphics allows us to move around the cloud of data points and look at it from any direction. Moreover, the computer can move around a ten-dimensional cloud of points almost as easily as a three-dimensional one. We cannot, of course, see ten dimensions. In fact, we do not directly see three dimensions on a piece of paper or video screen. In both cases, the computer presents a *projection* of the cloud of data onto the two-dimensional surface of the video screen. The computer can be instructed to compute and display changing projections, just as if we were moving around a many-dimensional cloud of data. Outliers and other important relations among the variables come clearly into view when they are scanned from the correct angle. To reduce the amount of time needed to discover the most meaningful viewing angles, the computer can even be programmed to search for interesting projections.

EXAMPLE ▶ *Studying Earthquake Locations*

Computer graphic displays of multidimensional data aid geologists who are studying a pattern of earthquakes (see Spotlight 6.2). For many years, the only view scientists had of earthquake epicenters was a standard map projection showing the two-dimensional location in latitude and longitude. Figure 6.13 shows earthquake epicenters near the Fiji Islands in the Pacific Ocean. Information about the depths of the epicenters below the earth's surface doesn't appear on these maps.

FIGURE 6.13 Earthquake epicenters near the Fiji Islands as seen from the earth's surface.

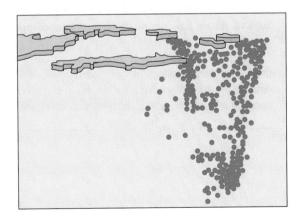

FIGURE 6.14 Earthquake epicenters after changing viewing angle to show also depth beneath the earth's surface.

Spotlight   *Visual Statistics: Eyeballing the Data*

6.2

Dr. Peter Huber, a statistician at Harvard University, comments on the computer's ability to display complex, multidimensional arrangements of data, giving new meaning to the term *descriptive statistics:*

Peter Huber

For most of the twentieth century, statistics focused on mathematical rigor and on small samples. In the past 10 years or so we have seen renewed interest in descriptive statistics, driven by the computer. It would be a mistake, however, to view this new emphasis as just a reaction of statistics to high tech. What we are witnessing now is a return to the descriptive statistics of the nineteenth century, completing unfinished business left over from that era. The computer makes it possible to do things that couldn't be done before.

In my view, descriptive statistics and mathematical statistics are complementary. In descriptive statistics, you see things, but you cannot test them in a formal way. In mathematical statistics, you may test for something preconceived, but you may overlook something you hadn't built into the test. So you simply have to do both so as to complement one with the other.

You also need subject specialists to complement the data analyst. In our experience, doing data analysis usually turns out to be a close conversational collaboration between the scientist (the one with the data) and the data analyst. They sit in front of the screen, discuss what they see, suggest the next action to be taken. Data analysis requires dynamic collaboration between the statistician and the subject-matter specialist.

The first step in any data analysis is data inspection, mainly to get familiar with the data and find extraordinary features. The next step is modification; enhance the picture by lines, maybe color groups, cluster and label selected points, maybe even fit some model to it. This leads to the most important step in data analysis: *comparison*. Without comparing things, you are not able to interact with your data. To help with interpretation, you almost always have to compare things—either several data sets or a data set and a model or different models for the same data set. Interpretation is the next step. And after interpretation, one usually has to begin another round of modeling. Very often the entire cycle starts again.

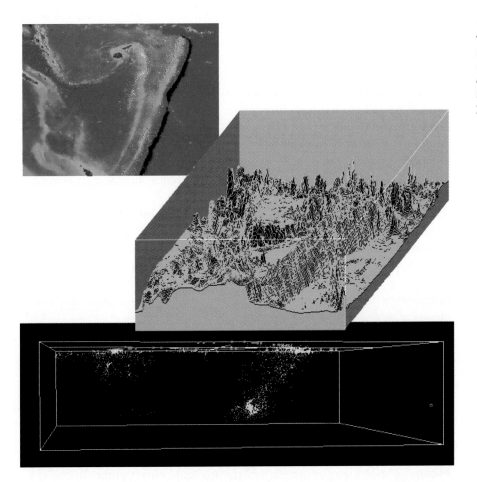

FIGURE 6.15 The Tonga Trench represented in color (above), in perspective (middle), and in three dimensions (below). Red indicates volcanoes and yellow, earthquakes.

A computer graphics system can present and manipulate a three-dimensional view of the epicenters that adds depth to latitude and longitude. This kind of picture permits geologists to examine earthquake patterns in light of plate tectonics, the geological theory of movements of vast plates that make up the earth's crust.

In the region of the Fiji Islands, a plate moving in from the east has bent beneath the plate on the west so that the eastern plate is diving straight down into the earth's mantle. The collision and bending of the plate causes many earthquakes in this region. The three-dimensional view that emerges as the computer continuously changes our viewing angle (see Figures 6.14 and 6.15) shows that earthquakes occur along the boundary between the plates deep inside the earth. The graphics system turns the image constantly, giving a much clearer view than a still picture provides. The detailed computer images even enable geologists to spot wrinkles and other features in the surfaces of the plates. ◆

REVIEW VOCABULARY

Boxplot   A graph of the five-number summary. A box spans the quartiles, with an interior line marking the median. Whiskers extend out from this box to the extreme high and low observations.

Distribution   The pattern of outcomes of a variable. The distribution describes what values the variable takes and how often each value occurs.

Dotplot   A graphical display of the distribution of a variable for small data sets. Each observation is represented by a dot above its value on a scale of the variable being measured.

Exploratory data analysis   The practice of examining data for unanticipated patterns or effects, as opposed to seeking answers to specific questions.

Five-number summary   A summary of a distribution of values consisting of the median, the upper and lower quartiles, and the largest and smallest observations.

Frequency   The number of times an outcome or group of outcomes occurs in a set of data.

Histogram   A graph of the frequencies of all outcomes (often divided into classes) for a single variable. The height of each bar is the frequency of the class of outcomes covered by the base of the bar. All classes should have the same width.

Least squares regression line   A line drawn on a scatterplot that makes the sum of the squares of the vertical distances of the data points from the line as small as possible. The regression line can be used to predict the response variable y for a given value of the explanatory variable x.

Mean   The ordinary arithmetic average of a set of observations. To find the mean, add all the observations and divide the sum by the number of observations summed.

Median   The midpoint of a set of observations. Half the observations fall below the median and half fall above.

Outlier   A data point that falls clearly outside the overall pattern of a set of data.

Quartiles   The first quartile of a distribution is the point with 25% of the observations falling below it; the third quartile is the point with 75% below it.

Response variable, explanatory variable   A response variable measures an outcome of a study. An explanatory variable attempts to explain the observed outcomes.

Scatterplot   A graph of the values of two variables as points in the plane. Each value of the explanatory variable is plotted on the horizontal axis and the corresponding value of the response variable on the vertical axis.

Skewed distribution   A distribution in which observations on one side of the median extend notably farther from the median than do observations on the other side. In a right-skewed distribution, the larger observations extend farther to the right of the median than the smaller observations extend to the left.

Standard deviation   A measure of the spread of a distribution about its mean as center. It is the square root of the average squared deviation of the observations from their mean.

Symmetric distribution   A distribution with a histogram or dotplot in which the part to the left of the median is roughly a mirror image of the part to the right of the median.

Variance   A measure of the spread of a distribution about its mean. It is the average squared deviation of the observations from their mean. The square root of the variance is the standard deviation.

SUGGESTED READINGS

CHAMBERS, JOHN M., WILLIAM S. CLEVELAND, BEAT KLEINER, AND PAUL A. TUKEY.   *Graphical Methods for Data Analysis,* Wadsworth, Belmont, Calif., 1983. A detailed survey of modern graphical methods, many requiring a computer for effective use.

CLEVELAND, WILLIAM S.   *The Elements of Graphing Data,* Wadsworth, Monterey, Calif., 1985. A careful study of the most effective elementary ways to present data graphically, with much sound advice on improving simple graphs.

MOORE, DAVID S.   *The Basic Practice of Statistics,* Freeman, New York, 1995. The first two chapters of this text provide a more extensive treatment of displaying and describing data for one and two variables. They cover the material of this chapter in more detail and present much new material on both technique and interpretation.

TUFTE, EDWARD R.   *The Visual Display of Quantitative Information,* Graphics Press, Cheshire, Conn., 1983. A beautifully printed book with classic examples such as Minard's work and suggestions for both statisticians and graphic artists.

VELLEMAN, PAUL F., and DAVID C. HOAGLIN.   Data analysis. In David C. Hoaglin and David S. Moore (eds.), *Perspectives on Contemporary Statistics,* Mathematical Association of America, Washington, D.C., 1992, pp. 19–39. A conceptual essay that presupposes knowledge of the basic techniques described in this chapter and in the text by Moore.

EXERCISES ▲ *Optional.* ▪ *Advanced.* ◆ *Discussion.*

Displaying Distributions

1.   Here are the number of days on which hail was observed at Evansville, Indiana, in each of 11 consecutive years:

4   5   3   2   4   2   0   5   0   1   1

Make a dotplot of these data. Are there any outliers or other unusual features?

2.   In an experiment on the effect of a drug on reaction time, a subject is asked to depress a button whenever a light flashes. Her reaction times for 11 trials are (in milliseconds)

96   101   102   138   93   99   107   93   95   100   100

Make a dotplot of these observations. Are there any outliers or other unusual features?

3.   The following histogram displays data on the hour at which the first flash of lightning was observed each day during a study in Colorado. Describe this distribution: Is it roughly symmetric or distinctly skewed? Where is the center? Are there any outliers or gaps?

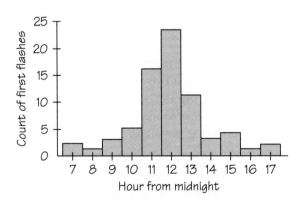

◆   4.   Members of a health maintenance organization (HMO) can make an unlimited number of visits to its member physicians for a fixed annual fee. The histogram at the top of the next page displays the distribution of the number of visits made by members of one HMO. (The vertical scale gives the proportion of the members in each class rather than the frequency.) Describe this distribution: Is it roughly symmetric or distinctly skewed? Are there any outliers or gaps?

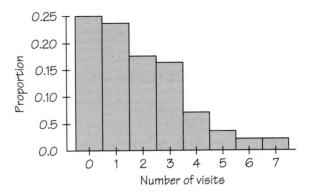

5.   In 1798 the English scientist Henry Cavendish measured the density of the earth in a careful experiment with a torsion balance. Here are his 23 repeated measurements of the same quantity (the density of the earth relative to that of water) made with the same instrument. (From S. M. Stigler, Do robust estimators work with real data? *Annals of Statistics* 5 [1977]: 1055–1078.)

5.36	5.62	5.27	5.46	5.53	5.57
5.29	5.29	5.39	5.30	5.10	5.79
5.58	5.44	5.42	5.75	5.34	5.63
5.65	5.34	5.47	5.68	5.85	

(a)   Make a histogram of these data.

(b)   Describe the distribution: Is it approximately symmetric or distinctly skewed? Are there gaps or outliers?

6.   A fisheries researcher compiled the following data on lengths of 6-year-old white female crappies (in millimeters):

217	230	220	221	225	223
219	217	225	228	234	222
231	222	220	222	222	223
225	214	221	233	227	234
223	225	253	220	213	224
235	283	210	218	235	231

(a)   The data range from 210 to 283 mm. Group them into 5 classes of width 15 mm starting with

$$210 \leq \text{length} < 225$$

as the leftmost class. Draw a well-labeled frequency histogram of the grouped data.

(b) Describe the distribution: Is it roughly symmetric or clearly skewed? Are there gaps or outliers?

Describing Distributions

7.   Return to the data on annual number of days with hail given in Exercise 1.

(a) Find the mean and the median number of hail days.
(b) Find the first and third quartiles of the number of hail days.

8.   Find the five-number summary of the reaction times given in Exercise 2.

9.   Here are the percentages of the popular vote won by the successful candidate in each of the presidential elections from 1948 to 1992:

Year	1948	1952	1956	1960	1964	1968	1972	1976	1980	1984	1988	1992
%	49.6	55.1	57.4	49.7	61.1	43.4	60.7	50.1	50.7	58.8	53.9	43.2

(a) Make a dotplot of these percents. Are there any outliers?
(b) What is the median percent of the vote won by the successful candidate in presidential elections?
(c) Call an election a landslide if the winner's percent falls at or above the third quartile. Find the third quartile. Which elections were landslides?

10.   A random sample of 12 Wisconsin farms showed the following soybean yields (bushels per acre):

41   28   34   36   26   44   39   32   40   35   36   33

(a) Make a dotplot of these data. Are there any outliers or other unusual features?
(b) Find the mean and median yield.
(c) Find the first and third quartiles of the yields.

11.   Find the standard deviation of the counts of days with hail given in Exercise 1. Rather than use a calculator, do the following:

(a) Make a table of the deviations and squared deviations of the observations from their mean, like the table in the example on page 234. Find the sum of the squared deviations and use this sum to obtain s.

(b) Make a dotplot of the data and mark on your plot the values of two of the deviations, one for an observation above the mean and the other for an observation below the mean.

◆ 12. Find the mean and standard deviation of the reaction times in Exercise 2. (Use a calculator.) Then find the mean and standard deviation when the outlier (138) is left out. What does your work show about the behavior of $\bar{x}$ and s? Explain why the mean is larger than the median.

◆ 13. Return to the data on lengths of fish given in Exercise 6.

(a) Find the five-number summary of this distribution. What range of lengths contains the middle 50% of the distribution?
(b) From the shape of the distribution, do you expect the mean to be larger than the median, smaller than the median, or about the same as the median? Find the mean and verify your expectation.
(c) Find the standard deviation. Based on the shape of the distribution, are $\bar{x}$ and s acceptable summary measures of center and spread?

◆ 14. Find the median and quartiles of Cavendish's measurements of the density of the earth in Exercise 5. Then give a five-number summary. How is the symmetry of the distribution reflected in the five-number summary?

15. The mean of the 23 measurements in Exercise 5 was Cavendish's best estimate of the density of the earth. Find this mean. Then find the standard deviation. (Because of the symmetry of the distribution, it can be summarized by $\bar{x}$ and s.)

◆ 16. A study of the size of jury awards in civil cases (such as injury, product liability, and medical malpractice) showed that the median award in Cook County, Illinois, was about $8000. But the mean award was about $69,000. Explain how this great difference between two measures of center can occur.

◆ 17. A news report says that "in the National Football League, average salaries have climbed from $79,000 in 1980 to $737,000 last season" (*Forbes,* December 19, 1994). That $737,000 is the mean salary. Is the median salary of National Football League players higher, lower, or about the same as the mean? Explain your answer.

18. Scores of adults on the Stanford-Binet IQ test have mean 100 and standard deviation 15. What is the variance of scores on this test?

19. Find the five-number summary for the home run data in Table 6.1 and make a boxplot of this distribution. What range contains the middle half of league-leading home run totals?

◆ 20.   Joe DiMaggio played center field for the Yankees for 13 years. He was succeeded by Mickey Mantle, who played for 18 years. Here are the number of home runs hit each year by DiMaggio:

29   46   32   30   31   30   21   25   20   39   14   32   12

and by Mantle:

13   23   21   27   37   52   34   42   31   40   54   30   15   35   19   23   22   18

Compute the five-number summary for each player, and make side-by-side boxplots of the home run distributions. What does your comparison show about the relative effectiveness of DiMaggio and Mantle as home run hitters?

Displaying Relations

◆ 21.   Manatees are large, gentle sea creatures that live along the Florida coast. Many manatees are killed or injured by power boats. Table 6.5 gives data on power boat registrations (in thousands) and the number of manatees killed by boats in Florida in the years 1977 to 1990.

TABLE 6.5	**Florida Powerboats and Manatee Deaths, 1977–1990**				
Year	Boats (thousands)	Manatees Killed	Year	Boats (thousands)	Manatees Killed
1977	447	13	1984	559	34
1978	460	21	1985	585	33
1979	481	24	1986	614	33
1980	498	16	1987	645	39
1981	513	24	1988	675	43
1982	512	20	1989	711	50
1983	526	15	1990	719	47

(a)   Which is the explanatory variable?

(b)   Make a scatterplot of these data.

(c)   Describe the form, direction, and strength of the overall pattern. Are there any outliers or other important deviations?

◆ 22.   The table at the top of the next page lists the retail prices of several consumer items in Washington, D.C., and in Japan. Make a scatterplot with the U.S. price as the explanatory variable. Describe the overall pattern of the relationship between U.S. and Japanese consumer prices. Are there any items with prices that

Item	Washington	Japan
First-run movie	$5.00	$13.00
Pound of rice	0.50	1.89
Big Mac hamburger	1.78	3.08
Electricity (kWh)	0.06	0.33
Gallon of gasoline	0.95	3.50
Cantaloupe	0.75	13.00
Dozen eggs	1.09	1.70
Subway	0.80	1.00
Quart of milk	0.55	1.48

SOURCE: *USA Today,* January 9, 1989.

fall clearly outside the overall pattern? If so, describe how these items differ from the overall pattern.

◆ 23. How does the fuel consumption of a car change as its speed increases? The accompanying table gives data for a British Ford Escort. Speed is measured in kilometers per hour, and fuel consumption is measured in liters of gasoline used per 100 kilometers traveled.

Fuel Consumption versus Speed for a Compact Car

Speed (km/h)	Fuel Used (liters/100 km)	Speed (km/h)	Fuel Used (liters/100 km)
10	21.00	90	7.57
20	13.00	100	8.27
30	10.00	110	9.03
40	8.00	120	9.87
50	7.00	130	10.79
60	5.90	140	11.77
70	6.30	150	12.83
80	6.95		

SOURCE: Based on T. N. Lam, Estimating fuel consumption from engine size, *Journal of Transportation Engineering*, 111 (1985): 339–357.

(a) Make a scatterplot. (Which is the explanatory variable?)
(b) Describe the form of the relationship. Explain why the form of the relationship makes sense.
(c) How would you describe the direction of this relationship?
(d) Is the relationship reasonably strong or quite weak? Explain your answer.

Regression Lines

24.   Continue your study of the manatee data from Exercise 21. Fit a line to your scatterplot by eye and draw the line on your graph. If power boat registrations remained at 716,000, predict the number of manatees that will be killed by boats.

25.   Here are four more years of manatee data, in the same form as in Exercise 21:

| 1991 | 716 | 53 | 1993 | 716 | 35 |
| 1992 | 716 | 38 | 1994 | 735 | 49 |

(a)   Plot manatee deaths against year for 1977 to 1994. Describe the pattern of change over time that you see.

(b)   In Exercise 24, you predicted manatee deaths in a year with 716,000 power boat registrations. In fact, power boat registrations remained at 716,000 for the next three years. Compare the mean manatee deaths in these years with your prediction from Exercise 24. How accurate was the prediction?

26.   Suppose that in some far future year 2 million power boats are registered in Florida. Extend the fitted line you found in Exercise 24 and use it to predict manatees killed. Explain why this prediction is very unreliable. (Using a fitted line to predict the response to an *x* value outside the range of the data used to fit the line is called *extrapolation*. Extrapolation often produces unreliable predictions.)

27.   We saw that the least squares regression line of gas consumed on degree-days for the home heating data of Table 6.3 is $y = 1.23 + 0.202x$. Use this line to predict the daily gas consumption for this home in a month averaging 20 degree-days per day and in a month averaging 40 degree-days per day.

◆   28.   Concrete road pavement gains strength over time as it cures. Highway engineers use regression lines to predict the strength after 28 days (when curing is complete) from measurements made after 7 days. Let *x* be strength (in pounds per square inch) after 7 days and *y* the strength after 28 days. One set of data gave the least squares regression line to be

$$y = 1389 + 0.96x$$

(a)   Explain in words what the slope 0.96 tells us about the curing of concrete.

(b) A test of some new pavement after 7 days shows that its strength is 3300 pounds per square inch. Predict the strength of this pavement after 28 days.

29. Table 6.6 shows the true number of calories in 10 common foods and the average number of calories estimated for these same foods in a sample of 3368 people.

(a) Make a scatterplot of these data, with true calories on the horizontal axis. Is there a general straight-line pattern? Which foods are outliers from the pattern?

(b) Fit a line by eye to the data (ignoring the outliers). If a food product contains 200 calories, what would you guess the general public's estimated calorie level for that food to be?

(c) The least squares regression line, computed without dropping the outliers, is

$$y = 58.6 + 1.30x$$

Here y = estimated calories and x = true calories. Draw this line on your scatterplot. (*Hint:* Use the equation to find y for two values of x, then plot the two (x, y) points and draw the line through them.) Use the least squares line to predict y when $x = 200$.

TABLE 6.6	True and Estimated Calories in 10 Common Foods	
Food	Guessed Calories	Correct Calories
8 oz whole milk	196	159
5 oz spaghetti with tomato sauce	394	163
5 oz macaroni with cheese	350	269
One slice wheat bread	117	61
One slice white bread	136	76
2-oz candy bar	364	260
Saltine cracker	74	12
Medium-size apple	107	80
Medium-size potato	160	88
Cream-filled snack cake	419	160

SOURCE: Wheat Industry Council, reported in *USA Today,* October 20, 1983.

The difference between your results in parts (b) and (c) reflects the influence of the outliers. It is often difficult to decide whether to include outliers in making a prediction.

Least Squares Regression (*Optional*)

30.   Calculate the least squares regression line of estimated calories *y* on true calories *x* from the data in Table 6.6. Verify that your result agrees with that given in Exercise 29c.

31.   Find the equation of the least squares regression line for the manatee data in Table 6.5. Use the equation to predict manatee deaths in a future year when 716,000 power boats are registered in Florida.

32.   The recipe for least squares regression will fit a line to any set of data on two variables, whether or not a line makes sense as a description of the data.

   (a)   Find the equation of the least squares regression line to predict gas mileage from speed for the car data in Exercise 23.
   (b)   Plot your line on the scatterplot from Exercise 23. Using a straight line for prediction is clearly a bad idea in this setting.

33.   Table 6.9 (see page 262) gives data on wine consumption and heart-attack death rates in 19 developed countries. Exercise 48 concerns these data. Calculate the least squares regression line of heart-disease death rate on wine consumption for the data in Table 6.9. Check that your result agrees with that given in Exercise 48.

Exploring Data on Several Variables

◆   34.   The table at the top of the next page gives data on the lean body mass (kilograms) and resting metabolic rate for 12 women and 7 men who are subjects in a study of obesity. The researchers suspect that lean body mass (that is, the subject's weight leaving out all fat) is an important influence on metabolic rate.

   (a)   Make a scatterplot of the data for the female subjects. Which is the explanatory variable?
   (b)   Does metabolic rate increase or decrease as lean body mass increases? What is the overall shape of the relationship? Are there any outliers?

Subject	Sex	Mass	Rate	Subject	Sex	Mass	Rate
1	M	62.0	1792	11	F	40.3	1189
2	M	62.9	1666	12	F	33.1	913
3	F	36.1	995	13	M	51.9	1460
4	F	54.6	1425	14	F	42.4	1124
5	F	48.5	1396	15	F	34.5	1052
6	F	42.0	1418	16	F	51.1	1347
7	M	47.4	1362	17	F	41.2	1204
8	F	50.6	1502	18	M	51.9	1867
9	F	42.0	1256	19	M	46.9	1439
10	M	48.7	1614				

◆ 35. Compare the distribution of lean body mass among the male subjects in Exercise 34 with the distribution for female subjects by making side-by-side boxplots. Describe what the plots show.

◆ 36. When observations on two variables fall into several categories, you can display more information in a scatterplot by plotting each category with a different symbol or a different color. Add the data for male subjects to your scatterplot in Exercise 34, using a different color or symbol than you used for females. Does the type of relationship you found for females hold for men also? How do the male subjects as a group differ from the female subjects as a group?

Chapter Exercises

◆ 37. Table 6.7 lists the percent of high school dropouts in each state. (More precisely, this is the percent of each state's residents of ages 16 through 19 who have not completed high school and are not currently enrolled in school.)

   (a) Make a histogram of these data.
   (b) Describe the overall shape of the distribution: It is roughly symmetric, clearly skewed to the right, or clearly skewed to the left? There is an outlier. Which state produced the outlier? Can you suggest an explanation for the outlier?

TABLE 6.7 Percentage of High School Dropouts, by State, 1991					
State	Percent	State	Percent	State	Percent
Alabama	12.6	Kentucky	13.0	North Dakota	4.3
Alaska	9.6	Louisiana	11.9	Ohio	8.8
Arizona	14.3	Maine	8.4	Oklahoma	9.9
Arkansas	10.9	Maryland	11.0	Oregon	11.0
California	14.3	Massachusetts	9.5	Pennsylvania	9.4
Colorado	9.6	Michigan	9.9	Rhode Island	12.9
Connecticut	9.2	Minnesota	6.1	South Carolina	11.9
Delaware	11.2	Mississippi	11.7	South Dakota	7.1
D.C.	19.1	Missouri	11.2	Tennessee	13.6
Florida	14.2	Montana	7.1	Texas	12.5
Georgia	14.1	Nebraska	6.6	Utah	7.9
Hawaii	7.0	Nevada	14.9	Vermont	8.7
Idaho	9.6	New Hampshire	9.9	Virginia	10.4
Illinois	10.4	New Jersey	9.3	Washington	10.2
Indiana	11.4	New Mexico	10.8	West Virginia	10.6
Iowa	6.5	New York	10.1	Wisconsin	6.9
Kansas	8.4	North Carolina	13.2	Wyoming	6.3

◆ 38. Make a boxplot of the school dropout data in Table 6.7. Is the box-plot inferior in any important way to the histogram you made in the previous exercise?

◆ 39. Do the southern states have higher school dropout rates than the rest of the country? Use Table 6.7 to answer this question. Decide which states you consider "southern." (You may want to look at a map.) Make side-by-side box-plots of school dropout percents for southern and nonsouthern states, and comment on the comparison of the two distributions.

40. A common criterion for detecting suspected outliers in a set of data is as follows:

1. Find the quartiles Q_1 and Q_3 and the *interquartile range, IQR =* $Q_3 - Q_1$. The interquartile range is the spread of the central half of the data.
2. Call an observation an outlier if it falls more than $1.5 \times IQR$ above the third quartile or below the first quartile.

Find the quartiles for the school dropout data in Table 6.7. Find the interquartile range *IQR*. Are there any outliers according to the $1.5 \times IQR$ criterion?

TABLE 6.8		Guinea Pig Survival Times (Days)							
43	45	53	56	56	57	58	66	67	73
74	79	80	80	81	81	81	82	83	83
84	88	89	91	91	92	92	97	99	99
100	100	101	102	102	102	103	104	107	108
109	113	114	118	121	123	126	128	137	138
139	144	145	147	156	162	174	178	179	184
191	198	211	214	243	249	329	380	403	511
522	598								

SOURCE: T. Bjerkedal, Acquisition of resistance in guinea pigs infected with different doses of virulent tubercle bacilli, *American Journal of Hygiene,* 72 (1960): 130–148.

◆ 41. Table 6.8 gives the survival times (in days) of 72 guinea pigs after they were infected by tubercle bacilli in a medical study. Make a histogram of these data. Is the survival-time distribution approximately symmetric or strongly skewed? Based on the shape of the distribution, would you prefer the five-number summary or $\bar{x}$ and s as a numerical description? Compute the numerical description you chose.

◆ 42. Find the mean and the median survival time for the guinea pigs in Table 6.8. Explain from the overall shape of the distribution the relationship between the two measures of center.

▩ 43. This is a standard deviation contest. You must choose four numbers from the whole numbers 0 to 10, with repeats allowed.

(a) Choose four numbers that have the smallest possible standard deviation.

(b) Choose four numbers that have the largest possible standard deviation.

(c) Is more than one choice possible in either part (a) or part (b)? Explain.

◆ 44. Choose a set of interesting data from the *Statistical Abstract of the United States* or an almanac (for example, populations of the states or per capita incomes of nations). Make a histogram of the data and describe the pattern and any outliers. Then give a numerical description of the data.

▩ 45. Colleges announce an "average" Scholastic Assessment Test score for their entering freshmen. Usually the college would like this "average" to be as high as possible. A *New York Times* article noted that "private colleges that buy lots of top students with merit scholarships prefer the mean, while open-enrollment public institutions like medians." Use what you know about the behavior of means and medians to explain these preferences.

◆ 46. A column in the business magazine *Forbes* (October 17, 1994) claimed that the median reader has a net worth of $789,000 and that "the average reader" has a net worth of $2,180,000. The "average" is the mean, and "net worth" is the total wealth that an individual possesses. Explain why the mean is so much larger than the median.

47. Give an example of a small set of data for which the mean is larger than the upper quartile.

◆ 48. There is some evidence that drinking moderate amounts of red wine reduces the risk of heart attacks. Table 6.9 gives data on wine consumption and deaths from heart disease in 19 developed countries as of 1989. Wine consumption is measured as liters of alcohol from drinking wine per person. The heart-disease death rate is deaths per 100,000 people.

(a) Make a scatterplot of these data arranged to show the possible influence of wine consumption on heart-disease deaths.
(b) The equation of the least squares regression line of heart-disease death rate on wine consumption is

$$y = 260.56 - 22.969x$$

Draw this line on your scatterplot.

TABLE 6.9	Wine Consumption and Heart-Disease Deaths, by Selected Countries				
Country	Alcohol from Wine	Heart-Disease Death Rate	Country	Alcohol from Wine	Heart-Disease Death Rate
Australia	2.5	211	Italy	7.9	107
Austria	3.9	167	Netherlands	1.8	167
Belg./Luxe.	2.9	131	New Zealand	1.9	266
Canada	2.4	191	Norway	0.8	227
Denmark	2.9	220	Spain	6.5	86
Finland	0.8	297	Sweden	1.6	207
France	9.1	71	Switzerland	5.8	115
Germany (West)	2.7	172	U.K.	1.3	285
Iceland	0.8	211	United States	1.2	199
Ireland	0.7	300			

SOURCE: *New York Times,* December 28, 1994.

(c) Use the regression line to predict the heart-disease death rate in a country where annual wine consumption is 5 liters of alcohol per person.

(d) Do these data come from an observational study or an experiment? Do you think the data give good reason to think that increasing the wine consumption of Americans (say, from 1.2 to 5 liters per person) would reduce the heart-disease death rate in the United States? Explain your answer.

◆ 49. Table 6.10 gives the average Scholastic Assessment Test (SAT) mathematics score for high school seniors in each state and also the percent of seniors in each state who take the examination, as of 1990. In some states, most

TABLE 6.10 Average SAT Scores, by State, 1990

State	Verbal Score	Math Score	% Taking SAT	State	Verbal Score	Math Score	% Taking SAT
Alabama	470	514	8	Montana	464	523	20
Alaska	438	476	42	Nebraska	484	546	10
Arizona	445	497	25	Nevada	434	487	24
Arkansas	470	511	6	New Hampshire	442	486	67
California	419	484	45	New Jersey	418	473	69
Colorado	456	513	28	New Mexico	480	527	12
Connecticut	430	471	74	New York	412	470	70
Delaware	433	470	58	North Carolina	401	440	55
D.C.	409	441	68	North Dakota	505	564	6
Florida	418	466	44	Ohio	450	499	22
Georgia	401	443	57	Oklahoma	478	523	9
Hawaii	404	481	52	Oregon	439	484	49
Idaho	466	502	17	Pennsylvania	420	463	64
Illinois	466	528	16	Rhode Island	422	461	62
Indiana	408	459	54	South Carolina	397	437	54
Iowa	511	577	5	South Dakota	506	555	5
Kansas	492	548	10	Tennessee	483	525	12
Kentucky	473	521	10	Texas	413	461	42
Louisiana	476	517	9	Utah	492	539	5
Maine	423	463	60	Vermont	431	466	62
Maryland	430	478	59	Virginia	425	470	58
Massachusetts	427	473	72	Washington	437	486	44
Michigan	454	514	12	West Virginia	443	490	15
Minnesota	477	542	14	Wisconsin	476	543	11
Mississippi	477	519	4	Wyoming	458	519	13
Missouri	473	522	12				

seniors who plan to attend college take the SAT. In other states, only students applying to selective out-of-state colleges take the SAT. Make a histogram of the SAT mathematics scores for the states. Describe the overall pattern of the distribution. Then explain why this pattern appears.

50.   The facts in Exercise 49 lead us to expect that the average SAT mathematics score in a state is influenced by what percent of high school seniors in that state take the SAT. Make an appropriate plot of the data from Table 6.10 to display the relationship between these two variables. What is the pattern of the relationship? Is the expectation correct? Is there a straight-line relationship?

◆   51.   Figure 6.16 is a scatterplot of the average SAT math score (y) versus the average SAT verbal score (x) for high school seniors in each of the 50 states. The data appear in Table 6.10. The plot shows a strong straight-line pattern. The least squares regression line for scores is

$$y = 27 + 1.03x$$

(a)   Do math SAT scores tend to be higher or lower than verbal SAT scores? How can you tell?

(b)   The average verbal SAT score in New York was 422. Use the regression line to predict New York's average math score. New York's actual math score was 466. What is the error (observed score minus predicted score)?

(c)   There is one outlier among the states. From the graph and Table 6.10, identify this state. Is this state's math score higher or lower than would be predicted from its verbal score? Can you suggest why this state might be an outlier?

FIGURE 6.16   Scatterplot of the average scores of high school seniors in each state on the Verbal and Mathematics sections of the SAT.

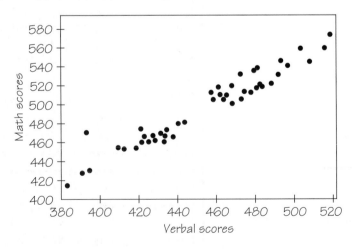

52. A study of sewage treatment measures the oxygen demand of decomposing solid wastes. If y is the logarithm of the oxygen demand (milligrams per minute) and x is the total solids (milligrams per liter of waste), measurements on 20 occasions give the following data:

x	7.2	7.8	7.1	6.4	6.4	5.1	5.9	5.3	5.0	5.0
y	1.56	0.9	0.75	0.72	0.31	0.36	0.11	0.11	-0.2	-0.15

x	4.8	4.4	4.3	3.7	3.9	3.6	4.4	3.3	2.9	2.8
y	0.0	0.0	-0.09	-0.22	-0.4	-0.15	-0.22	-0.4	-0.52	-0.05

(a) Make a scatterplot of these data. Is there an approximate straight-line relationship? Are there outliers?

(b) Draw a fitted line on your scatterplot by eye. Use your line to predict the log of the oxygen demand y when $x = 4$ milligrams per liter of solids.

WRITING PROJECTS

1 ▶ Part of analyzing data is to watch for implausible numbers. Here is part of a report on the problem of vacation cruise ships polluting the sea by dumping garbage overboard that appeared in *Conde Nast Traveler* magazine in June 1992:

> On a seven-day cruise, a medium-size ship (about 1,000 passengers) might accumulate 222,000 coffee cups, 72,000 soda cans, 40,000 beer cans and bottles, and 11,000 wine bottles.

Are these numbers plausible? Write a short essay arguing your position, with some arithmetic to back up your conclusions.

2 ▶ Table 6.7 records the school dropout rates for the states. Some states have higher rates than others. What factors do you think might account for this? For example, the high dropout rate for the District of Columbia suggests that a large urban population may contribute to a high state rate. Look in the *Statistical Abstract of the United States* for information you think is relevant. For example, the *Statistical Abstract* contains a table that gives the percent of each state's population that lives in metropolitan areas. (But "metropolitan

areas" cover suburbs along with central cities and small cities as well as large.) Write an essay suggesting some explanations, and accompany it with data and graphs that explore at least one of your suggestions.

3 ▸ Graphs good and bad fill the news media. Some publications, such as *USA Today*, make particularly heavy use of graphs to present data. Collect several graphs (at least five) from newspapers and magazines (not from advertisements). Use them as examples in a brief essay about the clarity, accuracy, and attractiveness of graphs in the news. You can find information on what makes good graphs in the books by Tufte and by Cleveland listed in Suggested Readings.

CHAPTER 7

PROBABILITY: THE MATHEMATICS OF CHANCE

Have you ever wondered how gambling, which is a recreation or an addiction for individuals, can be a business for the casino? A business requires predictable revenue from the service it offers, even when the service is a game of chance. Individual gamblers may win or lose. They can never say whether a day at Atlantic City or the local riverboat casino will turn a profit or a loss. But the casino itself does not gamble. Casinos are consistently profitable, and state governments make money both from running lotteries and from selling licenses for other forms of gambling.

It is a remarkable fact that the aggregate result of many thousands of chance outcomes can be known with near certainty. The casino need not load the dice, mark the cards, or alter the roulette wheel. It knows that in the long run each dollar bet will yield its five cents or so of revenue. It is therefore good business to concentrate on free floor shows or inexpensive bus fares to increase the flow of dollars bet. The flow of profit will follow.

Gambling houses are not alone in profiting from the fact that a chance outcome many times repeated is firmly predictable. For example, although a life insurance company does not know *which* of its policyholders will die next year, it can predict quite accurately *how many* will die. It sets its premiums by this knowledge, just as the casino sets its jackpots.

A phenomenon is called **random** if individual outcomes are uncertain but the long-term pattern of many individual outcomes is predictable.

FIGURE 7.1 Casino dice.

FIGURE 7.1 Casino dice.

To a statistician, "random" does not mean "haphazard." Randomness is a kind of order, an order that emerges only in the long run, over many repetitions. Many phenomena, both natural and of human design, are random. The life spans of insurance buyers and the hair color of children are examples of natural randomness. Indeed, quantum mechanics asserts that at the subatomic level the natural world is inherently random. Probability theory, the mathematical description of randomness, is essential to much of modern science.

Games of chance are examples of randomness deliberately produced by human effort. Casino dice (Figure 7.1) are carefully machined, and their drilled holes, called *pips,* are filled with material equal in density to the plastic body. This guarantees that the side with six pips has the same weight as the opposite side, which has only one pip. Thus, each side is equally likely to land upward. All the odds and payoffs of dice games rest on this carefully planned randomness.

Will it land heads or tails?

Statisticians and casino managers both rely on planned randomness, although statisticians use tables of random digits rather than dice and cards. The reasoning of statistical inference rests on planned randomness and on the mathematics of probability, the same mathematics that guarantees the profits of casinos and insurance companies. The mathematics of chance is the topic of this chapter.

WHAT IS PROBABILITY?

The mathematics of chance, the mathematical description of randomness, is called *probability theory.* Probability describes the predictable long-run patterns of random outcomes.

Toss a coin in the air. Will it land heads or tails? Sometimes it lands heads and sometimes tails. We cannot say what the next outcome will be. Perhaps you would argue that the coin looks balanced, so you think it has an equal chance of falling heads or tails on the next toss. Your personal probability of a head is then 1/2. Probability as the expression of personal opinion is a reasonable idea. But we have in mind a different meaning, one based on *observation* of random phenomena.

EXAMPLE ▶ *Ten Thousand Coin Tosses*

Suppose that we toss a coin not once but 10,000 times. John Kerrich, an English mathematician, actually did this while interned by the Germans during World War II. Figure 7.2 shows Kerrich's results. His first 10 tosses gave 4 heads, a proportion of 0.4. The proportion of heads increased to 0.5 after 20 tosses and to 0.57 after 30 tosses. Figure 7.2 shows how this proportion changes as more and more tosses are made. In a small number of tosses, the proportion of heads fluctuates—it is still essentially unpredictable. But many tosses produce a smoothing effect. After Kerrich had thrown his coin 5000 times, the proportion of heads was 0.507, or 50.7%. And in all 10,000 trials, he scored 5067 heads, again 50.7% of the total. After many trials, the proportion of heads settles down to a fixed number. This number is the probability of a head. ◆

> The **probability** of any outcome of a random phenomenon is the proportion of a very long run of trials in which the outcome occurs.

Probability answers the question, "What would happen if we did this many times?" The idea of probability uses what would happen in many trials to describe the uncertain outcome of a single trial. This "definition" of probability would not satisfy either a mathematician or a philosopher, but it fixes the idea in our minds. Strictly speaking, Kerrich's 0.507 only estimates the probability that his coin will come up heads. The proportion after 100,000 tosses would be closer to the true probability. In fact, we can't even say that Kerrich's coin is weighted in favor of heads. There is still enough unpredictability in the results of 10,000 tosses that 70 excess heads could easily occur even if heads and tails were equally probable.

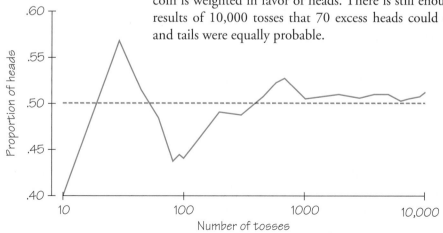

FIGURE 7.2 Proportion of heads versus number of tosses in Kerrich's coin-tossing experiment.

PROBABILITY MODELS

Gamblers have known for a long time that the fall of coins, cards, or dice stabilizes into definite patterns in the long run. France gave birth to the mathematics of probability in the seventeenth century when gamblers turned to mathematicians for advice (see Spotlight 7.1). The idea of probability rests on the observed fact that the average result of many thousands of chance outcomes can be known with near certainty. But a definition of probability as "long-run proportion" is vague. Who can say what "the long run" is? Instead, we give a mathematical description of *how probabilities behave,* based on our understanding of long-run proportions. This mathematical description of probability has the advantage that it applies equally well to probability thought of as personal assessment of chance. The same mathematics describes two different informal concepts of probability.

Our first task in assigning probabilities to outcomes is to list all the possible outcomes.

> The set of all possible outcomes of a random phenomenon is called the **sample space S.**

The sample space S can be very simple or very complex. When we toss a coin once, there are only two outcomes, heads and tails. So the sample space is $S = \{H, T\}$. If we draw a random sample of 1500 U.S. residents age 18 and over, as Gallup polls do, the sample space contains all possible choices of 1500 of the more than 190 million adults in the country. This S is extremely large. To count its members would require a number more than 8000 digits long! Each member of S is a possible Gallup poll sample, which explains the term *sample space.*

EXAMPLE ▶ *Rolling Dice*

Suppose that we roll a single die and observe the number of pips on the up face. The possible outcomes appear in Figure 7.3. We take the sample space, the set of these outcomes, to be

$$S = \{1, 2, 3, 4, 5, 6\}$$

What if we roll *two* dice, as many games of chance require? Figure 7.4 shows the possible combinations of up faces on the two dice in order. Think of the red die as being rolled first, then the green die. The sample space consists of these 36 outcomes.

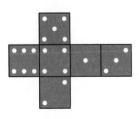

FIGURE 7.3 The possible outcomes for rolling one die.

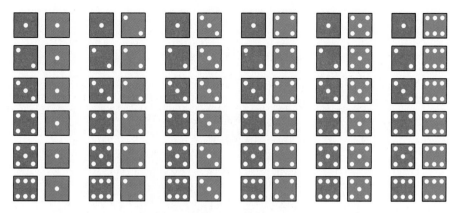

FIGURE 7.4 The possible outcomes for rolling two dice.

In craps and other games, all that matters is the *sum* of the pips on the up faces. Let's change the random outcomes we are interested in: roll two dice and count the pips on the up faces. Now there are only 11 possible outcomes, from a sum of 2 for rolling a double one through 3, 4, 5, and on up to 12 for rolling a double six. The sample space is now

$$S = \{2, 3, 4, 5, 6, 7, 8, 9, 10, 11, 12\}$$

As this example shows, it is important in choosing a sample space to be clear which random outcomes you will look for. Rolling two dice and recording the up faces in order is not the same as rolling two dice and recording only the sum of the two up faces. ◆

The next step in the mathematical description of chance is to assign probabilities to the outcomes that make up the sample space. There are many ways to assign probabilities, so it is convenient to start with some general rules that any assignment of probabilities to outcomes must obey. Let $P(s)$ stand for the probability of any outcome s in the sample space S. Here are the fundamental laws for assigning probabilities to outcomes:

LAW 1. Every probability $P(s)$ is a number between 0 and 1.
LAW 2. The sum of the probabilities $P(s)$ for all outcomes s in S is exactly 1.

These laws are based on our understanding of probability as long-run proportion. The first law reflects the fact that any proportion is a number between

Spotlight 7.1

The Mathematical Bernoullis

Jakob Bernoulli

Johann Bernoulli

Few families have made more contributions to mathematics than the Bernoullis of Basel, Switzerland. No fewer than seven Bernoullis, over three generations spanning the years between 1680 and 1800, were distinguished mathematicians. Five of them helped build the new mathematics of probability.

Jakob (1654–1705) and Johann (1667–1748) were sons of a prosperous Swiss merchant, but they studied mathematics against the will of their practical father. Both were among the finest mathematicians of their times, but it was Jakob who concentrated on probability. Several seventeenth-century mathematicians had started the study of games of chance, concentrating on counting outcomes to find chances. Jakob Bernoulli was the first to see clearly the idea of a long-run proportion as a way of measuring chance. He proved that if in a very large population (say, size N), K members have a property A, then the proportion of members of a sample of size n having property A must approach K/N (the probability of A) as the sample size n increases. This *law of large numbers*

helped to connect probability to the study of sequences of chance outcomes observed in human affairs.

Johann's son Daniel (1700–1782) and Jakob and Johann's nephew Nicholas (1687–1759) also studied probability. Nicholas saw that the pattern of births of male and female children could be described by probability. Despite his own rebellion against his father's strictures, Johann tried to make his son Daniel a merchant or a doctor. Daniel, undeterred, became yet another Bernoulli mathematician. He studied mainly the mathematics of flowing fluids (later applied to designing ships and aircraft) and elastic bodies. In the field of probability, he worked to fairly price games of chance and gave evidence for the effectiveness of inoculation against smallpox.

The Bernoulli family in mathematics, like their contemporaries the Bachs in music, is an unusual example of talent in one field appearing in successive generations. The Bernoullis' work helped probability to grow from its birthplace in the gambling hall to a respectable tool with worldwide applications.

0 and 1. If an outcome *never* occurs, it has probability 0; if it *always* occurs, the probability is 1; if it *sometimes* occurs, the proportion of trials producing the outcome is a number between 0 and 1. The second law says that some outcome must always occur, so that the sum of all their probabilities is 1. Laws 1 and 2 describe any legitimate assignment of probabilities to outcomes. An assignment that does not satisfy these laws does not make sense.

EXAMPLE ▶ *Rolling One Die*

Let's return to rolling a single die to see what a probability model looks like. The sample space (see Figure 7.3) is

$$S = \{1, 2, 3, 4, 5, 6\}$$

Any assignment of probabilities $P(1)$, $P(2)$, $P(3)$, $P(4)$, $P(5)$, and $P(6)$ to these six outcomes must assign numbers between 0 and 1 that add to exactly 1. If the die is a carefully made casino model, each face should have the same probability of landing upward. Then the six probabilities are all the same and have sum 1, so each must be 1/6. The probabilities are

$$P(1) = P(2) = P(3) = P(4) = P(5) = P(6)$$
$$= 1/6, \text{ or about } 0.167$$

Dice and other chance devices that are carefully made so as to have equally likely outcomes are often called *fair*. ◆

It is important to understand what the two laws for outcome probabilities accomplish and what they leave unsaid. They tell us which assignments of probabilities to outcomes make sense. But they do not tell us which assignment is correct, in the sense of accurately describing a real die. You must actually toss a die many times to find its correct probabilities. We are safe if we use the equal-probabilities model of the example for professional dice. But cheap dice with hollowed-out pips fall unequally. The 6 face is lightest and is located opposite the 1 face, which is heaviest. Thus, a cheap die might be described by a set of probabilities in which lighter faces are more likely to land upwards, such as

$$P(1) = 0.159 \quad P(4) = 0.166$$
$$P(2) = 0.163 \quad P(5) = 0.171$$
$$P(3) = 0.166 \quad P(6) = 0.175$$

These probabilities also satisfy Laws 1 and 2. They are legitimate, even if the dice are not.

Assigning probabilities to individual outcomes is not enough. We also want to assign probabilities to **events,** which are collections of outcomes. For example, what is the probability of rolling an odd number in one roll of a fair die? The odd outcomes are 1, 3, and 5. The proportion of rolls on which one of these numbers comes up must be the sum of the proportions on which each alone comes up. So thinking of probabilities as long-run proportions leads us to find the probability of any event by summing the probabilities of the outcomes that make up the event. In this case,

$$P(\text{outcome is odd}) = P(1) + P(3) + P(5)$$
$$= 3/6 = 0.5$$

We now have a complete *probability model* for rolling a fair die.

A **probability model** is a mathematical description of a random phenomenon consisting of two parts: a sample space S and a way of assigning probabilities to events.

Probability models for random phenomena with only finitely many possible outcomes have a simple form: assign probabilities to outcomes in a way that satisfies Laws 1 and 2, then find probabilities of events by adding up the probabilities of the outcomes that make up the event. You can state such a model by giving a table of the outcomes and their probabilities.

EXAMPLE ▶ *Household Size*

A household is a group of people living together, regardless of their relationship to each other. Sample surveys such as the Current Population Survey select a random sample of households. Here is the probability model for the number of people living in a randomly chosen American household:

Household size	1	2	3	4	5	6	7
Probability	.245	.323	.175	.155	.066	.023	.013

These probabilities are the proportions for all households in the country, and so give the probabilities that a single household chosen at random will have each size. (The very few households with more than 7 members are placed in the 7 group.) Check that this assignment of probabilities satisfies Laws 1 and 2.

The probability that a randomly chosen household has more than two members is

$$P(\text{size} > 2) = P(3) + P(4) + P(5) + P(6) + P(7)$$
$$= .175 + .155 + .066 + .023 + .013 = .432 \quad \blacklozenge$$

EQUALLY LIKELY OUTCOMES

A simple random sample gives all possible samples an equal chance to be chosen. Dealing from a well-shuffled deck gives all possible card hands an equal chance to be the hand you are dealt. When randomness is the product of human design, it is often the case that the outcomes in the sample space are all equally likely. Laws 1 and 2 force the assignment of probabilities in this case.

> If a random phenomenon has k possible outcomes, *all equally likely to occur,* then each individual outcome has probability $1/k$.

The probability of any event in the equally likely case is found, as usual, by adding the individual probabilities of the outcomes making up the event. Because each of these probabilities is the same $1/k$, we can add probabilities by just counting outcomes. Here is the rule:

> When all outcomes have equal probabilities, the probability of any event A is
>
> $$P(A) = \frac{\text{number of outcomes in } A}{\text{number of outcomes in } S}$$

EXAMPLE ▶ *Rolling Two Dice*

Roll two fair dice and record the pips on each of the two up faces. The sample space consists of the 36 outcomes pictured in Figure 7.4. Because of the balance of the dice, these outcomes are all equally likely. So each has probability 1/36.

What is the probability of rolling a 5? The event "roll a 5" contains the four outcomes

The probability is therefore 4/36, or about 0.111. What about the probability of rolling a 7? Look at Figure 7.4 and count six outcomes for which the sum of the pips is 7. The probability is therefore 6/36, or about 0.167. ◆

Be certain that you understand that the method of finding probabilities by counting outcomes applies *only* when all outcomes are equally likely. The *S* in Figure 7.4 does have equally likely outcomes. But if we choose to use the sample space for rolling two dice *and counting the pips,* we get

$$S = \{2, 3, 4, 5, 6, 7, 8, 9, 10, 11, 12\}$$

These outcomes do *not* have equal probabilities. We just saw, for example, that the probability of 5 is 4/36 and that the probability of a 7 is 6/36.

When outcomes are equally likely, finding probabilities leads to the study of counting methods, called **combinatorics.** Combinatorics is an important area of mathematics in its own right. Although we will not study combinatorics in detail, the following example uses the multiplication method that we called the *fundamental principle of counting* in Chapter 2.

EXAMPLE ▶ *Code Words*

A computer system assigns log-in identification codes to users by choosing three letters at random. All three-letter codes are therefore equally likely. What is the probability that the code assigned to you has no "x" in it?

First count the total number of code words. There are 26 letters that can occur in each position in the word. Any of the 26 letters in the first position can be combined with any of the 26 letters in the second position to give 26 × 26 choices. (This is true because the order of the letters matters, so that "ab" and "ba" are different choices.) Any of the 26 letters can then follow in the third position. The number of different codes is

$$26 \times 26 \times 26 = 17{,}576$$

Now count the number of code words that have no "x." These codes are made up of the other 25 letters. So there are

$$25 \times 25 \times 25 = 15{,}625$$

such codes. The probability that your code has no "x" is therefore

$$P(\text{no "x"}) = \frac{\text{number of codes with no "x"}}{\text{number of codes}}$$

$$= \frac{15{,}625}{17{,}576} = 0.889$$

Suppose that the computer is programmed to avoid repeated letters in the identification codes. Any of the 26 letters can still appear in the first position. But only the 25 remaining letters are allowed in the second position, so that there are 26×25 choices for the first two letters in the code. Any of these choices leaves 24 letters for the third position. The number of different codes without repeated letters is

$$26 \times 25 \times 24 = 15{,}600$$

Codes with no "x" are allowed one fewer choice in each position. There are

$$25 \times 24 \times 23 = 13{,}800$$

such codes. The probability that your code has no "x" is then

$$P(\text{no "x"}) = \frac{\text{number of codes with no "x"}}{\text{number of codes}}$$

$$= \frac{13{,}800}{15{,}600} = 0.885$$

Eliminating repeats slightly lowers your chance of avoiding an "x." ◆

The example makes use of two facts about counting that we often apply in finding probabilities:

A. Suppose we have a collection of n distinct items. We want to arrange k of these items in order, and the same item can appear several times in the arrangement. The number of possible arrangements is

$$n \times n \times \cdots \times n = n^k$$

B. Suppose we have a collection of n distinct items. We want to arrange k of these items in order, and any item can appear no more than once in the arrangement. The number of possible arrangements is

$$n \times (n-1) \times \cdots \times (n-k+1)$$

In the example, n (the number of letters available) is first 26, then 25, and k (the number of letters to be arranged to make a code) is 3. It is usually easier to think your way through the counting than to memorize the recipes.

EXAMPLE ▶ *How Many Orderings?*

A jury of 7 students is seated in a row of 7 chairs to judge a speaking competition. In how many orders can the students sit?

Because each chair holds only one student, no repeats are allowed. This is the case described by rule B with *n* and *k* both equal to 7. To think through the problem, proceed like this: any of the 7 students can sit in the first chair; then any of the 6 who remain can sit in the second chair; and so on. The number of arrangements is therefore

$$7 \times 6 \times 5 \times 4 \times 3 \times 2 \times 1 = 5040$$

(This number is often called 7!, read "seven factorial.") ◆

THE MEAN OF A RANDOM PHENOMENON

Suppose you are offered this choice of bets, each costing the same: bet A pays $10 if you win and you have probability 1/2 of winning, while bet B pays $10,000 and offers probability 1/10 of winning. You would very likely choose B even though A offers a better chance to win, because B pays much more if you win. It would be foolish to decide which bet to make just on the basis of the probability of winning. How much you can win is also important. When a random phenomenon has numerical outcomes, we are concerned with their amounts as well as with their probabilities.

What will be the average payoff of our two bets in many plays? Recall that the probabilities are the long-run proportions of plays on which each outcome occurs. Bet A produces $10 half the time in the long run and nothing half the time. So the average payoff should be

$$\left(\$10 \times \tfrac{1}{2}\right) + \left(\$0 \times \tfrac{1}{2}\right) = \$5$$

Bet B, on the other hand, pays out $10,000 on 1/10 of all bets in the long run. Bet B's average payoff is

$$\left(\$10{,}000 \times \tfrac{1}{10}\right) + \left(\$0 \times \tfrac{9}{10}\right) = \$1000$$

If you can place many bets, you should certainly choose B. Here is a general definition of the kind of "average outcome" we used to compare the two bets.

Suppose that the possible outcomes $s_1, s_2, \ldots, s_k$ in a sample space S are numbers, and that p_j is the probability of outcome s_j. The **mean** μ of this probability model is

$$\mu = s_1 p_1 + s_2 p_2 + \cdots + s_k p_k$$

Earlier, we met the mean $\bar{x}$, the average of n observations that we actually have in hand. The mean μ, on the other hand, describes the probability model rather than any one collection of observations. You can think of μ as a theoretical mean that says what average outcome we expect in the long run.

EXAMPLE ▶ *Mean Household Size*

The probability model for the number of people living in a randomly chosen household is

Household size	1	2	3	4	5	6	7
Probability	.245	.323	.175	.155	.066	.023	.013

The mean is therefore

$$\mu = (1)(.245) + (2)(.323) + (3)(.175) + (4)(.155) + (5)(.066)$$
$$+ (6)(.023) + (7)(.013)$$
$$= 2.595$$

In this case, the mean μ is the average size of all American households. If we took a random sample of, say, 100 households and recorded their sizes, we would call the average size for this sample $\bar{x}$. A second random sample would no doubt give a somewhat different value of $\bar{x}$. So $\bar{x}$ varies from sample to sample, but μ, which describes the distribution of probabilities, is a fixed number. ◆

The mean μ is an average outcome in two senses. The definition says that it is the average of the possible outcomes, not weighted equally but weighted by their probabilities. More likely outcomes get more weight in the average. An important fact of probability, the *law of large numbers,* says that μ is the average outcome in another sense as well.

Observe any random phenomenon having numerical outcomes with finite mean μ. According to the **law of large numbers,** as the random phenomenon is repeated a large number of times

- The proportion of trials on which each outcome occurs gets closer and closer to the probability of that outcome, and
- The mean $\bar{x}$ of the observed values gets closer and closer to μ.

These facts can actually be proved mathematically for any assignment of probabilities that satisfies Laws 1 and 2. The law of large numbers brings the study of basic probability back to a natural completion. We first observed that some phenomena are random in the sense of showing long-run regularity. Then we used the idea of long-run proportions to motivate the basic laws of probability. Those laws are mathematical idealizations that can be used without interpreting probability as proportion in many trials. Now the law of large numbers tells us that in many trials the proportion of trials on which an outcome occurs will always approach its probability.

SAMPLING DISTRIBUTIONS

Sampling, in a way, is a lot like gambling. Both rely on the deliberate use of chance. We want to apply probability to describe the results of sampling. At first glance, this is a formidable task. Suppose that we choose a simple random sample of size 100 from the more than 190 million adults in the United States. All possible samples are equally likely—that is the definition of simple random sampling. But there are an immense number of possible samples, so that finding probabilities by counting is not appealing. There are mathematical shortcuts, but there is also another way: rather than counting, we can actually choose a large number of samples and observe the outcomes. In practice, we program a computer to imitate (the formal word is *simulate*) drawing many samples. Let's try it.

EXAMPLE ▶ *A Sampling Experiment*

In Chapter 5, we looked at a Gallup poll that asked a sample of 1493 people "Are you afraid to go outside at night within a mile of your home because of crime?" Ask that question of a simple random sample of 100 adults; 48 say "Yes." That's 48% of the sample. Take another simple random sample of 100 adults; this time, 50% say "Yes." This is **sampling variability:** when we take repeated samples from the same population, the results will vary from sample to sample.

Figure 7.5a is a histogram of the percents who said "Yes" to the question in 1000 simple random samples from the same population. It shows the regular pattern of outcomes that is characteristic of random sampling. Now we can use the language of probability to describe this pattern.

One of the classes in the histogram covers the range

46% < percent saying "Yes" ≤ 48%

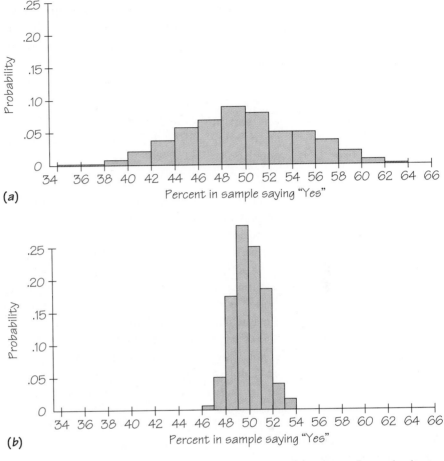

FIGURE 7.5   Sampling distributions that show the behavior of the percent of a sample who say "Yes" to an opinion poll question in many simple random samples from the same population. (a) Sample size 100. (b) Sample size 1493.

Exactly 70 of the 1000 samples had outcomes in this class. Because 1000 samples is a large number of repetitions of the random sampling, the proportion of outcomes that fall in this class is close to the probability of the class. So we estimate that the probability of getting an outcome greater than 46% but no larger than 48% is 70/1000, or 0.07. The height of the bar above that class in Figure 7.5a is 0.07.

Figure 7.5a is a new kind of histogram. The heights of the bars are the proportions of outcomes falling in each class, rather than the counts of outcomes. Because it shows the results of a large number of samples, the histogram presents the estimated probabilities for each class of sample outcomes. The heights of the bars add to 1 because they assign probabilities.   ◆

Statisticians call a number that is computed from a sample a **statistic.** The percent of our sample of 100 people who said "Yes" to the poll question is a statistic. The histogram in Figure 7.5a displays the sampling variability of this statistic by assigning probabilities to its possible values. These probabilities make up the *sampling distribution* of the statistic.

> The **sampling distribution** of a statistic is the distribution of values taken by the statistic in all possible samples of the same size from the same population.

Strictly speaking, the sampling distribution is the ideal pattern that would emerge if we looked at all possible samples of size 100 from our population. A distribution obtained from a fixed number of trials, like the 1000 trials in Figure 7.5a, is only an approximation to the sampling distribution. One of the uses of probability theory in statistics is to obtain exact sampling distributions without actually drawing many samples. The interpretation of a sampling distribution is the same, however, whether we obtain it by actual sampling or by the mathematics of probability.

Let's try a second sampling experiment. The Gallup poll asked 1493 people, not 100 people, about their fear of crime. We will take 1000 simple random samples of 1493 people. For each of these samples, calculate the percent who say that they are afraid to go out because of crime. Figure 7.5b displays the distribution of the 1000 sample percents, using the same scale as Figure 7.5a. This is the sampling distribution for this statistic.

EXAMPLE ▶ *Examining Sampling Distributions*

Let's apply our tools for describing distributions to the two sampling distributions illustrated in Figure 7.5. We will examine the *shape, center,* and *spread* of these distributions.

Both distributions share a distinctive shape. They are quite symmetric, with a single peak in the center. There are no outliers to disturb the pattern. Both are centered very close to 50%. In fact, the mean outcomes are 50.11 for samples of size 100 and 50.03 for samples of size 1493. The distribution of results for samples of size 1493 is much less spread out than the distribution for samples of 100 people—that is, the histogram in Figure 7.5b is taller and narrower than that in Figure 7.5a. The standard deviations of the 1000 sample results are 4.986 for the smaller samples and 1.289 for the larger samples.

In fact, the population from which we drew all of our samples contained exactly 50% who would say "Yes" to Gallup's question about fear of crime. The centers of the sampling distributions are very close to 50%. This reflects the lack of bias in simple random sampling. The spread of the sample results goes down as we take larger samples. So large samples usually give results close to the truth about the population. ◆

Our sampling experiment has both produced an approximate assignment of probabilities (without counting) and taught us a bit about how the sampling distribution behaves when we increase the size of the sample. Our goal is to learn enough about the mathematics of probability to get more exact results than sampling experiments provide. In the next chapter we'll learn specific recipes for the mean and standard deviation of the sampling distributions that Figure 7.5 approximates. The first step is to study the distinctive shape of these distributions. They are *normal distributions.* That is the topic to which we now turn.

NORMAL DISTRIBUTIONS

Although they differ in variability, the histograms in Figures 7.5a and 7.5b have similar shapes in other respects. Both are symmetric, with centers close to 50%. The tails fall off smoothly on either side, with no outliers. Suppose that we represent the shape of each histogram by drawing a smooth curve through the tops of the bars. If we do this carefully—using the actual probabilities of the outcomes rather than estimates from only 1000 samples—the two curves we obtain will be quite close to two members of the family of *normal curves.* The two normal curves appear in Figure 7.6.

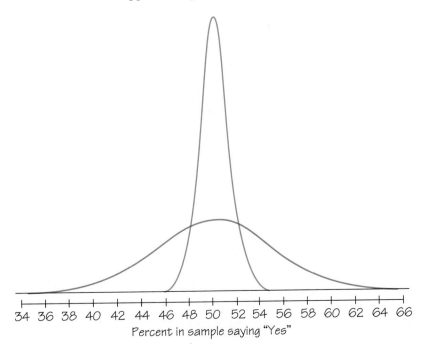

FIGURE 7.6 The normal curves that approximate the sampling distributions in Figure 7.5. The taller curve is for sample size 1493, the flatter curve for smaller samples of size 100. Each curve has area exactly 1 beneath it.

34 36 38 40 42 44 46 48 50 52 54 56 58 60 62 64 66
Percent in sample saying "Yes"

Normal curves introduce a new way of describing probabilities. We can describe an assignment of probabilities to the values of a statistic by a histogram like those of Figure 7.5. A histogram makes probability visible. The height of any bar is the probability of the outcomes spanned by the base of that bar. Because all bars have the same width, their area (height times width) is proportional to the probability. Normal curves can be thought of as approximations to a histogram of probabilities in which area is exactly equal to probability. Normal curves are easier to work with than histograms because many bars are replaced by a single smooth curve. Normal curves have the property that the total area under the curve is exactly 1, corresponding to the fact that all outcomes together have probability 1.

> A normal curve assigns probabilities to outcomes as follows: the probability of any interval of outcomes is the area under the normal curve above that interval. The total area under any normal curve is exactly 1.

EXAMPLE ▶ *Probability as Area*

Figure 7.7 is another drawing of the normal curve for the sampling distribution of Figure 7.5b. This curve assigns probabilities for the percent of a simple random sample of size 1493 who say "Yes" to Gallup's question about fear of crime.

The shaded area is the area under the normal curve between 50% and 52%. This area is 0.44. So the probability that between 50% and 52% of the people in a randomly chosen sample will say "Yes" is 0.44. ◆

Our original method of assigning probabilities was to give a probability to each individual outcome, then add these probabilities to get the probability of any event. Probability as area under a curve is the second important method of assigning probability, and is easier when there are many individual outcomes falling close together. Curves of different shapes describe different assignments of probability. We will emphasize the normal curves, because they describe probability in several important situations. An assignment of probabilities to outcomes by a normal curve is a **normal probability distribution.**

Figures 7.5 and 7.6 demonstrate that the sampling distribution of a sample proportion from a simple random sample is close to a normal distribution. This is not just a matter of artistic judgment. It is a mathematical fact, first proved by Abraham DeMoivre in 1718. Some other common statistics, such as the mean $\bar{x}$ of a large sample, also have sampling distributions that are approximately normal. A normal curve will not exactly describe a specific set of outcomes, such as our 1000 sample percentages. It is an idealized distribution that

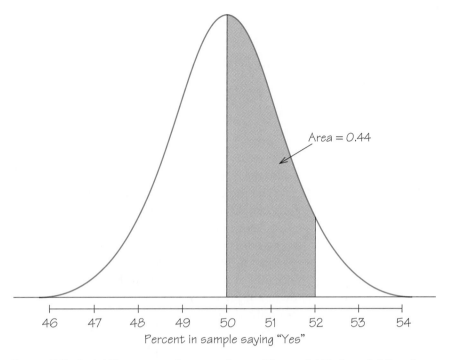

FIGURE 7.7   Probability as area under a normal curve. The area 0.44 is the probability of an outcome between 50 and 52.

is convenient to use and gives a good approximation to the actual distribution of outcomes.

There is a close connection between describing an assignment of probability to numerical outcomes and describing a set of data. Histograms can be used for both tasks. Similarly, smooth curves such as the normal curves can replace histograms for describing large sets of data as well as for assigning probabilities. Many sets of data are approximately described by normal distributions. The normal distributions therefore deserve more detailed study.

THE SHAPE OF NORMAL CURVES

Normal curves can be specified exactly by an equation, but we will be content with pictures like Figures 7.6 and 7.7. All normal curves are symmetric and bell-shaped, with tails that fall off rapidly. The center of the symmetric normal curve is the center of the distribution in several senses. It is the mean μ for the assignment of probabilities. It is also the median in the sense that half the probability (half the area under the curve) lies on each side of the center.

FIGURE 7.8   The mean
of a skewed distribution is
located further toward the
long tail than the median.

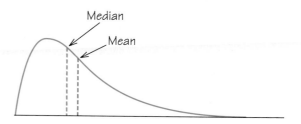

When probabilities are assigned as areas under a symmetric curve, the mean μ is also the median of the distribution.

The mean and median of a skewed distribution are not equal. Figure 7.8, for example, shows a right-skewed distribution. The right tail of the curve is much longer than the left. The prices of new houses are an example of a skewed distribution—there are many moderately priced houses and a few extravagantly priced mansions out in the right tail. Those mansions pull the mean, or average price, up, so that it is greater than the median. For example, the mean price of a new house in 1994 was $144,700, but the median price was only $123,000.

As we saw in Chapter 6, even the most cursory description of data on a single variable should include a measure of spread in addition to a measure of center or location. What about the spread of a normal curve? *Normal curves have the special property that their spread is completely measured by a single number, the standard deviation.* We learned in the last chapter how to calculate the standard deviation from a set of observations. For normal distributions, the standard deviation (like the mean) can be found directly from the curve.

The **mean** of a normal distribution lies at the center of symmetry of the normal curve.

To find **standard deviation** of a normal distribution, run a pencil along the normal curve from the center (the mean) outward. At first, the curve falls ever more steeply as you go out; farther from the mean it falls ever less steeply. The two points where the curvature changes are located one standard deviation on either side of the mean.

With a little practice you can locate the change-of-curvature points quite accurately. For example, Figure 7.9 shows the distribution of heights of American women ages 18 to 24. The shape of the curve is normal, with mean (and median) height $\mu = 64$ inches. The two change-of-curvature points are at 61.5 inches and 66.5 inches. The standard deviation of the distribution is the distance of either of these points from the mean, or 2.5 inches.

The usual notation for the standard deviation of a probability distribution is σ, the Greek lowercase sigma. Just as for the mean μ, it is possible to find σ for any distribution directly from the assignment of probabilities. We will not do this, but will content ourselves with being able to find σ for normal distributions by looking at the curves. Again, just as for the mean, we distinguish between s, the standard deviation of a given set of observations, and σ, the standard deviation of a probability distribution.

In Chapter 6, we often used the quartiles to indicate the spread of a distribution. Because the standard deviation completely describes the spread of any normal distribution, it tells us where the quartiles are. Here are the facts:

> The first quartile of any normal distribution is located 0.67σ below the mean; the third quartile is 0.67σ above the mean.

EXAMPLE ▶ *Heights of Young Women*

The distribution of heights of young women, shown in Figure 7.9, is approximately normal with mean $\mu = 64$ inches and standard deviation $\sigma = 2.5$ inches. The quartiles lie 0.67σ, or

$$(0.67)(2.5) = 1.7 \text{ inches}$$

on either side of the mean. The first quartile is $64 - 1.7$, or 62.3 inches. The third quartile is $64 + 1.7$, or 65.7 inches. Figure 7.10 marks the quartiles on the normal curve. They contain between them the middle 50% of women's heights. ◆

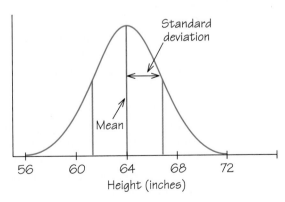

FIGURE 7.9 Locating the mean and standard deviation on a normal curve.

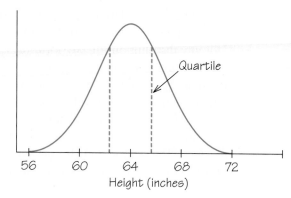

FIGURE 7.10   The quartiles of a normal distribution are located 0.67 standard deviation on either side of the mean.

The mean and standard deviation of normal curves have a special property: the shape of a normal distribution is completely specified by giving μ and σ. A measure of center and a measure of spread are not sufficient to determine the exact shape of most distributions of data, but the mean and standard deviation are enough when the distribution is normal. Changing the mean of a normal curve does not change its shape; it only moves the curve to a new location. Changing the standard deviation does change the shape. A normal curve with a smaller standard deviation is taller and narrower (has less spread) than one with a larger standard deviation. You can see this by comparing the two normal curves for our random sampling experiments in Figure 7.6. Both normal curves have the same mean, but the curve for samples of size 1493 has the smaller standard deviation.

THE 68–95–99.7 RULE

Another consequence of the fact that the mean and standard deviation completely specify a normal distribution is that all normal distributions are the same when we record observations in terms of how many standard deviations they lie from the mean. In particular, the probability that an observation falls within one, two, or three standard deviations of the mean is the same for all normal distributions. The probability of an outcome falling within one standard deviation on either side of the mean is 0.68. If we go out two standard deviations from the mean, the probability is 0.95. Finally, the probability of falling within three standard deviations of the mean is almost 1, or 0.997 to be exact. These facts can be derived mathematically from the equation of a normal curve. They are not true for distributions with other shapes.

Figure 7.11 illustrates these facts expressed in terms of percents. Together, we call them the *68–95–99.7 rule* for normal distributions.

According to the **68–95–99.7 rule,** in any normal distribution:

- 68% of the observations fall within one standard deviation of the mean.
- 95% of the observations fall within two standard deviations of the mean.
- 99.7% of the observations fall within three standard deviations of the mean.

Using the three numbers in the 68–95–99.7 rule, we can quickly derive helpful information about any normal distribution. More detailed information can be gleaned from tables of areas under the normal curves, but the 68–95–99.7 rule is adequate for our purposes.

EXAMPLE ▶ *Heights of Young Women*

The heights of American women between the ages of 18 and 24 are roughly normally distributed, with mean $\mu = 64$ inches and standard deviation $\sigma = 2.5$ inches. One standard deviation below the mean is $64 - 2.5$, or 61.5 inches. Similarly, one standard deviation above the mean is $64 + 2.5$, or 66.5 inches. The "68" part of the 68–95–99.7 rule says that about 68% of women are between 61.5 and 66.5 inches tall. Because two standard deviations are 5 inches, we know that 95% of young women are between $64 - 5$ and $64 + 5$, that is, between 59 and 69 inches tall. Almost all women have heights within three standard deviations of the mean, or between 56.5 and 71.5 inches. Very few women are 6 feet (72 inches) tall or over. ◆

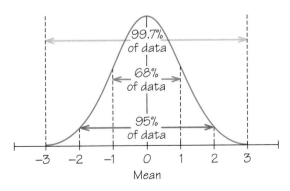

FIGURE 7.11   The 68–95–99.7 rule for normal distributions.

EXAMPLE ▶ *SAT Scores*

The distribution of scores on tests such as the Scholastic Assessment Test (SAT) is close to normal. SAT scores are adjusted so that the mean score is $\mu = 500$ and the standard deviation is $\sigma = 100$. This information allows us to answer many questions about SAT scores.

- *How high must a student score to fall in the top 25%?*
  The third quartile is $(0.67)(100) = 67$ points above the mean. So scores above 567 are in the top 25%.
- *What percent of scores fall between 200 and 800?*
  Scores of 200 and 800 are three standard deviations on either side of the mean. The 99.7 part of the 68–95–99.7 rule says that 99.7% of all scores lie in this range. (In fact, 200 and 800 are the lowest and highest scores that are reported on the SAT. The few scores higher than 800 are reported as 800.)
- *What percent of scores are above 700?*
  A score of 700 is two standard deviations above the mean. By the 95 part of the 68–95–99.7 rule, 95% of all scores fall between 300 and 700 and 5% fall below 300 or above 700. Because normal curves are symmetric, half of this 5% are above 700. So a score above 700 places a student in the top 2.5% of test-takers.

Sketching a normal curve with the points one, two, and three standard deviations from the mean marked can help you use the 68–95–99.7 rule. Figure 7.12 shows the distribution of SAT scores with the areas needed to find the percent of scores above 700. ◆

THE CENTRAL LIMIT THEOREM

The significance of normal distributions is explained in part by a key fact in probability theory known as the *central limit theorem*. This theorem says that the distribution of any random phenomenon tends to be normal if we average it over a large number of independent repetitions. The central limit theorem allows us to analyze and predict the results of chance phenomena if we average over many observations.

We have already seen the central limit theorem at work in our random sampling experiment. A single person drawn at random says either "Yes" or "No" to the opinion poll question. Only two outcomes are possible, and there is no normal curve in sight. However, the percent of "Yes" answers when 100 people are drawn at random roughly follows a normal distribution. You can think of the percent of "Yes" answers as an average of "Yes" or "No" over the

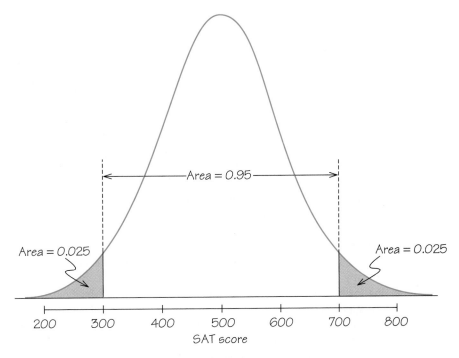

FIGURE 7.12   Using the 68–95–99.7 rule to find the percent of SAT scores that are above 700. For this normal curve, $\mu = 500$ and $\sigma = 100$.

100 people. When we sample 1493 people, the percent of "Yes" responses represents an average over a larger number of people and is even closer to a normal curve.

Our sampling experiment showed that samples of size 1493 have much less spread than samples of size 100. We describe spread by the standard deviation of the normal distribution of outcomes. The central limit theorem makes the relationship of standard deviation to sample size explicit. Here is a more exact statement:

The **central limit theorem** states that

- A sample mean or sample proportion from n trials on the same random phenomenon has a distribution that is approximately normal when n is large.
- The mean of this normal distribution is the same as the mean for a single trial.
- The standard deviation of this normal distribution is the standard deviation for a single trial divided by $\sqrt{n}$.

FIGURE 7.13   A gambler may win or lose at roulette, but in the long run the casino always wins.

$$P(\text{win } \$1) = 18/38$$
$$P(\text{lose } \$1) = 20/38$$

The mean outcome of a single bet on red is found in the usual way:

$$\mu = (1)\left(\tfrac{18}{38}\right) + (-1)\left(\tfrac{20}{38}\right)$$
$$= -\tfrac{2}{38} = -0.053$$

The law of large numbers says that the mean μ is the average outcome of a very large number of individual bets. In the long run, gamblers will lose (and the casino will win) an average of 5.3 cents per bet.   ◆

Just as when we ask only one person's opinion, there is no normal curve in sight when a gambler makes only one bet on red in roulette. But the central limit theorem ensures that the average outcome of many bets follows a distribution that is close to normal. Suppose that we place 50 bets in an evening's play. The mean outcome is the average winnings $\bar{x}$, the overall gain (or loss) divided by 50. If we win 30 and lose 20 times, the overall gain is $10, an average winnings of $\bar{x} = \$.20$ per bet. If we continue to gamble night after night,

FIGURE 7.14 The
distribution of winnings in
repeated bets on red or
black in roulette.

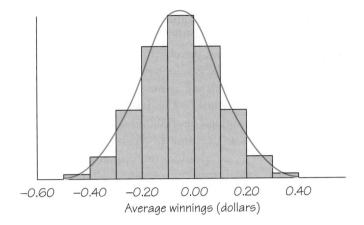

placing 50 bets each night, our average winnings per bet will vary from night to night. A histogram of these values will follow a normal distribution. Figure 7.14 shows the results of many trials of 50 bets each. The normal curve superimposed on the histogram is the distribution given by the central limit theorem in this case.

We know that the mean of the normal distribution in Figure 7.14 is the same as the mean of a single bet, -0.053. What is the standard deviation? The full spread of outcomes observed was -0.47 to 0.37, or 0.84 in all. By the 99.7 part of the 68–95–99.7 rule, the outcomes should span about three standard deviations on each side of the mean. The standard deviation is therefore about one-sixth of 0.84, or 0.14 (14 cents). Check this by locating the change-of-curvature points of the normal curve in the figure. From this combination of calculation and experiment we conclude that the average winnings in 50 bets follow approximately the normal distribution with mean -0.053 and standard deviation 0.14.

What will be the experience of a habitual gambler who places 50 bets per night? Almost all average nightly winnings will fall within three standard deviations of the mean, that is, between

$$-0.053 + (3)(0.14) = 0.367$$

and

$$-0.053 - (3)(0.14) = -0.473$$

The total winnings after 50 bets will therefore fall between

$$(0.367)(50) = 18.35$$

and

$$(-0.473)(50) = -23.65$$

The gambler may win as much as $18.35 or lose as much as $23.65. Gambling is exciting because the outcome, even after an evening of bets, is uncertain. It is possible to walk away a winner. It's all a matter of luck.

The casino, however, is in a different position. It doesn't want excitement, just a steady income. The house bets with all its customers—perhaps 100,000 individual bets on black or red in a week. The distribution of average customer winnings on 100,000 bets is very close to normal, and the mean is still the mean outcome for one bet, -0.053, a loss of 5.3 cents per dollar bet.

In addition, the central limit theorem says that the standard deviation of the distribution of average winnings decreases with the square root of the number of bets over which we are averaging. Now, 100,000 is 2000 times as much as 50. So the standard deviation of the casino's distribution (average winnings over 100,000 bets) is

$$\sqrt{2000} = 44.72$$

times as small as the standard deviation of the gambler's distribution (average over 50 bets). The gambler has standard deviation 0.14; the casino therefore has standard deviation

$$\frac{0.14}{44.72} = 0.003$$

There you have it. The individual gambler will experience wide variation in winnings; he or she gets excitement. The casino experiences very little variation; it has a business. Here is what the spread in the casino's average result looks like after 100,000 bets:

$$
\begin{aligned}
\text{Spread} &= \text{mean} \pm 3 \text{ standard deviations} \\
&= -0.053 \pm (3)(0.003) \\
&= -0.053 \pm 0.009 \\
&= -0.044 \text{ to } -0.062
\end{aligned}
$$

Because the casino covers so many bets, the standard deviation of the average winnings per bet becomes very narrow. And because the mean is negative, almost all outcomes will be negative. Thus, the gamblers' losses and the casino's winnings are almost certain to average between 4.4 and 6.2 cents for every dollar bet.

The gamblers who collectively placed those 100,000 bets will lose money. We are now in a position to estimate the probable range of their losses:

$$
\begin{aligned}
(-0.044)(100,000) &= -4400 \\
(-0.062)(100,000) &= -6200
\end{aligned}
$$

The gamblers are almost certain to lose—and the casino is almost certain to take in—between $4400 and $6200 on those 100,000 bets. What's more, we have seen from the central limit theorem that the more bets that are made, the narrower is the range of possible outcomes. That is how a casino can make a business out of gambling. The more money that is bet, the more accurately the casino can predict its profits.

REVIEW VOCABULARY

Central limit theorem   The average of many independent random outcomes is approximately normally distributed. When we average n independent repetitions of the same random phenomenon, the resulting distribution of outcomes has mean equal to the mean outcome of a single trial and standard deviation proportional to $1/\sqrt{n}$.

Combinatorics   The branch of mathematics that counts arrangements of objects.

Event   Any collection of possible outcomes of a random phenomenon. An event is a subset of the sample space.

Law of large numbers   As a random phenomenon is repeated many times, the mean $\bar{x}$ of the observed outcomes approaches the mean μ of the probability model.

Mean of a probability model   The average outcome of a random phenomenon with numerical values. When possible values $s_1, s_2, \ldots, s_k$ have probabilities $p_1, p_2, \ldots, p_k$, the mean is the average of the outcomes weighted by their probabilities, $\mu = s_1 p_1 + s_2 p_2 + \cdots + s_k p_k$.

Normal distributions   A family of probability models that assign probabilities to events as areas under a curve. The normal curves are symmetric and bell-shaped. A specific normal curve is completely described by giving its mean μ and its standard deviation σ.

Probability   A number between 0 and 1 that gives the long-run proportion of repetitions of a random phenomenon on which an event will occur.

Probability model   A sample space S together with an assignment of probabilities to events. If probabilities $P(s)$ are assigned to all outcomes s in S, they must satisfy two laws:

1.   For every outcome s, $0 \leq P(s) \leq 1$.
2.   The sum of the probabilities $P(s)$ over all outcomes s is exactly 1.

A probability model can also assign probabilities to events as areas under a curve. In this case, the total area under the curve must be exactly 1.

Random phenomenon A phenomenon is random if it is uncertain what the next outcome will be but each outcome nonetheless tends to occur in a fixed proportion of a very long sequence of repetitions. These long-run proportions are the probabilities of the outcomes.

Sample space A list of all possible outcomes of a random phenomenon.

Sampling distribution The distribution of values taken by a statistic when many random samples are drawn under the same circumstances. A sampling distribution consists of an assignment of probabilities to the possible values of a statistic.

Sampling variability The random variability in the value of a statistic (such as a sample mean or sample proportion) when random samples are drawn repeatedly from the same population.

68–95–99.7 rule In any normal distribution, 68% of the observations lie within one standard deviation on either side of the mean; 95% lie within two standard deviations of the mean; and 99.7% lie within three standard deviations of the mean.

Standard deviation of a probability model A measure of spread that is particularly appropriate for normal distributions. The standard deviation σ of a normal curve is the distance from the mean to the change-of-curvature points on either side.

Statistic A number computed from a sample, such as a sample mean or sample proportion. In random sampling, the value of a statistic will vary in repeated sampling.

SUGGESTED READINGS

MOSTELLER, FREDERICK, ROBERT E. K. ROURKE, AND GEORGE B. THOMAS. *Probability with Statistical Applications,* Addison-Wesley, Reading, Mass., 1970. A rich treatment of basic probability that requires only high school algebra but is somewhat sophisticated.

OLKIN, INGRAM, LEON J. GLESER, AND CYRUS DERMAN. *Probability Models and Applications,* Macmillan, New York, 1980. This book is distinguished by an emphasis on the use of probability to describe real phenomena and by outstanding examples of modeling. In level it falls between Mosteller et al. and Snell.

SNELL, J. LAURIE. *Introduction to Probability,* Random House, New York, 1988. A calculus-based text aimed at undergraduate mathematics majors. Recommended because of its excellent examples and historical remarks, and in particular because Snell makes good use of BASIC programs that are included in the text.

EXERCISES ▲ *Optional.* ■ *Advanced.* ◆ *Discussion.*

What Is Probability?

You can estimate an unknown probability by actually observing many repetitions of the random phenomenon in question. Exercises 1 to 4 produce rough estimates based on a small number of repetitions. You can see the random behavior in more detail by making a graph like Figure 7.2 rather than just reporting the final proportion of outcomes.

1. Toss a thumbtack on a hard surface 100 times. How many times did it land with the point up? What is the approximate probability of landing point up?

2. Hold a penny upright on its edge under your forefinger on a hard surface, then snap it with your other forefinger so that it spins for some time before falling. Based on 50 spins, what is the probability of heads?

3. Open your local telephone directory to any page and note whether the last digit of each of the first 100 telephone numbers on the page is odd or even. How many of the digits were odd? What is the approximate probability that the last digit of a telephone number is odd?

4. The table of random digits (Table 5.1 on p. 185) was produced by a random mechanism that gives each digit probability 0.1 of being a 0. What proportion of the first 200 digits in the table are 0s? This proportion is an estimate of the true probability, which in this case is known to be 0.1.

Probability Models

In each of Exercises 5 to 7, describe a reasonable sample space for the random phenomena mentioned. In some cases, more than one choice is possible.

5. Toss a coin 10 times.

   (a) Count the number of heads observed.
   (b) Calculate the percent of heads among the outcomes.
   (c) Record whether or not at least five heads occurred.

6. A female lab rat is about to give birth. You count the number of offspring in the litter. (We don't know how large rat litters can be, but you can set a reasonable upper limit if you want.)

7. Subjects in a clinical trial are assigned at random to either the new treatment group or the control group. For the next subject, you record treatment or control, male or female, and smoker or nonsmoker.

22.   Teachers in the Lost Valley Central School District are allowed up to 7 days of paid sick leave each year. Here is the distribution of the number of days of sick leave taken by the teachers last year. What is the mean number of days of sick leave that a teacher will take in a year?

Days taken	0	1	2	3	4	5	6	7
Percent of teachers	15	15	10	10	10	12	8	20

23.   In an experiment on the behavior of young children, each subject is placed in an area with five toys. The response of interest is the number of toys that the child plays with. Past experiments with many subjects have shown that the probability model for the number of toys played with is as follows.

Number of toys	0	1	2	3	4	5
Probability	.03	.16	.30	.23	.17	.11

(a)   What is the probability that a child will play with more than one toy during the experiment?

(b)   What is the mean number of toys a child will play with?

24.   An American roulette wheel has 38 slots numbered 0, 00, and 1 to 36. The ball is equally likely to come to rest in any of these slots when the wheel is spun. The slot numbers are laid out on a board on which gamblers place their bets. One column of numbers on the board contains a multiples of 3, that is, 3, 6, 9, . . . , 36. A gambler places a $1 column bet that pays out $3 if any of these numbers comes up.

(a)   What is the probability of winning?

(b)   What are the mean winnings for one play, taking into account the $1 cost of each play?

Sampling Distributions

◆ 25.   Return to Exercises 13 and 14 in Chapter 5 (pages 206–207). Working as a team with other students, draw 100 simple random samples of size 5 from this population. Compute the sample proportion $\hat{p}$ of females in each sample. What probability model, based on your experiment, describes the sampling distribution of $\hat{p}$? Make a histogram of this distribution. Describe the shape of this distribution. In particular, does it appear roughly normal? Then find the mean number of females in a sample.

◆ 26.   The following table contains the results of 100 repetitions of the drawing of a simple random sample of size 200 from a large lot of bearings, 10% of which do not conform to the specifications. The numbers in the table are the percents of nonconforming bearings in each sample of 200.

8.5	11.5	9	13.5	7.5	8.5	9	6.5	8	9
10	7.5	9	8	10.5	8.5	9	9.5	8	11.5
10	9	9	8.5	9.5	6.5	13.5	11	11.5	13
8.5	6.5	8	7	12	11	8	10.5	12	10.5
15	12	8.5	7	8	8	8.5	12	10.5	8
8.5	11.5	9	11.5	11	12	11.5	11.5	10	9.5
10	9	10	12.5	8	12	12	12	7.5	11
11	8	14	7.5	11	4.5	9.5	8	9.5	9.5
12.5	12	10	7.5	10.5	12.5	12	9.5	9.5	10
14	9	8.5	8.5	12.5	8.5	8.5	9	9.5	9

Give an estimated sampling distribution for the sample proportion in this situation by recording each outcome and the proportion of trials on which it occurred. Make a histogram of the distribution and describe its shape. Is the center close to 10%? Is the distribution roughly symmetric? Does it appear approximately normal? Find the mean outcome from your distribution. Is it close to 10%?

Normal Distributions

27.   The distribution of heights of adult American men is approximately normal, with mean 69 inches and standard deviation 2.5 inches. Draw a normal curve on which this mean and standard deviation are correctly located. (*Hint:* Draw the curve first, then mark the horizontal axis.)

28.   Using the normal distribution described in Exercise 27 and the 68–95–99.7 rule, answer the following questions about the heights of adult American men.

    (a)   What percent of men are taller than 74 inches?
    (b)   Between what heights do the middle 95% of American men fall?
    (c)   What percent of men are shorter than 66.5 inches?

29.   What are the quartiles of the distribution of heights of American men in Exercise 27?

30.   The figure that follows is a probability distribution that is not symmetric. The mean and median do not coincide. Which of the points marked is the mean of the distribution, and which is the median?

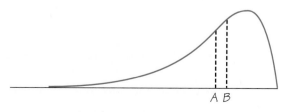

A B

31. Scores on the Wechsler Adult Intelligence Scale (a standard "IQ test") for the 20 to 34 age group are approximately normally distributed with $\mu = 110$ and $\sigma = 25$.

    (a) About what percent of people in this age group have scores above 110?

    (b) About what percent have scores above 160?

32. The army reports that the distribution of head circumference among soldiers is approximately normal with mean 22.8 inches and standard deviation 1.1 inches.

    (a) What percent of soldiers have head circumference greater than 23.9 inches?

    (b) What percent of soldiers have head circumference between 21.7 inches and 23.9 inches?

The Central Limit Theorem

33. A student makes a measurement in a chemistry laboratory and records the result in her lab report. When many students do this, the standard deviation of their individual measurements is $\mu = 10$ milligrams. Suppose the student repeats the measurement 3 times and records the mean $\bar{x}$ of her 3 measurements. What is the standard deviation $\sigma_{\bar{x}}$ of the mean result?

34. A student organization is planning to ask a sample of 50 students if they have noticed AIDS education brochures on campus. The sample percentage who say "Yes" will be reported. Their statistical advisor says that the standard deviation of this percentage will be about 7%. What would the standard deviation be if the sample contained 100 students rather than 50?

◆ 35. How many times must the student of Exercise 33 repeat the measurement to reduce the standard deviation of $\bar{x}$ to 5? Explain to someone who knows no statistics the advantage of reporting the average of several measurements rather than the result of a single measurement.

◆ 36. How large a sample is required in the setting of Exercise 34 to reduce the standard deviation of the percentage who say "Yes" from 7% to 3.5%? Explain to someone who knows no statistics the advantage of taking a larger sample in a survey of opinion.

Additional Exercises

37. Pick up a book and open to any page. Count the words in the first complete paragraph on that page and note how many of them begin with a vowel. (If the paragraph contains fewer than 100 words, include the next paragraph as well.) What do you estimate to be the probability that a word chosen at random from this book begins with a vowel?

38. A couple plans to have three children. What is the sample space S for each of the following random phenomena?

(a) Record the sex (M or F) of each child in order of birth.
(b) Record the number of girls.

39. Choose a student at random and record the number of dollars in bills (ignore change) that he or she is carrying. Give a reasonable sample space S for this random phenomenon. (We don't know the largest amount that a student could reasonably carry, so you will have to make a choice in stating the sample space.)

40. Here is the distribution of marital status for American women aged 25 to 29 years. If this is to be a legitimate probability model, what must be the probability that a woman in this age group is married?

Outcome	Single	Married	Widowed	Divorced
Probability	.288		.003	.076

◆ 41. A bridge deck contains 52 cards, four of each of the 13 face values ace, king, queen, jack, ten, nine, . . . , two. You deal a single card from such a deck and record the face value of the card dealt. Give an assignment of probabilities to these outcomes that should be correct if the deck is thoroughly shuffled. Give a second assignment of probabilities that is legitimate (that is, obeys Laws 1 and 2) but differs from your first choice. Then give a third assignment of probabilities that is *not* legitimate, and explain what is wrong with this choice.

◆ 42. Suppose that A and B are events that have no outcomes in common and thus cannot occur simultaneously. For example, in tossing three coins we could have $A = \{$First coin gives $H\}$ and $B = \{$First coin gives $T\}$. Starting from the fact that the probability of any event is the sum of the probabilities of the outcomes making up the event, explain why

$$P(A \text{ or } B \text{ occurs}) = P(A) + P(B)$$

must always be true for two such events.

■ 43. Automobile license plate numbers in Indiana consist of seven characters. The first three describe the county in which the car is licensed, while the last four are digits assigned at random. You are hoping for a plate on which these four digits are identical (like 7777). What is your probability of receiving such a plate?

■ 44. Automobile license plates in Hawaii consist of three letters followed by three digits.

(a)   How many different license plates are possible in Hawaii?

(b)   A visitor to Honolulu observes that all license plates seem to begin with one of E, F, G, or H. How many license plates are possible if all plates begin with one of these letters?

(c)   Suppose that a state allowed license plates consisting of any six letters or digits in any order. How many different license plates would then be possible?

■   45.   A monkey at a keyboard presses three keys and hits the letters a, g, and s in random order. How many possible three-letter "words" can the monkey type using only these letters? Which of these are meaningful English words? What is the probability that the word the monkey typed is meaningful?

46.   You are about to visit a new neighbor. You know that the family has four children, but you do not know their age or sex. Write down all possible arrangements of girls and boys in order from youngest to oldest, such as BBGG (the two youngest are boys, the two oldest girls). The laws of genetics say that all of these arrangements are equally likely.

(a)   What is the probability that the oldest child is a girl?

(b)   What is the probability that the family has at least three boys?

(c)   What is the probability that the family has at least three children of the same sex?

47.   A study selected a sample of fifth-grade pupils and recorded how many years of school they eventually completed. Based on this study we can give the following probability model for the years of school that will be completed by a randomly chosen fifth grader:

Years	4	5	6	7	8	9	10	11	12
Probability	.010	.007	.007	.013	.032	.068	.070	.041	.752

(a)   Verify that this is a legitimate probability model.

(b)   What outcomes make up the event "The student completed at least one year of high school"? (High school begins with the ninth grade.) What is the probability of this event?

(c)   What is the mean number of years of school completed?

48.   Keno is a common casino game. The house chooses 20 numbers between 1 and 80 at random and gamblers attempt to guess some of the numbers in advance. A bewildering variety of Keno bets are available. Here are some of the simpler Keno bets. Give the mean winnings for each.

(a)   A $1 bet on "Mark 1 number" pays $3 if the single number you mark is one of the 20 chosen; otherwise you lose your dollar.

(b) A $1 bet on "Mark 2 numbers" pays $12 if both your numbers are among the 20 chosen. The probability of this is about 0.06. Is Mark 2 a more or a less favorable bet than Mark 1?

49. Let us illustrate the idea of a sampling distribution in the case of a very small sample from a very small population. The population is the scores of 10 students on an exam:

Student	0	1	2	3	4	5	6	7	8	9
Score	82	62	80	58	72	73	65	66	74	62

The mean score in this population is 69.4. The sample is a simple random sample of size $n = 4$ drawn from the population. Because the students are labeled 0 to 9, a single random digit chooses one student for the sample.

(a) Use Table 5.1 to draw an SRS of size 4 from this population. Write the four scores in your sample and calculate the mean $\bar{x}$ of the sample scores. This statistic is an estimate of the population mean.

(b) Repeat this process 20 times. Make a histogram of your 20 values of $\bar{x}$. You are constructing the sampling distribution of $\bar{x}$. Is the center of your histogram close to 69.4?

(c) Twenty repetitions give a very crude approximation to the sampling distribution. Pool your work with that of other students—using different parts of Table 5.1—to obtain several hundred repetitions. Make a histogram of all the values of $\bar{x}$. Is the center close to 69.4? Is the shape approximately normal? This histogram is a better approximation to the sampling distribution.

50. The concentration of the active ingredient in capsules of a prescription painkiller varies according to a normal distribution with $\mu = 10\%$ and $\sigma = 0.2\%$.

(a) What is the median concentration? Explain your answer.

(b) What range of concentrations covers the middle 95% of all the capsules?

(c) What range covers the middle half of all capsules?

51. Answer the following questions for the painkiller in Exercise 50.

(a) What percent of all capsules have a concentration of active ingredient higher than 10.4%?

(b) What percent have a concentration higher than 10.6%?

52.   The length of human pregnancies from conception to birth varies according to a distribution that is approximately normal with mean 266 days and standard deviation 16 days.

(a)   Between what values do the lengths of the middle 95% of all pregnancies fall?

(b)   How short are the shortest 2.5% of all pregnancies?

53.   The *deciles* of a distribution are the points having 10% (lower decile) and 90% (upper decile) of the observations falling below them. The lower and upper deciles contain between them the central 80% of the data. The lower and upper deciles of any normal distribution are located 1.28 standard deviations on either side of the mean. What score is needed to place you in the top 10% of the distribution of SAT scores (normal with mean 500 and standard deviation 100)?

54.   Based on the information in Exercises 52 and 53, how short are the shortest 10% of human pregnancies?

WRITING PROJECTS

1 ▶   "France gave birth to the mathematics of probability in the seventeenth century when gamblers turned to mathematicians for advice." Some of the mathematicians in question were Pierre de Fermat and Blaise Pascal. Do some reading to learn more about the origins of probability and write a brief essay describing the role of Fermat and Pascal. (One good source is Carl B. Boyer, *A History of Mathematics,* Wiley, New York, 1991.)

2 ▶   State-run lotteries are common in the United States and in other countries, as Spotlight 7.2 suggests. Write a brief essay describing current state lotteries in the United States. How much money do they take in? How is the money that is kept used? What are the trends in the games offered? What other forms of gambling do revenue-hungry states license? (You can often find recent information about lotteries in the press. Consult, for example, the indexes to the *New York Times* in your library.)

3 ▶   Most people "overreact" to risks that have very low probability of occurring. The probability of dying from an airplane crash, a terrorist attack, or a tornado, for example, is extremely small. Yet public opinion and personal decisions often act as if these risks were as probable as death from an automobile accident or a heart attack. Write a brief essay describing how people assess risks, and what factors besides probability influence their actions. One reference is Richard J. Zeckhauser and W. Kip Vicusi, Risk within reason, *Science,* May 4, 1990, pp. 559–564.

CHAPTER 8

STATISTICAL INFERENCE

Inference is the process of reaching conclusions from evidence. Evidence can come in many forms. In a murder trial, evidence might be presented by the testimony of a witness, by a record of telephone conversations, or by analysis of blood samples. In the case of statistical inference, the evidence is provided by data. Informal statistical inference is often based on graphical presentation of data. Formal inference, the topic of this chapter, uses the language of probability to say how confident we are that our conclusion is correct.

Informal evidence is sometimes compelling. The gap in the histogram of Figure 6.4 (page 225), for example, demands an investigation of the inspection process. But in many cases it is difficult to reach a firm conclusion from informal evidence.

EXAMPLE ▶ *Was the Draft Lottery Unfair?*

Were men born late in the year more likely to draw low draft numbers in the first Vietnam-era draft lottery? The scatterplot (Figure 8.1) certainly does not show a strong relationship between birth date and draft number. Some clever data analysis, represented by adding the monthly medians to the plot, shows a trend toward lower numbers late in the year. The trend is not very strong, so we might well ask whether the outcome was simply due to chance rather than to systematic bias in the lottery. After all, any lottery will show some deviation from perfect uniformity due to the play of chance.

A calculation of probabilities shows, however, that a trend as strong as that in Figure 8.1 has probability less than 1 in 1000 in a truly random drawing. This calculation convinced everyone that the draft lottery was not a random drawing. ◆

FIGURE 8.1 Scatterplot of draft lottery numbers versus date of birth. The line on the plot connects the median draft number for each month. It appears that the lottery favored men born earlier in the year.

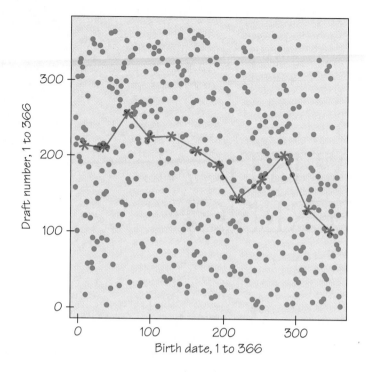

The purpose of formal statistical inference is to verify appearance by calculation. Statistical inference can be compared to an engineer's calculation of the load on a beam: we are more confident after the mathematics is done than we are if the engineer merely says that the beam looks large enough. Because we use random selection to produce data, the calculations behind statistical inference are calculations of probability. One of the most intriguing aspects of statistical inference is that *chance,* which we usually associate with uncertainty, is the ally rather than the enemy of confident conclusions.

EXAMPLE ▶ *Can We Trust an Opinion Poll?*

How can we trust the results of a random sample, knowing that a second sample would usually yield a different result? A Gallup poll of 1493 people finds that 45% are afraid to go out at night within a mile of their homes because of fear of crime. A second random sample would select a different 1493 people and give a result different from 45%. Knowing this, what can we say about the population of 190 million American adults on the basis of Gallup's sample?

A probability calculation tells us that 95% of Gallup's samples give a result within three percentage points of the truth about the population. We now realize that we are quite safe to believe that the truth for all adults lies between 42% and 48%. Because sample results will often miss by more than 1%, however, we can't confidently say that the truth lies between 44% and 46%. ◆

In both examples, a probability calculation answered the question, "What would happen if we did this many times?" In many random lotteries, only 1 in 1000 would give a trend as strong as that observed. In many Gallup poll samples, 95% would give a result within ± 3% of the truth for the population. This kind of probability statement is characteristic of statistical inference. Pay attention to the meaning of the probability that is announced along with the conclusions of any inference procedure. Understanding how probability is employed is the key to understanding statistical inference.

ESTIMATING A POPULATION PROPORTION

We will use a simplified version of the Gallup crime survey to introduce an important type of statistical inference. Like most national sample surveys, the Gallup poll uses a complex multistage sampling design. Suppose that we instead drew a *simple random sample* of 1500 adults and discovered that 675 of the people in this sample were afraid to go out at night because of crime. The **sample proportion** who stay home from fear of crime is

$$\hat{p} = \frac{675}{1500} = 0.45 = 45\%$$

We will call a sample proportion $\hat{p}$ (read as "p hat"), and we will always express sample proportions as percents.

The sample proportion $\hat{p} = 45\%$ refers to the 1500 people in this particular sample. We really want to know the *population proportion,* the percent (call it p) of all adult Americans who stay home at night for fear of crime. To discuss statistical inference intelligently, we must keep straight which numbers describe the sample and which describe the population.

> A number such as p that describes a population is called a **parameter.** A number such as $\hat{p}$ that is calculated from a sample is called a **statistic.**

It is easy to remember that *p*arameters belong to *p*opulations and *s*tatistics belong to *s*amples because the first letters agree. In an inference problem, parameters are usually unknown. We do not know, for example, the true proportion p of all adults who stay home at night for fear of crime. We use the statistic $\hat{p}$, which we know because we actually interviewed the sample, to estimate the unknown p. *Our goal is not simply to estimate p, but to say how accurate our* *estimate is.* To do this, we ask "What would happen if we took many samples? How close to the unknown *p* would the estimate $\hat{p}$ usually fall?"

To answer this question, we turn to the *sampling distribution* of $\hat{p}$. This is the distribution of values taken by the sample proportion as it varies from sample to sample in a large number of samples from the same population. We have simulated sampling distributions in earlier chapters. Now we want the mathematical facts. Here they are:

Choose a simple random sample of size *n* from a large population of which the percent *p* has some characteristic of interest. Let $\hat{p}$ be the percent of the sample having that characteristic. Then

- The sampling distribution of $\hat{p}$ is *approximately normal* and is closer to a normal distribution when the sample size *n* is large.
- The *mean* of the sampling distribution is exactly *p*.
- The *standard deviation* of the sampling distribution is

$$\sigma_{\hat{p}} = \sqrt{\frac{p(100 - p)}{n}}$$

To remind ourselves that this standard deviation belongs to the distribution of $\hat{p}$, we write it as $\sigma_{\hat{p}}$.

Figure 8.2 presents this sampling distribution as a normal curve. Both the center (mean) and spread (standard deviation) of this curve carry important statistical messages.

First, the mean of the curve is the true proportion *p* of people afraid to go out at night. This fact says that $\hat{p}$ has no bias or systematic error as an estimator of the unknown *p*. In repeated sampling our result will sometimes be high and sometimes low, but the long-run average result, the mean of the sampling distribution, will be correct. Of course, in practice we don't know the numerical value of the parameter *p*. But we now know that, whatever value *p* has, the observed values of the statistic $\hat{p}$ cluster around it as shown in Figure 8.2.

Being correct on the average is not enough. A good estimator must also be highly repeatable in the sense of giving nearly the same answer in repeated samples. Repeatability is described by the spread of the sampling distribution, as measured by its standard deviation. If we repeated the sampling many times, sending out waves of interviewers across the nation, each time we would get a value of the sample proportion $\hat{p}$ somewhere along the curve in Figure 8.2. How far from the true *p* these sample results lie depends on the standard devia-

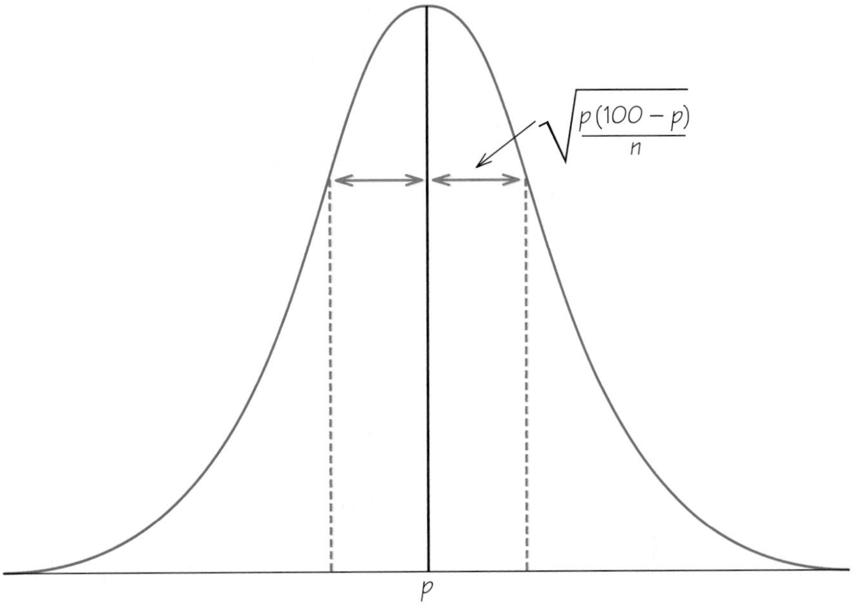

FIGURE 8.2 The sampling distribution of the sample proportion $\hat{p}$. It is approximately normal, with mean p and standard deviation $\sqrt{p(100 - p)/n}$.

tion $\sigma_{\hat{p}}$ of this normal curve. The standard deviation gets smaller as the sample size n gets larger. Our simulations in Chapter 7 (pages 280–283) showed this effect. What is new is the exact relationship between n and the standard deviation. *The standard deviation depends on the square root $\sqrt{n}$.* To cut the spread of the sampling distribution in half, we must take four times as many observations.

EXAMPLE ▶ *Sampling Distribution for the Crime Survey*

Suppose that in fact 40% of all adults fear to go out at night because of crime. That is, suppose that $p = 40\%$. Take a simple random sample of size $n = 1500$ people. In repeated samples, the sample percent $\hat{p}$ will vary according to a normal distribution with

$$\text{mean} = p = 40\%$$

$$\text{standard deviation } \sigma_{\hat{p}} = \sqrt{\frac{p(100 - p)}{n}}$$

$$= \sqrt{\frac{(40)(60)}{1500}} = \sqrt{1.6} = 1.265\%$$

In practice the value of the parameter p is unknown. Our calculations show that the standard deviation of $\hat{p}$ is small. So the sample percent $\hat{p}$ will usually lie quite close to p.

Suppose now that the truth about the population is $p = 50\%$ rather than 40%. The mean of the sampling distribution moves to 50%. The standard deviation changes to

$$\sigma_{\hat{p}} = \sqrt{\frac{p(100 - p)}{n}}$$

$$= \sqrt{\frac{(50)(50)}{1500}} = \sqrt{1.67} = 1.29\%$$

The standard deviation $\sigma_{\hat{p}}$ does not change very much when p changes. That is, when we take a sample of the same size from different populations, the center of the sampling distribution of $\hat{p}$ moves to the true p for each population, but the spread stays about the same.

The size of the sample is the major influence on the spread. Suppose that we took a sample of only $n = 375$ instead of 1500 people from a population for which $p = 40\%$. The mean of the distribution of $\hat{p}$ is still 40% — the sample size doesn't change the center of the sampling distribution. But the standard deviation increases to

$$\sigma_{\hat{p}} = \sqrt{\frac{p(100 - p)}{n}}$$

$$= \sqrt{\frac{(40)(60)}{375}} = \sqrt{6.4} = 2.53\%$$

Because the new sample size 375 is one-fourth of 1500, the new standard deviation 2.53% is twice as large as the previous result 1.265%. That's the $\sqrt{n}$ effect in action. ◆

CONFIDENCE INTERVALS

Our poll of 1500 people found that $\hat{p} = 45\%$. This is our best guess for the population percent p. How close to the true p is our guess likely to be? Well, $\hat{p}$

varies normally. The 95 part of the 68–95–99.7 rule says that $\hat{p}$ falls within two standard deviations of the true p (the mean of the sampling distribution) in 95% of all samples. So our guess based on this one sample is likely to be within two standard deviations, that is, within

$$2\sigma_{\hat{p}} = 2\sqrt{\frac{p(100 - p)}{1500}}$$

of the true p.

The catch is that this standard deviation depends on the unknown p. Fortunately, as the example demonstrated, $\sigma_{\hat{p}}$ changes only slowly as p changes, as long as p is not very close to either 0% or 100%. Because $\hat{p}$ is close to p, we simply substitute $\hat{p} = 45\%$ for the unknown p in the formula for the standard deviation. To indicate that the standard deviation is estimated rather than known exactly, we call it $\hat{\sigma}_{\hat{p}}$.

EXAMPLE ▶ *Estimated Standard Deviation for the Crime Survey*

We want to estimate the standard deviation of our observed sample proportion. The sample size was $n = 1500$, and for p we use the estimate $\hat{p} = 45\%$, based on our survey. The estimated standard deviation is

$$\hat{\sigma}_{\hat{p}} = \sqrt{\frac{(45)(55)}{1500}}$$

$$= \sqrt{1.65} = 1.285\% \quad ◆$$

Here, then, is our conclusion: in 95% of all samples, the sample proportion $\hat{p}$ will fall within 2×1.285, or about 2.6%, of the unknown population proportion p. We took one sample and got $\hat{p} = 45\%$. So we conclude that the p lies in the interval

45% ± 2.6%

or between 42.4% and 47.6%. We say that we are *95% confident* in this conclusion because we got the interval by calculating how close to p the sample proportion will lie in 95% of all samples. Our interval is a 95% *confidence interval* for estimating the unknown population proportion.

In mathematical terms, the probability is 0.95 that the sample proportion $\hat{p}$ will fall within ± 2.6% of the unknown true fraction p of all adults who are afraid to go out at night because of crime.

Figure 8.3 makes the idea clearer. The normal curve at the top of the figure is the sampling distribution of $\hat{p}$. As we take many samples, the actual values of $\hat{p}$ vary according to this distribution. The values of $\hat{p}$ observed in 25 samples appear as dots below the curve, together with the confidence intervals that extend out 2.6% on either side of the observed $\hat{p}$. The true population proportion p is marked by the vertical line. Although the intervals vary from sample to sample, all but one of these samples gave a confidence interval that covers the true p. To say that these are 95% confidence intervals is just to say that the interval covers the true p in 95% of all samples, and misses in only 5%. Be sure you understand that this 95% and 5% refer to what would happen if we continued to take samples forever. In a small number of samples, the number of confidence intervals that fail to cover the true p may be a bit more or less than 5% of the samples. In Figure 8.3, for example, one out of 25, or 4%, of the confidence intervals fails to contain p.

A 95% **confidence interval** is an interval obtained from the sample data by a method that in 95% of all samples will produce an interval containing the true population parameter.

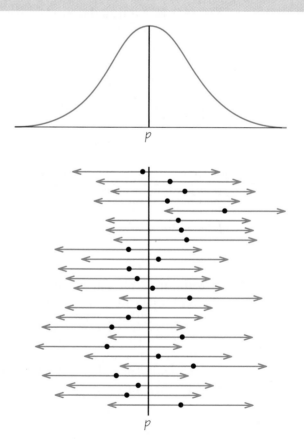

FIGURE 8.3 The behavior of 95% confidence intervals in repeated sampling. In the long run, 95% of all samples will produce intervals that contain the true value of p.

You can see in Figure 8.3 that a confidence interval from one particular sample can either hit or miss the unknown true parameter. We don't know whether our sample is one of the 95% that hit or one of the 5% that miss. To say that our interval 45% ± 2.6% is a 95% confidence interval means "We got this interval by a method that catches the true parameter 95% of the time."

We have now accomplished two things: we have seen what "95% confidence" means, and we have actually found a 95% confidence interval for estimating a population proportion. Here, as a summary, is the recipe for this interval:

A 95% confidence interval for a population proportion p, based on a simple random sample of size n, is

$$\hat{p} \pm 2\hat{\sigma}_{\hat{p}} = \hat{p} \pm 2\sqrt{\frac{\hat{p}(100 - \hat{p})}{n}}$$

Remember that both p and $\hat{p}$ are measured in percent. This recipe is only approximately correct, but is quite accurate when the sample size n is large.

EXAMPLE ▶ *Germination of Seeds*

We test a simple random sample of 100 seeds from a new lot for germination; 87 of the 100 germinate. The sample proportion that germinates is

$$\hat{p} = \frac{87}{100} = 0.87 = 87\%$$

The 95% confidence interval for estimating the proportion p of all seeds in the lot that will germinate is

$$\hat{p} \pm 2\sqrt{\frac{\hat{p}(100 - \hat{p})}{n}} = 87 \pm 2\sqrt{\frac{(87)(13)}{100}}$$
$$= 87 \pm 2\sqrt{11.31}$$
$$= 87\% \pm 6.7\%$$

We are 95% confident that between 80.3% and 93.7% of the entire lot of seeds will germinate.   ◆

The confidence interval depends on the size n of the sample; larger samples give shorter intervals. But the interval does *not* depend on the size of the population. This is true as long as the population is much larger than the sample. The confidence interval in the example works for a sample of 100 from a lot of 10,000 seeds as well as for a sample of 100 from a lot of 1,000,000 seeds. Put another way, what matters is how many seeds you examine, not what percent of the population you examine.

Any confidence interval has two essential pieces: the interval itself and the confidence level. The interval usually has the form

estimate $\pm$ margin of error

The estimate is a sample statistic (such as $\hat{p}$) that estimates the unknown parameter. The margin of error indicates how accurate this estimate is. In the germination example, the estimate is 87% and the margin of error is $\pm 6.7\%$.

The *confidence level* states how confident we are that our interval contains the true parameter. Although 95% confidence is common, you can hold out for higher confidence, such as 99%, or be satisfied with lower confidence, such as 90%. Our 95% confidence interval was based on the middle 95% of a normal distribution. A 99% confidence interval requires the middle 99% of the distribution, and so is wider (has a larger margin of error). Similarly, a 90% confidence interval is shorter than a 95% interval obtained from the same data. There is a trade-off between how closely we can pin down the parameter (the margin of error) and how confident we can be in the result.

Spotlight   *How the Poll Was Taken*

8.1

In February 1995, a New York Times/CBS News Poll carried out a survey of public opinion on issues before the new Congress. For example, 59% of the poll respondents said that they would prefer balancing the federal budget over cutting taxes. The *Times* published the poll results on February 28, 1995, along with a box titled "How the Poll Was Conducted." Here is part of that box. The methods and margins of error described by the *Times* are typical of national opinion polls.

The latest New York Times/CBS News Poll is based on telephone interviews conducted from last Wednesday through Saturday with 1,190 adults around the United States, excluding Alaska and Hawaii.

The sample of telephone exchanges called was selected by a computer. . . . For each exchange, the telephone numbers were formed by random digits, thus permitting access to both listed and unlisted numbers. Within each household, one adult was designated by a random procedure to be the respondent for the survey. . . .

In theory, in 19 cases out of 20 the results based on such samples will differ by no more than three percentage points in either direction from what would have been obtained by seeking out all American adults. . . .

In addition to sampling error, the practical difficulties of conducting any survey of public opinion may introduce other sources of error into the poll.

EXAMPLE ▶ *Understanding the News*

The results of opinion polls and other sample surveys are common in the news. News reports often give a margin of error, but rarely state a confidence level. (See Spotlight 8.1 for an exception.) A news report of our crime survey would say "The survey found that 45% of all Americans are afraid to go out at night because of crime. The margin of error in the survey is plus or minus 2.6 percentage points."

We need to know both the margin of error and the confidence level, because higher confidence requires a larger margin of error. In fact, there is an unspoken understanding in news releases: almost all public opinion polls announce the margin of error for 95% confidence. So if a story about an opinion poll gives a margin of error without a confidence level, you can usually assume 95%. ◆

The Bureau of Labor Statistics, on the other hand, chooses to announce the monthly unemployment rate at the 90% level of confidence. Basing its conclusions on the Current Population Survey of 60,000 households, the bureau says that the published unemployment rate is within ± 0.2% (two-tenths of 1 percent) of the figure it would get if it interviewed all workers. When the headlines announce a 7.9% unemployment rate, the bureau is saying—with 90% confidence—that between 7.7% and 8.1% of the labor force is out of work.

Opinion polls often have margins of error of about ± 3%. The much smaller margin of error for the announced unemployment rate is due to the much larger sample interviewed by the Current Population Survey. Larger samples give smaller margins of error at the same confidence level. However, the square root of n that appears in the calculations shows that in order to reduce our margin of error by half, we need a sample size four times bigger. To obtain a very small margin of error, the Current Population Survey takes the trouble to interview a sample of 60,000 people, compared with the Gallup poll's usual 1500. The Gallup poll can afford to be 3% off. The unemployment rate must be more exact because so many economic and political decisions depend on it.

ESTIMATING A POPULATION MEAN

The statistician's tool kit contains many different confidence intervals, matching the many different population parameters that we may wish to estimate. We have met the confidence interval for estimating a population proportion p. Now we want to estimate a population mean. We have regularly used the sample mean $\bar{x}$ of a sample of observations to describe the center of a set of data. Now we will use the sample mean $\bar{x}$ to estimate the unknown mean μ of the entire population from which the sample is drawn. We use μ, the symbol for the mean of a probability distribution, for the population mean because it is the mean of the distribution of the results of drawing a single observation at random from the population. The sample mean $\bar{x}$ is a statistic that will vary in repeated samples, while the population mean μ is a parameter, a fixed number. Fortunately, the new confidence interval for estimating μ is quite similar to the familiar confidence interval for estimating p, because both intervals are based on a normal sampling distribution.

EXAMPLE ▶ *Scholastic Assessment Test Scores*

How well would high school seniors do on the mathematics part of the Scholastic Assessment Test (SAT)? Although about a million students take the SAT each year, these students are planning to attend college and are not representative of the entire population of high school seniors. We therefore select a simple random sample of 500 seniors and administer the mathematics SAT to them. Their average score is $\bar{x} = 451$. What can we say about the mean score μ for the entire population, if we want to be 95% confident in our conclusion? ◆

We once again use a statistic—the mean $\bar{x}$ of our sample—to estimate an unknown parameter—the mean score μ for the entire population. To give a confidence interval, we need to know the sampling distribution of $\bar{x}$. The central limit theorem tells us that this distribution is close to normal. The mean of the sampling distribution is the same as the mean μ of the population from which we drew our sample. That is, the sample mean has no bias or systematic error as an estimator of the unknown μ. To find the standard deviation, we need to know something about the population. SAT scores for any large population follow a distribution that is close to normal. Moreover, the tests are arranged so that the standard deviation for the population used to develop the tests is $\sigma = 100$. The standard deviation σ of SAT scores may vary a bit among different populations of students. We will simplify our work by assuming that we know that $\sigma = 100$ for the population we are interested in.

Recall from Chapter 7 that the sampling distribution of the sample mean $\bar{x}$ has a standard deviation that decreases with the square root $\sqrt{n}$ of the sample size. Because individual SAT scores have a distribution that is close to normal, we don't even have to call on the central limit theorem. It's a fact that if the distribution of the population is normal, then the sampling distribution of $\bar{x}$ is also normal, no matter how small the sample is. Here is a summary of all of these facts about sample means:

Suppose that $\bar{x}$ is the mean of a simple random sample of size n drawn from a large population that has mean μ and standard deviation σ.

- The *mean* of the sampling distribution of $\bar{x}$ is μ.
- The *standard deviation* of the sampling distribution of $\bar{x}$ is $\sigma_{\bar{x}} = \sigma/\sqrt{n}$.
- If the distribution of the population is normal, then the sampling distribution of $\bar{x}$ is normal with mean μ and standard deviation $\sigma/\sqrt{n}$.
- The *central limit theorem* says that for any population with finite standard deviation σ, the sampling distribution of $\bar{x}$ gets close to the normal distribution with mean μ and standard deviation $\sigma/\sqrt{n}$ as the sample size n gets large.

Figure 8.4 shows the relation between the distribution of a single observation drawn from the population and the distribution of the mean of several (in this case, 10) observations. The mean of several observations is less variable than individual observations. Now we have in hand the facts we need to give a confidence interval for the mean SAT mathematics score for all high school seniors based on our sample of 500 students from this population.

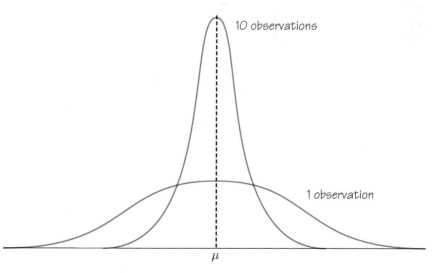

FIGURE 8.4 The sampling distribution of the sample mean $\bar{x}$ compared with the distribution of a single observation.

EXAMPLE ▶ *Estimating the Mean SAT Score*

The normal sampling distribution of $\bar{x}$ has mean equal to the unknown population mean μ. The standard deviation of the sampling distribution is

$$\sigma_{\bar{x}} = \frac{\sigma}{\sqrt{n}}$$

$$= \frac{100}{\sqrt{500}} = 4.47$$

By the 95 part of the 68–95–99.7 rule, $\bar{x}$ will fall within two standard deviations of μ in 95% of all samples. Two standard deviations is 2×4.47, or about 9 points. We observed $\bar{x} = 451$ in our sample. So we are 95% confident that the population mean μ lies in the interval

451 ± 9

or between 442 and 460. ◆

As in the case of the confidence interval for a population proportion p, we can give a recipe that summarizes our development. The confidence interval again has the form

estimate ± margin of error

where the estimate is now the sample mean $\bar{x}$.

Suppose that a population is described by a normal distribution with unknown mean μ and known standard deviation σ. Draw a simple random sample of size n from this population and calculate the sample mean $\bar{x}$. A 95% confidence interval for the population mean μ is

$$\bar{x} \pm 2\sigma_{\bar{x}} = \bar{x} \pm 2\frac{\sigma}{\sqrt{n}}$$

Because of the central limit theorem, this recipe is also approximately correct when the population is not described by a normal distribution, if our sample is large. Often in practice the standard deviation σ of the population is not known in advance. Then we must estimate σ by the standard deviation s of the sample. We will not concern ourselves with this situation.

Here is another example of estimating a population mean.

E X A M P L E ▶ *Estimating Dust in Coal Mines*

Because the mean of several observations is less variable than a single observation, it is good practice to take the average of several observations when accuracy is important. The amount of dust in the atmosphere of coal mines is measured by exposing a filter in the mine and then weighing the dust collected by the filter. The weighing is not perfectly precise. In fact, repeated weighings of the same filter will vary according to a normal distribution. The values that would be obtained in many weighings form the population we are interested in. The mean μ of this population is the true weight (that is, there is no bias in the weighing). The population standard deviation describes the precision of the weighing; it is known to be $\sigma = 0.08$ milligram (mg). Each filter is weighed three times and the mean weight is reported.

For one filter the three weights are

$$123.1 \text{ mg} \quad 122.5 \text{ mg} \quad 123.7 \text{ mg}$$

What is the 95% confidence interval for the true weight μ?

First compute the sample mean

$$\bar{x} = \frac{123.1 + 122.5 + 123.7}{3}$$

$$= \frac{369.3}{3} = 123.1 \text{ mg}$$

Then the 95% confidence interval is

$$\bar{x} \pm 2\frac{\sigma}{\sqrt{n}} = 123.1 \pm 2\frac{0.08}{\sqrt{3}}$$

$$= 123.1 \pm (2)(0.046) = 123.1 \pm .09$$

We are 95% confident that the true weight is between 123.01 mg and 123.19 mg. ◆

STATISTICAL PROCESS CONTROL

Statistical methods are widely used to gather social and economic information and in research on a wide variety of subjects. Most of our examples to this point, such as the Current Population Survey and the Physicians' Health Study, have illustrated these two types of applications of statistics. Statistics also contributes to the drive to improve the quality of manufactured products. Along with new technology such as robots and new management emphases such as cooperating with workers and suppliers, statistical ideas are an important part of any manufacturer's efforts to compete in the worldwide marketplace. In this section we look at one simple but important statistical tool for monitoring and improving quality, the control chart.

EXAMPLE ▶ *Monitoring Quality During Production*

AT&T Technologies manufactures the computerized electronic switches that interconnect our telephones. These switches are largely composed of complex electronic elements called circuit packs. AT&T needs efficient methods to check newly manufactured circuit packs for defects. The best strategy is to prevent defects by monitoring the manufacturing process to catch problems early rather than wait to inspect the finished switches (see Spotlight 8.2).

The 2000 electrical connections that attach components to the printed wiring board are all soldered at once as a conveyor carries the circuit pack through a wave of hot liquid solder. This wave-soldering operation is delicate. If the speed of the conveyor and the temperature of the solder are not just right, bad connections will appear both in the circuit pack and in our telephone conversations. AT&T therefore monitors the performance of the wave-soldering machine constantly and takes immediate action if something goes wrong.

An inspector takes a sample of five newly soldered circuit boards every hour and examines them carefully. The inspector calculates a number that expresses the quality of soldering for the board. A score of 100 represents the desired standard of quality. Lower scores represent poorer quality, while higher scores mean that the quality is better than the target. The inspector plots the sample mean $\bar{x}$ of the five quality scores on a **control chart.** The control chart is the key tool for keeping the process on target. ◆

There will always be some chance variation in the mean quality scores over time. Any industrial process will produce some variability. Constant fiddling with the process in response to small variations is unnecessary and wasteful. The purpose of the control chart is to help us distinguish the usual natural variation in the process from the added variation that indicates that the process has been disturbed. When unusual variation is spotted, we look for a specific cause. The disturbance may be caused by a new operator who hasn't been properly trained, for example, or by a malfunction in the machine.

Just plotting the mean quality scores against time can be helpful. We can see if there is a trend up or down, for example. We can check whether the last point plotted falls outside the pattern of the earlier points and so suggests that something has gone wrong. Figure 8.5 is a plot against time for the mean qual-

FIGURE 8.5 Plot of
sample mean quality scores
versus time. There appears
to be a downward shift
after sample number 10.

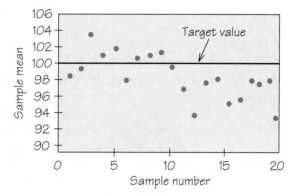

ity scores for 20 hourly samples. The graph appears to show that the level of
quality dropped shortly after sample number 10. The horizontal center line at
the target value 100 helps us see the trend. Once again, we would like to confirm this appearance by calculation. Adding the result of a simple calculation
turns the plot against time into a control chart.

When the wave-soldering machine is performing its task properly, the
quality index for individual circuit packs will vary according to a normal distribution. Let's suppose that we know from long experience that the mean of this
distribution should be 100 and the standard deviation 4. We are plotting the
mean $\bar{x}$ of five observations. What range of variation do we expect to see in the
values of $\bar{x}$? We know that the distribution of $\bar{x}$ in many samples is again
normal, with mean 100 and standard deviation

$$\sigma_{\bar{x}} = \frac{\sigma}{\sqrt{n}}$$

$$= \frac{4}{\sqrt{5}} = 1.79$$

By the 95 part of the 68−95−99.7 rule, 95% of all values of $\bar{x}$ will fall within
two standard deviations of the mean, that is, between

$$100 - (2)(1.79) = 96.42$$

and

$$100 + (2)(1.79) = 103.58$$

When we draw many samples, only 5% of the values of $\bar{x}$ will fall outside this
range if the process is operating undisturbed. In particular, only 2.5% (half of
5%) of all samples will give an $\bar{x}$ less than 96.42. A mean quality score this low

Spotlight
Check the Process Before the Product

8.2

Connie Moore

At its Oklahoma City plant, AT&T uses statistical process

Connie Moore, a process control engineer at AT&T, gives her views on ensuring product quality.

If we looked at 100% of the product it would take more time, more people, and would not give us any better information about the process. There was a time when industry thought that a quality control department's function was to in-spect quality at the end of the line. Now we know that the only reasonable philosophy is to build it right the first time.

I've been told that unless you measure how you're doing as you go along, you'll never know if you're done or if you succeeded. That's why I think that statistics and people are such an important combination. Statistics is the tool that tells us how we're doing as we go along and people are the force that drives us until we've succeeded.

is good evidence that something has gone wrong with the wave-soldering process.

Figure 8.6 is the control chart for the observations from Figure 8.5. The dashed line is the control limit 96.42 that indicates when action should be taken. The control chart shows convincingly that the quality of the process has deteriorated. Not only are the means for samples 12, 15, 16, and 20 below the control limit, but the last 11 means are all below the center line. In the long run, only half of these means should be below the center line if the process mean is really 100. It appears that the process quality shifted downward at about sample 9 or 10. In practice, the first out-of-control point at sample 12 would trigger an investigation to find and correct the cause of this trend. Here is a summary of the steps in constructing a control chart such as Figure 8.6:

Suppose that a process follows a normal distribution with mean μ and standard deviation σ when operating undisturbed. To monitor the process, take samples of size n at regular intervals. An $\bar{x}$ *control chart* for the process is a plot of the sample means $\bar{x}$ against time with a solid *center line* at μ and dashed *control limits* at $\mu - 2\sigma/\sqrt{n}$ and $\mu + 2\sigma/\sqrt{n}$.

There are many variations on the control chart idea. Although most control charts have both upper and lower control limits lying at equal distances above and below the center line, we drew only the lower limit in Figure 8.6. This is because only decreases in the quality index concern us in this example. Our control limits are placed two standard deviations out from the center line. It is more common in industry to place the control limits three standard deviations out in order to minimize the number of false alarms. Such limits contain 99.7% of all values of $\bar{x}$ if the process has not been disturbed. It is also common to keep control charts for statistics other than the sample mean $\bar{x}$.

More important than these details are the statistical ideas that are the basis for control charts. First, our goal is to distinguish expected from unexpected variation. Second, we use the normal sampling distribution of $\bar{x}$ and the 68–95–99.7 rule to specify the range of expected variation. Third, we combine this formal inference with a graph of the data that can be used in the factory by people with little statistical training.

Why sample only five circuit boards each hour? The purpose of statistical process control is not to check the function of the circuit packs; they will be rigorously tested when completed. Rather, the goal is to monitor the soldering process and correct any malfunctions quickly. It is not practical to check every circuit pack at every stage of manufacture. Instead, statistical sampling techniques give a quick and economical way to keep the process running smoothly.

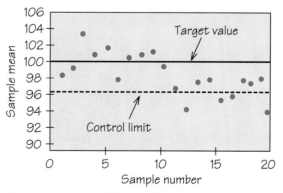

FIGURE 8.6 An $\bar{x}$ control chart for soldering quality.

Control charts based on samples keep down costs by catching malfunctions quickly, allowing a faulty process to be corrected immediately. This eliminates the need to repair or scrap products at the end of the assembly line.

EXAMPLE ▶ *Control Chart for a Machining Operation*

An important operation in producing cast aluminum aircraft frame parts is the machining of the raw casting. When the machining process is operating properly, a critical dimension of the parts varies according to a normal distribution with mean $\mu = 1.50$ centimeters (cm) and standard deviation $\sigma = 0.20$ cm. A sample of four parts is measured each hour in order to keep an $\bar{x}$ control chart. Here are the data for the past 20 hours.

Hour	1	2	3	4	5	6	7	8	9	10
$\bar{x}$	1.60	1.54	1.31	1.45	1.40	1.61	1.47	1.45	1.45	1.52

Hour	11	12	13	14	15	16	17	18	19	20
$\bar{x}$	1.49	1.66	1.60	1.57	1.73	1.68	1.58	1.60	1.57	1.73

The control chart appears in Figure 8.7. The center line is at $\mu = 1.50$. The control limits are

$$\mu \pm 2\frac{\sigma}{\sqrt{n}} = 1.50 \pm 2\frac{.20}{\sqrt{4}}$$
$$= 1.50 \pm .20$$
$$= 1.3 \text{ and } 1.7$$

The points for hours 15 and 20 are outside the control limits. Moreover, the last 9 points all lie above the center line. This is very unlikely to occur if the

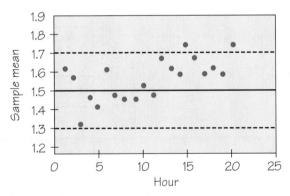

FIGURE 8.7 An $\bar{x}$ control chart for machining operation.

mean remains at 1.5, so it is additional evidence that some outside cause has disturbed the process. A common criterion is to look for a disturbing cause when 8 straight points fall on the same side of the center line. This criterion would lead to action at hour 19. In this case, the point out of control at hour 15 would already have called for action. ◆

THE PERILS OF DATA ANALYSIS

Statistical designs for collecting data may, like the Physicians' Health Study, involve experimentation. Or they may use sampling procedures such as those used in the Current Population Survey and also for process control. In both cases, we rely on randomization and the mathematics of probability to compute sampling distributions. From a sampling distribution we can obtain results that have known levels of confidence.

However, formal statistical inference, as reflected in levels of confidence, is secondary to well-designed data collection and to insight into the behavior of data. Inference can't correct basic flaws in the production of the data, such as the use of voluntary response samples. Moreover, the effects of *hidden variables* can make even an apparently clear inference misleading. We saw in Chapter 5 that a well-designed experiment can control for the effects of hidden variables. When an experiment is not possible, we may need to do some statistical detective work. Let's look at an example. Although the example is imaginary, it is based on a study of admissions to graduate programs at the University of California at Berkeley.

EXAMPLE ▶ *Admissions Discrimination?*

Metro University has several limited-enrollment courses that admit only some of the students who apply. There are complaints about discrimination in the admissions process. These complaints seem to be based on clear numerical evidence. Eighty men applied to limited-enrollment courses at Metro, and 35 were admitted. The percent of male applicants admitted was

$$\frac{35}{80} = 0.44 = 44\%$$

On the other hand, only 20 of the 60 women who applied were accepted. The success rate among women was

$$\frac{20}{60} = 0.33 = 33\%$$

TABLE 8.1	Metro Admissions	
	Men	Women
Admit	35	20
Deny	45	40
Total	80	60

Almost half the men, but only one-third of the women, were admitted.

A probability calculation shows that this difference is much larger than could reasonably be expected to occur simply by chance. That is, formal inference backs up the appearance that a systematically higher percent of men are being admitted. Are men being favored over women? ◆

The data in the example are displayed in a **two-way table** in Table 8.1. When variables simply place subjects into categories, such as male-female or admit-deny, we cannot draw a scatterplot to display the relationship between them. Instead we display the counts in a two-way table and see the relationship by comparing percents. Comparison of the percents of successful male and female applicants (44% and 33%) shows that men are more likely to be admitted. Does this difference reflect discrimination? Let's look more closely.

E X A M P L E ▶ *The Hidden Variable Strikes*

Our closer look breaks down the data according to which limited-enrollment course each individual applied for. There are only two such courses, organic chemistry, and history and sociology of the TV sitcom. Table 8.2 displays the counts for these two courses separately, as side-by-side two-way tables. Check that when you add the entries in these tables you get Table 8.1. These are the same 80 men and 60 women—we have just added information about another variable.

TABLE 8.2	Metro Admissions by Course				
Organic Chemistry			**TV Sitcom**		
	Men	Women		Men	Women
Admit	5	10	Admit	30	10
Deny	15	30	Deny	30	10
Total	20	40	Total	60	20

Table 8.2 shows that the organic chemistry course was hard to get into. Forty women applied to it, and only 10 got in. That's one-fourth of the female applicants. Twenty men also applied, and 5 were admitted. That's also one-fourth. So women and men were treated exactly alike. The sitcom course had more openings. Table 8.2 shows that 30 of the 60 men who applied were admitted, and that 10 of the 20 female applicants were accepted. That is, 50% of each group got in.

These numbers are remarkable. Each course appears to make no distinction between men and women. Yet the overall totals show that 44% of the men who applied were admitted, compared with 33% of the women. How can this be?

There is no mystery once we look at the data. Table 8.2 shows that 40 of the 60 women applied for the organic chemistry course and 60 of the 80 men signed up for the sitcom class. That is, most women signed up for the course that was hard to get into, and most men applied for the easier course. That's why fewer women than men were admitted. The hidden variable—which course a student applied for—explains the apparent inequity.   ◆

This example provides a warning about statistical evidence, especially when the data do not come from an experiment. Without walking through the example, we would not suspect that data showing equality in each of several cases can appear as evidence of inequality when the cases are lumped together. First appearances can be deceiving, even in statistics. Statistical evidence not based on experiments can show that an effect is present—more men than women get into limited courses at Metro—but does not show *why* the effect is present. At Metro, a hidden variable was responsible. In another case, an investigation might reveal discrimination. At a time when statistical evidence of all kinds is increasingly being used to formulate social policy and resolve legal disputes, it is crucial that we select, analyze, and interpret our data with great care.

Even if we produce data carefully and analyze them properly, we can't absolutely guarantee correct conclusions. There is always some chance, however small, that random selection will lead to a false conclusion. The strength of statistical inference is that the chance of a false conclusion is known and can be controlled by setting the confidence level as high as we think necessary.

Statistics does not produce proof. But in a world where proof is always wanting and most evidence is uncertain, statistical evidence is often the best evidence available.

REVIEW VOCABULARY

Confidence interval   An interval computed from a sample by a method that has a known probability of producing an interval containing the unknown parameter. This probability is called the *confidence level.* Confidence intervals usually have the form estimate $\pm$ margin of error.

Control chart   A graph showing the value of a statistic for successive samples (for example, one sample each hour or one sample each shift). The graph also contains a *center line* at the target value for the process parameter and *control limits* that the statistic will rarely fall outside of unless the process drifts away from the target. The purpose of a control chart is to monitor a process over time and signal when some unusual source of variation interferes with the process.

Parameter   A number that describes the population. In statistical inference, the goal is often to estimate an unknown parameter or make a decision about its value.

Sample mean   The mean (arithmetic average) $\bar{x}$ of the observations in a sample. The sample mean from a simple random sample is used to estimate the unknown mean μ of the population from which the sample was drawn.

Sample proportion   The proportion $\hat{p}$ of the members of a sample having some characteristic (such as agreeing with an opinion poll question). The sample proportion from a simple random sample is used to estimate the corresponding proportion p in the population from which the sample was drawn.

Statistic   A number that describes a sample. A statistic can be calculated from the sample data alone, and does not involve any unknown parameters of the population.

Two-way table   A table showing the frequencies (counts) or percentages of outcomes that are classified according to two variables (such as applicants classified by both sex and admission decision.)

SUGGESTED READINGS

FREEDMAN, DAVID, ROBERT PISANI, ROGER PURVES, AND ANI ADHIKARI. *Statistics,* 2nd ed., Norton, New York, 1991. Chapters 21 and 23 discuss sampling distributions and confidence intervals for proportions and means.

MOORE, DAVID S.   *The Basic Practice of Statistics,* Freeman, New York, 1995. Chapter 4 of this text discusses the sampling distributions of sample means and proportions and control charts. Chapter 5 presents the reasoning of inference in detail. Chapters 6 and 7 discuss the practical use of inference about proportions and means.

MOSES, LINCOLN E.   The reasoning of statistical inference. In David C. Hoaglin and David S. Moore (eds.), *Perspectives on Contemporary Statistics,* Mathematical Association of America, Washington, D.C., 1992, pp. 107–122. This broad essay on the nature of inference requires some knowledge of probability and should be read after one of the introductory texts just cited.

EXERCISES   ▲ *Optional.*   ■ *Advanced.*   ◆ *Discussion.*

Estimating a Population Proportion

Identify each of the boldface numbers in Exercises 1 to 3 as either a *parameter* or a *statistic.*

1.   A random sample of female college students has a mean height of **64.5** inches, which is greater than the **63**-inch mean height of all adult American women.

2.   A sample of students of high academic ability under 13 years of age was given the SAT mathematics examination, which is usually taken by high school seniors. The mean score for the females in the sample was **386,** whereas the mean score of the males was **416.**

3.   About **6%** of all U.S. households are without telephones, but another **28%** have unlisted numbers.

4.   Exercises 1 to 3 in Chapter 5 (page 204) describe the results of three samples. Find the sample proportion $\hat{p}$ for each sample, first as a decimal fraction and then as a percent.

5.   About 35% of residential telephones in the San Francisco area have unlisted numbers. A telephone sales organization uses random digit dialing to dial a random sample of 200 residential telephone numbers. The percent of the sample that are unlisted numbers is the statistic of interest. What are the mean and standard deviation of the sampling distribution of this statistic?

6.   In a midwestern state, 84% of the households have Christmas trees at holiday time. A sample survey asks a random sample of 400 households "Did you have a Christmas tree this year?" What is the sampling distribution of the percent who say "Yes"?

7.   The standard deviation $\sigma_{\hat{p}}$ of a sample proportion $\hat{p}$ varies with true value of the population proportion p. Fortunately, it does not vary greatly unless p is near 0% or 100%. Suppose that the size of the sample is $n = 1500$. Evaluate $\sigma_{\hat{p}}$ for $p = 30\%, 40\%, 50\%, 60\%,$ and 70%. Then evaluate $\sigma_{\hat{p}}$ for $p = 0\%, 10\%,$ and 20%. In which range does $\sigma_{\hat{p}}$ change most rapidly as p changes? Make a graph of $\sigma_{\hat{p}}$ against p.

Confidence Intervals

◆   8.   The report of a sample survey of 1500 adults says, "With 95% confidence, between 27% and 33% of all American adults believe that drugs are the most serious problem facing our nation's public schools." Explain to someone who knows no statistics what the phrase "95% confidence" means in this report.

◆   9.   A Gallup poll of a random sample of 1540 adults asked, "Do you happen to jog?" Fifteen percent answered "Yes." The news item stated that this poll has a 3% margin of error. Explain carefully to someone who knows no statistics what is meant by a "3% margin of error."

10.   Suppose that the poll in Exercise 9 had used a simple random sample of size 1540, of whom 15% answered "Yes." Give a 95% confidence interval for the percent of all adults who would have answered "Yes" if asked.

11.   The U.S. Forest Service is considering additional restrictions on the number of vehicles allowed to enter Yellowstone National Park. To assess public reaction, the Service asks a simple random sample of 150 visitors if they favor the proposal. Of these, 89 say "Yes." Give a 95% confidence interval for the proportion of all visitors to Yellowstone who favor the restrictions. Are you 95% confident that more than half are in favor? Explain your answer.

12.   *Organic Gardening* magazine once reported the results of a test to see whether mayonnaise jars would break when used for home canning. Here is their conclusion:

> The mayonnaise jars didn't do badly—only 3 out of 100 broke. Statistically this means you'd expect between 0% and 6.4% to break.

Verify this statistical statement by giving a 95% confidence interval.

Exercises 13 to 16 are based on the following situation. A news report says that a national opinion poll of 1500 randomly selected adults found that 43% thought they would be worse off during the next year. The news report went on to say that the margin of error in the poll result is ±3 percentage points with 95% confidence.

◆   13.   Which of the following sources of error are included in the poll's margin of error?

(a)   The poll dialed telephone numbers at random and so missed all people without phones.

(b)   Some people whose numbers were chosen never answered the phone in several calls.

(c)   There is chance variation in the random selection of telephone numbers.

◆ 14. Would a 90% confidence interval based on the poll results have a margin of error less than ±3 percentage points, equal to ±3 percentage points, or greater than ±3 percentage points? Explain your answer.

◆ 15. If the poll had interviewed 1000 persons rather than 1500 (and still found 43% believing they would be worse off), would the margin of error for 95% confidence be less than ±3 percentage points, equal to ±3 percentage points, or greater than ±3 percentage points? Explain your answer.

◆ 16. Suppose that the poll had obtained the outcome 43% by a similar random sampling method from all adults in New York State (population 18 million) instead of from all adults in the United States (population 260 million). Would the margin of error for 95% confidence be less than ±3 percentage points, equal to ±3 percentage points, or greater than ±3 percentage points? Explain your answer.

Estimating a Population Mean

17. A shipment of machined parts has a critical dimension that is normally distributed with mean 12 centimeters and standard deviation 0.01 centimeter. The acceptance sampling team measures a random sample of 25 of these parts. What is the sampling distribution of the sample mean $\bar{x}$ of the critical dimension for these parts?

18. The Acculturation Rating Scale for Mexican Americans (ARSMA) is a psychological test that evaluates the degree to which Mexican Americans have adapted to Anglo/English culture. The scores in a large population are normally distributed with mean 3.0 and standard deviation 0.8. A researcher gives the test to a random sample of 12 Mexican Americans. What is the sampling distribution of their average score?

19. Electrical pin connectors for use in computers are gold plated for better conductivity. The specified plating thickness is 0.001 inch. Due to variations in the plating process, the actual plating thickness on different pins has a normal distribution with mean 0.001 inch and standard deviation 0.0001 inch.

    (a) What range of plating thickness contains 95% of all pins?
    (b) Quality control samples of four pins are taken regularly during production. The plating thickness is measured and the sample mean of the four measurements is recorded on a control chart. What range of plating thickness contains 95% of the recorded sample means?

20. Scores on the American College Testing (ACT) college admissions examination for the reference population used to develop the test vary

normally with mean $\mu = 18$ and standard deviation $\sigma = 6$. The range of reported scores is 1 to 36.

(a)   What range of scores contains the middle 95% of all students in the reference population?
(b)   If the ACT scores of 25 randomly selected students are averaged, what range contains the middle 95% of the averages $\bar{x}$?

21.   Errors in careful measurements often have a distribution that is close to normal. Experience shows that the error in a surveying method varies when a measurement is repeated according to a normal distribution with mean 0 (that is, the procedure does not systematically overestimate or underestimate the true distance) and standard deviation 0.03 meter. A surveyor repeats each measurement three times and uses the mean of the three measurements as the final value. The error in this value is the mean $\bar{x}$ of the errors in the three individual measurements.

(a)   What is the distribution of the mean error $\bar{x}$ when the surveyor measures many distances?
(b)   Between what values do 95% of the errors fall?

22.   A laboratory scale is known to have a standard deviation of $\sigma = 0.001$ gram in repeated weighings. Suppose that scale readings in repeated weighings are normally distributed, with mean equal to the true weight of the specimen. Three weighings of a specimen give (in grams)

3.412   3.414   3.415

Give a 95% confidence interval for the true weight of the specimen. What are the estimate and the margin of error in this interval?

23.   An instrument in a chemistry laboratory measures the concentration of trace substances in specimens. When the instrument makes repeated measurements on the same specimen, the readings are known to vary normally with standard deviation $\sigma = 0.03$. It is customary to make three readings and use the sample mean as the final result. For a particular specimen, the readings are

53.12   53.08   53.17

Give a 95% confidence interval for the mean of the distribution of readings for this specimen. (This mean is the true concentration if the instrument has no bias.)

◆   24.   Find the margin of error for 95% confidence in Exercise 22 if the laboratory weighs each specimen 12 times rather than three times. Check that your result is half as large as the margin of error you found in Exercise 22. Explain why you knew without calculating that the new margin of error would be half as large.

25.   The Family Adaptability and Cohesion Evaluation Scales (FACES) is a psychological test that measures two different aspects of family behavior. One of these is "cohesion," which is the degree to which family members are emotionally connected to each other. Suppose it is known that the cohesion scores for adults vary normally with standard deviation $\sigma = 5$. A researcher administers FACES to a sample of 33 adults in families with a runaway teenager. The mean cohesion score in this sample is $\bar{x} = 36.9$. Give a 95% confidence interval for the mean cohesion (as rated by an adult) of families with runaway teenagers.

26.   A milk processor monitors the number of bacteria per milliliter in raw milk received for processing. A random sample of 10 one-milliliter specimens from milk supplied by one producer gives the following data:

    5370   4890   5100   4500   5260   5150   4900   4760   4700   4870

Suppose it is known that the bacteria count varies normally and that the standard deviation is $\sigma = 265$ per milliliter. Give a 95% confidence interval for the mean bacteria count per milliliter in this producer's milk.

27.   An automatic lathe machines shafts to specified diameters as part of a manufacturing operation. Due to small variations in the operation of the lathe, the actual diameters produced follow a normal distribution with standard deviation 0.0005 inch. The shafts now being produced are supposed to have a diameter of 0.75 inch. You measure 10 such shafts and find that they have a sample diameter of $\bar{x} = 0.7505$. Give a 95% confidence interval for the true mean diameter μ of the shafts being produced. Are you confident that the true mean is not 0.75 inch?

Statistical Process Control

28.   Give the center line and control limits for a control chart for means $\bar{x}$ of samples of size 4 in the gold-plating process described in Exercise 19. Use 2σ limits, as in the text example.

29.   The laboratory instrument of Exercise 23 is monitored by measuring a specimen with known concentration 50 each morning. The specimen is measured three times and an $\bar{x}$ control chart is kept. What should be the center line and 2σ control limits for this chart?

30.   It is common for laboratories to keep a control chart for a measurement process based on regular measurements of a standard specimen. Suppose that you are maintaining a control chart for the scale in Exercise 22 by weighing a 5-gram standard weight three times at regular intervals. What should be the center line of your chart? What are the control limits if you decide to use 3σ limits?

Use 3σ limits in the control charts of Exercises 31 through 33. That is, use control limits that are three standard deviations on either side of the mean μ. Look for runs of 8 or more consecutive observations on the same side of the center line as well as for individual points outside the control limits. These are common control chart signals used in industry. Notice that the center line and control limits are the same for all three charts.

31.   In the data set below are $\bar{x}$'s from samples of size 4 with $\mu = 101.5$ and $\sigma = 0.2$. Only random variation is present. Make an $\bar{x}$ chart of these data using the given μ and σ. Are any points out of control?

Sample	$\bar{x}$	Sample	$\bar{x}$	Sample	$\bar{x}$
1	101.627	8	101.458	15	101.429
2	101.613	9	101.552	16	101.477
3	101.493	10	101.463	17	101.570
4	101.602	11	101.383	18	101.623
5	101.360	12	101.715	19	101.472
6	101.374	13	101.485	20	101.531
7	101.592	14	101.509		

32.   The following set of $\bar{x}$'s for samples of size 4 illustrates the effect of a shift in the standard deviation. The process has $\mu = 101.5$ and $\sigma = 0.2$ when it is undisturbed. The first 10 samples are taken in this condition. Then the process is disturbed so that for the last 10 samples $\mu = 101.5$ and $\sigma = 0.3$. Make a control chart for these data using the undisturbed μ and σ. Are any points out of control? Is the increase in σ visible in any way on the chart?

Sample	$\bar{x}$	Sample	$\bar{x}$	Sample	$\bar{x}$
1	101.602	8	101.453	15	101.756
2	101.547	9	101.446	16	101.707
3	101.312	10	101.522	17	101.612
4	101.449	11	101.664	18	101.628
5	101.401	12	101.823	19	101.603
6	101.608	13	101.629	20	101.816
7	101.471	14	101.602		

33.   The following set of $\bar{x}$'s for samples of size 4 illustrates the effect of a steady drift in the mean of the population. The first 10 samples have $\mu = 101.5$ and $\sigma = 0.2$. Then the process is disturbed so that the last 10 samples have $\sigma = 0.2$ and μ increasing by 0.04 in each successive sample, reaching 101.7 at sample 15 and 101.9 at sample 20. Make an $\bar{x}$ chart for these data. Are any points out of control? Is the upward drift in μ visible in any way on the chart?

Sample	$\bar{x}$	Sample	$\bar{x}$	Sample	$\bar{x}$
1	101.458	8	101.695	15	101.896
2	101.618	9	101.351	16	101.634
3	101.507	10	101.555	17	101.632
4	101.494	11	101.453	18	101.824
5	101.533	12	101.258	19	101.968
6	101.334	13	101.557	20	101.783
7	101.547	14	101.484		

The Perils of Data Analysis

◆ 34.   The U.S. government publication *Science and Engineering Indicators, 1991* states that the median salary of newly graduated female engineers and scientists in 1991 was only 73% of the median salary for males. When the new graduates were broken down by field, however, the picture changed. Women earned at least 84% as much as men in every field of engineering and science. The median salary for women was higher than that of men in many engineering disciplines. How can women do nearly as well as men in every field, yet fall far behind men when we look at all young engineers and scientists?

◆ 35.   In a study of the effect of parents' smoking habits on the smoking habits of high school students, researchers interviewed students in eight high schools in Arizona. The results appear in the following two-way table.

	Student Smokes	Student Does Not Smoke
Both parents smoke	400	1380
One parent smokes	416	1823
Neither parent smokes	188	1168

SOURCE: S. V. Zagona, ed., *Studies and Issues in Smoking Behavior,* University of Arizona Press, Tucson, 1967, pp. 157–180.

Describe the association between the smoking habits of parents and their high school children by computing and comparing several percents. Then summarize the results in plain language.

◆ 36.   Here is a two-way table of degrees awarded by colleges and universities in 1995, by type of degree and the sex of the recipient. The counts are in thousands of people.

	Bachelor's	Master's	Professional	Doctorate
Male	548	182	44	25
Female	630	195	31	17

(a)   What percent of all bachelor's degrees were earned by women?
(b)   How many master's degrees were awarded?
(c)   What percent of all degrees earned by women were doctorates?
(d)   Summarize in words the relation between the level of degrees and the sex of the recipient. Back your summary by computing and comparing appropriate percents.

◆   37.   The following pair of two-way tables compares the batting records of two baseball players, Bill and Will. How well a batter hits may depend on whether the pitcher is a right-hander or a left-hander, so both the type of pitcher and the result (hit or out) are recorded for each time at bat.

BILL	**Right-Hander**		**Left-Hander**	
	Count	Percent	Count	Percent
Hits	40	40	80	20
Outs	60	60	320	80

WILL	**Right-Hander**		**Left-Hander**	
	Count	Percent	Count	Percent
Hits	120	30	10	10
Outs	280	70	90	90

(a)   Combine the information in these tables to make a two-way table of batter (Bill or Will) by outcome (hit or out) for all times at bat. Which player gets a hit a higher proportion of the time? (The proportion of at bats in which a player gets a hit is his batting average.)

(b)   Who has the higher batting average against right-handed pitching? Who has the higher batting average against left-handed pitching?

(c)   Explain carefully, as if talking to a skeptical baseball manager, how it is possible for one player to do better against both right-handers and left-handers and yet have a lower overall batting average. Which hitter would you prefer to have on your team?

◆ 38.   A community has two hospitals. Hospital A is a large medical center, and Hospital B is a fashionable spa for prosperous patients. An article in the local newspaper claims that a higher percent of surgery patients die at Hospital A than at Hospital B. The paper says that people who need surgery should go to Hospital B. The following pair of two-way tables looks at the data but adds information about a hidden variable, the condition of the patients before surgery.

HOSPITAL A	**Good Condition**		**Poor Condition**	
	Count	Percent	Count	Percent
Died	6		57	
Survived	594		1443	

HOSPITAL B	**Good Condition**		**Poor Condition**	
	Count	Percent	Count	Percent
Died	8		8	
Survived	592		192	

(a)   Fill in the percent columns in the tables. Use the percents to show that both a higher percent of patients in good condition and a higher percent of patients in poor condition survive at Hospital A than at Hospital B.

(b)   Combine the information in the two tables to make a single table of patient outcome (died or survived) by hospital (A or B). Show that, as the newspaper reported, a higher percent of patients survive at Hospital B.

(c)   Explain carefully, as if talking to a skeptical reporter, how Hospital A can have a poorer overall survival rate even though it does better than B for both classes of patients.

Additional Exercises

◆ 39.   A *New York Times* poll on women's issues interviewed 1025 women randomly selected from the United States, excluding Alaska and Hawaii. The poll found that 47% of the women said they do not get enough time for themselves.

(a) The poll announced a margin of error of ± 3 percentage points for 95% confidence in its conclusions. What is the 95% confidence interval for the percent of all adult women who think they do not get enough time for themselves?

(b) Explain to someone who knows no statistics why we can't just say that 47% of all adult women do not get enough time for themselves.

(c) Then explain clearly what "95% confidence" means.

40. Oxides of nitrogen (called NOX for short) emitted by cars and trucks are important contributors to air pollution. The amount of NOX emitted by a particular model varies from vehicle to vehicle. For one light truck model, NOX emissions vary with mean μ that is unknown and standard deviation $\sigma = 0.4$ gram per mile. You test a simple random sample of 50 of these trucks. The sample mean NOX level $\bar{x}$ estimates the unknown μ. You will get different values of $\bar{x}$ if you repeat your sampling.

(a) The sampling distribution of $\bar{x}$ is approximately normal. What are its mean and standard deviation?

(b) Sketch the normal curve for the sampling distribution of $\bar{x}$. Mark its mean and the values one, two, and three standard deviations on either side of the mean.

(c) According to the 68–95–99.7 rule, about 95% of all values of $\bar{x}$ lie within a distance m of the mean of the sampling distribution. What is m? Shade the region on the axis of your sketch that is within m of the mean.

(d) Whenever $\bar{x}$ falls in the region you shaded, the unknown population mean μ lies in the confidence interval $\bar{x} \pm m$. For what percent of all possible samples does this happen?

41. A simple random sample of students at Upper Wabash Tech is asked whether they favor limiting enrollment in crowded majors as a way of keeping the quality of instruction high. The student government suspects that the plan will be unpopular among freshmen, who have not yet been admitted to a major. Here are the responses for freshmen and seniors.

	Favor	Oppose
Freshmen	40	160
Seniors	80	20

(a) Give a 95% confidence interval for the percent of all freshmen who support the plan.

(b) Give a 95% confidence interval for the percent of all seniors who support the plan.

◆ 42. Ronald Reagan was president from 1981 to 1988. A Gallup poll of 1514 adults taken between July 30 and August 2, 1983, asked, "Do you approve of the way Ronald Reagan is handling his job as President?" Of these, 41% said "Yes."

(a) If the poll had used a simple random sample, what would have been the margin of error in a 95% confidence interval?

(b) The actual margin of error for a Gallup poll of this same size is $\pm 3\%$. Why does this not agree with your result in part (a)?

◆ 43. The Degree of Reading Power (DRP) is a test of the reading ability of children. Here are DRP scores for a sample of 44 third-grade students in a suburban school district.

40	26	39	14	42	18	25	43	46	27	19
47	19	26	35	34	15	44	40	38	31	46
52	25	35	35	33	29	34	41	49	28	52
47	35	48	22	33	41	51	27	14	54	45

(a) We expect the distribution of DRP scores to be close to normal. Make a dotplot or histogram of the distribution of these 44 scores and describe its shape.

(b) Suppose that the standard deviation of the population of DRP scores is known to be $\sigma = 11$. Give a 95% confidence interval for the mean score in the school district.

(c) Would you trust your conclusion from part (b) if these scores came from a single class in one school in the district? Why?

▨ 44. In the text we used the sampling distribution of $\hat{p}$ and the 68–95–99.7 rule to give a 95% confidence interval for a population proportion p.

(a) Explain carefully why

$$\hat{p} \pm \sqrt{\frac{\hat{p}(100 - \hat{p})}{n}}$$

is a 68% confidence interval for p.

(b) Give the recipe for a 99.7% confidence interval for p.

45.   Use the result of Exercise 44a and the data in Exercise 11 to give a 68% confidence interval for the percent of visitors to Yellowstone who support restricting the number of vehicles allowed into the park. Compare the width of the 68% interval with that of the 95% interval from Exercise 13 and explain the difference in plain language.

46.   Use the result of Exercise 44b and the data in Exercise 11 to give a 99.7% confidence interval for the percent of all adults who jog. Compare the width of the 99.7% interval with that of the 95% confidence interval from Exercise 10. What is the reason for the difference in widths?

47.   The upper and lower deciles of any normal distribution are located 1.28 standard deviations above and below the mean. (The lower decile is the point with probability 10% below it; the upper decile has probability 90% below it.)

(a)   Use this information to give a recipe for an 80% confidence interval for a population proportion p based on the sample proportion $\hat{p}$ that is accurate for large sample sizes n.
(b)   Give an 80% confidence interval for the proportion of visitors to Yellowstone favoring vehicle restrictions, using the data in Exercise 11.

48.   The upper and lower deciles of any normal distribution are located 1.28 standard deviations above and below the mean.

(a)   Use this information to give a recipe for an 80% confidence interval for the mean μ of a normal population based on the sample mean $\bar{x}$ of a simple random sample of size n.
(b)   Give an 80% confidence interval for the mean FACES cohesion score in Exercise 25.

WRITING PROJECTS

1 ▶   The margin of error announced for a sample survey takes into account the chance variation due to random sampling. In practice, survey results can be in error for other reasons. Some subjects can't be contacted, others lie or don't remember information, and the wording of the questions will influence the responses. These are the "practical difficulties" mentioned in Spotlight 8.1.

Write a brief discussion of the most important practical difficulties encountered in opinion polls and other surveys of human populations. You will want to read more on the subject. Some sources are Section 1.5 of *Statistics: Concepts and Controversies;* Section 19.6 of *Statistics* (see Suggested Readings in Chapter 5 for both books); and the article by P. E. Converse and M. W. Traugott, Assessing the accuracy of polls and surveys, *Science,* 234 (1986): 1094–1098.

2 ▸ There are many confidence intervals for use in settings other than the two described in this chapter. One common setting is to estimate the difference between *two* population proportions, p_1 and p_2. For example, p_1 could be the proportion of women and p_2 the proportion of men who stay home at night due to fear of crime. Read an account of the confidence interval for $p_1 - p_2$ in a statistical methods text (for example, in Section 7.2 of Moore's *Basic Practice of Statistics*). Then explain carefully how this new confidence interval arises from the same reasoning we have used: find an estimate, discover that the sampling distribution of the estimate is at least approximately normal, learn the standard deviation of this normal distribution, and go out two standard deviations from the estimate to get 95% confidence.

III

Y ou've seen bar-coded numbers on everything from milk cartons to books. And you've seen long numbers on everything from credit cards to airline tickets. Of course these numbers function as identification numbers just as social security numbers do and currency serial numbers do. However, because of inexpensive, fast, reliable computers, the identification numbers used by businesses today are more sophisticated than those of earlier years. Modern identification numbers have a built-in "check" to partially ensure that the numbers have been correctly entered into a computer or have been correctly scanned by an optical device. Identification numbers for people often contain a variety of personal information such as sex, date of birth, and portions of names.

Identification numbers are examples of codes. A **code** is a group of symbols that represent information. Codes existed thousands of years ago: hieroglyphics, the Greek alphabet, and Roman numerals. Many codes have

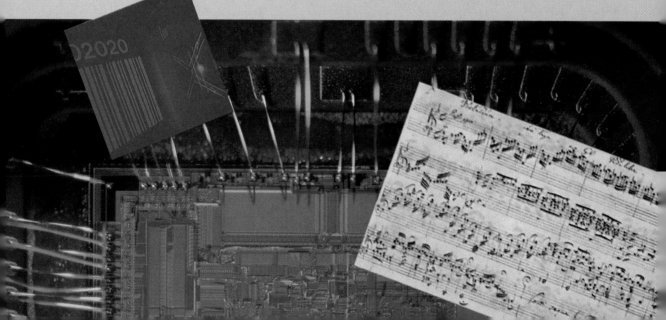

CODING
INFORMATION

been developed for a particular application: musical scores, the Morse code, and the "genetic code" used to describe the makeup of DNA. About 40 years ago an MIT graduate student named David Huffman invented a code that is now used in computers, high-definition television, modems, and even VCR Plus+ devices that automatically program a VCR (see Spotlight 10.5). In recent decades coding schemes have been invented that do more than simply code data. For example, data from space probes, signals from compact discs, and modem, fax machine, and high-definition television (see Spotlight 10.1) transmissions are coded so that errors in the data that occur during transmission can be corrected. Premium television services such as Home Box Office (HBO) and The Disney Channel code their signals so that only those with a decoding device receive the service.

In this part of the text we will examine some of the ways that information is coded.

CHAPTER 9

IDENTIFICATION NUMBERS

Modern identification numbers have at least two functions. Obviously, an identification number should unambiguously identify the person or thing to which it is associated. Not obvious is a "self-checking" aspect of the number.

Look at the ISBN printed on the back of this book. The number 0-7167-2841-9 distinguishes this book from all others. The last digit "9" is there solely to detect errors that may occur when the ISBN is entered into a computer. Look at the bottom of the airline ticket shown in Figure 9.1. Notice the letters "ck" (for "check") above the last digit of the stock control number and above the last digit of the document number. They also are there for the purpose of error detection. Grocery items, credit cards, overnight mail, magazines, personal checks, traveler's checks, soft-drink cans, automobiles, and many other items you encounter daily have identification numbers that code data and include check digits for error detection. In this chapter we examine some of the methods that are used to assign identification numbers and check digits.

FIGURE 9.1 Airline ticket with identification number 121399538132 and check digit 5; stock control number 2551016950 and check digit 6.

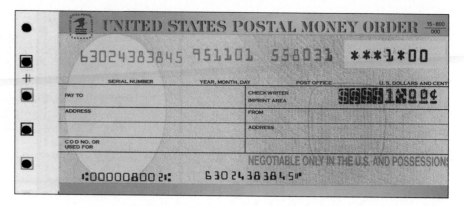

FIGURE 9.2 Money order with identification number 6302438384 and appended check digit 5. The check digit is the remainder upon dividing the sum of the digits by 9.

Let us begin by considering the U.S. Postal Service money order shown in Figure 9.2. The first 10 digits of the 11-digit number 63024383845 simply identify the money order. The last digit, 5, serves as an **error-detecting code** or mechanism. Let us see how this mechanism works. The eleventh (last) digit of a Postal Service money order number is the remainder obtained when the sum of the first 10 digits of the number is divided by 9. In our example the last digit is 5 because $6 + 3 + 0 + 2 + 4 + 3 + 8 + 3 + 8 + 4 = 41$ and the remainder when 41 is divided by 9 is 5.

Now suppose instead of the correct number, the number 63054383845 (an error in the fourth position) were entered into a computer programmed for error detection of money orders. The machine would divide the sum of the first 10 digits of the entered number, 44, by 9 and obtain a remainder of 8. Since the last digit of the entered number is 5 rather than 8, the entered number cannot be correct. This crude method of error detection will not detect the mistake of replacing a 0 with a 9, or vice versa. Nor will it detect the transposition of digits, such as 63204383845 instead of 63024383845 (the digits in positions three and four have been transposed). (Spotlight 9.1 shows what can happen as a result of a transposition error.)

American Express (see Figure 9.3) and VISA traveler's checks also utilize a check digit determined by division by 9. In these cases, the check digit is chosen so that the sum of the digits, including the check digit, is evenly divisible by 9.

EXAMPLE ▶ *The American Express Travelers Cheque*

The American Express Travelers Cheque with the identification number 387505055 has check digit 7 because $3 + 8 + 7 + 5 + 0 + 5 + 0 + 5 + 5 = 38$ and $38 + 7$ is evenly divisible by 9. ◆

FIGURE 9.3 Traveler's check with identification number 387505055 and check digit 7. The check digit is chosen so that the sum of all the digits including the check digit is evenly divisible by 9. Notice that the check digit is included as part of the identification number along the bottom but not included in the upper right corner. The number along the bottom is read by a computer.

The scheme used on airline tickets, Federal Express mail, UPS packages, and Avis and National rental cars assigns the remainder upon division by 7 of the number itself as the check digit (see Figure 9.1) rather than dividing the sum of the digits by 7. For example, the check digit for the number 540047 is 4 since $540047 = 7 \times 77149 + 4$. This method will not detect the substitution of 0 for a 7, 1 for an 8, 2 for a 9, or vice versa. However, unlike the Postal Service method, it will detect transpositions of adjacent digits with the exceptions of the pairs 0, 7; 1, 8; or 2, 9. For example, if 5400474 were entered into a computer as 4500474 (the first two digits are transposed), the machine would determine that the check digit should be 3 since $450047 = 7 \times 64292 + 3$. Because the last digit of the entered number is not 3, the error has been detected.

The scheme used on grocery products, the so-called *Universal Product Code (UPC)*, is more sophisticated. Consider the number 0 38000 00127 7 found on the bottom of a box of corn flakes. The first digit identifies a broad category of goods, the next five digits identify the manufacturer, the next five the product, and the last is a check. Suppose this number were entered into a computer as 0 58000 00127 7 (a mistake in the second position). How would the computer recognize the mistake?

The computer is programmed to carry out the following computation: add the digits in positions 1, 3, 5, 7, 9, 11 and triple the result; then add this tally to the sum of the remaining digits. If the result doesn't end with a 0, the computer knows the entered number is incorrect.

For the incorrect corn flakes number, we have $((0 + 8 + 0 + 0 + 1 + 7) \times 3) + (5 + 0 + 0 + 0 + 2 + 7) = (16 \times 3) + 14 = 62$. Since 62 doesn't end with 0, the error is detected. Notice that had we used the correct

Wrong Number: Demolition Crew Wrecks House at 415, Not 451

It was a case of mistaken identity. A transposed address that resulted in a bulldozer blunder.

City orders had called for demolition on Tuesday of the boarded-up house at 451 Fuller Ave. SE. . . .

But when the dust settled, 451 Fuller stood untouched. Down the street at 415 Fuller Ave. SE, only a basement remained.

Source: Doug Guthrie, *Grand Rapids Press*, December 5, 1990.

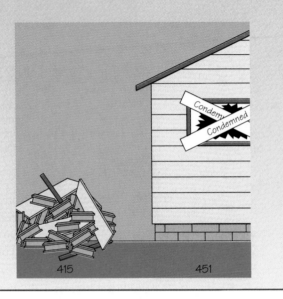

digit 3 in the second position instead of 5, the sum would have ended in a 0 as it should. This simple scheme detects *all* single position errors and about 89% of all other kinds of errors.

The U.S. banking system uses a variation of the UPC scheme that appends check digits to the numbers assigned to banks. Each bank has an eight-digit identification number $a_1 a_2 \cdots a_8$ together with a check digit a_9 so that a_9 is the last digit of $7a_1 + 3a_2 + 9a_3 + 7a_4 + 3a_5 + 9a_6 + 7a_7 + 3a_8$. The numbers 7, 3, and 9 used in this formula are called **weights.** The weights were carefully chosen so that all single-digit errors and most transposition errors are detected. (The use of different weights in adjacent positions permits the detection of most transposition errors.)

EXAMPLE ▶ *Bank Identification Number*

The First Chicago Bank has the number 071000013 on the bottom of all its checks (see Spotlight 9.2). The check digit 3 is the last digit of $7 \cdot 0 + 3 \cdot 7 + 9 \cdot 1 + 7 \cdot 0 + 3 \cdot 0 + 9 \cdot 0 + 7 \cdot 0 + 3 \cdot 1 = 33$. ◆

One of the most efficient error-detection methods is one used by all major credit card companies, as well as by many libraries, blood banks, photofinishing companies, German banks, and the South Dakota driver's license department. It is called **Codabar.** Say a bank intends to issue a credit card with the identification number 312560019643001. It must then add an extra digit for error detection. This is done as follows. Add the digits in positions 1, 3, 5, 7, 9, 11, 13, and 15 and double the result: $(3 + 2 + 6 + 0 + 9 + 4 + 0 + 1) \times 2 = 50$. Next, count the number of digits in positions 1, 3, 5, 7, 9, 11, 13, and 15 that exceed 4 and add this to the total. For our example, only 6 and 9 exceed 4, so the count is 2 and our running total is 52. Now add in the remaining digits: $52 + (1 + 5 + 0 + 1 + 6 + 3 + 0) = 68$.

The check digit is whatever is needed to bring the final tally to a number that ends with 0. Since $68 + 2 = 70$, the check digit for our example is 2. This digit is appended to the end of the number the bank issues for identification purposes. Errors in input data are detected by applying the same algorithm to the input, including the check digit. If the correct number is entered into a computer, the result will end in a zero. If the result doesn't end with a zero, a mistake has been made. The credit card shown in Figure 9.4 is reproduced from an ad promoting the Citibank VISA card. Notice that the check digit on the card is not valid since the Codabar algorithm yields

$$(4 + 2 + 0 + 1 + 3 + 5 + 7 + 9) \times 2 + 3$$
$$+ (1 + 8 + 0 + 2 + 4 + 6 + 8) + 0 = 94$$

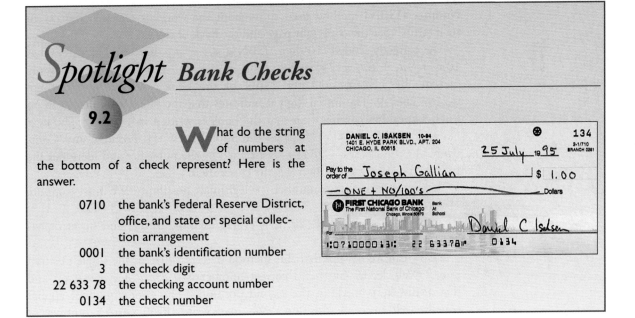

Spotlight *Bank Checks*

9.2

What do the string of numbers at the bottom of a check represent? Here is the answer.

0710	the bank's Federal Reserve District, office, and state or special collection arrangement
0001	the bank's identification number
3	the check digit
22 633 78	the checking account number
0134	the check number

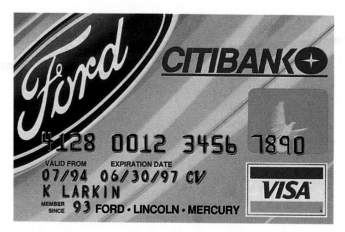

FIGURE 9.4　VISA card with an invalid Codabar number.

which does not end in 0. This method allows computers to detect 100% of single-position errors and about 98% of other common errors.

Besides detecting errors, the check digit offers partial protection against fraudulent numbers. A person who wanted to create a phony credit card, bank account number, or driver's license number would have to know the appropriate check digit scheme for the number to go unchallenged by the computer. (See Spotlight 9.3.)

Thus far we have not discussed any schemes that detect 100% of single errors and 100% of transposition errors. The **International Standard Book Number (ISBN)** method used throughout the world is one that detects all such errors. (See the copyright page and the back of this book.)

A correctly coded 10-digit ISBN $a_1 a_2 \cdots a_{10}$ has the property that $10a_1 + 9a_2 + 8a_3 + 7a_4 + 6a_5 + 5a_6 + 4a_7 + 3a_8 + 2a_9 + a_{10}$ is evenly divisible by 11. Consider the ISBN of the book you are now reading: 0-7167-2841-9. The initial digit 0 indicates that the book is published in an English-speaking country (*not* that the book is written in English). The next block of digits—7167—identifies the publisher, W. H. Freeman and Company. The third block—2841—is assigned by the publisher and identifies the particular book. The last digit 9 is the check digit.

Let us verify that this number is a legitimate possibility. We must compute $10 \cdot 0 + 9 \cdot 7 + 8 \cdot 1 + 7 \cdot 6 + 6 \cdot 7 + 5 \cdot 2 + 4 \cdot 8 + 3 \cdot 4 + 2 \cdot 1 + 9 = 220$. Since $220 = 11 \cdot 20$, it is evenly divisible by 11, and no error has been detected.

How can we be sure that this method detects 100% of the single-position errors? Well, let us say that a correct number is $a_1 a_2 a_3 a_4 a_5 a_6 a_7 a_8 a_9 a_{10}$ and that a mistake is made in the second position. (The same argument applies equally well in every position.) We may write this incorrect number as

Spotlight *Credit Card Fraud*

9.3

In 1994 a computer program designed to create credit card numbers began to appear on on-line computer services such as America Online and many electronic bulletin boards. Known as the Credit Master, the program uses legitimate bank codes and produces the correct check digit. Only 3 to 5% of the numbers the program produces actually correspond to active accounts. The program cannot produce the expiration date or the holder's name, which are often checked before spending is authorized. Credit card companies say that they know of no significant losses as the result of Credit Master. "We consider it as a threat, but the formula was never meant to be high-tech security screening," said Dennis Fiene, director of fraud control for VISA.

Source: Adapted from Ashley Dunn, "A pirate computer program builds credit card numbers," *New York Times,* March 19, 1995, p. 18.

$a_1 a_2' a_3 a_4 a_5 a_6 a_7 a_8 a_9 a_{10}$, where $a_2' \neq a_2$. Now in order for this error to go undetected, it must be the case that $10a_1 + 9a_2' + 8a_3 + 7a_4 + 6a_5 + 5a_6 + 4a_7 + 3a_8 + 2a_9 + a_{10}$ is evenly divisible by 11. Then, since both $10a_1 + 9a_2 + 8a_3 + 7a_4 + 6a_5 + 5a_6 + 4a_7 + 3a_8 + 2a_9 + a_{10}$ and $10a_1 + 9a_2' + 8a_3 + 7a_4 + 6a_5 + 5a_6 + 4a_7 + 3a_8 + 2a_9 + a_{10}$ are divisible by 11, so is their difference:

$$(10 \cdot a_1 + 9 \cdot a_2 + 8 \cdot a_3 + \cdots + 1 \cdot a_{10})$$
$$- (10 \cdot a_1 + 9 \cdot a_2' + 8 \cdot a_3 + \cdots + 1 \cdot a_{10}) = 9 \cdot (a_2 - a_2')$$

Because a_2 and a_2' are distinct digits between 0 and 9, their difference must be one of $\pm 1, \ldots, \pm 9$. Thus the only possibilities for the number $9 \cdot (a_2 - a_2')$ are $\pm 9, \pm 18, \pm 27, \pm 36, \pm 45, \pm 54, \pm 63, \pm 72, \pm 81, \pm 90$, and none of these is divisible by 11. So, a single-position error cannot go undetected.

Since this method, in contrast to the others we have described, detects all single-position errors and all transposition errors, why is it not used more? Well, it does have a drawback. Say the next title published by Freeman is to have 1910 for the third block. (All Freeman books begin with 0-7167-.) What check digit should be assigned? Call it a. Then $10 \cdot 0 + 9 \cdot 7 + 8 \cdot 1 + 7 \cdot 6 + 6 \cdot 7 + 5 \cdot 1 + 4 \cdot 9 + 3 \cdot 1 + 2 \cdot 0 + a = 199 + a$. Since the next integer after 199 that is divisible by 11 is 209, we see that $a = 10$. But appending 10 to the existing 9-digit number would result in an 11-digit number instead of a 10-digit one. This is the only flaw in the ISBN scheme. To avoid this flaw, publishers use an X to represent the check digit 10. As a result not all ISBNs consist solely of digits (some end with X). Publishers could avoid this inconsistency by simply refraining from using numbers that require an X.

At this point the reader might naturally ask, "Why are there so many different methods for achieving the same purpose?" Like many practices in the "real world," historical accident and lack of knowledge about existing methods seem to be the explanation.

Many identification numbers utilize both alphabetic and numerical characters. One of the most prevalent of these was developed in 1975 and is called **Code 39.** Code 39 permits the 26 uppercase letters A through Z and the digits 0 through 9. Because Code 39 has been chosen by the Department of Defense, the automotive companies, and the health industry for use by their suppliers it has become the workhorse of nonretail business.

A typical example of a Code 39 number is 210SA0162322ZAY. The last character is the "check." The check character is determined by assigning the letters A through Z the numerical values 10 through 35, respectively. The original number, composed of the digits 0 through 9 and letters A through Z, is now converted to a string $a_1, a_2, \ldots, a_{14}, a_{15}$, where the a_i are integers between 0 and 35. The check character a_{15} is chosen so that $15a_1 + 14a_2 + 13a_3 + \cdots + 2a_{14} + a_{15}$ is divisible by 36. Finally, a_{15} is converted to its alphabetic counterpart if it is greater than 9 (for example, 13 is converted to D).

E X A M P L E ▶ *Code 39 Number 210SA0162322ZA*

Let us examine the Code 39 method for the number 210SA0162322ZA. Here is how we determine the check character. First we convert the alphabetic characters to their numerical counterparts: 210*SA*0162322*ZA* → 2, 1, 0, 28, 10, 0, 1, 6, 2, 3, 2, 2, 35, 10. Then we compute

$$15 \cdot 2 + 14 \cdot 1 + 13 \cdot 0 + 12 \cdot 28 + 11 \cdot 10 + 10 \cdot 0 + 9 \cdot 1 + 8 \cdot 6$$
$$+ 7 \cdot 2 + 6 \cdot 3 + 5 \cdot 2 + 4 \cdot 2 + 3 \cdot 35 + 2 \cdot 10 = 30 + 14 + 0 + 336$$
$$+ 110 + 0 + 9 + 48 + 14 + 18 + 10 + 8 + 105 + 20 = 722$$

Now we select a_{15} so that $722 + a_{15}$ is divisible by 36. Since 722 divided by 36 has a remainder of 2 ($722 = 36 \cdot 20 + 2$), we choose a_{15} as 34. Finally, we convert 34 to Y. Thus the number becomes 210SA0162322ZAY. ◆

In many applications of Code 39 the seven special characters -, ., space, $, /, +, and % are permitted. These characters are assigned the numerical values

36 through 42, respectively. In these applications the check character is determined by the remainder upon division by 43 instead of 36.

THE ZIP CODE

Identification numbers occasionally **encode** geographical data. **ZIP codes,** social security numbers, and telephone numbers are prime examples. In 1963 the U.S. Postal Service numbered every American post office with a five-digit ZIP code. The numbers begin with zeros at the point farthest east—00601 for Adjuntas, Puerto Rico—and work up to nines at the point farthest west—99950 for Ketchikan, Alaska (see Figure 9.5). Here's what the five digits mean.

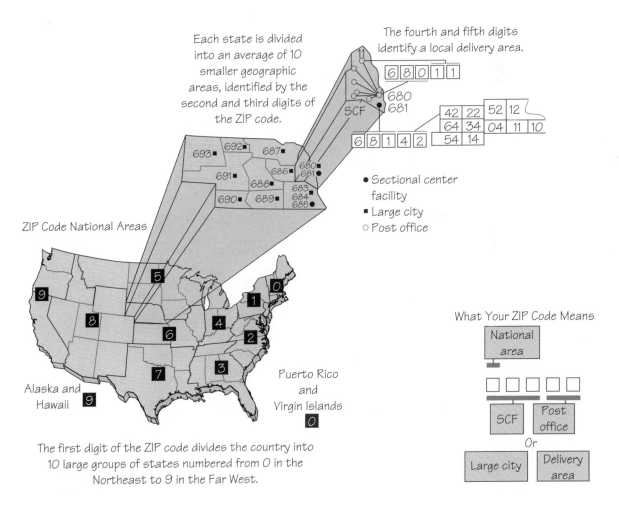

FIGURE 9.5   ZIP code scheme.

Let's use one of the ZIP codes for Lincoln, Nebraska, as an example:

68588

6   The first digit represents one of 10 geographical areas, usually a group of states. The numbers begin at points farthest east (0) and end at the points farthest west (9).

85   The second two digits, in combination with the first, identify a central mail-distribution point known as a sectional center. The location of a sectional center is based on geography, transportation facilities, and population density; although just four centers serve the entire state of Utah, there are six of them for New York City alone.

88   The last two digits indicate the town, or local post office. The order is often alphabetic for towns within a delivery area—for example, towns with names beginning with A usually have low numbers. (There are many exceptions to this, such as towns that came into existence after the ZIP code scheme was created.) In many cases the largest city in a region will be given the digits 01 and surrounding towns assigned succeeding digits alphabetically (see Figure 9.6).

FIGURE 9.6 Postal ZIP codes in Farmville, Virginia, area.

Andersonville	23911	Hampden-Sydney	23943
Boydton	23917	Kenbridge	23944
Buckingham	23921	Keysville	23947
Burkeville	23922	Lunenburg	23952
Charlotte Court House	23923	Meherrin	23954
Chase City	23924	Nottoway	23955
Clarksville	23927	Pamplin	23958
Crewe	23930	Phenix	23959
Cullen	23934	Prospect	23960
Darlington Heights	23935	Red Oak	23964
Dillwyn	23936	Rice	23966
Drakes Branch	23937	Skipwith	23968
Dundas	23938	Victoria	23974
Farmville	23901	Wylliesburg	23976
Green Bay	23942		

In 1983 the U.S. Postal Service added four digits to the ZIP code. When four digits are added after a dash—for example, 68588-1234—the number is called the **ZIP + 4 code.** Mail with the ZIP + 4 coding benefits from cheaper bulk rates, being easier to sort with automated equipment. It's also helpful for businesses that wish to sort the recipients of their mailings by geographical location. The first two numbers of the four-digit suffix represent a

Hand-held scanner reading the shipping bar code on a crate.

delivery sector, which may be several blocks, a group of streets, several office buildings, or a small geographical area. The last two numbers narrow the area further: They might denote one floor of a large office building, a department in a large firm, or a group of post office boxes.

For businesses that receive an enormous volume of mail the ZIP + 4 code permits automation of in-house mailroom sorting. For example, the first seven digits of all mail sent the University of Minnesota, Duluth, are 55812-24. The school has designated nine pairs of digits for the last two positions to direct the mail to the appropriate dormitory or apartment complex. Because of the enormous volume of mail addressed to the president, President Clinton has a secret ZIP + 4 code so that mail from his friends can be identified immediately.

BAR CODES

In modern applications bar codes and identification numbers go hand in hand. Bar coding is a method for automated data collection. It is a way to transmit information rapidly, accurately, and efficiently to a computer.

A **bar code** is a series of dark bars and light spaces that represent characters.

FIGURE 9.7 ZIP + 4 bar code.

> **NO POSTAGE
> NECESSARY
> IF MAILED
> IN THE
> UNITED STATES**

BUSINESS REPLY CARD
FIRST CLASS   PERMIT NO. 165   DOWNERS GROVE, ILLINOIS

POSTAGE WILL BE PAID BY ADDRESSEE

UNIVERSITY

SUBSCRIPTION

SERVICE

**1213 BUTTERFIELD ROAD
DOWNERS GROVE, IL 60515-9968**

Guard bar										Guard bar
Binary code	01100	11000	01010	00011	01010	10100	10100	01100	10010	00011
Digit code	6	0	5	1	5	9	9	6	8	1

Check digit

To **decode** the information in a bar code, a beam of light is passed over the bars and spaces via a scanning device, such as a hand-held wand or a fixed-beam device. The dark bars reflect very little light back to the scanner, whereas the light spaces reflect much light. The differences in reflection intensities are detected by the scanner and converted to strings of 0s and 1s that represent specific numbers and letters. Such strings are called a *binary coding* of the numbers and letters.

> Any system for representing data with only two symbols is a **binary code.**

ZIP Code Bar Code

The simplest bar code is the **Postnet code** used by the U.S. Postal Service and commonly found on business reply forms (see Figure 9.7). For a ZIP + 4 code there are 52 vertical bars of two possible lengths (long and short). The long bars at the beginning and end are called *guard bars* and together provide a frame for the remaining 50 bars. In blocks of five, the 50 bars within the guard bars represent the ZIP + 4 code and a tenth digit for error correction. Each block of five is composed of exactly two long bars and three short bars, according to the pattern shown in Figure 9.8.

FIGURE 9.8 The Postnet bar code.

Decimal Digit	Bar Code
1	ꞁꞁ‖‖
2	ꞁꞁ‖ꞁ‖
3	ꞁꞁ‖‖ꞁ
4	ꞁꞁꞁꞁ
5	ꞁꞁꞁꞁ
6	ꞁ‖‖ꞁꞁ
7	‖ꞁꞁꞁ
8	‖ꞁꞁ‖ꞁ
9	‖ꞁ‖ꞁꞁ
0	‖‖ꞁꞁꞁ

The tenth digit of a Postnet code number is a check digit chosen so that the sum of the nine digits of the ZIP + 4 code and the tenth one is evenly divisible by 10. That is, the check digit C for the ZIP + 4 code $a_1 a_2 \cdots a_9$ is the digit with the property that the sum $(a_1 + a_2 + \cdots + a_9 + C)$ ends with 0. For example, the ZIP + 4 code 80321-0421 has the check digit 9 since $8 + 0 + 3 + 2 + 1 + 0 + 4 + 2 + 1 = 21$ and $21 + 9 = 30$ ends with 0.

Because each digit is represented by exactly two long bars and three short ones, any error in reading or printing a single bar would result in a block of five with only one long bar or three long bars. In either case, the error is detected. (This is the reason behind the choice of five bars to code each digit rather than four bars. With five bars per digit, there are exactly 10 arrangements composed of two long bars and three short bars. Any misreading of a single bar in such a block is therefore recognizable, since it does not match any other of the blocks for the 10 digits.) And since the block location of the error is known, the check digit permits the correction of the error. Let's look at an example of an incorrectly printed bar code and see how the error is correctable.

The scanner ignores the guard bars at the beginning and the end and reads the remaining bars in blocks of five as shown below. (We have inserted dividing lines for readability.)

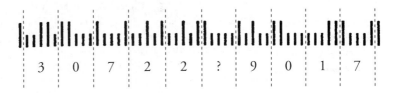

Since the sixth block has only one long bar, it is an incorrect one. To correct the error, the computer linked with the bar code scanner sums the remaining 9 digits to obtain 31. Since the sum of all 10 digits ends with 0, the correct value for the sixth digit must be 9.

Beginning in 1993, large organizations and businesses that wanted to receive reduced rates for ZIP + 4 bar-coded mail were required to use a 12-digit bar code called the *delivery-point bar code*. This code permits machines to sort a letter into the order it will be delivered by the carrier. (Mail for the first location on a mail route occurs first, mail for the second location on a route occurs second, and so on.)

The 12-digit bar code uses the Postnet bar scheme to code the 12-digit string composed of the 9-digit ZIP + 4 number followed by the last two digits of the street address or box number and a check digit chosen so that the sum of all 12 digits is evenly divisible by 10. For example, a letter addressed to 1738 Maple Street with ZIP + 4 code 55811-2742 would have the Postnet bar code for the digits 558112742384 (38 is from the street address and 4 is the check digit).

The UPC Bar Code

The bar code that we encounter most often is the **Universal Product Code (UPC).** The UPC was first used on grocery items in 1973 and has since spread to most retail products. The UPC bar code translates a 12-digit number into bars that can be quickly and accurately read by a laser scanner. The number has four components—two five-digit numbers sandwiched between two single digits—as shown in Figure 9.9.

FIGURE 9.9   UPC identification number 5 43000 21031 9. The initial 5 indicates the number is a manufacturer's coupon. The block 43000 identifies the manufacturer as Kraft General Foods. The block 21031 identifies the product. The last digit, 9, is a check digit.

Here is what the four components represent:

5   The first digit identifies the kind of product. For example, a 2 signals random weight items, such as cheese or meat; a 3 means drug and certain health-related products; a 4 means products marked for price reduction by the retailer; a 5 signals cents-off coupons (see Figure 9.9).

43000   The next five digits identify the manufacturer. This number is assigned by the Uniform Code Council in Dayton, Ohio.

21031   The next five digits are assigned by the manufacturer to identify the product, and can include size, color, or other important information (but not price).

9   The final digit is the check digit. This digit is often not printed, but it is always included in the bar code.

Each digit of the UPC code is represented by a space divided into seven modules of equal width, as illustrated in Figure 9.10. How these seven modules are filled depends on the digit being represented and whether the digit being represented is part of the manufacturer's number or the product number. In every case there are two light "spaces" and two dark bars of various thicknesses that alternate. A UPC code has on each end two long bars of one module thickness separated by a light space of one module thickness. These three modules are called the *guard bar patterns* (see Figure 9.11). The guard bar pat-

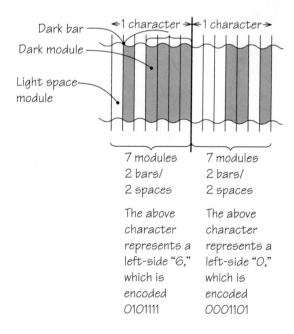

FIGURE 9.10 UPC bar coding for a left-side 6 and left-side 0.

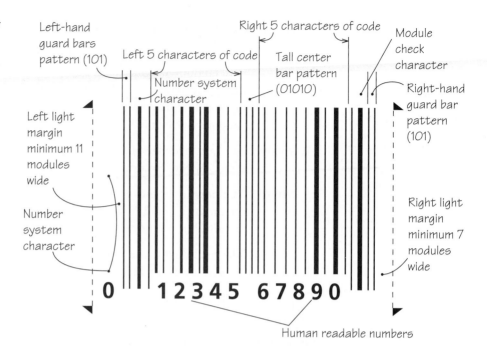

FIGURE 9.11 UPC bar code format.

Left-hand guard bars pattern (101)

Right 5 characters of code

Module check character

Left 5 characters of code

Number system character

Tall center bar pattern (01010)

Right-hand guard bar pattern (101)

Left light margin minimum 11 modules wide

Number system character

Right light margin minimum 7 modules wide

Human readable numbers

Entomologist Stephen Buchmann developed a reliable, inexpensive way to track bees using the same technology that supermarkets use to speed up the checkout lines and keep track of inventory. He glued bar code labels onto the backs of 100 bees and placed a laser scanner above the hive. In the past, researchers marked bees with paint or tags, but the monitoring of activity required the presence of a human observer.

terns define the thickness of a single module of each type. They are not part of the identification number. The manufacturer's number and the product number are separated by a center bar pattern consisting of the following five modules: a light space, a (long) dark bar, a light space, a (long) dark bar, and a light space (see Figure 9.11). The center bar pattern is not part of the identification number but merely serves to separate the manufacturer's number and product number. Figure 9.10 shows how the digits 6 and 0 in a manufacturer's number are coded.

Observe the following pattern in Figure 9.10: a light space of one module thickness, a dark bar of one module thickness, a light space of one module thickness, a dark bar of four module thickness. Symbolically such a pattern of light spaces and dark bars is represented as 0101111. Here each 0 means a one-module-thickness light space and each 1 means a one-module-thickness dark bar. Figure 9.12 illustrates how the thicknesses of the spaces and bars are translated into a binary code.

Table 9.1 shows the binary code for all digits. Notice that the code for the digits in the product number (the block of five digits on the right side) can be obtained from the code for the digits in the manufacturer's number (the block of digits on the left side), and vice versa, by replacing each 0 by a 1 and each 1 by a 0. Thus the code 0111011 for 7 in a manufacturer's number becomes 1000100 in the product number. Also notice that each manufacturer's number has an odd number of 1s, whereas each product number has an even number of 1s. This permits a computer linked with an optical scanner to determine

FIGURE 9.12 Translation of bars and space modules into binary code (see top of bars). The guard pattern defines a single module thickness for a bar and a space.

TABLE 9.1	**Binary UPC Coding**	
Digit	Manufacturer's Number	Product Number
0	0001101	1110010
1	0011001	1100110
2	0010011	1101100
3	0111101	1000010
4	0100011	1011100
5	0110001	1001110
6	0101111	1010000
7	0111011	1000100
8	0110111	1001000
9	0001011	1110100

whether the bar code was scanned left to right or right to left. (If the first block of digits has an even number of 1s for each digit, the scanning is being done right to left.) Thus scanning can be done in either direction without ambiguity.

Optional ▶ Encoding Personal Data

Consider this social security number: 189-31-9431. What information about the holder can be deduced from the number? Only that the holder obtained it in Pennsylvania (see Spotlight 9.4). Figure 9.13 shows an Illinois driver's license number: I225-1637-2133. What information about the holder can be deduced from this number? This time we can determine the date of birth, sex, and much about the person's name.

These two examples illustrate the extremes in coding personal data. The social security number has no personal data encoded in the number. It is entirely determined by the place and time it is issued, not the individual to

FIGURE 9.13   Illinois driver's license.

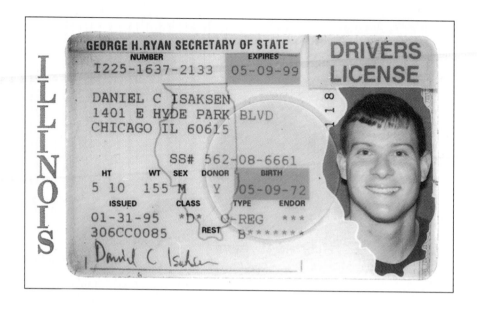

whom it is assigned. In contrast, in some states the driver's license numbers are entirely determined by personal information about the holders (see Spotlight 9.5). It is no coincidence that the unsophisticated social security numbering scheme predates computers. Agencies that have large data bases that include personal information such as names, sex, and dates of birth find it convenient to encode these data into identification numbers. Examples of such agencies are the National Archives (where census records are kept), genealogical research centers, the Library of Congress, and state motor vehicle departments.

There are many methods in use to encode personal data such as name, sex, and date of birth. Perhaps the most widely used application of these methods is to assign driver's license numbers in some states. Coding license numbers solely from personal data enables automobile insurers, government entities, and law enforcement agencies to determine the number from the personal data.

Many states encode the surname, first name, middle initial, date of birth, and sex by quite sophisticated schemes (see Figure 9.13).

In one scheme the first four characters of the license number are obtained by applying the **Soundex Coding System** to the surname as follows:

1. Delete all occurrences of h and w. (For example, Schworer becomes Scorer and Hughgill becomes uggill.)

2. Assign numbers to the remaining letters as follows:

   a, e, i, o, u, y → 0
   b, f, p, v → 1   l → 4
   c, g, j, k, q, s, x, z → 2   m, n → 5
   d, t → 3   r → 6

Spotlight 9.4

Social Security Numbers

The first three digits of social security numbers show where the number was applied for.

Changes in population have forced some numbers to be moved or assigned out of sequence over the years.

001–003	New Hampshire	387–399	Wisconsin	526–527 & 600–601	Arizona
004–007	Maine	400–407	Kentucky		
008–009	Vermont	408–415	Tennessee	528–529	Utah
010–034	Massachusetts	416–424	Alabama	530	Nevada
035–039	Rhode Island	425–428 & 587–588	Mississippi	531–539	Washington
040–049	Connecticut			540–544	Oregon
050–134	New York	429–432	Arkansas	545–573 & 602–626	California
135–158	New Jersey	433–439	Louisiana		
159–211	Pennsylvania	440–448	Oklahoma	574	Alaska
212–220	Maryland	449–467	Texas	575–576	Hawaii
221–222	Delaware	468–477	Minnesota	577–579	District of Columbia
223–231	Virginia	478–485	Iowa	580	Virgin Islands
232–236	West Virginia	486–500	Missouri	580–584 & 596–599	Puerto Rico
232, 237–246	North Carolina	501–502	North Dakota		
247–251	South Carolina	503–504	South Dakota	586	Guam
252–260	Georgia	505–508	Nebraska	586	American Samoa
261–267 & 589–595	Florida	509–515	Kansas	586	Philippines
		516–517	Montana	700–728	*through July 1, 1963, reserved for railroad employees*
268–302	Ohio	518–519	Idaho		
303–317	Indiana	520	Wyoming		
318–361	Illinois	521–524	Colorado		
362–386	Michigan	525 & 585	New Mexico		

Source: Social Security Administration.

3. If two or more letters with the same numerical value are adjacent, omit all but the first. (For example, Scorer becomes Sorer and uggill becomes ugil.)

4. Delete the first character of the original name if still present. (Sorer becomes orer.)

5. Delete all occurrences of a, e, i, o, u, and y.

6. Retain only the first three digits corresponding to the remaining letters; append trailing zeros if fewer than three letters remain; precede the three digits by the first letter of the surname.

Encoding New York Driver's License Numbers

Prior to September 1992, the state of New York utilized a complex algorithm based on an individual's name, and day, month, and year of birth to assign driver's license numbers. The first character of the number is the first character of the last name (J for Jones, S for Smith). The next 12 digits are determined by converting the second, third, fourth, and fifth letters of the last name, the first three of the first name, and the middle initial to the corresponding position on the alphabet (A converts to 1, B converts to 2, . . . , Z to 26, a blank to 0). Let us denote these alphabetic positions by $n_1, n_2, \ldots, n_8$. That is, n_1 is the alphabet position of the second letter of the last name, n_2 is the alphabet position of the third letter of the last name, and so on.

If the last name has 4 or more characters (so that $n_3 > 0$), the formula for the first 12 digits of the license number is

$$10{,}017{,}758{,}323 \cdot n_1 + 371{,}538{,}441 \cdot n_2$$
$$+ \ 13{,}779{,}585 \cdot n_3 + 510{,}355 \cdot n_4$$
$$+ \ 19{,}657 \cdot n_5 + 729 \cdot n_6 + 27 \cdot n_7$$
$$+ \ n_8 - 385{,}829{,}132$$

(When the resulting number has fewer than 12 digits, we fill in with 0s on the left.)

Applying this formula to Alvy J. Singer, we have

i	n	g	e	a	l	v	j
n_1	n_2	n_3	n_4	n_5	n_6	n_7	n_8
9	14	7	5	1	12	22	10

So the formula gives

$$10{,}017{,}758{,}323 \cdot 9 + 371{,}538{,}441 \cdot 14$$
$$+ \ 13{,}779{,}585 \cdot 7 + 510{,}355 \cdot 5$$
$$+ \ 19{,}657 \cdot 1 + 729 \cdot 12 + 27 \cdot 22$$

$$+ \ 10 - 385{,}829{,}132$$
$$= 095074571828$$

If the last name has 3 characters (for example, Lee), the term $-385{,}829{,}132$ must be replaced by $-385{,}318{,}778$. But if it has exactly 2 characters (for example, Ho), then $-385{,}829{,}132$ must be replaced by $-371{,}539{,}194$. Here is the formula for Julia Ho (no middle name):

o	—	—	—	j	u	l	—
n_1	n_2	n_3	n_4	n_5	n_6	n_7	n_8
15	0	0	0	10	21	12	0

$$10{,}017{,}758{,}323 \cdot 15 + 371{,}538{,}441 \cdot 0$$
$$+ \ 13{,}779{,}585 \cdot 0 + 510{,}355 \cdot 0$$
$$+ \ 19{,}657 \cdot 10 + 729 \cdot 21$$
$$+ \ 27 \cdot 12 + 0 - 371{,}539{,}194$$
$$= 1498950047854$$

In the cases where the first name consists of exactly one letter (for example, D. Denise Aiken, J. Thomas Bucker), we must add 26 to the number produced by the formula.

The next three digits of the driver's license number are determined by the day and month of birth, and the sex. For a male born on day d and month m we append $63m + 2d$ to the 12 digits computed from the name (July 13 gives $63 \cdot 7 + 2 \cdot 13$). For a female we append $63m + 2d + 1$. The last two digits of the number represent the year of birth (42 for 1942, 63 for 1963). So, for Alvy J. Singer, born on January 13, 1935, we have S09507457182808935. For Julia Ho, born on October 2, 1950, we have H1498950478563550. In cases where this algorithm yields identical numbers for more than one person, a "tie breaking" digit is inserted before the two digits representing the year of birth.

Figure 9.14 shows three examples.

What is the advantage of this method? It is an error-correcting scheme. Indeed, it is designed so that likely misspellings of a name nevertheless result in the correct coding of the name. For example, frequent misspellings of the name Erickson are Ericksen, Eriksen, Ericson, and Ericsen. Observe that all of these yield the same coding as Erickson. If a law enforcement official, a genealogical researcher, a librarian, or an airline reservation agent wanted to pull up the file from a data bank for someone whose name was pronounced "Erickson," the correct spelling isn't essential because the computer searches for records that are coded as E-625 for all spelling variations. (The Soundex system was designed for the U.S. Census Bureau when much census data were obtained orally; see Spotlight 9.6. Airlines use a somewhat different system called the *Davidson Consonant Code*.)

There are many schemes for encoding the date of birth and the sex in driver's license numbers. For example, the last five digits of Illinois and Florida driver's license numbers capture the year and date of birth as well as the sex. In Illinois, each day of the year is assigned a three-digit number in sequence beginning with 001 for January 1. However, each month is assumed to have 31 days. Thus, March 1 is given the number 063 since both January and February are assumed to have 31 days. These numbers are then used to identify the

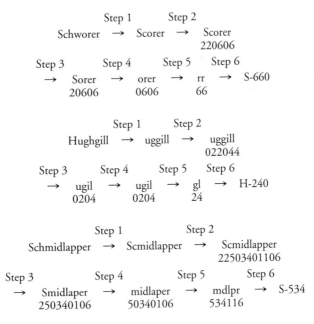

FIGURE 9.14 The Soundex Coding System.

Spotlight

Census Records at the National Archives

9.6

One of the best places to look for information pertaining to family history is the old censuses that are kept by the National Archives in Washington, D.C. By law, census records are open to the public 72 years after the census was taken. The 1920 census material was available to researchers beginning in 1992. The data from the 1880, 1900, 1910, and 1920 censuses (records from 1890 were destroyed by fire) were put on cards during the 1930s as a Works Progress Administration project. This information was coded using the Soundex system so that names that sound alike regardless of how they are spelled are grouped together. On old documents family names were so often misspelled—especially those that were not of British origin—that genealogists say several variations of a name may apply to one set of ancestors.

To look for a surname on the index, the researchers must work out the Soundex code. This code together with the state identifies a page

number on microfilm where the data are located. A typical census Soundex card is shown here. Note the Soundex code in the upper left-hand corner (B350).

B350					OHIO	
Bitton, George H.				VOL. _36_	E.D. _176_	
(HEAD OF FAMILY)				SHEET _8_	LINE _17_	
w _Feb._	_1853_	_47_	_England_		N.R.	
(COLOR) (MONTH)	(YEAR)	(AGE)	(BIRTHPLACE)		(CITIZENSHIP)	
Cuyahoga			Cleveland Twp.			
(COUNTY)			(N.C.D.)			
Cleveland			Howard		42	
(CITY)			(STREET)		(HOUSE NUMBER)	

OTHER MEMBERS OF FAMILY

NAME	RELATION-SHIP	BIRTH MONTH	BIRTH YEAR	AGE	BIRTH-PLACE	CITIZEN-SHIP
Bitton, P. Elizabeth	W	Aug.	1864	36	Ohio	
O. Adelaide	D	Nov.	1893	6	Ohio	

1900 CENSUS—INDEX
DEPARTMENT OF COMMERCE   U.S. GOVERNMENT PRINTING OFFICE
BUREAU OF THE CENSUS

month and day of birth of male drivers. For females, the scheme is identical except 600 is added to the number. The last two digits of the year of birth, separated by a dash (probably to obscure that fact that they represent the year of birth), are listed in the fifth and fourth positions from the end of the driver's license number. Thus, a male born on October 13, 1940, would have the last five digits 4-0292 ($292 = 9 \cdot 31 + 13$), whereas a female born on the same day would have 4-0892. The scheme to identify birth date and sex in Florida is the same as in Illinois except each month is assumed to have 40 days and 500 is added for women. For example, the five digits 4-9585 belong to a woman born on March 5, 1949. ◆

REVIEW VOCABULARY

Bar code   A code that employs bars and spaces to represent information.

Binary code   A coding scheme that uses two symbols, usually 0 and 1.

Codabar   An error-detection method used by all major credit card companies, many libraries, blood banks, and others.

Code   A group of symbols that represent information together with a set of rules for interpreting the symbols.

Code 39   An alphanumeric code that is widely used on nonretail items.

Decoding   Translating code into data.

Encoding   Translating data into code.

Error-detecting code   A code in which certain types of errors can be detected.

International Standard Book Number (ISBN)   A 10-digit identification number used on books throughout the world that contains a check digit for error detection.

Postnet code   The bar code used by the U.S. Postal Service for ZIP codes.

Soundex Coding System   An encoding scheme for surnames based on sound.

Universal Product Code (UPC)   A bar code and identification number that are used on most retail items. The UPC code detects 100% of all single-digit errors and most other types of errors.

Weights   Numbers used in the calculation of check digits.

ZIP code   A five-digit code used by the U.S. Postal Service to divide the country into geographical units to speed sorting of the mail.

ZIP + 4 code   The nine-digit code used by the U.S. Postal Service to refine ZIP codes into smaller units.

SUGGESTED READINGS

COLLINS, D. J., AND N. WHIPPLE.   *Using Bar Code,* Data Capture Institute, Duxbury, Mass., 1990. Contains extensive information on bar codes.

DAVIDSON, L.   Retrieval of misspelled names in an airline passenger record system, *Communications of the Association for Computing Machinery,* 5 (1962): 169–171. Describes the method used by airlines (the Davidson Consonant Code) to store and retrieve passenger names.

GALLIAN, J.   Assigning driver's license numbers, *Mathematics Magazine,* 64 (1992): 13–22. Discusses various methods used by the states to assign driver's license numbers. Several of these methods include check digits for error detection.

GALLIAN, J.   The mathematics of identification numbers, *College Mathematics Journal,* 22 (1991): 194–202. A comprehensive survey of check digit schemes that are associated with identification numbers.

GALLIAN, J., AND S. WINTERS.   Modular arithmetic in the marketplace, *American Mathematical Monthly,* 95 (1988): 548–551. A detailed analysis of the check digit schemes presented in this chapter. In particular, the error-detection rates for the various schemes are given.

HARMAN, C. K., AND R. ADAMS.   *Reading Between the Lines,* Helmers, Peterborough, N.H., 1989. Contains extensive information on bar codes.

PHILIPS, LAWRENCE.   Hanging on the Metaphone, *Computer Language,* 7 (December 1990): 39–43. Describes a sound-based retrieval algorithm that in some respects is superior to Soundex.

ROUGHTON, KAREN, AND DAVID A. TYCKOSEN.   Browsing with sound: Sound-based codes and automated authority control, *Information Technology and Libraries,* 4 (June 1985): 130–136. Explains the Soundex system and the Davidson Consonant Code and their uses. The Davidson Consonant Code is a sound-based retrieval algorithm used by airline passengers retrieval systems.

WAGNER, N., AND P. PUTTER.   Error detecting decimal digits, *Communications of the Association for Computing Machinery,* 32 (1989): 106–110. Describes the experience of two mathematicians hired by a large mail-order company to make recommendations for an error-correction scheme for the company's account numbers. They recommended a four-digit method.

EXERCISES   ▲ *Optional.*   ■ *Advanced.*   ◆ *Discussion.*

Postnet Codes

1.   Determine the ZIP + 4 code and check digit for each of the following Postnet bar codes.:

(a)   ‖⎮⎮⎕⎮⎕⎮‖⎕‖⎕‖⎕‖⎕⎮⎮⎕‖⎮⎕‖‖⎕⎮⎮‖⎮⎕⎮⎕‖⎕‖

(b)   ‖⎮‖⎕‖⎮⎮⎕⎮‖⎕⎮⎕‖‖⎕⎮‖⎮⎕⎮⎕‖⎕‖⎕‖⎕‖⎮‖‖

(c)   ‖⎕⎮⎕‖⎕⎮⎕⎮⎕⎮‖⎮⎕‖‖⎕⎮‖⎕⎮⎕‖⎕⎮⎕‖⎕⎮‖⎕⎕⎕⎕‖‖⎕‖‖‖

2.   Determine the ZIP + 4 code and check digit for each of the following Postnet bar codes:

(a) |₁₁₁||₁|₁|₁||₁₁₁||₁₁₁||₁₁₁|₁||₁₁₁||₁||₁||₁₁₁₁₁||₁||

(b) |₁||₁₁||₁₁₁||₁₁₁||₁₁₁||₁₁||₁₁||₁₁||₁₁₁||||₁₁||₁₁₁||₁||₁|

(c) |₁₁||₁₁₁|₁||₁₁||₁||₁₁||₁₁₁₁|||||₁₁|₁₁|₁|₁₁₁||₁₁₁||||₁||||

3.   In each Postnet bar code below, exactly one mistake occurs (that is, a long bar appears instead of a short one, or vice versa). Determine the correct ZIP code.

(a) |₁₁|₁|||₁₁₁|₁₁||₁₁|₁₁₁|₁||||₁₁₁|₁|₁₁₁||₁||₁₁₁|₁₁||₁||

(b) |₁|₁|₁||₁|₁|₁₁|₁₁||₁||₁₁|₁||₁|₁|₁|₁|₁|₁|₁||₁₁|₁|||||₁₁|

(c) |₁|₁₁₁|₁||||₁₁₁|₁₁||₁₁|₁|₁₁₁₁|₁₁₁|₁₁₁|||||₁₁₁₁₁₁||₁₁₁₁||

4.   Below is a 12-digit delivery-point bar code. Determine the ZIP + 4 number, the last two digits of the street address, and the check digit.

|₁|₁|₁₁|₁|₁|₁₁₁|₁₁₁₁₁||₁₁₁||₁₁|₁|₁||₁₁₁|₁₁|₁|₁|₁|₁|₁|₁₁|₁₁||₁₁₁|||

5.   Explain why any two errors in a particular block of five bars in a Postnet are always detectable. Explain why not all such errors can be corrected.

Identification Numbers

6.   Determine the check digit for a money order with identification number 7234541780.

7.   Determine the check digit for a money order with identification number 395398164.

8.   Suppose a money order with the identification number and check digit 21720421168 is erroneously copied as 27750421168. Will the check digit detect the error? Explain your reasoning.

9.   Determine the check digit for the United Parcel Service (UPS) identification number 873345672.

10.   Determine the check digit for the Avis rental car with identification number 540047.

11.   Determine the check digit for the airline ticket number 30860422052.

12.   Determine the check digit for the UPC number 05074311502.

13.   Determine the check digit for the UPC number 38137009213.

14.   Determine the check digit for the ISBN 0-669-33907.

15.   Determine the check digit for the ISBN 0-669-19493.

16.   Determine the check digit for the bank number 09100001.

17.   Determine the check digit for the bank number 09190204.

18.   Determine if the Master Card number 3541 0232 0033 2270 is valid.

19.   Use the Codabar scheme to determine the check digit for the number 300125600196431.

20.   Determine the check character for the Code 39 number 210SA0162305ZA. (Assume the code uses only 36 characters.)

21.   Determine the check character for the number 3050-0000 HEAD using the 43-character Code 39 scheme. (Be sure to include the hyphen and space as characters.)

22.   Change 173 into Postnet code.

23.   Is there any mathematical reason for a check digit to be at the end of an identification number? Explain your reasoning.

24.   Suppose the first block of a UPC bar code following the guard bar pattern a scanner reads is 1000100. Is the scanner reading left to right or right to left?

25.   For some products, such as soft-drink cans and magazines, an 8-digit UPC number called Version E is used instead of the 12-digit number. The method of calculating the eighth digit, which is the check digit, depends on the value of the seventh digit. Use the fact that the check digit a_8 for a UPC Version E identification number $a_1 a_2 a_3 a_4 a_5 a_6 a_7$, where a_7 is 0, 1, or 2, is chosen so that $a_1 + a_2 + 3a_3 + 3a_4 + a_5 + 3a_6 + a_7 + a_8$ is divisible by 10 to determine the check digit for the following Version E numbers:

(a)   0121690
(b)   0274551
(c)   0760022
(d)   0496580

26.   Use the fact that the check digit a_8 for a UPC Version E identification number $a_1 a_2 a_3 a_4 a_5 a_6 a_7$, where a_7 is 4, is chosen so that $a_1 + a_2 +$

$3a_3 + a_4 + 3a_5 + 3a_6 + a_8$ is divisible by 10 to determine the check digit for the following Version E numbers:

(a) 0754704
(b) 0774714
(c) 0724444

27. The ISBN 0-669-03925-4 is the result of a transposition of two adjacent digits not involving the first or last digit. Determine the correct ISBN.

28. Explain why the bank scheme will detect the error $751 \cdots \rightarrow 157 \cdots$, but the UPC scheme will not.

29. Suppose the check digit a_9 for bank checks were chosen to be the last digit of $3a_1 + 7a_2 + a_3 + 3a_4 + 7a_5 + a_6 + 3a_7 + 7a_8$ instead of the way described in the chapter. How would this compare with the actual check digit?

Encoding Personal Data (Optional)

30. Determine the Soundex code for Smith, Schmid, Smyth, and Schmidt.

31. Determine the Soundex code for Skow, Sachs, Lennon, Lloyd, Ehrheart, and Ollenburger.

32. Determine the last five digits of an Illinois driver's license number for a male born on July 18, 1942.

33. In Florida the last three digits of the driver's license number of a female with birth month m and birth date b are $40(m-1) + b + 500$. For both males and females, the fourth and fifth digits from the end give the year of birth. Determine the last five digits of a Florida driver's license number for a female born on July 18, 1942.

Additional Exercises

34. In Florida the last three digits of the driver's license number of a male with birth month m and birth date b are $40(m-1) + b$. For both males and females, the fourth and fifth digits from the end give the year of birth. Determine the dates of birth of people with the numbers whose last five digits are 42218 and 53953.

35. For driver's license numbers issued in New York prior to September 1992, the last two digits were the year of birth. The three digits preceding the year encoded the sex and the month and day of birth. For a woman with birth

month m and birth date b the three digits were $63m + 2b + 1$. For a man with birth month m and birth date b the three digits were $63m + 2b$. Determine the birth months, birth dates, and sexes of drivers with the three digits 248 and 601 preceding the year.

36. The state of Utah appends a ninth digit a_9 to an eight-digit driver's license number $a_1 a_2 \cdots a_8$ so that $9a_1 + 8a_2 + 7a_3 + 6a_4 + 5a_5 + 4a_6 + 3a_7 + 2a_8 + a_9$ is divisible by 10.

   (a) If the first eight digits of a Utah license number are 14910573, what is the ninth digit?

   (b) Suppose a legitimate Utah license number 149105767 is miscopied as 149105267. How would you know a mistake was made? Is there any way you could determine the correct number? Suppose you know the error was in the seventh position, could you correct the mistake?

   (c) If a legitimate Utah number 149105767 were miscopied as 199105767, would you be able to tell a mistake was made? Explain.

   (d) Explain why any transposition error involving adjacent digits of a Utah number would be detected.

37. Form all possible strings consisting of exactly three a's and two b's and arrange the strings in alphabetical order (for example, the first two possibilities are *aaabb* and *aabab*). Do you see any relationship between your list and the Postnet code?

38. Suppose the check digit a_{10} of ISBN numbers were chosen so that $a_1 + 2a_2 + 3a_3 + 4a_4 + 5a_5 + 6a_6 + 7a_7 + 8a_8 + 9a_9 + 10a_{10}$ is divisible by 11 instead of the way described in the chapter. How would this compare with the actual check digit?

39. The Canadian province of Quebec assigns a check digit a_{12} to an 11-digit driver's license number $a_1 a_2 \cdots a_{11}$ so that $12a_1 + 11a_2 + 10a_3 + 9a_4 + 8a_5 + 7a_6 + 6a_7 + 5a_8 + 4a_9 + 3a_{10} + 2a_{11} + a_{12}$ is divisible by 10. Criticize this method. Describe all single-digit errors that are undetected by this scheme. How does the transposition of two adjacent digits of a number affect the check digit of a number?

40. Examine the bar code on the back of six recently published books that begin with the digit 9. How does this bar code format differ from the UPC code used on grocery items? How is the UPC number related to the ISBN? How is the last digit of the UPC number (the check digit) determined?

41. Determine the New York State driver's license number for Leonard Zelig (no middle name) born on February 5, 1940, using the formula in Spotlight 9.5.

Building Regulations: 1985 Class 0
FINE ART WALLCOVERINGS LTD.
HOLMES CHAPEL, CHESHIRE

MADE IN ENGLAND
FABRIQUE EN ANGLETERRE

5 011419 194056

42. At the left is an actual identification number and bar code from a roll of wallpaper. What appears to be wrong with them? Speculate on the reason for the apparent violation of the UPC format.

43. The state of Washington encodes the last two digits of the year of birth into driver's license numbers (in positions 8 and 9) by subtracting the two-digit number from 100. For example, a person born in 1942 has 58 in positions 8 and 9, whereas a person born in 1971 has 29 in positions 8 and 9. Speculate on the reason for subtracting the birth year from 100.

44. Driver's license number assignment schemes that utilize personal data occasionally produce the same number for different people. Speculate about circumstances under which this is more likely to occur.

45. Consider a UPC number in which the digits 7 and 2 appear consecutively (that is, the number has the form $\cdots 72 \cdots$). Will the error caused by transposing these digits (that is, the number is taken as $\cdots 27 \cdots$) be detected? What if the digits 6 and 2 are transposed instead? State the general criterion for the detection of an error of the form

$$\cdots ab \cdots \longrightarrow \cdots ba \cdots$$

using the UPC scheme.

46. Apply the Soundex code to common ways to misspell your name. Do they give the same code as your name does?

47. The Canadian postal system has assigned each geographical region a six-character code composed of alternating letters and digits, such as P7B5E1 and K7L3N6. Discuss the advantages this scheme has over the five-digit ZIP code used in the United States.

WRITING PROJECTS

1 ▶ Prepare a report on coded information in your location. Possibilities for investigation include driver's license numbers in your state; student ID numbers and bar codes at your school; bar codes used by your school library and city library. Identify the coding schemes and, when possible, determine

whether a check digit is employed. Include samples. The Suggested Readings for this chapter contain information that will assist you.

2 ▶   Prepare a report on the driver's license coding schemes used by Minnesota, Michigan, Maryland, and Washington (the first three states use the same method). J. Gallian's "Assigning Driver's License Numbers" (see Suggested Readings) has the information you will need.

3 ▶   Imagine that you are employed by a small company that doesn't use identification numbers and bar codes for its employees or products. As requested by your boss, prepare a report discussing the various methods and make a recommendation.

4 ▶   The Davidson Consonant Code and Metaphone are two text-retrieval algorithms based on sound that are alternatives to Soundex. Prepare a report on either of these encoding schemes. The article by Roughton and Tyckosen and the one by Philips (see Suggested Readings) contain the information you will need.

TRANSMITTING INFORMATION

Data stored in computers may be modified by naturally occurring radiation; information transmitted from communication satellites and space probes is subject to a variety of electromagnetic interference; compact discs are corrupted by dust, dirt, scratches, and fingerprints; magnetic tapes deteriorate; and entry of data into computers by humans is subject to frequent error. In this chapter we illustrate a mathematical method for correcting errors. We also illustrate a way data can be coded to reduce transmission time and storage space, and ways to securely transmit secret messages.

BINARY CODES

Because of the way computers are built, in high-tech applications such as compact disc players, fax machines, high-definition television, modems, and signals sent back from space probes, data are represented as strings of 0s and 1s rather than the usual digits 0 through 9 and letters *A* through *Z*. Recall from Chapter 9 that a system for coding data with 0s and 1s is called a *binary code.* There are many binary codes in use. In this section we will illustrate one way binary codes can be devised so that errors can be corrected.

The idea behind error-correction schemes is simple and one you often use. To illustrate, suppose you are reading the employment section of a newspaper and you see the phrase "must have a minimum of bive years experience." Instantly you detect an error since "bive" is not a word in the English language. Moreover, you are fairly confident that the intended word is "five." Why so?

$S potlight$ The Ubiquitous Reed–Solomon Codes

10.1

One of the mathematical ideas underlying current error-correcting techniques for everything from computer hard disk drives to CD players was first introduced in 1960 by Irving Reed and Gustave Solomon. Reed–Solomon codes made possible the stunning pictures of the outer planets sent back by the space probes *Voyager 1* and 2. They make it possible to scratch a compact disc and still enjoy the music.

"When you talk about CD players and digital audio tape and now digital television, and various other digital imaging systems that are coming—all of those need Reed–Solomon [codes] as an integral part of the system," says Robert McEliece, a coding theorist at Caltech.

Why? Because digital information consists of 0s and 1s and a physical device that may occasionally confuse the two. *Voyager 2*, for example, was transmitting data at incredibly low power—barely a whisper—over billions of miles. Error-correcting codes are a kind of safety net, mathematical insurance against the vagaries of an imperfect material world.

In 1960, the theory of error-correcting codes was only about a decade old. Through the 1950s, a number of researchers began experimenting with a variety of error-correcting codes. But the Reed–Solomon paper, McEliece says, "hit the jackpot." "In hindsight it seems obvious," Reed recently said. However, he added, "coding theory was not a sub-

Irving Reed and Gustave Solomon
at the Jet Propulsion Laboratory in 1989 monitor the
encounter of *Voyager 2* with Neptune.

ject when we published the paper." The two authors knew they had a nice result; they didn't know what impact the paper would have.

Three decades later, the impact is clear. The vast array of applications, both current and pending, has settled the questions of the practicality and significance of Reed–Solomon codes. Billions of dollars in modern technology depend on ideas that stem from Reed and Solomon's original work.

Source: Adapted from an article by Barry Cipra with permission from *SIAM News*, January 1993, p. 1.

Because "five" is a word and it makes the phrase sensible. In other phrases, words such as "bike" or "give" might be sensible alternatives to "bive." Using the extra information provided by the context, we are often able to infer the intended meaning when errors occur.

Over the past 40 years mathematicians and engineers have devised highly sophisticated schemes to build extra information in messages composed of 0s and 1s that often permits one to infer the correct message even though the message may have been received incorrectly (see Spotlight 10.1). As a simple example, let's say that our message is 1001. We will build extra information into this message with the aid of the diagram in Figure 10.1. Begin by placing the four message digits in the four overlapping regions I, II, III, IV, with the digit in position 1 (starting at the left of the sequence) in region I, the digit in position 2 in region II, and so on. For regions V, VI, and VII, assign 0 or 1 so that the total number of 1s in each circle is even. See Figure 10.2.

We have now encoded our message 1001 using the diagram as 1001101. Now suppose that this encoded message is received as 0001101 (an error in the first position). How would we know an error was made? We place each digit from the received message in its appropriate region as in Figure 10.3.

Noting that in both circles A and B there is an odd number of 1s, we instantly realize that something is wrong, since the intended message had an even number of 1s in each circle. How do we correct the error? Since circles A and B have the wrong parity (parity refers to the oddness or evenness of a number; even integers have **even parity;** odd integers have **odd parity**) and C does not, the error is located in the portion of the diagram in circles A and B, but not in circle C; that is, region I (see Figure 10.4). Here we also see the advantage of

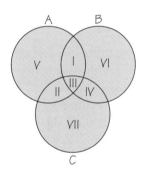

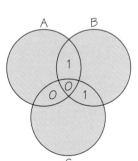

FIGURE 10.1 Diagram for message 1001.

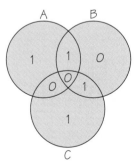

FIGURE 10.2 Diagram for encoded message 1001101.

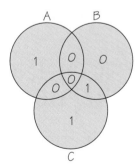

FIGURE 10.3 Diagram of received message 0001101.

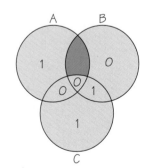

FIGURE 10.4 Circles A and B but not C have wrong parity.

TABLE 10.1	
Message	Code Word
0000	→ 0000000
0001	→ 0001011
0010	→ 0010111
0100	→ 0100101
1000	→ 1000110
1100	→ 1100011
1010	→ 1010001
1001	→ 1001101
0110	→ 0110010
0101	→ 0101110
0011	→ 0011100
1110	→ 1110100
1101	→ 1101000
1011	→ 1011010
0111	→ 0111001
1111	→ 1111111

using only 0s and 1s to encode data. If you have only two possibilities and one of them is incorrect, then the other one must be correct. Since the 0 in region I is incorrect, we know 1 is correct. This technique can be used to encode all 16 possible binary messages of length 4 as shown in the right column of Table 10.1. The encoded messages are called *code words*. The extra three digits appended to each string of length 4 provides the "extra information" that is sufficient to infer the intended four-digit message as long as the received seven-digit message has at most one error. If a received message has two or more errors, this method will not always yield the correct message.

ENCODING WITH PARITY-CHECK SUMS

In practice, binary messages consist of strings longer than four digits, and diagrams are too cumbersome for encoding them and decoding them. Rather, the messages are encoded by appending extra digits determined by the parity of various sums of certain portions of the messages. We illustrate this method for the 16 messages shown in the left column of Table 10.1. (See also Spotlight 10.2.)

Spotlight Neil Sloane

10.2

Neil Sloane
at work, wearing his famous "Codemart"
T-shirt (952 points in a sphere).

In the middle of Neil Sloane's office, which is in the center of AT&T Bell Laboratories, which in turn is at the heart of the Information Age, there sits a tidy little pyramid of shiny steel balls stacked up like oranges at a neighborhood grocery. Sloane has been pondering different ways to pile up balls of one kind or another for most of his professional life. Along the way he has become one of the world's leading researchers in the field of sphere packing, a field that has become indispensable to modern communications. Without it we might not have modems or compact discs or satellite photos of Neptune. "Computers would still exist," says Sloane. "But they wouldn't be able to talk to one another."

To exchange information rapidly and correctly, machines must code it. As it turns out, designing a code is a lot like packing spheres: both involve cramming things together into the tightest possible arrangement. Sloane, fittingly, is also one of the world's leading coding theorists, not least because he has studied the shiny steel balls on his desk so intently.

Here's how a code might work. Imagine, for example, that you want to transmit a child's drawing that used every one of the 64 colors found in a jumbo box of Crayola crayons. For transmission, you could code each of those colors as a number—say, the integers from 1 to 64. Then you could divide the image into many small units, or pixels, and assign a code to each one based on the color it contains. The transmission would then be a steady stream of those numbers, one for each pixel.

In digital systems, however, all those numbers would have to be represented as strings of 0s and 1s. Because there are 64 possible combinations of 0s and 1s in a six-digit string, you could handle the entire Crayola palette with 64 different six-digit "code words." For example, 000000 could represent the first color, 000001 the next color, 000010 the next, and so on.

But in a noisy signal two different code words might look practically the same. A bit of noise, for example, might shift a spike of current to the wrong place, so that 001000 looks like 000100. The receiver might then wrongly color someone's eyes. An efficient way to keep the colors straight in spite of noise is to add four extra digits to the six-digit code words. The receiver, programmed to know the 64 permissible combinations, could now spot any other combination as an error introduced by noise and it would automatically correct the error to the "nearest" permissible color.

In fact, says Sloane, "If any of those ten digits were wrong, you could still figure out what the right crayon was."

Source: Adapted from an article by David Berreby, *Discover,* October 1990.

Our goal is to take any binary string $a_1a_2a_3a_4$ and append three check digits $c_1c_2c_3$ so that any single error in any of the seven positions can be corrected. This is done as follows: choose

$$c_1 = 0 \text{ if } a_1 + a_2 + a_3 \text{ is even}$$
$$c_1 = 1 \text{ if } a_1 + a_2 + a_3 \text{ is odd}$$
$$c_2 = 0 \text{ if } a_1 + a_3 + a_4 \text{ is even}$$
$$c_2 = 1 \text{ if } a_1 + a_3 + a_4 \text{ is odd}$$
$$c_3 = 0 \text{ if } a_2 + a_3 + a_4 \text{ is even}$$
$$c_3 = 1 \text{ if } a_2 + a_3 + a_4 \text{ is odd}$$

The sums $a_1 + a_2 + a_3$, $a_1 + a_3 + a_4$, and $a_2 + a_3 + a_4$ are called **parity-check sums.** They are so named because their function is to guarantee that the sum of various components of the encoded message is even. Indeed, c_1 is defined so that $a_1 + a_2 + a_3 + c_1$ is even. (Recall that this is precisely how the value in region V was defined.) Similarly, c_2 is defined so that $a_1 + a_2 + a_4 + c_2$ is even, and c_3 is defined so that $a_2 + a_3 + a_4 + c_3$ is even.

Let us revisit the message 1001 we considered in Figure 10.1. Then $a_1a_2a_3a_4 = 1001$ and

$$c_1 = 1 \text{ since } 1 + 0 + 0 \text{ is odd}$$
$$c_2 = 0 \text{ since } 1 + 0 + 1 \text{ is even}$$

and

$$c_3 = 1 \text{ since } 0 + 0 + 1 \text{ is odd}$$

So, because $c_1c_2c_3 = 101$, we have $1001 \rightarrow 1001101$.

Now how is the intended message determined from a received encoded message? This process is called **decoding.** Say, for instance, that the message 1000, which has been encoded using parity check sums as $u = 1000110$, is received as $v = 1010110$ (an error in the third position). We simply compare v with each of the 16 code words (that is, the possible correct messages) in Table 10.1 and decode it as the one that differs from v in the fewest positions. (Put another way, we decode v as the code word that agrees with v in the most positions). In the situation that there is more than one code word that differs from v in the fewest positions, we do not decode. To carry out this comparison it is convenient to define the distance between two strings of equal length.

The **distance between two strings** of equal length is the number of positions in which the strings differ.

TABLE 10.2

v	1010110	1010110	1010110	1010110	1010110	1010110	1010110	1010110
code word	0000000	0001011	0010111	0100101	1000110	1100011	1010001	1001101
distance	4	5	2	5	1	4	3	4
v	1010110	1010110	1010110	1010110	1010110	1010110	1010110	1010110
code word	0110010	0101110	0011100	1110100	1101000	1011010	0111001	1111111
distance	3	4	3	2	5	2	6	3

For example, the distance between $v = 1010110$ and $u = 1000110$ is 1, since they differ in only one position (the third). In contrast, the distance between 1000110 and 0111001 is 7, since they differ in all seven positions. Thus our decoding procedure is simply to decode any received message v as the code word v' that is "nearest" to v in the sense that among all distances between v and code words, the distance between v and v' is a minimum. (If there is more than one possibility for v', we do not decode.) Table 10.2 shows the distance between $v = 1010110$ and all 16 code words. From this table we see that v will be decoded as u, since it differs from u in only one position while it differs from all others in the table in at least two positions. This method is called nearest-neighbor decoding.

> The **nearest-neighbor decoding** method decodes a received message as the code word that agrees with the message in the most positions.

Assuming that errors occur independently, the nearest-neighbor method decodes each received message as the one it most likely represents.

The scheme we have just described was first proposed in 1948 by Richard Hamming, a mathematician at Bell Laboratories. (See Spotlight 10.3.) It is one of an infinite number of codes that are called the *Hamming codes*.

Strings of 0s and 1s obtained from all possible k-tuples of 0s and 1s by appending extra 0s and 1s using parity-check sums, as illustrated earlier, are called binary linear codes. The strings with the appended digits are called code words.

> A **binary linear code** consists of words composed of 0s and 1s obtained from all possible k-tuple messages by using parity-check sums to append check digits to the messages. The resulting strings are called **code words**.

Spotlight   *Richard Hamming*

10.3

Richard W. Hamming was born in Chicago, Illinois, on February 11, 1915. He graduated from the University of Chicago with a B.S. degree in mathematics. In 1939, he received an M.A. degree in mathematics from the University of Nebraska, and in 1942, a Ph.D. in mathematics from the University of Illinois.

During the latter part of World War II, Hamming was at Los Alamos, where he was involved in computing atomic bomb designs. In 1946, he joined Bell Telephone Laboratories, where he worked in mathematics, computing, engineering, and science.

When Hamming arrived at Bell Laboratories, the Model V computer there had over 9000 relays and over 50 pieces of teletype apparatus. It occupied about 1000 square feet of floor space and weighed some 10 tons. (In computing power it equaled some of today's hand-held calculators.) The input was entered in the machine via a punched paper tape, which had two holes per row. Each row was read as a unit. The sensing relays would prevent further computation if more or less than two holes appeared in a given row. Similar checks were used in nearly every step of a computation. If such a check failed when the operating personnel were not present, the problem had to be rerun. This inefficiency led Hamming to investigate the possibility of automatic error correction. Many years later he said to an interviewer:

Richard Hamming

Two weekends in a row I came in and found that all my stuff had been dumped and nothing was done. I was really aroused and annoyed because I wanted those answers and two weekends had been lost. And so I said, "Damn it, if the machine can detect an error, why can't it locate the position of the error and correct it?"

In 1950, Hamming published his famous paper on error-correcting codes, resulting in the use of Hamming codes in modern computers and a new branch of information theory.

Source: Adapted from T. Thompson, *From Error-Correcting Codes Through Sphere Packing to Simple Groups* (Washington, D.C.: Mathematical Association of America, 1983).

You should think of a binary linear code as a set of *n*-tuples where each *n*-tuple is composed of two parts: the message part, consisting of the original *k*-digit messages, and the remaining check digit part.

The longer the messages are, the more check digits are required to correct errors. For example, binary messages consisting of six digits require four check digits to ensure that all messages with one error can be decoded correctly.

Where there is no possibility of confusion, it is customary to denote an n-tuple $(a_1, a_2, \ldots, a_n)$ more concisely as $a_1 a_2 \cdots a_n$, as we did in Table 10.1.

Given a binary linear code, how can we tell if it will correct errors and how many errors it will detect? It is remarkably easy. We examine all the code words to find one that has the fewest number of 1s excluding the code word consisting entirely of 0s. Call this minimum number of 1s in any nonzero code word the weight of the code and denote it by t.

> The **weight of a code** is the minimum number of 1s that occur among all nonzero code words of that code.

If t is odd, the code will correct any $(t - 1)/2$ or fewer errors; if t is even, the code will correct any $(t - 2)/2$ or fewer errors. If we prefer simply to detect errors rather than to correct them (as is often the case in applications), the code will detect any $t - 1$ or fewer errors.

Applying this test to the code in Table 10.1 we see that the weight is 3, so it will correct any $(3 - 1)/2 = 1$ error or it will detect any $3 - 1 = 2$ errors. Be careful here. We must decide *in advance* whether we want our code to correct single errors or detect double errors. It can do whichever we choose, but not both. If we decide to detect errors, then we will not decode any message that was not among our original list of encoded messages (just as "bive" is not a word in the English language). Instead, we simply note that an error was made and, in most applications, request a retransmission. An example of this occurs when a bar code reader at the supermarket detects an error and it does not emit a sound (in effect, requesting a rescanning). On the other hand, if we decide to correct errors, we will decode any received message as its nearest neighbor.

Here is an example of another binary linear code. Let the set of messages be {000, 001, 010, 100, 110, 101, 011, 111} and append three check digits c_1, c_2, and c_3 using

$$c_1 = 0 \text{ if } a_1 + a_2 + a_3 \text{ is even}$$
$$c_1 = 1 \text{ if } a_1 + a_2 + a_3 \text{ is odd}$$
$$c_2 = 0 \text{ if } a_1 + a_3 \text{ is even}$$
$$c_2 = 1 \text{ if } a_1 + a_3 \text{ is odd}$$
$$c_3 = 0 \text{ if } a_2 + a_3 \text{ is even}$$
$$c_3 = 1 \text{ if } a_2 + a_3 \text{ is odd}$$

For example, if we take $a_1 a_2 a_3$ as 101, we have

$$c_1 = 0 \text{ since } 1 + 0 + 1 \text{ is even}$$
$$c_2 = 0 \text{ since } 1 + 1 \text{ is even}$$
$$c_3 = 1 \text{ since } 0 + 1 \text{ is odd}$$

So we encode 101 by appending 001, that is, $101 \rightarrow 101001$. The entire code is shown in Table 10.3.

Since the minimum number of 1s of any nonzero code word is 3, this code will either correct any single error or detect any double error, whichever we choose.

Spotlight *Jessie MacWilliams*

10.4

Jessie MacWilliams

An important contributor to coding theory was Jessie MacWilliams. Born in 1917 in England, she received a B.A. in 1938 and a M.A. degree in 1939 from Cambridge University and then came to the United States to study at Johns Hopkins University. After a year at Johns Hopkins, she went to Harvard, where she studied a second year. In 1955, with three children aged 13, 11, and 9, MacWilliams became a programmer at Bell Labs, where she learned about coding theory. Although she made a major discovery about codes while a programmer, she could not obtain a promotion to a math research position without a Ph.D. degree. She completed some of the requirements for the Ph.D. while working full time at Bell Labs and looking after her family. She then returned to Harvard for a year (1961–1962) and finished her degree. Interestingly, both MacWilliams and her daughter Ann were studying mathematics at Harvard at the same time.

MacWilliams returned to Bell Labs where she remained until her retirement in 1983. While at Bell Labs she made many contributions to the subject of error-correcting codes, including *The Theory of Error-Correcting Codes*, written jointly with Neil Sloane—a book that is still a leader in the field. One of her results of great theoretical importance is known as the "MacWilliams identity." She died on May 27, 1990, at the age of 73.

TABLE 10.3	
Message	Code Word
000	→ 000000
001	→ 001111
010	→ 010101
100	→ 100110
110	→ 110011
101	→ 101001
011	→ 011010
111	→ 111100

It is natural to ask how the method of appending extra digits with parity-check sums enables us to detect or even correct errors. Error detection is obvious. Think of how a computer spell checker works. If you type "bive" instead of "five," the spell checker detects the error because the string "bive" is not on its list of valid words. On the other hand, if you type "give" instead of "five," the spell checker will not detect the error since "give" is a valid word.

Our error-detection scheme works the same way, except that if we add extra digits to ensure that our code words differ in many positions, say t positions, then even $t - 1$ mistakes will not convert one code word into another code word. And if every pair of code words differs from each other in at least three positions, we can correct any single error since the incorrect received word will differ from the correct code word in one position, but it will differ from all others in two or more positions. Thus, in this case, the correct word is the "nearest neighbor." So the role of the parity-check sums is to ensure that code words differ in many positions. For example, consider the code in Table 10.1. The messages 1000 and 1100 differ in only the second position. But the two parity-check sums $a_1 + a_2 + a_3$ and $a_2 + a_3 + a_4$ will guarantee that encoded words for these messages will have different values in positions 5 and 7 as well as position 2. It is the job of mathematicians to discover the appropriate parity-check sums to correct several errors in long, complicated codes.

Data Compression

Binary linear codes are fixed-length codes. In a fixed-length code each code word is represented by the same number of digits (or symbols). In contrast, the Morse code (see Figure 10.5), designed for the telegraph, is a **variable-length** code. That is, a code in which the number of symbols for each code word may vary.

FIGURE 10.5 Morse code.

A	.—	N	—.
B	—...	O	———
C	—.—.	P	.——.
D	—..	Q	——.—
E	.	R	.—.
F	..—.	S	...
G	——.	T	—
H	....	U	..—
I	..	V	...—
J	.———	W	.——
K	—.—	X	—..—
L	.—..	Y	—.——
M	——	Z	——..

Notice that in the Morse code the letters that occur most frequently have the shortest coding, whereas the letters that occur the least frequently have the longest coding. By assigning the code in this manner, telegrams could convey more information per line than would be the case for fixed-length codes or a randomly assigned variable-length coding of the letters. The Morse code is an example of data compression. Figure 10.6 shows a typical frequency distribution for letters in English-language text material.

Data compression is the process of encoding data so that the most frequently occurring data are represented by the fewest symbols.

FIGURE 10.6 A widely used frequency table for letters in normal English usage.

	A	B	C	D	E	F	G	H	I	J	K	L	M
Percentage:	8	1.5	3	4	13	2	1.5	6	6.5	0.5	0.5	3.5	3

	N	O	P	Q	R	S	T	U	V	W	X	Y	Z
Percentage:	7	8	2	0.25	6.5	6	9	3	1	1.5	0.5	2	0.25

Let us illustrate the principles of data compression with a simple example. Biologists are able to describe genes by specifying sequences composed of the four letters A, T, G, and C, which represent the four nucleotides adenine, thymine, guanine, and cytosine, respectively. One way to encode a sequence such as AAACAGTAAC in fixed-length binary form would be to encode the letters as

$$A \longrightarrow 00 \quad C \longrightarrow 01 \quad T \longrightarrow 10 \quad G \longrightarrow 11$$

The corresponding binary code for the sequence AAACAGTAAC is then

00000001001110000001

On the other hand, if we knew from experience that the hierarchy of occurrence of the letters is A, C, T, and G (that is, A occurs most frequently, C second most frequently, and so on), and that A occurs much more frequently than T and G together, the most efficient binary encoding would be

$$A \longrightarrow 0 \qquad C \longrightarrow 10 \qquad T \longrightarrow 110 \qquad G \longrightarrow 111$$

For this encoding scheme the sequence AAACAGTAAC is encoded as

0001001111100010

Notice that this binary sequence has 20% fewer digits than our previous sequence, in which each letter was assigned a fixed length of 2 (16 digits versus 20 digits). However, to realize this savings, we have made decoding more difficult. For the binary sequence using the fixed length of two symbols per character, we decode the sequence by taking the digits two at a time in succession and converting them to the corresponding letters. For the compressed coding, we can decode by examining the digits in groups of three.

E X A M P L E ▶ *Decode 0001001111100010*

Consider the compressed binary sequence 0001001111100010. Look at the first three digits: 000. Since our code words have one, two, or three digits and neither 00 nor 000 is a code word, the sequence 000 can only represent the *three* code words 0, 0, and 0. Now look at the next three digits: 100. Again, since neither 1 nor 100 is a code word, the sequence 100 represents the *two* code words 10 and 0. The next three digits, 111, can only represent the code word 111 since the other three code words all contain at least one 0. Next consider the sequence 110. Since neither 1 nor 11 is a code word, the sequence 110 can only represent 110 itself. Continuing in this fashion, we can decode the entire sequence to obtain AAACAGTAAC.

The following observation can simplify the decoding process for compressed sequences. Note that 0 only occurs at the end of a code word. Thus each time you see a 0, it is the end of the code word. Also, because the code words 0, 10 and 110 end in a 0, the only circumstances under which there are three consecutive 1s is when the code word is 111. So, to quickly decode a compressed binary sequence using our coding scheme, insert a comma after every 0 and after every three consecutive 1s. The digits between the commas are code words. ◆

$\mathcal{S}potlight$ *David Huffman*

10.5

Large networks of IBM computers use it. So do high-definition televisions, modems, and a popular electronic device that takes the brainwork out of programming a videocassette recorder. All these digital wonders rely on the results of a 40-year-old term paper by an MIT graduate student—a data compression scheme known as Huffman encoding.

In 1951 David Huffman and his classmates in an electrical engineering graduate course on information theory were given the choice of a term paper or a final exam. For the term paper, Huffman's professor had assigned what at first appeared to be a simple problem. Students were asked to find the most efficient method of representing numbers, letters, or other symbols using binary code. Huffman worked on the problem for months, developing a number of approaches, but none that he could prove to be the most efficient. Finally, he despaired of ever reaching a solution and decided to start studying for the final. Just as he was throwing his notes in the garbage, the solution came to him. "It was the most singular moment of my life," Huffman says. "There was the absolute lightning of sudden realization. It was my luck to be there at the right time and also not have my professor discourage me by telling me that other good people had struggled with the problem," he says. When presented with his student's discovery, Huffman recalls, his professor exclaimed: "Is that all there is to it!"

"The Huffman code is one of the fundamental ideas that people in computer science and data communications are using all the time," says Donald Knuth of Stanford University. Although others have used Huffman's code to help make mil-

David Huffman

lions of dollars, Huffman's main compensation was dispensation from the final exam. He never tried to patent an invention from his work and experiences only a twinge of regret at not having used his creation to make himself rich. "If I had the best of both worlds, I would have had recognition as a scientist, and I would have gotten monetary rewards," he says. "I guess I got one and not the other."

But Huffman has received other compensation. A few years ago an acquaintance told him that he had noticed that a reference to the code was spelled with a lowercase "H." Remarked his friend to Huffman, "David, I guess your name has finally entered the language."

Source: Adapted from an article by Gary Stix, *Scientific American,* September 1991, pp. 54, 58.

EXAMPLE ▶ *Code AGAACTAATTGACA and Decode the Result*

Recall: A → 0, C → 10, T → 110, and G → 111. So

AGAACTAATTGACA ⟶ 0111001011000110110111 0100

To decode the encoded sequence we insert commas after every 0 and after every occurrence of 111 and convert to letters:

0,111,0, 0,10,110,0, 0,110,110,111,0,10,0
A, G, A, A,C, T, A,A, T, T, G, A,C,A ◆

Modern data compression schemes were first invented in the 1950s (see Spotlight 10.5). They are now routinely used by modems and fax machines for data transmissions and by computers for data storage. In many cases data compression results in a savings of up to 50% on telephone charges or storage space.

CRYPTOGRAPHY

Thus far we have discussed ways in which data can be encoded to detect errors or correct errors in transmission. In many situations there is also a desire for security against unauthorized interpretation of coded data (that is, a desire for secrecy). The process of disguising data is called **encryption. Cryptology** is the study of methods to make and break secret codes. Access to computers is controlled by **passwords,** which are stored in encrypted form in the computers. Banking transactions, military transmissions, and intelligence information are encrypted. (Among the earliest methods for encrypting information is one attributed to Julius Caesar, who used it to communicate with his soldiers. See Exercise 14.) Premium television services such as HBO, Showtime, and The Disney Channel also have a need to prevent their television signals to local cable operators and satellite dish subscribers from being received free by satellite dish owners.

The method used by these services involves a monthly password, a subscriber sequence called a key, and the addition of binary sequences. We add two binary sequences $a_1 a_2 \cdots a_n$ and $b_1 b_2 \cdots b_n$ as follows:

$$
\begin{array}{r}
a_1 a_2 \cdots a_n \\
+\ b_1 b_2 \cdots b_n \\
\hline
c_1 c_2 \cdots c_n
\end{array}
$$

where $c_i = 0$ if $a_i = b_i$ and $c_i = 1$ if $a_i \neq b_i$. Equivalently, $c_i = 0$ if $a_i + b_i$ is 0 or 2 and $c_i = 1$ if $a_i + b_i$ is 1. (Add a_i and b_i in the ordinary way but replace 2 by 0.)

The latest and most intricate ciphering machine is being explained by Major William F. Friedman, Chief of Signal Intelligence (1930s)

E X A M P L E ▶ *Sum of Binary Sequences*

$$
\begin{array}{r}
11000111 \\
+\ 01110110 \\
\hline
10110001
\end{array}
\qquad
\begin{array}{r}
00111011 \\
+\ 01100101 \\
\hline
01011110
\end{array}
\qquad
\begin{array}{r}
10011100 \\
+\ 10011100 \\
\hline
00000000
\end{array}
$$

The data security method we describe hinges on the fact that the sum of two binary sequences $a_1 a_2 \cdots a_n + b_1 b_2 \cdots b_n = 00 \cdots 0$ if and only if the sequences are identical.

Beginning in 1984, HBO scrambled its signal. To unscramble the signal, a cable system operator or dish owner who pays a monthly fee has to have a password that is changed monthly. The password is transmitted along with the scrambled signal. Although HBO uses binary sequences of length 56, we will illustrate the method with sequences of length 8.

Let us say that the password for this month is p. Each subscriber of the service is assigned a sequence uniquely associated with him or her called a **key.** Let us say that the list of keys issued by HBO to its customers is $k_1, k_2, \ldots$. HBO transmits the password p, and the encrypted sequences $k_1 + p, k_2 + p, \ldots$ (that is, one sequence for each authorized user). A microprocessor in each subscriber's decoding box adds its key, say k_i, to each of the encrypted sequences. That is, it calculates $k_i + (k_1 + p)$, $k_i + (k_2 + p)$, ... As it does so, the microprocessor compares each of these calculated sequences with the correct password p. When one of the sequences matches p, the microprocessor will unscramble the signal. Notice that the correct password p will be produced precisely when k_i is added to $k_i + p$, since $k_i + (k_i + p) = (k_i + k_i) + p = 00 \cdots 0 + p = p$ and $k_j + (k_j + p) \neq p$ when $k_j \neq k_i$. (That is, key k_i "unlocks" the encrypted sequence $k_i + p$ and no other.) If a subscriber with key k_i fails to pay the monthly bill, HBO can terminate the service by not transmitting the sequence $k_i + p$ the next month. ◆

EXAMPLE ▶ *Encryption and Decoding*

Suppose the password for this month is $p = 10101100$ and your key is $k = 00111101$. One of the sequences transmitted by HBO is $k + p$:

$$\begin{array}{r} 00111101 \\ + \ 10101100 \\ \hline 10010001 \end{array}$$

Your decoder box adds your key $k = 00111101$ to each of the sequences received. Eventually, it finds the sequence obtained by adding the password to your key (namely, $p + k = 10010001$) and calculates

$$\begin{array}{r} 00111101 \\ + \ 10010001 \\ \hline 10101100 \end{array}$$

to obtain the password p. Once the password has been found the decoder descrambles the signal.

One might suspect that a computer hacker could find the password by simply trying a large number of possible keys until one "unlocks" the password. But with sequences of length 56 there are 2^{56} possible keys, of which fewer than a million are used by HBO to scramble its monthly password. The number 2^{56} is so large (it exceeds 72 quadrillion), however, that even if one tries a billion possible keys, the chance of finding one that works is essentially 0. ◆

Optional ▶ **Public Key Cryptography**

In the mid-1970s Ron Rivest, Adi Shamir, and Len Adleman devised an ingenious method that permits each person who is to receive a secret message to publicly tell how to scramble messages sent to him or her. And even though the method used to scramble the message is known publicly, only the person for whom it is intended will be able to unscramble the message.

Before presenting this idea, we first introduce a method of counting that you often use. For example, if it is now September, what month will it be 25 months henceforth? Of course, you answer October but the interesting fact is that you didn't arrive at the answer by starting with September and counting off 25 months. Instead, without even thinking about it, you simply observed that $25 = 12 \cdot 2 + 1$ and you added one month to September. Similarly, if it is now Wednesday, you know that in 23 days it will be Friday. This time, you arrived at your answer by noting that $23 = 7 \cdot 3 + 2$, so you added 2 days to Wednesday instead of counting off 23 days. Likewise, if your electricity is off for 26 hours, you know you must advance your clock 2 hours, since $26 = 2 \cdot 12 + 2$. Surprisingly, this simple idea has numerous important applications in mathematics and computer science.

Before describing the method for transmitting messages secretly, it is convenient to introduce a notation for the kind of arithmetic described in the previous paragraph. For any positive integers a and n we write a mod n (read: "a modulo n" or just "a mod n") to be the remainder when a is divided by n. Thus,

$$3 \bmod 2 = 1 \text{ since } 3 = 1 \cdot 2 + 1$$
$$6 \bmod 2 = 0 \text{ since } 6 = 2 \cdot 3 + 0$$
$$4 \bmod 3 = 1 \text{ since } 4 = 1 \cdot 3 + 1$$
$$15 \bmod 3 = 0 \text{ since } 15 = 5 \cdot 3 + 0$$
$$12 \bmod 10 = 2 \text{ since } 12 = 1 \cdot 10 + 2$$
$$37 \bmod 10 = 7 \text{ since } 37 = 3 \cdot 10 + 7$$
$$98 \bmod 85 = 13 \text{ since } 98 = 1 \cdot 85 + 13$$
$$342 \bmod 85 = 2 \text{ since } 342 = 4 \cdot 85 + 2$$
$$62 \bmod 85 = 62 \text{ since } 62 = 0 \cdot 85 + 62$$

Arithmetic involving mod n is called **modular arithmetic.** One rule of modular arithmetic we will need is

$$(ab) \bmod n = ((a \bmod n)(b \bmod n)) \bmod n$$

This rule allows you to replace integers greater than or equal to n with integers less than n to simplify calculations. You should think of the rule as saying, "Mod before you multiply."

E X A M P L E ▶ *Modular Arithmetic*

$$(17 \cdot 23) \bmod 10 = ((17 \bmod 10)(23 \bmod 10)) \bmod 10$$
$$= (7 \cdot 3) \bmod 10 = 21 \bmod 10 = 1$$

$$(22 \cdot 19) \bmod 8 = ((22 \bmod 8)(19 \bmod 8)) \bmod 8$$
$$= (6 \cdot 3) \bmod 8 = 18 \bmod 8 = 2$$

$$(100 \cdot 8) \bmod 85 = ((100 \bmod 85) \cdot (8 \bmod 85)) \bmod 85$$
$$= (15 \cdot 8) \bmod 85 = 120 \bmod 85 = 35 \quad ◆$$

We now describe the Rivest, Shamir, and Adleman method by way of a simple example that nevertheless illustrates the essential features of the method. Say we wish to send the message "IBM." We convert the message to digits by replacing A by 1, B by 2, . . . , and Z by 26. So the message IBM becomes 9213. The person to whom the message is to be sent has picked two primes p and q, say, $p = 5$ and $q = 17$ (recall that a *prime* is an integer greater than 1 whose only divisors are 1 and itself), and a number r that has no divisors in common with the least common multiple m of $(p - 1) = 4$ and $(q - 1) = 16$ other than 1, say, $r = 3$, and published $n = pq = 85$ and r in a public directory. The receiver also must find a number s so that $r \cdot s = 1 \bmod m$ (this is where knowledge of p and q is necessary). That is, $3 \cdot s = 1 \bmod 16$. This number is 11. (The number s can be found by calculating successive powers of $r \bmod m$; when 1 is reached, the previous power of r is s. In our example we have

$$3 \bmod 16 = 3, \quad 3^2 \bmod 16 = 9, \quad 3^3 \bmod 16 = 11, \quad 3^4 \bmod 16 = 1$$

so $s = 3^3 \bmod 16 = 11$.)

We consult this directory to find n and r, then send the "scrambled" numbers $9^3 \bmod 85$, $2^3 \bmod 85$, and $13^3 \bmod 85$ rather than 9, 2 and 13 and the receiver will unscramble them. Thus we send

$$9^3 \bmod 85 = 49$$
$$2^3 \bmod 85 = 8$$

and

$$13^3 \bmod 85 = 72$$

Now the receiver must take the numbers he or she receives, 49, 8, and 72, and convert them back to 9, 2, and 13 by calculating $49^{11} \bmod 85$, $8^{11} \bmod 85$, and $72^{11} \bmod 85$.

The calculation of $49^{11} \bmod 85$ can be simplified as follows:[1]

$$49 \bmod 85 = 49$$
$$49^2 \bmod 85 = 2401 \bmod 85 = 21$$
$$49^4 \bmod 85 = 49^2 \cdot 49^2 \bmod 85 = 21 \cdot 21 \bmod 85 = 441 \bmod 85$$
$$= 16 \bmod 85$$
$$49^8 \bmod 85 = 49^4 \cdot 49^4 \bmod 85 = 16 \cdot 16 \bmod 85 = 1$$

---

[1] To determine $49^2 \bmod 85$ with a calculator, enter 49×49 to obtain 2401, then divide 2401 by 85 to obtain 28.247058. Finally, enter $2401 - (28 \times 85)$ to obtain 21.

So $49^{11} \bmod 85 = (49^8 \bmod 85)(49^2 \bmod 85)(49 \bmod 85)$

$$= (1 \cdot 21 \cdot 49) \bmod 85$$
$$= 1029 \bmod 85$$
$$= 9 \bmod 85$$

Thus, the receiver has correctly determined the code for "I." The calculations for $8^{11} \bmod 85$ and $72^{11} \bmod 85$ are left as exercises for the reader. Notice that without knowing how pq factors, one cannot find the least common multiple of $p - 1$ and $q - 1$ (in our case, 16), and therefore the s that is needed to determine the intended message.

The procedure just described is called the **RSA public key encryption scheme** in honor of Rivest, Shamir, and Adleman, who discovered the method. The method is practical and secure because there exist efficient methods for finding very large prime numbers (say about 100 digits long) and for multiplying large numbers, but no one knows an efficient algorithm for factoring large integers (say about 200 digits long).

The algorithm is summarized below. (In practice, the messages are not sent one letter at a time. Rather, the entire message is converted to decimal form with A represented by 01, B by 02, . . . , and a space by 00. The message is then broken up into blocks of uniform size and the blocks are sent. See Step 2 under Sender below.)

Receiver

1. Pick very large primes p and q and compute $n = pq$.
2. Compute the least common multiple of $p - 1$ and $q - 1$; let us call it m.
3. Pick r so that it has no divisors in common with m other than 1 (any such r will do).
4. Find s so that $rs = 1$ modulo m (there is always exactly one such s between 1 and m).
5. Publicly announce n and r, but keep p, q, and s secret.

Sender

1. Convert the message to a string of digits.
2. Break up the message into uniformly sized blocks of digits, appending 0s in the last block if necessary; call them M_1, M_2, . . . , M_k. For example, for a string such as 2105092315, we would use $M_1 = 2105$, $M_2 = 0923$, and $M_3 = 1500$.
3. Check to see that the greatest common divisor of each M_i and n is 1. If not, n can be factored and the code is broken. (In practice, the primes p and q are so large that they exceed all M_i, so this step may be omitted.)
4. Calculate and send $R_i = M_i^r \bmod n$.

Receiver

1. For each received message R_i, calculate $R_i^s \bmod n$.
2. Convert the string of digits back to a string of characters.

Why does this method work? It works because of a basic property of modular arithmetic and the choice of r. It so happens that the number m has the property that for each x having no common divisors with n except 1, we have $x^m = 1 \bmod n$. So, because each message M_i has no common divisors with n except 1, and r was chosen so that $rs = 1 + mt$ for some t, we have modulo n.

$$R_i^s = (M_i^r)^s = M_i^{rs} = M_i^{1+mt} = M_i(M_i^m)^t = M_i 1^t = M_i \quad \blacklozenge$$

REVIEW VOCABULARY

Binary linear code   A code consisting of words composed of 0s and 1s obtained by using parity-check sums to append check digits to messages.
Code words   Words from a binary linear code.
Cryptography   The study of how to make and break secret codes.
Data compression   The process of encoding data so that the most frequently occurring data are represented by the fewest symbols.
Decoding   The process of translating received data into code words.
Distance between two messages   The distance between two messages is the number of positions in which they differ.
Encryption   The process of encoding data to protect against unauthorized interpretation.
Even parity   Even integers are said to have even parity.
Key   A string used to decode data.
Modular arithmetic   Addition and multiplication involving modulo n.
Nearest-neighbor decoding   A method that decodes a received message as the code word that agrees with the message in the most positions.
Odd parity   Odd integers are said to have odd parity.
Parity-check sums   Sums of digits whose parities determine the check digits.
Password   A word used to encode data.
RSA public key encryption   A method of encoding that permits each person to announce publicly the means by which secret messages are to be sent to him or her.
Variable-length code   A code in which the number of symbols for each code word may vary.
Weight of a code   The minimum number of 1s that occur among all nonzero code words of a code.

SUGGESTED READINGS

DENEEN, L.   Secret encryption with public keys. *UMAP Journal,* 8 (1987): 9–29. Describes several ways in which modular arithmetic can be used to code secret messages, from a simple scheme used by Julius Caesar to the RSA public key encryption method.

KAHN, DAVID.   *Codebreakers: The Story of Secret Writing,* Macmillan, New York, 1967. A monumental, illustrated history.

KORHEIM, ALAN G.   *Cryptography: A Primer,* Wiley, New York, 1981. An introduction to cryptography.

MCELIECE, R.   The reliability of computer memories, *Scientific American,* 252 (1985): 88–95. Discusses why and how error-correcting codes are employed in computer memories.

RICHARDS, I.   The invisible prime factor, *American Scientist,* 70 (1982):176–179. Explains how elementary number theory and modular arithmetic can be used to test whether an integer is prime and how prime numbers can be used to create secret codes that are extremely difficult to break.

THOMPSON, T.   *From Error-Correcting Codes Through Sphere Packing to Simple Groups,* Mathematical Association of America, Washington, D.C., 1983. Chapter 1 of this award-winning book gives a fascinating historical account of the origins of error-correcting codes.

EXERCISES   ▲ *Optional.*   ■ *Advanced.*   ◆ *Discussion.*

Binary Codes

1.   Use the diagram method shown in Figures 10.1 and 10.2 to verify the code words in Table 10.1 for the messages 0101, 1011, and 1111.

2.   Use the diagram method to decode the received messages 0111011 and 0100110.

3.   Find the distance between each of the following pairs of words:

   (a)   11011011 and 10100110
   (b)   01110100 and 11101100

4. Referring to Table 10.1, use the nearest-neighbor method to decode the received words 0000110 and 1110100.

5. If the code word 0110010 is received as 1001101, how is it decoded using the diagram method?

6. Suppose a received word has the diagram arrangement shown below:

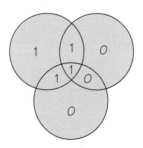

What can we conclude about the received word?

Parity-Check Sums

7. Determine the binary linear code that consists of all possible three digit messages with three check digits appended using the parity-check sums $a_2 + a_3$, $a_1 + a_3$, and $a_1 + a_2$. (That is, $c_1 = 0$ if $a_2 + a_3$ is even; $c_1 = 1$ if $a_2 + a_3$ is odd, and similarly for c_2 and c_3.)

8. Let C be the code

{0000000, 1110100, 0111010, 0011101,

1001110, 0100111, 1010011, 1101001}

What is the error-correcting capability of C? What is the error-detecting capability of C?

9. Find all code words for binary messages of length 4 by adding three check digits using the parity-check sums $a_2 + a_3 + a_4$, $a_2 + a_4$, and $a_1 + a_2 + a_3$. Will this code correct any single error?

10. Consider the binary linear code

$C = \{00000, 10011, 01010, 11001, 00101, 10110, 01111, 11100\}$

Use nearest-neighbor decoding to decode 11101 and 01100. If the received word 11101 has exactly one error, can you determine the intended code word? Explain your reasoning.

11.   Construct a binary linear code using all eight possible binary messages of length 3 and appending three check digits using the parity-check sums $a_1 + a_2$, $a_2 + a_3$, and $a_1 + a_2 + a_3$. Decode each of the received words

001001, 011000, 000110, 100001

by the nearest-neighbor method.

12.   Add the following pairs of binary sequences:

(a)   10111011 and 01111011
(b)   11101000 and 01110001

13.   All binary linear codes have the property that the sum of two code words is another code word. Use this fact to determine which of the following sets cannot be a binary linear code:

(a)   {0000, 0011, 0111, 0110, 1001, 1010, 1100, 1111}
(b)   {0000, 0010, 0111, 0001, 1000, 1010, 1101, 1111}
(c)   {0000, 0110, 1011, 1101}

Cryptography

14.   The *Caesar cipher* encrypts messages by replacing each letter of the alphabet with the letter shown beneath it

ABCDEFGHIJKLMNOPQRSTUVWXYZ
DEFGHIJKLMNOPQRSTUVWXYZABC

Use the Caesar cipher to encrypt the message RETREAT. Determine the intended message corresponding to the encrypted message DWWDFN.

▲ 15.   Use the RSA scheme with $p = 5$, $q = 17$, and $r = 3$ to determine the numbers sent for the message VIP.

▲ 16.   Use the RSA scheme with $p = 5$, $q = 17$, and $r = 3$ to decode the received numbers 52 and 72.

▲ 17. In the RSA scheme with $p = 5$, $q = 17$, and $r = 5$, determine the value of s.

▲ 18. Why can't we use the RSA scheme with $p = 7$ and $q = 11$ and $r = 3$?

▲ 19. Assume that the letters of the alphabet have been converted to integers according to the correspondence $A \rightarrow 0$, $B \rightarrow 1$, $C \rightarrow 2$, . . . , $Z \rightarrow 25$. Write a formula that describes the result of applying the Caesar cipher (see Exercise 14) to the integer x. (*Hint:* Use modular arithmetic.)

Data Compression

▨ 20. Suppose we code a five-symbol set {*A, B, C, D, E*} into binary form as follows:

$$A \longrightarrow 0, B \longrightarrow 10, C \longrightarrow 110, D \longrightarrow 1110, \text{ and } E \longrightarrow 1111$$

Convert the sequence to *AEAADBAABCB* into binary code. Determine the sequence of symbols represented by the binary code 01000110100011111110.

▨ 21. Use the code in Exercise 20 to convert the sequence *EABAADABB* into binary code. Determine the sequence of letters represented by the binary code 00100011001111101110.

▨ 22. Devise a variable-length binary coding scheme for a six-symbol set {*A, B, C, D, E, F*}. Assume that *A* is the most frequently occurring symbol, *B* is the second most frequently occurring symbol, and so on.

23. Judging from the Morse code, what are the three most frequently occurring consonants in English text material? What is the most frequently occurring vowel?

24. In English, the letter H occurs more often than D, G, K, and W, but in Morse code, H has a longer code than D, G, K, and W. Speculate on the reason for this apparent violation of data compression principles.

25. Explain why the Morse code must include a space after each letter but fixed-length codes do not.

Additional Exercises

26.   Let $v = a_1 a_2 \ldots a_n$ and $u = b_1 b_2 \ldots b_n$ be binary sequences. Explain why the number of 1s in $v + u$ is the same as the distance between u and v.

27.   Extend the code words listed in Table 10.1 to eight digits by appending a 0 to words of even weight and a 1 to words of odd weight. What is the error-detecting and error-correcting capability of the new code?

28.   Suppose the weight of a binary linear code is 6. How many errors can the code correct? How many errors can the code detect?

29.   How many code words are there in a binary linear code that has all possible messages of length 5 with three check digits appended?

30.   Explain why no binary linear code with three message digits and three check digits can correct all possible double errors.

31.   A *ternary* code is formed by starting with all possible strings of a fixed length comprised of 0s, 1s, and 2s and appending extra digits that are also 0s, 1s, or 2s. Form a ternary code by appending to each message $a_1 a_2$ the check digits $c_1 c_2$ using:

$c_1 = 0$ if $a_1 + a_2$ is 0 or 3
$c_1 = 1$ if $a_1 + a_2$ is 1 or 4
$c_1 = 2$ if $a_1 + a_2$ is 2
$c_2 = 0$ if $2a_1 + a_2$ is 0, 3 or 6
$c_2 = 1$ if $2a_1 + a_2$ is 1 or 4
$c_2 = 2$ if $2a_1 + a_2$ is 2 or 5

32.   Use the ternary code in Exercise 31 and the nearest-neighbor method to decode the received word 1211.

33.   Suppose a ternary code is formed by starting with all possible strings of length 4 composed of 0s, 1s, and 2s and appending two extra digits that are also 0s, 1s, and 2s. How many code words are there in this code? How many possible received words are there in this code?

▲ 34.   For each part below explain how modular arithmetic can be used to answer the question.

(a) If today is Wednesday, what day of the week will it be in 16 days?
(b) If a clock (with hands) indicates that it is now 4 o'clock, what will it indicate in 37 hours?

(c) If a military person says it is now 0400, what time would it be in 37 hours? (Instead of A.M. and P.M., military people use 1300 for 1 P.M., 1400 for 2 P.M., and so on.)

(d) If it is now July 20, what day will it be in 65 days?

(e) If the five-digit odometer of an automobile reads 97,000 miles now, what will it read in 12,000 miles?

WRITING PROJECTS

1 ▶ Prepare a report on cryptography. Discuss at least three methods of encryption. Discuss the interface between computers and cryptography.

2 ▶ Prepare a report on applications of modular arithmetic. Explain the calculation of the check digits described in Exercises 7, 9, 11, and 31 with modular arithmetic. Use modular arithmetic to describe the error-detection schemes used in Chapter 9.

3 ▶ Prepare a report on the early history of error-correcting codes. The reference by Thompson (see Suggested Readings) has the information you will need.

A revolution currently taking place in the field of mathematics is its application to the study of human beings—their behavior, values, interactions, conflicts, and decision making, as well as their interface with modern technology and various institutions. This revolution could eventually prove to be as far reaching as the turning of mathematics to study physical objects and their motion some three centuries ago. As mathematics and computers play an increasingly important role in understanding our social institutions, a new profession is emerging devoted to thinking mathematically about human affairs.

In particular, decision making is being influenced profoundly by modern mathematics, and several particularly mathematical subjects have been created primarily to assist in arriving at sound decisions. While many aspects involved in arriving at a decision are cultural rather than quantitative, there are many ingredients of contemporary decision making

SOCIAL CHOICE
and DECISION MAKING

that are mathematical in nature; some of these are addressed by mathematical models that are developed in Part IV.

In Chapter 11 we discuss the important problem of social choice. How does a group of individuals, each with his or her own set of values, select one outcome from a list of possibilities? This problem arises frequently in a democratic society, and even in authoritarian institutions where decisions are made by more than one person. While "majority rule" is a good system for deciding an election involving just two candidates, there is no perfect way of deciding an election when three or more candidates are running. Group decision making is often a strategic encounter, and citizens need to be aware of the difficulties that can arise when some participants have an incentive to manipulate the outcome.

In Chapter 12 we consider decision-making bodies in which the individual voters or parties do not have equal power. In particular, we will look at weighted voting systems such as the electoral college, stockholders

in a corporation, or political parties in a national assembly, in which the voters cast different numbers of votes. While the notion of power is central in political science, it is typically difficult to quantify. We will find that a voter's power in such a system may not be proportional to the number of votes that he or she is entitled to cast. We describe two well-known indices for measuring power in weighted voting systems which will enable us to assess the fairness of weighted voting systems.

A general theme of Part IV concerns the idea of fairness in decision making. In Chapter 13 this becomes most explicit. Here we describe some fair-decision schemes in which a group of individuals with different values can be assured of each receiving what he or she views as a fair share when dividing up objects like cakes or the good in an estate. An important theme of this chapter is finding procedures that produce "envy-free" allocations, in which each person gets a largest portion (as he or she values the cake or other goods) and hence does not envy anybody else.

In Chapter 14 we discuss the apportionment problem, which is to round a set of fractions to whole numbers while preserving their sum; of course, the sum of the original fractions must be a whole number to start. Apportionment problems occur when resources must be allocated in integer quantities—for instance, when college administrators allocate faculty positions to each department.

The most important apportionment problem is the allocation of seats in the U.S. House of Representatives to the 50 states. It is this problem that initiated the study of apportionment, when President George Washington vetoed the first congressional apportionment bill.

Chapter 15 introduces the mathematical field called game theory, which describes situations involving two or more decision makers having different goals. Game theory provides a collection of models to assist in the analysis of conflict and cooperation. It prescribes optimal strategies for games of total conflict in which one player's gain is equal to the other player's loss. It also provides insights into more cooperative situations in which players are trying to coordinate their choices, as well as encounters of partial conflict that involve aspects of both competition and cooperation. Particular games, including those known as Prisoners' Dilemma and Chicken, provide us with insights into certain social paradoxes that we routinely meet in our daily lives.

In Chapter 16 we introduce an extension of game theory, called the theory of moves, which is designed to model strategic situations in which players think beyond the immediate consequences of departing from an outcome. Instead, they think about the consequences that such a move might trigger in terms of a possible countermove on the part of the other player, counter-counter moves, and so on. Besides modeling the more farsighted thinking of decision makers, this theory allows for the possibility that some players may have greater power than others and shows what effects this greater power has on outcomes.

CHAPTER 11

SOCIAL CHOICE: THE IMPOSSIBLE DREAM

The basic question of *social choice*, of how groups can best arrive at decisions, has occupied social philosophers and political scientists for centuries. Indeed, voting is a subject that lies at the very heart of representative government and participatory democracy.

Social choice theory attempts to address the problem of finding good procedures that will turn individual preferences for different candidates—or *alternatives*, as they are often called—into a single choice by the whole group. The goal is to find such procedures that will result in an outcome that "reflects the will of the people."

This search for good voting systems, as we shall see, is plagued by a variety of counterintuitive results and disturbing outcomes. In fact, it turns out that one can prove (mathematically) that no one will ever find a completely satisfactory voting system for three or more alternatives.

The elections with which we are most familiar—those at the national political level—typically involve only two candidates. Ignoring the electoral college, such elections pose little difficulty in theory or practice: each voter casts a vote for one of the two candidates, and majority wins. We will begin our discussion of voting systems with this two-alternative case.

On the other hand, there are real-world situations where elections must be held to choose a single winner among three or more candidates, as in the presidential election of 1992 in which George Bush, Bill Clinton, and Ross Perot were the candidates. The procedure used in that election was—ignoring the

Clinton, Perot, and Bush at the second presidential debate, in Richmond, Virginia (1992).

electoral college—*plurality voting,* wherein a ballot is a choice for one of the three, and the candidate with the most votes wins.

Plurality voting, however, is not the only method that can be used to elect a single candidate from a choice of three or more, and we will investigate several other methods in this chapter. Most of these methods (the one called *approval voting* is the exception) use a ballot in which a voter provides a rank ordering of the candidates (without ties) indicating the order in which he or she prefers the candidates.

A ballot consisting of such a rank ordering of candidates (which we often picture as a vertical list with the most preferred candidate on top and the least preferred on the bottom) is called a **preference list,** or, more completely, an **individual preference list,** since it is a statement of the preferences of one of the individuals who is voting.

Ballots that are preference lists allow each voter to make a much clearer statement of his or her preferences than do ballots that correspond to a single vote for a single candidate. Such ballots (preference lists) are already used in a wide range of applications, such as rating football teams and scoring track meets.

We present four particular methods of choosing a winner when there are three or more alternatives. For each method, we illustrate the failure of some natural property that one would like to have satisfied by any voting system that is being used. Finally, we face the fact that the unfortunate shortcomings of the particular procedures we have presented are largely unavoidable. There are difficulties with elections involving three or more candidates that are simply insurmountable.

ELECTIONS WITH ONLY TWO ALTERNATIVES

When choosing between two alternatives, the first type of voting to suggest itself is **majority rule:** each voter indicates a preference for one of the two candidates, and the candidate with the most votes wins. Majority rule has at least three desirable properties:

1. All voters are treated equally. That is, if any two voters were to exchange (marked) ballots prior to submitting them, the outcome of the election would be the same.
2. Both candidates are treated equally. That is, if a new election were held and every voter were to reverse his or her vote, then the outcome of the previous election would be reversed as well.
3. If a new election were held and a single voter were to change his or her ballot from being a vote for the loser of the previous election to being a vote for the winner of the previous election, and everyone else voted exactly as before, then the outcome of the new election would be the same as the outcome of the previous election.

It is easy to devise voting systems for two alternatives in which these fail, but each such voting system quickly reveals its undesirability. For example, condition 1 is not satisfied by a *dictatorship* (whereby all ballots except that of the dictator are ignored); condition 2 is not satisfied by *imposed rule* (whereby candidate X wins regardless of who votes for whom); and condition 3 is not satisfied by *minority rule* (whereby the candidate with the fewest votes wins).

But maybe there are voting systems in the two-alternative case that are superior to majority rule in the sense of satisfying the three properties just listed *and* some other properties that we might also wish to have satisfied. This, however, turns out not to be the case. In 1952, Kenneth May proved the following:

> If the number of voters is odd, and we are interested only in voting systems that never result in a tie, then majority rule is the *only* voting system for two alternatives that satisfies the three conditions just listed.

This is an important and elegant result. **May's theorem** says that, for two alternatives, our work in finding a good voting system is done.

ELECTIONS WITH THREE OR MORE ALTERNATIVES: PROCEDURES AND PROBLEMS

In sharp contrast with the case of two alternatives is the situation in which there are three or more candidates. Here, we find no shortage of procedures that suggest themselves and that seem to represent perfectly reasonable ways to choose a winner from among three or more alternatives. Closer inspection, however, reveals shortcomings with all of these. We illustrate this with a consideration of four well-known procedures. Additional procedures (and additional shortcomings) can be found in the exercises.

In what follows, we assume that a ballot is an individual preference list. We allow ties in the election result and assume that, in the real world, either the number of voters is so large that ties in the election result will virtually never occur or that they can be broken by some kind of random device.

Political rally in
Washington, D.C.

Plurality Voting and the Condorcet Winner Criterion

In **plurality voting,** only first-place votes are considered. Thus, while we will consider plurality voting in the context of preference lists, a ballot might just as well be a single vote for a single candidate. The candidate with the most votes wins, even though this may be considerably fewer than one-half the total votes cast. This is perhaps the most common system in use today. It is how we chose Bill Clinton over George Bush and Ross Perot in 1992.

E X A M P L E ▶ *Plurality Voting and the 1980 U.S. Senate Race in New York State*

Plurality voting (and one of its drawbacks) is illustrated by the 1980 U.S. Senate race in New York among Alfonse D'Amato (a conservative), Elizabeth Holtzman (a liberal), and Jacob Javits (also a liberal). Reasonable estimates (based largely on exit polls) suggest that voters ranked the candidates according to the following table.

22%	23%	15%	29%	7%	4%
D	D	H	H	J	J
H	J	D	J	H	D
J	H	J	D	D	H

◆

Plurality voting led to D'Amato winning with 22% + 23% = 45% of the vote to 15% + 29% = 44% for Holtzman, and 7% + 4% = 11% for Javits. In this example, however, Holtzman was what is called a **Condorcet winner:** she would have defeated each of the other candidates in a head-to-head (i.e., a "one-on-one") election. That is, Holtzman would have defeated D'Amato in a two-candidate contest

15% + 29% + 7% = 51% (for Holtzman)

to

22% + 23% + 4% = 49% (for D'Amato)

and she would have defeated Javits in a two-candidate contest

22% + 15% + 29% = 66% (for Holtzman)

to

23% + 7% + 4% = 34% (for Javits)

A voting procedure is said to satisfy the **Condorcet winner criterion** provided that, for every possible sequence of preference lists, either (1) there is no Condorcet winner (as is often the case), or (2) there is a Condorcet winner (which, if it exists, is always unique) and it is the unique winner of the election.

The Condorcet winner criterion is certainly a property that one would like to see satisfied. However, the D'Amato–Holtzman–Javits election shows that plurality voting fails to satisfy the Condorcet winner criterion.

Perhaps a more fundamental drawback of plurality voting is the extent to which the ballots provide no opportunity for a voter to express any preferences except for naming his or her top choice. No use is made, for example, of the fact that a candidate may be no one's first choice, but everyone's close second choice.

Finally, there is yet another shortcoming of plurality voting. It is subject to what is called *manipulability:* there are elections in which it is to a voter's advantage to submit a ballot that misrepresents his or her true preferences. For example, in the presidential election of 1992, many voters who ranked Ross Perot over George Bush and Bill Clinton chose to vote for Bush or Clinton rather than "throw away" their vote on a candidate who they felt had no chance. As we move on to consider voting systems that make more significant use of a voter's rank ordering of the candidates, we will see that manipulability becomes even more of an issue.

The Borda Count and Independence of Irrelevant Alternatives

In many elections that use preference lists as ballots, the goal is to arrive at a final group rank ordering of all the contestants that best expresses the desires of the electorate. The purpose is not only to determine the winner, say, the class valedictorian, but also to arrive at who finished second, third, and so on, as in the case of one's rank in his or her senior class. In other applications, such as an election to a hall of fame, the first few finishers each receive the award, while the remaining nominees are also-rans.

One common mechanism for achieving this objective is to assign points to each voter's rankings and then to sum these for all voters to obtain the total points for each candidate. If there are 10 candidates, for example, then we could assign 10 points to each first-place vote for a given candidate, 9 points for each second-place vote, 8 for each third, and so forth. The candidate with the highest total number of points is the winner. Subsequent positions are assigned to those with the next highest tallies.

A voting method that assigns points in a descending manner to each voter's subsequent ranking and then sums these points to arrive at a group's final ranking is called a *rank method.* The special case in which there are n alternatives with each first place vote worth $n - 1$ points, each second place vote worth $n - 2$ points, and so on down to each last place vote worth zero points is known as the **Borda count.**

Rank methods other than the Borda count are not uncommon. For example, a track meet can be thought of as an "election" in which each event is a "voter" and each of the schools competing is a "candidate." If the order of finish in the 100-meter dash is School A, School B, School C, School D, then points are often awarded to each school as follows: 5 points for first place, 3 for second place, 2 for third place, and 1 for fourth place.

Rank methods using point assignments similar to what we just described are also used in sports polls. Our next example, however, uses the special case in which we are most interested: the Borda count.

EXAMPLE ▶ *The Borda Count and a Football Poll*

Suppose that a poll by 25 sports announcers is used to rank the football teams from among the following four universities: Miami (of Florida), Notre Dame, Penn State, and Southern California. They elect to assign 3 points to each announcer's first choice, 2 points to a second, 1 to a third, and 0 for a fourth. (Thus, they are using the Borda count.) There turn out to be 24 possible rankings, but assume that only the 5 in the following table appear.

| | **Number of Announcers** | | | | | |
Rank	8	6	5	4	2	Points
First	Mi	ND	PS	SC	ND	3
Second	ND	Mi	Mi	PS	PS	2
Third	PS	SC	ND	Mi	SC	1
Fourth	SC	PS	SC	ND	Mi	0

We calculate the total number of points for each team as follows:

Mi: $(3)(8) + (2)(6) + (2)(5) + (1)(4) + (0)(2) = 50$
ND: $(2)(8) + (3)(6) + (1)(5) + (0)(4) + (3)(2) = 45$
PS: $(1)(8) + (0)(6) + (3)(5) + (2)(4) + (2)(2) = 35$
SC: $(0)(8) + (1)(6) + (0)(5) + (3)(4) + (1)(2) = 20$

The resulting ranking is (1) Miami, 50 points; (2) Notre Dame, 45; (3) Penn State, 35: and (4) Southern California, 20. ◆

The Borda count certainly seems to be a reasonable way to choose a winner from among several alternatives (or to arrive at a group ranking of the alternatives). It also has its shortcomings, however, one of which is the failure of a property known as *independence of irrelevant alternatives.*

To describe this property, suppose that an election yields one alternative (call it *A*) as a winner and another alternative (call it *B*) as a nonwinner. Suppose that a new election is now held and that, although some of the voters may have changed their preference lists, no one who had previously ranked *A* over *B* changed his or her list so as now to have *B* over *A*.

If this new election were to yield *B* as a winner, then *B*—in terms of the outcome of the election—would have moved up to at least a tie with *A* on the basis of ballot changes involving alternatives *other than A or B*. One could argue that these other alternatives ought to be irrelevant to the question of whether *A* is more desirable than *B* or *B* is more desirable than *A*.

A voting system is said to satisfy **independence of irrelevant alternatives (IIA)** if it is impossible for an alternative *B* to move from nonwinner status to winner status unless at least one voter reverses the order in which he or she had *B* and the winning alternative ranked.

The following illustration shows that the Borda count fails to satisfy independence of irrelevant alternatives. Suppose the initial 5 ballots are as here:

| | Number of Voters | | |
Rank	3	2	Points
First	*A*	*C*	2
Second	*B*	*B*	1
Third	*C*	*A*	0

We calculate the total number of points for each alternative as follows:

A: (2)(3) + (0)(2) = 6
B: (1)(3) + (1)(2) = 5
C: (0)(3) + (2)(2) = 4

The winner is *A* (with 6 points), and *B* is a nonwinner (with 5 points). But now suppose that the two voters (on the right) change their ballots by moving *C* down between *A* and *B*. The lists then become

	Number of Voters		
Rank	3	2	Points
First	A	B	2
Second	B	C	1
Third	C	A	0

For this new election, we calculate the total number for each alternative:

A: $(2)(3) + (0)(2) = 6$
B: $(1)(3) + (2)(2) = 7$
C: $(0)(3) + (1)(2) = 2$

The Borda count therefore now yields B as the winner (with 7 points). Thus, B has gone from being a nonwinner to being a winner, even though no one changed his or her mind about whether B is preferred to A, or vice versa. This shows that the Borda count does not satisfy independence of irrelevant alternatives.

The results of these two elections also illustrate the problem of manipulability. Suppose the initial 5 ballots, reproduced below, represent the true preferences of the voters.

	Number of Voters		
Rank	3	2	Points
First	A	C	2
Second	B	B	1
Third	C	A	0

If all the voters are **sincere** (submit ballots that represent their true preferences), then A wins, as we saw, with 6 points. But A is the least preferred candidate of the two voters on the right. Hence, if they vote **strategically** (submitting ballots that do not represent their true preferences), then the ballots actually cast might be as follows:

	Number of Voters		
Rank	3	2	Points
First	A	B	2
Second	B	C	1
Third	C	A	0

Thus, the winner would be candidate *B* (who is preferred to candidate *A* by both of the voters who voted strategically).

Much more on manipulability will be said later in the chapter, but let us note for now a famous remark of Jean-Charles de Borda (1733–1799), made when a colleague pointed out how easily his Borda count can be manipulated: "My scheme is only intended for honest men!"

Sequential Pairwise Voting and the Pareto Condition

In our voting-theoretic context, an **agenda** will be understood to be a listing (in some order) of the alternatives. This listing is not to be confused with any of the preference lists, and, to avoid confusion, we will present agendas as horizontal lists and continue to present preference lists vertically.

> **Sequential pairwise voting** starts with an agenda and pits the first alternative against the second in a one-on-one contest. The winner (or both, if they tie) then moves on to confront the third alternative in the list, one-on-one. Losers are deleted. This process continues throughout the entire agenda, and those remaining at the end are the winners.

For a given sequence of individual preference lists, the particular agenda chosen can greatly affect the outcome of the election. In fact, Exercise 17 at the end of the chapter presents a sequence of ballots with the property that any one of the candidates can be made the winner by a suitable choice of agenda. This notwithstanding, we will see later in the chapter that sequential pairwise voting arises naturally in the legislative process.

EXAMPLE ▶ *Sequential Pairwise Voting*

Assume we have four alternatives and that the agenda is *A, B, C, D*. Consider the following sequence of three preference lists:

	Number of Voters		
Rank	1	1	1
First	*A*	*C*	*B*
Second	*B*	*A*	*D*
Third	*D*	*B*	*C*
Fourth	*C*	*D*	*A*

The first one-on-one pits *A* against *B*, and *A* wins by a score of 2 to 1 (meaning that two of the voters—the two on the left—prefer *A* to *B*, and one of the voters prefers *B* to *A*). Thus *B* is eliminated, and *A* moves on to confront *C*. Since *C* wins this one-on-one (by a score of 2 to 1), *A* is eliminated. Finally *C* takes on *D*, and *D* wins by a score of 2 to 1. Thus, *D* is the winner. ◆

There is something very troubling about the outcome of the preceding example, especially if you are candidate *B*. *Everyone* prefers *B* to *D*!

> Sequential pairwise voting fails to satisfy what is called the **Pareto condition:** if everyone prefers one alternative (in this case, *B*) to another alternative (*D*), then this latter alternative (*D*) is not among the winners.

The Hare System and Monotonicity

The social choice procedure known as the *Hare system* was introduced by Thomas Hare in 1861, and is also known by names such as the "single transferable vote system." In 1862, John Stuart Mill described the Hare system as being "among the greatest improvements yet made in the theory and practice of government." Today the system is used to elect public officials in Australia, Malta, the Republic of Ireland, and Northern Ireland.

> The **Hare system** proceeds to arrive at a winner by repeatedly deleting alternatives that are "least preferred" in the sense of being at the top of the fewest preference lists. If a single alternative remains after all others have been eliminated, it alone is the winner. If two or more alternatives remain and all of these remaining alternatives would be eliminated in the next round (because they all have the same number of first place votes), then these alternatives are declared to be tied for the win.

EXAMPLE ▶ *The Hare System*

Suppose we have the following sequence of preference lists:

	Number of Voters			
Rank	7	5	4	1
First	*A*	*C*	*B*	*D*
Second	*D*	*A*	*C*	*B*
Third	*B*	*B*	*D*	*A*
Fourth	*C*	*D*	*A*	*C*

Alternative D has 1 first-place vote, B has 4 first-place votes, C has 5 first-place votes, and A has 7 first-place votes. Thus, alternative D, having the fewest, is eliminated in round one. The sequence of preference lists now becomes

	Number of Voters			
Rank	7	5	4	1
First	A	C	B	B
Second	B	A	C	A
Third	C	B	A	C

Alternatives C and B now each have 5 first-place votes (because B gained one with D's elimination), while alternative A has 7 first-place votes. Thus, B and C are eliminated in round two because they now have the fewest first place votes. Alternative A is the only one remaining, and so it wins the election. ◆

In the preceding example, suppose that the voter on the far right moves alternative A up on his list. Let's look at the new election. Notice that, even though A won the last election, the only change we are making in ballots for the new election is one that is favorable to A. The preference lists for the new election are as follows:

	Number of Voters			
Rank	7	5	4	1
First	A	C	B	D
Second	D	A	C	A
Third	B	B	D	B
Fourth	C	D	A	C

If we apply the Hare system again, D is still eliminated in round one. But now only B is eliminated in round two, because B only has 4 first-place votes (to 5 for C and 8 for A). Thus, after these first two rounds, the preference lists are as follows:

	Number of Voters			
Rank	7	5	4	1
First	A	C	C	A
Second	C	A	A	C

We now have A on top of 8 lists and C on top of 9 lists. Thus, at stage three, A (our previous winner!) is eliminated and C is the winner of this new election.

Clearly, this is once again quite counterintuitive. Alternative *A* won the original election, the only change in ballots made was one favorable to *A* (and no one else), and then *A* lost the next election.

> This example shows that the Hare system does not satisfy **monotonicity:** if an alternative is a winner, and a new election is held in which the only ballot change made is for some voter to exchange that winning alternative with the one immediately above it on his or her ballot, then the original winner should remain a winner.

The fact that the Hare system does not satisfy monotonicity is considered by many—and with good reason—to be a glaring defect. Moreover, the previous example can be used to show that the Hare system, like the other voting systems we have looked at, can be manipulated (see Exercise 18).

Summary

We have introduced four voting procedures (plurality, the Borda count, sequential pairwise voting, and the Hare system), and four desirable properties of voting systems (the Condorcet winner criterion, independence of irrelevant alternatives, the Pareto condition, and monotonicity). We showed that each of the four voting procedures failed to satisfy at least one of the four properties. What about the others? While space prevents a complete discussion, the following table summarizes the results.

	CWC	IIA	Pareto	Mono
Plurality	No	No	Yes	Yes
Borda count	No	No	Yes	Yes
Sequential pairs	Yes	No	No	Yes
Hare system	No	No	Yes	No

INSURMOUNTABLE DIFFICULTIES: FROM PARADOX TO IMPOSSIBILITY

All four of the voting procedures that we have discussed turn out to be flawed in one way or another. You, the reader, may well ask at this point why we don't simply present *one* voting method for the three-alternative case that has all the desirable properties we want to have satisfied. This is, after all, exactly what we did for the two-alternative case.

The answer to this question is extremely important. The difficulties in the three-alternative case are not, in any way, tied to a few particular voting methods that we happen to present in a text such as this (or that we choose to use in the real world). The fact is, there are difficulties that will be present *regardless* of what voting method is used, and this applies even to voting methods not yet discovered. We discuss such difficulties in what follows.

The Voting Paradox of Condorcet

The Marquis de Condorcet (1743–1794) may have been the first to realize that serious difficulties can *always* arise in elections where there are three or more alternatives. The word that is typically used in connection with Condorcet's observation is "paradox." In general, the word *paradox* is applied whenever there is a situation in which apparently logical reasoning leads to an outcome that seems impossible. Quite often, a situation that is first described as being paradoxical comes, in the fullness of time, to be seen as simply a fact that reveals the extent to which our previous intuition was flawed.

Condorcet considered the following set of three preference lists and found that they indeed lead to a situation that seems paradoxical:

	Number of Voters		
Rank	1	1	1
First	A	B	C
Second	B	C	A
Third	C	A	B

If we view society as being broken down into thirds, with one-third holding each of Condorcet's preference lists, then society certainly seems to favor A to B (two-thirds to one-third) and B to C (again, two-thirds to one-third). Thus, we would expect society to prefer A to C. That is, we would expect the relation of social preference to be *transitive:* if A is "better than" B, and B is "better than" C, then surely A is "better than" C. But exactly the opposite is true. Society not only fails to prefer A to C but, in fact, rather strongly prefers C to A (i.e., by a two-thirds to one-third margin)! With, say, 10 alternatives, a similar phenomenon can occur with "two-thirds" replaced by "90%."

The fact that two-thirds of society can prefer A to B, two-thirds prefer B to C, and two-thirds prefer C to A is known as **Condorcet's voting paradox.**

Some authors apply the phrase "voting paradox" to any sequence of ballots in which there is no Condorcet winner. Either way, Condorcet's voting paradox is truly one of the cornerstones of modern social choice theory. We will make use of it in the remaining sections of this chapter.

Manipulability

As we have seen, there are voting situations in which one fares better by voting for less preferred alternatives rather than more preferred alternatives. This, however, is just one form of insincerity that arises within voting theory. For another, consider the following.

EXAMPLE ▶ *Bogus Amendments*

Diversionary amendments, when strategically introduced, can mislead some voters into acting against their own interest. To see how this can happen, assume that three representatives *A*, *B*, and *C* each have the choice of voting in favor of or against a new bill *N*. Voting against this new law means that the old law *O* will prevail. Assume that two of the three voters do prefer the proposed bill *N* over the existing law *O*, as indicated in this table of preferences:

	Voter		
	A	*B*	*C*
First choice	*N*	*N*	*O*
Second choice	*O*	*O*	*N*

In a direct comparison between the outcomes *N* and *O*, *N* will win by a vote of 2 to 1. Nevertheless, voter *C* may attempt to defeat *N* by the following maneuver. He proposes to modify the new bill *N* with an amended version called *M*, chosen so that *A* prefers *M* most of all, and *B* prefers *M* least of all. This may be done, perhaps, by merely shifting some of the proposed reward in bill *N* from *B* to *A*.

For *C*'s strategy to work, he must behave (i.e., vote) as if *O* is still his first choice, but that he likes *M* better than *N*. This may require deception on the part of *C*.

What effect will this have? Standard practice requires amendments to be considered, one at a time, before the main motion is brought to the floor. Thus, the voting method we find in use here is sequential pairwise voting with the agenda

M   N   O

Moreover, the sequence of preference lists is the following (which the reader will want to compare with those in the voting paradox of Condorcet):

	Voter		
	A	B	C
First choice	M	N	O
Second choice	N	O	M
Third choice	O	M	N

When voting between M and N, M wins over N by 2 to 1. At the second step, O beats M by 2 to 1. Thus, voter C has tricked the others into defeating N and maintaining the status quo, O. Voter A should have noticed this tactic and resisted the temptation to initially vote for the fleeting amendment M. In reality, however, C could have publicized amendment M in A's district, making it difficult for her to vote no, even if she saw C's trick (which is familiar to virtually all legislators). ◆

In the course of a long, complex agenda and heated debate, we must be continuously on guard to avoid being manipulated into voting against our own long-range interests. Those designing the agenda can often rig it in their own favor. For example, one contingent might stack up a larger number of popular outcomes and pit them against a *single* highly desired one in an attempt to eliminate this single outcome at an early stage of the agenda. As a general rule of thumb, it's best to enter the more preferred outcomes at a later stage of the agenda. The chances of survival may increase when there are fewer competing alternatives and fewer remaining votes to be taken.

It would be nice, of course, to have a voting system that is not vulnerable to this kind of deceptive strategy. This turns out to be too much to ask. In the early 1970s, Allan Gibbard and Mark Satterthwaite independently proved that in a context similar to the ones we have considered, there is no social choice procedure, except a dictatorship, that completely avoids situations wherein it is better for at least one of the voters to mark his or her ballot in an insincere way. Thus, in terms of strategy, honesty may not be the best policy.

Impossibility

Nothing in the remarkable body of work produced by Nobel laureate Kenneth Arrow of Stanford University is as well known or widely acclaimed as the result known as **Arrow's impossibility theorem** (see Spotlight 11.1). How, though, can one mathematically prove that it is *impossible* to find a voting system that satisfies certain properties?

Our goal here is to consider a version of Arrow's theorem and at least to sketch the argument behind it. This version is taken from the 1995 text cited in Suggested Readings, and uses stronger hypotheses than Arrow's original theorem.

The framework for our present considerations will be the same one with which we have been working. Ballots will be preference lists (without ties), and the outcome of an election will be either a single alternative (the winner) or a group of alternatives (tied for the win). We *do* demand of any social choice procedure that it definitely produce at least one winner when confronted by any sequence of preference lists.

Recall that the first two examples of social choice procedures that we considered were plurality voting and the Borda count. Moreover, we showed that plurality voting failed to satisfy the Condorcet winner criterion (CWC), and that the Borda count failed to satisfy independence of irrelevant alternatives (IIA). The theorem we want to prove here is the following:

> There does not exist, and never will exist, *any* social choice procedure that satisfies both the CWC and IIA.

More specifically, we claim that if a "voting rule" of some kind were to be found that satisfied both the CWC and IIA, then, when confronted by the three preference lists occurring in the voting paradox of Condorcet, this voting rule would *fail* to produce a winner (and thus not be a social choice procedure in the sense that we are using the phrase). Let's see why this is true.

The argument really comes in three separate, but extremely similar, pieces—one for each of the three alternatives. Piece 1 argues that alternative *A* can't be among the winners; piece 2 that *B* can't be among the winners; and piece 3 that *C* can't be among the winners. We'll do piece 1 and leave the others for the interested reader. The sequence of preference lists that we are considering is the following:

	Number of Voters		
Rank	1	1	1
First	*A*	*B*	*C*
Second	*B*	*C*	*A*
Third	*C*	*A*	*B*

Spotlight *Kenneth J. Arrow*

11.1

For centuries, mathematicians have been in search of a perfect voting system. Finally, in 1951, economist Kenneth Arrow proved that finding an absolutely fair and decisive voting system is impossible. Arrow is the Joan Kenney Professor of Economics, as well as a professor of operations research, at Stanford University. In 1972, he received the Nobel Memorial Prize in Economic Science for his outstanding work in the theory of general economic equilibrium. His numerous other honors include the 1986 von Neumann Theory Prize for his fundamental contributions to the decision sciences. He has served as president of the American Economic Association, the Institute of Management Sciences, and other organizations. Dr. Arrow talks about the process by which he developed his famous impossibility theorem and his ideas on the laws that govern voting systems:

Kenneth Arrow

My first interest was in the theory of corporations. In a firm with many owners, how do the owners agree when they have different opinions, for example, about the prospects of the company? I was thinking of stockholders. In the course of this, I realized that there was a paradox involved—that majority voting can lead to cycles. I then dropped that discussion because I was frustrated by it.

I happened to be working with The RAND Corporation one summer about a year or two later. They were very interested in applying concepts of rationality, particularly of game theory, to military and diplomatic affairs. That summer, I felt not like an economist but instead like a general social scientist or a mathematically oriented social scientist. There was tremendous interest in game theory, which was then new.

Someone there asked me, "What does it mean in terms of national interest?" I said, "Oh, that's a very simple matter," and he said, "Well, why don't you write us a little memorandum on the subject." Trying to write that memorandum led to a sharper formulation of the social-choice question, and I realized that I had been thinking of it earlier in that other context.

I think that society must choose among a number of alternative policies. These policies may be thought of as quite comprehensive, covering a number of aspects: foreign policy, budgetary policy, or whatever. Now, each individual member of the society has a preference, or a set of preferences, over these alternatives. I guess that you can say one alternative is better than another. And these individual preferences have a property I call *rationality* or *consistency,* or more specifically, what is technically known as *transitivity:* if I prefer *a* to *b,* and *b* to *c,* then I prefer *a* to *c.*

Imagine that society has to make these choices among a set. Each individual has a preference ordering, a ranking of these alternatives. But we really want society, in some sense, to give a ranking of these alternatives. Well, you can always produce a ranking, but you would like it to have some properties. One is that, of course, it be responsive in some sense to the individual rankings. Another is that when you finish, you end up with a real ranking, that is, something that satisfies these consistency, or transitivity, properties. And a third condition is that when choosing between a number of alternatives, all I should take into account are the preferences of the individuals among those alternatives. If certain things are possible and some are impossible, I shouldn't ask individuals whether they care about the impossible alternatives, only the possible ones.

It turns out that if you impose the conditions I just stated, there is no method of putting together the individual preferences that satisfies all of them.

The whole idea of the axiomatic method was very much in the air among anybody who studied mathematics, particularly among those who studied the foundations of mathematics. The idea is that if you want to find out something, to find the properties, you say, "What would I like it to be?" [You do this] instead of trying to investigate special cases. And I was really accustomed to this approach. Of course, the actual process did involve trial and error.

But I went in with the idea that there was some method of handling this problem. I started out with some examples. I had already discovered

that these led to some problems. The next thing that was reasonable was to write down a condition that I could outlaw. Then I constructed another example, another method that seemed to meet that problem, and something else didn't seem very right about it. Then I had to postulate that we have some other property. I found I was having difficulty satisfying all of these properties that I thought were desirable, and it occurred to me that they couldn't be satisfied.

After having formulated three or four conditions of this kind, I kept on experimenting. And lo and behold, no matter what I did, there was nothing that would satisfy these axioms. So after a few days of this, I began to get the idea that maybe there was another kind of theorem here, namely, that there was no voting method that would satisfy all the conditions that I regarded as rational and reasonable. It was at this point that I set out to prove it. And it actually turned out to be a matter of only a few days' work.

It should be made clear that my impossibility theorem is really a theorem [showing that] the contradictions are possible, not that they are necessary. What I claim is that given any voting procedure, there will be some possible set of preference orders for individuals that will lead to a contradiction of one of these axioms.

But you say, "Well, okay, since we can't get perfection, let's at least try to find a method that works well most of the time." Then when you do have a problem, you don't notice it as much. So my theorem is not a completely destructive or negative feature any more than the second law of thermodynamics means that people don't work on improving the efficiency of engines. We're told you'll never get 100% efficient engines. That's a fact— and a law. It doesn't mean you wouldn't like to go from 40% to 50%.

Our starting point, however, will be to ask what our hypothetical voting rule must do when confronted by a slightly different sequence of preference lists:

	Number of Voters		
Rank	1	1	1
First	*A*	*C*	*C*
Second	*B*	*B*	*A*
Third	*C*	*A*	*B*

Here, alternative *C* is clearly a Condorcet winner, and thus it must be the unique winner of the election contested under our hypothetical voting rule. Therefore, *C* is a winner and *A* is a nonwinner (for *this* sequence of preference lists).

However, because our hypothetical voting rule satisfies independence of irrelevant alternatives, we know that alternative *A* will remain a nonwinner as long as no one reverses his or her ordering of *A* and *C*. But to arrive at the preference lists from the voting paradox, we can move *B* (the alternative that is irrelevant to *A* and *C*) up one slot in the second voter's list.

Thus, because of IIA, we know that alternative *A* is a nonwinner when our voting rule is confronted by the preference lists from the voting paradox of Condorcet. This is one-third of the argument. As we mentioned before, similar arguments (see Exercise 21) show that *B* and *C* are also nonwinners when our voting rule is confronted by the preference lists from the voting paradox of Condorcet. Therefore, we have shown that no voting system which is guaranteed to produce at least one winner can satisfy both CWC and IIA.

A BETTER APPROACH? APPROVAL VOTING

Elections in which there are only two candidates present no problem. Majority rule is, as we have seen, an eminently successful voting system in both theory and practice. If there are three or more candidates, however, the situation changes quite dramatically. While several voting systems suggest themselves (plurality, the Borda count, sequential pairwise voting, and the Hare system), each fails to satisfy one or more desired properties (the Condorcet winner criterion, independence of irrelevant alternatives, the Pareto condition, and monotonicity). Manipulability is an ever-present problem. Moreover, when all is said and done, Arrow's impossibility theorem says that any search for an ideal voting system of the kind we have discussed is doomed to failure.

Where does this leave us? More than intellectual issues are at stake here: over 550,000 elected officials serve in approximately 80,000 governments in the United States. Whether it is a small academic department voting on the best senior thesis or a democratic country electing a new leader, multicandidate elections will be contested in one way or another. If there is no perfect voting system — and perhaps not even a best voting system (whatever that may mean; that is, best in what way?) — what can we do?

Perhaps the answer is that different situations lend themselves to different voting systems, and what is required is a judicious blend of common sense with an awareness of what the mathematical theory has to say. For example, while both the Hare system and the Borda count are subject to manipulability, it is clearly easier to manipulate the latter (recall the quote of Jean-Charles de Borda on page 420). Thus, people may tend to vote more sincerely, rather than strategically, if the Hare system is used instead of the Borda count. This may be a consideration when choosing a voting system for a faculty governance system, for example.

For national political elections, there are also practical considerations. The kind of ballot we are considering (an individual preference list) is certainly more complicated than the ballots we now employ, and preference lists cannot be used with existing voting machines. There is, however, a voting system that avoids the practical difficulties caused by the type of ballot being used that has much else to commend it. It is called *approval voting*.

Under **approval voting,** each voter is allowed to give one vote to as many of the candidates as he or she finds acceptable. No limit is set on the number of candidates for whom an individual can vote. Voters show disapproval of other candidates simply by not voting for them.

The winner under approval voting is the candidate who receives the largest number of approval votes. This approach is also appropriate in situations where more than one candidate can win, for example, in electing new members to an exclusive society such as the National Academy of Sciences or the Baseball Hall of Fame.

Approval voting was proposed independently by several analysts in the 1970s. Probably the best-known official elected by approval voting today is the secretary-general of the United Nations. In the 1980s, several academic and professional societies initiated the use of approval voting. Examples include the Institute of Electrical and Electronics Engineers (IEEE), with about 400,000 members, and the National Academy of Sciences. In Eastern Europe and some former Soviet republics, approval voting has been used in the form wherein one disapproves of (instead of approving of) as many candidates as one wishes.

MALKEVITCH, JOSEPH, AND WALTER MEYER.   *Graphs, Models and Finite Mathematics,* Prentice Hall, Englewood Cliffs, N.J., 1974. In chapter 10 there is an introduction to the problem of voting, including a discussion of the properties desired of any voting method (Arrow's axioms).

MERRILL, SAMUEL III.   *Making Multicandidate Elections More Democratic,* Princeton University Press, Princeton, N.J., 1988. This is a well-written treatment of voting from quite a practical point of view.

NURMI, HANNU.   *Compairing Voting Systems,* D. Reidel, Dordrecht, 1987. This monograph provides an excellent treatment, at a somewhat more technical level, of the topics dealt with in this chapter.

SARRI, DONALD G.   *The Geometry of Voting,* Springer-Verlag, New York, 1994. This monograph provides an advanced treatment of voting that focuses on ranking methods like the Borda count.

TAYLOR, ALAN D.   *Mathematics and Politics: Strategy, Voting, Power, and Proof,* Springer-Verlag, New York, 1995. Chapters 5 and 10 give an expanded treatment of the topics considered here, with proofs included. It is also intended for nonmajors.

EXERCISES   ▲ *Optional.*   ■ *Advanced.*   ◆ *Discussion.*

Elections with Only Two Alternatives

1.   In a few sentences, explain why a dictatorship (the voting procedure for two alternatives that is described on page 413) satisfies conditions (2) and (3) on page 413, but not (1).

2.   In a few sentences, explain why imposed rule (the voting procedure for two alternatives that is described on page 413) satisfies conditions (1) and (3) on page 413, but not (2).

3.   In a few sentences, explain why minority rule (the voting procedure for two alternatives that is described on page 413) satisfies conditions (1) and (2) on page 413, but not (3).

■   4.   Find (or invent) a voting rule for two alternatives that satisfies

   (a)   condition (1) on page 413, but neither (2) nor (3).
   (b)   condition (2) on page 413, but neither (1) nor (3).
   (c)   condition (3) on page 413, but neither (1) nor (2).

Elections with Three or More Alternatives: Procedures and Problems

5. (Everyone wins.) Consider the following set of preference lists:

			Number of Voters				
Rank	3	1	1	1	1	1	1
First	A	A	B	B	C	C	D
Second	D	B	C	C	B	D	C
Third	B	C	D	A	D	B	B
Fourth	C	D	A	D	A	A	A

Note that the first list is held by three voters, not just one. Calculate the winner using

   (a) plurality voting
   (b) the Borda count
   (c) the Hare system
   (d) sequential pairwise voting with the agenda *A, B, C, D*

6. Consider the following set of preference lists:

		Number of Voters			
Rank	2	2	1	1	1
First	C	D	C	B	A
Second	A	A	D	D	D
Third	B	C	A	A	B
Fourth	D	B	B	C	C

Calculate the winner using

   (a) plurality voting
   (b) the Borda count
   (c) the Hare system
   (d) sequential pairwise voting with the agenda *B, D, C, A*

7. Consider the following set of preference lists:

	Number of Voters					
Rank	2	2	1	1	1	1
First	A	E	A	B	C	D
Second	B	B	D	E	E	E
Third	C	D	C	C	D	A
Fourth	D	C	B	D	A	B
Fifth	E	A	E	A	B	C

Calculate the winner using

    (a) plurality voting
    (b) the Borda count
    (c) the Hare system
    (d) sequential pairwise voting with the agenda *B, D, C, A, E*

8. Consider the following set of preference lists:

	Number of Voters				
Rank	1	1	1	1	1
First	A	B	C	D	E
Second	B	C	B	C	D
Third	E	A	E	A	C
Fourth	D	D	D	E	A
Fifth	C	E	A	B	B

Calculate the winner using

    (a) plurality voting
    (b) the Borda count
    (c) the Hare system
    (d) sequential pairwise voting with the agenda *A, B, C, D, E*

9. Consider the following set of preference lists:

	Number of Voters				
Rank	2	2	1	1	1
First	A	B	A	C	D
Second	D	D	B	B	B
Third	C	A	D	D	A
Fourth	B	C	C	A	C

Calculate the winner using

    (a)   plurality voting
    (b)   the Borda count
    (c)   the Hare system
    (d)   sequential pairwise voting with the agenda B, D, C, A

10.   Consider the following set of preference lists:

	Number of Voters				
Rank	2	2	1	1	1
First	C	E	C	D	A
Second	E	B	A	E	E
Third	D	D	D	A	C
Fourth	A	C	E	C	D
Fifth	B	A	B	B	B

Calculate the winner using

    (a)   plurality voting
    (b)   the Borda count
    (c)   the Hare system
    (d)   sequential pairwise voting with the agenda A, B, C, D, E

11.   In a few sentences, explain why plurality voting satisfies

    (a)   the Pareto condition
    (b)   monotonicity

12.   In a few sentences, explain why the Borda count satisfies

    (a)   the Pareto condition
    (b)   monotonicity

13.   In a few sentences, explain why sequential pairwise voting satisfies

    (a)   the Condorcet winner criterion
    (b)   monotonicity

14.   In a few sentences, explain why the Hare system satisfies the Pareto condition.

23.   How many different ways can a voter

   (a)   rank 3 choices when ties are not allowed, but *incomplete* rankings can be submitted (e.g., a first choice without giving a second or third choice)?
   (b)   rank 3 choices when ties are allowed and *complete* rankings are required?

24.   Ten board members vote by approval voting on eight candidates for new positions on their board as indicated in the following table. An **X** indicates an approval vote. For example, voter 1, in the first column, approves of candidates *A, D, E, F*, and *G*, and disapproves of *B, C*, and *H*.

Candidates	Voters 1	2	3	4	5	6	7	8	9	10
A	X	X	X			X	X	X		X
B		X	X	X	X	X	X	X	X	
C			X					X		
D	X	X	X	X	X		X	X	X	X
E	X		X		X		X		X	
F	X		X	X	X	X	X	X		X
G	X	X	X	X	X			X		
H		X		X		X		X		X

   (a)   Which candidate is chosen for the board if just one of them is to be elected?
   (b)   Which candidates are chosen if the top four are selected?
   (c)   Which candidates are elected if 80% approval is necessary and at most four are elected?
   (d)   Which candidates are elected if 60% approval is necessary and at most four are elected?

25.   To be elected to the Baseball Hall of Fame a player must be retired for five years and receive a vote from 75% of some 420 actual voters. (A few eligible voters often do not cast a ballot.) In the election for January 1993 there were 423 voters, and the top five finishers (and their number of votes) were

Reggie Jackson (396)
Phil Niekro (278)
Orlando Cepeda (252)
Tony Perez (233)
Steve Garvey (176)

(a)   Who was elected in 1993?

(b)   How many more votes would Perez have needed to have been elected?

(c)   What percent of the voters who did not vote for Garvey would have had to change and vote for him in order for Garvey to have been elected?

▲   (d)   Is there any way in which a voter can vote in an insincere manner to help or hurt the chances of some player?

26.   A player remains on the ballot for the Baseball Hall of Fame for 15 years provided he receives 5% of the votes cast each year. Some other players (and their votes) in the 1993 election were

Mickey Lolich (43)
Thurman Munson (40)
Rusty Staub (32)
Bill Maddock (19)
Ron Cey (8)

Which of these five players meets the 5% cutoff criterion (for the 423 votes cast) for remaining on the ballot for the 1994 election?

27.   Consider the following set of preference lists:

	Number of Voters						
Rank	1	1	1	1	1	1	1
First	C	D	C	B	E	D	C
Second	A	A	E	D	D	E	A
Third	E	E	D	A	A	A	E
Fourth	B	C	A	E	C	B	B
Fifth	D	B	B	C	B	C	D

Calculate the winner using

(a)   plurality voting
(b)   the Borda count
(c)   sequential pairwise voting with the agenda *A, B, C, D, E*
(d)   the Hare system

28.   An interesting variant of the Hare system was proposed by the psychologist Clyde Coombs. It operates exactly as does the Hare system, but instead of deleting alternatives with the fewest first-place votes, it deletes those with the most last-place votes.

(a)   Use the Coombs procedure to find the winner if the ballots are as in Exercise 27.

(b) Show that for two voters and three alternatives, it is possible to have ballots that result in one alternative winning if the Coomb's procedure is used and a tie between the other two if the Hare system is used.

29. The 45 members of a school's football team vote on three nominees, *A, B,* and *C,* by approval voting for the award of "most improved player" as indicated in the following table. An **X** indicates an approval vote.

Nominee	Number of Voters							
	7	8	9	9	6	3	1	2
A	X			X	X		X	
B		X		X		X	X	
C			X		X	X	X	

(a) Which nominee is selected for the award?

(b) Which nominee gets announced as runner-up for the award?

(c) Note that two of the players "abstained," that is, approved of none of the nominees. Note also that one person approved of all three of the nominees. What would be the difference in the outcome if one were to "abstain" or "approve of everyone"?

WRITING PROJECTS

1 ▶ In the 1992 presidential election, the final results were as follows:

Candidates	Number of Votes	Percentage of Votes
Clinton	43,727,625	43
Bush	38,165,180	38
Perot	19,236,411	19

Making reasonable assumptions about voters' preference schedules, discuss how the election might have turned out under the different voting methods discussed in this chapter.

2 ▶ Frequently in presidential campaigns, the winner of the first few primaries is given front-runner status that can lead to the nomination of his or her party. Frequently there are several candidates running in early primaries such as New Hampshire. Consider a recent election (for example, the 1996 Republican primaries), and discuss how the nominating process might have proceeded through the campaign if approval voting had been used to decide primary winners.

WEIGHTED VOTING SYSTEMS

I n some voting situations, the "one person, one vote" principle does not apply. For example, when the shareholders of a public corporation elect a board of directors, each shareholder is entitled to one vote per share owned. Shareholders who own relatively large numbers of shares usually have greater influence in such an election than the small shareholders do.

A *weighted voting system* is a decision-making procedure in which the participants have varying numbers of votes. Examples of such systems include shareholder elections and the election of the president of the United States by the electoral college (see Spotlight 12.1). Some legislative bodies have such strong party discipline that each legislator always votes as dictated by his or her party. These legislatures are weighted voting systems in which the participants are the political party organizations, each of which is entitled to a number of votes equal to the size of its delegation in the legislature.

The *power* of a participant in a weighted voting system can be roughly defined as the ability of the participant to influence a decision. There are several ways to measure mathematically the power of a participant in a weighted voting system. We will study two ways, the *Banzhaf power index* and the *Shapley–Shubik power index*. Either of these indices provides a much more accurate measure of a participant's power than the number of votes that the participant is entitled to cast.

Delegates at the 1992
Republican National
Convention.

HOW WEIGHTED VOTING WORKS

In 1958, the Board of Supervisors of Nassau County, New York, consisted of six supervisors from five municipalities. Two of the supervisors were elected at large from the city of Hempstead, which had more than half of the county's population. To compensate for the unequal populations of the municipalities, the supervisors were given weighted votes, as described in Table 12.1.

The total number of votes assigned to the supervisors was 30, and a simple majority (16 votes) was required to pass a measure. Since the two Hempstead supervisors controlled 18 votes between them, they would have had the power to pass any measure without consulting their colleagues from the smaller municipalities. However, the Nassau County Charter contained a provision requiring that any measure must have the support of supervisors from two different municipalities in order to pass. This provision complicates our analysis, so we will ignore it for now and return to it later.

TABLE 12.1	**Weighted Voting, Nassau County Board of Supervisors, 1958**
Municipality	Number of Votes
Hempstead ⎱ Hempstead ⎰	⎰ 9
North Hempstead	7
Oyster Bay	3
Glen Cove	1
Long Beach	1
Total	30

To pass a measure, the two Hempstead supervisors can vote together, or one of the Hempstead supervisors can vote with the North Hempstead supervisor. If one of the Hempstead supervisors should sponsor a bill, he or she will quickly find out that it is not worthwhile to lobby the supervisors of Oyster Bay, Glen Cove, and Long Beach to obtain their support. Between them, these supervisors have only five votes, so even if they added their votes to the sponsor's nine votes, the total would be only 14—not enough to pass the measure. If the three supervisors from the smaller communities joined the North Hempstead supervisor, their votes would total only 12. No bill can pass without the support of at least two of the Hempstead and North Hempstead supervisors, and if it has the support of two of these supervisors, it will pass without the help of anyone else.

In this situation, the three supervisors from Oyster Bay, Glen Cove, and Long Beach have no voting power. They might influence the decision-making process by serving on committees, by introducing bills, and by participating in the debate, but they are disenfranchised by the board's voting system. A voter whose vote will never be needed to pass any measure, or to defeat any measure, is called a **dummy.**

Notation for Weighted Voting

A **weighted voting system** is a system that consists of a set of voters V_1, V_2, V_3, . . . , V_n, who are assigned **weights,** $w(V_1)$, $w(V_2)$, $w(V_3)$, . . . , $w(V_n)$. Each voter's weight is the number of votes that he or she is allowed to cast. The total number q of votes necessary to pass a measure is called the **quota** of the system.

$\mathcal{S}potlight$ *The Electoral College*

12.1

In a presidential election the voters select representatives to the electoral college, who are called *electors*. The number of electors allotted to a state is equal to the size of its congressional delegation; thus a state with one congressional district gets three electors, since it has one representative in the U.S. House of Representatives and two senators. A state with 25 representatives and two senators would be entitled to 27 electors.

Each state requires its electors to vote as a bloc for the candidate who received a plurality in the state's general election. Thus the electoral college is a weighted voting system with 51 participants (the states and the District of Columbia). The weights range from 3 to 54, and the quota is a simple majority of 270.

A third-party candidate (such as H. Ross Perot in 1992) can acquire a significant percentage of the popular vote, and prevent anyone from receiving a majority. However, third-party candidates rarely get any electoral votes, so one of the major-party candidates usually gets a majority in the electoral college. The system doesn't always work as planned, though. In 1876 Samuel J. Tilden received a *majority* of the popular vote, but lost by one vote in the electoral college to Rutherford B. Hayes.

The shorthand notation

$$[q: w(V_1), w(V_2), \ldots, w(V_n)]$$

is used to describe the weighted voting system with n voters $V_1, V_2, \ldots, V_n$, with weights $w(V_1), \ldots, w(V_n)$ and with quota q. Thus, the voting system used by the Nassau County Board of Supervisors in 1958 is expressed as

$$[q: w(H_1), w(H_2), w(N), w(B), w(G), w(L)] = [16:9, 9, 7, 3, 1, 1]$$

where H_1 and H_2 are the two Hempstead supervisors, N is the North Hempstead supervisor, and B, G, and L are the Oyster Bay, Glen Cove, and Long Beach supervisors, respectively.

A set of voters who have joined together to vote in favor of an issue, or to oppose an issue, is called a **coalition.** The coalition may consist of all the voters or any subset of the voters. It may consist of just one voter, or it may even be *empty.* For example, if the voting body is unanimously in favor of a motion, then the coalition opposing the motion is empty.

A coalition of voters in favor of a measure is a **winning coalition** if the sum of its weights equals or exceeds the quota q. A **blocking coalition** is a subset of voters opposing a motion, with enough votes to defeat it.

If the voting system is to reach an unambiguous decision, it is important not to permit two winning coalitions in opposition to each other. For this reason, we will require $q > \frac{1}{2}w$, where w is the sum of the weights of all participants in the voting system. Thus, if a coalition is winning, it must have more than half of the votes. The voters who did not join the coalition will then have less than half of the votes, and cannot possibly win.

In a voting system with total weight w and quota q, any coalition with weight more than $w - q$ is a blocking coalition. For example, in the Nassau County Board of Supervisors of 1958, $w = 30$, $q = 16$, and $w - q = 14$; so any coalition with more than 14 votes is a blocking coalition.

Winning coalitions always have enough votes to block a measure, since they have more than half the votes. However, there can be blocking coalitions whose votes total less than the quota. For a simple example, consider a voting system with four voters, each with one vote, and a quota of three votes. Any coalition of two voters opposing a measure is a blocking coalition, although these voters could not pass any measure they favored without being joined by a third voter.

A jury in a criminal trial might have 12 members. A decision to convict or to acquit must be unanimous, so a winning coalition requires all 12 jurors. If the jury cannot agree on a verdict, a mistrial is declared and the prosecution has the right to demand a new trial. Therefore, in this case every coalition with at least one member is a blocking coalition.

If the voting weight of one voter is greater than or equal to the quota for passing a measure, then that voter is called a **dictator.** The other voters are then dummies, since their votes will have no effect. For example, if one stock-

Stock certificates.

holder owns 51% of the shares in a corporation, then he or she is a dictator and controls the business of the corporation.

A voter who has enough votes to block any measure is said to have **veto power.** Of course, a dictator automatically has veto power, but it is possible to have veto power without being a dictator. For example, we have just seen that every juror in a criminal trial has veto power.

EXAMPLE ▶ *Weighted Voting Systems*

1. Consider a small corporation owned by two people, *A* and *B*, who possess 60% and 40% of the stock, respectively. If measures are allowed to pass by a simple majority, we express this voting system as

    $$[q: w(A), w(B)] = [51:60, 40]$$

    In this example, shareholder *A* is a dictator, and *B* is a dummy.

2. Let us examine a second company, with three shareholders, *A*, *B*, and *C*, who hold 49%, 48%, and 3% of the stock, respectively. Thus

    $$[q: w(A), w(B), w(C)] = [51:49, 48, 3]$$

    There is no dictator; indeed, this company is more "democratic" than one might expect. Any coalition of two or more shareholders has a simple majority, so the power is equally divided among the three shareholders.

3. A third company has shareholders *A*, *B*, *C*, and *D*. Shareholders *A*, *B*, and *C* each own 26% of the stock, while *D* holds the remaining 22%. The voting system for this corporation is

    $$[q: w(A), w(B), w(C), w(D)] = [51:26, 26, 26, 22]$$

    A measure will pass if it gains the support of two of the shareholders *A*, *B*, and *C*. Although *D*'s share of the company is not much less than the shares of any of the other three shareholders, *D* is a dummy. The power in this company is equally divided among *A*, *B*, and *C*.

These examples show that the relationship between voting weight and voting power is not simple. We can be sure that if a voter *A* has more votes than another voter *B*, then *B* is not more powerful than *A*, and that voters with the same weights will have the same voting power. However, the second example shows that two voters may be equally powerful, even if one has many more votes than the other. On the other hand, the first and third examples show that a dummy voter may have almost as many votes as a voter who has considerable power. ◆

THE BANZHAF POWER INDEX

We have seen from our examples that participants in a weighted voting system cannot take their fraction of the total vote as a meaningful indication of their share of voting power. Power is the ability to win. An individual can appear frequently on the winning side, however, without being powerful. For example, if a professional sports team usually wins, that does not mean that all members of the team can demand high salaries. The high-salaried players are those who are crucial to winning. Similarly, the real significance of a vote is whether it is essential to victory.

One reasonable measure of voting power is the frequency with which a participant's vote is critical for winning coalitions. This measure is a count of the number of different ways that the participant alone can turn defeat into victory, or vice versa.

> A **critical voter** is a voter who belongs to a winning or blocking coalition, and who can single-handedly cause that coalition to lose by changing his or her vote.
>
> A voter's **Banzhaf power index** is the number of distinct winning coalitions in which the participant is a critical voter, plus the number of distinct blocking coalitions in which he or she is a critical voter.

Since every winning coalition would become a blocking coalition if all of its members reversed their votes, some voters may be counted as critical voters twice in the same coalition, once for being critical voters in a winning coalition and again for being critical voters in a blocking coalition.

The Banzhaf power index was developed in 1965 by an attorney, John F. Banzhaf III (see Spotlight 12.2), in an analysis of weighted voting that was provocatively entitled "Weighted Voting Doesn't Work." Because the Banzhaf power index is a relatively new concept, terminology has not become standardized yet. For example, in some of the literature cited at the end of this chapter, terms such as *pivotal voter* and *swing voter* are used in place of *critical voter*.

EXAMPLE ▶ *A Three-Person Committee*

A committee has a chairperson A, with two votes, and two other members, B and C, each of whom has one vote. The quota for passing a measure is three votes. We can express this weighted voting system as

$$[q : w(A), w(B), w(C)] = [3 : 2, 1, 1]$$

Spotlight *Power Indices*

12.2

Lloyd S. Shapley   **John F. Banzhaf III**   **Martin Shubik**

The first widely accepted numerical index for assessing power in voting systems was the Shapley–Shubik index, developed in 1954 by a mathematician, Lloyd S. Shapley, currently at the University of California, Los Angeles, and an economist, Martin Shubik, of Yale University. A particular voter's power as measured by this index is proportional to the number of different *permutations* (or orderings) of the voters in which he or she has the potential to cast the pivotal vote—the vote that first turns from losing to winning.

The Banzhaf power index was introduced in 1965 by John F. Banzhaf III, a law professor at George Washington University and well-known consumer advocate. This index is the one most often cited in court rulings, perhaps because Banzhaf brought several cases to court and continues to file *amicus curiae* briefs when courts evaluate weighted voting systems. A voter's Banzhaf index is the number of different possible voting *combinations* in which he or she casts a critical vote—a vote in favor of a motion that is necessary for the motion to pass, or a vote against a motion that is essential for its defeat.

The winning coalitions are all those whose weights sum to 3 or 4: $\{A, B\}$, $\{A, C\}$, and $\{A, B, C\}$.

The chairperson A has veto power, and so is a critical voter in each of the three winning coalitions. This means that if she defects from any of these coalitions, it becomes a **losing coalition.** For instance, if A votes yes with B and C, the measure passes with four votes. When A switches to a no vote, the measure fails to pass since there are only two yes votes. She is also a critical voter in three blocking coalitions: $\{A\}$, $\{A, B\}$, and $\{A, C\}$. She is not a critical voter in the blocking coalition {*A, B, C*}, since {*B, C*}, with a total weight of 2, would still block if she defected. The members *B* and *C* have equal power. Neither is a critical voter in the winning coalition {*A, B, C*}, since both would have to defect to turn that coalition into a losing one. Member *B* is a critical voter in {*A, B*} as a winning coalition, since *A* cannot pass a motion by herself, but not a critical voter in {*A, B*} as a blocking coalition, since *A* can veto a motion by herself. Similarly, *C* is a critical voter in {*A, C*}, as a winning coalition, but not as a blocking coalition. Finally, both *B* and *C* are critical voters in the blocking coalition {*B, C*}. Thus, *A* is a critical voter in three winning coalitions and three blocking coalitions, and so has a Banzhaf power index of 6. The committee members *B* and *C* each have Banzhaf power indices of 2, counting one critical winning and one critical blocking vote each. According to the Banzhaf model, *A* has three times as much power as *B* (or *C*), even though her vote has only twice the weight.

To summarize, the Banzhaf index for this voting system is (6, 2, 2). Notice that each of the voters in the example is a critical voter in equal numbers of winning coalitions and blocking coalitions. These are not coincidences. If a voter who is a critical voter defects from a winning coalition to join the opposing coalition, the opposing coalition becomes a blocking coalition, and the same voter is now a critical voter in the blocking coalition. If a critical voter defects from a blocking coalition, that coalition would no longer have enough weight to block, and the opposing coalition would win. Thus, every voter is a critical voter in exactly the same number of blocking coalitions as he or she is a critical voter in winning coalitions. Knowing this, we can determine a participant's Banzhaf index simply by counting the winning coalitions in which he or she is a critical voter and doubling the result. ◆

Computing the Banzhaf Power Index

In a weighted voting system with no more than four voters, it is not difficult to calculate the Banzhaf power index by a brute force method. We merely list all of the theoretically possible ways that the participants can vote; that is, all the different **combinations** of yes and no votes. If there are n voters, there will be 2^n such combinations (see "How to Count Combinations," on page 456). Thus, with three voters, there are eight combinations, and with four voters, sixteen combinations. We examine each combination to identify the critical voters: the voters who would change the outcome if they alone switched their votes. Each voter's Banzhaf power index is the number of critical votes cast by that voter, when all 2^n voting combinations have been considered.

Let us use the brute force method to recalculate the Banzhaf power index of the voting system

$$[q : w(A), w(B), w(C)] = [3 : 2, 1, 1]$$

TABLE 12.2 Combinations of Votes in the Three-Person Committee

Member:	A	B	C			
Weight:	2	1	1			
Combinations				Votes	Pass	Fail
Y	Y	Y		4	P	
Y	Y	N		3	P	
Y	N	Y		3	P	
Y	N	N		2		F
N	Y	Y		2		F
N	Y	N		1		F
N	N	Y		1		F
N	N	N		0		F

for the three-person committee that was presented earlier. Table 12.2 lists the eight combinations of voters, according to whether they vote yes (Y) or no (N). Whether the issue will pass (P) or fail (F) is indicated in the outcome columns to the right of the combinations. If the issue will pass, the coalition that voted Y is a winning coalition; if the issue fails the voters who voted N form a blocking coalition.

Each row of the table is examined to see which of the voters are critical voters. This means checking each yes and no vote in each combination to determine whether a switch of that one vote will change the result.

For example, the first combination

A	B	C	
Y	Y	Y	P

results in passing the measure by a unanimous vote. If the voter A changes her vote from yes to no

A	B	C	
Y	Y	Y	P
↓			
N	Y	Y	F

then the outcome changes to F. We will indicate that A is a critical voter in this combination by circling the Y that indicates her yes vote:

A	B	C	
Ⓨ	Y	Y	P

On the other hand, if only voter B switches his vote from yes to no in the first combination, the result remains the same: the issue passes, 3 to 1:

A	B	C	
Y	Y	Y	P
Y	N	Y	P

Therefore B isn't a critical voter in this voting combination. For the same reason, voter C is also not a critical voter in this combination.

Now let us consider the second combination:

A	B	C	
Y	Y	N	P

If voter A changes her vote from yes to no, the measure will no longer pass:

A	B	C	
Ⓨ	Y	N	P
N	Y	N	F

Furthermore, if voter B changes his vote, the measure will be defeated:

A	B	C	
Y	Ⓨ	N	P
Y	N	N	F

Therefore A and B are critical voters in the second combination. Voter C is not a critical voter, since if she changes her vote, the outcome will not change:

A	B	C	
Y	Y	N$\downarrow$	P
Y	Y	Y	P

We proceed in the same way with each row of Table 12.2, determining each critical voter and circling the corresponding Y or N. Table 12.3 summarizes the result of calculations. Counting the number of circles in each voter's column, we find that the Banzhaf power index is (6, 2, 2).

EXAMPLE ▶ *A Corporation with Four Shareholders*

Consider a weighted voting system

$$[q: w(A), w(B), w(C), w(D)] = [51:40, 30, 20, 10]$$

This could represent four shareholders, A, B, C, and D in a corporation, owning 40%, 30%, 20%, and 10% of the stock, respectively. A simple majority (taken here as 51%) is necessary to pass a measure.

TABLE 12.3 The Critical Voters in Each Combination of Votes in the Three-Person Committee

Member:	A	B	C	Votes	Pass	Fail
Weight:	2	1	1			
	Combinations					
	Ⓨ	Y	Y	4	P	
	Ⓨ	Ⓨ	N	3	P	
	Ⓨ	N	Ⓨ	3	P	
	Y	Ⓝ	Ⓝ	2		F
	Ⓝ	Y	Y	2		F
	Ⓝ	Y	N	1		F
	Ⓝ	N	Y	1		F
	N	N	N	0		F
Number of critical votes	6	2	2			

TABLE 12.4 The Critical Voters in Each Combination of Votes in the Four-Stockholder Corporation

Stockholder:	A	B	C	D			
Percent Ownership:	40	30	20	10			
	Combinations				Votes	Pass	Fail
	Y	Y	Y	Y	100	P	
	Ⓨ	Y	Y	N	90	P	
	Ⓨ	Ⓨ	N	Y	80	P	
	Ⓨ	Ⓨ	N	N	70	P	
	Ⓨ	N	Ⓨ	Y	70	P	
	Ⓨ	N	Ⓨ	N	60	P	
	Y	Ⓝ	Ⓝ	Y	50		F
	Y	Ⓝ	Ⓝ	N	40		F
	N	Ⓨ	Ⓨ	Ⓨ	60	P	
	Ⓝ	Y	Y	Ⓝ	50		F
	Ⓝ	Y	Ⓝ	Y	40		F
	Ⓝ	Y	N	N	30		F
	Ⓝ	Ⓝ	Y	Y	30		F
	Ⓝ	N	Y	N	20		F
	N	N	N	Y	10		F
	N	N	N	N	0		F
Number of critical votes	10	6	6	2			

We will find the Banzhaf power index by listing the $2^4 = 16$ distinct combinations of yes (Y) and no (N) for the shareholders. Each combination is shown in one of the rows of Table 12.4. The total percent of yes votes for each combination is indicated at its right. The issue passes (P) or fails (F), depending on whether or not the percent of yes votes meets the quota of 51%. Each vote in each combination must be examined to determine whether or not it is a critical vote.

Will the change of this vote alter the result? If it will, the vote is a critical vote and it is circled. By counting the number of critical votes in each shareholder's column, we find that the Banzhaf power index is (10, 6, 6, 2). Notice that while B and C own different amounts of stock, they have the same voting power.

Again, we have counted each critical voter in winning and in blocking coalitions. We could have made the process more efficient by counting only the critical votes in winning coalitions or, alternatively, we could have counted only the critical votes in blocking coalitions.

To restrict our attention to the winning coalitions, consider only the rows that show the measure passing. A voter is a critical voter in a voting combination given by one of these rows if (1) the voter voted Y, and (2) the measure would fail if the voter switched his or her vote to N. In Table 12.4, counting either the critical votes in winning coalitions, or the critical votes in blocking coalitions, will yield half the Banzhaf index: (5, 3, 3, 1). ◆

How to Count Combinations

How do we know that for n voters there is a total of 2^n voting combinations? To answer this question, we will use the multiplication principle. Each voter has two options: to vote either yes or no, and the voters are independent of each other. Therefore the number of ways n voters can cast their votes is

$$\underbrace{2 \times 2 \times 2 \times \cdots \times 2}_{n \text{ factors}} = 2^n$$

The number of combinations with exactly k yes votes is called C_k^n. When speaking, people often refer to C_k^n as "n choose k." Another common notation for C_k^n is $\binom{n}{k}$.

If $n = 5$ and $k = 1$, C_k^n is the number of ways of getting one yes vote in a committee of five; thus $C_1^5 = 5$. To calculate C_2^5, we must determine the number of ways a coalition of two voters can form in the same committee. Name the committee members A, B, C, D, and E. The two-member coalitions involving A are $\{A, B\}$, $\{A, C\}$, $\{A, D\}$, and $\{A, E\}$. Similarly, there are four two-member coalitions involving each of the other members, for a total of 5×4 coalitions. However we counted each coalition twice! For example, $\{A, B\}$ was counted both as a coalition involving A and as a coalition involving B. We can correct this multiple counting by dividing the number of coalitions that we counted by two, to get $C_2^5 = 10$.

To calculate C_6^n we would ask, "How many combinations consist of 6 yes votes and $n - 6$ no votes?" Of course, the answer is "none" if $n < 6$, so we will assume $n \geq 6$. If we decide to keep track of the order in which the six yes voters cast their votes, then there are n voters who could cast the first yes vote. The first voter cannot vote again, so there are $n - 1$ voters who could cast the second yes vote. Similarly, there are $n - 2$ voters who could cast the third yes vote, $n - 3$ who could cast the fourth, $n - 4$ who could cast the fifth, and $n - 5$ who could cast the sixth. Again, by the multiplication principle, there are

$$n \times (n - 1) \times (n - 2) \times (n - 3) \times (n - 4) \times (n - 5)$$

ways that exactly six voters could vote yes, *if we keep track of the order in which they vote.*

To correct for multiple counting now, we must divide by the number of times that each six-member coalition was counted. This is the number of ways that the voters in the coalition can be put in order. In each coalition, there are six voters who could have been first, and when the first is known, five who could have been second, and so on. There are $6 \times 5 \times 4 \times 3 \times 2 \times 1 = 720$ different orderings, so that if we want to find the number of ways that there could be six yes votes without reference to order, we must divide by $6 \times 5 \times 4 \times 3 \times 2 \times 1$ to get

$$C_6^n = \frac{n \times (n-1) \times (n-2) \times (n-3) \times (n-4) \times (n-5)}{6 \times 5 \times 4 \times 3 \times 2 \times 1}$$

combinations with six yes votes. There is nothing special about the number six here; the number of combinations with k yes votes is given by a similar formula.

> The symbol C_k^n refers to the number of ways that k voters can be selected from a set of n voters. The formula for C_k^n is
>
> $$C_k^n = \frac{n \times (n-1) \times (n-2) \times \cdots \times (n-k+1)}{k \times (k-1) \times (k-2) \times \cdots \times 1}$$
>
> An easy way to remember this formula is that both the numerator and the denominator have k (the number of yes votes) factors; the factors of the numerator start with n (the total number of voters) and count down, while the factors of the denominator start with k and count down. Many scientific calculators have keys for calculating C_k^n.

Efficient counting methods make it possible to compute the Banzhaf power index of large weighted voting systems. The method of counting combinations applies to systems in which most of the voters have equal weights, as in the following example. When there are many different weights, more advanced counting methods must be used. For example, in 1992 the electoral college of the United States had 21 different weights, ranging from 3 to 54 votes. The brute force method is completely impractical: the participants are the 50 states and the District of Columbia, so there are $2^{51} = 2,251,799,813,685,248$ combinations to examine. This would take over 70 years at one million combinations per second. However, it is possible to compute the Banzhaf index of the electoral college in a few minutes with a home computer. Table 12.5 displays the Banzhaf index, which was calculated by using *generating functions,* an

---

TABLE 12.5 The Electoral College*

States	Electoral Votes	Nominal Power (%)	Banzhaf Power (%)
CA	54	10.04	11.14
NY	33	6.13	6.20
TX	32	5.95	6.00
FL	25	4.65	4.63
PA	23	4.28	4.25
IL	22	4.09	4.06
OH	21	3.90	3.87
MI	18	3.35	3.30
NJ	15	2.79	2.75
NC	14	2.60	2.56
GA, VA	13	2.42	2.38
IN, MA	12	2.23	2.19
MO, TN, WA, WI	11	2.04	2.01
MD, MN	10	1.86	1.82
AL, LA	9	1.67	1.64
AZ, CO, CT, KY, OK, SC	8	1.49	1.46
IA, MS, OR	7	1.30	1.28
AR, KS	6	1.12	1.09
NE, NM, UT, WV	5	0.93	0.91
HI, ID, ME, NV, NH, RI	4	0.74	0.73
AK, DE, DC, MT, ND, SD, VT, WY	3	0.56	0.55

* A state's nominal power is the number of electoral votes that it has, expressed as a percentage of the total number of electoral votes for the nation (538). A state's Banzhaf power is the Banzhaf power index, expressed as a percentage of the total number of critical votes for all states in all winning and blocking coalitions (9,426,404,631,750,950 critical votes in all).

---

advanced counting method described in the article by Paul J. Affuso and Steven J. Brams cited in Suggested Readings in this chapter.

EXAMPLE ▶ *A Seven-Person Committee*

A certain committee has a chairperson with three votes and six ordinary members, each of whom casts one vote. The quota for passing a measure is five. If we let C represent the chairperson and $M_1, \ldots, M_6$ represent the ordinary members, then the weighted system used by the committee is

$$[q: w(C), w(M_1), w(M_2), w(M_3), w(M_4), w(M_5), w(M_6)]$$
$$= [5:3, 1, 1, 1, 1, 1, 1]$$

We will calculate the Banzhaf power index for each person in the committee. An ordinary member such as M_6 will be a critical voter in winning coalitions consisting of M_6 and exactly four other ordinary members, and in winning coalitions consisting of the chair C, M_6, and exactly one other ordinary member. These coalitions have exactly five votes. A coalition with fewer than five votes would be losing, and in a coalition with more than five votes, M_6 would not be a critical voter.

There are C_4^5 ways to assemble a five-vote coalition consisting of M_6 and four of the five other ordinary members, and C_1^5 ways to choose one of $M_1 - M_5$ to join M_6 and C to form a minimal winning coalition. Since

$$C_4^5 = \frac{5 \times 4 \times 3 \times 2}{4 \times 3 \times 2 \times 1} = 5$$

and $C_1^5 = 5$, M_6 casts critical votes in 10 winning coalitions. The Banzhaf power index of M_6 (and each of the other ordinary members) is therefore 20: 10 for the winning coalitions and 10 for the blocking coalitions.

The chair C is a critical voter in any winning coalition consisting of C and 2, 3, or 4 ordinary members. If C is joined by 5 or all 6 ordinary members, then C would not be a critical voter. There are C_2^6 ways to choose 2 ordinary members to join the chair, C_3^6 ways to choose 3, and C_4^6 ways to choose 4. The number of winning coalitions in which C is a critical voter is thus

$$C_2^6 + C_3^6 + C_4^6 = \frac{6 \times 5}{2 \times 1} + \frac{6 \times 5 \times 4}{3 \times 2 \times 1} + \frac{6 \times 5 \times 4 \times 3}{4 \times 3 \times 2 \times 1} = 50$$

and the Banzhaf power index of C is 100. According to the Banzhaf model, the chair has five times the power of an ordinary member in this committee. ◆

EQUIVALENT VOTING SYSTEMS

The weights in a weighted voting system are to determine which coalitions are winning and which are losing. But if there are just two voters, A and B, how many really different voting systems are there? We can agree that the empty coalition ({ }) is always a losing coalition, and that the unanimous coalition ({A, B}) must be a winning coalition. Therefore there are only three distinct voting systems involving A and B: in the first, unanimous consent is required for each measure, so the only winning coalition is {A, B}. In the second, A is the dictator, and the winning coalitions are {A} and {A, B}. The third voting system has winning coalitions {B} and {A, B}; B is the dictator. Of course, there is an infinite number of ways that we can assign weights to the voters and a quota for passing measures in this two-voter system. However, there are only three ways to distribute the voting power: A as dictator, B as dictator, or consensus rule.

Two voting systems involving the same numbers of voters are called **equivalent** if there is a way for the voters in one system to exchange places with the voters in the other system without changing the winning coalitions. Every coalition that was winning before the switch should be winning after the switch, and every coalition that was losing before the switch should be losing after the switch.

For example, the weighted voting systems $[q: w(A), w(B)] = [50:49, 1]$ and $[q: w(C), w(D)] = [4:3, 3]$ are equivalent because in each system, unanimous support is required to pass a measure. If A and C change places, and B and D change places, the distribution of power will be the same.

Now let's consider the voting systems $[q: w(A), w(B)] = [2:2, 1]$ and $[q: w(A), w(B)] = [5:3, 6]$. In the first, A is a dictator, while in the second, B dictates. The winning coalitions for the first system, $\{A\}$ and $\{A, B\}$, correspond to the winning coalitions $\{B\}$, $\{A, B\}$ of the second if A and B exchange places, so the systems are equivalent. This shows that "equivalent" does not mean "the same." There is a distinction between the system where A dictates to B, and the system where B dictates to A. They are equivalent because each has a dictator.

Every voting system with only two voters is either equivalent to a system with a dictator or to one that requires consensus. The number of distinct systems increases as the number of voters increases. Table 12.6 lists all five of the distinct three-voter systems. Each of these systems can be presented as a weighted voting system, and suitable weights are given in the table. These systems can be transformed into four-voter systems by adding a dummy to each. There are nine additional four-voter systems in which each voter has power, for a total of 14 distinct systems with four voters.

TABLE 12.6	Voting Systems with Three Participants		
System	Minimal Winning Coalitions	Weights	Banzhaf Index
Dictator	$\{A\}$	$[3:3, 1, 1]$	$(8, 0, 0)$
Clique	$\{A, B\}$	$[4:2, 2, 1]$	$(4, 4, 0)$
Majority	$\{A, B\}, \{A, C\}, \{B, C\}$	$[2:1, 1, 1]$	$(4, 4, 4)$
Chair veto	$\{A, B\}, \{A, C\}$	$[3:2, 1, 1]$	$(6, 2, 2)$
Unanimous	$\{A, B, C\}$	$[3:1, 1, 1]$	$(2, 2, 2)$

Finding the Weights

A country called Atlantis is governed by a king, a council with three members, and a parliament with five members. All measures are considered first by parliament. Any measure that is approved by a vote of at least three members of parliament is sent to the council, which can either accept the measure by a majority vote or reject and kill the measure. Any measure that is passed by the parliament and the council is sent to the king, who may sign the measure into law or veto it. Although this voting system is not a weighted voting system, it is still possible to analyze the winning coalitions. Every winning coalition must include the king, at least two councilors, and at least three members of parliament.

Atlantis is governed by a voting system that is not equivalent to any weighted voting system. To see this, notice that a coalition consisting of the king, two councilors, and three members of parliament is a winning coalition. If a councilor leaves this coalition and a member of parliament joins, there will be the king, only one councilor, and four members of parliament—a losing coalition. In weighted voting systems, losing coalitions always have less total weight than winning coalitions. Therefore, if the Atlantis government were a weighted voting system, the councilor who left would have to have a greater weight than the member of parliament who joined. However, suppose that a member of parliament defects from the original winning coalition and is replaced by a councilor. Exactly the same reasoning shows that the member of parliament should have the greater weight. Since we can't have it both ways, we have to conclude that Atlantis is not governed by a weighted voting system.

Many voting systems are not presented as weighted systems, but are equivalent to weighted systems. Here are two examples.

EXAMPLE ▶ *Australia*

Australia has six states: New South Wales (N), Northern Territory (T), Queensland (Q), South Australia (S), Victoria (V), and Western Australia (W). Some national decisions are made by the following system: the states are each given one vote, and the federal government (G) has two votes. The total number of votes is eight, an even number, which makes ties possible. By law, all ties are settled in favor of the federal government. Minimal winning coalitions thus consist of the federal government with two states, or any coalition of five of the six states. The tie-breaking provision prevents this voting system from being a valid weighted voting system with these weights.

However, if we assign the federal government three votes, leave the states with one vote each, and set the quota at five, we will have an equivalent weighted voting system. If five states are aligned against the government, the extra vote will bring the government's coalition to a total of four, and the five states will still win, 5–4. On the other hand, if the government has two states on its side, the additional vote will give the government's coalition the victory, 5–4, without needing to break a tie. This weighted voting system

$$[q:w(G), w(N), w(T), w(Q), w(S), w(V), w(W)]$$
$$= [5:3, 1, 1, 1, 1, 1, 1]$$

is the same as that of the seven-person committee that we analyzed on page 458. The Banzhaf power index is therefore (100, 20, 20, 20, 20, 20, 20), and the government has five times the power of an individual state. ◆

EXAMPLE ▶ *The Nassau County Board of Supervisors in 1958*

Table 12.1 (page 445) gives the weighted vote allotted to each supervisor in the 1958 Nassau County Board. To prevent Hempstead's two supervisors from assuming a dictatorship, the county's charter then contained a provision saying no measure shall be adopted without the support of supervisors from at least two municipalities. With this provision, the county's voting system is not correctly presented as a weighted voting system, since the two Hempstead supervisors alone form a losing coalition with 18 votes, while one Hempstead supervisor can team up with the North Hempstead supervisor to win with 16 votes.

Let us compute the Banzhaf index of this voting system. The two Hempstead supervisors will be referred to as H_1 and H_2. The other supervisors are North Hempstead (N), Oyster Bay (B), Glen Cove (G), and Long Beach (L). To have enough votes to win, a coalition must contain H_1 and H_2, or one Hempstead supervisor and N. Furthermore, $\{H_1, N\}$ and $\{H_2, N\}$ are winning coalitions, but $\{H_1, H_2\}$ is not. Therefore, the **minimal winning coalitions** are as follows:

$$\{H_1, N\} \quad \{H_2, N\} \quad \{H_1, H_2, B\} \quad \{H_1, H_2, G\} \quad \text{and} \quad \{H_1, H_2, L\}$$

The supervisor H_1 will be a critical voter in any winning coalition that

(i)   Contains $\{H_1, N\}$ and does not contain H_2, or
(ii) Contains $\{H_1, H_2\}$ and at least one of B, G, or L, but not N.

The set $\{B, G, L\}$, has $2^3 = 8$ subsets, including $\{\}$. The union of any one of these subsets with $\{H_1, N\}$ will give a coalition of type (i); thus there are eight coalitions of this type. We can form seven coalitions of type (ii) by taking the union of $\{H_1, H_2\}$ with any *nonempty* subset of $\{B, G, L\}$. It follows that H_1 casts $8 + 7 = 15$ critical votes in winning coalitions. Remembering that there is an equal number of blocking coalitions in which H_1 is a critical voter, we find that the Banzhaf index of H_1 is 30. Of course, H_2 has the same amount of power.

Spotlight 12.3 A Mathematical Quagmire

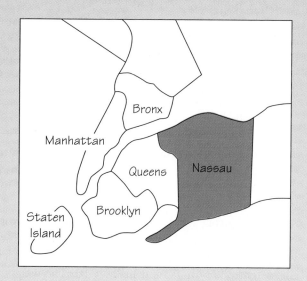

In a 1965 law review article, John F. Banzhaf III analyzed the weighted voting system used by the Board of Supervisors of Nassau County, New York. The article inspired legal action against several elected bodies that employ weighted voting systems.

The first legal challenge to weighted voting was to invalidate the voting system of the Washington County, New York, Board of Supervisors. In its decision, the New York Court of Appeals drew a corollary from Banzhaf's work that provided a way to fix weighted voting systems: each supervisor's *Banzhaf power index,* rather than his or her voting weight, should be proportional to the population of the district that he or she represents.

In the Washington County decision, the court observed that expert opinion and detailed com-

puter analyses would be needed to justify any weighted voting system, and predicted that the courts would eventually be dragged into a "mathematical quagmire."

A series of five lawsuits, spanning 25 years, challenged weighted voting in the Nassau County Board of Supervisors. These cases proved to be the mathematical quagmire that the appeals court had feared. The courts attempted to force Nassau County to comply with the Washington County decision. Although the county made a sincere attempt to do so, every voting system that it devised faced a new legal challenge. With conflicting expert testimony, the U.S. District Court finally ruled in 1993 that weighted voting was inherently unfair.

Banzhaf's law review article, which initially drew attention to weighted voting in Nassau County, was aptly titled "Weighted Voting Doesn't Work."

The North Hempstead supervisor will be a critical voter in the winning coalition $\{H_1, H_2, N\}$. In addition, N is a critical voter in all winning coalitions formed by assembling N with any subset of $\{B, G, L\}$ ($\{\}$ is allowed) and exactly one of H_1 and H_2. There are eight subsets of $\{B, G, L\}$, so that makes eight winning coalitions with N, H_1, and not H_2, and the same number with N, H_2, and not H_1. Counting $\{H_1, H_2, N\}$, this makes a total of 17 winning coalitions in which N casts a critical vote. Including the blocking coalitions, the Banzhaf index of N is $2 \times 17 = 34$.

The supervisors B, G, and L each have a Banzhaf index of 2, since each is a critical voter only when he or she forms a winning coalition with the two Hempstead supervisors or a blocking coalition by joining forces with the rest of the county against the two Hempstead supervisors. The Banzhaf index for this voting system is therefore

(30, 30, 34, 2, 2, 2)

The provision that prevents $\{H_1, H_2\}$ from being a winning coalition has two side effects. One was intentional: B, G, and L now have some voting power. The second side effect, making N more powerful than H_1 or H_2, may have been unintended.

Is this voting system *equivalent* to a weighted voting system? The Banzhaf index gives a clue to the answer. Since N is more powerful than H_1 or H_2, N's vote must have more weight. Let us increase N's weight to one vote more than the weight of H_1 or H_2 and make it 10. The weights of the other supervisors will be unchanged. Table 12.7 shows that the new weighted voting system will have the right winning coalitions if the quota is changed to 19. The table lists all of the minimal winning coalitions, and with their new weights, each has at least 19 votes. Furthermore, $\{H_1, H_2\}$, which is supposed to be a losing

---

TABLE 12.7 Minimal Winning Coalitions, Nassau County Supervisors, 1958

Coalition	Weights						Total
	H_1	H_2	N	B	G	L	
$\{H_1, N\}$	9		10				19
$\{H_2, N\}$		9	10				19
$\{H_1, H_2, B\}$	9	9		3			21
$\{H_1, H_2, G\}$	9	9			1		19
$\{H_1, H_2, L\}$	9	9				1	19

coalition, has only 18 votes—less than the quota. Thus, the weighted voting system

$$[q: w(H_1), w(H_2), w(N), w(B), w(G), w(L)] = [19:9, 9, 10, 3, 1, 1]$$

is equivalent to the voting system actually used by the Nassau County Board of Supervisors in 1958. ◆

The following example shows how to calculate for a voting system that we know is not equivalent to any weighted voting system.

EXAMPLE ▶ *A King's Power*

We will calculate the Banzhaf index for each member of the government of Atlantis (see page 461). Let us start with the king. Since he has veto power, the king is a critical voter in every winning coalition. There are $C_2^3 + C_3^3 = 4$ ways to form a majority coalition (which would have two or three members) in the council, and

$$C_3^5 + C_4^5 + C_5^5 = 10 + 5 + 1 = 16$$

ways to form a three-, four-, or five-member majority coalition in the parliament. By the multiplication principle, there are $4 \times 16 = 64$ winning coalitions, and thus the Banzhaf power index of the king, counting the blocking coalitions as well as the winning coalitions, is $2 \times 64 = 128$.

Now let us determine the Banzhaf power index for a councilor. He or she will be a critical voter in any coalition that includes the king, one other councilor, and a majority of parliament. There are $C_1^2 = 2$ ways to choose the other councilor, and, as before, 16 ways to choose a majority coalition in parliament. A councilor is therefore a critical voter in $2 \times 16 = 32$ winning coalitions and 32 blocking coalitions for a Banzhaf power index of 64.

A member of parliament will be a critical voter in any coalition that includes the king, a majority of the council, and exactly two other members of parliament. There are four ways to get a majority coalition in the council (this was determined when we found the Banzhaf power index of the king) and $C_2^4 = 6$ ways to choose the other two members of parliament. By the multiplication principle, each member of parliament is a critical voter in $4 \times 6 = 24$ winning coalitions, 24 blocking coalitions, and has a Banzhaf power index of 48. The Banzhaf power index for the voting system as a whole is therefore

$$(128, 64, 64, 64, 48, 48, 48, 48, 48)$$

By this measure, the king is twice as powerful as a councilor and $2\frac{2}{3}$ times as powerful as a member of parliament. ◆

THE SHAPLEY–SHUBIK POWER INDEX

In some political situations, coalitions are built one voter at a time. The most important voter in the sequence is the one who turns the coalition from a losing coalition into a winning coalition. In 1954, a power index based on this idea was introduced by Lloyd Shapley and Martin Shubik (see Spotlight 12.2). To calculate the index, one considers *permutations* of voters.

A **permutation** of voters is an ordering of all of the voters in a voting system.

For our purposes, a permutation represents a spectrum of opinion on some issue. For example, suppose that the issue is animal rights. Here the spectrum might range from a voter who would outlaw the sale of cow's milk to a voter who would legalize cockfighting. If an animal rights bill is being drafted, it must be written so as to receive the support of a coalition with enough votes to meet the quota.

The first voter in a permutation whose vote would make the coalition a winning coalition (if he or she could be induced to join) is called the **pivotal voter** in that permutation. Each permutation has exactly one pivotal voter.

If a bill is drafted so that it secures the support of the pivotal voter in the permutation corresponding to the spectrum of opinion on the issue, then the bill should pass.

Like the Banzhaf power index, computation of the Shapley–Shubik power index is basically a counting problem. The first step is to determine the number of permutations of a set of n voters. There are n voters who could be first on a list. When the first voter is identified, there are $n - 1$ left who could be the second. As we form the list, the number of voters available for the next position decreases until finally only one voter is left to be last on the list. According to the multiplication principle, the number of permutations is the product of the consecutive numbers from n down to 1, which is called the **factorial** of n:

$$n! = n \times (n - 1) \times (n - 2) \times \cdots \times 2 \times 1$$

TABLE 12.8	Permutations and Pivotal Voters for the Three-Person Committee				
Permutations			Weights		
A	Ⓑ	C	2	$\underline{3}$	4
A	Ⓒ	B	2	$\underline{3}$	4
B	Ⓐ	C	1	$\underline{3}$	4
B	C	Ⓐ	1	2	$\underline{4}$
C	Ⓐ	B	1	$\underline{3}$	4
C	B	Ⓐ	1	2	$\underline{4}$

Consider the three-person committee in which the chair has veto power, with the weighted voting system

$$[q : w(A), w(B), w(C)] = [3 : 2, 1, 1]$$

There are $3! = 1 \times 2 \times 3 = 6$ permutations of the members A, B, and C. Table 12.8 displays all six permutations. Next to each permutation, the total weight of the first voter, of the first two voters, and of all three voters is shown. The first number in this sequence to exceed the quota is underlined, and the corresponding pivotal voter's symbol is circled. We see that A is pivotal in four permutations, while B and C are each pivotal in one.

The **Shapley–Shubik power index** of a voter is the fraction of the permutations in which that voter is pivotal.

In the three-person committee, the Shapley–Shubik index for A is $\frac{4}{6}$, and B and C each have a Shapley–Shubik index of $\frac{1}{6}$. According to the Shapley–Shubik model, the chairperson of this committee, A, has four times as much voting power as an ordinary member. Recall that the Banzhaf power index for the same committee was $(6, 2, 2)$, so according to the Banzhaf model, A is three times as powerful as B or C.

How to Compute the Shapley–Shubik Power Index

For voting systems with small numbers of voters, the Shapley-Shubik power index can be calculated by making a list of all the voting permutations and identifying the pivotal voter in each.

EXAMPLE ▶ *The Corporation with Four Shareholders*

Let us calculate the Shapley–Shubik power index for the corporation whose shareholders *A, B, C,* and *D* own 40%, 30%, 20%, and 10% of the stock, respectively. This weighted voting system is presented as

$$[q: w(A),\ w(B),\ w(C),\ w(D)] = 51:40,\ 30,\ 20,\ 10]$$

There are 4! = 24 permutations to consider; they are shown in Table 12.9. In ten of the permutations, *A* is the pivotal voter; *B* and *C* are each pivotal voters

TABLE 12.9	Permutations and Pivotal Voters for the Four-Person Corporation									
Permutations				Weights				Pivot		
A	Ⓑ	*C*	*D*	40	<u>70</u>	90	100		B	
A	Ⓑ	*D*	*C*	40	<u>70</u>	80	100		B	
A	Ⓒ	*B*	*D*	40	<u>60</u>	90	100			C
A	Ⓒ	*D*	*B*	40	<u>60</u>	70	100			C
A	*D*	Ⓑ	*C*	40	50	<u>80</u>	100		B	
A	*D*	Ⓒ	*B*	40	50	<u>70</u>	100			C
B	Ⓐ	*C*	*D*	30	<u>70</u>	90	100	A		
B	Ⓐ	*D*	*C*	30	<u>70</u>	80	100	A		
B	*C*	Ⓐ	*D*	30	50	<u>90</u>	100	A		
B	*C*	Ⓓ	*A*	30	50	<u>60</u>	100			D
B	*D*	Ⓐ	*C*	30	40	<u>80</u>	100	A		
B	*D*	Ⓒ	*A*	30	40	<u>60</u>	100			C
C	Ⓐ	*B*	*D*	20	<u>60</u>	90	100	A		
C	Ⓐ	*D*	*B*	20	<u>60</u>	70	100	A		
C	*B*	Ⓐ	*D*	20	50	<u>90</u>	100	A		
C	*B*	Ⓓ	*A*	20	50	<u>60</u>	100			D
C	*D*	Ⓐ	*B*	20	30	<u>70</u>	100	A		
C	*D*	Ⓑ	*A*	20	50	<u>60</u>	100		B	
D	*A*	Ⓑ	*C*	10	50	<u>80</u>	100		B	
D	*A*	Ⓒ	*B*	10	50	<u>70</u>	100			C
D	*B*	Ⓐ	*C*	10	40	<u>80</u>	100	A		
D	*B*	Ⓒ	*A*	10	40	<u>60</u>	100			C
D	*C*	Ⓐ	*B*	10	30	<u>70</u>	100	A		
D	*C*	Ⓑ	*A*	10	30	<u>60</u>	100		B	

in six; and D is the pivotal voter in two permutations. Therefore the Shapley–Shubik index for this voting system is

$$\left(\frac{10}{24}, \frac{6}{24}, \frac{6}{24}, \frac{2}{24} \right)$$

The Banzhaf power index gives the same ratios of power in this case. In most cases, however, the two indices do not agree, and they may not even be approximately the same. ◆

The combinatorial explosion makes it impractical to list all of the permutations if there are more than four voters. Computer programs that will run through all $5! = 120$ permutations of five voters and $6! = 720$ permutations of six voters are available, but for more than 10 voters, even a computer would have trouble counting all of the permutations, considering that 11! is almost 40 million.

To calculate the Shapley–Shubik power index without listing the permutations, we can use two facts: first, voters whose weights are equal are interchangeable and thus will have the same Shapley–Shubik power indices, and second, the sum of the Shapley–Shubik power indices of all voters is equal to 1. For example, consider an n-voter weighted voting system in which all voters have equal weight. All voters will then have the same Shapley–Shubik power index, $1/n$. The next simplest case is one where all of the voters but one have equal weights (as in the Australia example).

EXAMPLE ▶ *The Shapley–Shubik Index of the Australian Voting System*

Recall that this is a seven-voter system in which the federal government has three votes and each of the six states has one; the quota for passing a measure is five. There are $7! = 5040$ voting permutations to consider, so we will consider groups of permutations rather than one permutation at a time. Each group will be identified by the position occupied by the federal government. Thus, the first group would be *GSSSSSS*, in which the government is first. Counting from the left, we see that votes are accumulated in the sequence 3,4,5,6,7,8,9, and that the third participant (a state) is the pivotal voter. In the next group, *SGSSSSS,* the vote accumulation sequence is 1,4,5,6,7,8,9; again, a state is the pivotal voter. The government is the pivotal voter in groups 3, 4, and 5: *SSGSSSS, SSSGSSS,* and *SSSSGSS,* with vote accumulation sequences 1,2,5,6,7,8,9, 1,2,3,6,7,8,9, and 1,2,3,4,7,8,9, respectively. In the final two groups, *SSSSSGS* and *SSSSSSG,* the vote accumulation sequences will be 1,2,3,4,5,8,9 and 1,2,3,4,5,6,9, respectively, and a state will be the pivotal voter again.

Each of the seven groups that we have considered has 6! permutations, since there are 6! orderings for the states. Because the groups are of equal size, each has $\frac{1}{7}$ of the permutations. The government is the pivotal voter in all permutations in groups 3, 4, and 5, and in none of the permutations in the other groups. Therefore, the Shapley–Shubik power index for the government is $\frac{3}{7}$.

The total Shapley–Shubik power index for the states is the fraction of power not held by the federal government: $\frac{4}{7}$. Since each of the six states has the same amount of power, the Shapley–Shubik index for each state is $\frac{4}{7} \div 6 = \frac{2}{21}$.

The Shapley–Shubik index for this weighted voting system is therefore

$$\left(\frac{3}{7}, \frac{2}{21}, \frac{2}{21}, \frac{2}{21}, \frac{2}{21}, \frac{2}{21}, \frac{2}{21} \right)$$

Since $\frac{3}{7} \div \frac{2}{21} = \frac{9}{2} = 4\frac{1}{2}$, the Shapley–Shubik model indicates that the government has four and one-half times as much power as an individual state. This is in close agreement with the Banzhaf model, which held that the government was five times as powerful as an individual state.   ◆

Comparison of the Banzhaf and Shapley–Shubik Indices

Deciding which power index best describes the distribution of power in a particular voting system is a subjective judgment. The heart of the issue is the distinction between permutations and combinations. A particular voting permutation reflects the range of opinion concerning an issue, while a voting combination tells us who is for and who is against a measure. The Shapley–Shubik index of a voter is the fraction of the voting permutations in which that voter is the pivotal voter. In a decision-making body, it is natural that some participants will frequently be found at extreme ends of voting permutations, while others will more frequently occupy middle positions. Those who are less inclined to extreme positions are pivotal voters more frequently than those with extreme views, and so have more power. However, the purpose of a power index is to measure the distribution of power that is built into the system, not the power that some voters acquire as a consequence of their political views. The Shapley–Shubik index is appropriate if we believe that on most issues before the voting body, there is a one-dimensional spectrum of opinion. The Banzhaf index does not attempt to model the dynamics of the legislative process, and it best reflects the situation when the participants operate unpredictably without consulting one another.

In many legislatures, the political dynamic is too complex to be accurately modeled by either index. When there are many points of view to consider, the opinions that are represented cannot be strictly ordered between two extremes. In these cases, the Banzhaf and Shapley–Shubik indices provide measurements of voting power from two points of view.

SYSTEMS WITH LARGE NUMBERS OF VOTERS

Analysis of voting systems with voters numbering in the thousands or millions is possible only if all of the voters, with only a few exceptions, are equally powerful.

EXAMPLE ▶ *A Corporation with 9001 Shareholders*

A corporation has one shareholder who owns 1000 shares of the total outstanding stock and 9000 shareholders, each of whom has one share. How much voting power does the big shareholder have? Our calculation will assume that all shareholders are fully participating in the process (an unusual occurrence), and we will determine the Shapley–Shubik index. The number of voting permutations—9001!—is more than astronomical. With over 31,000 digits, it would occupy more than seven pages of this book if written out.

There are 10,000 shares in this corporation, and in any voting permutation, the voter with share 5001 is the pivotal voter. Thus the big shareholder will occupy the pivotal position if he is preceded by at least 4001 but not more that 5000 little shareholders. Let us partition the voting permutations into 9001 groups, according to the location of the big shareholder. We see that the big shareholder is the pivotal voter in all permutations in groups 4002 through 5001, or 1000 groups out of the 9001. Since the groups are of equal size, his Shapley–Shubik index is $\frac{1000}{9001}$. The remaining $\frac{8001}{9001}$ power goes to the 9000 small shareholders.

This is one instance where the Banzhaf and Shapley–Shubik power indices disagree. We will see that the Banzhaf power index gives almost 100% of the power to the 10% shareholder, in contrast to the approximately 11% granted to that shareholder by the Shapley–Shubik power index.

It is not feasible to compute the Banzhaf index for the 9001-shareholder corporation, but we can approximate it by using some ideas about probability that were developed in Chapter 7. ◆

Optional ▶ **Interpreting the Power Indices as Probabilities**

In the Shapley–Shubik model, the measure of the power of a voter A is

$$\frac{\text{number of permutations in which } A \text{ is the pivot}}{\text{total number of permutations}}$$

Thus, if we consider each permutation to be equally likely, a voter's Shapley–Shubik power index is the probability that he or she will be the pivotal voter.

To interpret the Banzhaf power index in terms of probability, make the assumption that all voting combinations are equally likely. This would be the case if each voter decided which way to vote by tossing a coin. Then the probability that A will be a critical voter in a winning or blocking coalition is equal to

$$\frac{\text{number of voting combinations in which } A \text{ is a critical voter}}{\text{total number of voting combinations}}$$

With n voters, there are 2^n voting combinations in all; therefore the probability that A will be a critical voter is obtained by dividing A's Banzhaf power index by 2^n. ◆

EXAMPLE ▶ *Estimating the Banzhaf Power Index of the 9001-Shareholder Corporation*

In the 9001-shareholder corporation, one shareholder has 1000 shares of stock, and the other 9000 shareholders have one share each. We will approximate the Banzhaf power index of a shareholder A by approximating the probability that

---

Spotlight
12.4

Large or Small: Which States Are Favored in the Electoral College System?

The Electoral College, operating under the general ticket or unit rule ("winner take all") method, is, in my estimation, the most unfair, inaccurate, uncertain, and undemocratic institution of all.

—Senator Karl Mundt (R.–South Dakota)

What did Senator Mundt have in mind? With three electoral votes and a population of about 600,000, South Dakota has one vote for each 200,000 of its citizens. If every state were as well represented in the electoral college, there would have to be about 1250 electors, instead of 538, and 150 of them would be from California.

Senator Mundt made the statement quoted above after reading an article by John F. Banzhaf III entitled "One Man, 3.312 . . . Votes." This article

asserted that a voter's probability of casting a critical vote in a general election with n voters is inversely proportional to the square root of n. A voter in a large state has less chance of casting a critical vote than does a voter in a small state, but swings a greater number of electoral votes. Surprisingly, this square root rule implies that a voter in California has almost three times as much influence as a voter in South Dakota. This is based on California's nominal voting power of 54 electoral votes. When California's actual Banzhaf power index is used, the imbalance is even greater.

Presidential political campaigns have confirmed these observations. Per capita campaign spending is much higher in states with large populations than it is in the less populous states.

A will be a critical voter, under the assumption that each shareholder lets a coin toss decide his or her vote. The big shareholder will be a critical voter unless his coalition is joined by fewer than 4001 small shareholders (which would make it a losing, nonblocking coalition), or if his coalition is joined by at least 5001 small shareholders (which would make it a blocking or winning coalition with no critical voters).

Suppose that 9000 people each toss a coin, and we count the number of heads. The expected number of heads is $\frac{1}{2} \times 9000 = 4500$, and the standard deviation (see Chapter 7) is $\sqrt{\frac{1}{2} \times \frac{1}{2} \times 9000}$ which is approximately 50. By the 68–95–99.7 rule, 68% of the time there will be 4500 ± 50 heads, or between 4450 and 4550 heads. Ninety-five percent of the time there will be between 4400 and 4600 heads, and 99.7% of the time, between 4350 and 4650 heads. It is almost certain that there will be between 4000 and 5000 heads and thus that the big shareholder will be a critical voter. Under these circumstances, however, a small shareholder has practically no chance of being a critical voter, since he or she would have to be in a coalition with either exactly 5000 other small shareholders, or exactly 4000 other small shareholders and the big shareholder. The big shareholder is a critical voter in almost all of the voting combinations, and small shareholders are critical voters in practically no voting combinations. ◆

REVIEW VOCABULARY

Banzhaf power index A numerical measure of power for participants in a voting system. A participant's Banzhaf index is the number of winning or blocking coalitions in which he or she is a critical voter.

Blocking coalition A set of participants in a voting system that can prevent a measure from passing by voting against it.

C_k^n A set with n elements has C_k^n subsets with k elements. This number, referred to as "n choose k," is given by the formula

$$C_k^n = \frac{n \times (n-1) \times \cdots \times (n-k+1)}{k \times (k-1) \times \cdots \times 1}.$$

Remember that the numerator is the product of k numbers starting with n and counting down; the denominator is the product of k numbers starting with k and counting down.

Coalition A set consisting of some, all, or none of the participants in a voting system.

Combination A partitioning of a set into a subset and its complementary subset; for example, a list of voters indicating the vote of each on an issue.

There is a total of 2^n combinations in an n-element set, and C_k^n combinations in which the subset has k elements and its complement has $n - k$ elements.

Critical voter   A member of a winning coalition whose vote is essential for the coalition to win, or a member of a blocking coalition whose vote is essential for the coalition to block.

Dictator   A participant in a voting system who can pass any issue even if all other voters oppose it, and block any issue even if all other voters approve it.

Dummy   A participant in a voting system who has no power. A dummy is never a critical voter in any winning or blocking coalition, and is never the pivotal voter in any permutation.

Equivalent voting systems   Two voting systems are equivalent if there is a way for all of the voters of the first system to exchange places with the voters of the second system and preserve all winning (and losing) coalitions.

Factorial   If n is a positive integer, the factorial of n, denoted $n!$, is the product of all the positive integers less than or equal to n. It is usually a big number: $10!$ is a seven-digit number, $9000!$ is a seven-page number.

Losing coalition   A coalition that does not have the voting power to get its way.

Minimal winning coalition   A winning coalition that will become losing if any member defects. Each member is a critical voter.

Permutation   A specific ordering from first to last of the elements of a set; for example, an ordering of the participants in a voting system.

Pivotal voter   The first voter in a permutation who, with his or her predecessors in the permutation, will form a winning coalition. Each permutation has one and only one pivotal voter.

Quota   The minimum number of votes necessary to pass a measure in a weighted voting system.

Shapley–Shubik power index   A numerical measure of power for participants in a voting system. A participant's Shapley–Shubik index is the number of permutations of the voters in which he or she is the pivotal voter, divided by the number of permutations ($n!$ if there are n participants).

Veto power   A voter has veto power if no issue can pass without his or her vote. A voter with veto power is a one-person blocking coalition.

Weight   The number of votes assigned to a voter in a weighted voting system, or the total number of votes of all voters in a coalition.

Weighted voting system   A voting system in which the participants can have different numbers of votes. It can be represented as $[q : w(A_1), w(A_2), \ldots, w(A_n)]$, where $A_1, \ldots, A_n$ are the voters, $w(A_1), \ldots, w(A_n)$ represent the numbers of votes held by these voters, and q is the quota necessary to win.

Winning coalition   A set of participants in a voting system who can pass a measure by voting for it.

SUGGESTED READINGS

AFFUSO, PAUL J., AND STEVEN J. BRAMS. Power and size: A new paradox, *Theory and Decision*, 7 (1976): 29–56. This paper explains the use of generating functions to calculate the Banzhaf and Shapley–Shubik power indices.

BANZHAF, JOHN F. III. Weighted voting doesn't work, *Rutgers Law Review*, 19 (1965): 317–343. The author defines the Banzhaf index, and uses it to show that the weighted voting system in use by the Nassau County Board of Supervisors was unfair.

BANZHAF, JOHN F. III. One man, 3.312 . . . votes: A mathematical analysis of the electoral college, *Villanova Law* Review, 13 (1968): 304–332. The author shows that voters residing in the more populous states have more power to influence the outcome of a presidential election. The same issue has commentaries on Banzhaf's analysis (pages 333–346). Another commentary appears in the same *Review,* 14 (1968): 86–96.

BRAMS, STEVEN J. *Game Theory and Politics,* Free Press, New York, 1975. Chapter 5 treats the Shapley–Shubik and Banzhaf indices.

BRAMS, STEVEN J. *Paradoxes in Politics: An Introduction to the Nonobvious in Political Science,* Free Press, New York, 1976.

BRAMS, STEVEN J. *The Presidential Election Game,* Yale University Press, New Haven, Conn., 1978.

BRAMS, STEVEN J., W. F. LUCAS, AND P. D. STRAFFIN, JR., EDS. *Political and Related Models. Modules in Applied Mathematics,* vol. 2, Springer-Verlag, New York, 1983. Chapters 9–11 are devoted to measuring power in weighted and other types of voting systems. The Banzhaf and Shapley–Shubik indices are the focus of chapters 9 and 11; chapter 10 is about an index based on counting minimal winning coalitions.

GOLDBERG, SAMUEL. *Probability in Social Science,* Birkhäuser, Boston, 1983. Chapter 1 is devoted to the Shapley–Shubik index.

IANNUCCI V. *BOARD OF SUPERVISORS OF WASHINGTON COUNTY.* 20 N.Y. 2d 244, 251, 229 N.E. 2d 195, 198, 282 N.Y.S. 2d 502, 507 (1967). This code will help a law librarian find this case for you. It opened a "mathematical quagmire."

LAMBERT, JOHN P. Voting games, power indices, and presidential elections, *UMAP Journal,* 9 (1988): 213–267.

LUCAS, WILLIAM F. *Fair Voting: Weighted Votes for Unequal Constituencies,* COMAP: HistoMAP Module 19, Lexington, Mass., 1992. An introduction to the power indices with emphasis on the historical aspects.

MORRIS V. BOARD OF ESTIMATE. 489 U.S. 688 (1989). In this case, the U.S. Supreme Court ruled that the Banzhaf square root rule was not a realistic measure of a citizen's voting power and could not be used to justify weighted voting in the New York City Board of Estimate.

SHUBIK, MARTIN, ED. *Game Theory and Related Approaches to Social Behavior,* Wiley, New York, 1964. This book reprints Shapley and Shubik's 1954 paper, in which their index was introduced, as chapter 9. Chapter 10, originally published in 1960 and 1962, is an analysis of the electoral college by I. Mann and L. S. Shapley. Chapter 11 is a study of power in the U.S. Congress based on the Shapley–Shubik index, originally published in 1956 by R. D. Luce and A. A. Rogow.

TAYLOR, ALAN D. *Mathematics and Politics: Strategy, Voting Power, and Proof,* Springer-Verlag, New York, 1995. Chapter 4 covers weighted voting systems and their analysis using the Shapley–Shubik and Banzhaf indices. It has no mathematical prerequisites, but it does include carefully written logical arguments that must be carefully read.

EXERCISES ▲ *Optional.* ■ *Advanced.* ◆ *Discussion.*

How Weighted Voting Works

◆ 1. Will a blocking coalition turn into a winning coalition if every voter in the coalition votes Y? Consider the following examples:

(a) A committee with 9 members, each with one vote, where majority rules.

(b) A committee with 12 members, each with one vote, where majority rules.

(c) A jury with 9 members in a criminal trial, where a unanimous decision is necessary to convict or to acquit.

◆ 2. Is it possible to have a weighted voting system in which more votes are required to block a measure than to pass a measure?

3. For each of the following weighted voting systems, list (i) all winning coalitions containing the voter *A*; (ii) all blocking coalitions containing the voter *A*; (iii) all losing coalitions containing the voter *A*; (iv) all dummy voters.

    (a) $[q: w(A), w(B)] = [51:52, 48]$.
    (b) $[q: w(A), w(B), w(C)] = [2:1, 1, 1]$.
    (c) $[q: w(A), w(B), w(C)] = [3:2, 2, 1]$.
    (d) $[q: w(A), w(B), w(C)] = [8:5, 4, 3]$.
    (e) $[q: w(A), w(B), w(C), w(D)] = [51:45, 43, 8, 4]$.
    (f) $[q: w(A), w(B), w(C), w(D)] = [51:28, 27, 26, 19]$.
    (g) $[q: w(A), w(B), w(C), w(D)] = [16:10, 10, 10, 1]$.
    (h) $[q: w(A), w(B), w(C), w(D), w(E)] = [21:10, 10, 10, 10, 1]$.

◆   4. A comparison of the voting systems in parts (g) and (h) of Exercise 3 reveals a paradox. Explain it.

5. A *minimal winning coalition* is a winning coalition that will be a losing coalition if any one of its members deserts it. Describe the minimal winning coalitions for the weighted voting system

$$[q: w(A), w(B), w(C), w(D)] = [51:30, 25, 25, 21]$$

The Banzhaf Power Index

6. (a)   List the 16 possible combinations of how four voters, *A*, *B*, *C*, and *D*, can vote either yes (Y) or no (N) on an issue.
    (b)   List the 16 subsets of the set {*A, B, C, D*}.
◆   (c)   How do the lists in parts (a) and (b) correspond to each other?
    (d)   In how many of the combinations in part (a) is the vote
        (i)   4 Y to 0 N?
        (ii)   3 Y to 1 N?
        (iii) 2 Y to 2 N?

7. For the voting system in Exercise 5, list all winning coalitions in which

    (a)   *A* is a critical voter.
    (b)   *B* is a critical voter.

8. Calculate the Banzhaf index for the voting system in Exercise 5.

9. Calculate the Banzhaf index for each of the weighted voting systems in Exercise 3.

10.   Calculate the Banzhaf index for the weighted voting system $[q : w(A), w(B), w(C), w(D)] = [6 : 4, 3, 2, 1]$.

◆   11.   Explain why $C_k^n = C_{n-k}^n$. Do not use the formula; give your explanation in terms of the definition "C_k^n is the number of voting combinations for a set of n voters in which k voters vote Y."

12.   Calculate the following:

     (a) C_3^6
     (b) C_{100}^{50}
     (c) C_3^{10}
     (d) C_7^{10}

13.   Calculate the following:

     (a) C_4^6
     (b) C_2^{100}
     (c) C_{98}^{100}
     (d) C_5^{10}

14.   Before being declared unconstitutional by a federal district court in 1993, the weighted voting system of the Nassau County Board of Supervisors was changed several times. The weights in use since 1958 were as follows:

Year	$[q : w(H_1), w(H_2), w(N), w(B), w(G), w(L)]$
1958	$[16 : 9, 9, 7, 3, 1, 1]$
1964	$[58 : 31, 31, 21, 28, 2, 2]$
1970	$[63 : 31, 31, 21, 28, 2, 2]$
1976	$[71 : 35, 35, 23, 32, 2, 3]$
1982	$[65 : 30, 28, 15, 22, 6, 7]$

where H_1 is the presiding supervisor, always from Hempstead, H_2 is the second supervisor from Hempstead, and N, B, G, and L are the supervisors from North Hempstead, Oyster Bay, Glen Cove, and Long Beach, respectively.

◆   (a)   From 1970 on, more than a simple majority was required to pass any measure. Give an argument in favor of this policy from the viewpoint of a supervisor who would benefit from it, and an argument against the policy from the viewpoint of a supervisor who would lose some power.

   (b)   In which years were some supervisors dummy voters?

   (c)   Suppose that the two Hempstead supervisors always vote together. In which years are some of the supervisors dummy voters?

   (d)   Assume that the two Hempstead supervisors always agree, so that the board is in effect a five-voter system. Determine the Banzhaf index of this system in each year.

   (e)   Table 12.10 gives the 1980 census for each municipality, the number of votes assigned to each supervisor, and the Banzhaf index for each supervisor in 1982. Do you think the voting scheme is fair?

15.   The Nassau County Board of Supervisors had a higher quota for votes that require a two-thirds majority. In 1982, this quota was 72.

   (a)   Assume that the two Hempstead supervisors vote together, and determine the Banzhaf index for each municipality.

   (b)   Is there any justification for using the quota 72 to represent a two-thirds majority? Is the distribution of power proportional to population (see Table 12.10)?

TABLE 12.10   **Nassau County Board of Supervisors, 1982**

Supervisor From	Population	Number of Votes	Banzhaf Power Index	
Quota			65	72
Hempstead (Presiding)	738,517	30	30	26
Hempstead		28	26	22
North Hempstead	218,624	15	18	18
Oyster Bay	305,750	22	22	18
Glen Cove	24,618	6	2	2
Long Beach	43,073	7	6	6
Totals	1,321,582	108	104	92

Equivalent Voting Systems

16. Consider a four-person voting system with voters A, B, C, and D. The winning coalitions are

$$\{A, B, C, D\}, \ \{A, B, C\}, \ \{A, B, D\}, \ \{A, C, D\}, \text{ and } \{A, B\}$$

(a) List the minimal winning coalitions.
(b) Determine the Banzhaf power index for this voting system.
(c) Find an equivalent weighted voting system.

17. A committee has a chairperson and six ordinary members. It uses majority rule, except that the chairperson is only allowed to vote when it is necessary to break a tie. Give an equivalent weighted voting system for the committee.

18. The committee in Exercise 17 is reduced in size; now there are just five ordinary members. The rules are the same. Give an equivalent weighted voting system for the committee.

◆ 19. A five-member committee has the following voting system. The chairperson can pass or block any motion that she supports or opposes, provided that at least one other member is on her side. Show that this voting system is equivalent to the weighted voting system

$$[q\!:\!w(C), \ w(M_1), \ w(M_2), \ w(M_3), \ w(M_4)] = [4\!:\!3, \ 1, \ 1, \ 1, \ 1].$$

20. Calculate the Banzhaf index for the weighted voting system in Exercise 19.

21. Which of the following voting systems are equivalent to weighted voting systems? Find the weights and quota for those that are.

(a) A committee of three faculty and the dean. To pass a measure, at least two faculty members and the dean must vote yes.
(b) A committee of three faculty, the dean, and the provost. To pass a measure, two faculty, the dean, and the provost must vote yes.
(c) A four-member faculty committee and a three-member administration committee vote separately on each issue. The measure passes if it receives the support of a majority of each of the committees.

22. Calculate the Banzhaf index of each of the voting systems in Exercise 21.

◆ 23. How many *distinct* (nonequivalent) voting systems with four voters can you find? Systems that have dummies don't count. The challenge is to find all nine. (*Hint:* Each voting system can be specified by listing all of the minimal winning coalitions. If there are no dummies, each voter belongs to at least one minimal winning coalition. Remember that two winning coalitions cannot be disjoint, if the system is to be decisive.)

The Shapley–Shubik Power Index

24. For the voting system in Exercise 5, list all permutations of the voters in which

(a) *A* is the pivotal voter.
(b) *B* is the pivotal voter.

25. Calculate the Shapley–Shubik index for the system in Exercise 5.

26. Calculate the Shapley–Shubik index for the weighted voting system in Exercise 19.

27. Calculate the Shapley–Shubik index of each of the voting systems in Exercise 21.

Systems with Large Numbers of Voters

28. A corporation has 120 shares of stock outstanding. There are 80 shareholders who own one share each, and one shareholder who owns 20 shares. To pass an issue, owners representing 61 shares must vote yes. Determine the Shapley–Shubik index of each shareholder.

29. Estimate the power of the shareholders of the corporation in Exercise 28, as measured by the Banzhaf index.

Miscellaneous Problems

◆ 30. The vice president of the United States is allowed to break ties in the U.S. Senate. How does his or her Banzhaf power index compare with that of an individual senator?

31.   Determine the Shapley–Shubik power index for the four-person voting system described in Exercise 16.

32.   A corporation has four shareholders and a total of 100 shares. The quota for passing a measure is the votes of shareholders owning 51 or more shares. The number of shares owned are as follows:

A	48 shares
B	23 shares
C	22 shares
D	7 shares

All transactions must be in whole numbers of shares; sales of fractional shares are not permitted.

(a)   List the minimal winning coalitions.

(b)   How many shares can A sell, without changing the set of minimal winning coalitions, to one of the other shareholders, or to E, who currently owns none of the stock? (Notice that A may be able to sell more to D than to B, etc.)

(c)   How many shares can D sell without changing the set of minimal winning coalitions, to B, C, or E? Again, it is conceivable that D would be able to sell more to one stockholder than to another.

(d)   How many shares can D sell to B, C, or E without becoming a dummy?

(e)   How many shares can B sell to C without changing the set of minimal winning coalitions?

33.   Which of the following voting systems is equivalent to the voting system in use by the corporation in Exercise 32?

(a) $[q: w(A), w(B), w(C)] = [3:1, 1, 1, 1]$

(b) $[q: w(A), w(B), w(C)] = [3:2, 1, 1, 1]$

(c) $[q: w(A), w(B), w(C)] = [5:3, 1, 1, 1]$

(d) $[q: w(A), w(B), w(C)] = [5:3, 2, 1, 1]$

(e) $[q: w(A), w(B), w(C)] = [5:3, 2, 2, 2]$

34.   Determine the Banzhaf and Shapley–Shubik power indices for the corporation in Exercise 32.

◆   35.   Explain why $C_k^n = 0$ when $k > n$.

36. A nine-member committee has a chairperson and eight ordinary members. A motion can pass if and only if it has the support of the chairperson and at least two other members, or if it has the support of all eight ordinary members.

    (a) Find an equivalent weighted voting system.
    (b) Determine the Banzhaf power index.
    (c) Determine the Shapley–Shubik power index.

37. Consider the $2m$-person voting system in which each participant has one vote, and a simple majority wins. In the notation for weighted voting systems, this system can be expressed as

$$[q:w(V_1), \ldots, w(V_{2m})] = [m + 1:1, \ldots, 1]$$

Assume that all voting combinations are equally likely. What is the probability that a voter will be a critical voter, when $m = 1, 2, 3, 4, 5, 6,$ or 7?

38. The New York City Board of Estimate consists of the mayor, the comptroller, the city council president, and the presidents of each of the five boroughs. It employed a voting system in which the city officials each had two votes, and the borough presidents each had one; the quota to pass a measure was six. This voting system was declared unconstitutional by the U.S. Supreme Court in 1989.[1]

    (a) Describe the minimal winning coalitions.
    (b) Determine the Banzhaf power index.

39. Here is a proposed weighted voting system for the New York City Board of Estimate that is based on the populations of the boroughs (see Exercise 38):

$$[q:w(M), w(C), w(P), w(K), w(H), w(Q), w(X), w(S)]$$
$$= [71:35, 35, 35, 11.3, 7.3, 9.6, 6.0, 1.8]$$

Find a simpler system of weights that yields an equivalent voting system.

40. The United Nations Security Council has five permanent members: China, France, Russia, the United Kingdom, and the United States, and 10 other members that serve 2-year terms. To resolve a dispute not involving a member of the Council, nine votes, including the votes of each of the permanent members, is required. (Thus each permanent member has veto power.)

---

[1] *Morris v. Board of Estimate.*

◆ (a) Show that this voting system is equivalent to the weighted voting system in which each permanent member has 7 votes, each ordinary member has 1 vote, and the quota is 39.

▨ (b) Compute the Banzhaf index for the Security Council.

▨ (c) Compute the Shapley–Shubik index for the Security Council. (This is harder than computing the Banzhaf index.)

◆ (d) Which index is most appropriate for measuring power in the Security Council?

▲ 41. The ABC College Student Senate is a five-member body with simple majority rule; each member has one vote. Two members of the Senate, A and B, have a pact to vote together.

(a) In how many voting combinations do A and B cast the same vote?

(b) Say that $\{A, B\}$ is a critical voter in a voting combination if the result of the vote would be reversed if both A and B defected. In how many of the voting combinations in part (a) is $\{A, B\}$ a critical voter?

(c) In how many voting combinations in part (a) would Senator C, who is not involved in a pact, be a critical voter?

(d) How does A's pact with B affect the probability that A will be a critical voter? How does it affect C's chance of being a critical voter?

▲ 42. Senators A and B have quarreled in the ABC College Student Senate (see Exercise 41). In the future, they will never be on the same side of *any* issue.

(a) In how many voting combinations will A and B be on opposite sides?

(b) In how many of the voting combinations in part (a) is A a critical voter?

(c) In how many of the voting combinations in part (a) is C a critical voter?

(d) How does the quarrel affect the distribution of power, as compared to independent voting?

▲ 43. This problem is like Exercise 41, in that two members of the Student Senate, *A* and *B*, have formed an alliance. However, this senate has only four members.

(a) Show that the voting system in effect is

$$[q: w(\{A, B\}), w(C), w(D)] = [3:2, 1, 1]$$

and compute the Banzhaf index.

(b) Senators *C* and *D* quarrel, and resolve to oppose each other on every issue. Assuming that *A* and *B* maintain their alliance, determine the Banzhaf index, and the probability that each voter (*{A, B}* is considered to be a single voter here) will be a critical voter.

WRITING PROJECTS

1 ▶ The most important weighted voting system in the United States is the electoral college (see Spotlights 12.1 and 12.4). Three alternate methods to elect the president of the United States have been proposed:

- *Direct election.* The electoral college would be abolished, and the candidate receiving a plurality of the votes would be elected. Most versions of this system include a runoff election or a vote in the House of Representatives in cases where no candidate receives more than 40% of the vote.
- *District system.* In each congressional district, and in the District of Columbia, the candidate receiving the plurality would select one elector. Furthermore, in each state, including the District of Columbia, the candidate receiving the plurality would receive two electors. In effect, the unit rule would be retained for the District of Columbia and for states with a single congressional district. Larger states would typically have electors representing both parties.
- *Proportional system.* Each state and the District of Columbia would have fractional electoral votes assigned to each candidate in proportion to the number of popular votes received. Under this system, President

Clinton, who received 126,054 popular votes out of 287,580 cast in the District of Columbia in 1992, would have received

$$\frac{126,054}{287,580} \times 3 = 1.31498$$

of the District's 3 electoral votes. Obviously, there would be no actual electors involved in the process.

Should the present electoral college, operating under the unit rule, be replaced by one of these systems? A starting point to answer this question is the article by John Banzhaf III, "One Man, 3.312 . . . Votes" (1968). Another reference is "The a Priori Voting Strength of the Electoral College," by I. Mann and L. S. Shapley, in the anthology edited by Martin Shubik (1964). *The Presidential Election Game,* by Steven Brams (1978), contains useful references to Senate hearings on electoral college reform.

2 ▶   Choose an organization that uses weighted voting, and explain the organization's reasons for using weighted voting. If you can, compute the Banzhaf and Shapley–Shubik indices for the system. If they differ significantly in their allocation of power, which, if either, represents the true balance of power best?

CHAPTER 13

FAIR DIVISION

"Gimme the Plaza, the jet, and $150 million, too."

—Headline, *New York Post,* February 13,
1990, reporting Ivana Trump's divorce
settlement demands of husband, Donald

Though hardly typical, the 1991 divorce ending the 13-year marriage of Donald and Ivana Trump was one of an estimated 1,187,000 divorces in the United States that year. Few of them involved distributing marital property that included an 118-room mansion, a 282-foot yacht, and a small airline, as was the case in the Trumps' divorce. Many, however, gave rise to the same difficulties—and opportunities—in deciding exactly who gets what in a breakup.

How are divisions of assets in a divorce handled today? The answer often seems to be, "Not very well," even in states like New York where there are domestic relation laws regarding equitable distribution of assets when a marriage dissolves. The point is that an asset that is of considerable value to one party (for reasons ranging from personal circumstances to sentimentality) may be of little interest to the other.

Thus, a property settlement is perhaps the only aspect of a divorce that provides an opportunity for some kind of win-win situation. If the two parties place different values on the items to be distributed (or the issues to be resolved), then there should be settlements (allocations) that leave each party feeling that he or she has been met more than halfway.

Divorces are one example of what is called a **fair-division problem.** In general, a fair-division problem consists of *n* individuals, called **players,** and either a single object (like a cake) or some set of objects (like the property in a divorce). The *n* players partition the object or set of objects into *n* disjoint pieces S_1, S_2, . . . , S_n. A **fair-division procedure** or **fair-division scheme** is a method for solving a fair-division problem in which each player has a way to realize a share that he or she considers fair in his or her own value system.

These considerations bring us to the two questions we address in this chapter:

1. Have real-world problems such as divorce settlements led to the development of new mathematics that has potential applications in such contexts?

Spotlight 13.1 Fifty Years of Cake Cutting

Will this cake be divided fairly?

The modern era of cake cutting began with the investigations of the Polish mathematician Hugo Steinhaus during World War II. His research, and that of dozens of others over the past half century, involved dealing with two fundamental difficulties. First, allocation schemes that work in the context of two or three players often do not generalize easily to the context of four or more players. Second, procedures that yield envy-free allocations are considerably harder to obtain than procedures that yield proportional allocations.

The mathematics inspired by these two difficulties over the past 50 years constitutes a rather

2. If so, is the mathematics of interest in its own right, in the sense of involving nontrivial questions and answers that shed light on the fundamental issue of fairness?

We begin this chapter with a procedure developed within the last few years called the *adjusted winner procedure*. To describe this procedure we revisit the Trump divorce and analyze it using this new procedure, as was initially done by Catherine Duran (see Suggested Readings in this chapter).

elegant corner of the large and important area of fair division. Steinhaus's investigations in the 1940s led to his observation that there is a rather natural extension of divide-and-choose to the case of three players. This is the "lone-divider scheme" described on page 499. Steinhaus's method was generalized to an arbitrary number of players by Harold W. Kuhn of Princeton University in 1967.

Unable to extend his scheme from three to four players, Steinhaus proposed the problem to some Polish colleagues. Two of them, Stefan Banach and Bronislaw Knaster, solved this problem in the mid-1940s by producing the "last-diminisher scheme" described on page 500.

In addition to the schemes devised by Banach, Knaster, and Kuhn, there are other well-known constructive procedures for obtaining a proportional allocation among four or more players. One of these is due to A. M. Fink of Iowa State University and appears in Exercise 23.

Another constructive procedure of note, although different in flavor from the others, is the 1961 recasting by Lester E. Dubins and Edwin H. Spanier of the University of California at Berkeley of the last-diminisher method as a "moving-knife scheme" (illustrated in Exercise 25). The trade-off here involves giving up the "discrete" nature of the last-diminisher method in exchange for the conceptual simplicity of the moving knife.

Although the existence of an envy-free allocation (even for four or more players) was known to

Steinhaus in the 1940s, the first *constructive* procedure for producing an envy-free allocation among three players was not found until around 1960. At that time, John L. Selfridge of Northern Illinois University and, later but independently, John H. Conway of Princeton University found the elegant scheme presented on pages 504–506. Although never published by either, the scheme was quickly and widely disseminated by Richard K. Guy of the University of Calgary and others; eventually it appeared in several treatments of the problem by different authors.

In 1980, a moving-knife procedure for producing an envy-free allocation among three players was found by Walter R. Stromquist of Daniel Wagner Associates. Shortly thereafter, another scheme, capable of being recast as a moving-knife solution of the three-player case, was found by a law professor at the University of Virginia, Saul X. Levmore, and a former student of his, Elizabeth Early Cook.

In 1992, Steven J. Brams, a political scientist at New York University, and Alan D. Taylor, a mathematician at Union College, succeeded in finding a constructive procedure for producing an envy-free allocation among four or more players. In 1994, Brams, Taylor, and William S. Zwicker (also from Union College) found a moving-knife solution to the four-person envy-free problem. No moving-knife scheme is known that will produce an envy-free allocation among five or more players.

Then, turning to inheritances, we describe an old allocation scheme that was discovered by the Polish mathematician Bronislaw Knaster during World War II.

Bridging the gap between schemes with obvious real-world potential, such as divorce and inheritance procedures, and schemes that address fundamental mathematical questions of fair division (as do the procedures treated later in this chapter) is the ancient two-person scheme known as divide-and-choose. An application of this scheme to the Law of the Sea Treaty is described.

Divide-and-choose sets the stage for the mathematical investigations of fair division that have gone on for the past half-century. These investigations have often been phrased within the metaphor of "cake cutting." We present four cake-cutting schemes. The first two of these—found by Steinhaus and Banach–Knaster in the 1940s—yield allocations that are "proportional," meaning that each player receives what he or she perceives to be at least his or her fair share of the cake. The last two of these—found by Selfridge–Conway in 1960 and Brams–Taylor in 1992—yield allocations that are "envy-free" in the sense that each player receives what he or she perceives to be a piece at least tied for largest.

THE ADJUSTED WINNER DIVORCE PROCEDURE

In this section we present a recently developed scheme, called the *adjusted winner procedure,* for handling property settlements in a divorce or in an inheritance involving only two heirs. Generalizations of the adjusted winner scheme to three or more parties are somewhat less satisfactory, and so we will introduce a different, and much older, scheme in the next section to handle such cases. This older scheme, however, has the drawback that it requires each party to have an adequate bankroll with which to work.

EXAMPLE ▶ *The Trump Divorce*

Although Ivana had initially estimated Donald's assets to be in the $5 billion range, public disclosures by Donald later revealed his financial instability, and by early 1991, both Donald and Ivana were willing to attempt an out-of-court settlement.

Of course, we have no way of knowing the exact values and states of mind of Donald and Ivana Trump at the time of these negotiations, and we can only offer a rough approximation of the actual items involved, as some assets were

Trump Tower, Midtown
Manhattan.

taken back by the bank in default proceedings. Child custody, we should note, had already been settled to the satisfaction of both.

For the sake of a fairly realistic illustration, let us take as the marital assets the following: a 45-room mansion in Greenwich, Connecticut; the 118-room Mar-a-Lago mansion in Palm Beach, Florida; an apartment in the Trump Plaza; a 50-room Trump Tower triplex; and just over a million dollars in cash and jewelry.

The starting point of the adjusted winner scheme is to have each party (independently and simultaneously) distribute 100 points over the items in a way that reflects their relative worth to that party. In practice, this is at best a daunting task, and one that may require a considerable amount of assistance from a trained facilitator. For our example—given what we know—let's assume that Donald and Ivana used the following point assignments:

Point Allocations		
Marital Asset	Donald	Ivana
Connecticut estate	10	38
Palm Beach mansion	40	20
Trump Plaza apartment	10	30
Trump Tower triplex	38	10
Cash and jewelry	2	2

Some of the reasons behind the point totals are as follows. In the four marital agreements signed by the Trumps over the years, Ivana had always received the Connecticut estate, indicating that it was probably worth more to her than to him. The Palm Beach mansion was purely a vacation home to Ivana, but represented an important business opportunity for Donald. The Trump Tower apartment was home to Ivana and the children (thus explaining its relative importance to her), while Donald was living at the triplex (thus explaining its relative importance to him). The cash and jewelry were not much of an issue (but one we need in order to illustrate fully the adjusted winner procedure).

The adjusted winner procedure now allocates the property as follows:

1. Each party is initially given each asset for which he or she placed more points than the other party. Thus, Donald initially receives the Palm Beach mansion (40 of his points) and the Trump Tower triplex (38 of his points), while Ivana initially receives the Connecticut estate (38 of her points) and the Trump Plaza apartment (30 of her points). Notice that Donald now has $40 + 38 = 78$ of his points and Ivana only has $38 + 30 = 68$ of her points. The issue upon which they placed the same number of points (the cash and jewelry, on which each placed 2 points) now goes to Ivana, because she has fewer points so far. This increases her point total to $68 + 2 = 70$.

2. We now start transferring assets from Donald to Ivana until their point totals are equalized. The order in which this is done—that is, which assets get transferred before which others—is extremely important and determined as follows.

The assets which Donald currently has are arranged, from left to right, so that the fractions

$$\frac{\text{Donald's point value of the asset}}{\text{Ivana's point value of the asset}}$$

increase (or stay the same) as we scan from left to right. For our example, the fractions for the two items Donald has are

40/20 (Palm Beach)   38/10 (triplex)

We now transfer, in the preceding order, assets (or fractions thereof) from Donald to Ivana until equality of points is achieved.

If we were to transfer the Palm Beach mansion (worth 40 points to Donald and 20 points to Ivana) completely to Ivana, then she would have far more points than Donald: $70 + 20 = 90$ of her points to $78 - 40 = 38$ of his points.

Hence, we want to find what fraction—call it x—of the mansion Donald should retain, while giving the rest—namely, $1 - x$—to Ivana. For example, if Donald retains $\frac{4}{5}$ of the mansion, then Ivana will get $1 - \left(\frac{4}{5}\right) = \frac{1}{5}$ of the mansion. The points Donald receives from his fraction of the mansion will be x times 40, while the points Ivana receives from her fraction of the mansion will be $1 - x$ times 20. Thus, to equalize points we want x to satisfy the following:

$$38 + 40x = 70 + 20(1 - x)$$

That is, Donald receives 38 of his points from the Trump Tower triplex, and 40 times x of his points from his fraction x of the Palm Beach mansion. Ivana receives 70 of her points from the Connecticut estate, the Trump Plaza apartment, and the cash and jewelry, while she receives 20 times $1 - x$ of her points from her fraction $1 - x$ of the Palm Beach mansion.

Solving for x yields the following:

$$38 + 40x = 70 + 20 - 20x$$
$$38 + 40x = 90 - 20x$$
$$60x = 52$$
$$x = 52/60$$

With $x = \frac{52}{60}$, the number of his points that Donald receives is

$$38 + 40 \times (52/60) = 38 + 2080/60 \approx 38 + 34.7 = 72.7$$

Similarly, the number of her points that Ivana receives is

$$70 + 20 \times (8/60) = 70 + 160/60 \approx 70 + 2.7 = 72.7$$

Thus, equality of points is achieved when Donald retains $\frac{52}{60}$ (about 87%) ownership of the Palm Beach mansion, and Ivana gets the remaining $\frac{8}{60}$ (about 13%) ownership.   ◆

In point of fact, the actual settlement reached by Ivana and Donald Trump was extremely close to that produced by the adjusted winner procedure: Donald received the Trump Tower triplex, and Ivana received the Connecticut estate, the Trump Plaza apartment, and the cash and jewelry. And what about the Palm Beach mansion that gets split 87–13 by the adjusted winner procedure? In reality, Ivana was awarded use of it for one month a year as a vacation home—not too far off the kind of split we came up with here.

Having seen how the adjusted winner procedure works, one must now ask the following question: Exactly what is it about the allocation produced by this scheme that would make one want to use it? The answer is given by the following theorem (whose proof can be found in the 1996 monograph by Brams and Taylor cited in Suggested Readings):

> **Theorem:** For two parties, the **adjusted winner procedure** produces an allocation, based on each player's assignment of 100 points over the items to be divided, that has the following properties:
>
> 1. The allocation is **equitable:** this means that both players receive the same number of points.
> 2. The allocation is **envy-free:** this means that neither player would be happier with what the other received.
> 3. The allocation is **Pareto-optimal:** this means that no other allocation, arrived at by any means, can make one party better off without making the other party worse off.

Economists consider Pareto optimality to be an extremely important property. The fact that the adjusted winner procedure produces an allocation that is efficient in this sense leads one to hope that it can and will play a future role in real-world dispute resolution.

THE KNASTER INHERITANCE PROCEDURE

The adjusted winner procedure can be applied in the case of an inheritance if there are only two heirs. For *more than two heirs,* there is quite a different scheme, first proposed by Bronislaw Knaster in 1945. It has a drawback, though, in that it requires the heirs to have a large amount of cash at their disposal.

EXAMPLE ▶ *A Four-Person Inheritance*

Suppose (for the moment) that there is just one object—a house—and four heirs—Bob, Carol, Ted, and Alice. Knaster's scheme begins with each heir bidding (simultaneously and independently) on the house. Assume, for example, that the bids are

Bob	Carol	Ted	Alice
$120,000	$200,000	$140,000	$180,000

Carol, being the high bidder, is awarded the house. Her fair share, however, is only one-fourth of the $200,000 she thinks the house is worth, and so she places $150,000 (which is three-fourths of the $200,000 she bid) into a temporary "kitty."

Each of the other heirs now withdraws from the kitty his or her fair share, that is, one-fourth of his or her bid. Thus

Bob withdraws	$120,000/4 = $30,000
Ted withdraws	$140,000/4 = $35,000
Alice withdraws	$180,000/4 = $45,000

Thus, from the $150,000 kitty, a total of $30,000 + $35,000 + $45,000 = $110,000 is withdrawn, and each of the four heirs now feels that he or she has the equivalent of one-fourth of the estate. Moreover, there is a $40,000 surplus ($150,000 kitty − $110,000 withdrawn), which is now divided equally among the four heirs (so each receives an additional $10,000). The final settlement is

Bob	Carol	Ted	Alice
$40,000	house − $140,000	$45,000	$55,000

This illustrates Knaster's procedure for the simple case in which there is only one object. But what do we do if our same four heirs have to divide an estate consisting of, say, a house (as before), a cabin, and a boat? The easiest answer is to handle the estate one object at a time (proceeding for each object as we just did for the house). To illustrate, assume that our four heirs submit the following bids:

	Bob	Carol	Ted	Alice
House	$120,000	$200,000	$140,000	$180,000
Cabin	60,000	40,000	90,000	50,000
Boat	30,000	24,000	20,000	20,000

We have already settled the house. Let's handle the cabin the same way. Thus, Ted is awarded the cabin based on his high bid of $90,000. His fair share is one-fourth of this, so he places three-fourths of $90,000 (which is $67,500) into the kitty.

Bob withdraws from the kitty $60,000/4 = $15,000. Carol withdraws $40,000/4 = $10,000, and Alice withdraws $50,000/4 = $12,500. Thus, from the $67,500 kitty, a total of $15,000 + $10,000 + $12,500 = $37,500 is withdrawn.

The surplus left in the kitty is thus $30,000, and this is again split equally ($7500 each) among the four heirs. The final settlement on the cabin is

Bob	Carol	Ted	Alice
$22,500	$17,500	cabin − $60,000	$20,000

If we were now to do the same for the boat (we leave the details to the reader), the corresponding final settlement would be

Bob	Carol	Ted	Alice
boat − $20,875	$7625	$6625	$6625

Putting the three separate analyses (house, cabin, and boat) together, we get a final settlement of

Bob: boat + ($40,000 + $22,500 − $20,875 = $41,625)
Carol: house + (− $140,000 + $17,500 + $7625 = − $114,875)
Ted: cabin + ($45,000 − $60,000 + $6625 = − $8375)
Alice: $55,000 + $20,000 + $6625 = $81,625. ◆

Notice that in the final settlement, Carol gets the house, but she must put up $114,875 in cash (and Ted gets the cabin, but he must put up $8375 in cash). This cash is then disbursed to Bob and Alice. In practice, Carol's having this amount of cash available may be a real problem. This is the key drawback to Knaster's procedure. Nevertheless, Knaster's procedure shows again that whenever some participants have different evaluations of some objects, there is an allocation in which everyone obtains more than a fair share.

DIVIDE-AND-CHOOSE

There are vast mineral resources under the seabed, all of which, one might argue, should be available to both the developed and the developing countries. In the absence of some kind of agreement, however, what is to prevent the developed countries from mining all of the most promising tracts before the developing countries have reached a technological level where they can begin their own mining operations? Such an agreement went into effect on November 16, 1994, with 159 signatories (including the United States). It was called **the Convention of the Law of the Sea,** and it protects the interests of the developing countries by means of the following fair-division procedure.

Whenever a developed country wants to mine a portion of the seabed, that country must propose a division of the portion into two tracts. An international mining company called the Enterprise, funded by the developed countries but representing the interests of the developing countries through the International Seabed Authority, then chooses one of the two tracts to be reserved for later use by the developing countries.

> The preceding rules constitute a fair-division procedure known as **divide-and-choose:** one party divides the object into two parts in any way that he desires, and the other party chooses whichever part she wants.

As a fair-division procedure, the origins of divide-and-choose go back at least 2800 years to Hesiod's *Theogeny.* The Greek gods Prometheus and Zeus had to divide a portion of meat. Prometheus began by placing the meat into two piles, and Zeus selected one.

Actually, a fair-division procedure consists of both rules and strategies, and all we have described so far are the rules of divide-and-choose. But the strategies here are quite obvious: the divider makes the two parts equal in his estimation, and the chooser selects whichever piece she feels is more valuable.

Rules and strategies differ from each other in the following sense: a referee could determine if a rule were being followed, even without knowing the preferences of the players. Strategies represent choices of how players follow the rules, given their individual preferences (and any other knowledge and/or goals they may have).

The strategies on which we will focus in our discussion of fair-division procedures are those that require no knowledge of the preferences of the other players and yet provide some kind of minimal degree of satisfaction even in the face of collusion by the other players. For example, the strategies just given for divide-and-choose guarantee each player a piece that he or she would not wish to trade for that received by the other.

There are, to be sure, other strategic considerations that might be relevant. For example, in divide-and-choose, would you rather be the divider or the chooser? The answer, given our assumptions that nothing is known of the preferences of the others, is to be the chooser. However, if you knew the preferences of your opponent (and her value of spite), then you might want to be the divider.

As a final comment on strategic considerations, we need only look to the origins of the well-known expression "the lion's share." It comes from one of Aesop's fables, as reported by Todd Lowry in *Archaeology of Economic Ideas* (1987, p. 130):

It seems that a lion, a fox, and an ass participated in a joint hunt. On request, the ass divides the kill into three equal shares and invites the others to choose. Enraged, the lion eats the ass, then asks the fox to make the division. The fox piles all the kill into one great heap except for one tiny morsel. Delighted at this division, the lion asks, "Who has taught you, my very excellent fellow, the art of division?" to which the fox replies, "I learnt it from the ass, by witnessing his fate."

CAKE-DIVISION SCHEMES: PROPORTIONALITY

The modern era of fair division in mathematics began in Poland during World War II (see Spotlight 13.1). At this time, Hugo Steinhaus asked what is, in retrospect, the obvious question: What is the "natural" generalization of divide-and-choose to three or more people? The metaphor that has been used in this context, going back at least to the English political theorist James Harrington (1611–1677), is a cake. We picture different players valuing different parts of the cake differently because of concentrations of certain flavors or depth of frosting. (Don't, however, think of a layer cake, because that is a context in which the difficult "envy-free" questions we ask can be answered by simply dividing each layer equally among the players.) Thus, we ask the following:

Can one devise a **cake-division scheme** for n players—that is, a procedure that the players can use to allocate a cake among themselves (no outside arbitrators)—so that each player has a strategy that will guarantee her a piece with which she is "satisfied," even in the face of collusion by the others?

As we have seen, divide-and-choose is a cake-division scheme for two players, if by "satisfied" we mean either "thinks his piece is of size or value at least one-half," or "does not want to trade what he received for what anyone else received." These two notions of satisfaction are so important that we define them precisely.

A cake-division scheme will be called **proportional** if each player's strategy guarantees him a piece of size or value at least $1/n$ in his own estimation. It will be called **envy-free** if each player's strategy guarantees him a piece he considers to be at least tied for largest.

It turns out that for $n = 2$, a scheme is envy-free if and only if it is proportional; that is, for $n = 2$, the two notions of fair division are exactly the same. For $n > 2$, however, all we can say is that an envy-free scheme is automatically proportional. For example, if a three-person allocation is not proportional, then one player (call him Bob), thinks that he received less than one-third. Bob then feels that the other two are sharing more than two-thirds between them, and thus that at least one of the two (call her Carol) must have more than one-third. But then Bob will envy Carol, and so the allocation is not envy-free. Since all nonproportional allocations fail to be envy-free, it follows that if an allocation *is* envy-free, then it must be proportional.

Many schemes that are proportional, however, fail to be envy-free as we will soon show. Thus, proportional schemes are fairly easy to come by, but envy-free schemes are fairly hard to come by.

E X A M P L E ▶ *The Steinhaus Proportional Procedure for Three Players (Lone Divider)*

Given three players—Bob, Carol, and Ted—we have Bob divide the cake into three pieces, call them X, Y, and Z, each of which he thinks is of size or value exactly one-third. Let's speak of Carol as "approving of a piece" if she thinks it is of size or value at least one-third. Similarly, we will speak of Ted as "approving of a piece" if the same criterion applies. Notice that both Carol and Ted must approve of at least one piece.

If there are distinct pieces, say, X and Y, with Carol approving of X and Ted approving of Y, then we give the third piece, Z, to Bob (and, of course, X to Carol and Y to Ted), and we are done. The problem case is where both Carol and Ted approve of only one piece and it is the *same* piece.

Thus, let's assume that Carol and Ted approve of only piece X, and hence (of more importance to us) both *disapprove* of piece Z. Let XY denote the result of putting piece X and piece Y back together to form a single piece. Notice that both Carol and Ted think that XY is at least two-thirds of the cake, since both disapprove of Z. Thus, we can give Z to Bob and let Carol and Ted use divide-and-choose on XY. Since half of two-thirds is one-third, both Carol and Ted are guaranteed a proportional share (as is Bob, who approved of all three pieces). ◆

The method just described, which guarantees proportional shares but is not necessarily envy-free and is sometimes called the **lone-divider method,** was discovered by Steinhaus around 1944. Unfortunately, it does not extend easily to more than three players. It was left to Steinhaus's students, Stefan Banach and Bronislaw Knaster, to devise a method for more than three players. Picking up where Steinhaus left off (and traveling in quite a different direction), they devised the proportional scheme that today is referred to as the **last-diminisher method.** Like the lone-divider method it is proportional but not envy-free. We illustrate it for the case of four players (Bob, Carol, Ted, and Alice), and we include both the rules and the strategies that guarantee each player his or her fair share.

EXAMPLE ▶ *The Banach–Knaster Proportional Procedure for Four or More Players (Last Diminisher)*

Bob cuts from the cake a piece that he thinks is of size one-fourth, and hands it to Carol. If Carol thinks the piece handed her is larger than one-fourth, she trims it to size one-fourth in her estimation, places the trimmings back on the cake, and passes the diminished piece to Ted. If Carol thinks the piece handed her is of size at most one-fourth, she passes it unaltered to Ted.

Ted now proceeds exactly as did Carol, trimming the piece to size one-fourth if he thinks it is larger than this and passing it (diminished or unaltered) on to Alice. Alice does the same, but, being the last player, simply holds onto the piece momentarily instead of passing it to anyone.

Notice that everyone now thinks the piece is of size at most one-fourth, and the last person to trim it (or Bob, if no one trimmed it) thinks the piece is of size exactly one-fourth. Thus, the procedure now allocates this piece to the last person who trimmed it (and to Bob if no one trimmed it).

Assume for the moment that it was Ted who trimmed the piece last, and so he takes this piece and exits the game. Bob, Carol, and Alice all think that at least three-fourths of the cake is left, and so they can start the process over with (say) Bob beginning by cutting a piece from what remains that he thinks is one-fourth of the original cake. Carol and Alice are both given a chance to trim it to size one-fourth in their estimation, and again, the last one to trim it takes that piece and exits the game. The two remaining players both think that at least half the cake is left, and so they can use divide-and-choose to divide it between themselves and thus be assured of a piece that is of size at least one-fourth in their estimation. ◆

For a concrete illustration of the last diminisher method for the simple case where there are only three players (Bob, Carol, and Ted), suppose that all three players view the cake as having 18 units of "value," with each unit of value represented by a small square. Suppose, however, that the players value various parts of the cake differently (or that Bob views the cake as being perfectly rectangular, whereas Carol and Ted see it as skewed in opposite ways). We represent this pictorially as follows:

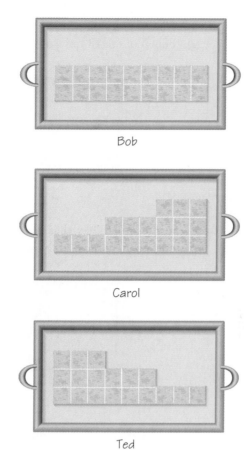

Bob

Carol

Ted

In step 1, Bob cuts from the cake a piece—call it *A*—that he considers to be of size or value one-third (because there are only three players this time). We'll assume he does this by making a vertical cut as follows:

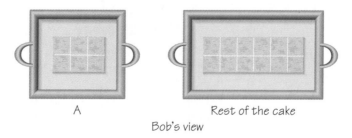

A

Rest of the cake

Bob's view

From Carol's point of view (or value system), the piece *A* appears to contain only 3 of the 18 units of value:

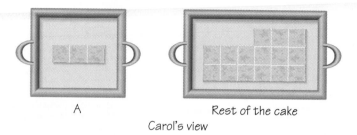

A   Rest of the cake

Carol's view

Because Carol thinks that *A* represents less than one-third of the cake, she passes *A* unaltered to Ted. Now Ted sees *A* as follows:

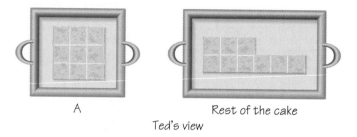

A   Rest of the cake

Ted's view

Thus, Ted thinks that *A* represents one-half of the cake (9 out of 18 units of value), and so he will trim it to what he thinks is one-third (6 out of 18 units of value). Let's assume that he does this with another vertical cut, with the trimmed version of *A* now called A'.

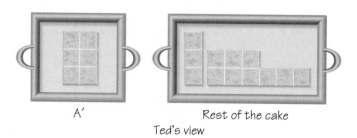

A'   Rest of the cake

Ted's view

Everyone has now had a chance to diminish the piece *A* that Bob initially cut from the cake, and so A' goes to the last person to trim it, namely, Ted. So Ted takes A' (which he thinks is of size $\frac{6}{18} = \frac{1}{3}$) and exits the game. Notice that what is left is seen by Bob and Carol as follows:

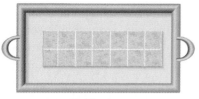

Bob's view
(14 units of value)

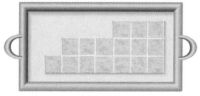

Carol's view
(16 units of value)

The final step has Bob and Carol use divide-and-choose. Note that if Bob is the divider and elects to make a vertical cut (halving the middle column of squares), then Carol will see the division as leaving 6 squares to the left of the cut and 10 squares to the right of the cut. She will thus choose the piece to the right of the cut. Hence, the final allocation finds Ted thinking he has $\frac{6}{18}$ of the cake, Bob thinking he has $\frac{7}{18}$ of the cake, and Carol thinking she has $\frac{10}{18}$ of the cake.

CAKE-DIVISION SCHEMES: THE PROBLEM OF ENVY

Divide-and-choose has a property that neither of the last two procedures possesses: it can assure each player of a piece of cake he or she considers the largest or tied for the largest. In the case of only two players, this means that each player can get what he or she perceives to be at least half the cake, no matter what the other player does. Thus, divide-and-choose is an envy-free procedure.

Steinhaus's $n = 3$ proportional procedure (the lone-divider method) is not envy-free. For example, consider the case where Carol and Ted both find one piece unacceptable (and this piece is given to Bob). Carol and Ted will not envy each other when one divides and the other chooses, but Bob may think that this is not a 50–50 split. Indeed, if Bob divided the cake initially into what he thought was three equal pieces, an unequal split of the remaining two-thirds of the cake by Carol and Ted means that Bob will prefer the larger of these two pieces to the one-third he got. Consequently, Bob will envy the person who got this larger piece.

Neither is the last-diminisher method envy-free. For example, if Bob initially cuts a piece of cake of size one-fourth, and no one else trims it, then Bob receives this piece and exits the game. If Carol is the one to make the next initial cut, she may well cut a piece from the cake that she thinks is of size one-fourth, but that Bob thinks is of size considerably more than one-fourth. But Bob is out of the game. Thus, if Ted and Alice think this piece is of size less than one-fourth, then Carol receives it, and so Bob will envy Carol.

Nevertheless, there do exist cake-division schemes that are envy-free. We present two of these in what follows.

EXAMPLE ▶ *The Selfridge–Conway Envy-Free Procedure for Three Players*

We start with a cake and three people. The point we wish to arrive at is an envy-free allocation of the entire cake among the three people in a finite number of steps. This task may seem formidable; however, quite often in mathematics, an important part of solving a problem involves breaking the problem into identifiable parts. In this case, let us call our starting point *A*, and the final point we wish to reach *C*. Now let us identify an appropriate in-between point *B* that makes going from *A* to *C*—via *B*—more manageable. Our in-between point *B* is the following:

Point B: Getting a constructive procedure that gives an envy-free allocation of *part* of the cake.

Can we constructively obtain three pieces of cake, whose union may not be the whole cake, which can be given to the three people so that each thinks he or she received a piece at least tied for largest? This turns out to be quite easy, with the solution due to John Selfridge and John Conway, who arrived at it independently around 1960. The following process and strategies do the trick:

1. Player 1 cuts the cake into three pieces he considers to be the same size. He hands the three pieces to player 2.
2. Player 2 trims at most one of the three pieces so as to create at least a two-way tie for largest. Setting the trimmings aside, player 2 hands the three pieces (one of which may have been trimmed) to player 3.

3. Player 3 now chooses, from among the three pieces, one that he considers to be at least tied for largest.

4. Player 2 next chooses, from the two remaining pieces, one that she considers to be at least tied for largest, with the proviso that if she trimmed a piece in step 2, and player 3 did not choose this piece, then she must now choose it.

5. Player 1 receives the remaining piece.

Let us reconsider the five steps of this trimming procedure to assure ourselves that each player experiences no envy. Recall that player 1 cuts the cake into three pieces, and player 2 trims one of these three pieces. Now player 3 chooses, and, as the first to choose, he certainly envies no one. Player 2 created a two-way tie for largest, and at least one of these two pieces is still available after player 3 selects his piece. Hence, player 2 can choose one of the tied pieces she created and will envy no one. Finally, player 1 created a three-way tie for largest and, because of the proviso in step 4, the trimmed piece is not the one left over. Thus, player 1 can choose an untrimmed piece and therefore will envy no one.

So far we have gone from point A to point B: starting with a cake and three players, we have constructively obtained (in finitely many steps) an envy-free allocation of all of the cake, except the part T that player 2 trimmed from one of the pieces. We will now describe how T can be allocated among the three players in such a way that the resulting allocation of the whole cake is envy-free. (This is the rest of the Selfridge–Conway procedure.)

The key observation for the $n = 3$ case is that player 1 will not envy the player who received the trimmed piece, even if that player were to be given all of T. Recall that player 1 created a three-way tie and received an untrimmed piece. The union of the trimmed piece and the trimmings yields a piece that player 1 considers to be exactly the same size as the one he received. Thus, assume that it is player 3 who received the trimmed piece (it could as well be player 2). Then player 1 will not envy player 3, however T is allocated.

The next step ensures that neither player 2 nor player 3 will envy another player when it comes time to allocate T. Let player 2 cut T into three pieces she considers to be the same size. Let the players choose which of the three pieces they want in the following order: player 3, player 1, player 2.

To see that this yields an envy-free allocation, notice that player 3 envies no one, because he is choosing first. Player 1 does not envy player 2, because he is choosing ahead of her; and player 1 does not envy player 3 because, as pointed out earlier, player 1 will not envy the player who received the trimmed piece. Finally, player 2 envies no one, because she made all three pieces of T the same size.

Hence, for $n = 3$, the **Selfridge–Conway procedure** will give an envy-free allocation of all the cake except T, followed by an allocation of T that gives an envy-free allocation of all the cake. ◆

A naive attempt to generalize to $n = 4$ what we have done for $n = 3$ would proceed as follows: we would begin by having player 1 cut the cake into four pieces he considers to be the same size. Then we would have players 2 and 3 trim some pieces (but how many?) to create ties for the largest. Finally, we would have the players choose from among the pieces—some of which would have been trimmed—in the following order: player 4, player 3, player 2, player 1.

This approach fails because player 1 could be left in a position of envy. In order to understand how the approach could fail, consider how many pieces player 3 might have to trim in order to create a sufficient supply of pieces tied for largest so that he is guaranteed to have one available when it is his turn to choose. Player 3 might have to trim one piece to create a two-way tie for largest. Player 2 might need to trim two pieces to create a three-way tie for largest (since, if there were only a two-way tie for largest, player 3 might further trim one of these pieces and player 4 might choose the other). This leaves player 1 in a possible position of envy, because we could have a situation where player 2 trims two pieces and player 3 trims a third piece, and player 4 then chooses the only untrimmed piece. If this happens, player 1, by being forced to choose a trimmed piece, will definitely envy player 4.

All is not lost, however, since there are slight modifications of the Selfridge–Conway procedure that will work for arbitrary n. We describe one such modification for the case $n = 4$.

E X A M P L E ▶ *The 1992 Envy-Free Procedure for Four or More Players (The Trimming Procedure)*

With four or more players, the new idea we introduce is to have the first player cut the cake into more pieces than there are players. For simplicity, we illustrate this with four players.

1. Player 1 cuts the cake into *five* pieces she considers to be the same size. She hands the five pieces to player 2.

2. Player 2 trims at most two of the five pieces so as to create at least a three-way tie for largest. Setting the trimmings aside, player 2 hands the five pieces—one or two of which may have been trimmed—to player 3.

3. Player 3 trims at most one of the five pieces she has been handed so as to create at least a two-way tie for largest. This, of course, may involve further trimming of a piece that player 2 already trimmed in step 2.

She sets her trimmings aside with those of player 2, handing the further altered collection of five pieces to player 4.

4. Player 4 now chooses from among the five pieces, some of which may have been trimmed by player 2 and/or player 3, a piece that he considers to be at least tied for largest. (The remaining steps now reverse the order of initial play.)

5. Player 3 chooses next, from among the four remaining pieces, a piece that she considers to be at least tied for largest, with the proviso that if she trimmed a piece in step 3, and player 4 did not choose this piece, then player 3 must choose it now.

6. Player 2 chooses next, from among the three remaining pieces, a piece that he considers to be at least tied for largest, with the proviso that if he trimmed a piece or pieces in step 2, and one of these is still available, then he must now choose such a piece.

7. Player 1 now chooses, from the remaining two pieces, one that was not trimmed.

This time, check the procedure on your own to be certain that it achieves an envy-free allocation of part of the cake. Notice, however, that for $n = 4$ we not only have the trimmings from steps 2 and 3 left over but also one of the five pieces, perhaps now trimmed, with which we started.

Thus, we have again gone from point A to point B: starting with a cake and n players, we have constructively obtained (in finitely many steps) an envy-free allocation of part of the cake. Our next step (before arriving at C) is a fairly small one:

Point B': Getting a constructive procedure that gives an envy-free allocation of all the cake, but using *infinitely* many steps.

The intuition behind the infinite scheme is simple: one applies the **trimming procedure,** a scheme that yields an envy-free allocation of part of the cake over and over again, with each application yielding an envy-free allocation of part of what was left over from the previous application. Eventually the whole cake is allocated.

The next question for the $n = 4$ case is how to get from point B' to point C, where the whole cake is allocated in a finite number of steps. That is, how do we make the infinite scheme finite? Our approach mirrors what we did in the $n = 3$ case: if two players have different preferences, then we arrange the partial-allocation scheme so that each player thinks he received a piece of cake strictly larger than the other player.

Eventually, when the size of the crumb is sufficiently small (in the eyes of these two players), then neither will care if the other player should get the whole crumb (because both will think his piece is already larger than the other's piece, even with the crumb added to it). This is just a glimpse of what is needed; full details of the trimming procedure can be found in the article by Brams and Taylor in Suggested Readings. ◆

Thus, there is a solution to the problem of envy-freeness that obviates the need for an endless process of finer and finer divisions. Practically speaking, however, it is only necessary to know that the trimmings become progressively smaller; then one can stop this procedure when what remains no longer matters much to the players.

APPLYING THE TRIMMING PROCEDURE TO INDIVISIBLE GOODS

Although we have used the metaphor of cake cutting throughout our earlier discussion on the problem of envy, the idea of successive trimming is nonetheless applicable to problems of fair division other than parceling out the last crumbs of a cake. The main practical problem in applying the trimming procedure is that many fair-division problems involve goods that cannot be divided up at all, much less trimmed in fine amounts. Such goods are said to be *indivisible*.

Although the trimming procedure, as such, is not applicable to allocation problems involving indivisible goods, it can be adapted to such problems under certain conditions. The key condition is that there be a sufficient quantity of more divisible goods, like small items or, even better, money, which can be trimmed in lieu of the indivisible good.

Take, for example, the problem of dividing up an estate, in which a house is the single big item. Assume there are four heirs, but only one heir thinks the house is worth one-fifth of the estate (with the others thinking it is worth less). If this person is the one to make the initial division, and if, in addition, he knows that the other heirs do not value the house so highly, he can begin by dividing up the estate into five pieces, with one piece being just the house. If none of the other three heirs thinks this indivisible piece has to be trimmed on the first round—even after other trimmings are made—then the house can, in effect, be "reserved" for the heir who thinks it is the most valuable piece.

This example illustrates how one player's knowledge of the preferences of the others need not always be exploitative but can, instead, facilitate the search for a solution. It will not always be apparent, however, precisely what information players should reveal and what they should hide (as is true in most negotiations). But because the trimming procedure has certain safeguards built in—

The division of Berlin after World War II.

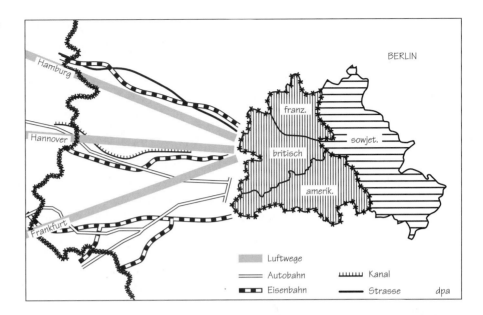

A checkpoint at the Berlin Wall.

in particular, allocating in stages in addition to ensuring envy-freeness in each—players probably can afford to be more open about their preferences than if these safeguards were absent.

It is interesting to recall that when the Allies agreed in 1944 to partition Germany into sectors after World War II (first stage), they at first did not reach agreement about what to do with Berlin. Subsequently, they decided to partition Berlin itself into sectors (second stage), even though this city fell 110 miles within the Soviet sector. Berlin was simply too valuable a "piece" for the Western Allies (Great Britain, France, and the United States) to cede to the Soviets, which suggests how, after a leftover piece is trimmed off, it can be subsequently divided under the trimming procedure.

Yet, what if a large piece like Berlin is not divisible? In the settlement of an estate, this might be the house, as we suggested earlier, which may be worth half the estate to the claimants. In this situation, there may be no alternative but to sell this big item and use the proceeds to make the remaining estate more liquid or, in our terms, "trimmable."

EXAMPLE ▶ *Dividing Up an Estate*

To illustrate the trimming procedure in the case of an estate, assume the estate is composed of six items. Four heirs have valuations for each item that are indicated by points that sum to 100 for each in Table 13.1. Notice that all heirs consider H (the house) to be worth half the estate.

Assume no heir has sufficient resources to pay off the other three to get the house. Accordingly, the heirs agree to sell the house on the open market. Suppose they get exactly 50 for it, which is what they agree it is worth. (If they get less, say, 40, this would change their totals to 90 but would not affect the trim-

TABLE 13.1 Fair Division of an Estate by Trimming

Item	Heirs			
	Bob	Carol	Ted	Alice
1. House (H)	50	50	50	50
2. Boat (B)	20	10	10	10
3. Car (C)	10	20	10	10
4. Furniture (F)	10	10	10	10
5. Piano (P)	10	0	10	10
6. Art (A)	0	10	10	10
Total point valuation of estate	100	100	100	100

ming process in a fundamental way.) After the sale, H is replaced by 50 (divisible) points rather than a single indivisible item.

We start with Bob (it could be any of the heirs) dividing the estate into five parts among which he is indifferent. For example, Bob could chose a division of the estate into the following five parts, all of which are worth 20 points to him. (Note that the 50 points replacing the house are divided unequally among the five parts to bring each up to value 20 in Bob's eyes.)

Bob: B $C + 10$ $F + 10$ $P + 10$ $A + 20$

If Carol goes next, she must create at least a three-way tie for largest. Because she initially assigned 20 points to C and 10 points to A, $C + 10$ and $A + 20$ will be the largest parts for her (each is worth 30), and she will trim each by 10 to create a tie with her next-largest item ($F + 10$), which is worth 20. We underscore to indicate a tie for largest.

Carol: B $\underline{C}$ $\underline{F + 10}$ $P + 10$ $\underline{A + 10}$

If Ted goes next, he must create at least a two-way tie for largest. But because $F + 10$, $P + 10$, and $A + 10$ are all worth 20 to him, he need do no trimming, but his ties are different from Carol's:

Ted: B C $\underline{F + 10}$ $\underline{P + 10}$ $\underline{A + 10}$

Now Alice must choose a part from the five that she considers to be at least tied for largest. This part will be $F + 10$, $P + 10$, or $A + 10$, because her preferences are the same as Ted's. Assume she chooses $A + 10$, and Ted next chooses $P + 10$. Then Carol has two remaining pieces, $F + 10$ and C, that she considers tied for largest. But since C is the result of her trimming a piece (namely, $C + 10$), Carol must take it (C) because it is available. Finally, assume Bob chooses B. Then the first-stage allocations to heirs Bob, Carol, Ted, and Alice are, respectively, B, C, $P + 10$, and $A + 10$, leaving $F + 10$ and the 20 trimmed by Carol for the second stage of the procedure.

Now all the heirs value F equally (10), but none can divide $F + 30$ ($F + 10 + 20$) into five equal parts as Bob did at the beginning of the first stage of the procedure. Thus, we can see that there could be a problem of indivisibility at the second stage in some instances, which would necessitate selling other items to provide proceeds that can be divided in later stages.

In our example, however, there is no such necessity, because $F + 30$ can be exactly divided into four equal parts for the four heirs:

$\underline{F}$ $\underline{10}$ $\underline{10}$ 10

Adding these four equal parts, in this order, to the previous first-stage envy-free

Divide-and-Choose

10.   Suppose that Bob, Carol, and Ted view a cake as pictured in the example illustrating the Banach–Knaster scheme (see page 501). Assume that all cuts that will be made are vertical.

(a)   If Bob and Carol use divide-and-choose to divide the cake between them, how large a piece will each receive (assuming they follow the suggested strategies that go with divide-and-choose)?

(b)   If Carol and Ted use divide-and-choose to divide the cake between them, how large a piece will each receive (assuming they follow the suggested strategies that go with divide-and-choose)?

11.   If you and another person are using divide-and-choose to divide something between you, would you rather be the divider or the chooser? (Assume that neither of you knows anything about the preferences of the other.)

12.   Suppose that Bob and Carol view a cake as pictured in the example illustrating the Banach–Knaster scheme (see page 501). Assume that all cuts that will be made are vertical. Assume that Bob and Carol know how each other values the cake, and that neither is spiteful. Suppose they are to divide the cake using the *rules*, but not necessarily the *strategies*, of divide-and-choose.

(a)   Is Bob better off being the divider or the chooser? Why?

(b)   Discuss this in relation to Exercise 11.

Cake-Division Schemes: Proportionality

13.   Suppose that players 1, 2, and 3 view a cake as follows:

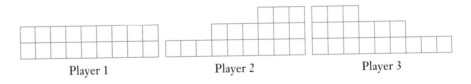

Player 1   Player 2   Player 3

Notice that each player views the cake as having 18 square units of area (or value). Assume that each player regards a piece as acceptable if and only if it is at least $\frac{18}{3} = 6$ square units of area (his or her "fair share"). Assume also that all cuts made correspond to vertical lines.

(a) Provide a total of three drawings to show how each player views a division of the cake by player 1 into three pieces he or she considers to be the same size or value. Label the pieces *A*, *B*, and *C*.

(b) Identify two of these pieces that player 2 finds acceptable, and two that player 3 finds acceptable.

(c) Show that a feasible assignment of fair pieces can be achieved by letting the players choose in the following order: player 3, player 2, player 1. Indicate how many square units of value each player thinks he or she received. Is there any other order in which players can choose pieces (in this example) that also results in a feasible assignment?

14. Suppose that players 1, 2, and 3 view a cake as follows:

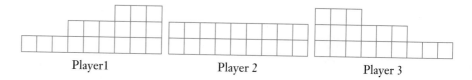

Player1   Player 2   Player 3

(a) Provide a total of three drawings to show how each player views a division of the cake by player 1 into three pieces he or she considers to be the same size or value. Label the pieces *A*, *B*, and *C*. (We are still assuming that all cuts correspond to vertical lines, so this will require a cut along a vertical center line of some of the squares.)

(b) Show that neither player 2 nor player 3 finds more than one of the three pieces acceptable (with "acceptable" defined as in Exercise 13).

(c) Identify a single piece that player 2 and player 3 agree is *not* acceptable. (There are actually two such pieces; for definiteness, find the one on the right.)

(d) Assume that players 2 and 3 give the piece from part (c) to player 1. Suppose they reassemble the rest and players 2 and 3 divide it between themselves using divide-and-choose (with a single vertical cut). Determine what size piece each of the three players will think he or she received (1) if player 2 divides and player 3 chooses, and (2) if player 3 divides and player 2 chooses.

15.   Suppose players 1, 2, and 3 view a cake as in Exercise 14. Illustrate the last-diminisher method (still restricting attention to vertical cuts and, furthermore, assuming that the piece potentially being diminished is a piece off the left side of the cake) by following steps (a)–(h) below:

(a)   Draw a picture showing the third of the cake (6 squares) that player 1 will slice off the cake.

(b)   Determine if player 2 will pass or further diminish this piece. If he or she would further diminish it, make a new drawing.

(c)   Determine if player 3 will pass or further diminish this piece. If he or she would further diminish it, make a new drawing.

(d)   Determine who receives the piece cut off the cake and what size or value he or she thinks it is. (Actually, we *knew* what size the person receiving this first piece would think it was, assuming he or she followed the prescribed strategy. How did we know this?)

(e)   Finish the last-diminisher method using divide-and-choose on what remains, with the lowest-numbered player who remains doing the dividing.

(f)   Redo step (e) with the other player doing the dividing.

(g)   Redo step (e), but with the last two players using the last-diminisher method directly, instead of divide-and-choose (with the order as in step (e)).

(h)   Redo step (g) with the order reversed.

16.   Suppose players 1, 2, and 3 view the cake as in Exercise 14. Illustrate the envy-free procedure for $n = 3$ (yielding an allocation of part of the cake) by following steps (a)–(c) below. Again, restrict attention to vertical cuts.

(a)   Provide a total of three drawings to show how each player views a division of the cake by player 1 into three pieces he or she considers to be the same size or value. Label the pieces *A*, *B*, and *C*. (This is the same as Exercise 14a.)

(b)   Redraw the picture from player 2's view, and illustrate the trimming of piece *A* that he or she would do. Label the trimmed piece *A'* and the actual trimmings *T*.

(c)   Indicate which piece each player would choose (and what he or she thinks its size is) if the players choose in the following order: player 3, player 2, player 1 according to the envy-free procedure on pages 504–505. Does the proviso in step 4 come into play here?

17. Apply the remainder of the Selfridge–Conway procedure from pages 505–506 to what was obtained in Exercise 16 by completing (a)–(c) below:

(a) Draw a picture of T from each player's view.

(b) The procedure calls for the player (other than player 1) who did not receive the trimmed piece to divide T into three pieces he or she considers to be the same size. Here, that would be player 2. Illustrate this division, and label the pieces X, Y, Z.

(c) Indicate which parts of T (and the sizes or values) the players will choose when they go in the following order: player 3, player 1, player 2.

18. Consider the trimming procedure for four people described on pages 506–507. Explain why each player experiences no envy.

19. In generalizing the $n = 4$ envy-free procedure to arbitrary n, player 1 cut the cake into $2^{n-2} + 1$ pieces. For $n = 4$, we had $2^{n-2} + 1 = 5$. Suppose that $n = 5$. Then $2^{n-2} + 1 = 9$. Determine how many pieces each player must trim to make the procedure work for $n = 5$.

▲ 20. Suppose that $n = 5$, and we begin by having player 1 cut the cake into 16 pieces of the same size. Suppose player 2 creates an eight-way tie, then player 3 creates a four-way tie, and, finally, player 4 creates a two-way tie. Let the players now choose in the following order: player 5, player 4, player 3, player 2, player 1. Show that no provisos about choosing trimmed pieces are needed to ensure envy-freeness.

▲ 21. Consider the envy-free procedure for $n = 4$, wherein player 1 cuts the cake into five pieces he considers to be the same size. Since he (player 1) gets an untrimmed piece, he thinks the size of the leftover L_1 from the first stage is at most 4/5 of the original cake. (He may think it as little as 1/5 if the other players did no trimming, leaving only one of the five pieces he cut for the second stage.) If we now do the same thing with L_1, then he will think the size of the leftover L_2 from the second stage is at most $(4/5)(4/5) = 16/25 = .64$ of the cake. How large a portion of the cake will he think the maximum size of the eleventh leftover is? Is the size of this leftover less than 1/10 of the cake?

Applying the Trimming Procedure to Indivisible Goods

22.   Suppose that four heirs have the following valuations for each of seven items in an estate, indicated by points that sum to 100 for each.

	Heirs			
Item	I	II	III	IV
Money (M)	40	40	40	40
Boat (B)	15	10	20	15
Car (C)	10	15	10	10
Furniture (F)	10	10	10	14
Piano (P)	17	10	10	5
Art (A)	5	15	10	10
Dog (D)	3	0	0	6

Use the trimming procedure to find an envy-free settlement of the estate among the four heirs by completing steps (a)–(n) below.

(a)   Show how heir I can divide the estate into five parts that she considers to be of equal value. (Although there is more than one way to do this, choose the way that combines the art and dog, together with some money, to make one part, and then combine the car with some money to make another part.)

(b)   Write down heir II's values of the five parts handed to her by heir I.

(c)   Indicate how heir II can trim two of the parts to create a three-way tie for most valuable part. Make the trimming consist of money if possible.

(d)   Write down heir III's value of the five parts handed to her (after the trimming) by heir II.

(e)   Indicate how heir III can trim one of the parts to create a two-way tie for most valuable part.

(f)   Indicate which part heir IV will now choose (and how many points of value she thinks it is worth).

(g)   Indicate which part heir III will now choose (and how many points of value she thinks it is worth). Does the proviso about choosing a piece you trimmed if one is available come into play here?

(h)   Indicate which part heir II will now choose (and how many points of value she thinks it is worth). Assume she really does not want the dog.

(i)   Indicate which part heir I will now choose (and how many points of value she thinks it is worth).

(j)   In dividing up the trimmings and the piece not chosen, assume the art is sold for 10 units of money. Show that this yields:

	Heirs			
Item	I	II	III	IV
Money (M)	32	32	32	32
Dog (D)	3	0	0	6

(k)   Show how heir I can divide the estate into five parts that she considers to be of equal value.

(l)   Explain why neither heir II nor heir III will do any trimming.

(m)   Indicate which part each will choose when they go in the order: heir IV, heir III, heir II, heir I.

(n)   What is the obvious thing to do with what is left?

Additional Exercises

23.   The Banach–Knaster last-diminisher method is not the only well-known cake-division scheme that yields a proportional allocation for any number of players. There is also one due to A. M. Fink (sometimes called the *lone-chooser method*). For three players (Bob, Carol, and Ted) it works as follows:

(i)   Bob and Carol divide the cake into two pieces using divide-and-choose.

(ii)   Bob now divides the piece he has into three parts that he considers to be the same size. Carol does the same with the piece she has.

(iii)   Ted now chooses whichever of Bob's three pieces that he (Ted) thinks is largest, and Ted chooses whichever of Carol's three pieces that he thinks is largest.

(iv)   Bob keeps his remaining two pieces, as does Carol.

    (a)   Explain why Ted thinks he is getting at least one-third of the cake.

    (b)   Explain why Bob and Carol each think they are receiving at least one-third of the cake.

    (c)   Explain why, in general, this scheme is not envy-free.

24.   In A. M. Fink's scheme (described in Exercise 23), suppose that a fourth person (Alice) comes along after Bob, Carol, and Ted have already divided the cake among themselves so that each of the three thinks he or she has a piece of size at least one-third. Mimic what was done in the three-person case to obtain an allocation among the four that is proportional. (*Hint:* Begin by having Bob, Carol, and Ted divide the pieces they have into a certain number—how many?—of equal parts.)

25.   There is a "moving-knife" version of the Banach–Knaster scheme that appears in the Dubins–Spanier paper in Suggested Readings. To describe it, we picture the cake as being rectangular, and the procedure beginning with a referee holding a knife along the left edge as illustrated below.

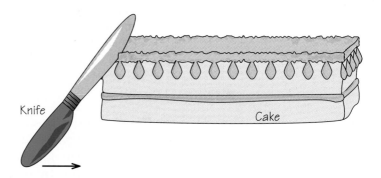

Assume, for the sake of illustration, that there are four players (Bob, Carol, Ted, and Alice). The referee starts moving the knife from left to right over the cake (keeping it parallel to the position in which it started) until one of the players (assume it is Bob) calls "cut." At this time, a cut is made, and the piece to the left of the knife is given to Bob, and he exits the game. The knife starts moving again, and the process continues. The strategies are for each player to call cut whenever it would yield him or her a piece of size at least one-fourth.

    (a)   Explain why this procedure produces an allocation that is proportional.

    (b)   Explain why the resulting allocation is not, in general, envy-free.

    (c)   Explain why, if you are not the first player to call cut, there is a strategy different from the one suggested that is never worse for you, and sometimes better.

26.   There is a two-person moving-knife cake-division scheme due to A. K. Austin that leads to each player receiving a piece of cake that he or she considers to be of size *exactly* one-half. It begins by having one of the two players (Bob) place two knives over the cake, one of which is at the left edge, and the other of which is parallel to the first and placed so that the piece between the knives (*A* in the picture below) is of size exactly one-half in Bob's estimation.

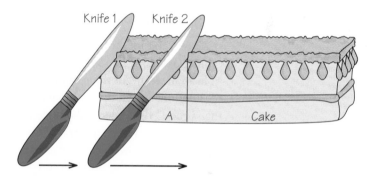

If Carol agrees this is a 50–50 division, we are done. Otherwise, Bob starts moving both knives to the right—perhaps at different rates—so that the piece between the knives remains of size one-half in his eyes. Carol calls stop at the point when she also thinks the piece between the two knives is of size exactly one-half.

    (a)   If the knife on the right were to reach the right-hand edge, where would the knife on the left be?

    (b)   Explain why there definitely *is* a point where Carol thinks the piece between the two knives is of size exactly one-half. (Hint: If Carol thinks the piece is too small at the beginning, what will she think of it at the end?)

WRITING PROJECTS

1 ▶   It turns out that there is no way to extend the adjusted winner procedure to three or more players. That is, there are point assignments by three players to three objects so that no allocation satisfies the three desired properties of equitability (equal points), envy-freeness, and Pareto optimality. On the other hand, there are separate procedures that will realize any two of the three properties. Thus, trade-offs must be made, and these may depend on the circumstances. Discuss your feelings regarding the relative importance of the three properties, and circumstance which may affect the choice of which two of the three properties one might wish to have satisfied.

2 ▶   One of the most important differences between the three-person and the *n*-person envy-free procedures is that the latter procedure may take more than two stages. And, of course, the more stages there are, the more cuts and trimmings that may be necessary. Do you consider this a serious practical problem, or is it mainly a theoretical problem? Why?

3 ▶   Besides the division of an estate among heirs, or the allocation of cabinet positions to political parties in a coalition government, what other possible applications of the trimming procedure can you think of? Give examples of situations not in the text in which it is "bad" things like chores—rather than "good" things like cake (for most of us, anyway)—that you might want to allocate in an envy-free way? (*Optional:* In applying the trimming procedure to bad things, how might you create ties for worst with, say, next-worst by adding things to the worst piece?)

CHAPTER 14

APPORTIONMENT

THE APPORTIONMENT PROBLEM

The delegates who wrote the U.S. Constitution in 1787 (see Figure 14.1) established a House of Representatives, which "shall be apportioned among the several states within this union according to their respective Numbers . . . " (Article I, Section 2). The Constitution does not specify the total number of representatives. In 1790, it was decided that there would be 105 representatives, and the number steadily increased until 1910, when Congress permanently fixed the House size at 435 seats.

The apportionment problem arises because it is unlikely that any state's fair share of the House will be a whole number. For example, in the census of 1790, the population of the United States was found to be 3,615,920. The most populous state was Virginia, with 630,560 people, while the least populous was Delaware, with 55,540. With 105 seats in the House, Virginia's fair share would be

$$\frac{\text{population of Virginia}}{\text{population of the United States}} \times 105$$

or

$$\frac{630,560}{3,615,920} \times 105 = 18.310$$

Delaware's share would be

$$\frac{55,540}{3,615,920} \times 105 = 1.613$$

FIGURE 14.1   The signing of the U.S. Constitution.

Should these numbers be rounded to give Delaware 2 representatives and Virginia 18? Virginians would find this inequitable, since each of their 18 representatives would have to account for 35,031 constituents, while each of the 2 Delaware representatives would represent 27,770 people.

The **apportionment problem** is to round a set of fractions so that their sum is maintained at its original value. The rounding procedure must not be an arbitrary one, but one that can be applied consistently. Any such rounding procedure is called an **apportionment method.**

We will discuss several apportionment methods in this chapter, and our focus will be on how to implement the methods and on criteria for deciding which is the most appropriate method to use.

Apportionment problems occur in many contexts besides the U.S. House of Representatives. For example, many countries are governed by parliaments in which each party is allotted a number of seats determined by the number of votes the party received in a general election. If a party obtains, say, 41.23% of the votes in a national election for a parliament with 120 seats, its fair share is 41.23% of 120, or 49.476. The number of individual representatives assigned to this party—its apportionment—must be a whole number, such as 49, 50, or some other integer value.

An apportionment problem that has nothing to do with politics occurs when a school or college schedules classes. Suppose a small high school has one math teacher who teaches all geometry, precalculus, and calculus sections. She has time to teach a total of five sections. The enrollments are 52 for geometry, 33 for precalculus, and 15 for calculus. How many sections of each course should be scheduled? The total number of students is 100, and 52% of them are taking geometry. Therefore geometry's share is 52% of the 5 sections available, or 2.6 sections. The precalculus share is 33% of 5, or 1.65 sections, and the calculus share is 15% of 5, or 0.75 section. The apportionment will be the actual numbers of sections taught, which have to be whole numbers.

The House of Representatives is the best-known and most frequently studied case of political apportionment. Some half-dozen different apportionment methods have been seriously considered for use by the Congress. In addition to the *Hill–Huntington method* that is currently used, three other methods have been implemented in the past: the methods of *Thomas Jefferson, Alexander Hamilton,* and *Daniel Webster.*

Although many apportionment problems do not involve the House of Representatives, the terminology that we will use in discussing them will refer to *states, populations,* and a *House size.* In a course-scheduling problem, the states would represent the subjects, the populations would be the numbers of students enrolled in each subject, and the House size would be the total number of sections to be taught.

Let n be the number of states, and let the populations of these states be denoted

$$p_1, p_2, \ldots, p_i, \ldots, p_n$$

We will denote the House size by h. The total population will be denoted p; thus

$$p = p_1 + p_2 + \cdots + p_n$$

If we divide p by h, we obtain the **population of the average congressional district,** which we will denote by c, so that

$$c = \frac{p}{h}$$

A state's fair share, or *quota,* is the number of congressional districts necessary to accommodate its population. It can be found by dividing the state's population by the average district population c, so that

$$q_i = \frac{p_i}{c}$$

is the quota for state i.

> In an apportionment problem, the **quota** is the exact share that would be allocated if a whole number were not required.

An apportionment is given by n integers

$$a_1, a_2, \ldots, a_i, \ldots, a_n,$$

with a_i representing the number of representatives apportioned to state i. Ideally, the numbers a_i should be as close as possible to the quotas q_i. Furthermore the total number of representatives has to be equal to the House size:

$$a_1 + a_2 + \cdots + a_n = h$$

In choosing an apportionment method, we must decide what we mean by the phrase "the numbers a_i should be as close as possible to the quotas q_i."

As one might expect, apportionment involves the rounding of numbers. We will consider several ways to do this. Given a number q, $\lfloor q \rfloor$ denotes the integer part of q; that is, the fractional part is simply discarded and q is rounded down. With this stingy rounding, $\lfloor 7.00001 \rfloor = 7$, $\lfloor 7 \rfloor = 7$, and $\lfloor 6.99999 \rfloor = 6$. The notation for rounding up to the next integer is $\lceil q \rceil$. With this generous rounding, $\lceil 7.00001 \rceil = 8$, but $\lceil 7 \rceil = 7$. Finally, we will use $\langle q \rangle$ to denote the rounding of q to the nearest integer.[1] Thus, $\langle q \rangle = \lfloor q \rfloor$ if the fractional part of q is less than 0.5, and $\langle q \rangle = \lceil q \rceil$ if the fractional part of q is greater than or equal to 0.5. Thus, $\langle 7.49999 \rangle = 7$, but $\langle 7.5 \rangle = 8$.

THE HAMILTON METHOD

The first apportionment method considered by Congress was the *method of largest fractions,* or the *method of Alexander Hamilton.*

> With the **Hamilton method,** state i receives either its **lower quota,** which is the integer part of its quota (in the notation just introduced, $\lfloor q_i \rfloor$), or its **upper quota,** $\lceil q_i \rceil$. The states that receive their upper quotas are those whose quotas have the largest fractional parts.

---

[1] Surprisingly, there is no commonly accepted notation for this type of rounding, even though it is preferred for most purposes. The notations $\lfloor q \rfloor$ and $\lceil q \rceil$ are standard.

The apportionment method of largest fractions, also known as the Hamilton method, was named for Alexander Hamilton.

Implementing the Hamilton method is a three-step procedure. First, we calculate each state's quota. Second, we assign to each state its lower quota of representatives. Unless each quota is an integer, the total number of seats assigned at this point will be less than the House size h, and this leaves a number of additional seats to be apportioned. The third step is to distribute these additional seats, one each, to those states whose quotas have the largest fractional parts.

It is possible that a tie will be encountered, in which the quotas of two states have identical fractional parts. All apportionment methods are vulnerable to occasional ties, but in practice, ties rarely occur when large populations are involved.

The Hamilton method is straightforward, but it provoked the first presidential veto in U.S. history when George Washington rejected an apportionment bill that was based on the Hamilton method. Washington's objection appears to have been that the fractions of seats gained by some states in the last phase of the Hamilton apportionment were not proportional to the states' populations.

EXAMPLE ▶ *Hamilton's Apportionment of the House of Representatives*

In 1790, there were 15 states, and the House had 105 seats. Table 14.1 displays the calculations leading to Alexander Hamilton's proposed apportionment.

The total population was 3,613,920, so the average congressional district population was $c = 3{,}613{,}920 \div 105 = 34{,}418$. The quotas shown in the table are obtained by dividing each state's population by c.

The third column of Table 14.1 shows the quotas for the states, and the fourth column, the lower quotas. We see that if each state were given its lower quota, only 97 seats would have been apportioned. The remaining 8 seats go to the 8 states whose quotas had the largest fractional parts. In column 5 of the table, the states are ranked in decreasing order according to the *fractional parts* of their quotas; the top 8 states receive an additional seat. The Hamilton apportionments appear in the sixth column (labeled a_i).   ◆

President Washington's veto prevented the Hamilton method from being used in 1792, but it was adopted by Congress in 1850 and remained in use until 1900. The half century of experience with the Hamilton method revealed a paradox.

TABLE 14.1		Apportioning the House of Representatives by the Hamilton Method			
State	Population	Quota	Lower Quota	Rank	Apportionment
Virginia	630,560	18.310	18	10	18
Massachusetts	475,327	13.803	13	5	14
Pennsylvania	432,879	12.570	12	8	13
North Carolina	353,523	10.266	10	11	10
New York	331,589	9.629	9	6	10
Maryland	278,514	8.088	8	14	8
Connecticut	236,841	6.877	6	4	7
South Carolina	206,236	5.989	5	2	6
New Jersey	179,570	5.214	5	12	5
New Hampshire	141,822	4.118	4	13	4
Vermont	85,533	2.484	2	9	2
Georgia	70,835	2.057	2	15	2
Kentucky	68,705	1.995	1	1	2
Rhode Island	68,446	1.988	1	3	2
Delaware	55,540	1.613	1	7	2
Totals	3,615,920	105.000	97	—	105

Paradoxes of the Hamilton Method

A *paradox* is a fact that seems obviously false. The first Hamilton apportionment paradox, called the *Alabama paradox,* was discovered in 1881. As part of the reapportionment procedure mandated by the Constitution every 10 years, the Census Bureau had supplied Congress with a table of congressional apportionments for a range of different House sizes from 275 to 350, based on the 1880 census. The table revealed a strange phenomenon.

With a 299-seat House, Alabama's quota was 7.646. The fractional part ranked 20th of the 38 states, and that was just enough to give Alabama its upper quota. Illinois and Texas, with quotas of 18.64 and 9.64, respectively, ranked below Alabama and received their lower quotas. The next column of the table, corresponding to a House size of 300 seats, showed changes in the apportionments for these three states: Alabama now had 7 seats instead of 8, while Illinois and Texas received increased apportionments of 19 seats 10 seats, respectively. Alabama had *lost* a seat as a result of an increase in the House size. This happened because the Illinois and Texas quotas increased more than Alabama's quota did. With a 300-seat House, Alabama's quota increased to 7.671, Illinois's to 18.702, and Texas's to 9.672. Since the fractional parts of the quotas for Illinois and Texas were larger, those states were given their upper quotas, and Alabama was left with its lower quota.

The **Alabama paradox** occurs when a state loses a seat as the result of an increase in the House size.

EXAMPLE ▶ *A Mathematics Department Meets the Alabama Paradox*

A mathematics department assigns teaching assistants to cover recitations for college algebra, calculus I, calculus II, calculus III, and contemporary mathematics. There are 30 teaching assistants available, and the enrollments are given in Table 14.2. The department will use the Hamilton method to apportion the teaching assistants to the five subjects. The total number of students enrolled is 750, so the average section size will be $750 \div 30 = 25$ students. The quotas shown in the table are derived by dividing the course enrollments by 25. The lower quotas add up to 27, so the three courses whose quotas have the largest fractional parts, calculus I and III and contemporary mathematics, are entitled to their upper quotas.

After these calculations are finished, the graduate school authorizes the department to hire an additional teaching assistant. The apportionment calculations are repeated; with 31 teaching assistants the average section size will be

TABLE 14.2   Apportioning 30 Teaching Assistants

Course	Enrollment	Quota	Lower Quota	Rank	Apportionment
College algebra	188	7.52	7	4	7
Calculus I	142	5.68	5	2	6
Calculus II	138	5.52	5	5	5
Calculus III	64	2.56	2	3	3
Contemporary mathematics	218	8.72	8	1	9
Totals	750	30.00	27	—	30

$750 \div 31 = 24.1935$. Table 14.3 displays the rest of the calculations. The calculus III instructor will be dismayed, since the increased number of teaching assistants has caused her course to *lose* one section. While the average section size has decreased slightly, calculus III sections have increased from 21 students to 32 students! ◆

The Alabama paradox was the cause of the abandonment of the Hamilton method. In 1901, tables of apportionment for all House sizes between 350 and 400 were prepared by the Census Bureau, using the Hamilton method and the 1900 census figures. Colorado, a Populist state, received two seats with a House size of 357, and three seats for House sizes 350–356 *and* 358–400. When the House Apportionment Committee submitted a bill making the House size equal to 357, it was accused in debate of manipulating the apportionment to Colorado's disadvantage. The bill was rejected, and another apportionment method (the method of Daniel Webster) was chosen to apportion the House.

TABLE 14.3   Apportioning 31 Teaching Assistants

Course	Enrollment	Quota	Lower Quota	Rank	Apportionment
College algebra	188	7.771	7	2	8
Calculus I	142	5.869	5	1	6
Calculus II	138	5.704	5	3	6
Calculus III	64	2.645	2	4	2
Contemporary mathematics	218	9.011	9	5	9
Totals	750	31.000	28	—	31

The Alabama paradox was a serious problem in the nineteenth century, because Congress increased the House size with every reapportionment (this was done to prevent any state from losing any seats it had had before the reapportionment). Now that the House size is permanently fixed at 435 members, one might think it safe to return to the Hamilton method.

A second paradox, the *population paradox,* is a much worse problem, and could present itself even with a fixed House size. Suppose that there have been many errors in the census, and when they are all corrected, the apportionment of the House has to be recalculated based on the new data. It turns out that your state was originally undercounted; but even so, the new apportionment based on a larger population for your state results in a loss of one seat. Would you then be surprised to find that another state, which had originally been overcounted (and thus the correct population was lower than the one on which the original apportionment was based) *gained* a seat? This has never happened, of course, but with the Hamilton method of apportionment, it could.

In Exercise 3 we consider this situation in an imaginary country with four states and a 100-seat legislature, apportioned by the Hamilton method. The census corrections result in substantial increases in population for the three largest states, and a slight loss of population for the smallest state. In the recalculated apportionments, the middle two states each lose one seat, while the largest and smallest states each gain a seat. Thus, two states that gained population saw their apportionments decrease, while one state that lost population gained a seat.

> The **population paradox** is that one state's apportionment can decrease and another state's apportionment can increase, although the first state had gained population, and the second state had lost population.

The explanation of these paradoxes was the basis for George Washington's objection to the Hamilton method. The fractional parts of the quotas, which determine the states that get additional seats, are not proportional to the populations.

DIVISOR METHODS

The Jefferson Method

President Washington could have vetoed the Hamilton apportionment bill on constitutional grounds, because Article I, Section 3, of the Constitution requires that each congressional district have a population of at least 30,000.

Thomas Jefferson favored a method of apportionment biased in favor of states with large populations.

With only 55,540 people, Delaware's population was too small for the two congressional districts assigned to it by the bill. Washington turned to Thomas Jefferson for a more acceptable method of apportionment.

The **Jefferson method** set a minimum district population d. Each state's population was divided by the number d, and the state's apportionment was to be the integer part of the quotient. In other words, all fractional seats were discarded. Because each state's population is divided by a fixed number d to obtain the apportionment, d is called the *divisor*, and the Jefferson method is a *divisor method*. The divisor taken by Jefferson was $d = 33,000$. His state, Virginia, had a population of 630,560, according to the 1790 census. Since

$$\frac{630,560}{33,000} = 19.108$$

Jefferson apportioned to Virginia 19 seats, discarding the fractional part, 0.108. Jefferson's bill was passed by Congress and signed into law by the president in time to apportion the House for the 1794 election.

A **divisor method** of apportionment determines each state's apportionment by dividing its population by a common divisor d and rounding the resulting quotient. Divisor methods differ in the rule used to round the quotient.

With the Jefferson method, no state can have a district with a population less than d, because a state's apportionment is determined by calculating how many d person districts will fit in that state and discarding any remainders. Thus, as long as $d \geq 30,000$, the requirement that congressional districts have a population of at least 30,000 will be satisfied.

To achieve a House size of 105, the divisor had to be carefully chosen. For example, $d = 30,000$ would have resulted in larger apportionments for several states and a House size of 112, while $d = 36,000$ would have decreased several apportionments, and the House size would have been 91.

EXAMPLE ▶ *Jefferson's Apportionment*

Table 14.4 displays the apportionment according to the 1790 census by the Jefferson method, including the determination of the divisor d. The first four columns of the table are identical to the first four columns of Table 14.1. We start by putting $d = c$, the average district population, as the divisor. The quotients appearing in column 3 are the quotas, and the lower quotas appear in column 4. Each state's lower quota, which we denote n_i, is its tentative apportionment. We see immediately that too few seats have been apportioned.

TABLE 14.4	Apportioning the House of Representatives by the Jefferson Method					
State	Population	Quota	n_i	Critical Divisors		a_i
Virginia	630,560	18.310	18	33,187	31,528	19
Massachusetts	475,327	13.803	13	33,951	31,688	14
Pennsylvania	432,879	12.570	12	33,298	30,920	13
North Carolina	353,523	10.266	10	32,138		10
New York	331,589	9.629	9	33,158	30,144	10
Maryland	278,514	8.088	8	30,946		8
Connecticut	236,841	6.877	6	33,834	29,605	7
South Carolina	206,236	5.989	5	34,372	29,462	6
New Jersey	179,570	5.214	5	29,928		5
New Hampshire	141,822	4.118	4	28,364		4
Vermont	85,533	2.484	2	28,511		2
Georgia	70,835	2.057	2	23,611		2
Kentucky	68,705	1.995	1	34,352	22,902	2
Rhode Island	68,446	1.988	1	34,223	22,815	2
Delaware	55,540	1.613	1	27,770		1
Totals	3,615,920	105.000	97	—	—	105

To apportion more seats, it is necessary to choose a divisor smaller than c. This is done by calculating the *critical divisor* for each state.

> The **critical divisor** for a state is the divisor nearest to c that will cause that state's apportionment to increase, if an increase is necessary to fill the house; or to decrease, if too many seats have been apportioned already.

The critical divisors are shown in the fifth column of Table 14.4. To see how they were calculated, consider the case of Virginia, with a population of 630,560 and a tentative apportionment of 18. A divisor that will apportion one more seat to Virginia must satisfy the equation

$$\left\lfloor \frac{630,560}{d} \right\rfloor = 18 + 1$$

There is a range of numbers that will satisfy this equation, all less than c. The largest is $d = \frac{630,560}{18 + 1} = 33,187$, so Virginia's critical divisor is 33,187. The critical divisors for the remaining states are computed in the same way: divide the state's population by $n_i + 1$, where n_i is the tentative apportionment. Thus, for the Jefferson method, the critical divisor is

$$d_i = \frac{p_i}{n_i + 1}$$

Additional seats are apportioned to states in the order of their critical divisors. When a state's tentative apportionment increases, its critical divisor must be recalculated before another seat is apportioned. Thus, South Carolina was the recipient of the first additional seat, bringing its tentative apportionment to 6. The new critical divisor, obtained by dividing the population of South Carolina by 7 to get 29,642, is entered in the next column of Table 14.4. The second additional seat goes to Kentucky since its critical divisor is now the largest. The process continues until all 105 seats have been apportioned.

Figure 14.2 shows how the total number of seats apportioned increases as the divisor decreases. When the divisor passes the critical divisor of some state, that state's apportionment (and thus the total) increases by 1. Eight states will have received increased apportionments when the critical divisor of New York is reached. Therefore New York's critical divisor, 33,158, could be used to obtain the correct apportionment. Since we have already obtained the apportionment, it is not necessary to do the arithmetic. The final apportionments are shown in column 7 (labeled a_i) of Table 14.4.

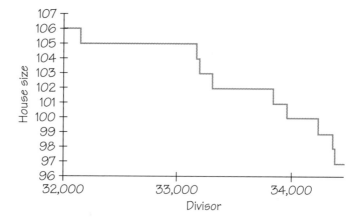

FIGURE 14.2 House size as determined by the divisor. The Jefferson method applied to data from the 1790 census. The vertical jumps in the graph occur at the critical divisors.

If the 1790 House size had been 106, North Carolina would have received the next seat, since Table 14.4 shows its critical divisor to be the largest. The 107th and 108th seats go to Massachusetts and Virginia, respectively. ◆

As with the Hamilton method, ties are possible with the Jefferson method. They can occur only when two states have identical critical divisors, a rare occurrence when dealing with large populations.

With the Jefferson method, no state can receive less than its lower quota as its apportionment, since the lower quota is the starting point. However, a state can be given more than its upper quota. This phenomenon occurred for the first time in the apportionment based on the 1820 census, which recorded that New York had a population of 1,368,775. The total population of the United States was found to be 8,969,878. Since the House had 213 members, the average district population was $c = 8{,}969{,}878 \div 213 = 42{,}112$. New York's quota was therefore $q = 1{,}368{,}775 \div 42{,}112 = 32.503$. The Hamilton method would have apportioned 33 seats to New York, but the Jefferson method, using the divisor $d = 39{,}900$, gave New York $\lfloor \frac{1{,}368{,}775}{39{,}900} \rfloor = 34$ seats. When the House was reapportioned following the 1830 census, the Jefferson method awarded New York 40 seats, although its quota was only 38.593.

An apportionment method is said to satisfy the **quota condition** if in every situation each state's apportionment is equal to either its lower quota or its upper quota. It only takes one example like the 1820 and 1830 apportionments of the House of Representatives to show that the Jefferson method does not satisfy the quota condition. In fact, if the House had continued to use the Jefferson method, it would have violated the quota condition in every apportionment since 1850. The states receiving seats in excess of their upper quotas would always be the largest states, because the Jefferson method is biased in favor of large states.

With the Hamilton method, each state starts with its lower quota, and some states are given their upper quotas to fill the House. There is no way for a state to receive less than its lower quota, or more than its upper quota, so the Hamilton method does satisfy the quota condition. This was obvious to the Congress in 1850, so it based its apportionment on the Hamilton method.[2]

The Jefferson method is not troubled by the paradoxes of the Hamilton method. Consider the Alabama paradox, in which a state loses a seat when the House size is increased. With any apportionment method, the apportionments of some states must be increased when the size of the House is increased. The Jefferson method increases the number of seats apportioned by using a smaller divisor. Since the apportionment for each state is obtained by rounding down the quantity p_i/d, and each such quantity increases when the divisor decreases, no apportionment will decrease.

The population paradox is also impossible with the Jefferson method. Suppose that the population of state A increases, and the population of state B decreases. If A loses a seat, then the divisor d must have increased proportionally more than A's population. But then the population of B divided by d will also be less than it originally was, so B cannot gain a seat.

Congress has never used a method of apportionment that satisfies the quota condition and avoids the paradoxes. It would seem to be desirable to have such a method, and in the 1970s, the mathematicians Michel L. Balinski and H. Peyton Young set out to find one. They succeeded in finding a method, which they called the *quota method,* that satisfies the quota condition (as the Hamilton method does) and avoids the Alabama paradox (as the Jefferson method does). However, the population paradox remained. Balinski and Young subsequently proved that the only apportionment methods that are free of the population paradox are the divisor methods. Since it is also known that every divisor method is capable of violating the quota condition, Balinski and Young have proved an impossibility theorem like the one of Kenneth Arrow (discussed in Chapter 11). No apportionment method is both free of paradoxes and satisfies the quota condition.

The Webster Method

The **Webster method** is the divisor method that employs the round to the nearest integer rule, rounding up when the fractional part is greater than or equal to $\frac{1}{2}$ and rounding down when the fractional part is less than $\frac{1}{2}$. In the notation introduced in page 530, the apportionment for state i is $\langle p_i/d \rangle$.

---

[2] The origins of the Hamilton method had been forgotten in 1850, and the method was named for Congressman Samuel Vinton, who had rediscovered the method.

Statesman and orator
Daniel Webster
(1782–1852), who
developed a divisor method
for apportioning the U.S.
House of Representatives.

The Webster and Jefferson methods share some strengths and defects. For example, both are immune to the Alabama and population paradoxes, and neither satisfies the quota condition. However, in many respects, Webster's is a better method than Jefferson's. While the Jefferson method obviously favors the large states, the Webster method has been shown to be neutral, favoring neither the large nor the small states. Furthermore, while the Jefferson method almost always gives some state more than its upper quota, the Webster method would never have violated the quota condition in any of the 21 congressional apportionments that have occurred since 1790.

EXAMPLE ▶ *Apportioning Classes*

A high school has five mathematics teachers, who can teach a total of 17 sections. The subjects, with their enrollments, are algebra I, 155; geometry, 124; algebra II, 158; precalculus, 85; and calculus, 55. We will use the Webster method to decide how many sections to allot to each course.

The first step is to calculate the average class size. The total enrollment is 577, so there will be an average of $c = 577 \div 17 = 33.941$ students per section. Table 14.5 displays the rest of the calculation. Column 3 gives the quota of sections for each subject. The quota is obtained by dividing the number of students enrolled in the subject by the average section size (c), just as in the

TABLE 14.5	**Apportioning Classes by the Webster Method**					
Subject	Enrollment	Quota	n_i	Critical Divisors		a_i
Algebra I	155	4.56	5	34.445	44.286	4
Geometry	124	3.65	4	35.429		4
Algebra II	158	4.65	5	34.725		5
Precalculus	85	2.50	3	34.000	56.667	2
Calculus	55	1.62	2	36.667		2
Totals	577	17.00	19	—	—	17

Hamilton and Jefferson calculations. However, in column 4, we put the *rounded* quota $\langle q_i \rangle$ instead of the lower quota.

If the sum of the entries in column 4 had been equal to 17, our job would be finished: the apportionments would be these rounded quotas. In this example, the sum turns out to be 19, so we must reduce the number of sections apportioned by 2.

The procedure is to find a *larger* divisor. For subject i, the critical divisor d_i is the divisor that is just large enough to reduce the apportionment of subject i by 1 must satisfy the equation

$$\frac{p_i}{d_i} = n_i - 0.5$$

Thus, $d_i = \frac{p_i}{n_i - 0.5}$.

The critical divisors appear in column 5. The smallest critical divisor belongs to precalculus, so that divisor will reduce the precalculus apportionment by 1 without affecting the other subjects. We reduce the precalculus apportionment to 2 sections and recalculate the critical divisor as $85 \div (2 - 0.5) = 56.667$. We compare critical divisors again to find the second apportionment to reduce. Now the smallest critical divisor belongs to algebra I. The final apportionment is shown in column 7 of Table 14.5. ◆

We have seen that, unlike the Jefferson method, it may be necessary to take away seats when using the Webster method. If the total apportionment must be reduced, always take a seat from the state with the smallest critical divisor. When increasing the total apportionment, always give a seat to the state with the largest critical divisor.

For example, if the sum of the rounded quotas in column 4 of Table 14.5 had been less than 17, a smaller divisor would be needed. To obtain it, the critical divisors in column 5 would be calculated by the formula

$$d_i = \frac{p_i}{n_i + 0.5}$$

The subject with the largest critical divisor would receive an additional section, and its critical divisor would be recalculated. If another section had to be apportioned, it would go the the subject that now had the largest critical divisor.

As with the Jefferson method, ties can occur with the Webster method in the unlikely event that two states have identical critical divisors. For example, suppose state 1 has a population of 500,000 and $n_1 = 2$, and state 2 has a population of 300,000 with $n_2 = 1$. If it is necessary to increase the apportionment, we would calculate $d_1 = 500,000 \div 2.5 = 200,000$ and $d_2 = 300,000 \div 1.5 = 200,000$. The two states would have the same priority for receiving the next seat, and there would be a tie, unless some other state had a greater critical divisor.

The Hill–Huntington Method

The **Hill–Huntington method** is a divisor method that has been used to apportion the U.S. House of Representatives since 1940. Like the Jefferson and Webster methods, the apportionment is obtained by dividing each state's population by a divisor and then rounding the resulting quotient to obtain a whole number. Recall that when we used the Jefferson method, all quotients p_i/d were rounded down to $\lfloor p_i/d \rfloor$, while with the Webster method the quotients were rounded to the nearest whole number $\langle p_i/d \rangle$. This is essentially the only difference between those two methods. The Hill–Huntington method also follows the same procedure, but it has a more complicated way of rounding.

To start, we need to know what is meant by the *geometric mean* of two numbers A and B. These numbers are not allowed to be negative, but 0 is allowed. Imagine a rectangle whose sides have lengths A and B. The area would be the product AB. The geometric mean of A and B is the length S of the side of a square with the same area as this rectangle. Since the square's area will be S^2, we have $S^2 = AB$, so $S = \sqrt{AB}$.

The Hill–Huntington rounding of a number q is equal to $\lfloor q \rfloor$ if $q < \sqrt{\lfloor q \rfloor \lceil q \rceil}$, which is the geometric mean of $\lfloor q \rfloor$ and $\lceil q \rceil$. Otherwise, it is equal to $\lceil q \rceil$. For example, suppose that $q = 7.485$. Jefferson and Webster would round q down to 7. Hill and Huntington would calculate the geometric mean of $\lfloor q \rfloor = 7$ and $\lceil q \rceil = 8$ to get $\sqrt{7 \times 8} = 7.48331. \ldots$ With their rounding method, one rounds up if q is greater than or equal to the geometric mean, and rounds down if q is less. Thus, Hill–Huntington would round 7.485 up to get 8. There is no standard notation for Hill–Huntington rounding, so we will use $\langle\langle q \rangle\rangle$ to denote this of rounding a number q.

The calculations follow the general plan of the Jefferson and Webster methods. Starting with the average district size as the divisor, round each state's quota the Hill–Huntington way to obtain a first tentative apportionment. If the sum of the tentative apportionments is equal to the House size, the job is finished. If not, a list of critical divisors must be constructed, each critical divisor being chosen to be just sufficient to change the corresponding state's apportionment by one seat in the desired direction. If n_i is the number of seats apportioned tentatively to a state, and the total apportionment is too small, the critical divisor for that state is

$$d_i = \frac{p_i}{\sqrt{n_i(n_i + 1)}}$$

When this divisor is used, the quotient for the state will be $p_i/d_i = \sqrt{n_i(n_i + 1)}$, which is the cutoff number for rounding up to $n_i + 1$. If the total apportionment is too large

$$d_i = \frac{p_i}{\sqrt{n_i(n_i - 1)}}$$

should be used, since then p_i/d_i would be equal to the cutoff number for rounding down to $n_i - 1$.

As with the Webster method, a seat is taken from the state with the smallest critical divisor if it is necessary to reduce the total apportionment, and a seat is given to the state with the largest critical divisor if it is necessary to increase the total number of seats apportioned. If further adjustments are necessary, the critical divisor of the state whose apportionment was altered is recomputed, and the process is repeated.

If $p_i/d < 1$, so that $\lfloor p_i/d \rfloor = 0$ and $\lceil p_i/d \rceil = 1$, the cutoff for rounding p_i/d up is $\sqrt{0 \times 1} = 0$, so $\langle\langle p_i/d \rangle\rangle = 1$. Thus a zero apportionment is impossible with the Hill–Huntington method.

With the Hill–Huntington, ties are less likely than they are with other methods. This is an advantage of the method over most other methods of apportionment, but not a decisive one, because ties are rare with any method.

EXAMPLE ▶ *Apportionment of the 1790 Congress Revisited*

The calculations leading to the Hill–Huntington apportionment of the Congress are given in Table 14.6. Notice the similarity between Tables 14.6 and 14.4. To aid in the process of rounding, Hill–Huntington style, the fourth column of the table, labeled G.M., shows the geometric mean of the lower and upper quotas. Column 5 gives the rounded quotas. These add up to 106, more than the House size of 105, so we have to choose a slightly larger divisor. Column 6 gives the critical divisors, calculated by the formula $d_i = p_i/\sqrt{n_i(n_i - 1)}$. The smallest critical divisor, Pennsylvania's, will give the correct apportionment. It isn't necessary to actually carry out the divisions. Penn-

TABLE 14.6	Apportioning the House of Representatives by the Hill–Huntington Method					
State	Population	Quota	G.M.	n_i	Critical Divisor	a_i
Virginia	630,560	18.310	18.493	18	36,047	18
Massachusetts	475,327	13.803	13.491	14	35,234	14
Pennsylvania	432,879	12.570	12.490	13	34,659	12
North Carolina	353,523	10.266	10.488	10	37,265	10
New York	331,589	9.629	9.487	10	34,953	10
Maryland	278,514	8.088	8.485	8	37,218	8
Connecticut	236,841	6.877	6.481	7	36,546	7
South Carolina	206,236	5.989	5.477	6	37,654	6
New Jersey	179,570	5.214	5.477	5	40,154	5
New Hampshire	141,822	4.118	4.472	4	40,941	4
Vermont	85,533	2.484	2.449	3	34,919	3
Georgia	70,835	2.057	2.449	2	50,088	2
Kentucky	68,705	1.995	1.414	2	48,582	2
Rhode Island	68,446	1.988	1.414	2	48,398	2
Delaware	55,540	1.613	1.414	2	39,273	2
Totals	3,615,920	105.000	—	106	—	105

sylvania will lose a seat, and the other apportionments will remain the same. The final apportionments are given in column 7 (labeled a_i). ◆

If Tables 14.1, 14.4, and 14.6 are compared, we will find that the three apportionments are different. That is why politicians find the choice of apportionment methods so important.

WHICH DIVISOR METHOD IS THE BEST?

All divisor methods work by dividing each state's population by a common divisor d and rounding the quotients in some consistent way. Thus, the Jefferson method rounds all of the quotients p_i/d downward to $\lfloor p_i/d \rfloor$, the Webster method rounds to the nearest whole number $\langle p_i/d \rangle$, and the Hill–Huntington method rounds p_i/d to $\langle\langle p_i/d \rangle\rangle$.

Although divisor methods differ only in their rounding rules, they can produce different apportionments. Which divisor method is the fairest? We

Spotlight

A Legal Challenge to Apportionment

14.1

In 1991, the Census Bureau reported the new apportionment that will be in effect for the congressional elections in the years 1992–2000. Several states lost representatives: New York lost 3, and Ohio and Pennsylvania lost 2 apiece. Montana, whose apportionment decreased from 2 to 1, sustained the greatest percentage loss, and Montana sued to restore the lost seat. As precedents, Montana referred to the two famous cases, *Baker* v. *Carr* and *Wesberry* v. *Sanders,* in which the U.S. Supreme Court required legislative and congressional district boundaries to be drawn so as to make district populations equal.

Montana argued that the correct apportionment would be the one that met the *Baker* and *Wesberry* criterion of having districts as nearly equal in population as possible, and asked the Court to require the Census Bureau to recompute the apportionments using the Dean method, which minimizes differences in district populations. This would have resulted in the transfer of a congressional seat from Washington to Montana.

In *U.S. Department of Commerce* v. *Montana,* the Supreme Court unanimously rejected Montana's claim. The opinion of the Court, written by Justice Stevens, pointed out that *intra*state districts, which were the subject of the *Baker* and *Wesberry* cases, could be equalized in population by drawing district boundaries correctly. Since congressional districts can't cross state lines, some inequity is inevitable in congressional apportionment. The opinion conceded that there were alternatives to the Hill–Huntington method, but concluded that the choice of apportionment method was best left to Congress.

can try to find out by measuring the *inequity* resulting from each apportionment method. This can be done by making comparisons of congressional district populations.

> If state i is apportioned a_i seats, and its population is p_i, then its **district population** is p_i / a_i.

In an ideal apportionment, each state would have the same district population, but this ideal is impossible to achieve in practice. For example, census data for 1990 show that the district populations ranged from 543,105, for Washington, to 803,655, for Montana (see Spotlight 14.1).

District population is not the only measure that can be used. Another is *representative share*.

> Suppose that state i has a_i seats and population p_i. The quotient a_i / p_i is called the **representative share;** it represents the share of a congressional seat given to each citizen of that state.

It may seem that district population and representative share are two sides of the same coin. After all, it is true that

$$\text{representative share} = \frac{1}{\text{district population}}$$

However, an apportionment that minimizes differences in representative share may not minimize differences in district population, and vice versa.

It can be shown that the apportionment method that gives the most equitable apportionment in terms of representative share is the Webster method. Another divisor method, the *Dean method* (not studied in this text), minimizes differences in district population. Huntington's research on divisor methods led him to suggest that it would be best to compare either district populations or representative shares by considering relative differences, rather than absolute differences between them.

> Given two positive numbers A and B, with $A > B$, the **absolute difference** is $A - B$, and the **relative difference** is the quotient $\frac{A - B}{B} \times 100\%$.

$Spotlight$ *Mathematics and Politics: A Strange Mixture*

14.2

Walter F. Willcox

Edward V. Huntington

The first American to consider apportionment from a theoretical point of view was Walter Willcox (1861–1964), who strongly advocated the Webster method and had computed the apportionment of 1900. His arguments convinced the Congress to use the Webster method again in 1910. In 1911, Joseph Hill, a statistician at the Census Bureau, proposed the Hill–Huntington method, with the strong endorsement of Edward V. Huntington, a mathematics professor at Harvard.

In 1920, the two methods were in competition. There were significant differences in the apportionments determined by the two methods, and the result was Washington gridlock: no apportionment bill passed during the decade, and the 1910 apportionments were retained throughout the 1920s. In preparation for the 1930 census results, the National Academy of Sciences formed a committee to determine whether either method was biased in favor of large or small states.

In 1929, the committee reported that the Hill–Huntington method was the more neutral (this conclusion was disproved in 1980 by Balinski and Young).

The 1930 census was remarkable in that the apportionments calculated by the Webster method were the same as the Hill–Huntington apportionments. The House was therefore reapportioned, but the method used could be claimed to be either one of the competing methods. The coincidence was almost repeated in the 1940 census, but there was one difference. The Hill–Huntington method gave the last seat to Arkansas, while Webster's method gave it to Michigan (see pages 549–550). At the time, Michigan was a predominantly Republican state, and Arkansas was in the Democratic column. The vote on the apportionment bill split strictly along party lines, with Democrats supporting the Hill–Huntington method, and Republicans voting for the Webster method. Since the Democrats had the majority, the Hill–Huntington method became the law.

For any two states, it turns out that the *relative difference* in district populations is equal to the relative difference in representative share (see Exercise 19). Therefore an apportionment method that minimizes *relative* difference in representative shares will also minimize the relative difference in district populations. The Hill–Huntington method gives the apportionment in which the relative difference in representative shares (or district populations) is as small as possible.

EXAMPLE ▶ *Inequities in the 77th Congress*

In 1940, Michigan had a population of 5,256,106 and was apportioned 17 seats in the House of Representatives. Therefore, each citizen of Michigan had a representative share of

$$\frac{17}{5,256,106} = 0.000003234 \text{ seat}$$

or 3.234 microseats (a microseat is one-millionth of a seat).

To calculate representative shares in microseats, divide the state's apportionment by its population, *expressed in millions.* Arkansas, with a population of 1,949,387, or 1.949,387 million, received 7 seats, so each citizen of that state had a representative share of 7/1.949,387 = 3.591 microseats. The absolute difference in representative share between Arkansas and Michigan was therefore

$$3.591 - 3.234 = 0.357 \text{ microseats}$$

If a seat had been taken from Arkansas and given to Michigan, then the representative share for a Michigander would have been 18/5.256,106 = 3.425 microseats, while each Arkansan would have been left with a representative share of 3.078 microseats.

Now it is Michigan that has the larger representative share, but the absolute difference

$$3.425 - 3.078 = 0.347 \text{ microseats}$$

is less than it was before the transfer was made. In terms of absolute difference in representative share, it would have been more equitable to have given Arkansas 6 seats and Michigan 18 seats. This would be the Webster apportionment, since that method minimizes absolute differences in representative share.

Spotlight 14.2 explains how this example may have convinced Congress to prefer the Hill–Huntington method to the Webster method.

The 1940 apportionment was done with the Hill–Huntington method, which minimizes *relative* rather than absolute differences in representative share.

If Michigan had 18 seats and Arkansas had 6 in the 77th Congress, then the relative difference in representative shares would be found by subtracting the smaller representative share (Arkansas's) from the larger (Michigan's), and expressing the result as a percentage of the smaller representative share. Thus, the relative difference would have been

$$\frac{3.425 - 3.078}{3.078} \times 100\% = 11.27\%$$

in Michigan's favor. The actual apportionment of 17 seats for Michigan and 7 for Arkansas gave representative shares of 3.591 microseats for Arkansas and 3.234 for Michigan. The relative difference was

$$\frac{3.591 - 3.234}{3.234} \times 100\% = 11.02\%$$

Since the relative inequity was less when Michigan had 17 seats and Arkansas had 7, this was the preferred apportionment with the Hill–Huntington method. ◆

Since each divisor method can lead to a slightly different apportionment, one way to decide which apportionment method to use is to decide, in a political debate, which type of inequity should be minimized. Challenges to apportionments have followed this approach (see Spotlight 14.1). Since there are valid arguments to be made for using absolute difference in representative share, absolute difference in district population, or relative difference in either district population or representative share as the best measure of inequity, this debate will never end.

Another approach to choosing an apportionment method is to consider *bias* in favor of large or small states. The Jefferson method, for example, favors the more populous states. It rounds all quotas downward, and in doing so, a small state is likely to lose a greater percentage of its quota than a large state will.

In 1980, Balinski and Young proved that the Webster method of apportionment is the only divisor method that is completely unbiased toward small or large states. The Hill–Huntington method uses a more generous rounding rule than the Webster method does, and has a slight bias in favor of small states.

REVIEW VOCABULARY

$\lfloor q \rfloor$ The integer part of a number q; for example, $\lfloor \pi \rfloor = 3$.

$\lceil q \rceil$ The result of rounding a number up to the next integer; for example, $\lceil \pi \rceil = 4$.

$\langle q \rangle$ The result of rounding a number q in the usual way: round down if the fractional part of q is less than 0.5, and round up otherwise. For example, $\langle \frac{86}{57} \rangle = 2$. This notation is not standard, and is used only in this text.

$\langle\langle q \rangle\rangle$ The result of rounding a number q the Hill–Huntington way: round down if q is less than the geometric mean of $\lfloor q \rfloor$ and $\lceil q \rceil$, and round up otherwise. For example, to compute $\langle\langle 2.45 \rangle\rangle$, calculate the geometric mean of $\lfloor 2.45 \rfloor = 2$ and $\lceil 2.45 \rceil = 3$, which is $\sqrt{2 \times 3} \approx 2.449$. Since $2.45 > 2.449$, we round up: $\langle\langle 2.45 \rangle\rangle = 3$.

Absolute difference   The absolute difference of two numbers is obtained by subtracting the smaller number from the larger number.

Alabama paradox   An apportionment method suffers the Alabama paradox if it is possible for some state to lose a representative solely because the size of the House is increased.

Apportionment method   A systematic way of computing solutions of apportionment problems.

Apportionment problem   To round a list of fractions to integers in a way that preserves the sum of the original fractions.

Average district population   The total population divided by the House size.

Critical divisor   The critical divisor for a state is the largest divisor that will cause the state's apportionment to increase, if an increase is necessary to fill the House; or it is the smallest divisor that will cause a state's apportionment to decrease, if a decreased apportionment is needed.

District population   A state's population divided by its apportionment.

Divisor method   One of many apportionment methods in which the apportionments are determined by dividing the populations of the states by a number d, called the *divisor*, and rounding the resulting quotients to adjacent integer values. Divisor methods differ in their rounding rules. The methods of Jefferson, Webster, and Hill–Huntington are divisor methods.

Hamilton method   An apportionment method advocated by Alexander Hamilton. This method assigns to each state either its lower quota or its upper quota. The states that receive their upper quotas are those whose quotas have the largest fractional parts.

Hill–Huntington method An apportionment method named for the statistician Joseph Hill and the mathematician Edward Huntington. This divisor method minimizes relative differences in both representative shares and district populations. It is based on the "Hill–Huntington" way of rounding, so a state's apportionment is $\langle\langle p_i/d \rangle\rangle$.

Jefferson method An apportionment method invented by Thomas Jefferson. It is a divisor method that rounds all fractions downward, and a state's apportionment is $\lfloor p_i/d \rfloor$.

Lower quota The integer part $\lfloor q_i \rfloor$ of a state's quota q_i.

Population paradox If changes in population cause one state's apportionment to increase, and another's to decrease, although the first state's population had decreased and the second state's population had increased, the population paradox has occurred. This paradox is possible with all apportionment methods *except* divisor methods.

Quota A state's quota in an apportionment problem is the number of seats it would receive if fractional seats could be awarded. The quota for state i is $q_i = p_i/c$, where p_i is the population of the state and c is the average district population.

Quota condition An apportionment method satisfies the quota condition if in every situation each state's apportionment is equal to either its lower quota or its upper quota. All divisor methods fail in some cases to satisfy the quota condition.

Relative difference The relative difference between two positive numbers is obtained by subtracting the smaller number from the larger, and expressing the result as a percentage of the smaller number. Thus, the relative difference of 120 and 100 is 20%.

Representative share A state's representative share is the state's apportionment divided by its population. It is intended to represent the amount of influence a citizen of that state would have on his or her representative.

Upper quota If a state's quota is q_i, its upper quota is $\lceil q_i \rceil$.

Webster method A divisor method of apportionment invented by Congressman Daniel Webster. It is based on rounding fractions the usual way, so that the apportionment for state i is $\langle p_i/d \rangle$. The Webster method minimizes differences of representative share between states.

SUGGESTED READINGS

BALINSKI, M. L., AND H. P. YOUNG. *Fair Representation: Meeting the Ideal of One Man, One Vote,* Yale University Press, New Haven, Conn., 1982. In the 1970s, Balinski and Young analyzed apportionment methods in depth. Their point of view was to postulate the desirable properties of an apportionment method as axioms and to deduce from the axioms the characteristics of the best method. This book combines an account of the history of apportionment of the U.S. House of Representatives with the results of their research.

COMMONWEALTH OF MASSACHUSETTS v. MOSBACHER, 785 Federal Supplement 230 (District of Massachusetts 1992). This opinion concerns a suit by Massachusetts to increase its representation. The Commonwealth argued that the Bureau of Census did not fairly assign federal employees who are stationed abroad to their home states, but this part of the opinion is not of interest to us. However, Massachusetts also claimed that the Hill–Huntington method was an unfair method of apportionment, and asked the Court to replace that method with the Webster method. The discussion of this portion of the claim is Section D of the opinion, and starts on page 253. The Federal Supplement is available in law libraries.

ERNST, LAWRENCE R. Apportionment methods for the House of Representatives and the court challenges, *Management Science,* 40 (1994): 1207–1227. Ernst, who wrote briefs for the government in both the *Montana* and the *Massachusetts* cases, reviews the apportionment problem, and the arguments in favor of and against each of the divisor methods. The article includes a summary of the arguments used by both sides in the two court cases.

LUCAS, W. F. The apportionment problem. In S. J. Brams, W. F. Lucas, and P. D. Straffin, Jr. (eds.), *Political and Related Models,* Springer-Verlag, New York, 1983, pp. 358–396. An introduction to apportionment, written at a somewhat more advanced level than the presentation in this text.

U.S.. DEPARTMENT OF COMMERCE v. MONTANA, 112 Supreme Court 1415 (1992). This opinion, available in any law library, gives the grounds for rejecting the Montana suit to replace the Hill–Huntington method with the Dean method.

YOUNG, H. PEYTON. *Equity,* Princeton University Press, Princeton, N.J., 1994. Chapter 3 covers apportionment and focuses on which apportionment method is the most equitable.

EXERCISES ▲ *Optional.* ■ *Advanced.* ◆ *Discussion.*

The Hamilton Method

1. A country has a parliament with 577 seats. In an election, the Democratic Socialists receive 323,829 votes, the Social Democrats, 880,702 votes; the Christian Democrats, 5,572,614 votes; the Greens, 1,222,498 votes; and the Communists, 111,224 votes. The number of seats won by each party is to be proportional to the number of votes cast in its favor. Calculate the quota for each party and apportion the seats by the Hamilton method.

2. A very small country has three states, with populations of 59,000, 76,000, and 14,000. Use the Hamilton method to apportion the seats of the 35-seat National Legislature. Repeat the calculation for 36, 37, 38, 39, and 40 seats. Does the Alabama paradox occur?

3. A country has four states, *A, B, C,* and *D.* Its House of Representatives has 100 members, and by law, the Hamilton method is used to apportion it. When a census is taken, there are some systematic errors, and the census has to be repeated. The accompanying table gives the original and corrected figures.

State	Original Census	Corrected Population
A	5,525,381	5,657,564
B	3,470,152	3,507,464
C	3,864,226	3,885,693
D	201,205	201,049
Totals	13,060,962	13,251,770

(a) Apportion the House using the original census.
(b) Reapportion, using the corrected populations.
(c) Explain how this is an example of the population paradox.

4. Suppose that a country has three states, with populations 254,000, 153,000, and 103,000, respectively. The legislature has 102 seats. Show that if the Hamilton method is used to apportion seats, a tie will result. How would you suggest breaking the tie?

Divisor Methods

◆ 5. Show that for any number q, $\langle q \rangle = \lfloor q + 0.5 \rfloor$.

6. A small high school has one mathematics teacher who can teach a total of five sections. The subjects that she teaches, and their enrollments, are as follows: geometry, 43; algebra, 42; calculus, 12. Apportion sections to the subjects using the Hamilton, Jefferson, and Webster methods.

7. Repeat Exercise 6 using the following enrollments: geometry, 76; algebra, 19; calculus, 20.

8. Suppose that three states have populations 1,000,000, 2,000,000, and 2,500,000, respectively. The legislature has 26 seats. Show that if the

Jefferson method is used to apportion the legislature, then a tie will result. How would you break the tie?

9. A three-state country has a legislature with 36 seats. The states have populations of 155,000, 105,000, and 100,000. Show that a tie will result if the seats are apportioned by the Webster method.

10. A country has a 100-seat parliament with one major party, the National party, and 10 splinter parties, In a recent election, the National party received 87.85% of the vote. The splinter parties received the following percentages: 1.26, 1.25, 1.24, 1.23, 1.22, 1.21, 1.20, 1.19, 1.18, and 1.17.

◆   (a) What is the best apportionment for the parliament?
    (b) Compute the apportionments according to the methods of Hamilton, Jefferson, and Webster.
    (c) Do any of the methods in part (b) violate the quota condition? That is, does some party receive either more than its upper quota or less than its lower quota?

11. A country with a 100-seat parliament has an election in which one party captures 92.15% of the vote. Five splinter parties receive the following percentages of the vote: 1.59, 1.58, 1.57, 1.56, and 1.55. Answer all of the questions in Exercise 10.

12. Suppose that the Jefferson method is used to apportion the House according to the 1990 census. Decide the contest between Massachusetts and Oklahoma for the last seat available to them. The population figures are given in Exercise 21. (*Hint:* Start with the assumption that Massachusetts already has 10 seats and Oklahoma has 5 and compare their critical divisors.)

Which Divisor Method Is Best?

13. Determine the relative difference between the numbers 5 and 7.

14. Jim is 72 inches tall and Alice is 65 inches tall. What is the relative difference in their heights?

15. Professor Roe's salary is $68,000; she just received a $4000 raise. What is the relative difference between her present salary and her salary before the raise?

16. In the 1991 apportionment of Congress, the average congressional district in Pennsylvania had a population of 567,843. The corresponding figure for New Jersey was 596,049.

    (a) Which state is the more favored in this apportionment?
    (b) What is the relative difference in the district sizes?

17.   According to the 1990 census, the population of Ohio was 10,887,325; Ohio was apportioned 19 House seats. The population of Kansas was 2,485,600, and Kansas received 4 House seats.

(a)   Determine the average congressional district sizes for these states.

(b)   Determine the relative difference in these district sizes.

(c)   Suppose a seat were transferred from Ohio to Kansas, giving Ohio 18 seats and Kansas 5. What would the relative difference in district sizes now be?

18.   Table 14.7 shows the Hill–Huntington apportionments for several states, based on the 1970 census.

(a)   Which state has the largest district size and which has the smallest?

(b)   If a seat were transferred from the state with the smallest district size to the one with the largest, would the inequity be less (as measured by absolute difference in district size)?

(c)   If a seat were transferred from the state with the smallest district size to the one with the largest, would the inequity be less (as measured by relative difference in district size)?

(d)   Citizens of which state have the largest representative share? The smallest?

(e)   If a seat were transferred from the state with the smallest representative share to the one with the largest, would the inequity be less (as measured by absolute difference in representative share)?

(f)   Which state would benefit from a change of apportionment method, and which method would that state prefer?

TABLE 14.7	Some 1970 Apportionments	
State	Population	Apportionment
California	20,098,863	43
Connecticut	3,050,693	6
Montana	701,573	2
Oregon	2,110,810	4
South Dakota	673,247	2

◆ 19. Let the populations of states A and B be p_A and p_B, respectively. The apportionments will be a_A and a_B. Assuming that district populations for state A are larger than district populations in state B, show that the relative difference in district populations is

$$\frac{p_A a_B - p_B a_A}{p_B a_A} \times 100\%$$

Also show that this expression is equal to the relative difference in representative share. Hence the relative difference in district populations is equal to the relative difference in representative shares.

◆ 20. Which divisor method would be the most appropriate for apportioning sections to classes according to class enrollments, as in the senior high school example?

21. In *Massachusetts* v. *Mosbacher,* Massachusetts contested its 1991 apportionment, claiming a systematic census undercount of Massachusetts residents living abroad. Another issue in the suit was the claim by Massachusetts that the Hill–Huntington method of apportionment is unconstitutional, because it does not reflect the "one person, one vote" principle as well as the Webster method does. Massachusetts sought an additional House seat that had been awarded to Oklahoma. Would Massachusetts have gained a seat from Oklahoma if the Webster method had been used to apportion the House of Representatives in 1991? Use the following populations and Hill–Huntington apportionments:

State	Population	Apportionment
Massachusetts	6,029,051	10
Oklahoma	3,145,585	6

The Hill–Huntington Method

22. Find the geometric mean of each pair of numbers: (a) $0, 1$; (b) $1, 2$; (c) $2, 3$; (d) $3, 4$.

23. Use the Hill–Huntington method to apportion sections to the three mathematics courses in the small high school. The enrollment data are in Exercise 6.

◆ 24. (a) Show that for any positive numbers A and B, the geometric mean is less than the arithmetic mean,[3] except when $A = B$; then the two means are equal. (*Hint:* Show that a right triangle can be formed, in which the length of the hypotenuse is the arithmetic mean, and one of the legs is the geometric mean.)

(b) Show that the Hill–Huntington way of rounding is more generous than the Webster way.

25. Suppose that the governor of Kansas believes that the population of his state was undercounted. What increase in population would be large enough to entitle Kansas to take a seat from Ohio, if the apportionment is by the Hill–Huntington method? The data needed for this problem are given in Exercise 17.

26. Ties can occur when the Hill–Huntington method is used to apportion, if two states have identical populations, or if the number of states is larger than the House size. Is there any other way that ties can occur with this method?

Additional Exercises

◆ 27. Here is an apportionment method that should please everyone! Just give each state its upper quota.

(a) Show that the House size will be more than the planned House size h.

(b) Do you think California would be enthusiastic about this method, or would that state prefer to give each state its *lower* quota?

28. A country has five states, with populations 5,576,330, 1,387,342, 3,334,241, 7,512,860, and 310,968. Its House of Representatives is apportioned by the Hamilton method.

(a) Calculate the apportionments for House sizes of 82, 83, and 84. Does the Alabama paradox occur?

(b) Repeat the calculations for House sizes of 89, 90, and 91.

29. A country has six states with populations 27,774; 25,178; 19,947; 14,614; 9225; and 3292. Its House of Representatives has 36 seats. Find the apportionment using the methods of Hamilton, Jefferson, Webster, and Hill–Huntington.

---

[3] The arithmetic mean of A and B is equal to $(A + B)/2$.

30. A country that is governed by a parliamentary democracy has two political parties, the Liberals and the Tories. The number of seats awarded to a party is supposed to be proportional to the number of votes it receives in the election. Suppose the Liberals receive 49% of the vote. If the total number of seats in parliament is 99, how many seats do the Liberals get with the Hamilton method? With the Webster method? With the Jefferson method?

◆ 31. A country with a parliamentary government has two parties that capture 100% of the vote between them. Each party is awarded seats in proportion to the number of votes received.

    (a) Show that the Webster and Hamilton methods will always give the same apportionment in this two-party situation.

    (b) Show that the Alabama and population paradoxes cannot occur when the Hamilton method is used to apportion seats between two parties or states.

    (c) Show that the Webster method satisfies the quota condition when the seats are apportioned between two parties or states.

    (d) Will the Jefferson and Hill–Huntington methods also yield the same apportionments as the Hamilton method?

32. If the 1790 Congress were apportioned by the Webster method, would the result differ from the Hill–Huntington method (see Table 14.6)?

33. The following apportionment method was invented by Congressman William Lowndes of South Carolina in 1822. Lowndes starts, as Hamilton does, by giving each state its lower quota. But where Hamilton apportions the remaining seats to the states whose quotas have the largest fractional parts—in other words, the states for which the *absolute difference* between q_i and $\lfloor q_i \rfloor$ is greatest—Lowndes gives the extra seats to the states where the *relative* difference between q_i and $\lfloor q_i \rfloor$ is greatest, raising as many as necessary to their upper quotas to fill the House.

◆   (a) Would this method be more beneficial to states with large populations or small populations, as compared with the Hamilton method?

◆   (b) Does the Lowndes method satisfy the quota condition?

◆   (c) Would there be any trouble with paradoxes with the Lowndes method?

    (d) Use the method to apportion the 1790 House of Representatives.

▨ 34. The Hill–Huntington method, based on the 1990 census, apportioned 9 seats to Washington, based on a population of 4,887,941. The figures for Massachusetts are given in Exercise 21. How much would have to be added to the population of Massachusetts to entitle that state to take a seat from Washington, if apportionment is done by the Hill–Huntington method? With the Webster method?

■ 35.   Let q_1, q_2, . . . , q_n be the quotas for n states in an apportionment problem, and let the apportionments assigned by some apportionment method be denoted a_1, a_2, . . . , a_n. The *absolute deviation* for state i is defined to be $|q_i - a_i|$; it is a measure of the amount by which the state's apportionment differs from its quota. The *maximum absolute deviation* is the largest of these numbers. Show that the Hamilton method always gives the least possible maximum absolute deviation.

WRITING PROJECTS

1 ▶   Does the Hill–Huntington method best reflect the intentions of the founding fathers, as these intentions were set down in the Constitution and in the debate during the 1787 Constitutional Convention? Good sources of information here include the following publications listed under Suggested Readings: *Fair Representation,* by Balinski and Young; *Equity,* by H. Peyton Young; "Apportionment Methods," by Lawrence Ernst; and the two court opinions, *Massachusetts* v. *Mosbacher* and *U.S. Department of Commerce* v. *Montana.* This writing project requires that you state your answer to the question and make a case for it.

2 ▶   Suppose that in 1990, Congress reverted to its nineteenth-century habit of increasing the House size with every reapportionment so that no state would have a decrease in the size of its delegation. How many seats would have been added to the House, and which states would get them? (*Warning:* The apportionments of some states might *increase* as a result of using this method.) As the first step of this project, look up the populations and apportionments for the 50 states in an almanac.

CHAPTER 15

GAME THEORY: THE MATHEMATICS OF COMPETITION

Conflict has been a central theme throughout human history and literature. It arises whenever two or more individuals, with different values, compete to try to control the course of events. *Game theory* uses mathematical tools to study situations involving both conflict and cooperation. Its study was greatly stimulated by the publication in 1944 of the monumental *Theory of Games and Economic Behavior* by John von Neumann and Oskar Morgenstern (see Spotlight 15.1).

The *players* in a game, who may be people, organizations, or even countries, choose from a list of options available to them—that is, courses of action they might take—that are called **strategies.** The strategies chosen by the players lead to *outcomes,* which describe the consequences of their choices. We assume that the players have *preferences* for the outcomes: they like some more than others.

Game theory analyzes the **rational choice** of strategies—that is, how players select strategies to obtain preferred outcomes. Among areas to which game theory has been applied are bargaining tactics in labor-management disputes, resource-allocation decisions in political campaigns, military choices in international crises, and the use of threats by animals in habitat acquisition and protection.

S*potlight* *Historical Highlights*

15.1

John von Neumann

Oskar Morgenstern

As early as the seventeenth century, such outstanding scientists as Christiaan Huygens (1629–1695) and Gottfried W. Leibniz (1646–1716) proposed the creation of a discipline that would make use of the scientific method to study human conflict and interactions. Throughout the nineteenth century, several leading economists created simple mathematical examples to analyze particular examples of competitive encounters. The first general mathematical theorem in this subject was proved by the distinguished logician Ernst Zermelo (1871–1956) in 1912. It stated that any finite game with perfect information, such as checkers or chess, has an optimal solution in *pure* strategies; that is, no randomization or secrecy is necessary. A game is said to have *perfect information* if at each stage of the play, every player is aware of all past moves by himself and others as well as all future choices that are allowed. This theorem is an example of an *existence theorem:* it demonstrates that there must exist a best way to play such a game, but it does not provide a detailed plan for playing a complex game, like chess, to achieve victory.

The famous mathematician F. E. Émile Borel (1871–1956) introduced the notion of a *mixed,* or randomized, strategy when he investigated some elementary duels around 1920. The fact that every two-person, *zero-sum* game must have optimal mixed strategies and an *expected value* for the game was proved by John von Neumann (1903–1957) in 1928. Von Neumann's result was extended to the existence of equilibrium outcomes in mixed strategies for multiperson games that are either *constant-sum* or *variable-sum* by John F. Nash, Jr. (1931–   ), in 1951.

Modern game theory dates from the publication in 1944 of *Theory of Games and Economic Behavior* by the Hungarian-American mathematician John von Neumann and the Austrian-American economist Oskar Morgenstern (1902–1977). They introduced the first general model and solution concept for multiperson *cooperative games,* which are primarily concerned with coalition formation (economic cartels, voting blocs, and military alliances) and the resulting distribution of gains or losses. Several other suggestions for a "solution" to such games have since been proposed. These include the value concept of Lloyd S. Shapley (1923–   ), which relates to fair allocation and economic prices and serves as well as an index of voting power (see Chapter 12).

The French artist Georges Mathieu designed a medal for the Paris Musée de la Monnaie in 1971 to honor game theory. It was the seventeenth medal to "commemorate 18 stages in the development of Western consciousness." The first medal was for the Edict of Milan in A.D. 313. Game theory also has a mascot, the tiger, arising from the Princeton University tiger and the Russian abbreviation of the term "game theory" (ТЕОРИЯ ИГР).

Unlike the subject of *individual* decision making, which researchers in psychology, statistics, and other disciplines study, game theory analyzes situations in which there are at least two players, who may find themselves in conflict because of different goals or objectives. The outcome depends on the choices of *all* the players. In this sense decision making is *collective,* but this is not to say that the players necessarily cooperate when they choose strategies. Indeed, many strategy choices are noncooperative, such as those between combatants in warfare or competitors in sports. In these encounters, the adversaries' objectives may be at cross-purposes: a gain for one means a loss for the other. But in many activities, especially in economics and politics, there may be joint gains that can be realized from cooperation.

Most interactions probably involve a delicate mix of cooperative and noncooperative behavior. In business, for example, firms in an industry cooperate to gain tax breaks even as they compete for shares in the marketplace.

In the next two sections we present several simple examples of two-person games of **total conflict,** in which what one player wins the other player loses, so cooperation never benefits the players. We distinguish two different kinds of solutions to such games. Next we analyze two well-known games of **partial conflict,** in which the players can benefit from cooperation but may have strong incentives not to cooperate. We then turn to the analysis of a larger three-person voting game, in which we show how to eliminate undesirable strategies in stages. Finally, we offer some general comments on solving matrix games and discuss different applications of game theory.

TWO-PERSON TOTAL-CONFLICT GAMES: PURE STRATEGIES

For some games with two players, determining the best strategies for the players is straightforward. We begin with such a case.

EXAMPLE ▶ *A Location Game*

Two young entrepreneurs, Henry and Lisa, plan to locate a new restaurant at a main-route intersection in the nearby mountains. They agree on all aspects of the restaurant except one. Lisa likes low elevations, whereas Henry wants greater heights—the higher, the better. In this one regard, their preferences are diametrically opposed. What is better for Henry is worse for Lisa, and likewise what is good for Lisa is bad for Henry.

The layout for their location problem is shown in Figure 15.1. Observe that three routes, Avenue A, Boulevard B, and County Road C (blue lines), run in an east-west direction, and that three highways, numbered 1, 2, and 3

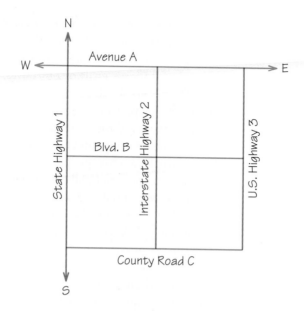

Figure 15.1 The road map for the location example.

(red lines), run in a north-south direction. Table 15.1 shows the altitudes at the nine corresponding intersections; the same information (which is in thousands of feet) is shown in three dimensions in Figure 15.2.

To maximize the number of customers, Henry and Lisa agree that the restaurant should be at a location where one of the three routes intersects one of the three highways. But they cannot agree on which intersection, so they decide to turn their decision into the following competitive game: Henry will select one of the three routes, A, B, or C, and Lisa will simultaneously choose one of the three highways, 1, 2, or 3.

Henry is pessimistic and considers the lowest altitude along each of the Routes A, B, and C. These are the numbers 4, 5, and 2, which are the respective *row minima*, indicated in the right-hand column of Table 15.2. He notes that the highest of these values is 5. By choosing the corresponding route, B, Henry can guarantee himself an altitude of at least 5000 feet.

TABLE 15.1	Heights (in thousands of feet) of the Nine Intersections		
		Highways	
Routes	1	2	3
A	10	4	6
B	6	5	9
C	2	3	7

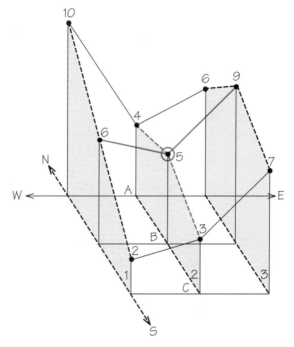

FIGURE 15.2 Three-dimensional road map showing Henry's and Lisa's selections.

The number 5 in the right-hand column is referred to as the **maximin,** which is circled in Table 15.2. It is the maximum value of the minimum numbers in the three rows in the table. The strategy that corresponds to the maximin (for Henry, Route B) is called his **maximin strategy.**

TABLE 15.2 The Heights in Table 15.1, with the Row Minima and Column Maxima

	Routes	Lisa Highways 1	2	3	Row Minima
	A	10	4	6	4
Henry	B	6	5	9	⑤
	C	2	3	7	2
	Column maxima	10	⑤	9	

Lisa likewise does a worst-case analysis and lists the highest—for her, the worst—elevations for each highway. These numbers, 10, 5, and 9, are the column maxima and are listed in the bottom line of Table 15.2. From Lisa's point of view, the best of these outcomes is 5. If she picks Interstate Highway 2, then she is assured of an elevation of no more than 5000 feet.

> The number 5 in the bottom line of Table 15.2 is referred to as the **minimax,** which is circled in the table. It is the minimum value of the maximum numbers in the three columns. The strategy that corresponds to the minimax (for Lisa, Highway 2) is called her **minimax strategy.**

To summarize, Henry has a strategy that will ensure the height is 5 or higher, and Lisa has a strategy that will ensure the height is 5 or lower. The height of 5 at the intersection of Route B and Highway 2 is, simultaneously, the lowest value along Boulevard B and the highest along Interstate Highway 2. In other words, the maximin and the minimax are both equal to 5 for the location game.

> When the maximin and the minimax are the same, the resulting outcome is called a **saddlepoint.**

The reason for the term *saddlepoint* should be clear from the saddle-shaped payoff surface shown in Figure 15.2. The middle point on a horse saddle is simultaneously the lowest point along the spine of the horse and the highest point between the rider's legs. In our example, one might also think of the saddlepoint as a mountain pass: as one drives through the pass, the car is at a high point on a highway (in the north-south direction) and at a low point on a route (in the east-west direction).

The resolution of this contest is for Henry to pick B and Lisa to pick 2. This puts them at an elevation of 5, which is simultaneously the maximin and the minimax.

> If a game has a saddlepoint (5 in our example), it gives the **value** of the game. Players can guarantee at least this value by choosing their maximin and minimax strategies (B for Henry and 2 for Lisa).

There is no need for secrecy in a game with a saddlepoint. Even if Henry were to reveal his choice of B in advance, Lisa would be unable to use this knowledge to exploit him. In fact, both players can use the height information in our example to compute the optimal strategy for their opponent as well as for themselves. In games with saddlepoints, players' worst-case analyses lead to the best *guaranteed* outcome—in the sense that each player can ensure that he or she does not do worse than a certain amount (5 in our example)—and may do better (if the opponent deviates from a maximin or minimax strategy).   ◆

Another well-known game with a saddlepoint is tic-tac-toe. Two players alternately place an × or an ○, respectively, in one of the nine unoccupied spaces in a 3 × 3 grid. The winner is the first player to have three ×'s, or three ○'s, in either the same row, the same column, or along a diagonal.

An explicit list of all strategies for either the first- or second-moving player in tic-tac-toe is long and complicated, because it specifies a complete plan for all possible contingencies that can arise. For the first-moving player, for example, a strategy might say "put an × in the middle square, then an × in the corner if your opponent puts an ○ in a noncorner position, etc." While, initially, young children find this game interesting to play, before long they discover that each player can always prevent the other player from winning by forcing a tie, making the game quite boring. All strategies that force a tie, it turns out, are a saddlepoint in tic-tac-toe.

E X A M P L E ▶ *The Restricted-Location Game*

Assume in our location game that Henry and Lisa are informed by the county officials that it is against the law to locate a restaurant on either Boulevard B or Interstate Highway 2. These two choices, which provided our earlier solution, are now forbidden. The resulting location game without these two strategies is given in Table 15.3 (with payoffs again expressed in thousands of feet).

As before, Henry and Lisa can each do a worst-case analysis. Henry is worried about the minimum number in each row, and Lisa is concerned with the

TABLE 15.3	Heights Without Boulevard B and Interstate Highway 2	
	Highways	
Routes	1	3
A	10	6
C	2	7

TABLE 15.4 Heights in Table 15.3, with Row Minima and Column Maxima

		Lisa		
		Highways		
	Routes	1	3	Row Minima
Henry	A	10	6	⑥
	C	2	7	2
	Column maxima	10	⑦	

maximum number in each column. These are listed in the right column and bottom row, respectively, in Table 15.4.

Henry sees from the row that his maximin is 6, so he can guarantee a height of 6000 feet or more by choosing Route A. Likewise, Lisa observes that her minimax is 7, so she can keep the elevation of the restaurant down to 7000 feet or less by selecting Highway 3. There is a gap of 1 ($= 7 - 6$) between the maximin and minimax. When the maximin is less than the minimax, as in this case, then a game does *not* have a saddlepoint, but it does have a value (described in the next section).

If Henry does play his maximin strategy, Route A, and Lisa plays her minimax strategy, Highway 3, then the resulting payoff is 6. However, Henry may be motivated to gamble in this case by playing his other strategy, Route C; if Lisa sticks to her conservative strategy, Highway 3, then his payoff is 7. Henry will have gained one unit (1000 feet), going from 6 to 7.

This is, however, a risky move. If Lisa suspected it, she might counter by selecting Highway 1. The payoff would then be 2, the best for Lisa and the worst for Henry. So Henry's gamble to gain one unit (6 to 7) has the risk that he might lose 4 units (6 to 2).

But then there is no incentive for Lisa to play her nonminimax strategy (that is, to play Highway 1) in this restricted-location game if she believes Henry, in turn, move to his maximin strategy (Route A) leading to a payoff of 10. This is worse than 6 from her viewpoint.   ◆

In two-player games that have saddlepoints, like our original 3 × 3 location game and tic-tac-toe, each player can calculate the maximin and minimax strategies for both players before the game is even played. Once the solution has been determined by either mathematical analysis or practical experience (as was probably true of tic-tac-toe), there may be little interest in actually playing the game.

But this is decidedly not the case for much more complex games, like chess, whose solution has not yet been determined—and is unlikely to be in the foreseeable future. Even though computers are able to beat world champions on occasion, the computer's winning moves will not necessarily be optimal against those of *all* other opponents. Nevertheless, we know that chess, like tic-tac-toe, has a saddlepoint, but we do not know whether it yields a win for white, a win for black, or a draw.

Unlike chess, many games, like the 2 × 2 restricted-location game, do not have an outcome that can always be guaranteed. These games, which include poker, involve uncertainty and risk. One desires in such games not to have one's strategy detected in advance, because this information can be exploited by an opponent. It is no surprise, then, that poker players are told to keep a "poker face," revealing nothing about their likely moves. But this advice is not very helpful in telling the players what actually to do in the game, such as how many cards to ask for in draw poker.

We will show that there are optimal ways to play two-person total-conflict games without a saddlepoint so as not to reveal one's choices. But their solution is by no means as straightforward as that of games with a saddlepoint.

TWO-PERSON TOTAL-CONFLICT GAMES: MIXED STRATEGIES

Probably most competitive games do not have a saddlepoint, like that we found in our first location-game example. Rather, as is illustrated in our restricted-location game, in which the maximin and minimax were not the same, players must try to keep secret their strategy choices, lest their opponent use this information to his or her advantage.

In particular, players must take care to *conceal* the strategy they will select until the encounter actually takes place, when it is too late for the opponent to alter his or her choice. If the game is repeated, a player will want to *vary* his or her strategy in order to surprise the opponent.

In parlor games like poker, players often use the tactic of *bluffing*. This tactic involves a player's sometimes raising the stakes when he has a low hand so that opponents cannot guess whether or not his hand is high or low—and may, therefore, miscalculate whether to stay in or drop out of the game (a player would prefer opponents to stay in when he has a high hand and drop out when he has a low hand). In military engagements, too, secrecy and even deception are often crucial to success.

In many sporting events, a team tries to surprise or mislead the opposition. A pitcher in baseball will not signal the type of pitch he or she intends to throw in advance, varying the type throughout the game to try to keep the batter off balance. In fact, we next consider a confrontation between a pitcher and batter in more detail.

EXAMPLE ▶ *A Duel Game*

Assume that a particular baseball pitcher can throw either a blazing fastball or a slow curve into the strike zone and so has two strategies: *fast* (denoted by *F*) and *curve* (*C*). The pitcher faces a batter who attempts to guess, before each pitch is thrown, whether it will be a fastball or a curve, giving the batter also two strategies: guess *F* and guess *C*. Assume that the batter has the following batting averages, which are known by both players.

- .300 if the batter guesses fast (*F*) and the pitcher throws fast (*F*)
- .200 if the batter guesses fast (*F*) and the pitcher throws curve (*C*)
- .100 if the batter guesses curve (*C*) and the pitcher throws fast (*F*)
- .500 if the batter guesses curve (*C*) and the pitcher throws curve (*C*)

A player's batting average is the number of times he hits safely divided by his number of times at bat. If a batter hit safely 3 times out of 10, for example, his average would be .300.

This game is summarized in Table 15.5. We see from the right-hand column in the table that the batter's maximin is .200, which is realized when he selects his first strategy *F*. Thus, the batter can "play it safe" by always guessing a fastball, which will result in his batting .200, hardly enough for him to remain on the team.

We see from the bottom row of the table that the pitcher's minimax is .300, which is obtained when he throws fast (*F*). Note that the batter's maximin of .200 is less than the pitcher's minimax of .300, so this game does not have a saddlepoint. There is a gap of .100 (= .300 − .200) between these two numbers.

TABLE 15.5	Batting Averages in a Baseball Duel			
		Pitcher		Row Minima
		F	C	
Batter	F	.300	.200	(.200)
	C	.100	.500	.100
	Column maxima	(.300)	.500	

The pitcher and the batter use mixed strategies.

Each player would like to play so as to win as much for himself of the .100 payoff in the gap as possible. That is, the batter would like to average more than .200, whereas the pitcher wants to hold the batter down to less than .300. ◆

A Flawed Approach

If the batter and pitcher in our example consider how they might outguess each other, they might reason along the following lines:

1. *Pitcher* (to himself): If I choose strategy *F*, I hold the batter down to .300 (the minimax) or less. However, the batter is likely to guess *F* because it guarantees him at least .200 (his maximin), and it actually provides him with .300 against my *F* pitch. In this case, the batter wins all the .100 payoff in the gap.

2. *Batter* (to himself): I can figure out that the pitcher is reasoning as in step 1 and will try to surprise me with *C*. So I should fool him and guess *C*. I would thus average .500, which will show him up for trying to gamble and outguess me!

3. *Pitcher* (to himself): But if the batter is thinking as in step 2, that is, of guessing *C*, I, on second thought, should really throw *F*. This will lead to an average of only .200 for the batter and teach him to not try to outguess me!

This type of cyclical reasoning can go on forever: I think that he thinks that I think that he thinks. . . . . It provides no resolution of the players' decision problem.

Clearly, there is no pitch, or guess, that is best in all circumstances. Nevertheless, both the pitcher and the batter *can* do better, but not by trying to anticipate the choices of each other. The answer to their problem lies in the notion of a *mixed strategy.*

A Better Idea

The play of many total-conflict games requires an element of surprise, which can be realized in practice by making use of a mixed strategy.

> A **mixed strategy** is a particular randomization over a player's strategies (which henceforth we call **pure strategies**—the definite options a player can choose—to distinguish them from mixed strategies). Each one of the player's pure strategies is assigned some probability, indicating the relative frequency with which the pure strategy will be played. The specific pure strategy that will be used in any given play of the game is selected by some appropriate probabilistic mechanism or random device.

Note that a pure strategy is a special case of a mixed strategy, with the probability of 1 assigned to just one pure strategy. When a player resorts to a mixed strategy, the resulting outcome of the game is no longer predictable in advance. Rather, it must be described in terms of the probabilistic notion of an *expected value.*

> If each of the n payoffs s_1, s_2, . . . , s_n will occur with the probabilities p_1, p_2, . . . , p_n, respectively, then the average, or **expected value E,** is given by
>
> $$E = p_1 s_1 + p_2 s_2 + \cdots + p_n s_n$$
>
> We assume that the probabilities sum to 1 and that each probability p_i is never negative. That is, we assume that $p_1 + p_2 + \cdots + p_n = 1$, and $p_i \geq 0$ $(i = 1, 2, . . . , n)$.

To see how mixed strategies and expected values are used in the analysis of games, we turn to what is perhaps the simplest of all competitive games without a saddlepoint.

EXAMPLE ▶ *Matching Pennies*

In matching pennies, each of two players simultaneously shows either a head *H* or a tail *T*. If the two coins match, with either two heads or two tails, then the first player (Player I) receives both coins (a win of 1 for Player I). If the coins do not match, that is, if one is an *H* and the other is a *T*, then the second player (Player II) receives the two coins (a loss of 1 for Player I). These wins and losses for Player I are shown in Table 15.6.

> The game in Table 15.6 is described by a **payoff matrix.** The rows and columns correspond to the strategies of the two players, and the numerical entries give the payoffs to Player I when these strategies are chosen.

TABLE 15.6	Wins and Losses for Player I in Matching Pennies		
		Player II	
		H	*T*
Player I	*H*	1	− 1
	T	− 1	1

Although the entries in our earlier tables for the location game also gave payoffs, they were not monetary, as here.

The two rows in Table 15.6 correspond to Player I's two pure strategies, *H* and *T*, and the two columns to Player II's two pure strategies, also *H* and *T*. The numbers in the table are the corresponding winnings for Player I and losses for Player II. If two *H*'s or two *T*'s are played, Player I wins 1 from Player II. When both an *H* and a *T* are played, Player I pays out 1 to Player II.

It is fruitless for one player to attempt to outguess the other in this game. They should instead resort to mixed strategies and use expected values to estimate their likely gains or losses.

The best thing for Player I to do is randomly to select *H* half the time and *T* half the time. This mixed strategy can be expressed as

$$(p_H, p_T) = (p_1, p_2) = (1 - p, p) = \left(\tfrac{1}{2}, \tfrac{1}{2}\right)$$

Note that the probabilities of choosing $H (1 - p)$ and $T (p)$ do indeed sum to 1, as required; in particular, when $p = \tfrac{1}{2}$, $1 - p = 1 - \tfrac{1}{2} = \tfrac{1}{2}$.

This mixture can be realized in practice by the flip of the coin. Player I's resulting expected value is

$$E_H = \tfrac{1}{2}(1) + \tfrac{1}{2}(-1) = 0$$

whenever Player II plays H (first column of Table 15.6), and

$$E_T = \tfrac{1}{2}(-1) + \tfrac{1}{2}(1) = 0$$

whenever Player II plays T (second column).

> Player I's average outcome of 0 is the **(mixed-strategy) value** of the game; unlike the use of this notion in games with a saddlepoint, the value here can only be realized by the use of mixed strategies.

The value of 0 is really an expected value and so must be understood in a statistical sense. That is, in a given play of the game, Player I will either win 1 or lose 1. However, his or her expectation over many plays of this game is 0. The optimal mixed strategy for Player II is likewise a 50–50 mix of H and T, which also leads to an expectation of 0, making the game fair.

> A game is **fair** if its value is 0 and, consequently, it favors neither player when at least one player uses an *optimal* (mixed) strategy — one that guarantees that the resulting payoff is the best that this player can obtain against all possible strategy choices (pure or mixed) by an opponent.

Player II gains nothing by knowing that Player I is using the optimal mixed strategy $\left(\tfrac{1}{2}, \tfrac{1}{2}\right)$. However, Player I must not reveal to Player II whether H or T will be displayed *in any given play* of the game before Player II makes his or her own choice of H or T. Even without this information, if Player II knew that Player I was using a particular *nonoptimal* mixed strategy $(p_1, p_2) = (1 - p, p)$, where $p \neq \tfrac{1}{2}$ — that is, not choosing a 50–50 mixture between H and T — then Player II could take advantage of this knowledge and increase his or her average winnings over time to something greater than the value of 0. (See Exercise 14.) ◆

E X A M P L E ▶ *Nonsymmetrical Matching*

In this game, Players I and II can again show either heads H or tails T. When two H's appear, Player II pays \$5 to Player I. When two T's appear, Player II pays \$1 to Player I. When one H and one T are displayed, then Player II collects $3 from Player I. Note that although the sum of Player I's gains ($5 + $1 = $6) when there are two H's or two T's, and the sum of Player II's gains ($3 + $3 = $6) otherwise, are the same, the game is nonsymmetrical.

> A **nonsymmetrical** two-person total-conflict game is one in which the row player's gains ($5 and $1 in our example) are different from the column player's gains (always $3), except when there is a tie (not possible in our example, but possible in chess). In matching pennies, the payoff for winning is the same for each player, so the game is *symmetrical.*

TABLE 15.7 Payoffs for Player I in a Nonsymmetrical Matching Game

		Player II	
		H	T
Player I	H	5	− 3
	T	− 3	1

The game just described is given by the payoff matrix in Table 15.7, which shows the payoffs that Player I receives from Player II. A worst-case analysis, like that which solved our initial location game, is of little help here. Player I may lose $3 whether he plays H or T, making his maximin − 3. Player II can keep her losses down to $1 by always playing T (and thus avoiding the loss of $5 when two H's appear), so Player II's minimax is 1. However if Player II chooses T and Player I knows this, then Player I will also play T and collect $1 from Player II. Can Player II do better than lose $1 in each play of the game?

Consider the situation where Player I uses a mixed strategy $(p_H, p_T) = (1 − p, p)$, which involves playing H with probability $1 − p$ and playing T with probability p, where $0 \le p \le 1$. Against Player II's pure strategy H, Player I's expected value is

$$E_H = (5)(1 − p) + (− 3)p = 5 − 8p$$

Against Player II's pure strategy T, Player I's expectation is

$$E_T = (− 3)(1 − p) + 1p = − 3 + 4p$$

These two linear equations in the variable p are depicted in Figure 15.3. Note that the four points where these two lines intersect the two vertical lines, $p = 0$ and $p = 1$, are the four payoffs appearing in the payoff matrix.

The point at which the lines given by E_H and E_T intersect can be found by setting $E_H = E_T$, yielding

$$5 - 8p = -3 + 4p$$
$$-12p = -8$$

so $p = \frac{2}{3}$. To the left of $p = \frac{2}{3}$, $E_H < E_T$, and to the right $E_T > E_H$; at $p = \frac{2}{3}$, $E_H = E_T$. If Player I chooses $(p_H, p_T) = (1 - p, p) = (\frac{1}{3}, \frac{2}{3})$, he can ensure

$$E_H = 5 - 8(2/3) = E_T = -3 + 4(2/3) = -1/3$$

regardless of what Player II does.

In other words, Player I's optimal mixed strategy is to pick H and T with probabilities $\frac{1}{3}$ and $\frac{2}{3}$, respectively, which gives Player I an expected value of $-\frac{1}{3}$. As can be seen from Figure 15.3, $-\frac{1}{3}$ is the highest expected value that Player I can guarantee against *both* strategies H and T of Player II. Although H yields Player I a higher expected value for $p < \frac{2}{3}$, and T yields him a higher expected value for $p > \frac{2}{3}$, Player I's choice of $(p_H, p_T) = (\frac{1}{3}, \frac{2}{3})$ protects him against an expected loss greater than $-\frac{1}{3}$, which neither of his pure strategies does (each may produce a maximum loss of -3).

A similar calculation for Player II results in the same optimal mixed strategy $(\frac{1}{3}, \frac{2}{3})$ and expected value $-\frac{1}{3}$. But because the payoffs for Player II are losses, $-\frac{1}{3}$ means that she gains $\frac{1}{3}$ on the average.

This game is therefore unfair, even though the sum of the amounts ($6) that Player I might have to pay Player II when he loses is the same as the sum that Player II might have to pay Player I when she loses. Interestingly, it is Player II, who will win an average of $33\frac{1}{3}$ cents each time the game is played, who is favored, even though she may have to pay more to Player I when she loses (a maximum of $5) than Player I will ever have to pay her (a maximum of $3).   ◆

FIGURE 15.3   Solution to the nonsymmetrical matching pennies.

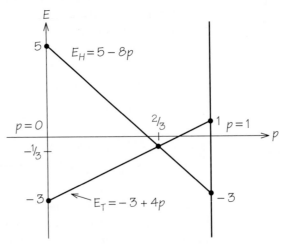

The symmetrical and nonsymmetrical matching games are examples of what are called *zero-sum games*.

A **zero-sum game** is one in which the payoff to one player is the negative of the corresponding payoff to the other, so the sum of the payoffs to the two players is always zero. These games can be completely described by a payoff matrix, in which the numbers represent the payoffs to Player I, while their negatives are the payoffs to Player II.

Zero-sum games are total-conflict games, in which what one player wins the other loses. But not all total-conflict games are zero-sum—in particular, the sum of the payoffs could be some other constant. Nevertheless, the strategic nature of these latter games is the same as that of zero-sum games: what one player wins the other player still loses. This was true in our location game, in which Henry's payoff was greater the higher the altitude, and Lisa's greater the lower the altitude.

Scoring in professional chess tournaments usually assigns a payoff of 1 for winning, 0 for losing, and $\frac{1}{2}$ to each player for a tie, making the sum of the payoffs to the two players always 1. Such games are called **constant-sum,** which can readily be converted to zero-sum games. Thus, chess could as well be scored -1 for a loss, $+1$ for a win, and 0 for a tie, making the constant 0 in this case. Although constant-sum and zero-sum games have the same strategic nature, constant-sum games are a more general class because the constant need not be zero.

The solution in the symmetrical version of matching pennies illustrated how mixed strategies of $\left(\frac{1}{2}, \frac{1}{2}\right)$ guarantee each player the value of 0, but we did not give a *solution technique* for finding optimal mixed strategies. In the nonsymmetrical version of matching pennies, by contrast, we illustrated a procedure that can be applied to *every* payoff matrix in which each player has only two strategies.

We must use more complex methods, which we will not describe here, to find mixed-strategy solutions when one or both players have more than two strategies. However, one should always check first to see whether a game has a saddlepoint before employing any solution method for finding optimal mixed strategies.

In our next example, which is the earlier duel between the pitcher and the batter given by the 2 × 2 payoff matrix in Table 15.4, we already showed that there is no saddlepoint. Thus, the solution will necessarily be in mixed strategies.

EXAMPLE ▶ *The Duel Game Revisited*

In Table 15.8, we add probabilities, which we explain next, to Table 15.4, where F indicates fastball and C indicates curve ball. The pitcher should use a mixed strategy $(p_1, p_2) = (p_F, p_C) = (1 - p, p)$. The probabilities $1 - p$ and p (where $0 \leq p \leq 1$) are indicated below the matrix and under the corresponding strategies, F and C, for the pitcher. If the pitcher plays a mixed strategy $(1 - p, p)$ against the two pure strategies, F and C, for the batter, he realizes the respective expected values:

$$E_F = (.3)(1 - p) + .2p = .3 - .1p$$
$$E_C = (.1)(1 - p) + .5p = .1 + .4p$$

As in the nonsymmetrical matching-pennies game, the solution to this game occurs at the intersection of the two lines given by E_F and E_C. Setting the equations of these lines equal to each other yields $p = .4$, giving $E_F = E_C = E = .260$.

Thus, the pitcher should use his optimal mixed strategy that selects F with probability $1 - p = 3/5$ and C with probability $p = 2/5$. This choice will hold the batter down to a batting average of .260, which is the value of the game. We stress that .260 is an average and must be interpreted in a statistical manner. It says that about one time in four the batter will get a hit, but not what will happen on any particular time at bat.

Assume that the batter uses a mixed strategy $(q_1, q_2) = (q_F, q_C) = (1 - q, q)$, as indicated to the right of the game matrix in Table 15.8. This mixed strategy, when played against the pitcher's pure strategies, F and C, results in the respective expected values:

$$E_F = (.3)(1 - q) + .1q = .3 - .2q$$
$$E_C = (.2)(1 - q) + .5q = .2 + .3q$$

The intersection of these two lines occurs at the point $q = .2$, giving $E_F = E_C = E = .260$. The batter's optimal mixed strategy is, therefore, $(q_F, q_C) = (\frac{4}{5}, \frac{1}{5})$, which gives him the same batting average of .260.   ◆

TABLE 15.8	A Baseball Duel with Probabilities		
		Pitcher	
		F	C	
Batter	F	.300	.200	$1 - q$
	C	.100	.500	q
		$1 - p$	p	

We have seen that the outcome of .260, which is the value of the game, occurs when either the pitcher selects his optimal mixed strategy $\left(\frac{3}{5}, \frac{2}{5}\right)$ or the batter selects his optimal mixed strategy $\left(\frac{4}{5}, \frac{1}{5}\right)$. This particular result holds true for every two-person zero-sum game; it is the fundamental theorem for such games and is known as the *minimax theorem*.

> The **minimax theorem** guarantees that there is a unique game value (.260 in our example), and an optimal strategy for each player, so that either player alone can realize at least this value by playing this strategy, which may be pure or mixed.

While our previous examples illustrate this theorem, they are not a proof of it, which can be found in advanced game theory texts.

PARTIAL-CONFLICT GAMES

The 2×2 matrix games presented so far have been total-conflict games: one player's gain was equal to the other player's loss. Although most parlor games, like chess or poker, are games of total conflict, and therefore constant-sum, most real-life games are surely not. (Elections, in which there is usually a clear-cut winner and one or more losers, probably come as close to being games of total conflict as we find in the real world.) We will consider two games of partial conflict, in which the players' preferences are not diametrically opposed, that have often been used to model many real-world conflicts.

> Games of partial conflict are **variable-sum games,** in which the sum of payoffs to the players at the different outcomes varies.

There is some mutual gain to be realized by both players if they can cooperate in partial-conflict games, but this may be difficult to do in the absence of either good communication or trust. When these elements are lacking, we are in the realm of noncooperative games. *Noncooperative games* are games in which no binding agreement is possible or can be enforced. Even if communication is allowed in such games, there is no assurance that a player can trust an opponent to choose a particular strategy that he or she promised to select.

In fact, the players' self-interests may lead them to make strategy choices that yield both lower payoffs than they could have achieved by cooperating. Two partial-conflict games illustrate this problem, which is partly resolved by different rules of play that will be described in Chapter 16 on the theory of moves.

EXAMPLE ▶ *Prisoners' Dilemma*

Prisoners' Dilemma is a two-person variable-sum game. It provides a simple explanation of the forces at work behind arms races, price wars, and the population problem. In these and other similar situations, the players can do better by cooperating. But there may be no compelling reasons for them to do so, such as credible threats of retaliation for not cooperating. The term *Prisoners' Dilemma* was first given to this game by Princeton mathematician Albert W. Tucker (1905–1994) in 1950.

Before formally defining this game, we introduce it through a story, which involves two persons, accused of a crime, who are held incommunicado. Each has two choices: to maintain his or her innocence, or to sign a confession accusing the partner of committing the crime.

Now it is in each suspect's interest to confess, that is, to implicate the partner and thereby try to receive a reduced sentence. Yet when both suspects confess, they ensure a bad outcome—namely, they are both found guilty. What is good for the prisoners as a pair—to deny having committed the crime, leaving the state with insufficient evidence to convict them—is frustrated by their pursuit of their own individual rewards.

Prisoners' Dilemma, as we already noted, has many applications, but we will use it here to model a recurrent problem in international relations: arms races between antagonistic countries, which earlier included the superpowers but now include such countries as India and Pakistan and Israel and some of its Arab neighbors.

For simplicity, assume there are two nations, Red and Blue. Each can independently select one of two policies:

    A: Arm in preparation for a possible war (noncooperation).
    D: Disarm, or at least negotiate an arms-control agreement
         (cooperation).

There are four possible outcomes:

    (D, D):   Red and Blue disarm, which is *next best* for both because, while advantageous to each, it also entails certain risks.
    (A, A):   Red and Blue arm, which is *next worst* for both, because they spend needlessly on arms and are comparatively no better off than at *(D, D)*.
    (A, D):   Red arms and Blue disarms, which is *best for Red* and *worst for Blue,* because Red gains a big edge over Blue.
    (D, A):   Red disarms and Blue arms, which is *worst for Red* and *best for Blue,* because Blue gains a big edge over Red.

TABLE 15.9	The Outcomes in an Arms Race, as Modeled by Prisoners' Dilemma		
		Blue	
		A	D
Red A		Arms race	Favors red
D		Favors blue	Disarmament

This situation can be modeled by means of the matrix in Table 15.9, which gives the possible outcomes that can occur. Here, Red's choice involves picking one of the two rows, whereas Blue's choice involves picking one of the two columns.

We assume that the players can rank the four outcomes from best to worst, where 4 = best, 3 = next best, 2 = next worst, and 1 = worst. Thus, the higher the number, the greater the payoff, but these payoffs are only **ordinal:** they indicate an ordering of outcomes from best to worst, but they say nothing about the *degree* to which a player prefers one outcome over another. To illustrate, if a player despises the outcome that he or she ranks 1 but sees little difference among the outcomes ranked 4, 3, and 2, the "payoff distance" between 4 and 2 will be less than that between 2 and 1, even though the numerical difference between 4 and 2 is greater.

The ordinal payoffs to the players for choosing their strategies of A and D are shown in Table 15.10, where the first number in the pair indicates the payoff to the row player (Red), and the second number the payoff to the column player (Blue). Thus, for example, the pair (1, 4) in the second row and first column signifies a payoff of 1 (worst outcome) to Red and a payoff of 4 (best outcome) to Blue. This outcome occurs when Red unilaterally disarms while Blue continues to arm, making Blue, in a sense, the winner and Red the loser.

TABLE 15.10	Ordinal Payoffs in an Arms Race, as Modeled by Prisoners' Dilemma		
		Blue	
		A	D
Red A		(2, 2)	(4, 1)
D		(1, 4)	(3, 3)

Let us examine this strategic situation more closely. Should Red select strategy A or D? There are two cases to consider, which depend on what Blue does:

- If Blue selects A: Red will receive a payoff of 2 for A and 1 for D, so it will choose A.
- If Blue selects D: Red will receive a payoff of 4 for A and 3 for D, so it will choose D.

In both cases, Red's first strategy (A) gives it a more desirable outcome than its second strategy (D). Consequently, we say that A is Red's **dominant strategy,** because it is always advantageous for Red to choose A over D.

In Prisoners' Dilemma, A dominates D for Red, so we presume that a rational Red would choose A. A similar argument leads Blue to choose A as well, that is, to pursue a policy of arming. Thus, when each nation strives to maximize its own payoffs independently, the pair is driven to the outcome (A, A), with payoffs of $(2, 2)$. The better outcome for both, (D, D), with payoffs of $(3, 3)$, appears unobtainable when this game is played noncooperatively.

The outcome (A, A), which is the product of dominant strategy choices by both players in Prisoners' Dilemma, is a *Nash equilibrium.*

When no player can benefit by departing unilaterally (i.e., by itself) from its strategy associated with an outcome, the strategies of the players constitute a **Nash equilibrium.** (Technically, while it is the set of strategies that define the equilibrium, the choice of these strategies leads to an outcome that we shall also refer to as the equilibrium.)

Note that in Prisoners' Dilemma, if either player departs from (A, A), the payoff for the departing player who switches to D drops from 2 to 1 at (D, A) and (A, D). Not only is there no benefit from departing, but there actually is a loss, with the D player punished with its worst payoff of 1. These losses would presumably deter each nation from moving away from the Nash equilibrium of (A, A), assuming the other nation sticks to A.

Even if both nations agreed in advance jointly to pursue the socially beneficial solution, (D, D), the $(3, 3)$ outcome is unstable. This is because if either nation alone reneges on the agreement and secretly arms, it will benefit, obtaining its best payoff of 4. Consequently, each nation would be tempted to go back on its word and select A. Especially if nations have no great confidence in the trustworthiness of their opponents, they would have good reason to try to protect themselves against defection from an agreement.

Prisoners' Dilemma is a two-person variable-sum game in which each player has two strategies, cooperate or defect. Defect dominates cooperate for both players, even though the mutual-defection outcome, which is the unique Nash equilibrium in the game, is worse for both players than the mutual-cooperation outcome.

Note that if 4, 3, 2, and 1 in Prisoners' Dilemma were not just ranks but numerical payoffs, their sum would be $2 + 2 = 4$ at the mutual-defection outcome and $3 + 3 = 6$ at the mutual-cooperation outcome. At the other two outcomes, the sum, $1 + 4 = 5$, is still different, illustrating why Prisoners' Dilemma is a variable-sum game.

In real life, of course, people often manage to escape the noncooperative Nash equilibrium in Prisoners' Dilemma. Either the game is played within a larger context, wherein other incentives are at work, such as cultural norms that prescribe cooperation (though this is just another way of saying that defection from (D, D) is not rational, rendering the game not Prisoners' Dilemma), or the game is played on a repeated basis—it is not a one-shot affair—so players can induce cooperation by setting a pattern of rewards for cooperation and penalties for noncooperation.

In a repeated game, factors like reputation and trust may play a role. Realizing the mutual advantages of cooperation in costly arms races, players may inch toward the cooperative outcome by slowly phasing down their acquisition of weapons over time, or even destroying them (the United States and Russia have begun doing exactly this). They may also initiate other productive measures, such as improving their communication channels, making inspection procedures more reliable, writing agreements that are truly enforceable, or imposing penalties for violators when their violations are detected (as may be possible by reconnaissance or spy satellites).

Prisoners' Dilemma illustrates the intractable nature of certain competitive situations that blend conflict and cooperation. The standoff that results at the Nash equilibrium of (2, 2) is obviously not as good for the players as that which they could achieve by cooperating—but they risk a good deal if the other player defects.

The fact that the players must forsake their dominant strategies to achieve the (3, 3) cooperative outcome (see Table 15.10) makes this outcome a difficult one to sustain in one-shot play. On the other hand, assume that the players can threaten each other with a policy of tit-for-tat in repeated play: "I'll cooperate on each round unless you defect, in which case I will defect until you start cooperating again." If these threats are credible, the players may well shun their defect strategies and try to establish a pattern of cooperation in early rounds, thereby fostering the choice of (3, 3) in the future. Alternatively, they may look ahead, in a manner that will be described in Chapter 16, to try to stabilize (3, 3). ◆

EXAMPLE ▶ *Chicken*

Let us look at one other two-person game of partial conflict, known as *Chicken*, which also can lead to troublesome outcomes. Two drivers approach each other at high speed. Each must decide at the last minute whether to swerve to the right or not swerve. Here are the possible consequences of their actions:

1. Neither driver swerves, and the cars collide head-on, which is the worst outcome for both because they are killed (payoff of 1).
2. Both drivers swerve—and each is mildly disgraced for "chickening out"—but they do survive, which is the next-best outcome for both (payoff of 3).
3. One of the drivers swerves and badly loses face, which is his next-worst outcome (payoff of 2), whereas the other does not swerve and is perceived as the winner, which is her best outcome (payoff of 4).

These outcomes and their associated strategies are summarized in Table 15.11.

If both drivers persist in their attempts to "win" with a payoff of 4 by not swerving, the resulting outcome will be mutual disaster, giving each driver his or her worst payoff of 1. Clearly, it is better for both drivers to back down and each obtain 3 by not swerving, but neither wants to be in the position of being intimidated into swerving (payoff of 2) when the other does not (payoff of 4).

Notice that neither player in Chicken has a dominant strategy. His or her better strategy depends on what the other players does: swerve if the other does not, don't swerve if the other player swerves, making this game's choices highly interdependent, which is characteristic of many games. The Nash equilibria in

TABLE 15.11	Payoffs in a Driver Confrontation, as Modeled by Chicken		
		Driver 2	
		Swerve	Not Swerve
Driver 1	Swerve	(3, 3)	(2, 4)
	Not swerve	(4, 2)	(1, 1)

Chicken, moreover, are (4, 2) and (2, 4), suggesting that the compromise of (3, 3) will not be easy to achieve because both players will have an incentive to deviate from it in order to try to be the winner.

Chicken is a two-person variable-sum game in which each player has two strategies: to swerve to avoid a collision or not to swerve and possibly cause a collision. Neither player has a dominant strategy. The compromise outcome, in which both players swerve, and the disaster outcome, in which both players do not, are not Nash equilibria; the other two outcomes, in which one player swerves and the other does not, are Nash equilibria.

In fact, there is a third Nash equilibrium in Chicken, but it is in mixed strategies, which can only be computed if the payoffs are not ranks, as we have assumed here, but numerical values. Even if the payoffs were numerical, however, we can show that this equilibrium is always worse for both players than the cooperative (3, 3) outcome. Moreover, it is implausible that players would sometimes swerve and sometimes not—randomizing according to particular probabilities—in the actual play of this game, compared with either trying to win outright or reaching a compromise.

The two pure-strategy Nash equilibria in Chicken suggest that, insofar as there is a "solution" to this game, it is that one player will succeed when the other caves in to avoid the mutual-disaster outcome. But there certainly are real-life cases in which a major confrontation was defused and a compromise of sorts was achieved in Chicken-type games. This fact suggests that the one-sided solution given by the two pure-strategy Nash equilibria may not be the only pure-strategy solution. We will pursue this subject further in Chapter 16, where we allow for more farsighted thinking than standard game theory provides, which can yield (3, 3) as an equilibrium outcome in Chicken as well as in Prisoners' Dilemma.

International crises, labor-management disputes, and other conflicts in which escalating demands may end in wars, strikes, and other catastrophic outcomes have been modeled by the game of Chicken. But Chicken, like Prisoners' Dilemma, is only one of the 78 essentially different 2×2 ordinal games, in which each player can rank the four possible outcomes from best to worst. (Other examples of such games are given in Chapter 16.)

Chicken and Prisoners' Dilemma, however, are especially disturbing, because the cooperative (3, 3) outcome in each is not a Nash equilibrium. Unlike a constant-sum game, in which the losses of one player are offset by the gains of the other, *both* players can end up doing badly—at (2, 2) in Prisoners' Dilemma and (1, 1) in Chicken—in these variable-sum games. ◆

Spotlight 15.2
The 1994 Nobel Memorial Prize in Economics

John C. Harsanyi

John F. Nash

The Nobel Memorial Prize in Economics was awarded to three game theorists in 1994, marking the fiftieth anniversary of the publication of von Neumann and Morgenstern's *Theory of Games and Economic Behavior* (see Spotlight 15.1, p. 562). The recipients were as follows:

• *John C. Harsanyi* (1920–   ) of the University of California, Berkeley, a Hungarian-American who emigrated from Hungary to Australia in 1950 and then to the United States in 1956. He is well known for extending game theory to the study of ethics and how societal institutions, each of whose members' satisfaction can be measured against that of others, choose among alternatives. His other major contribution was to give a precise definition

to "incomplete information" in games in which players may be thought of as different types, and probabilities assigned to each type. His analysis of such games is applicable to the modeling of many real-life conflicts in which there are severe constraints on the information that players have about each other. Harsanyi was trained in both mathematics and economics, and also has strong interests in philosophy.

• *John F. Nash* (1928–   ) of Princeton University, an American mathematician who did path-breaking work in both noncooperative game theory (the "Nash equilibrium" is named after him) and cooperative game theory, especially on bargaining, in which axioms or assumptions are specified and a unique solution that satis-

Reinhard Selten

fies these axioms is derived. Nash obtained his results in the early 1950s, when he was only in his 20s, after which he became mentally ill and was unable to work. Fortunately, he has made a remarkable recovery and has now resumed research.

- *Reinhard Selten* (1930–) of the University of Bonn, a German mathematician who proposed significant refinements in the concept of a Nash equilibrium that help to distinguish those that are most plausible in games (often there are many such equilibria, which creates a selection problem). Some of his work on equilibrium selection was done in collaboration with Harsanyi. Selten is also noted for pioneering work on developing game-theoretic models in evolutionary biology. He is an advo-

cate of experimental testing of game-theoretic solutions to determine those that are most likely to be chosen by human subjects, and using these empirical results to refine the theory.

Other contemporary game theorists have also made important advances in the theory, including Lloyd S. Shapley (1923–), an American mathematician who proposed the cooperative game-theoretic solution concept known as the "Shapley value." He extended this work with Robert J. Aumann (1930–), a mathematician who was born in Germany, emigrated as a child to the United States before World War II, and then moved to Israel. Aumann also gave a precise formulation to "common knowledge," developing some of its consequences in games, and proposed the notion of a "correlated equilibrium," which has been helpful in understanding how players coordinate their choices in games.

Game theory has provided important theoretical foundations in economics, starting with microeconomics but now extending to macroeconomics and international economics. It also has been increasingly applied in political science, especially in the study of voting, elections, and international relations. In addition, game theory has contributed major insights in biology, particularly in understanding the evolution of species and conditions under which animals—humans included—fight each other for territory or act altruistically. It has also illuminated certain fields in philosophy, including ethics, philosophy of religion, and political philosophy, and inspired many experiments in social psychology.

LARGER GAMES

We have shown how to compute optimal pure and mixed strategies, and the values ensured by using them, in 2×2 constant-sum games. In 2×2 variable-sum games, we focused on Nash equilibria as a solution concept in Prisoners' Dilemma and Chicken, but we found that this notion of a stable outcome did not justify the choice of cooperative strategies in either of these games.

We turn next to a somewhat larger game, in which there are three players, each of whom can choose among three strategies, which is technically a $3 \times 3 \times 3$ game. In this game, we eliminate certain undesirable strategies, but in stages, to arrive at a Nash equilibrium that seems quite plausible.

If one of the three players has a dominant strategy in the $3 \times 3 \times 3$ game, we suppose this player will choose it. The game can thereby be reduced to a 3×3 game between the other two players, since the player with the dominant strategy can be assumed already to have made his choice. (Of course, if no player has a dominant strategy in a three-person game, it cannot be reduced in this manner to a two-person game.)

Now the 3×3 game is not one of total conflict, so the minimax theorem, guaranteeing players the value in a two-person zero-sum game, is not applicable. Even if the game were zero-sum, the fact that we assume the players can only rank outcomes, but not assign numerical values to them, prevents their calculating optimal mixed strategies in it.

The problem in finding a solution to the 3×3 game is not a lack of Nash equilibria. Rather, there are too many! So the question becomes which, if any, are likely to be selected by the players. Specifically, is one more appealing than the others? The answer is "yes," but it requires extending the idea of dominance, discussed in the previous section, to its successive application in different stages of play.

EXAMPLE ▶ *The Paradox of the Chair's Position—And an Escape*

The $3 \times 3 \times 3$ game we analyze involves voting, illustrating the applicability of game theory to politics. There is also a tie-in to the analysis of weighted voting in Chapter 12, because one of the players (the chair in the voting body), while not having more votes than the others, can break ties. This would seem to make the chair more powerful, in some sense, than the other players.

As we shall see, however, rather than making the chair more powerful—as we earlier measured voting power—the possession of a tie-breaking vote backfires, preventing the chair from obtaining a preferred outcome. However, we do indicate a possible escape for the chair from this unenviable position.

The power indices described in Chapter 12 do not take into account the preferences of the differently weighted players—all combinations (Banzhaf index) or permutations (Shapley–Shubik index) were assumed to be equally likely. But as we shall show next, the greater resources that a player has do not always translate into greater **power,** by which we mean the ability of a player to obtain a preferred outcome.

Now one would suppose that the chair of a voting body, if he or she has a tie-breaking vote in addition to a regular vote, would have more power than other members. Yet, under circumstances to be spelled out shortly, the chair may actually be at a disadvantage relative to the other members.

THE PARADOX: To illustrate this problem, suppose there is a set of three voters, $V = \{X, Y, Z\}$, and a set of three alternatives, $A = \{x, y, z\}$, from which the voters choose. Assume that voter X prefers x to y to z, indicated by xyz; voter Y's preference is yzx, and voter Z's is zxy. These preferences give rise to a *Condorcet voting paradox* (discussed in Chapter 11), because the social ordering, according to majority rule, is *intransitive:* although a majority (voters X and Z) prefer x to y, and a majority (voters X and Y) prefer y to z, a majority (voters Y and Z) prefer z to x. So there is no *Condorcet winner*—an alternative that would beat all others in separate pairwise contests. Instead, every alternative can be beaten by one other.

Assume that the voting procedure used by the three voters, who choose from among the three alternatives, is the **plurality procedure,** under which the alternative with the most votes wins. If there is a three-way tie (there can never be a two-way tie if there are three voters), we assume the chair X can break the tie, giving the chair what would appear to be an edge over the other two voters, Y and Z.

To begin, assume that voting is sincere.

> Under **sincere voting,** every voter votes for his or her most-preferred alternative, without taking into account what the other players might do.

In this case, X will prevail by being able to break the tie in favor of x. However, X's apparent advantage disappears if voting is "sophisticated" (to be defined shortly), as we shall demonstrate.

To see why, first note that X has a dominant strategy of "vote for x": it is never worse and sometimes better than her other two strategies, whatever the other two voters do. Thus, if the other two voters vote for the same alternative, it wins, and X cannot do better than vote sincerely for x, so voting sincerely is never worse. On the other hand, if the other two voters disagree, X's tie-

FIGURE 15.4
Sophisticated voting, given
X chooses "vote for *x*." The
dominated strategies of
each voter are crossed out
in the first reduction,
leaving two (undominated)
strategies for *Y* and one
(dominant) strategy for *Z*.
Given these eliminations,
Y would then eliminate
"vote for *y*" in the second
reduction, making *z* the
sophisticated outcome.

FIRST REDUCTION

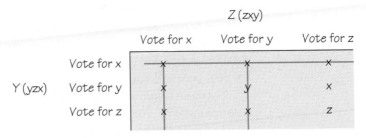

SECOND REDUCTION

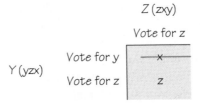

breaking vote (along with her regular vote) for *x* will be decisive in *x*'s selection, which is *X*'s best outcome.

Given the dominant choice of *x* on the part of *X*, *Y* and *Z* face the strategy choices shown in Figure 15.4. *Y* has one, and *Z* has two, **dominated strategies,** which are never better and sometimes worse than some other strategy, whatever the other two voters do. For example, observe that "vote for *x*" by *Y* always leads to his worst alternative, *x*. The dominated strategies are crossed out in the top matrix in Figure 15.4.

This leaves *Y* with two *undominated* strategies that are neither dominant nor dominated; "vote for *y*" and "vote for *z*." "Vote for *y*" is better than "vote for *z*" if *Z* chooses *y* (leading to *y* rather than *x*), whereas the reverse is the case if *Z* chooses *z* (leading to *z* rather than *x*). By contrast, *Z* has a dominant strategy of "vote for *z*," which leads to outcomes at least as good and sometimes better than his other two strategies.

If voters have complete information about each other's preferences, then they can perceive the situation in terms of the top matrix in Figure 15.4 and eliminate the dominated strategies that are crossed out (first reduction). The elimination of these strategies gives the bottom matrix in Figure 15.4. Then *Y*, choosing between "vote for *y*" and "vote for *z*" in this matrix, would cross out "vote for *y*" (second reduction), now dominated because that choice would result in *x*'s winning due to the chair's tie-breaking vote. Instead, *Y* would choose "vote for z," ensuring z's election, which is Z's best outcome, but only the next-best outcome for Y. In this manner z, which is not the first choice of a majority and could in fact be beaten by y in a pairwise contest, becomes the sophisticated outcome.

> The successive elimination of dominated strategies by voters—insofar as this is possible—beginning in our example with X's choice of x in favor of y and z, is called **sophisticated voting.**

Sophisticated voting results in a Nash equilibrium, because none of the three players can do better by departing from his or her sophisticated strategy when the other two players choose theirs. This is clearly true for X, because x is her dominant strategy; given X's choice of x, z is dominant for Z; and given these choices by X and Z, z is dominant for Y. These "contingent" dominance relations, in general, make sophisticated strategies a Nash equilibrium.

Observe, however, that there are four other Nash equilibria in this game. First, the choice of each of x, y, or z by all three voters are all Nash equilibria, because no single voter's departure can change the outcome to a different one, much less a better one, for that player. In addition, the choice of x by X, y by Y, and x by Z—resulting in x—is also a Nash equilibrium, because no voter's departure would lead to a better outcome for him or her.

In game-theoretic terms, sophisticated voting produces a different and smaller game in which some formerly undominated strategies in the larger game become dominated in the smaller game. The removal of such strategies, sometimes in several successive stages, in effect enables sophisticated voters to determine what outcomes eventually *will* be chosen by eliminating those outcomes that definitely *will not* be chosen. Voters can thereby ensure that their worst outcomes will not be chosen by successively removing dominated strategies, given the presumption that other voters do likewise.

> How does sophisticated voting affect the chair's presumed extra voting power? The chair's tie-breaking vote is not only not helpful, but it is positively harmful: it guarantees that X's worst outcome (z) will be chosen if voting is sophisticated! This situation, in which the chair's tie-breaking vote hurts rather than helps the chair, is called the **paradox of the chair's position.**

Given this unfortunate state of affairs for the chair, we might ask whether a chair, or the largest voting faction in a voting body comprising three factions—none of which commands a majority—has any recourse. It would appear not: the sophisticated outcome, z, is supported by both Y and Z, which no voting strategy of X can upset.

AN ESCAPE: A chair is often in the unique position, after the other voters have already committed themselves, of being the last voter to have to make a strategy choice. However, this position does not furnish a ready solution to the chair's problem if voting is truly sophisticated, for sophisticated voting implies that voters act upon both their own preferences and a knowledge of the preferences of the other voters. Therefore, the order of voting is immaterial: all voters can predict sophisticated choices beforehand and act accordingly. Thus, even a chair's (unexpected) deviation from a sophisticated strategy cannot effect a better outcome for it.

But now assume that the chair, by virtue of her position, can obtain information about the preferences of the other two voters, but they cannot obtain information about her preference. Further, assume that each of the two regular members is informed of the other's preference. If voting is to be sophisticated, the chair's preference must be made known to the regular member; however, the chair is not compelled to tell the truth. The question is: Can a chair, by announcing a preference different from her true preference, induce a more-preferred sophisticated outcome?

Recall that the chair, having a tie-breaking vote, will always have a dominant strategy, which happens to be her sincere choice in our example, if voting is sophisticated. Indeed, the other voters need only know (and believe) her announced first choice, and not her complete preference scale, to determine what her sophisticated strategy will be.

A **deception strategy** on the part of the chair is any *announced* most-preferred alternative that differs from her sincere choice. Her deception is **tacit** if she chooses the strategy she announces, thereby not revealing her deception.

Tacit deception will be profitable for the chair if it induces an outcome that the chair prefers to the sophisticated outcome, based on her sincere preferences.

In our example, the chair (X), by announcing her first choice to be y instead of x, can induce the sophisticated outcome y, which she would prefer to z. This can be seen from the reductions shown in Figure 15.5. Note that the first-reduction matrix gives the outcomes as Y and Z *perceive* them, after X's deceptive announcement of y as her first choice (it does not matter in what order she ranks x and z below y).

FIRST REDUCTION

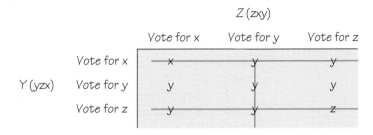

SECOND REDUCTION

		Z (zxy)	
		Vote for x	Vote for z
Y (yzx)	Vote for y	y	y

FIGURE 15.5 Tacit deception outcome, given X chooses "vote for y." The dominated strategies of each voter are crossed out in the first reduction, leaving two (undominated) strategies for Z and one (dominant) strategy for Y. Given these eliminations, Z's two undominated strategies, both yielding y, remain undominated in the second reduction—so no strategies are in fact crossed out. The (manipulated) outcome is y, whichever of Z's two remaining strategies he chooses, making y the tacit-deception outcome. If X actually voted for x after falsely announcing a first choice of y, the (manipulated) outcome would be x; however, X's deception would be revealed.

Y's elimination of dominated strategies "vote for x" and "vote for z," and Z's elimination of dominated strategy "vote for y," give the second-reduction matrix shown in Figure 15.5, which cannot be reduced further since Z's two remaining strategies both yield y. Thus, tacit deception, by changing the outcome from z to y, is profitable for X (as well as Y).

Suppose now that X, as the chair, actually chooses "vote for x" after announcing her (insincere) preference for y. Then the sophisticated outcome she induces will be x, her first preference. In other words, the chair can induce her most-preferred outcome by announcing a bogus preference for y and, contrary to her announcement, voting sincerely in the end. Because y, the tacit-deception outcome, is not a Nash equilibrium, X can benefit by voting for x, contrary to her announced preference for y.

> Such **revealed deception** involves announcing a deception strategy but then voting contrary to the announcement.

Recall that an insincere announcement by the chair improves her position somewhat, inducing her next-best outcome (y), when deception remains tacit.

However, when this announcement is followed instead by the chair's vote for her most-preferred alternative (x)—flouting this announcement—then the chair can achieve her best outcome (x).

Of course, revealed deception becomes apparent after the vote, unless it is secret, and probably cannot be used very frequently. If it were, the chair's announcements would quickly lose credibility and thus their inducement value.

The deception-strategy game we have described for the chair can also be played by a regular member if he or she is privy to information that the chair and the other regular members are not. (The results of such special knowledge will not duplicate those for the chair, however, because the chair has an extra resource—her tie-breaking vote.) We shall not carry this analysis further, though, because our main purpose has been to demonstrate that there is a resolution of sorts to the paradox of the chair's position. It requires, however, that the information available to some players in the game be restricted, which has the effect of endowing one player (the chair) with still greater resources.

This, it must be admitted, is a rather deceptive way out of a problem that seems genuine. If voting is sophisticated, the chair, despite the added weight of her position, will not necessarily enjoy greater control over outcomes than the other members. In fact, the reverse might be the case, as the paradox demonstrates.

Clearly, power defined as control over outcomes is not synonymous with power defined as control over resources (e.g., a tie-breaking vote or simply more votes). The strategic situation facing voters intervenes and may cause them to reassess their sincere strategies in light of the additional resources that a chair possesses. In so doing, they may be led to "gang up" against the chair—that is, to vote in such a way as to undermine the impact of her extra resources—handing the chair a worse outcome than she would have achieved without them. These resources in effect become a burden to bear, not power to wield.

We stress that Y and Z do not form a coalition against X in the sense of coordinating their strategies and agreeing to act together in a cooperative game, wherein binding agreements are possible. Rather, they behave as isolated individuals; at most they could be said to form an "implicit coalition." Such a coalition does not imply even communication between its members but simply agreement based on their common perceived strategic interests. ◆

USING GAME THEORY

Solving Matrix Games

Given any payoff matrix, the first thing we ask is whether it is zero-sum (or constant-sum). If so, we check to see whether it has a saddlepoint by determining the minimum number of each row and the maximum number of each column, as we did in several earlier examples. If the maximum of the row minima (maximin) is equal to the minimum of the column maxima (minimax), then the game has a saddlepoint. The resulting value, and the corresponding pure strategies, provide a solution to the game.

This value will appear in the payoff matrix as the smallest number in its row and the largest in its column. In the 3 × 3 location game, this number was 5 (5000 feet).

Like our voting game, dominated strategies can successively be eliminated in the 3 × 3 location game. Thus, Route B dominates Route C, and Highway 2 dominates Highway 3; having made these eliminations, Highway 2 dominates Highway 1; having made this elimination, Route B dominates Route A. Thus, Highway 2 and Route B survive the successive eliminations, yielding the saddlepoint of 5. The successive-elimination procedure therefore provides an alternative method for finding the saddlepoint in this 3 × 3 location game. Unfortunately, it does not work to find the saddlepoint in *all* two-person zero-sum games bigger than 2 × 2.

Recall that instead of eliminating dominated strategies in the 3 × 3 location game, we eliminated Route B and Highway 2, which dominated other strategies, to obtain the 2 × 2 restricted-location game in Table 15.3. In this game, there were no dominated strategies and, hence, no saddlepoint.

If a two-person zero-sum game does not have a saddlepoint, which was the case not only in the restricted-location game but also for matching pennies, the nonsymmetrical matching game, and the baseball duel, the solution will be in mixed strategies. To find the optimal mix in a 2 × 2 game, we calculate the expected value to a player from choosing its first strategy with probability p and its second with probability $1 - p$, assuming that the other player chooses its first pure strategy (yielding one expected value) and its second pure strategy (yielding another expected value).

Setting these two expected values equal to each other yields a unique value for p that gives the optimal mix, $(1 - p, p)$, with which the player should choose its first and second strategies. Substituting the numerical solution of p back into either expected-value equation gives the value of the game, which each player can guarantee for itself whatever strategy its opponent chooses.

Several general algorithms have been developed since 1945 to find mixed-strategy solutions to large constant-sum games. This work has mostly been done in the field of linear programming, using such algorithms as the simplex method of G. B. Dantzig and the more recent method of N. K. Karmarkar (see Chapter 4).

In variable-sum games, we also begin by successively eliminating dominated strategies, if there are any. The outcomes that remain do not depend on the numerical values we attach to them but only on their ranking from best to worst by the players, as illustrated in the three-person voting example.

Care must be taken in interpreting this solution, however. It began with the choice of a dominant strategy by the chair (X) — and her elimination of her two dominated strategies. Presuming these eliminations, Y and Z were then able to eliminate their own dominated strategies in the first reduction, and Y in turn eliminated a dominated strategy in the second reduction, leading finally to the outcome z, supported by Y and Z.

This solution is a fairly demanding one, because it assumes considerable calculational abilities on the part of the players. Less demanding, of course, is that players simply choose their dominant strategies, as is possible in Prisoners' Dilemma, but of course they may not have such strategies.

In the game of Chicken, for example, neither player has a dominant (or dominated) strategy, so the game cannot be reduced. In such situations, we ascertain what outcomes are Nash equilibria. There are two (in pure strategies) in Chicken, suggesting that the only stable outcomes in this game occur when one player gives in and the other other does not. In Prisoners' Dilemma, by comparison, the choice by the players of their dominant strategies singles out the mutual-defection outcome as the unique Nash equilibrium, which is worse for both players than the cooperative outcome.

In both Chicken and Prisoners' Dilemma, there seems no good reason for the choice of the (3, 3) cooperative outcome, at least if each game is played only once, because this outcome is not a Nash equilibrium. In Chapter 16, we shall describe different rules, and a different rationale, for cooperation in these games.

Practical Applications

The element of surprise, as captured by mixed strategies, is essential in many encounters. For example, mixed strategies are used in various inspection procedures and auditing schemes to deter potential cheaters; by making inspection or auditing choices random, they are rendered unpredictable.

Investigators or regulatory agencies monitor certain accounts as well as take various actions to check for faults, errors, or illegal activities. The investigators include bank auditors, customs agents, insurance investigators, and quality-control experts. The National Bureau of Standards is responsible for monitoring the accuracy of measuring instruments and for maintaining reliable standards. The Nuclear Regulatory Agency demands an accounting of dangerous nuclear material as part of its safeguards program. The Internal Revenue Service attempts to identify those cheating on taxes.

Military or intelligence services may wish to intercept a weapon hidden among many decoys or plant a secret agent disguised to look like a respectable individual. Because it is prohibitively expensive to check the authenticity of each and every possible item or person, efficient methods need to be used to check for violations. Both optimal detection and optimal concealment strate-

Soviet General Secretary Mikhail Gorbachev and U.S. President Ronald Reagan sign the Intermediate Range Nuclear Forces Treaty in Washington, D.C., on December 8, 1987.

gies can be modeled as a game between an inspector trying to increase the probability of detection, and a violator trying to evade detection.

Some of these games are constant-sum: the violator "wins" when the evasion is successful and "loses" when it is not. On the other hand, cheating on arms-control agreements may well be variable-sum if both the inspector and the cheater would prefer that no cheating occur to there being cheating and public disclosure of it. The latter could be an embarrassment to both sides, especially if it undermines an arms-control agreement both sides wanted and the cheating is not too significant.

We alluded earlier to the strategy of bluffing in poker, which is used to try to keep the other player or players guessing about the true nature of one's hand. The optimal probability with which one should bluff can be calculated in a particular situation (see Exercise 17). Besides poker, bluffing is common in many bargaining situations, whereby a player raises the stakes (e.g., labor threatens a strike in labor-management negotiations), even if it may ultimately have to back down if its "hand is called."

Perhaps the greatest value of game theory is the framework it provides for understanding the rational underpinnings of conflict in the world today. As a case in point, a confrontation over the budget between the Democratic President Bill Clinton and the Republican Congress resulted in the shutdown of part of the federal government on two occasions between November 1995 and

January 1996. Many government workers were frustrated in not being able to do their jobs, even though they knew they would be paid for not working, not to mention the many citizens either greatly hurt or substantially inconvenienced by the shutdown.

Viewed as a game of Chicken, in which each side wanted to get its way not only for the moment but also to establish a precedent for the future, this conflict was not so foolish as it might seem at first glance. As another example, the constant price wars among the airlines suggest competitors caught up in a Prisoners' Dilemma, in which they all suffer from lower fares but cannot avoid their dominant strategies of not cooperating, perhaps to try to seize a quick advantage or hurt the competition even more (and possibly even eliminate a competitor). All in all, we believe that game theory offers fundamental insights into conflicts at all levels, especially its *seemingly* irrational features which, on second look, are often not ill-conceived.

REVIEW VOCABULARY

Chicken   A two-person variable-sum symmetric game in which each player has two strategies: to swerve to avoid a collision, or not to swerve and cause a collision if the opponent has not swerved. Neither player has a dominant strategy; the compromise outcome, in which both players swerve, is not a Nash equilibrium, but the two outcomes in which one player swerves and the other does not are Nash equilibria.

Constant-sum game   A game in which the sum of payoffs to the players at each outcome is a constant, which can be converted to a zero-sum game by an appropriate change in the payoffs to the players that does not alter the strategic nature of the game.

Deception strategy   A player's announcement of a false preference to induce other players to choose strategies favorable to the deceiver.

Dominant strategy   A strategy that is sometimes better and never worse for a player than every other strategy, whatever strategies the other players choose.

Dominated strategy   A strategy that is sometimes worse and never better for a player than some other strategy, whatever strategies the other players choose.

Expected value E   If each of the n payoffs $s_1, s_2, \ldots, s_n$ occurs with respective probabilities $p_1, p_2, \ldots, p_n$, then the expected value E is

$$E = p_1 s_1 + p_2 s_2 + \cdots + p_n s_n$$

where $p_1 + p_2 + \cdots + p_n = 1$ and $p_i \geq 0$ $(i = 1, 2, \ldots, n)$.

Fair game   A zero-sum game is fair when the (expected) value of the game, obtained by using optimal strategies (pure or mixed), is zero.

Maximin   In a two-person zero-sum game, the largest of the minimum payoffs in each row of a payoff matrix.

Maximin strategy   In a two-person zero-sum game, the pure strategy of the row player corresponding to the maximin in a payoff matrix.

Minimax   In a two-person zero-sum game, the smallest of the maximum payoffs in each column of a payoff matrix.

Minimax strategy   In a two-person zero-sum game, the pure strategy of the column player corresponding to the minimax in a payoff matrix.

Minimax theorem   The fundamental theorem for two-person constant-sum games, stating that there always exist optimal pure or mixed strategies that enable the two players to guarantee the value of the game.

Mixed strategy   A strategy that involves the random choice of pure strategies, according to particular probabilities. A mixed strategy of a player is optimal if it guarantees the value of the game.

Nash equilibrium   Strategies associated with an outcome such that no player can benefit by choosing a different strategy, given that the other players do not depart from their strategies.

Nonsymmetrical game   A two-person constant-sum game in which the row player's gains are different from the column player's gains, except when there is a tie.

Ordinal game   A game in which the players rank the outcomes from best to worst.

Paradox of the chair's position   This paradox occurs when being chair (with a tie-breaking vote) hurts rather than helps the chair if voting is sophisticated.

Partial-conflict game   A variable-sum game in which both players can benefit by cooperation, but they may have strong incentives not to do so.

Payoff matrix   A rectangular array of numbers. In a two-person game, the rows and columns correspond to the strategies of the two players, and the numerical entries give the payoffs to the players when these strategies are selected.

Plurality procedure   A voting procedure in which the alternative with the most votes wins.

Power   The ability of a player to induce a preferred outcome.

Prisoners' Dilemma   A two-person variable-sum symmetric game in which each player has two strategies, cooperate or defect. Cooperate dominates defect for both players, even though the mutual-defection outcome, which is the unique Nash equilibrium in the game, is worse for both players than the mutual-cooperation outcome.

Pure strategy   A course of action a player can choose in a game, which does not involve randomized choices.

Rational choice   A choice that leads to a preferred outcome.
Revealed deception   Involves falsely announcing a strategy to be dominant but subsequently choosing another strategy—inconsistent with one's announcement—that reveals one's deception.
Saddlepoint   In a two-person constant-sum game, the payoff that results when the maximin and the minimax are the same, which is the value of the game. The saddlepoint has the shape of a saddle-shaped surface and is also a Nash equilibrium.
Sincere voting   Voting for one's most-preferred alternative in a situation.
Sophisticated voting   Involves the successive elimination of dominated strategies by voters.
Strategy   One of the courses of action a player can choose in a game; strategies are mixed or pure, depending on whether they are selected in a randomized fashion (mixed) or not (pure).
Tacit deception   Involves falsely announcing a strategy to be dominant, and subsequently choosing this strategy—consistent with one's announcement—thereby not revealing one's deception.
Total-conflict game   A zero-sum or constant-sum game, in which what one player wins the other player loses.
Value   In a two-person zero-sum game, if there is a saddlepoint, this is the value; otherwise, it is the expected payoff resulting when the players choose their optimal mixed strategies.
Variable-sum game   A game in which the sum of the payoffs to the players at the different outcomes varies.
Zero-sum game   A constant-sum game in which the payoff to one player is the negative of the payoff to the other player, so the sum of the payoffs to the players at each outcome is zero.

SUGGESTED READINGS

AUMANN, ROBERT J., AND SERGIU HART, EDS.   *Handbook of Game Theory with Economic Applications,* Elsevier, Amsterdam, 1992 (vol. 1), 1994 (vol. 2). A comprehensive treatment of game theory and its applications, developed in long chapters written by leading experts. A third (and final) volume is forthcoming.

BAIRD, DOUGLAS G., ROBERT H. GERTNER, AND RANDAL C. PICKER.   *Game Theory and the Law,* Harvard University Press, Cambridge, Mass., 1994. A good treatment of how game theory informs various branches of the law.

BINMORE, KEN.   *Fun and Games: A Text on Game Theory,* Heath, Lexington, Mass., 1992. This best-selling text is a provocative introduction to game theory at an elementary-intermediate level.

BRAMS, STEVEN J.   *Biblical Games: A Strategic Analysis of Stories in the Old Testament,* MIT Press, Cambridge, Mass., 1980. About 20 stories of conflict and intrigue in the Hebrew Bible are modeled as simple games, in many of which God is a player. By and large, Brams argues, the players made rational choices.

BRAMS, STEVEN J.   *Negotiation Games: Applying Game Theory to Bargaining and Arbitration,* Routledge, New York, 1990. Game-theoretic models of negotiation have been developed in several disciplines; this book provides a survey of different models and applications.

BRAMS, STEVEN J.   *Superpower Games: Applying Game Theory to Superpower Conflict,* Yale University Press, New Haven, Conn., 1985. The superpower conflict, as it existed during the cold war (roughly from 1945 to 1990), has evaporated since the demise of the Soviet Union, but the models developed in this book to study deterrence, arms races, and the verification of arms-control agreements are relevant to conflicts between other countries today.

DIXIT, AVINASH, AND BARRY NALEBUFF.   *Thinking Strategically: The Competitive Edge in Business, Politics, and Everyday Life,* Norton, New York, 1991. A best-selling popular treatment of applications of game theory, with many stimulating examples.

GARDNER, ROY.   *Game Theory for Business and Economics,* Wiley, New York, 1995. A good introduction to game theory and its business and economic applications, with many examples.

GIBBONS, ROBERT.   *Game Theory for Applied Economists,* Princeton University Press, Princeton, N.J., 1993. There are a number of texts on game-theoretic models in economics, many at an advanced level; this is an intermediate-level text that emphasizes applications.

LUCE, R. DUNCAN, AND HOWARD RAIFFA.   *Games and Decisions: Introduction and Critical Survey,* Wiley, New York, 1957; Dover, New York, 1989. This venerable survey of most of early game theory presents two-person constant-sum and variable-sum games in chapters 4 and 5. Several different algorithms for solving the zero-sum case are given in appendix 6; the minimax theorem, and its equivalence to the "duality theorem" in linear programming, are given in appendixes 2 and 5.

MCDONALD, JOHN.   *The Game of Business,* Doubleday, New York, 1975; Anchor, New York, 1977. A superb collection of cases in which elementary tools of game theory are used to explicate some classic battles in the business world.

MORROW, JAMES D.   *Game Theory for Political Scientists,* Princeton University Press, Princeton, N.J., 1994. Develops tools of game theory and discusses game-theoretic models used in political science.

POUNDSTONE, WILLIAM. *Prisoner's Dilemma: John von Neumann, Game Theory, and the Puzzle of the Bomb,* Doubleday, New York, 1992. A history of game theory as well as a biography of its mathematician founder, including his views on the use of nuclear weapons.

SIGMUND, KARL. *Games of Life: Explorations in Ecology, Evolution, and Behavior,* Oxford University Press, Oxford, 1993. Game theory is placed in the broader context of evolutionary models in biology and related fields in a very readable account.

TAYLOR, ALAN D. *Mathematics and Politics: Strategy, Voting, Power and Proof,* Springer-Verlag, New York, 1995. An elementary but sophisticated treatment of how game theory and social choice theory illuminate the study of politics in areas as different as voting and crisis escalation.

WEIBULL, JÖRGEN. *Evolutionary Game Theory,* MIT Press, Cambridge, Mass., 1995. Applications of game theory in biology, especially the study of evolution, have grown rapidly, and this is an advanced, up-to-date assessment.

WILLIAMS, JOHN D. *The Compleat Strategyst: Being a Primer on the Theory of Games of Strategy,* McGraw-Hill, New York, 1954, rev. 1966; Dover, New York, 1986. This gem, which contains many simple illustrations, is a humorous primer on two-person zero-sum games.

EXERCISES   ▲ *Optional.*   ■ *Advanced.*   ◆ *Discussion.*

Total-Conflict Games

Consider the following five two-person zero-sum games, wherein the payoffs represent gains to the row Player I and losses to the column Player II:

1. $\begin{bmatrix} 6 & 5 \\ 4 & 2 \end{bmatrix}$ 2. $\begin{bmatrix} 0 & 3 \\ -5 & 1 \\ 1 & 6 \end{bmatrix}$ 3. $\begin{bmatrix} -2 & 3 \\ 1 & -2 \end{bmatrix}$

4. $\begin{bmatrix} 13 & 11 \\ 12 & 14 \\ 10 & 11 \end{bmatrix}$ 5. $\begin{bmatrix} -10 & -17 & -30 \\ -15 & -15 & -25 \\ -20 & -20 & -20 \end{bmatrix}$

(a)   Which of these games have saddlepoints?

(b)   Find the maximin strategy of Player I, the minimax strategy of Player II, and the value for those games given in part (a).

(c)   List dominated strategies in these games that the players should avoid because the resulting payoffs are worse than those for some alternative strategy.

Solve the following three games of batter-versus-pitcher in baseball, wherein the pitcher can throw one of two pitches and the batter can guess either of these two pitches. The batter's batting averages are given in the game matrix.

6.

		Pitcher	
		Fastball	Curve
Batter	Fastball	.300	.200
	Curve	.100	.400

7.

		Pitcher	
		Fastball	Knuckleball
Batter	Fastball	.500	.200
	Knuckleball	.200	.300

8.

		Pitcher	
		Blooperball	Knuckleball
Batter	Blooperball	.400	.200
	Knuckleball	.250	.250

9.   A businessman has the choice of either not cheating on his income tax or cheating and making $1000 if not audited. If caught cheating, he will pay a fine of $2000 in addition to the $1000 he owes. He feels good if he does not cheat and is not audited (worth $100). If he does not cheat and is audited, he evaluates this outcome as − $100 (for the lost day). Viewing the game as a two-person zero-sum game between the businessman and the tax agency, what are the optimal mixed strategies for each player and the value of the game?

10.   When it is third down and short yardage to go for a first down in American football, the quarterback can decide to run the ball or pass it. Similarly, the other team can commit itself to defend more heavily against a run or

a pass. This can be modeled as a 2×2 matrix game, wherein the payoffs are the probabilities of obtaining a first down. Find the solution of this game.

		Defense	
		Run	Pass
Offense	Run	.5	.8
	Pass	.7	.2

11.   You have the choice of either parking illegally on the street or else parking in the lot and paying $16. Parking illegally is free if the police officer is not patrolling, but you receive a $40 parking ticket if she is. However, you are peeved when you pay to park in the lot on days when the officer does not patrol, and you are willing to assess this outcome as costing $32 ($16 for parking plus $16 for your time, inconvenience, and grief). It seems reasonable to assume that the police officer ranks her preferences in the order (1) giving you a ticket, (2) not patrolling with you parked in the lot, (3) patrolling with you in the lot, and (4) not patrolling with you parked illegally.

   (a)   Describe this as a matrix game, assuming that you are playing a zero-sum game with the officer.
   (b)   Solve this matrix game for its optimal mixed strategies and its value.
   (c)   Discuss whether it is reasonable or not to assume that this game is zero-sum.
   (d)   Assuming that you play this parking game each working day of the year, how do you implement an optimal mixed strategy?

12.   Describe how a pure strategy for a player in a matrix game can be considered as merely a special case of a mixed strategy.

13.   (a)   Describe in detail *one* pure strategy for the player who moves first in the game of tic-tac-toe. (This strategy must tell how to respond to all possible moves of the other player.) (*Hint:* You may wish to make use of the symmetry in the 3×3 grid in this game; that is, there is one "center" box, four "corner" boxes, and four "side" boxes.)
   (b)   Is your strategy optimal in the sense that it will guarantee the first player a tie (and possibly a win) in the game?

14.   In the matching-pennies example, consider the case where Player I favors heads H over tails T. For example, assume that Player I plays H three-fourths of the time and T only one-fourth of the time—a nonoptimal mixed strategy. What should Player II do if he knows this?

15. Assume in the nonsymmetrical matching example that Player II is using the nonoptimal mixed strategy $(p, 1 - p) = (\frac{1}{2}, \frac{1}{2})$; that is, he is playing H and T with the same frequency. What should Player I do in this case if she knows this?

16. You plan to manufacture a new product for sale next year, and you can decide to make either a small quantity, in anticipation of a poor economy and few sales, or a large quantity, hoping for brisk sales. Your expected profits are indicated in the following table:

		Economy	
		Poor	Good
Quantity	Small	$500,000	$300,000
	Large	$100,000	$900,000

If you want to avoid risk and believe that the economy is playing an optimal mixed strategy against you in a two-person zero-sum game, then what is your optimal mixed strategy and the resulting expected value? Discuss some alternative ways that you may go about making your decision.

17. Consider the following miniature poker game with two players, I and II. Each antes $1. Each player is dealt either a high card H or a low card L, with probability one-half. Player I then folds or bets $1. If Player I bets, then Player II either folds, calls, or raises $1. Finally, if II raises, I either folds or calls.

Most choices by the players are rather obvious, at least to anyone who has played poker: if either player holds H, that player always bets or raises if he or she gets the choice. The question remains of how often one should bluff— that is, continue to play while holding a low card in the hope that one's opponent also holds a low card.

This poker game can be represented by the following matrix game, wherein the payoffs are the *expected* winnings for Player I (depending upon the random deal) and the dominated strategies have been eliminated:

		Player II (when holding L)		
		Folds	Calls	Raises
Player I (when holding L)	Folds initially	− .25	0	.25
	Bets first and folds later	0	0	− .25
	Bets first and calls later	− .25	− .25	0

(a) Are there any strategies in this matrix game that a player should avoid playing?

(b) Solve this game.

(c) Which player is in the more favored position?

(d) Should one ever bluff?

Partial-Conflict Games

Consider the following three two-person variable-sum games. Discuss the players' possible behavior when these games are played in a noncooperative manner (i.e., with no prior communication or agreements). The first payoff is to the row player; the second, to the column player.

18.

	Player II	
Player I	(4, 4)	(1, 3)
	(3, 1)	(2, 2)

19. Battle of the sexes:

		She buys a ticket for:	
		Boxing	Ballet
He buys a ticket for:	Boxing	(4, 3)	(2, 2)
	Ballet	(1, 1)	(3, 4)

20.

	Player II	
Player I	(2, 1)	(4, 2)
	(1, 4)	(3, 3)

Larger Games

21. For the preferences of the players given in the text—*xyz* for *X* (chair), *yzx* for *Y*, and *zxy* for *Z*—verify that the strategy choices of *x* by *X*, *y* by *Y*, and *x* by *Z* are a Nash equilibrium. Does this equilibrium seem to you defensible as the social choice by the voters? Under what circumstances might the voters choose these strategies rather than their sophisticated strategies?

22. What is the sophisticated outcome, and the sophisticated strategies of voters, if the preferences of *X* (chair), *Y*, and *Z* are *xyz*, *yxz*, and *zyx*, respec-

tively? What are the other Nash equilibria in this game? Can tacit or revealed deception help the chair?

23.   Assume that the preferences of the three voters are the same as those in Exercise 22, but *X* does not have a tie-breaking vote. What is the sophisticated outcome? Can tacit or revealed deception help *X*?

24.   Assume that the preference of *Y* in Exercise 22 changes from *yxz* to *yzx*, but the preferences of the other two voters remain the same. When there is no chair, show that all three possible outcomes can occur under sophisticated voting. Can tacit or revealed deception ensure *X*, *Y*, or *Z*'s best outcome if that voter has a tie-breaking vote?

Additional Exercises

Consider the following three two-person zero-sum games, wherein the payoffs represent gains to the row Player I and losses to the column Player II:

25. $\begin{bmatrix} 3 & 6 \\ 5 & 4 \end{bmatrix}$ 26. $\begin{bmatrix} -1 & 3 \\ 2 & 0 \end{bmatrix}$ 27. $\begin{bmatrix} 6 & 5 & 6 & 5 \\ 1 & 4 & 2 & -1 \\ 8 & 5 & 7 & 5 \\ 0 & 2 & 6 & 2 \end{bmatrix}$

(a)   Which of these games have saddlepoints?
(b)   Find the optimal strategy for Player I and for Player II, and the value for those games given in part (a).
(c)   List dominated strategies in these games that the players should avoid because the resulting payoffs are worse than those for some alternative strategy.

28.   Consider the game played between the opposing goalie and a soccer player who, after a penalty, is allowed a free kick. The kicker can elect to kick toward one of the two corners of the net, or else aim for the center of the goal. The goalie can decide to commit in advance (before the kicker's kick) to either one of the sides, or else remain in the center until he sees the direction of the kick. This two-person zero-sum game can be represented as follows, wherein the payoffs are the probability of scoring a goal:

		Goalie		
		Breaks left	Remains center	Breaks right
	Kicks left	.5	.9	.9
Kicker	Kicks center	1	0	1
	Kicks right	.9	.9	.5

If we assume that decisions between the left or right side are made symmetrically (i.e., with equal probabilities), then this game can be represented by a 2×2 matrix as follows, where $.7 = \left(\frac{1}{2}\right)(.5) + \left(\frac{1}{2}\right)(.9)$:

		Goalie	
		Remains center	Breaks side
Kicker	Kicks center	0	1
	Kicks side	.9	.7

Find the optimal mixed strategies for the kicker and goalie and the value of this game.

29. (a) Describe in detail *one* pure strategy for the player who moves second in the game of tic-tac-toe.
    (b) Is your strategy in part (a) optimal in the sense that it will guarantee the second player a tie (and possibly a win) in the game?

▲ 30. Find a two-person zero-sum game with a saddlepoint in which the successive elimination of dominated strategies does *not* lead to the saddlepoint (unlike the 3×3 location game; you may restrict yourself to 3×3 games).

31. On an overcast morning, deciding whether to carry your umbrella can be viewed as a game between yourself and nature as follows:

		Weather	
		Rain	No rain
You	Carry umbrella	Stay dry	Lug umbrella
	Leave it home	Get wet	Hands free

Let's assume that you are willing to assign the following numerical payoffs to these outcomes, and that you are also willing to make decisions on the basis of expected values (that is, average payoffs):

$$(\text{Carry umbrella, rain}) = -2$$
$$(\text{Carry umbrella, no rain}) = -1$$
$$(\text{Leave it home, rain}) = -5$$
$$(\text{Leave it home, no rain}) = 3$$

(a) If the weather forecast says there is a 50% chance of rain, should you carry your umbrella or not? What if you believe there is a 75% chance of rain?

(b) If you are conservative and wish to protect against the worst case, what pure strategy should you pick?

(c) If you are rather paranoid and believe that nature will pick an optimal strategy in this two-person zero-sum game, then what strategy should you choose?

(d) Another approach to this decision problem is to assign payoffs to represent what your *regret* will be after you know nature's decision. In this case, each such payoff is the best payoff you could have received under that state of nature, minus the corresponding payoff in the previous table.

		Weather	
		Rain	No rain
You	Carry umbrella	$0 = (-2) - (-2)$	$4 = 3 - (-1)$
	Leave it home	$3 = (-2) - (-5)$	$0 = 3 - 3$

What strategy should you select if you wish to minimize your maximum possible regret?

Consider the two following two-person variable-sum games. Discuss the players' possible behavior when these games are played in a noncooperative manner (i.e., with no prior communication or agreements). The first payoff is to the row player, the second to the column player.

32.

	Player II	
Player I	(2, 4)	(4, 3)
	(1, 2)	(3, 1)

33.

	Player II	
Player I	(3, 4)	(2, 3)
	(1, 2)	(4, 1)

■   34.   Under a voting system called approval voting, a voter can vote for as many alternatives as he or she wishes. (If there are three alternatives, the only undominated strategies of a voter under approval voting are to vote for his or her best, or two best, choices). If voters X, Y, and Z have paradox-of-voting preferences of xyz, yzx, and zxy, and X is the chair, show that x is the sophisticated outcome, obtained by all voters' voting for their two best choices.

▲   35.   Show by example that approval voting is not immune to the paradox of the chair's position.

WRITING PROJECTS

1 ▶   In tennis, one player often prefers to play from the baseline while her opponent prefers a serve-and-volley game (i.e., likes to come to the net). The baseline player attempts to hit passing shots. This player has a choice of hitting "down the line" or "crosscourt." The net player must often guess correctly which direction the ball will go in order to cover the shot. Formulate this situation as a duel game and discuss appropriate strategies for the players.

2 ▶   Consider a conflict that you, personally, had—with a parent, a boss, a girlfriend or boyfriend, or some other acquaintance—in which each of you had to make a choice without being sure of what the other person would do. What strategies did you seriously consider adopting, and what options do you think the other person considered? What plausible outcomes do you think each set of strategy choices would have led to? How would you rank these outcomes from best to worst, and how do you think the other player would have ranked them? Analyze the resulting game, and state whether you think you and the other person made optimal choices. If not, what upset your or the other person's rationality?

3 ▶   It is sometimes argued that game theory does not take account of the (irrational?) emotions of people, such as anger, jealousy, or love. What is your opinion about this question? Give an example, real or hypothetical, that supports your position, paying particular attention to whether the players acted consistently with, or contrarily to, their preferences.

CHAPTER 16

THEORY OF MOVES: A DYNAMIC APPROACH TO GAMES

The **theory of moves (TOM),** while based on game theory (see Chapter 15), makes major changes in its rules. These changes make game theory a more dynamic theory. In particular, by postulating that players think ahead not just to the immediate consequences of making moves but also to the consequences of countermoves to these moves, counter-countermoves, and so on, it extends the strategic analysis of conflict into the more distant future.

TOM also elucidates the role that different kinds of power can have in conflicts. In addition, it shows how misinformation, perhaps caused by misperceptions or deception, may affect player choices and game outcomes.

TOM has been applied to a number of different strategic situations in politics, economics, sociology, fiction, and the Bible, among other areas. In this chapter we will mention some of these that can be modeled by simple 2×2 matrix games of the kind already studied in Chapter 15. As was the case for several games analyzed in Chapter 15, we make the assumption that players can only rank outcomes from best to worst, not associate precise numerical payoffs with them. Thus, mixed strategies and expected-value calculations, which depend on quantitative values and probabilities, are not used in TOM. By comparison, TOM throws into bold relief what players can obtain through sequences of moves, based only on pure strategies.

Chess players apply the reasoning of TOM when they think ahead to consequences of moves, countermoves, and so on.

We will describe how TOM helps farsighted players resolve the dilemmas they face in games like Prisoners' Dilemma and Chicken. But first we develop the main ideas of TOM for a less well-known game, called Success, which we use to model the Samson and Delilah story in the Bible. We then consider other games—including games with more than two players or more than two strategies for each player—and ask what TOM says about their strategic characteristics. We relate the analysis at several points to real-life cases in politics and other areas of application.

GAME THEORY REVISITED

In Chapter 15 we used payoff matrices to describe *games in strategic form.* In these games, the row and column players' choices of strategies led to an outcome, from which each player received a payoff. These strategy choices were assumed to be simultaneous. In this chapter, however, we shall use a "game tree" to analyze the *sequential* choices players can make, as occurs when first you move, then I move, and so on, which are called *games in extensive form.*

Game trees can be drawn in different ways. For the three-person game that we illustrate at the end of the chapter, the tree is upside down, branching out at the bottom rather than at the top. But for the two-person games that make up most of this chapter, we indicate the sequential choices of players along a line, going from left to right. This sideways tree provides an economical representation of moves within a 2 × 2 payoff matrix, as does a related arrow diagram that illustrates cycling within the matrix.

Although TOM starts off with a payoff matrix, it does not assume that players choose strategies simultaneously. Rather, players look ahead and plan their moves, based on rules of play that enable them to make sequential choices.

We begin by applying the rules of standard game theory to a game called Success. Next we use the rules of TOM to show how the rational outcome of this game depends on where play starts and who, if anybody, has certain kinds of power.

EXAMPLE ▶ *The Game of Success*

Consider the game shown in Figure 16.1, which we call Success. Notice that Row (R) has two strategies, s_1 and s_2, and Column (C) also has two strategies, t_1 and t_2, making Success, in appearance, a 2 × 2 game much like Prisoners' Dilemma and Chicken. Unlike the latter two games, however, R's two strategies do not lead to the same payoffs as C's two strategies, which makes Success an **asymmetric game.**

For example, if R chooses s_1, he obtains payoffs of either 2 or 4, depending on what C does. Because neither of C's two strategies, t_1 or t_2, leads to these payoffs (4 and 1 if she chooses t_1, 2 and 3 if she chooses t_2), the strategic consequences of play are different for the two players.

Like Prisoners' Dilemma and Chicken, we assume Success is an *ordinal game:* the payoffs indicate only an ordering of outcomes from best to worst (4 = best, 3 = next best, 2 = next worst, and 1 = worst). As before, the higher the rank, the better the payoff, but the ranks do not indicate whether a player prefers, say, 4 to 3 more than 2 to 1, or vice versa.

FIGURE 16.1 The game of Success according to standard game theory.

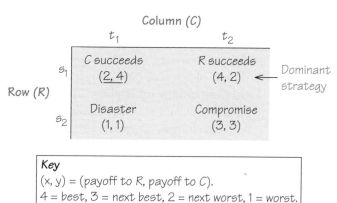

Key
$(x, y) = $ (payoff to R, payoff to C).
4 = best, 3 = next best, 2 = next worst, 1 = worst.
Nash equilibrium underscored.

Assume R chooses s_1 and C chooses t_1 in the Figure 16.1 payoff matrix. The resulting outcome is that shown in the upper left-hand corner of the matrix, with a payoff of 2 to R and 4 to C, or next worst for R and best for C. As shorthand verbal descriptions of these outcomes, we call $(2, 4)$ "C succeeds," $(4, 2)$ "R succeeds," $(3, 3)$ "Compromise," and $(1, 1)$ "Disaster."

Consider what standard game theory, in which players are assumed to make simultaneous strategy choices, tells us about this game. (If the players' choices are not literally simultaneous, game theory assumes they are made independently of each other, so neither R nor C knows the other's choice when each makes his or her own choice.) First consider what strategy is rational for R to choose. If C selects t_1, R has a choice between $(2, 4)$ and $(1, 1)$ in the first column; his payoff will be 2 if he chooses s_1 and 1 if he chooses s_2. By comparison, if C chooses t_2, R has a choice between $(4, 2)$ and $3, 3)$ in the second column; his payoff will be 4 if he chooses s_1 and 3 if he chooses s_2.

Clearly, R is better off choosing s_1 regardless of the strategy that C chooses $(t_1$ or $t_2)$, which, as we showed in Chapter 15, makes s_1 a *dominant* strategy over s_2. By contrast, R's strategy of s_2 is *dominated* by s_1, because it always leads to worse payoffs than s_1, whichever strategy C chooses.

C, on the other hand, does not have a dominant strategy in Success. Her better strategy depends on R's strategy choice: if R chooses s_1, C is better off choosing t_1 because she prefers $(2, 4)$ to $(4, 2)$ in the first row; but if R chooses s_2, C is better off choosing t_2 because she prefers $(3, 3)$ to $(1, 1)$ in the second row. In Chicken, recall from Chapter 15, neither player has a dominant strategy, whereas both players have dominant strategies in Prisoners' Dilemma.

We assume Success to be a game of **complete information,** meaning that both players have full knowledge of the rules and each other's payoffs as well as their own. Therefore, C will know that R's dominant strategy is s_1. Because s_1 is always better than s_2 for R, C can surmise that R will choose s_1. Given that R chooses s_1, it is rational for C to choose t_1, yielding $(2, 4)$ as the rational outcome of Success.

Curiously, this outcome is only R's next worst (2), though R is the player with the dominant strategy. C, the player without a dominant strategy, obtains her best outcome (4). Nevertheless, $(2, 4)$ has a strong claim to be called *the* solution of Success. Not only is it the product of one player's (R's) dominant strategy and the other player's (C's) best response to this dominant choice, but it is also the unique Nash equilibrium, as was $(2, 2)$ in Prisoners' Dilemma.

Recall from Chapter 15 that a *Nash equilibrium* is an outcome—or, more precisely, the strategies associated with this outcome, which are said to be "in equilibrium"—from which neither player would unilaterally depart because he or she would do worse doing so. Thus, if R chooses s_1 and C chooses t_1, giving $(2, 4)$, R will not switch to s_2 because he would do worse at $(1, 1)$; and C will not switch to t_2 because she would do worse at $(4, 2)$. Hence, $(2, 4)$ is stable in the sense that, once chosen, neither player would have an incentive to switch to a different strategy, given that the other player does not switch.

This is not true in the case of the other three outcomes in Success. From $(4, 2)$, C can do better by departing to $(2, 4)$; from $(3, 3)$, R can do better by departing to $(4, 2)$; and from $(1, 1)$, either player can do better by departing, R to $(2, 4)$ and C to $(3, 3)$.

In the latter case, if *both* players switched their strategies in an effort to scramble away from the mutually worst outcome of $(1, 1)$, they would end up at $(4, 2)$, which also is better for both. Indeed, because $(4, 2)$ is R's best outcome, R would be the player who would most welcome a double departure; next most welcome would be a departure by C alone to $(3, 3)$; and least welcome a departure by just himself to $(2, 4)$.

C would not particularly welcome a double departure, obtaining only her next-worst payoff of 2. Like R, she would prefer that her adversary make the first move from $(1, 1)$, because R's departure would yield $(2, 4)$, whereas C's departure yields $(3, 3)$. There are other games, as we shall see later, in which the opposite is true: each player would prefer to be the *first* to depart from an inferior outcome rather than wait for his or her adversary to make the first move. ◆

Standard game theory, by assuming that the players choose strategies simultaneously, does not raise questions about the rationality of moving or departing from outcomes (beyond an immediate departure). In fact, however, most real-life games do not start *with* simultaneous strategy choices and then ask whether the resulting outcome is stable in the sense of being a Nash equilibrium. Rather, play begins with the players already *in* some state. The question then becomes whether or not they would benefit from staying in this state or moving, given the possibility that a move will set off a series of subsequent moves by the players. In the next section we discuss how the rules of TOM alter the players' calculations of stable outcomes in Success.

To get some perspective on our analysis of Success and other games, we note here that there are 78 distinct 2×2 ordinal games that are structurally distinct in the sense that no interchange of the players, their strategies, or any combination of these can transform one of these games into any other. These games represent *all* the different configurations of ordinal payoffs in which two players, each with two strategies, may find themselves embedded. (See Rapoport et al. in Suggested Readings for how the number 78 was determined.)

Success is only one such configuration; Prisoners' Dilemma and Chicken are two others. The rules of play we describe next apply to all 78 2×2 games. These rules can be extended to larger games, as we will illustrate later.

RULES OF TOM

The founders of game theory, John von Neumann and Oskar Morgenstern, defined a **game** to be "the totality of rules of play which describe it." The rules of TOM apply to all ordinal games between two players, each of whom has two strategies. The first four **rules of play** of TOM are as follows:

1. Play starts at an **initial state,** given at the intersection of the row and column of a payoff matrix (i.e., one of the four entries in a 2×2 payoff matrix).
2. Either player can unilaterally switch his or her strategy (i.e., make a **move**), thereby changing the initial state into a new state, in the same row or column as the initial state. The player who switches is called player 1 (*P1*).
3. Player 2 (*P2*) can respond by unilaterally switching his or her strategy, thereby moving the game to a new state.
4. The alternating responses continue until the player (*P1* or *P2*) whose turn it is to move next chooses not to switch his or her strategy. When this happens, the game terminates in a **final state,** which is the **outcome** of the game.

Note that the sequence of moves and countermoves is *strictly alternating*. First, say, *R* moves, then *C* moves, and so on, until one player stops, at which point the state reached is final and therefore the outcome of the game. We assume that no payoffs accrue to players from being in a state unless it becomes the outcome (which could be the initial state if the players choose not to move from it).

To assume otherwise would require that payoffs be numerical values, rather than ordinal ranks, which players can accumulate as they pass through states. But in most real-life games, payoffs cannot easily be quantified and summed across the states visited; moreover, the big reward in many games depends overwhelmingly on the final state reached, not on how it was reached. In politics, for example, the payoff for most politicians is not in campaigning, which is arduous and costly, but in winning.

Rule 1 differs radically from the corresponding rule of play in standard game theory, in which players simultaneously choose strategies in a matrix game that determines the outcome. Instead of starting with strategy choices, TOM assumes that players are already *in* some state at the start of play and receive payoffs from this state *only if they stay*. Based on these payoffs, they decide, individually, whether or not to change this state in order to try to do better.

To be sure, some decisions are made collectively by players, in which case it would be reasonable to say that they choose strategies from scratch, either simultaneously or by coordinating their choices. But if, say, two countries are coordinating their choices, as when they agree to sign a treaty, the important strategic question is what individualistic calculations led them to this point. The formality of jointly signing the treaty is simply the culmination of their negotiations and does not reveal the move-countermove process that preceded the signing. It is precisely these negotiations, and the calculations underlying them, that TOM is designed to uncover.

To continue this example, the parties who sign the treaty were in some prior state, from which both desired to move—or, perhaps, only one desired to move and the other could not prevent this move from happening. Eventually they may arrive at a new state (e.g., after treaty negotiations) in which it is rational for both countries to sign the treaty that has been negotiated.

Put another way, almost all outcomes of games that we observe have a history. TOM seeks to explain strategically the progression of (temporary) states that lead to a (more permanent) outcome. Consequently, play of a game starts in an initial state, at which players accrue payoffs only if they remain in that state so that it becomes the final state, or outcome, of the game.

If they do not remain in the initial state, they still know what payoffs they would have accrued had they stayed; hence, they can make a rational calculation of the advantages of staying or moving. They move precisely because they calculate that they can do better by switching states, anticipating a better outcome when the move-countermove process finally comes to rest. The game is different, but not the payoff matrix, when play starts in a different state.

Rules 1–4 say nothing about what *causes* a game to end, but only when: termination occurs when a "player whose turn it is to move next chooses not to switch his or her strategy" (rule 4). But when is it rational not to continue moving, or not to move at all from the initial state?

To answer this question partially, TOM postulates a **termination rule:**

5. If play returns to the initial state, the initial state becomes the outcome.

We will illustrate shortly how rational players, starting from some initial state, can predict what the outcome will be. The initial state will *not* be the outcome if the players find it rational to terminate play before returning to it. On the other hand, if after *P1* moves, it is rational for play of the game to cycle back to the initial state, then rule 5 says that there will be no further movement. After all, what is the point of continuing the move-countermove process if play will, once again, return to "square one," given that the players receive no payoffs along the way (i.e., before the outcome is reached)?

At this point, we make rule 5 only provisional. An alternative rule (5') that allows for cycling will be considered later (along with "moving power" as a way to break cycles).

A final rule of TOM is needed to ensure that *both* players take into account each other's calculations before deciding to move from the initial state. We call this rule the **two-sidedness rule:**

6. Each player takes into account the consequences of the other player's **rational choices,** as well as his or her own, in deciding whether or not to move from the initial state or any subsequent state. If it is rational for one player to move and the other player not to move from the initial state, then the player who moves takes **precedence:** his or her move overrides the player who stays, so the outcome will be induced by the player who moves.

Later we will show that if both players want to move first, or both want the other player to move first, this conflict can be resolved by "order power."

EXAMPLE ▶ *Applying TOM to Success*

If players have complete information, they can look ahead and anticipate the consequences of their moves and thereby decide whether or not to move from the initial state or any subsequent states reached. We next show how they can use backward induction to make this decision.

Backward induction is a reasoning process in which players, working backward from the last possible move in a game, anticipate each other's rational choices.

To illustrate backward induction, consider again Success in Figure 16.2 (the payoffs in brackets, just below the payoffs in parentheses, will be defined shortly). We show next the progression of moves, starting from each of the four possible initial states of Success and cycling back to this state, and indicate where rational players will terminate play:

1. Initial state (2, 4). If R moves first, the counterclockwise progression of moves from (2, 4) back to (2, 4) — with the player (R or C) who makes the next move shown below each state in the alternating sequence — is as follows (see Figure 16.2):

	State 1 R		State 2 C		State 3 R		State 4 C		State 1
R starts:	(2, 4)	→	(1, 1)	→	(3, 3)	→|	(4, 2)	→	(2, 4)
Survivor:	(3, 3)		(3, 3)		(3, 3)		(2, 4)		

Below the progression of states we indicate the survivor at each state.

> The **survivor** is the payoff selected at each state as the result of backward induction. It is determined by working backward, after a cycle has been completed and play returns to the initial state (state 1).

Assume the players' alternating moves have taken them counterclockwise in Success from (2, 4) to (1, 1) to (3, 3) to (4, 2), at which point C must decide whether to stop at (4, 2) or complete the cycle and return to (2, 4). Clearly, C prefers (2, 4) to (4, 2), so (2, 4) is listed as the survivor below (4, 2): because C *would* move the process back to (2, 4) should she reach (4, 2), the players know that if the move-countermove process reaches this state, the outcome will be (2, 4).

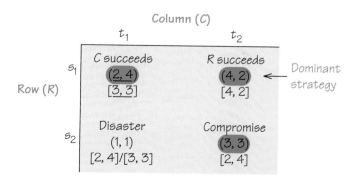

Key

(x, y) = (payoff to R, payoff to C).

[x, y] = [payoff to R, payoff to C] in anticipation game.

4 = best, 3 = next best, 2 = next worst, 1 = worst.

Nash equilibrium in original game and anticipation game underscored.

Nonmyopic equilibria (NMEs) highlighted in blue.

FIGURE 16.2   The game of Success according to the theory of moves.

Knowing this, would R at the prior state, $(3,3)$, move to $(4,2)$? Because R prefers $(3,3)$ to the survivor at $(4,2)$—namely, $(2,4)$—the answer is no. Hence, $(3,3)$ becomes the survivor when R must choose between stopping at $(3,3)$ and moving to $(4,2)$—which, as we just showed, would become $(2,4)$ once $(4,2)$ is reached.

At the prior state, $(1,1)$, C would prefer moving to $(3,3)$ than stopping at $(1,1)$, so $(3,3)$ again is the survivor if the process reaches $(1,1)$. Similarly, at the initial state, $(2,4)$, because R prefers the previous survivor, $(3,3)$, to $(2,4)$, $(3,3)$ is the survivor at this state as well.

The fact that $(3,3)$ is the survivor at initial state $(2,4)$ means that it is rational for R initially to move to $(1,1)$, and C subsequently to move to $(3,3)$, where the process will stop, making $(3,3)$ the rational choice if R has the opportunity to move first from initial state $(2,4)$. That is, after working *backward* from C's choice of completing the cycle or not from $(4,2)$, the players can reverse the process and, looking *forward*, determine that it is rational for R to move from $(2,4)$ to $(1,1)$, and C to move from $(1,1)$ to $(3,3)$, at which point R will stop the move-countermove process at $(3,3)$.

Notice that R does better at $(3,3)$ than at $(2,4)$, where he could have terminated play at the outset, and C does better at $(3,3)$ than at $(1,1)$, where she could have terminated play, given that R is the first to move. We indicate that $(3,3)$ is the consequence of backward induction by underscoring this state in the progression; it is the state at which **stoppage** of the process occurs. In addition, we indicate that it is not rational for R to move on from $(3,3)$ by the vertical line blocking the arrow emanating from $(3,3)$, which we refer to as **blockage:** a player will always stop at a blocked state, wherever it is in the progression. Stoppage occurs when blockage occurs for the *first* time from some initial state, as we illustrate next.

If C can move first from $(2,4)$, backward induction shows that $(2,4)$ is the last survivor, so $(2,4)$ is underscored when C starts. Consequently, C would *not* move from the initial state, where there is blockage (and stoppage), which is hardly surprising since C receives her best payoff in this state:

	C		R		C		R	
C starts:	$\underline{(2,4)}$	$\rightarrow\!\mid$	$(4,2)$	$\rightarrow\!\mid$	$(3,3)$	$\rightarrow$	$(1,1)$	$\rightarrow$	$(2,4)$
Survivor:	$(2,4)$		$(4,2)$		$(2,4)$		$(2,4)$	

As when R has the first move, $(2,4)$ is the first survivor, working backward from the end of the progression, and is also preferred by C at $(3,3)$. But then, because R at $(4,2)$ prefers this state to $(2,4)$, $(2,4)$ is temporarily displaced as the survivor. It returns as the last survivor, however, because C at $(2,4)$

prefers it to $(4, 2)$. Nonetheless, there are circumstances in which a player does better by departing from a best state, paradoxical as this may seem (see Exercises 7–9).

Thus, the first blockage and, therefore, stoppage occur at $(2, 4)$, but blockage occurs subsequently at $(4, 2)$ if, for any reason, stoppage does not terminate moves at the start. In other words, if C moved initially, R would then be blocked. Hence, blockage occurs at two states when C starts the move-countermove process, whereas it occurs only once when R has the first move.

The fact that the rational choice depends on which player has the first move—$(3, 3)$ is rational if R starts, $(2, 4)$ if C starts—leads to a conflict over what outcome will be selected when the process starts at $(2, 4)$. However, because it is not rational for C to move from the initial state, R's move takes precedence, according to rule 6, and overrides C's decision to stay. Consequently, when the initial state is $(2, 4)$, the outcome will be $(3, 3)$.

2. Initial state (4, 2). The progressions, survivors, stoppages, other blockages, and outcome from this state are as follows:

	R	C	R	C	
R starts:	$\underline{(4, 2)}$ →|	$(3, 3)$ →	$(1, 1)$ →	$(2, 4)$ →|	$(4, 2)$
Survivor:	$\overline{(4, 2)}$	$(2, 4)$	$(2, 4)$	$(2, 4)$	

	C	R	C	R	
C starts:	$\underline{(4, 2)}$ →|c	$(2, 4)$ →	$(1, 1)$ →	$(3, 3)$ →	$(4, 2)$
Survivor:	$\overline{(4, 2)}$	$(4, 2)$	$(4, 2)$	$(3, 3)$	
Outcome:	$(4, 2)$				

Clearly, when $(4, 2)$ is the initial state, there is no conflict between R and C about staying there. Yet, while neither player has an incentive to move from $(4, 2)$, each player's reasons for stoppage are different. If R starts, there is blockage at the start, whereas if C starts, there will be cycling back to $(4, 2)$.

Because cycling is no better for C than not moving, we assume that C will stay at $(4, 2)$, which we indicate by c (for "cycling") following the arrow at $(4, 2)$. This might be interpreted as a special kind of blockage: while rule 5 allows play to return to the initial state after one complete cycle, whence it terminates, there is no benefit to C from doing so. Consequently, we assume that C will not move initially from $(4, 2)$, simply to cycle once. Because R also will not move, there will be a consensus on the part of both players of staying at $(4, 2)$.

3. Initial state (3, 3). The progressions, survivors, stoppages, other blockages, and outcome from this state are as follows:

	R	C	R	C	
R starts:	$\underline{(3,3)}$ $\rightarrow\mid c$	$(4,2)$ $\rightarrow$	$(2,4)$ $\rightarrow$	$(1,1)$ $\rightarrow$	$(3,3)$
Survivor:	$\overline{(3,3)}$	$(3,3)$	$(3,3)$	$(3,3)$	

	C	R	C	R	
C starts:	$(3,3)$ $\rightarrow$	$(1,1)$ $\rightarrow$	$\underline{(2,4)}$ $\rightarrow\mid$	$(4,2)$ $\rightarrow\mid$	$(3,3)$
Survivor:	$(2,4)$	$(2,4)$	$\overline{(2,4)}$	$(4,2)$	

Outcome:	$(2,4)$

As from initial state $(2,4)$, there is a conflict. If R starts, $(3,3)$ is the rational choice, but if C starts, $(2,4)$ is. But because C's move takes precedence over R's staying, the outcome is that which C can induce—namely, $(2,4)$.

4. Initial state (1, 1). The progressions, survivors, stoppages, other blockages, and outcome from this state are as follows:

	R	C	R	C	
R starts:	$(1,1)$ $\rightarrow$	$\underline{(2,4)}$ $\rightarrow\mid$	$(4,2)$ $\rightarrow\mid$	$(3,3)$ $\rightarrow\mid$	$(1,1)$
Survivor:	$(2,4)$	$\overline{(2,4)}$	$(4,2)$	$(3,3)$	

	C	R	C	R	
C starts:	$(1,1)$ $\rightarrow$	$\underline{(3,3)}$ $\rightarrow\mid$	$(4,2)$ $\rightarrow$	$(2,4)$ $\rightarrow$	$(1,1)$
Survivor:	$(3,3)$	$\overline{(3,3)}$	$(2,4)$	$(2,4)$	

Outcome:	Indeterminate—$(2,4)/(3,3)$, depending on whether R or C starts.

Unlike the conflicts from initial states $(2,4)$ and $(3,3)$, it is rational for *both* players to move from initial state $(1,1)$. But, as we showed earlier, each player would prefer that the other player be *PI*, because

- R's initial move induces $(2,4)$, C's preferred state; and
- C's initial move induces $(3,3)$, R's preferred state.

Presumably, each player will try to hold out longer at $(1,1)$, hoping that the other will move first. Because neither player's move takes precedence according to the rules of play, neither rational choice can be singled out as *the* outcome. Hence, when play starts at $(1,1)$, we classify the state as **indeterminate**—either $(2,4)$ or $(3,3)$ can occur, depending on which player *P1* is; we write this state as $(2,4)/(3,3)$. Because the choice of first mover is not specified by the rules of play, indeterminacy is a consequence of TOM.

Typically, this kind of indeterminacy is characterized by bargaining, wherein each player tries to hold off being the first to make concessions. Although both players would benefit at either $(2,4)$ or $(3,3)$ over $(1,1)$, there is greater benefit to each in having the other player move first.

> A player has **order power** if he or she can determine the order of moves from an indeterminate state.

Thus, if R has order power, he can force C to depart first from $(1,1)$ and induce his preferred state of $(3,3)$; if C has order power, she can induce $(2,4)$ as the state. Note, however, that the state that R most prefers, $(4,2)$, is unattainable from $(1,1)$—it can occur only if the process starts at $(4,2)$.

To summarize, each of the initial states goes into the following final determinate states, or outcomes—except when there is a conflict, as there is from $(1,1)$, and neither player's move takes precedence, according to rule 6, and neither player wants to move first from $(1,1)$:

$$(2,4) \rightarrow (3,3); \quad (4,2) \rightarrow (4,2); \quad (3,3) \rightarrow (2,4); \quad (1,1) \rightarrow (2,4)/(3,3). \quad \blacklozenge$$

> The outcomes into which each state goes are **nonmyopic equilibria (NMEs).** They are the consequence of both players' looking ahead and anticipating where, from each of the initial states, the move-countermove process will end up.

We hasten to add that once reached, a player may have an incentive to leave an NME. Thus, NMEs might be better thought of as "reachable outcomes" from a state, rather than equilibria, because they may not be stable once reached. Only $(4,2)$ is stable in the sense that, if it is the initial state, neither player would depart from it. Although this is an argument for considering $(4,2)$ to be "more stable" than the other NMEs in Success, we shall not distinguish among the NMEs in a game. Perhaps the best way to think about NMEs is as states (i) at which players would stay or (ii) to which they would migrate, but if the latter, not necessarily stay there.

In Success the players can end up at every state except $(1,1)$, which therefore is not an NME. Note that from each initial state except $(1,1)$ the NMEs are unique. From $(1, 1)$ the NMEs are $(2, 4)$ or $(3, 3)$, depending on whether C or R has order power and can thereby dictate who moves first from this initial state (recall that each player would like to go second).

The move-countermove process can be interpreted as a bargaining process in which, starting at the initial state, a player can choose not to move (i.e., to accept an offer) or to move (i.e., reject an offer). If a player chooses to move, the other player can then terminate the game by accepting the offer, or continue it by moving to an adjacent offer, which may in turn be accepted or rejected, and so on.

If this process did not stabilize, the initial offer (i.e., first move) would not be worth making. But every 2×2 game contains at least one NME, because from each initial state there is an outcome (perhaps indeterminate) of the move-countermove process. If this outcome is both determinate and the same from every initial state, then it is the only NME; otherwise, there is more than one NME (Success has three).

In the offer-counteroffer interpretation, then, there is at least one offer that will always be accepted, so the process always stabilizes. But this may occur at more than one outcome if there is more than one NME, as in Success. In this situation, there might be a kind of positioning game played over the choice of an initial state, which we can analyze in terms of an anticipation game.

> An **anticipation game** is simply the game resulting from the substitution of the NMEs into which each of these states goes (in brackets in Figure 16.2) for the original payoffs at each of the four states (in parentheses in Figure 16.2).

When we apply standard game theory to the anticipation game, s_1 is a dominant strategy for R: if C chooses t_1, $[3, 3]$ is at least as good for R as $[2, 4]/[3, 3]$; if C chooses t_2, $[4, 2]$ is better for R than $[2, 4]$. By contrast, C's strategies are undominated, but anticipating that R will choose s_1, C's best response is t_1, yielding $[3, 3]$. Thus, if the players believe that they are choosing only initial states rather than outcomes when they choose their strategies, their choices of s_1 and t_1 will start them out at $(2, 4)$, whence a move by R and a countermove by C will, according to TOM, bring them to the NME of $(3, 3)$.

Players in most real-life games, whether they do or do not anticipate making moves from some initial state, rarely choose strategies simultaneously, as we argued earlier. Rather, they find themselves in some state, or status quo point, from which they consider moving. In Success, as we have seen, moves and countermoves can lead the players to three different NMEs, which is the maximum number that can occur in a 2×2 strict ordinal game; the minimum, as already noted, is one. Most 2×2 games have either one or two NMEs; in fact, Chicken and Success are the *only* 2×2 games (of the 78) that have three NMEs.

In these games, in particular, *where* play starts matters, which the unique $(2,4)$ Nash equilibrium in Success masks. To show how TOM gives insight into player choices in such games, consider the story of Samson and Delilah from the Book of Judges in the Old Testament.

INTERPRETING TOM

In using the game of Success to model the conflict between Samson and Delilah, we seek to show that Samson's behavior was *not* irrational, despite his ample later troubles. On the contrary, we will argue that his moves and Delilah's were entirely rational in Success.

To provide evidence that Samson and Delilah had the preferences of the players in Success, we quote extensively from the Bible. We suggest in the end, however, that it is possible that Delilah's preferences were different from those of the row player in Success—but this does not make a difference for rational play of the game.

Samson and Delilah, played by Victor Mature and Hedy Lamarr, 1949.

EXAMPLE ▶ *Samson and Delilah*

The background to the story is as follows: After aiding the flight of the Israelites from Egypt and delivering them into the promised land of Canaan, God became extremely upset by their recalcitrant ways and punished them severely:

> The Israelites again did what was offensive to the LORD, and the LORD delivered them into the hands of the Philistines for forty years. (Judg. 13:1)

But a new dawn appears at the birth of Samson, which is attended to by God and whose angel predicts: "He shall be the first to deliver Israel from the Philistines" (Judg. 13:5).

Samson developed a reputation as a ferocious warrior of inhuman strength. This served him well as judge (leader) of Israel for 20 years, but then he fell in love with a Philistine woman named Delilah. Apparently, Samson's love for Delilah was not reciprocated. Rather, Delilah was more receptive to serving as bait for Samson for a suitable payment. The lords of the Philistines made her a proposition:

> Coax him and find out what makes him so strong, and how we can overpower him, tie him up, and make him helpless; and we'll each give you eleven hundred shekels of silver. (Judg. 16:5)

After agreeing, Delilah asked Samson: "Tell me, what makes you so strong? And how could you be tied up and be made helpless?" (Judg. 16:6). Samson replied: "If I were to be tied with seven fresh tendons that had not been dried, I should become as weak as an ordinary man" (Judg. 16:7).

After Delilah bound Samson as he had instructed her, she hid men in the inner room and cried, "Samson, the Philistines are upon you!" (Judg. 16:9). Samson's lie quickly became apparent:

> Whereat he pulled the tendons apart, as a strand of tow [flax] comes apart at the touch of fire. So the secret of his strength remained unknown.
>
> Then Delilah said to Samson, "Oh, you deceived me; you lied to me! Do tell me how you could be tied up." (Judg. 16:9–10)

Twice more Samson lied to Delilah about the source of his strength, and she became progressively more frustrated by his deception. In exasperation, Delilah exclaimed:

> "How can you say you love me, when you don't confide in me? This makes three times that you've deceived me and haven't told me what makes you so strong." Finally, after she had nagged him and pressed him constantly, he was wearied to death and he confided everything to her. (Judg. 16:15–17)

The secret, of course, was Samson's long hair. When he told his secret to Delilah, she had his hair shaved off while he slept. The jig was then up when he was awakened:

> For he did not know that the LORD had departed him. The Philistines seized him and gouged out his eyes. They brought him down to Gaza and shackled him in bronze fetters, and he became a mill slave in the prison. After his hair was cut off, it began to grow back. (Judg. 16:20–22)

Thus is a slow time bomb set ticking. The climax approaches when Samson is summoned by the Philistines to a great celebration:

> They put him between the pillars. And Samson said to the boy who was leading him by the hand, "Let go of me and let me feel the pillars that the temple rests upon, that I may lean on them." Now the temple was full of men and women; all the lords of the Philistines were there, and there were some three thousand men and women on the roof watching Samson dance. Then Samson called to the LORD, "O Lord GOD! Please remember me, and give me strength just this once, O God, to take revenge of the Philistines if only for one of my two eyes." (Judg. 16:25–28)

Samson, his strength now restored, avenged himself on his captors in an unprecedented biblical reprisal (by a human being, not God) that sealed both his doom and the Philistines':

> He embraced the two middle pillars that the temple rested upon, one with his right arm and one with his left, and leaned against them; Samson cried, "Let me die with the Philistines!" and he pulled with all his might. The temple came crashing down on the lords and on all the people in it. Those who were slain by him as he died outnumbered those who had been slain by him when he lived. (Judg. 1:29–30)

There is irony, of course, in this reversal of roles, whereby the victim becomes the victor—and a victim as well. We do not suggest, however, that Samson planned for his own mutilation and ridicule only to provide himself with the later opportunity to retaliate massively against the Philistines. Perhaps this was in God's design, as foretold by the angel at Samson's birth, and in God's "seeking a pretext against the Philistines" (Judg. 14:4).

Although Samson's betrayal of the secret of his strength may seem stupid, it was entirely consistent with his previous behavior and apparent preferences. To put it bluntly, Samson was a man of carnal desires: he had lusted after several women before meeting Delilah, and Delilah was not the first to whose charms he fell prey. He would not withhold information if the right woman was around to wheedle it out of him. While Samson could fight the Philistines like a fiend, he could readily be disarmed by women he desired.

FIGURE 16.3 Samson and Delilah, as modeled by Success and Variation. The payoffs of Success are to the left of the slashes and the payoffs of Variation are to the right.

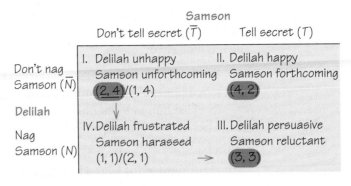

Key
(x, y) = (payoff to Delilah, payoff to Samson).
4 = best, 3 = next best, 2 = next worst, 1 = worst.
Nash equilibrium underscored.
Nonmyopic equilibria (NMEs) highlighted in blue.
Arrows indicate progression of states from (2, 4)/(1, 4) to NME of (3, 3).

The payoff matrix of the game we posit that Samson played with Delilah is shown in Figure 16.3. (Ignore for now the payoffs to the right of the slashes in the first column of Figure 16.3.) Samson's desire having been kindled, Delilah could trade on it either by nagging Samson for the secret of his strength (N) or not nagging him ($\overline{N}$) and hoping it would come out anyway. Samson, in turn, could either tell (T) the secret of his strength or not tell it ($\overline{T}$). Consider the consequences of each pair of strategy choices, starting from the upper left-hand state and moving in a clockwise direction:

I.   *Delilah unhappy, Samson unforthcoming: (2, 4).* The next-worst state for Delilah, because Samson withholds his secret, though she is not frustrated in an unsuccessful attempt to obtain it; the best state for Samson, because he keeps his secret and is not harassed.

II. *Delilah happy, Samson forthcoming: (4, 2).* The best state for Delilah, because she learns Samson's secret without making a pest of herself; the next-worst state for Samson, because he gives away his secret without being under duress.

III. *Delilah persuasive, Samson reluctant: (3, 3).* The next-best state for both players, because though Delilah would prefer not to nag (if Samson tells) and Samson would prefer not to succumb (if Delilah does not nag), Delilah gets her way when Samson tells; and Samson, under duress, has a respectable reason (i.e., Delilah's nagging) for telling.

IV. *Delilah frustrated, Samson harassed: (1, 1).* The worst state for both players, because Samson does not get peace of mind, and Delilah is frustrated in her effort to learn Samson's secret.

The Figure 16.3 game just described is Success and starts in state I at (2, 4), when Delilah chooses $\overline{N}$ and Samson chooses $\overline{T}$ during their period of getting acquainted. These strategy choices are consistent with Delilah's choosing her dominant strategy, and Samson his best response to this strategy, giving the Nash equilibrium of (2, 4), according to standard game theory.

The standard theory offers no explanation of why the players would ever move to a nonequilibrium outcome. But this is precisely what they do. Delilah switches to N, putting the players in state IV at (1, 1), and Samson responds with T, leading to state III at (3, 3), neither of which is a Nash equilibrium. By contrast, TOM leads to a different prediction when the initial state is I at (2, 4). From this state, TOM predicts the outcome to be (3, 3). Although (2, 4) and (4, 2) are also NMEs, they do not arise unless play starts elsewhere.

Thus, TOM leads to a unique prediction of (3, 3) if play starts in state I, which is the actual outcome of the story. (Arguably, the outcome is state II, after Delilah has stopped nagging, which is rational because Samson cannot respond by hiding his secret once it is already out; we discuss such infeasible moves later.) Doubtless, Samson did not anticipate having his eyes gouged out and being derided as a fool before the Philistines when he responded to Delilah's nagging by revealing his secret. On the other hand, because he was later able to kill thousands of Philistines at the same time that he ended his own humiliation, the resolution of this story can plausibly be seen as next best for Samson.

Although he surrendered his secret to the treacherous Delilah, Samson apparently never won her love, which seems to be the thing he most wanted. In fact, Delilah's decisive argument in coaxing the truth from Samson was that because he did not confide in her, he did not love her. What better way was there for Samson to counter this contention, and prove his love, than to comply with her request, even if it meant courting not just Delilah but disaster itself?

As for Delilah's preferences, it is hard to quarrel with the assumption that her two best states were associated with Samson's choice of T. What is less certain, however, is the order of preferences she held for her two worst states. Contrary to the Success representation in Figure 16.3, Delilah might have preferred to nag Samson than not had he in the end denied her his secret. For even though she would have failed to discover Samson's secret, Delilah would perhaps have felt less badly after having tried than if she had made no effort at all.

If this is the case, then 2 and 1 for Delilah would be interchanged in Success, giving a different game matrix. But, like Success, it is possible to show that this game, which we call Variation and whose payoffs are to the right of the slashes in the first column of Figure 16.3, yields $(3,3)$ as the NME when play starts in state I at $(1,4)$. Hence, this reordering of Delilah's preferences would not affect the rational outcome, according to TOM: thinking nonmyopically, Delilah would still switch to N, and Samson in turn would switch to T. Thus, even if Samson had only incomplete information about how Delilah ordered her two worst states, his play would not be affected. ◆

To summarize, where players start in a game, including the unique dominant-strategy Nash equilibria, may not be where they end up, according to TOM. However, the restriction that rules 5 and 6 place on the ability of players to cycle in games may not always be descriptive of a situation. Therefore, we next consider alternatives to these rationality rules (namely, rules 5′ and 6′) that permit cycling around a matrix and, in addition, allow players to terminate cycles through the exercise of "moving power."

CYCLIC GAMES AND MOVING POWER

Recall that the rules of play of TOM say nothing about what causes a game to end, only when. Rule 5, which forbids continual cycling, provides one answer. But this ban on cycling may not be realistic, as many protracted conflicts (e.g., the Arab-Israeli conflict), in which the protagonists have revisited the past again and again, make unmistakably clear.

To capture the cyclic aspect of certain conflicts and give players the ability to make choices in which they repeat themselves (why they may want to do so will be considered shortly), we define a subclass of the 78 games in which cycling is possible by defining a subclass of games in which it is *not*. Rule 5′ provides a sufficient condition for cycling not to occur:

> 5′. If at any state in the move-countermove process a player whose turn it is to move next receives his or her best payoff (i.e., 4), that player will not move from this state.

Rule 5′, in fact, prevents cycling in 42 of the 78 distinct 2×2 games. We call the remaining 36 games cyclic.

> A **cyclic game** is one in which, when the game cycles either clockwise or counterclockwise, neither player ever receives a best payoff (4) when it is his or her turn to move next.

Consider the circumstances under which players, who know not only their own payoffs but also the payoffs of their opponents, would have an incentive to cycle to try to outlast an opponent. By "outlasting" we mean that one (stronger) player can force the other (weaker) player to stop the move-countermove process at a state where the weaker player has the next move. Forcing stoppage at such a state involves the exercise of moving power.

> If one player (*P1*) has **moving power,** he or she can force the other player (*P2*) to stop, in the process of cycling, at one of the two states at which *P2* has the next move.

The state at which *P2* will stop is that which *P2* prefers. Because of the change of rules that now allow for cycling, moving power is very different from order power. For example, it may not always benefit the player who has it compared with the player who does not have it.

Rule 5′ specified what players would *not* do, namely, move from a best (4) state when it was their turn to move. However, this rule did not say anything about *where* cycling would stop, which the exercise of moving power determines by enabling the player who possesses it to break the cycle of moves.

Rule 6′ ensures there will be termination.

> 6′. At some point in the cycling, *P2* must stop.

This is not to say that *P1* will always exercise his or her moving power. In some games, as we shall see, it is rational for *P1* to terminate play, even though *P1* can always force *P2* to stop first.

(E X A M P L E) ▶ *Moving Power in Success and Alternative*

Moving power is **effective** in a cyclic game if the outcome that a player can implement with this power is better for him or her than the outcome that the other player can implement. To illustrate when moving power is effective, consider again the game of Success. The arrows in Figure 16.4a (ignore for now the distinction between the single and double arrows) illustrate the cyclicity of Success in a counterclockwise direction: starting at the upper right state, C benefits by moving from $(4, 2)$ to $(2, 4)$; R does not benefit by moving from $(2, 4)$ to $(1, 1)$ but departs from a 2, not a 4, state (so does not violate rule 5′); C benefits by moving from $(1, 1)$ to $(3, 3)$; and R benefits by moving from $(3, 3)$ to $(4, 2)$. Because no player, when it is his or her turn to move, ever departs from his or her best (i.e., 4) state, Success is cyclic.

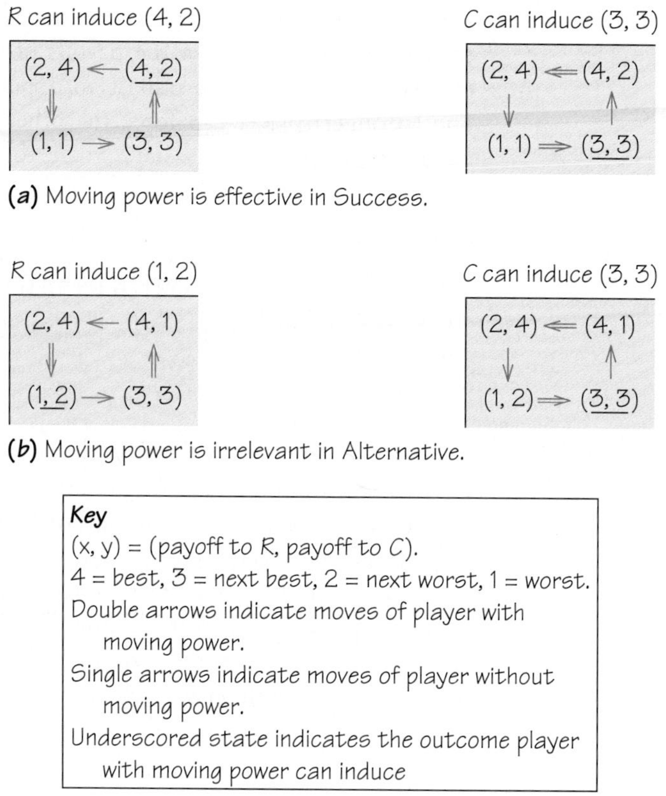

FIGURE 16.4   Moving power in two cyclic games.

R can induce (4, 2)

$$(2, 4) \leftarrow (\underline{4, 2})$$
$$\Downarrow \qquad \Uparrow$$
$$(1, 1) \rightarrow (3, 3)$$

C can induce (3, 3)

$$(2, 4) \Leftarrow (4, 2)$$
$$\downarrow \qquad \uparrow$$
$$(1, 1) \Rightarrow (\underline{3, 3})$$

(a) Moving power is effective in Success.

R can induce (1, 2)

$$(2, 4) \leftarrow (4, 1)$$
$$\Downarrow \qquad \Uparrow$$
$$(\underline{1, 2}) \rightarrow (3, 3)$$

C can induce (3, 3)

$$(2, 4) \Leftarrow (4, 1)$$
$$\downarrow \qquad \uparrow$$
$$(1, 2) \Rightarrow (\underline{3, 3})$$

(b) Moving power is irrelevant in Alternative.

Key
(x, y) = (payoff to R, payoff to C).
4 = best, 3 = next best, 2 = next worst, 1 = worst.
Double arrows indicate moves of player with
    moving power.
Single arrows indicate moves of player without
    moving power.
Underscored state indicates the outcome player
    with moving power can induce

To show what outcome R can implement if he has moving power, which might be thought of as greater stamina or endurance—in the sense that he can continue moving when the other must eventually stop—let his moves (vertical, as illustrated on the left side of Figure 16.4a) be represented by double arrows. C, whose (horizontal) moves are represented by single arrows, must stop in the cycling at either $(1, 1)$ or $(4, 2)$, from where her single arrows emanate that indicate she has the next move. Since she would prefer to stop at $(4, 2)$ rather than $(1, 1)$, R can implement his best outcome of $(4, 2)$ if he has moving power. On the other hand, if C has moving power (right side of Figure 16.4a), she can force R to stop at either $(2, 4)$ or $(3, 3)$, whence his single arrows emanate that indicate he has the next move. Since R would prefer to stop at $(3, 3)$ rather than $(2, 4)$, C can implement her next-best outcome of $(3, 3)$ if she has moving power. Thus, the possession of moving power benefits the player who possesses it—compared with the other player's possession of it—so it is effective in Success.

This is not the case in a game in which 1 and 2 are interchanged for C in Success, which we will refer to as the game Alternative. Applying the same reasoning to Alternative, shown in Figure 16.4b, we see that R can implement only $(1,2)$ — C prefers this to $(4,1)$, the other state where she moves from — but C can implement $(3,3)$ — R prefers this to $(2,4)$, the other state he moves from. Since R also prefers $(3,3)$ to $(1,2)$, moving power is not effective: R cannot implement a better outcome when he has it than when C has it. Instead, moving power is irrelevant in Alternative, because it would be in R's interest to stop at $(3,3)$, even if he has moving power, rather than to force C to stop at $(1,2)$.

More generally, moving power is **irrelevant** when the outcome one player can implement is better for both. Unlike Variation (see Figure 16.3), in which a switch of 1 and 2 by one player (R, or Delilah) does not change the outcome of Success when play commences at the upper left-hand state — $(3,3)$ remains the NME from this state — a switch of 1 and 2 by C, yielding Alternative, does change the effects of moving power in Success.

Moving power obviously constrains the freedom of choice of the player who does not possess it, because the player who possesses it can force the other player to stop. As we have seen in Alternative, R has no reason to exercise this kind of power, preferring to stop himself at $(3,3)$ in the move-countermove process, which renders his possession of moving power in this game irrelevant. On the other hand, the possession of moving power in Success by either R or C helps him or her obtain a better outcome than if the other player possessed it, rendering moving power effective in this game. ◆

In many real-world conflicts, there may be no clear recognition of which, if either, player has moving power. In fact, there may be a good deal of misinformation. For example, if both players believe they can hold out longer, cycling is likely to persist until one player succeeds in demonstrating greater strength or both players are exhausted by the repeated cycling. The latter may well have occurred in the Egyptian-Israeli conflict between 1948 and 1979, as well as similar recent conflicts (e.g., in South Africa and Northern Ireland). Although Egypt and Israel fought five wars in this period (1948, 1956, 1967, 1969–70, and 1973), at great cost to both sides, it still required considerable pressure from the United States to achieve the 1978 Camp David accords that paved the way for the signing of a peace treaty between Egypt and Israel in 1979. Likewise, similar outside pressure was exerted in the South African and Northern Ireland conflicts to induce settlements, as it was in the former Yugoslavia to induce the warring sides to sign a peace treaty in November 1995 after four years of bitter conflict that cost some 250,000 lives.

We offer a note of caution in interpreting cycling and the exercise of moving power. While Egypt and Israel cycled in and out of war for a generation, cycling in games like that of Samson and Delilah may, though theoretically possible, not be feasible. In particular, although Delilah was readily able to switch from being a seductress to being a nag, and Samson from being mum about the secret of his strength to revealing it, once Samson's secret was out, he

Israel's late Prime Minister Yitzhak Rabin (*left*) with PLO Chairman Yasir Arafat. Egyptian President Hosni Mubarak (*center*) hosted the historic PLO-Israeli peace accord in May 1994.

could not retract it, especially in light of the fact that Delilah tested every explanation, true or false, that he gave for his strength.

Thus, cycling in the Samson and Delilah game is not **feasible:** it contradicts what a reasonable interpretation of the strategies in this game permit. Whereas it is permissible for Samson to move from $\overline{T}$ to T (see Figure 16.3), a reverse switch is hard to entertain in the context of the story, though not necessarily in other interpretations of Success. For example, consider the original interpretation of this game given in Figure 16.1, in which $(3,3)$ is "Compromise." Starting from this state, one can readily imagine situations in which C switches from t_2 to t_1, plunging the players into "Disaster," hoping that this move will drive R to switch from s_2 to s_1, yielding "C succeeds," with payoffs of $(2,4)$ to the players.

In using TOM to model a strategic situation, the analyst must be sensitive to what strategy changes are feasible and infeasible. Also, even if moves are rational according to the rules, the state to which a game moves may, if feasible, be one from which a player cannot emerge to move on. As a case in point, assume C, starting at $(3,3)$ in the Figure 16.1 game, switches from t_2 to t_1, changing the state to $(1,1)$. Given that this move is feasible (which we assumed was not the case in the Samson and Delilah story), can R subsequently move on to $(2,4)$?

We suggest that the answer to this question depends on what interpretation one gives to "Disaster" at $(1,1)$. If this state means, say, nuclear war, nobody may survive in order to be able to "move on," making $(1,1)$, figuratively speaking, a black hole. On the other hand, if C's move to $(1, 1)$ creates a severe crisis, like the Cuban missile crisis of 1962, R may be able to respond, as the Soviet Union did, by withdrawing its nuclear weapons from Cuba, which abated this crisis.

To conclude, we have shown that, depending on the rules, the starting point, or who possesses order or moving power, can matter. Nevertheless, not all rational moves may be feasible, either because changing strategies or moving from certain states violates the interpretation of a game.

RETURN TO PRISONERS' DILEMMA AND CHICKEN

In Figure 16.5 we show Prisoners' Dilemma and Chicken, in which C = cooperation and $\overline{C}$ = noncooperation. As we did in the game of Success in Figure 16.2, we also show in brackets the anticipation games for each of these games, giving the NMEs into which each state goes, according to rules 1–6. Note that Prisoners' Dilemma has two NMEs: starting at $(2, 2)$, the players would remain stuck at Conflict; but from $(4, 1)$ or $(1, 4)$ they would go into Compromise, or stay there if they start out at $(3, 3)$. Hence, the dilemma arises only if the status quo state is Conflict. This is a somewhat more auspicious view of this game, at least for nonmyopic players, than the standard game theory solution (i.e., Conflict) suggested, unless there is repeated play that enables the players to use a strategy like tit-for-tat to induce cooperation (Chapter 15).

TOM also indicates that Compromise is an NME in Chicken if the players start out there. Order power comes into play if the players commence at either $(4, 2)$ or $(2, 4)$, and one player is advantaged: if the advantaged player has order power at these states, he or she can ensure $(3, 3)$ as the NME. This consequence of TOM is interesting, because it suggests that the advantaged player in each of these states should move to his or her *less* preferred state of $(3, 3)$, lest the other player induce his or her best outcome by moving the process through $(1, 1)$ to the NME of $(2, 4)$ or $(4, 2)$, respectively, from $(4, 2)$ and $(2, 4)$. Finally, if the initial state is Disaster, then the possession of order power enables the player who possesses it to force the other player to move away first (i.e., "chicken out"), leading to the nonmover's best state (i.e., 4).

In real-life games that can be modeled by Prisoners' Dilemma or Chicken, Compromise has sometimes been achieved, it seems, because the players were nonmyopic. For example, even before the collapse of the Soviet Union in the late 1980s, the superpowers agreed to certain limitations in their arms race, which was frequently modeled as a Prisoners' Dilemma. In the Cuban missile

FIGURE 16.5   Prisoners'
Dilemma and Chicken.

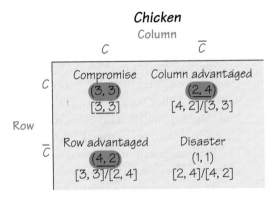

Prisoners' Dilemma

Chicken

Key
(x, y) = (payoff to Row, payoff to Column).
[x, y] = [payoff to Row, payoff to Column] in
    anticipation game.
4 = best, 3 = next best, 2 = next worst, 1 = worst.
C = cooperation; $\overline{C}$ = noncooperation.
Nash equilibria in original game and anticipation
    game underscored.
Nonmyopic equilibria (NMEs) highlighted in blue.

crisis of 1962, which was often modeled as a game of Chicken (notice that the payoffs to the players in the four states are the same as those in Success, but the configuration is different), a compromise was reached when the Soviets withdrew their missiles from Cuba and the United States promised not to invade the island in the future (an invasion of Cuba, with U.S. support, had been unsuccessfully attempted in 1961 at the infamous Bay of Pigs).

Assume rules 5′ and 6′ are operative. Then moving power is not defined in either Prisoners' Dilemma or Chicken because these games are not cyclic: whether the players cycle clockwise or counterclockwise, a state is reached in which a player whose turn it is to move next receives his or her best payoff and so would not move, which precludes cycling (rule 5′). In Prisoners' Dilemma, for example, if the move-countermove process is clockwise, the column player at some point would have to move from $(4, 1)$ to $(3, 3)$, whereas if it is counterclockwise, the row player would have to move from $(1, 4)$ to $(3, 3)$. Hence, while order power is applicable in Chicken, moving power is defined in neither this game nor Prisoners' Dilemma.

LARGER GAMES

So far we have applied TOM only to 2×2 games. But in this concluding section we turn to a game with more than two players and in which each player has more than two strategies.

EXAMPLE ▶ *A Truel*

A **truel** is like a duel, except that there are three players. Each player can either fire, or not fire, his or her gun at either of the other two players. We assume the goal of each player is, first, to survive and, second, to survive with as few other players as possible. Each player has one bullet and is a perfect shot, and no communication (e.g., to pick out a common target) that results in a binding agreement with other players is allowed, making the game noncooperative. We will discuss the answers that standard game theory, on the one hand, and TOM, on the other, give to what it is optimal for the players to do in the truel.

According to standard game theory, *at the start of play, each player fires at one of the other two players, killing that player.*

Why will the players all fire at each other? Because their own survival does not depend an iota on what they do. Since they cannot affect what happens to themselves but can only affect how many others survive (the fewer the better, according to the postulated secondary goal), they should all blaze away at each other. (Even if the rules of the play permitted shooting oneself, the primary goal of survival would preclude committing suicide.) In fact, the players all have dominant strategies to shoot at each other, because whether or not a player survives—we will discuss shortly the probabilities of doing so—he or she does at least as well shooting an opponent.

The game, and optimal strategies in it, would change if (i) the players were allowed more options, such as to fire in the air and thereby disarm themselves, or (ii) they did not have to choose simultaneously, and a particular order of play were specified. Thus, if the order of play were *A*, followed by *B* and *C* choosing simultaneously, followed by any player with a bullet remaining choosing, then *A* would fire in the air and *B* and *C* would subsequently shoot each other. (*A* is no threat to *B* or *C*, so neither of the latter will fire at *A*; on the other hand, if *B* or *C* did not fire immediately at the other, each might not survive to get in the last shot, so they both fire.) Thus, *A* will be the sole survivor. In 1992, a modified version of this scenario was played out in late-night television programming among the three major television networks, with ABC's effectively going first with "Nightline," its well-established news program, and CBS and NBC dueling on which host, David Letterman or Jay Leno, to choose for their entertainment shows. Regardless of their ultimate choices, ABC "won" when CBS and NBC were forced to divide the entertainment audience.

To return to the original game, the players' strategies of all firing have two possible consequences: either one player survives (even if two players fire at the same person, the third must fire at one of them, leaving only one survivor), or no player survives (if each player fires at a different person). In either event, there is no guarantee of survival. In fact, if each player has an equal probability of firing at one of the two other players, the probability that any particular player will survive is only .25.

The reason is that if the three players are *A*, *B*, and *C*, *A* will be killed when either *B* fires at him or her, *C* does, or both do. The only circumstance in which *A* will survive is if *B* and *C* fire at each other, which gives *A* one chance in four. Although this calculation implies that one of *A*, *B*, or *C* will survive with probability .75 if all outcomes are equally likely, more meaningful for each player is the low .25 individual probability of survival.

According to TOM, *no player will fire at any other, so all will survive.*

At the start of the truel, all the players are alive, which satisfies their primary goal of survival, though not their secondary goal of surviving with as few others as possible. Now assume that *A* contemplates shooting *B*, thereby reducing the number of survivors. But looking ahead, *A* knows that by firing first and killing *B*, he or she will be defenseless and be immediately shot by *C*, who will then be the sole survivor.

It is in *A*'s interest, therefore, not to shoot anybody at the start, and the same logic applies to each of the other players. Hence, everybody will survive, which is a happier outcome than that given by game theory's answer, in which everyone's primary goal is not satisfied—or, quantitatively speaking, satisfied only 25% of the time. ◆

The purpose of TOM, however, is not to produce "happier" outcomes but to provide a plausible model of a strategic situation that mimics what people might actually think and do in such a situation. We believe that the players in the truel, artificial as this kind of shoot-out may seem, would be motivated

Jay Leno   David Letterman   Ted Koppel

to think ahead, given the dire consequences of their actions. Following the reasoning of TOM, therefore, they would hold their fire, knowing that if one fired first, he or she would be the next target.

In Figure 16.6 we show this logic somewhat more formally with a **game tree,** in which A has three strategies, as indicated by the three branches that sprout from A: not shoot $(\overline{S})$, shoot B $(S \rightarrow B)$, or shoot C $(S \rightarrow C)$. The latter two branches, in turn, give survivors C and B, respectively, two strategies: not shoot $(\overline{S})$ or shoot A $(S \rightarrow A)$.

We assume that the players rank the outcomes as follows, which is consistent with their primary and secondary goals: $4 =$ best (lone survivor), $3 =$ next best (survivor with one other), $2 =$ next worst (survivor with two others), and $1 =$ worst (nonsurvivor). These payoffs are given for ordered triples (A, B, C); thus $(3, 3, 1)$ indicates the next-best payoffs for A and B and the worst payoff for C.

Note that play necessarily terminates when there is only one survivor, as is the case at $(1, 1, 4)$ and $(1, 4, 1)$. To keep the tree simple, we assume that play also terminates when either A initially or B or C subsequently chooses $\overline{S}$, giving outcomes of $(2, 2, 2)$, $(3, 3, 1)$, and $(3, 1, 3)$, respectively. Of course, we could allow the two or three surviving players in the latter cases to make subsequent choices in an extended game tree, but this example is meant only to illustrate the analysis of a game tree, not be the definitive statement on truel possibilities (more will be explored in the exercises).

As in 2×2 strategic-form games using TOM, we work backwards in extensive-form games, starting the analysis at the bottom of the tree. (By "bottom" we mean where play terminates; because this is where the tree branches out, the tree looks upside down in Figure 16.6.) Thus, because C prefers $(1, 1, 4)$ to $(3, 1, 3)$, we indicate that C would not choose $\overline{S}$ by "cutting" this branch with a scissors; similarly, B would not choose $\overline{S}$. Cutting a branch here is analogous to the blockage of a move in a 2×2 game.

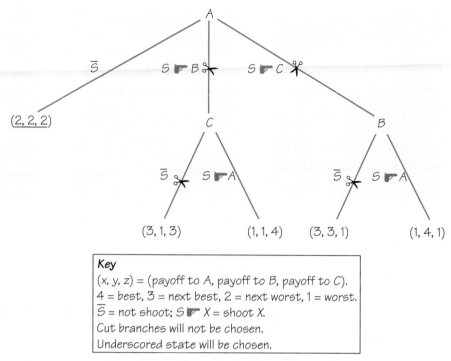

Key
(x, y, z) = (payoff to A, payoff to B, payoff to C).
4 = best, 3 = next best, 2 = next worst, 1 = worst.
$\overline{S}$ = not shoot; S ☞ X = shoot X.
Cut branches will not be chosen.
Underscored state will be chosen.

FIGURE 16.6 A game tree of a truel.

Moving up to the next level, A would know that if he or she chose $S \to B$, $(1, 1, 4)$ would be the outcome; if he or she chose $S \to C$, $(1, 4, 1)$ would be the outcome, making each, in our earlier terminology, the survivor from the bottom level. Choosing between these two outcomes and $(2, 2, 2)$, A would prefer the latter, so we cut the two branches, $S \to B$ and $S \to C$. Hence, A would choose $\overline{S}$, terminating play with nobody's shooting anybody else.

This, of course, is the conclusion we reached earlier, based on the reasoning that if A shot either B or C, he or she would end up dead, too. Because we could allow each player, like A, to choose among his or her three initial strategies in a $3 \times 3 \times 3$ game, and subsequently make moves and countermoves from the initial state (if feasible), the foregoing analysis applies to all players.

Underlying the completely different answers of game theory and TOM is a change in the rules of play that TOM introduces, namely, that players do not have to fire simultaneously at the start of play in the $3 \times 3 \times 3$ game. In standard game theory, however, if the players do not fire simultaneously, an order of play is posited, in which case the player to move first—and then the later players—would not fire, given that play continues until all bullets are expended or nobody chooses to fire.

To be sure, this is the same answer as that given by TOM. The point we stress here is that the standard theory does not raise the question of which order of play the players—if, thinking ahead, they could make this choice—would adopt, given their goals.

TOM, by contrast, leaves open the order of play by asking of each player: given your present situation (all alive), and the situation you anticipate will ensue if you fire first, should you do so? Because each player prefers living to the state he or she would bring about by being the first to shoot (certain death), none shoots. This analysis suggests that truels might be more effective than duels in preventing the outbreak of conflict.

We will not try to develop this argument into a more general model. The main point is that TOM introduces into a payoff matrix a look-ahead approach to game-theoretic analysis, which requires radical changes in the usual rules of play. In particular, these changes require the comparison of the past or present (initial state) with the future (final state)—perhaps several steps ahead—to which the moves and countermoves may transport the players. TOM also allows for the exercise of power (order or moving) when there is an asymmetry in the abilities of the players, which, as we saw, can also affect the outcome.

Unfortunately, there is no surefire way to determine which set of rules, or what kinds of power, are most applicable in a given situation. Our aim in this chapter has been to show that there are alternatives to standard game theory, particularly in looking at the *processes* by which outcomes are chosen, thereby making the analysis more dynamic.

REVIEW VOCABULARY

Anticipation game A game, described by a payoff matrix, whose entries, which are given in brackets, are the nonmyopic equilibria (NMEs) into which each state of the original game goes.

Asymmetric game A game in which the row player's strategies do not lead to the same payoffs as the column player's strategies.

Backward induction A reasoning process in which players, working backward from the last possible move in a game, anticipate each other's rational choices.

Blockage Occurs when it is not rational, based on backward induction, for a player to move from a state.

Complete information Each player knows the rules of the game, the preferences of every player for all possible states, and which, if either, player has order or moving power.

Cyclic game A 2 × 2 ordinal game in which, when the game cycles either clockwise or counterclockwise, neither player ever receives a best payoff when it is his or her turn to move next.

Effective moving power Moving power is effective when possessing it induces a better outcome for a player than when the other player possesses it.

Feasibility A move is feasible if it can plausibly be interpreted as possible in the situation being modeled.

Final state A final state is the state induced after all rational moves and countermoves (if any) from the initial state have been made, making it the outcome of the game.

Game A game is the totality of the rules that describe it.

Game tree A symbolic tree, based on the rules of play of a game, in which the vertices, or nodes, of the tree represent choice points, and the branches represent alternative courses of action that the players can select.

Indeterminate state A state in which the outcome induced depends on which player moves first (in which case order power is effective).

Initial state The state from which play commences.

Irrelevant moving power Moving power is irrelevant when the outcome induced by one player is better for both players than the outcome that the other player can induce.

Move A player's switch from one strategy to another in the payoff matrix of a strategic-form game.

Moving power In a cyclic game, the ability of one player to continue moving when the other player must eventually stop; a player exercises moving power in order to try to implement a preferred outcome.

Nonmyopic equilibrium (NME) A state to which rational players would move (or not move), anticipating all possible rational moves and countermoves from some initial state.

Order power The ability of a player to determine the order of moves in which the players depart from an indeterminate initial state in order to ensure a preferred outcome.

Outcome The final state of a game, from which no player would choose to move and at which the players receive their payoffs.

Precedence Occurs when the outcome induced by the player who moves overrides the outcome induced by the player who stays.

Rational choice A choice that leads to a preferred outcome, based on the rules of the game.

Rules of play Describe the possible choices of the players at each stage of play.

State An entry in a payoff matrix from which the players can move. Play of a game starts at an initial state and terminates at a final state, or outcome.

Stoppage Occurs when blockage occurs for the first time from some initial state.

Survivor The state that is selected at any stage as the result of backward induction.

Termination rule   Prescribes that play will terminate after one complete cycle.

Theory of moves (TOM)   A dynamic theory that describes optimal strategic choices in games in which the players, thinking ahead, can make moves and countermoves from an initial state.

Truel   The analogue of a duel, in which each of three players can fire or not fire his or her gun at either of the other two players.

Two-sidedness rule   Describes how a player determines whether or not to move from a state, based on the other player's rational choices as well as his or her own.

SUGGESTED READINGS

BRAMS, STEVEN J.   *Superior Beings: If They Exist, How Would We Know? Game-Theoretic Implications of Omniscience, Omnipotence, Immortality, and Incomprehensibility,* Springer-Verlag, New York, 1983. An application of game theory, and an early version of the theory of moves, to questions in the philosophy of religion, such as ascertaining the existence of a superior being and explaining the problem of evil in the world, based on an analysis of 2 × 2 ordinal games.

BRAMS, STEVEN J.   *Theory of Moves,* Cambridge University Press, Cambridge, 1994. The main source of the theory in this chapter; it also contains diverse applications, ranging from literature to theology.

RAPOPORT, ANATOL, MELVIN GUYER, AND DAVID GORDON.   *The 2 × 2 Game,* University of Michigan Press, Ann Arbor, 1976. A comprehensive review and classification of the 78 distinct 2 × 2 strict ordinal games, including the means by which the number 78 was determined.

TAYLOR, ALAN D.   *Mathematics and Politics: Strategy, Voting, Power, and Proof,* Springer-Verlag, New York, 1995. A mathematics textbook in which game theory, theory of moves, and social choice theory are used to model power, voting, and conflict and escalation processes.

EXERCISES   ▲ *Optional.*   ■ *Advanced.*   ◆ *Discussion.*

Game Theory Revisited

◆   1.   Define an ordinal game to be one of total conflict in which the best state (4) for one player is the worst state (1) for the other player, and the next-best state (3) for one player is the next-worst state (2) for the other. What relationship do these games have to zero-sum games?

2.   The three different total-conflict games with $(4, 1)$, $(1, 4)$, $(3, 2)$, and $(2, 3)$ as states are as follows:

$$
\begin{array}{|cc}
(2, 3) & (4, 1) \\
(1, 4) & (3, 2)
\end{array}
\qquad
\begin{array}{|cc}
(3, 2) & (4, 1) \\
(2, 3) & (1, 4)
\end{array}
\qquad
\begin{array}{|cc}
(2, 3) & (4, 1) \\
(3, 2) & (1, 4)
\end{array}
$$

Find the dominant strategies and Nash equilibria in each (if they exist).

Rules of TOM

3.   Show that two of the total-conflict games in Exercise 2 have one NME and the third has two NMEs. What, if any, relationship is there between the Nash equilibria and NMEs in the total-conflict games?

4.   Both Success and Chicken have as states $(3, 3)$, $(2, 4)$, $(4, 2)$, and $(1, 1)$, but these two games have very different properties. Find one other *symmetric* 2×2 game (like Chicken) with these four states, and one other *asymmetric* game (like Success). Are there any other games that have different configurations of these four states?

5.   Show that the other symmetric game in Exersise 4 has one NME, and the other asymmetric game has two NMEs. Which of the NMEs in these games coincide with Nash equilibria?

Interpreting TOM

◆   6.   An alternative game called Variation was suggested for Delilah's preferences in Figure 16.3. Does Variation seem more reasonable to you than Success as a model of the Samson and Delilah story? Why?

7.   The following "mugging game" has been proposed to model the conflict between a mugger and a victim:

		Mugger	
		Use force (F)	Don't use force ($\overline{F}$)
	Resist (R)	I. Fight $(2, 2)$ $[3, 4]$	II. Mugger fails $(4, 1)$ $[2, 2]$
Victim	Don't resist ($\overline{R}$)	II. Involuntary submission $(1, 3)$ $[3, 4]$	III. Voluntary submission $(3, 4)$ $[3, 4]$

◆   Do the ordinal payoffs in this game seem plausible to you? If not, propose a more plausible alternative game.

8.   Verify that the NMEs in the mugging game from each state are those shown in its anticipation game. If play of the game starts at state II, does it seem reasonable that, because the victim would *not* move from (4, 1) according to TOM, the mugger would move to (2, 2)?

◆   9.   Assume that the victim anticipates the (2, 2) outcome in the mugging game, starting at state II. Argue that it would be reasonable for him or her to take the initiative and move to (3, 4), making Voluntary submission rather than Fight the outcome from (4, 1). *Note:* In fact, TOM postulates a "two-sidedness convention" to cover this and six other 2 × 2 games. If one player (the victim in the mugging game) can induce a better state for *both* players [at (3, 4)] by moving rather than staying—instead of forcing the other player (the mugger) to move [to (2, 2)]—then that player (the victim) will move, even if he or she would otherwise prefer to stay [at (4, 1)], to induce a better outcome [(3, 4) rather than (2, 2)]. When amended by the two-sided convention, TOM says that (3, 4) rather than (2, 2) is the NME from (4, 1), making the mugging game a one-NME rather than a two-NME game.

Cyclic Games and Moving Power

10.   Show that the mugging game is cyclic in a counterclockwise direction. Is moving power effective or irrelevant in this game?

◆   11.   It was shown that moving power is irrelevant in Alternative (Figure 16-4b), leading to (3, 3), whichever player possessed it. But since (2, 4) as well as (3, 3) is an NME in Alternative, it would appear *not* to be in C's interest to exercise her moving power in order to try to get her preferred NME of (2, 4). Does this seem paradoxical?

▪   12.   If a 2 × 2 strict ordinal game is cyclic, show that it can cycle in only one direction (clockwise or counterclockwise, but not both).

Return to Prisoners' Dilemma and Chicken

13.   If (2, 4) or (4, 2) is the initial state in Chicken, show by backward induction that the player receiving 4 would not move, whereas the player receiving 2 would move the process from (2, 4) to (4, 2) or from (4, 2) to (2, 4), making (4, 2) and (2, 4) the NMEs from (2, 4) and (4, 2), respectively. Knowing this, would the 4-player at (2, 4) or (4, 2) have an incentive to "beat the 2-player to the punch" in order to move the process immediately to (3, 3)—instead of not moving and suffering a 2-outcome if the 2-player moved first? Does this reasoning agree with backward induction?

Larger Games

14.   Extend the game tree of the truel in Figure 16.6 to allow the additional possibility that if *A* does not shoot initially, then *B* has the choice of shooting or not shooting *C*. Will *A*, in fact, not shoot initially, and will *B* then shoot *C*?

15.   Extend the game tree in Exercise 14 to still another level to allow the possibility that if *A* does not shoot initially, and *B* shoots *C*, then *A* has the choice of shooting or not shooting *B*. What will happen in this case?

16.   Change Exercise 15 to allow for the possibility that if *A* does not shoot initially, and *B* shoots *or does not shoot C*, then *A* has the choice of shooting or not shooting *B*. What will happen in this case?

◆ 17.   What general conclusions would you draw in light of your answers to Exercises 14, 15, and 16?

Extensions to Threats

18.   Say that a player has a *threat* if he or she can choose a strategy that leads to the two worst states (1 and 2) of the other player. Show that both players have threats in Prisoners' Dilemma and Chicken. Which, if either, player has a threat in Success, Variation, Alternative, and the mugging game?

Additional Exercises

19.   Show that there are four different "almost" total-conflict games with states $(4, 1)$, $(1, 4)$, $(3, 3)$, $(2, 2)$. (If you should find more than four games, check whether interchanging players, their strategies, or both can transform one game into another, so that they are not strategically different.) Find the dominant strategies and Nash equilibria in each (if they exist).

▲ 20.   Find the NMEs in the "almost" total-conflict games in Exercise 19.

◆ 21.   In Shakespeare's *Macbeth,* Lady Macbeth badgers and cajoles Macbeth into murdering King Duncan. Argue that Success can be used to model their conflict—or show that another game better mirrors the preferences of the two players, and analyze the latter game.

22.   Show that if the players commenced play from the upper left state of either Success or Variation, it would make no difference which game was the "true" game—the players in both games would move to the lower right state of $(3, 3)$. Does TOM make the same prediction if the initial state is $(1, 1)$ in Success and $(2, 1)$ in the Variation?

23.   Show that Variation is cyclic and that moving power is effective in this game. Is it plausible to assume that moving power would be exercised in Variation if this game is used to model Samson and Delilah?

■   24.   Show that no symmetric 2 × 2 strict ordinal game is cyclic (but *not* by checking all possible symmetric games—there are a total of 12).

25.   Indicate the game that results if the row player's preferences are those in Prisoners' Dilemma, and the column player's preferences are those in Chicken. Show that this game has only one NME, which favors the player with Prisoners' Dilemma preferences. Does it seem strange to you that this game has only one NME, whereas Prisoners' Dilemma has two and Chicken has three?

■   26.   If a fourth player is added to the original truel, show that each player will have an incentive to shoot another (as in a duel).

◆   27.   If the players in Prisoners' Dilemma and Chicken say that they will carry out their threats if and only if the other player does not choose his or her cooperative strategy initially, and these threats are believed by both players, will (3, 3) be the outcome in both games? Might threats like this help in ameliorating, rather than aggravating, conflict in such games?

WRITING PROJECTS

1 ▶   Quentin Tarantino's films *Reservoir Dogs* (1992) and *Pulp Fiction* (1994) both have truels, but the choices that the characters make in each are completely different. Does TOM offer any insight into why?

2 ▶   Model a conflict in a Bible story, a work of fiction, or a real-life situation as a 2 × 2 strict ordinal game, comparing the results predicted by game theory and by TOM. Do you think these theories take proper account of the feelings or emotions of the characters, especially when their actions appear "irrational"?

3 ▶   Strikes and threatened strikes in different professional sports have become common, even as the salaries of professional athletes have skyrocketed. Explain why this has happened, using either standard game theory or TOM to model the conflict between the athletes and the team owners. Does either theory suggest how strikes might be avoided and stable settlements reached?

PART V

athematics is the study of patterns and relationships. It can be used to characterize the spiral growth of a sunflower's seeds, measure increase in populations, and calculate the effects of travel near the speed of light.

Mathematicians search for and classify numerical, geometrical, and even abstract patterns. In these chapters we follow some of those searches, concentrating on geometrical patterns, but also looking at what geometry can express about some numerical patterns. Examining the underlying patterns helps explain why some of the objects in the world around us have the shapes that they do and trains us to recognize the same patterns as they arise in contemporary problems.

ON SIZE and SHAPE

We investigate patterns in some BIG things: King Kong, tall trees, high mountains, large populations, the universe, and even symmetries that extend infinitely in all directions. We look at how an animal's size can greatly influence its form. We explore how savings accounts are similar to biological populations, and how changes in interest rates may lead to extinction of species.

Intertwined with size, shape is also a theme for these chapters. We examine why the shape of an animal must change as the animal grows. We ask: What is the shape of the universe? We analyze what shapes crystals can have. We enjoy the patterned beauty of African crafts and the prints of M. C. Escher.

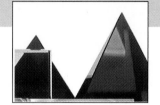

GROWTH AND FORM

Fantasy films have made us familiar with assorted giant creatures, including King Kong, Godzilla, and the 50-foot-high grasshoppers in *The Beginning of the End.* We also find supergiants in literature, such as the giant of "Jack and the Beanstalk," Giant Pope and Giant Pagan of *The Pilgrim's Progress,* and the Brobdingnagians of *Gulliver's Travels.*

Much as we appreciate those stories, even from an early age we don't really believe in monsters and giants. But could such beings ever exist? What problems would their enormous size cause them? How would they have to adapt in order to cope? (See Figure 17.1.)

Every species survives by adapting to its environment. In particular, it faces the **problem of scale:** how to adapt and survive at the different sizes from the beginning of life to the final size of a mature adult. For example, consider the giant panda, which ranges from barely 1 lb at birth to 275 lb in adulthood. A baby panda is at risk of being crushed by its mother; an adult panda needs to eat a great deal of food.

As a contrasting example, consider the horse. If a newborn foal weighed as little as a newborn panda, the foal would be too small to keep up with the moving herd and could not survive. An adult horse weighs much more than a panda and has to consume much more food; but the horse can move much more quickly and cover great distances, to take advantage of wide-ranging sources of sustenance.

There have been large land mammals (mammoths) and huge sea mammals (the blue whale)—not to mention the dinosaurs. But the tallest humans have been only 9 to 10 feet tall; the largest mammoth was 16 feet at the shoulder

Figure 17.1   Could King Kong actually exist?

(about twice as tall as an elephant); and even the tallest dinosaur, *Supersaurus*, stood only 40 feet high.

But what about supergiants and utterly huge monsters? That they have never existed suggests that there are physical limits to size. In fact, with a few simple principles of geometry, we can show not only that lizards and apes of such size are impossible, but also that none of the living beings and objects in our world could exist, unchanged in shape, on a vastly different scale, larger or smaller.

GEOMETRIC SIMILARITY

The powerful mathematical idea that we use is *geometric similarity*.

> Two objects are **geometrically similar** if they have the same shape, regardless of the materials of which they are made.

Similar objects need not be of the same size, but measurements of corresponding distances on the two objects must be proportional. For example, when a photo is enlarged, it is enlarged by the same factor in both the horizontal and vertical directions—in fact, in any direction whatever (such as a diagonal). We call this enlargement factor the *linear scaling factor*.

FIGURE 17.2 Two geometrically similar photographs.

FIGURE 17.2 Two geometrically similar photographs.

> The **linear scaling factor** of two geometrically similar objects is the ratio of a length of any part of the second to the corresponding part of the first.

In the photos in Figure 17.2, the linear scaling factor is 3; the enlargement is three times as wide and three times as high as the original. In fact, every pair of points goes to a new pair of points three times as far apart as the original ones.

We notice that the enlargement can be divided into $3 \times 3 = 9$ rectangles, each the size of the original. Hence, the enlargement has $3 \times 3 = 3^2 = 9$ times the area of the original. More generally, if the linear scaling factor is some general number M (not necessarily 3), the resulting enlargement has an area $M \times M = M^2$ ("M squared") times the area of the original.

> *The* area *of a scaled-up object goes up with the* square *of the linear scaling factor.*

What about enlarging three-dimensional objects? If we take a cube and enlarge it by a linear scaling factor of 3, it becomes three times as long, three times as high, and three times as deep as the original (see Figure 17.3).

What about volume? The enlarged cube has three layers, each with 3×3 = 9 little cubes, each the same size as the original. Thus, the total volume is $3 \times 3 \times 3 = 3^3 = 27$ times as much as the original cube.

> *The* volume *of a scaled-up object goes up with the* cube *of the linear scaling factor.*

Thus, for an object enlarged by a linear scaling factor of M, the enlargement has M^3 ("M cubed") $= M \times M \times M$ times the volume of the original. Like the relationship between surface area and M^2, this relationship holds even for irregularly shaped objects, such as science fiction monsters.

We observe, however, that the area of each face (side) of the enlarged cube is $3^2 = 9$ times as large as that of a face of the original cube, just as the area of the photo enlarged by a factor of 3 has nine times the area of the original. Since this fact is true for all six faces, the total surface area of the enlarged cube is nine times as much as the total surface area of the original cube.

More generally, for objects of any shape, the total *surface area* of a scaled-up object goes up with the *square* of the linear scaling factor. Thus, the surface area of an object scaled up by a factor of M is M^2 times the surface area of the original; this feature holds true even for irregular shapes.

Before we discuss scaling real three-dimensional objects, you should understand the pitfalls of the language for describing increases and decreases.

FIGURE 17.3   Cube (b) is made by enlarging cube (a) by a factor of 3.

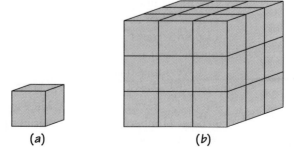

(a)   (b)

THE LANGUAGE OF GROWTH, ENLARGEMENT, AND DECREASE

In 1976, the average price of a home in Madison [Wisconsin] was $38,323 — about 108 percent less than in 1988.

—*Madison Business*, March 1991, p. 38.

House prices in Madison rose substantially, but we explore why this is an incorrect and confusing way to say so. In 1988 the average price was $80,000, which you can verify was 2.08 times the 1976 average price. What the author meant to say, in correct language, is that the 1988 price was 208% *of* the 1976 price, or that the 1988 price was 108% *more than* the 1976 price.

"x% of A" or "x% as large as" means $\frac{x}{100} \times A$.

"x% more than A" means A plus x% of A, in other words, $(1 + \frac{x}{100}) \times A$. Saying that A has "increased by x%" means the same thing.

"x% less than A" means A minus x% of A, in other words, $(1 - \frac{x}{100}) \times A$. Saying that A has "decreased by x%" means the same thing.

So, to say "108% less than" $80,000 would mean $80,000(1 - \frac{108}{100}) = $80,000(1 - 1.08) = $80,000(-0.08) = -$6400$. Clearly this is not what the author intended to say.

The terms "of," "times," and "as much as" refer to *multiplication* of the original amount, while the terms "more," "larger," and "greater" refer to *adding* to the original amount. For instance, "five times as much" means the same as "four times more than" (the original plus four times as much in addition). Similarly, the relationship of the original amount to the larger amount can be expressed in multiplicative terms ("one-fifth as much," "20% as much") or in subtractive terms ("four-fifths less than," "80% less than").

These two ways of expressing change are similar in their phrasing, and people (even people in the media) often say "five times more than" when they mean "five times as much" (or even "five times less than" when they mean "one-fifth as much"). All you can do is be aware of the potential confusion, try to figure out what was meant, and be careful in your own expression. In particular, avoid using both "times" and "more" together.

Finally, in discussions of percent we need to distinguish *percent* from percentage *points:* if support for the president has decreased from 60% to 30%, it has dropped 30 *percentage points* but decreased 50% (because the drop of 30 percentage points is 50% of the original 60 percentage points).

EXAMPLE ▶ *What About Those Homes?*

Returning to the Madison home prices, how can we state correctly what the author was trying to say? The 1976 figure, $38,323, is about 0.48 times the 1988 figure of $80,000 (38,323/80,000 = 0.48), or 48% of $80,000; so the writer could say "about 48% of what it is in 1988" or "about 52% less than in 1990."

What if the writer wanted to use the 1976 figure as a base (i.e., as the 100% for the calculation)? The 1988 price is about 2.08 times the 1976 price, so the 1988 price is "208% of," or "108% more than," the 1976 price.

Caution: The dollar comparisons here can be misleading, since they do not take into account that a dollar was worth less in purchasing power in 1988 than in 1976, because of inflation. Shortly we will explore the "numerical similarity" of dollars in different years and show how to take inflation into account.

Another caution: What does the author mean by "average," the mean or the median (see pages 227–229)? Prices of houses are skewed, with a small number of very expensive houses (as noted on page 286), and the mean may be much larger than the median. Government statisticians and economists usually use the median. The Madison housing article notes that the median price in 1976 was $34,000, so the $38,323 "average" must be the mean.   ◆

NUMERICAL SIMILARITY

House prices in different years are not directly comparable, in part because of inflation—a dollar today is not worth the same as a dollar in 1976. However, based on measures of inflation, we can determine the equivalent today of a 1976 price, or how much a 1976 dollar would be worth today.

The official measure of inflation is the Consumer Price Index (CPI), prepared by the Bureau of Labor Statistics. We describe and use here the CPI-U, the index for all urban consumers, which covers about 80% of the U.S. population and is the index of inflation that is usually referred to in newspaper and magazine articles.

Each month, the Bureau of Labor Statistics determines the average cost of a "market basket" of goods, including food, housing, transportation, clothing, and other items. It compares this cost to the cost of the same (or comparable) goods in a base period. The base period used to construct the CPI-U is

1982–1984. The index for 1982–1984 is set to 100, and the CPI-U for other years is calculated by using the proportion

$$\frac{\text{CPI for other year}}{100} = \frac{\text{cost of market basket in other year}}{\text{cost of market basket in base period}}$$

For example, the cost of the market basket in 1976 (in 1976 dollars) was 0.569 times the cost in 1982–1984 (in 1982–1984 dollars), so the CPI for 1976 is 100 × 0.569, or 56.9.

Table 17.1 shows the average CPI for each year from 1913 through 1995, with estimates for 1996 and 1997. This table can be used to convert the cost of an item in dollars for one year to what it would cost in dollars in a different year, using the proportion

$$\frac{\text{cost in year A}}{\text{cost in year B}} = \frac{\text{CPI for year A}}{\text{CPI for year B}}$$

TABLE 17.1   U.S. Consumer Price Index (1982–1984 = 100)

—	—	1931	15.2	1951	26.0	1971	40.5	1991	136.2
—	—	1932	13.7	1952	26.6	1972	41.8	1992	140.3
1913	9.9	1933	13.0	1953	26.7	1973	44.4	1993	144.5
1914	10.0	1934	13.4	1954	26.9	1974	49.3	1994	148.2
1915	10.1	1935	13.7	1955	26.8	1975	53.8	1995	152.4
1916	10.9	1936	13.9	1956	27.2	1976	56.9	1996 (est)	156.3
1917	12.8	1937	14.4	1957	28.1	1977	60.6	1997 (est)	160.5
1918	15.1	1938	14.1	1958	28.9	1978	65.2		
1919	17.3	1939	13.9	1959	29.1	1979	72.6		
1920	20.0	1940	14.0	1960	29.6	1980	82.4		
1921	17.9	1941	14.7	1961	29.9	1981	90.9		
1922	16.8	1942	16.3	1962	30.2	1982	96.5		
1923	17.1	1943	17.3	1963	30.6	1983	99.6		
1924	17.1	1944	17.6	1964	31.0	1984	103.9		
1925	17.5	1945	18.0	1965	31.5	1985	107.6		
1926	17.7	1946	19.5	1966	32.4	1986	109.6		
1927	17.4	1947	22.3	1967	33.4	1987	113.6		
1928	17.1	1948	24.1	1968	34.8	1988	118.3		
1929	17.1	1949	23.8	1969	36.7	1989	124.0		
1930	16.7	1950	24.1	1970	38.8	1990	130.7		

Note: This is the CPI-U index, which covers all urban consumers, about 80% of the U.S. population. Each index is an average for all cities for the year. The basis for the index is the period 1982–1984, for which the index was set equal to 100.

EXAMPLE   ▶ *The Price of a House and the Value of a Dollar*

We convert the average cost of a Madison house in 1976 dollars into a price in 1997 dollars.

We see from Table 17.1 that the CPI for 1976 is 56.9 and the CPI for 1997 is estimated to be 160.5. The average cost of a Madison house in 1976 was $38,323. Using the proportion, we have

$$\frac{\text{cost in 1997}}{\text{cost in 1976}} = \frac{\text{CPI for 1997}}{\text{CPI for 1976}}$$

or

$$\frac{\text{cost in 1997}}{\$38,323} = \frac{160.5}{56.9}$$

so that

$$\text{cost in 1997} = \$38,323 = \frac{160.5}{56.9} = \$38,323 \times 2.821 = \$108,099$$

To convert from 1997 dollars to 1976 dollars, we multiply by 1/2.821, or 0.355.

The 2.821 is the *scaling factor* for converting 1976 dollars to 1997 dollars. What we are observing is a proportion, or *numerical similarity*, between 1976 dollars and 1997 dollars, analogous to the geometrical similarity that we studied earlier.   ◆

We now move from scaling up one-dimensional dollars and two-dimensional photographs toward examining similarities between three-dimensional objects, including animals, buildings, and trees. Scaling three-dimensional objects involves considering the scaling of physical quantities such as distance, weight, area, and volume. Before we consider the possibility of a King Kong or a building ten times as tall as the Sears Tower in Chicago, we need to discuss the units in which quantities are measured.

MEASURING LENGTH, AREA, VOLUME, AND WEIGHT

We start with a brief introduction to the common units in which various physical quantities are measured, together with a handy table of scaling factors (often called *conversion factors*) and examples of how to convert successfully from one system of units to another.

U.S. Customary System

You are probably familiar with the common units of the *U.S. Customary System* of measurement and their abbreviations. But please examine Table 17.2 and pay close attention to the systematic way to convert from one unit to another and to the expression of approximate numbers in scientific notation. The symbol $\approx$ means "is approximately equal to."

TABLE 17.2   Units of the U.S. Customary System

Distance:

1 mile (mi) = 1760 yards (yd) = 5280 feet (ft) = 63,360 in.
1 yard (yd) = 3 feet (ft) = 36 in.
1 foot (ft) = 12 inches (in.)

Area:

$$
\begin{aligned}
1 \text{ square mile} &= 1 \text{ mi} \times 1 \text{ mi} = 5280 \text{ ft} \times 5280 \text{ ft} \\
&= 27{,}878{,}400 \text{ ft}^2 \approx 28 \times 10^6 \text{ ft}^2 \\
&= 63{,}360 \text{ in.} \times 63{,}360 \text{ in.} \\
&= 4{,}014{,}489{,}600 \text{ in.}^2 \approx 4 \times 10^9 \text{ in.}^2 \\
&= 640 \text{ acres} \\
1 \text{ acre} &= 43{,}560 \text{ ft}^2
\end{aligned}
$$

Volume:

$$
\begin{aligned}
1 \text{ cubic mile} &= 1 \text{ mi} \times 1 \text{ mi} \times 1 \text{ mi} \\
&= 5280 \text{ ft} \times 5280 \text{ ft} \times 5280 \text{ ft} \\
&= 147{,}197{,}952{,}000 \times \text{ft}^3 \\
&\approx 147 \times 10^9 \times \text{ft}^3 \\
&= 63{,}360 \text{ in.} \times 63{,}360 \text{ in.} \times 63{,}360 \text{ in.} \\
&\approx 2.5 \times 10^{14} \text{ in.}^3
\end{aligned}
$$

1 U.S. gallon (gal) = 4 U.S. quarts (qt) = 231 in.^{3}, exactly

Weight:

1 ton (t) = 2000 pounds (lb)

Metric System

With the notable exception of the United States, almost every country of the world uses a different measurement system in science, industry, and commerce—

the metric system. It was first proposed in France by Gabriel Mouton, Vicar of Lyons, in 1670 and was adopted in France in 1795. The fundamental unit of length, the *meter,* was originally defined to be one ten-millionth of the distance from the North Pole to the equator, as measured on the meridian through Paris. Later, the meter was redefined as the distance between two lines marked on a platinum-iridium bar kept at the International Bureau of Weights and Measures, near Paris, when the bar is kept at a temperature of 0°C. Finally, in 1960 the meter was redefined in terms of a standard reproducible in any laboratory, namely, 1,650,763.73 times the wavelength of the orange-red light emitted by atoms of the gas krypton-86 when an electrical charge is passed through them.

All other units of length, area, and volume are *defined* in terms of the meter; for example, a centimeter is a hundredth of a meter. The metric unit of weight, the *kilogram,* is defined as the weight of a platinum-iridium standard.

Table 17.3 lists the units of the metric system.

TABLE 17.3   Units of the Metric System

Distance:

$$1 \text{ meter (m)} = 100 \text{ centimeters (cm)}$$
$$1 \text{ kilometer (km)} = 1000 \text{ meters (m)}$$
$$= 100{,}000 \text{ centimeters (cm)} = 1 \times 10^5 \text{ cm}$$

Area:

$$1 \text{ square meter (m}^2) = 1 \text{ m} \times 1 \text{ m}$$
$$= 100 \text{ cm} \times 100 \text{ cm} = 10{,}000 \text{ (cm}^2) = 1 \times 10^4 \text{ cm}^2$$
$$1 \text{ hectare (ha)} = 10{,}000 \text{ m}^2$$

Volume:

$$1 \text{ liter (l)} = 1000 \text{ cm}^3 = 0.001 \text{ m}^3$$
$$1 \text{ cubic meter (m}^3) = 1 \text{ m} \times 1 \text{ m} \times 1 \text{ m}$$
$$= 100 \text{ cm} \times 100 \text{ cm} \times 100 \text{ cm}$$
$$= 1{,}000{,}000 \text{ cm}^3 = 1 \times 10^6 \text{ cm}^3 \text{ (or cc)}$$

Weight:

$$1 \text{ kilogram (kg)} = 1000 \text{ grams (g)}$$

Converting Between Systems

What are the scaling factors between the U.S. Customary System and the metric system? Since 1960, the fundamental units of the U.S. Customary System,

---

> **TABLE 17.4** **Conversions Between the U.S. Customary System and the Metric System**
>
> *Distance:*
>
> $\quad$ 1 in. = 2.54 cm, exactly
> $\quad\quad$ 1 ft = 12 in. = 12 × 2.54 cm
> $\quad\quad\quad$ = 30.48 cm = 0.3048 m, exactly
> $\quad$ 1 yd = 0.9144 m, exactly
> $\quad$ 1 mi = 5280 ft = 5280 × 30.48 cm
> $\quad\quad\quad$ = 160,934.4 cm, exactly ≈ 1.61 km
>
> $\quad$ 1 cm ≈ 0.3937 in. ≈ 0.4 in.
> $\quad\,$ 1 m ≈ 39.37 in. ≈ 3.281 ft
> $\quad$ 1 km ≈ 0.621 mi
>
> *Area:*
>
> $\quad$ 1 hectare (ha) = 2.47 acres
>
> *Volume:*
>
> $\quad$ 1 cubic meter (m^3) = 1000 liters
> $\quad\quad\quad\quad\quad\quad\quad\quad$ = 264.2 U.S. gallons
> $\quad\quad\quad\quad\quad\quad\quad\quad$ = 35.31 ft^3
> $\quad\quad\quad$ 1 liter (l) = 1000 cm^3
> $\quad\quad\quad\quad\quad\quad\quad\quad$ = 1.057 U.S. quarts (qt)
>
> *Weight:*
>
> $\quad$ 1 lb = 0.45359237 kg, exactly
> $\quad$ 1 kg ≈ 2.205 lb

the yard (for length) and the pound (for weight), have been *defined* in terms of metric units, so that we have

$\quad$ 1 yd = 0.9144 m, exactly

$\quad$ 1 lb = 0.45359237 kg, exactly

The scaling factors for other units are shown in Table 17.4.

$\quad$ We illustrate how to convert a measurement in one unit to the corresponding measurement in a different unit.

Comparison of a meter stick with a yardstick, a liter with a quart, and a kilogram with a pound.

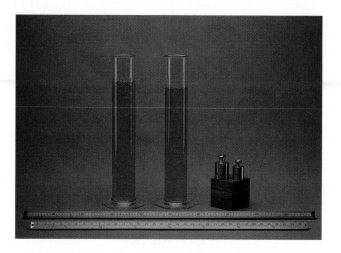

E X A M P L E ▶ *What's That in Feet?*

A foreign student tells his American student friends that he is 175 cm tall. They naturally ask how much that is in feet and inches.

We approach the conversion by using the scaling factor that 1 cm = 0.393701 in.:

$$175 \text{ cm} = 175 \times 1 \text{ cm} \approx 175 \times 0.393701 \text{ in.}$$

$$\approx 68.9 \text{ in.} = 68.9 \text{ in.} \times \frac{1 \text{ ft}}{12 \text{ in.}} \approx \frac{68.9}{12} \text{ ft} \approx 5.74 \text{ ft}$$

However, since we normally give height in feet and a whole number of inches, the height is

$$68.9 \text{ in.} = 5 \times (12 \text{ in.}) + 8.9 \text{ in.} = 5 \text{ ft} + 8.9 \text{ in.} \approx 5 \text{ ft } 9 \text{ in.}$$

Another way to approach the problem is by means of a proportion, similar to the one used with the CPI:

$$\frac{\text{height in inches}}{\text{height in cm}} = \frac{\text{length of 1 inch in inches}}{\text{length of 1 inch in cm}} = \frac{1 \text{ in.}}{2.54 \text{ cm}}$$

so that

$$\text{height in inches} = \text{height in cm} \times \frac{1 \text{ in.}}{2.54 \text{ cm}}$$

$$= 175 \text{ cm} \times \frac{1 \text{ in.}}{2.54 \text{ cm}} \approx 68.9 \text{ in.} \quad \blacklozenge$$

SCALING REAL OBJECTS

Real three-dimensional objects are made of matter, which has volume and mass. **Mass** is the aspect of matter that is affected by forces, according to physical laws. For example, your mass reacts to the gravitational force of the earth by staying close to it (and your mass exerts an equal force on the earth that tends to keep the earth close to you). We perceive the mass of an object when we try to move it (as in throwing a ball). When we try to lift an object, we perceive its mass as weight, due to the gravitational force that the earth exerts on it. As we will see, gravity exerts an enormous effect on the size and shape that objects and beings can assume.

Suppose that the two cubes in Figure 17.3 are made of steel, and that the first is 1 ft on a side and the second is 3 ft on a side. For each, its bottom face supports the weight of the entire cube. **Pressure** is force per unit area, so the pressure exerted on the bottom face by the weight of the cube is equal to the weight of the cube divided by the area of the bottom face, or

$$P = \frac{W}{A}$$

A cubic foot of steel weighs about 500 lb (we say it has a **density** of 500 lb per cubic foot).

The first cube weighs 500 lb and has a bottom face with area 1 ft^2, so the pressure exerted on this face is 500 lb/ft^2.

The second cube is 3 feet on a side. The area of the bottom face has increased with the square of the linear scaling factor, so it is $3^2 \times 1$ ft$^2 = 9$ ft^2. As we learned earlier in this chapter, volume goes up with the cube of the linear scaling factor. So this larger cube has a volume of $3^3 \times 1 = 27$ ft^3. Because both cubes are made of the same steel, the larger cube has 27 times as much steel as the smaller; hence it weighs 27 times as much as the smaller cube, or 27×500 lb $= 13,500$ lb.

When we divide this weight by the area of the bottom face (9 ft^2), we find that the pressure exerted on the bottom face is 1500 lb/ft^2, or three times the pressure on the bottom face of the original cube. This makes sense because over each 1 ft^2 area stands 3 ft^3 of steel. In general, if the linear scaling factor for the cube is M, the pressure on the bottom face is M times as much.

E X A M P L E ▶ *What About a 10-Foot Cube?*

If we scale the original cube of steel up to a cube 10 ft on a side, then the dimensions are

10 ft $\times$ 10 ft $\times$ 10 ft

The total volume is

$$V = \text{length} \times \text{width} \times \text{height}$$
$$= 10 \text{ ft} \times 10 \text{ ft} \times 10 \text{ ft} = 1000 \text{ ft}^3$$

The weight of the cube is

$$W = V \times \text{density}$$
$$= 1000 \text{ ft}^3 \times 500 \text{ lb/ft}^3 = 500{,}000 \text{ lb}$$

The area of the bottom face is

$$A = \text{length} \times \text{width}$$
$$= 10 \text{ ft} \times 10 \text{ ft} = 100 \text{ ft}^2$$

The pressure on the bottom face is

$$P = \frac{W}{A} = \frac{500{,}000 \text{ lb}}{100 \text{ ft}^2} = 5000 \text{ lb/ft}^2$$

This is 10 *times*—not "10 times *more* than"—the pressure on the bottom face of the original 1-foot cube. ◆

At some scale factor, the pressure on the bottom face will exceed the steel's ability to withstand that pressure—and the steel will deform under its own weight. That point for steel is reached for a cube about 3 miles on a side—the pressure exerted by the cube's weight exceeds the resistance to crushing (ability to withstand pressure, or **crushing strength**) of steel, which is about 7.5 million lb/ft². Since a mile is 5280 ft, a 3-mile-long cube of steel would be more than 15,000 times as long as the original 1-foot cube; that is, the linear scaling factor is more than 15,000. The pressure on the bottom face of the cube would therefore be more than 15,000 times as much as for the 1-foot cube, or $15{,}000 \times 500 \text{ lb/ft}^2 = 7.5$ million lb/ft².

EXAMPLE ▶ *What About the Petronas Towers?*

At 1462 ft (445 m), the twin Petronas Towers in Kuala Lumpur, Malaysia, are the tallest buildings in the world, if we don't count radio and television antennas. What is the pressure on the bottom of their walls?

The towers are made of reinforced concrete, which weighs about 160 lb/ft³. Over each square foot of bottom surface of one of their walls stands 1462 ft³ of reinforced concrete, which weighs $1462 \times 160 = 234{,}000$ lb. The pressure at the bottom of the wall, from the wall's weight alone, is 234,000 lb/ft². That's not counting the contents of the tower, which also must be supported!

Could we have a Super Petronas Tower 10 times as high? The bottom of its walls would have to support 2.4 million lb/ft^2. The crushing strength of reinforced concrete is about 8.5 million lb/ft^3, which would leave some safety margin.   ◆

Sorry, No King Kongs

Unfortunately, the resistance of bone to crushing is not nearly as great as that of steel. This fact helps to explain why there couldn't be any King Kongs (unless they were made of steel!). A King Kong scaled up by a factor of 20 would weigh $20^3 = 8000$ times as much. Though the weight increases with the cube of the linear scaling factor, the ability to support the weight—as measured by the cross-sectional area of the bones, like the area of the bottom face of the cube in Figure 17.3—increases only with the square of the linear scaling factor.

These simple consequences of the geometry of scaling apply not only to supermonsters but also to other objects, such as trees and mountains.

EXAMPLE ▶ *How Tall Can a Tree Be?*

Galileo suggested that no tree could grow taller than 300 feet (see Spotlight 17.1). The world's tallest trees are giant sequoias, which grow only on the west coast of the United States, and hence were unknown to Galileo. They grow to 360 feet (Figure 17.4).

FIGURE 17.4 Even these giant sequoias may grow no taller than their form and materials allow.

S*potlight* *Galileo and the Problem of Scale*

17.1

The Italian physicist and astronomer Galileo Galilei (1564–1642) was the first to describe the problem of scale, in 1638, in his *Dialogues Concerning Two New Sciences* (in which he also discussed the idea of the earth revolving around the sun):

You can plainly see the impossibility of increasing the size of structures to vast dimensions either in art or in nature; likewise, the impossibility of building ships, palaces, or temples of enormous size in such a way that their oars, yards, beams, iron-bolts, and, in short, all their other parts will hold together; nor can nature produce trees of extraordinary size because the branches would break down under their own weight, so also would it be impossible to build up the bony structures of men, horses, or other animals so as to hold together and perform their normal functions if these animals were to be increased enormously in height; for this increase in height can be accomplished only by employing a material which is harder and stronger than usual, or by enlarging the size of the bones, thus changing their shape until the form and appearance of the animals suggest a monstrosity.

To illustrate briefly, I have sketched a bone whose natural length has been increased three times and whose thickness has been multiplied until, for a correspondingly large animal, it would per-

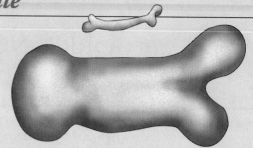

One bone, with another three times as long and thick enough to perform the same function in a scaled-up animal.

form the same function which the small bone performs for its small animal. From the figures shown here you can see how out of proportion the enlarged bone appears. Clearly then if one wishes to maintain in a great giant the same proportion of limb as that found in an ordinary man he must either find a harder and stronger material for making the bones, or he must admit a diminution of strength in comparison with men of medium stature; for if his height be increased inordinately he will fall and be crushed under his own weight. Whereas, if the size of a body be diminished, the strength of that body is not diminished in proportion; indeed the smaller the body the greater its relative strength. Thus a small dog could probably carry on his back two or three dogs of his own size; but I believe that a horse could not carry even one of his own size.

Translated by Henry Crew and Alfonso De Salvo, and published by Macmillan, 1914, and Northwestern University, 1946.

What can limit the height of a tree? If the roots do not adequately anchor it, a tall tree can blow over. (This, in fact, happened in 1990 to the world's tallest tree, the Dyerville Giant, a giant sequoia in Humboldt Redwoods State Park in California.) The tree could buckle or snap under its own weight and the force of a strong wind. The wood at the bottom will begin to crush if there is too much weight above. Finally, there is a limit to how far the tree can lift water and minerals from the roots to the leaves.

Could a tree be a mile high? To make an easy but rough estimate of the pressure at the base of the tree due to gravity, let's model the tree as a perfectly vertical cylinder. Over each square foot at the bottom, there is 5280 ft^3 of cells of wood, which we may think of as a column of water. To calculate how much that weighs, we first translate 1 ft^3 into metric measurement:

$$1 \text{ ft}^3 = (12 \text{ in.})^3 = (12 \times 2.54 \text{ cm})^3 = 28{,}316 \text{ cm}^3$$

A reason to convert to cubic centimeters is the convenient fact that water weighs just about exactly 1 gram per cubic centimeter. Now, 1 ft^3 of water weighs about 28,300 g = 28.3 kg = 28.3 $\times$ 2.20 lb $\approx$ 62 lb. Consequently, 5280 ft^3 of water weighs 5280 $\times$ 62 lb $\approx$ 327,000 lb, so the pressure on the bottom layer would be about 327,000 lb/ft^2.

Actually, freshly cut wood weighs only about *half* as much as water, so the pressure on the bottom layer would be half this figure, or 164,000 lb/ft^2. This is still an overestimate, however, since we assumed that the tree does not taper. A tree that tapers steadily looks like an elongated cone; as we will see when we model a mountain, a cone of the same radius and height as a cylinder has only one-third of the cylinder's volume. Using the more realistic cone model, the pressure at the bottom of the tree is one-third of 164,000 lb/ft^2, or 55,000 lb/ft^2.

A biological organism needs a safety factor of at least 2 to 4 times the minimum physical limits for its processes, so a tree a mile high would need from 110,000 to 220,000 lb/ft^2 of upward pressure for water and minerals. Tension in the string of water molecules from root to leaf ranges from 80,000 to 3.2 million lb/ft^2, for different kinds and heights of trees, so this consideration does not rule out mile-high trees.

However, at more than about 500 lb/in.^{2} = 70,000 lb/ft^2, the bottom of the tree would begin to crush under the weight above. On this point, the mile-high tree is barely feasible, with little margin of safety.

These considerations suggest that trees a mile high might be *physically* possible, but others suggest a lower maximum height. In addition, there are also biological considerations. The taller the tree, the greater the area from which it must draw water and minerals, for which nearby trees also compete. Moreover, for a tree to grow very tall, it would have to live for a very long time. Evolution and time may select against extremely tall trees; or maybe, for no reason at all, they have just never evolved. ◆

E X A M P L E ▶ *How High Can a Mountain Be?*

Gravity and the physical characteristics of wood limit the height of trees. Gravity also limits the height of mountains. Mountains differ in composition and shape, and some assumptions about those features are necessary to do any calculating. We want to make realistic assumptions that make it easy to estimate how high a mountain can be. In effect, we build a simple mathematical model of a mountain.

Let's suppose that the mountain is made entirely of granite, a common material in many mountains, and let's assume that the granite has uniform density. Granite weighs 165 lb/ft^3 and has a crushing strength of about 4 million lb/ft^2.

In the interests of both realism and simplicity, we assume that the mountain is in the shape of a cone whose width at the base is the same as its height. Let's model Mount Everest: the tallest earth mountain, it is about 6 miles high. The base, then, is a circle with a distance across (or diameter) of 6 miles. The radius of the circle is half the diameter, so the model Everest has a radius of 3 miles measured at the base (Figure 17.5). Since we are taking such a round number for the height of Everest, we record as significant only the first two digits of the results of the calculations.

What does the model Everest weigh? The relevant formula is

$$\text{weight} = \text{density} \times \text{volume}$$

We already know the density of granite (165 lb/ft^3), so to find the weight we need the formula for the volume of a cone of radius r and height h:

$$\text{volume} = \frac{1}{3}\,\pi r^2 h$$

For Everest, the radius is 3 miles and the height is 6 miles; π (pi) is about 3.14. Using those values in the formula, we find that the model Everest has a volume of about 57 cubic miles.

To find the weight of 57 cubic miles of granite, we need to convert units, since the density is given in pounds per cubic foot. Let's convert to units of feet:

$$
\begin{aligned}
1\ \text{mi}^3 &= 1\ \text{mi} \times 1\ \text{mi} \times 1\ \text{mi} \\
&= 5280\ \text{ft} \times 5280\ \text{ft} \times 5280\ \text{ft} \\
&\approx 1.5 \times 10^{11}\ \text{ft}^3
\end{aligned}
$$

Thus

$$57\ \text{mi}^3 \approx 57 \times 1.5 \times 10^{11}\ \text{ft}^3 \approx 8.6 \times 10^{12}\ \text{ft}^3$$

So we have

$$
\begin{aligned}
\text{weight of mountain} &= 165\ \text{lb/ft}^3 \times 8.6 \times 10^{12}\ \text{ft}^3 \\
&\approx 1.4 \times 10^{15}\ \text{lb} \\
&\approx 1.4\ \text{quadrillion lb}
\end{aligned}
$$

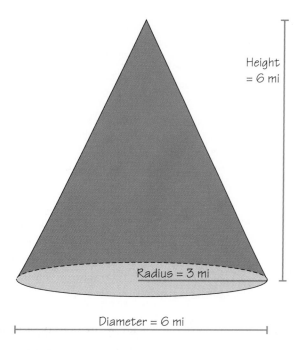

FIGURE 17.5   Model of Mt. Everest as a cone of granite.

Now that we know the weight of the mountain, we want to find out what the pressure is on the base of the cone and compare that with the crushing strength of granite. (Everest is standing, so if our model is any good, that pressure will be below the crushing strength.) Physics tells us that the weight of the mountain is spread evenly over the base of the cone (though we are oversimplifying the geology underlying mountains). Since

$$\text{pressure} = \frac{\text{weight}}{\text{area}}$$

we need to calculate the area of the base of the cone. The shape is a circle, and the familiar formula

$$\text{area} = \pi r^2$$

gives an area of 28 square miles for a radius of 3 miles.

Once again, we need to convert units in order to express the pressure in pounds per square foot, the units in which the crushing strength is expressed. We get

$$\text{area} = 28 \text{ mi}^2 = 28 \times 5280 \text{ ft} \times 5280 \text{ ft} \approx 8 \times 10^8 \text{ ft}^2$$

Then

$$\text{pressure} = \frac{\text{weight}}{\text{area}}$$

$$= \frac{1.4 \times 10^{15} \text{ lb}}{8 \times 10^8 \text{ ft}^2}$$

$$= 1.8 \times 10^6 \text{ lb/ft}^2$$

This number is below the crushing strength of granite, 4×10^6 lb/ft^2, with a safety factor of about 2.

For a mountain to come close to the limitation of the crushing strength of granite, it would have to be only about twice as high as Everest, or about 10 miles high. Other physical considerations suggest a maximum height of at most 15 miles. That no present mountains are that high may be a consequence of the earth's high amount of volcanic activity and the structural deformation of the earth's crust. ◆

What about mountains made of other materials—glass, ice, wood, old cars? They couldn't be nearly as high; the pressure would cause glass to flow, ice to melt, and old cars to compact. What about mountains on another planet? Their potential height depends on the gravity of the planet.

SOLVING THE PROBLEM OF SCALE

A large change in scale forces a change in either materials or form. A major manifestation of the scaling problem is the tension between weight and the need to support it. For example, a real building or machine must differ from a scale model: the balsa wood or plastic of the model would never be strong enough to use for the real thing, which must use aluminum, steel, or reinforced concrete. So one way to compensate for the problem of scale is to use stronger materials in the scaled-up object.

Another way to compensate is to redesign the object so that its weight is better distributed. Let's go back to the original cube. It supports all its weight on its bottom face. In the version scaled up by a factor of 3, each small cube of the bottom layer has a bottom face that is supporting that cube's weight plus the weight of the other two cubes piled on top of it.

Let's redesign the scaled-up cube, concentrating for simplicity only on the front face, with its nine small cubes. We take the three cubes on top and move them to the bottom, alongside the three already there. We take the three cubes on the second level, cut each in half, and put a half cube over each of the six

FIGURE 17.6 Nine small cubes rearranged to support greater weight.

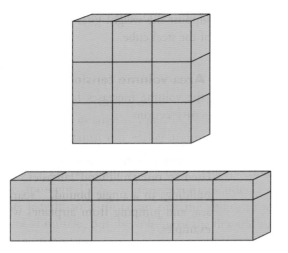

ground-level cubes (see Figure 17.6). We have the same volume and weight that we started with, but now there is less pressure on the bottom face of each small cube. Of course, the new design is not geometrically similar to the object that we started with—it's no longer a cube. We have solved the scaling problem by changing the proportions and redistributing the weight.

We observe in nature both strategies for adaptation to scaling: change of materials and change of form. Smaller animals generally do not have bony internal skeletons; larger animals generally do. Animals made of similar materials but differing greatly in size, such as a mouse and an elephant, most certainly differ in shape. If a mouse were scaled up to the size of an elephant, its legs could no longer support it; it would need the disproportionately thicker legs of the elephant. It would also need the elephant's thick hide to contain its tissue.

Some dinosaurs, like *Supersaurus* (which weighed 30 tons), had special adaptations to lighten their weight, such as hollow bones, just as some birds have. (Hollow bones also turn out to be stronger, a paradox that Galileo analyzed. Of two bones of the same weight and length, the hollow one is wider across at its midpoint, because of the air it contains; and the greater the width, the greater the resistance to fracture.)

FALLS, DIVES, JUMPS, AND FLIGHTS

The need to support weight can be thought of as a tension between volume and area. As an object is scaled up, its volume and weight go up together, as long as the density remains constant (for example, no air bubbles introduced into the steel to make it into a Swiss cheese!). At the same time, the ability to

Shakespeare painted Richard III as a humpbacked Machiavellian monster. Did Richard have an advantage in armored combat because he was short? That suggestion was made some years ago by one of the leading modern historians of the Tudor era, Garrett Mattingly.

Between a short man and a tall man, height increases by the linear dimension—from 5 feet 2 inches, say, to 6 feet—while the surface of the body increases as the square. Since it's the surface of the body that the armorer must plate with steel, the armor of a short warrior, like Richard, would be lighter than a tall warrior's by a lot more than the few inches' difference in height would indicate. So Richard's notorious deadliness in battle would have been possible at least in part because his armor, while protecting him as well as the big man's, left him less encumbered.

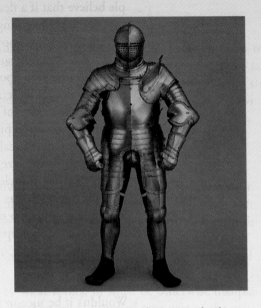

Did wearing armor give an advantage to the shorter warrior?

Of course, ostriches are not just scaled up sparrows, nor are eagles. The larger flying birds have disproportionately larger wings than a sparrow, to keep the wing loading down. The largest animal ever to have taken to the air was *Quetzalcoatlus northropi*, a flying reptile of 65 million years ago, with a wingspan of 36 feet and a weight of about 100 pounds.

You have to stay up, you have to keep moving—and you have to get up there. Here basic aerodynamics imposes further limits. Paleontologists originally thought that *Quetzalcoatlus northropi* weighed 200 pounds and had a 50-foot wingspan. Even though that works out to just about the same wing loading as for 100 pounds and a wingspan of 36 feet, other considerations from aerodynamics show that at 200 pounds, the reptile wouldn't have been able to get off the ground. ◆

After the lecture, someone said to Mattingly that he had grasped the right idea—but by the wrong end. Muscle power, the listener claimed, is a matter of bulk—and physical volume goes up by the cube, whereas the surface to be protected goes up by the square. So the large warrior should have more strength left over than the little guy after putting on his armor. And the large warrior, swinging a bigger club, can deliver a far more punishing blow—because the momentum of the club depends on its weight, which goes up with its volume, which means by the cube. Richard was at a terrible disadvantage.

But wait a minute, a second listener said. That's true about the club—but not about the muscles. The strength of a muscle is proportional not to its bulk but to the area of its cross section. And since the cross section of muscles increases by the square, just as the surface of the body does, the big guy, plated out, has no more, or less, advantage over the little guy than if both were naked.

But hang on, a third person interjected—an engineer. That's right about the muscles, but it's not right about the armor. The weight of the armor increases not simply with the increase in the surface area that it must cover but slightly faster. There must be reinforcing ribs. Or else the metal must be significantly thicker overall. So maybe Richard had an advantage.

Armor was made as thin as possible. Thickness, reinforcement, and structural stiffening were concentrated where opponents' weapons were likely to hit. From these strong, shaped places, the metal tapered away, until the sheet steel was as thin as the lid of a coffee can at the sides of the rib cage beneath the arms, or across the fingers, or at the cheek of a helmet.

Source: Horace F. Judson, *The Search for Solutions* (Baltimore: Johns Hopkins University Press, 1987), pp. 54–56.

KEEPING COOL (AND WARM)

Area-volume tension is also crucial to an animal's maintenance of thermal equilibrium. Both warm-blooded and cold-blooded animals gain or lose heat from the environment in proportion to body surface area. A warm-blooded animal usually is losing heat; its basal metabolism, or rate of food intake needed to maintain body heat, depends primarily on the amount of its surface area, the temperature of its environment, and the insulation provided by its coat or skin. Other factors being equal, a scaled-up mammal scales up its food consumption by *surface area* (proportional to the square of the linear scaling factor), *not by volume* (proportional to its cube). For example, a mouse eats about half of its weight in food every day, while a human consumes only about one-fiftieth of its own weight.

Mammals regulate their metabolism and maintain a constant internal body temperature. Cold-blooded animals, such as alligators or lizards, have a somewhat different problem. They absorb heat from the environment for energy, but they must also dissipate any excess heat to keep their temperature below unsafe levels. The amount of heat that must be gained or lost is proportional to total volume, because the entire animal must be warmed or cooled. But the heat is exchanged through the skin, so the rate is proportional to surface area.

Dimetrodon was a large mammal-like reptile that roamed present-day Texas and Oklahoma 280 million years ago (see Figure 17.7). *Dimetrodon* had a great "sail" or fan on its back. As an individual grew, and as the species evolved, the sail grew. But it did not grow according to *geometric similarity*, the kind of growth we refer to as *proportional growth*.

> **Proportional growth** is growth according to geometric similarity: the length of every part of the organism enlarges by the same linear scaling factor.

Instead, the area of *Dimetrodon*'s sail grew precisely in proportion to the volume of the animal, a fact that strongly suggests to paleontologists that the

FIGURE 17.7   *Dimetrodon* may have evolved a sail to absorb and dissipate heat efficiently.

sail was a temperature-regulating organ that was able to absorb or radiate heat. So, an individual *Dimetrodon* twice as long would have eight ($= 2^3$) times as much weight and volume and also a sail with eight times as much area. If it had grown according to geometric similarity, the sail would have been twice as high and twice as wide, and hence would have had only four times as much sail area. Larger specimens of *Dimetrodon* didn't look quite like scaled-up smaller ones; we would say that the sail grew disproportionately large compared to the rest of the animal.

Dimetrodon was a large animal, but the need for heat regulation is even more acute for smaller animals. Like human babies, small animals can lose heat quickly, because of their high ratio of surface area to volume. Leading paleontologists now believe that birds (most of whom are quite small) evolved from dinosaurs and that feathers are modified reptilian scales. Though not a prevailing view, it has been hypothesized that the wings of birds and insects evolved originally not for flight but as temperature control devices.

Some scientists have speculated that African Pygmies are small in part because a small body is better able to lose heat in the hot, humid climate of the Ituri Forest where Pygmies live. Other scientists have suggested that ancestors of human beings began walking on two legs in part to keep cool in a hot climate. Walking upright exposes much less area of the body to the rays of the sun than walking on all fours and also reduces the amount of water needed by about one-half.

SIMILARITY AND GROWTH

Although a large change of scale forces adaptive changes in materials or form, within narrow limits—perhaps up to a factor of 2—creatures can grow according to a law of similarity, that is, they can grow proportionally, so that their shape is preserved. A striking example of such growth is that of the chambered nautilus *(Nautilus pompilius)*. Each new chamber that is added onto the nautilus shell is larger than but geometrically similar to the previous chamber, and the shape of the shell as a whole—an *equiangular,* or *logarithmic,* spiral—remains the same (see Figure 17.8).

Most living things grow over the course of their lives by a factor greater than 2. We've seen with *Dimetrodon* that a big specimen was not just a scaled-up small one. Nor is a human adult simply a scaled-up baby. Relative to the length of the body, a baby's head is much larger than an adult's. The arms of the baby are disproportionately shorter than an adult's. In the growth from baby to adult, the body does not scale up as a whole. But different parts of the body scale up, each with a different scale factor. That is, a baby's eyes grow at

FIGURE 17.8 A chambered nautilus shell.

FIGURE 17.9 Modeling the changes in shape of a human head from infancy to adulthood.

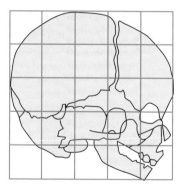

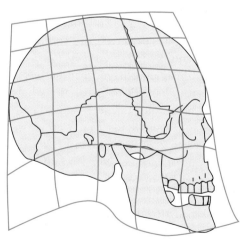

one rate to perhaps twice their original size, while the arms grow at another rate, to about four times their original size.

Although the laws for growth can be much more complicated than proportional growth (or even the allometric growth that we discuss in the next section), more sophisticated mathematics—for example, differential geometry, the geometry of curves and surfaces—permits analysis of complex and interlocking scalings. For a model of the process in which a baby's head changes shape to grow into an adult head, we can use graph paper: first, we put a picture of the baby's skull on graph paper, then we determine how to deform the grid until the pattern matches an adult skull (see Figure 17.9 and Spotlight 17.3). The same idea lies at the heart of computerized "morphing," the process in which the face of one film character can be made to change smoothly into the face of another, with different scalings for different parts of the face.

O p t i o n a l ▶ Allometry

If we measure the arm length or head size for humans of different ages and compare these measurements with body height, we observe that humans do not grow proportionally, that is, in a way that maintains geometric similarity. The head of a newborn baby may be one-third of the baby's length, but an adult's head is usually close to one-seventh of the individual's height. The arm, which at birth is one-third as long as the body, is by adulthood closer to two-fifths as long (see Figure 17.10a).

Ordinary graphing provides a way to test for differential growth. We can plot body height on the horizontal axis and arm length on the vertical axis (see Figure 17.10b). A straight line would indicate proportional growth, that is, according to geometric similarity; and we do get a straight line from age 9 months (0.75 years) or so on up. Up to age 9 months or so, we get a curve, which indicates that the ratio of arm length to height does not remain constant over the first year.

Is there an orderly law by which we can relate arm length to height? Let's plot again, this time using a different scale. For this **logarithmic scale,** we mark off equal units, as usual. But instead of labeling the marked points with 0, 1, 2, 3, etc., we label them with the corresponding powers of 10: $10^0 = 1$, $10^1 = 10$, $10^2 = 100$, $10^3 = 1000$, etc., which are also called **orders of magnitude.** Plotting a point on such a scale is not easy, since the point midway between 1 and 10 is not 5.5, but instead is closer to 3. Special graph paper (available in most college bookstores) marks smaller divisions and makes it easier to plot; paper marked with log scales on both axes is called **log-log paper,** while **semilog paper** has a logarithmic scale on just one axis. Also, many computer plotting packages can produce logarithmic scales.

Figure 17.10 (a) The proportions of the human body change with age. (b) A graph of human body growth on ordinary graph paper. The numbers shown beside the points indicate the age in years; they correspond to the stage of human development shown in part a. (c) A graph of human body growth on log-log paper.

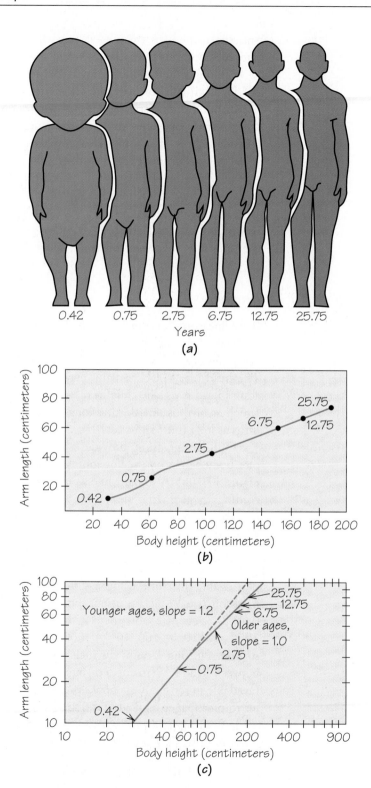

Spotlight

17.3

Helping to Find Missing Children

Photograph at age 10.

Age progression to age 15.

Lekeasha E. Gordon
Reported missing in 1992.

It can be valuable to be able to predict what a developing organism will look like in the future. For example, what does a child look like now who was kidnapped six years ago, at age 3?

At the National Center for Missing and Exploited Children (NCMEC) in Arlington, Virginia, a computer and a more sophisticated version of the graph-paper technique are used to answer such questions. Computer age-progression specialists scan photographs of both the missing child at age 3 and an older sibling or a biological parent at age 9 into a computer. Then the face of the 3-year-old is stretched, depending on age, to reflect craniofacial growth and merged with the image of the sibling or parent at 9 years old. The result is a rough idea of what the missing child may look like. As mathematicians and biologists refine their models of how faces change over time, this technique will improve. It may even become possible to gain an idea of how a child may look at age 40 or 65.

We could use a logarithmic scale for either height or arm length, or for both. Using logarithmic scales for both, as in Figure 17.10c, the data plot closely to a straight line. Looking carefully, we can discern two different straight lines: a steeper one that fits early development (we will see shortly that it has slope 1.2), and a less steep one (with slope 1.0) that fits development after 9 months of age.

The change from one line to another after 9 months indicates a change in pattern of growth. The pattern after 9 months, characterized by the straight line with slope 1, is indeed proportional growth (sometimes called **isometric growth**). For the pattern before 9 months, we know from the slope (1.2) being greater than 1 that arm length is increasing relatively faster than height. That earlier growth also follows a definite pattern, called *allometric growth*.

> **Allometric growth** is the growth of the length of one feature at a rate proportional to a power of the length of another.

We have seen that in geometric scaling, area grows according to the square (second power) and volume according to the cube (third power) of length, so we can say that they grow allometrically with length.

If we denote arm length by y and height by x, a straight-line fit on log-log paper corresponds to the algebraic relation

$$\log_{10} y = B + a \log_{10} x$$

where a is the slope of the line and B is the point where the graph crosses the vertical axis. If we raise 10 to the power of each side, we get

$$y = bx^a$$

where $b = 10^B$. This equation describes a **power curve:** y is a constant multiple of x raised to a certain power.

We can find approximate values for a for each of the two lines in the log-log plot from the coordinates of the points at the ends of the line. Those are the points for ages 0.42, 0.75, and 25.75. The observations and the corresponding logarithms are:

Age	Height	Log (height)	Arm Length	Log (arm length)
0.42	30.0	1.48	10.7	1.03
0.75	60.4	1.78	25.1	1.40
25.75	180.8	2.26	76.9	1.89

The slope for the line from age 0.42 to age 0.75 is the vertical change over the horizontal change, both converted to log units:

$$\frac{\log 25.1 - \log 10.7}{\log 60.4 - \log 30.0} = \frac{1.40 - 1.03}{1.78 - 1.48} = \frac{0.37}{0.30} \approx 1.2$$

The slope for the line from age 0.75 to age 25.75 is

$$\frac{\log 76.9 - \log 25.1}{\log 180.8 - \log 60.4} = \frac{1.89 - 1.40}{2.26 - 1.78} = \frac{0.49}{0.48} \approx 1.0$$

So $a = 1.2$ up to 9 months, and $a = 1.0$ after 9 months. Up to 9 months, arm length grows according to (height)$^{1.2}$; after 9 months, arm length grows according to $(\text{height})^{1.0}$. We get $y = bx^{1.0}$, which is a linear relationship describing proportional growth, that is, growth according to geometric similarity. On ordinary graph paper, proportional growth appears as a straight line, allometric growth as a curve. On log-log paper, both patterns appear as straight lines.

The technique of allometry has been used in the last few years by paleontologists to determine that all of the six known specimens identified as the earliest fossil bird *Archaeopteryx* are indeed from the same species, and that the minute fossil and puzzling fish known as *Palaeospondylus* (found only in Scotland) is probably just the larval stage of some better known fish.   ◆

CONCLUSION

We have examined the problem of scale and noted that a large change in scale forces a change in either materials or form. A particular instance of the problem of scale is area-volume tension, and we have seen how an animal's size and geometric shape affect its abilities to move and to keep itself warm or cool.

In this chapter we have explored the limitations on life imposed by dwelling in three dimensions. In Chapter 20 we will see that dimensionality also imposes surprising limits on artistic creativity in devising patterns.

REVIEW VOCABULARY

Allometric growth   A pattern of growth in which the length of one feature grows at a rate proportional to a power of the length of another feature.

Area-volume tension   A result of the fact that as an object is scaled up, the volume increases faster than the surface area and faster than areas of cross sections.

Crushing strength   The maximum ability of a substance to withstand pressure without crushing or deforming.

Density   Weight per unit volume.

Geometrically similar   Two objects are geometrically similar if they have the same shape, regardless of the materials of which they are made. They need not be of the same size. Corresponding linear dimensions must have the same factor of proportionality.

Isometric growth   Proportional growth.

Linear scaling factor   The number by which each linear dimension of an object is multiplied when it is scaled up or down; that is, the ratio of the length of any part of one of two geometrically similar objects to the length of the corresponding part of the second.

Logarithmic scale A scale on which equal divisions correspond to powers of 10.

Log-log paper Graph paper on which both the vertical and the horizontal scales are logarithmic scales, that is, the scales are marked in orders of magnitude 1, 10, 100, 1000, . . . , instead of 1, 2, 3, 4, ...

Mass The aspect of matter that is affected by forces, according to physical laws.

Orders of magnitude Powers of 10.

Power curve A curve described by an equation $y = bx^a$, so that y is proportional to a power of x.

Pressure Weight divided by area.

Problem of scale As an object or being is scaled up, its surface and cross-sectional areas increase at a different rate from its volume, forcing adaptations of materials or shape.

Proportional growth Growth according to geometric similarity: the length of every part of the organism enlarges by the same linear scaling factor.

Semilog paper Graph paper on which only one of the scales is a logarithmic scale.

Wing loading Weight supported divided by wing area.

SUGGESTED READINGS

CAMPBELL, R. B. Hercules' height, *UMAP Journal*, 5 (1984): 265–269. Pythagoras calculated the height of Hercules (who must have been a legend already in Pythagoras's time!) by assuming proportional scaling.

DEWDNEY, A. K. *200% of Nothing: An Eye-Opening Tour Through the Twists and Turns of Math Abuse and Innumeracy*, Wiley, New York, 1993.

DIAMOND, JARED M. Why are pygmies small? *Nature*, 354 (1991): 111–112.

DUDLEY, BRIAN A. C. *Mathematical and Biological Interrelations*, Wiley, New York, 1977. Excellent and gentle extended introduction to graphing, scale factors, and logarithmic plots.

FLASPOHLER, DAVID C., FRANK MASTRIANNA, AND RICHARD PULSKAMP. *The Consumer Price Index: What Does It Mean?*, 2nd ed. UMAP Modules in Undergraduate Mathematics and Its Applications: Module 639. COMAP, Inc., Lexington, Mass., 1994. Reprinted in *UMAP Modules: Tools for Teaching, 1993,* COMAP, Inc., Lexington, Mass., 1994.

GOULD, STEPHEN JAY. The origin and function of bizarre structures: Antler size and skull size in the "Irish elk," *Megaloceros giganteus. Evolution,* 28 (1974): 191–220. Logarithmic plots solve a longstanding mystery.

GOULD, STEPHEN JAY.   Size and shape. In *Ever Since Darwin,* Norton, New York, 1977, chapter 21.

HALDANE, J. B. S.   On being the right size. In *Possible Worlds and Other Papers,* Harper, New York, 1928. Reprinted in James R. Newman (ed.), *The World of Mathematics,* vol. 2, Simon & Schuster, New York, 1956, pp. 952–957. Also reprinted in John Maynard Smith (ed.), *On Being the Right Size and Other Essays by J. B. S. Haldane,* Oxford University Press, Oxford, 1985, pp. 1–8. Succinctly surveys area-volume tension, flying, the size of eyes, and even the best size for human institutions.

HILDEBRANDT, STEFAN, AND ANTHONY J. TROMBA.   *Mathematics and Optimal Form,* Scientific American Library, New York, 1985.

HOUCK, MARILYN A., JACQUES A. GAUTHIER, AND RICHARD E. STRAUSS.   Allometric scaling in the earliest fossil bird, *Archaeopteryx lithographica, Science,* January 12, 1990, 195–198.

HUXLEY, JULIAN.   The size of living things. In *Man Stands Alone,* Harper Brothers, New York, 1942.

HUXLEY, JULIAN.   *Problems of Relative Growth,* Methuen, London, 1932. Reprint Dover, New York, 1972. Many semilog and log-log plots of biological relations.

MCMAHON, T. A., AND J. T. BONNER.   *On Size and Life,* Scientific American Library, New York, 1983. Astonishingly beautiful and informative book on the effects of size and shape on living things.

MINEYEV, ANATOLY.   Trees worthy of Paul Bunyan: Why do trees grow so tall — but no taller? *Quantum,* 4 (January/February 1994): 4–10.

STEVENS, PETER S.   *Patterns in Nature,* Atlantic Monthly Press, Boston, 1974. Splendid treatment of the problem of scale and other physical phenomena in nature: flows, meanders, branching, trees, soap films, cracking, and packing.

THOMPSON, D'ARCY.   *On Growth and Form,* Cambridge University Press, Cambridge, 1917, 1961. "A discourse on science as though it were a humanity" (J. T. Bonner); this was the first book to describe in quantitative terms the processes of growth and shaping of biological forms.

THOMSON, KEITH STEWART.   The puzzle of *Palaeospondylus. American Scientist,* 80 (1992): 216–219.

TREFIL, JAMES S.   What would a giant look like? In *The Unexpected Vista: A Physicist's View of Nature,* chapter 10, pp. 156–171, Scribner, New York, 1983. Explanation of the effects of scaling up. In Trefil's illustration on p. 162, however, the eyes of the giants are unrealistically large.

WENT, F. W.   The size of man. *American Scientist,* 56 (1968): 400–413. Demonstrates a schism between the macroworld and the molecular world, by comparing human life with that of an ant.

WGBH EDUCATIONAL FOUNDATION AND PEACE RIVER FILMS.   *Nova: The Shape of Things,* 1985. Distributed by Vestron Video, Box 4000, Stamford, CT 06907.

ZHERDEV, A.   Horseflies and flying horses: Questions of scale in the animal kingdom. *Quantum,* 4 (May/June 1994): 32–37, 59–60.

EXERCISES

▲   *Optional.*   ■   *Advanced.*   ◆   *Discussion.*

Most of the exercises require a calculator; one that offers square roots will suffice.

Geometric Similarity

1.   Suppose you are printing photographs from negatives of so-called 35-millimeter film, whose frames are just about 1 inch by $1\frac{1}{2}$ inches (the actual size is 24 mm (millimeters) by 36 mm).

(a)   First you make some contact prints, which are exactly the same size as the negatives. What is the scaling factor of a contact print?

(b)   One enlargement you want to make is to be three times as high and three times as wide as the negative. What is the linear scaling factor for this print? How does its area compare with the area of the negative?

(c)   What is the scaling factor for a 4-by-6 print? What is the area of the print?

(d)   The size of so-called 3-by-5 prints can vary, depending on whether the print has a border or not; a common size is about $3\frac{1}{16}$ by $4\frac{19}{32}$ inches. Is this print geometrically similar to the negative?

(e)   The cost of photographic paper is very nearly proportional to the area of the paper. Suppose you are comparing the cost of getting 3-by-5 enlargements versus 4-by-6 enlargements, and let's assume for the sake of simplicity that the prints are exactly 3 in. by 5 in. and 4 in. by 6 in. The smaller prints cost 17 cents each, and the larger cost 50 cents each. From what you know about scaling, what can you say about the relative cost of the two kinds of prints?

(f) Based on the amount of paper used, what would you expect a 7-by-10 print to cost, considering the cost of the 3-by-5 prints in part (e)? Considering the cost of the 4-by-6 prints in part (e)?

2. The area of a circle of radius r is πr^2; expressed in terms of the diameter, $d = 2r$, the area is $\frac{1}{4}\pi d^2$. If we apply a linear scaling factor M to the diameter of a circle, then — as in the case of the square we considered in the text — the area of the scaled circle changes with M^2, the square of the linear scaling factor. A natural application of this idea, of course, is to your local pizza parlor and the prices on its menu. The actual prices at the pizza restaurant closest to Beloit College are $6, $7, $7.95, and $8.95, respectively, for small (10-inch), medium (12-inch), large (14-inch), and extra large (16-inch) cheese pizzas.

(a) What is the linear scaling factor for an extra large pizza compared to a small one?

(b) How many times as large in area is the extra large pizza compared to the small one?

(c) How much pizza does each size give per dollar? What "hidden" assumptions are you making about how the pizzas are scaled up?

(d) The corresponding prices for a pizza with "the works" are $8.25, $9.75, $11.95, and $13.95. Is there any size of these for which you get more pizza per dollar than some size of the cheese pizzas?

3. Toy trains, sometimes called model trains, come in various sizes or gauges. Not all toy trains are exact scale models of real trains, but some are.

(a) HO-gauge toy trains are usually built to an exact scale of 1 to 87, meaning that a part one foot long on the real train measures one eighty-seventh of a foot on the toy train. What is the linear scaling factor of an HO-gauge toy train?

(b) How does the volume of a real boxcar compare with the volume of an HO-gauge scale model?

(c) O-gauge toy trains are built to a scale of approximately $\frac{1}{4}$ inch to a foot, meaning that a part 1 ft long on the real train measures $\frac{1}{4}$ in. on the toy train. (In fact, O-gauge trains tend to be a little shorter than exact scale would demand, and their wheels are oversized compared to exact scale.) What is the linear scaling factor of an O-gauge toy train?

4.   Doll houses and their furnishings are customarily built to a scale of exactly 1 in. to 1 ft, meaning that an item 1 ft long in a real house is 1 in. long in a doll house.

   (a)   What is the linear scaling factor for a doll house?
   (b)   If a doll house were made of the same materials as a real house, how would their weights compare?

5.   Two geometric figures are *similar* if they have the same shape but not necessarily the same size. Indicate whether the geometric figures described below are always, sometimes, or never similar:

   (a)   Two squares
   (b)   Two isosceles triangles
   (c)   Two equilateral triangles
   (d)   Two pentagons
   (e)   Two regular pentagons
   (f)   Two rectangles
   (g)   A square and a rectangle
   (h)   Two circles
   (i)   A regular pentagon and a regular hexagon
   (j)   Two angles

6.   Identify each of the following statements as either true or false:

   (a)   Every polygon is similar to itself.
   (b)   If polygon *A* is similar to polygon *B* and polygon *B* is similar to polygon *C*, then polygon *A* is similar to polygon *C*.
   (c)   Corresponding interior angles of similar polygons are congruent.

7.   One of the famous problems of Greek antiquity was the *duplication of the cube*. Our knowledge of the history of the problem comes down to us from the third century B.C. from Eratosthenes of Cyrene, who is famous for his estimate of the circumference of the earth. According to him, the citizens of Delos were suffering from a plague. They consulted the oracle, who told them that to rid themselves of the plague, they must construct an altar to a particular god that would be geometrically similar to the existing one but double the volume.

   (a)   How would the volume of the new altar compare with the old if each of its linear dimensions were doubled?

(b)   What should the linear scaling factor be for the new altar? (The problem intended by the oracle was to construct with straight-edge and compasses a line segment equal in length to this particular linear scaling factor. Not until the nineteenth century was the task shown to be impossible. Eratosthenes relates that the Delians interpreted the problem in this sense, were perplexed, and asked Plato about it. Plato told them that the god didn't really want an altar of double the volume but wished to shame them for their "neglect of mathematics and their contempt for geometry.")

8.   The declining purchasing power of the dollar and the short life of a dollar bill in circulation suggest the desirability of abolishing the dollar bill in favor of a dollar coin. The Susan B. Anthony dollar coin of 1979–1980 was a failure with the U.S. public, who found it too small and light. Suppose you have been put in charge of designing a new dollar coin that is to be made of the same material as the current U.S. 25-cent piece and weigh four times as much. A quarter can be described geometrically as a circular cylinder approximately $\frac{15}{16}$ in. in diameter and $\frac{1}{16}$ in. thick. Since your new dollar should weigh four times as much, it needs to have four times the volume of a quarter. (You may find it helpful that the formula for the volume of a cylinder is $\pi \times (\text{diameter}/2)^2 \times \text{height}$.)

(a)   A member of your public advisory panel suggests that the requirements will be fulfilled if you just double the diameter and double the thickness. What do you tell this individual, in the most diplomatic terms?

(b)   If you go along with the member's suggestion to double the diameter, how thick does the coin need to be?

(c)   Another member of the board feels that doubling the diameter would produce a coin too large to be convenient and proposes instead that you scale up the quarter proportionally (she took a course from an earlier edition of this book). What would the dimensions be for this new dollar?

The Language of Growth, Enlargement, and Decrease

9.   Criticize the following statement, which appears on sacks of the product, and write a correct version:

"Erin's Own Irish sphagnum moss peat. It enriches your soil and makes your growing easier. Compressed to $2\frac{1}{2}$ times normal volume."

10.   Criticize the following claims, which were cited in the *New York Times* of 9/25/87 and 10/21/87:

(a)   A new dental rinse "reduces plaque on teeth by over 300%."
(b)   An airline working to decrease lost baggage has "already improved 100% in the last six months."
(c)   "If interest rates drop from 10% to 5%, that is a 100% reduction."

Numerical Similarity

11.   Here are some problems on using the Consumer Price Index (CPI) (see Table 17.1):

(a)   I bought my first LP record in 1965, at list price, for $4.98. How much would that be in 1997 dollars? How does that compare with the list price of a CD today?
(b)   My father bought a Royal portable typewriter in 1940 for $40 (I have the sales slip). What would be the equivalent price in 1997 dollars? How does that compare to the cost of a portable typewriter today?
(c)   My first-semester college mathematics book cost $10.75 in 1962. What would be the equivalent price in 1997 dollars? How does that compare to what you paid for this book? (My book had black and white text and figures, with no photographs, color or otherwise.)
(d)   In 1970, before the OPEC oil embargo, gasoline cost about 25 cents per gallon. In 1974, after the embargo, it cost about 70 cents per gallon. What would be the equivalent prices in 1997 dollars? How do they compare to the price of gasoline today?

12.   From the CPI table (see Table 17.1), you can determine the rate of inflation from one year to the next. For example, you find the rate of inflation from 1991 to 1992 by subtracting the two index numbers and dividing by the earlier one: $(140.6 - 136.2)/136.2 = 0.032 = 3.2\%$. Similarly, knowing the rate of inflation, you can compute one index number from another. (Thus, from learning the rate of inflation from newspapers, you can add entries to the CPI table for years past 1997.)

(a)   What was the rate of inflation from 1980 to 1981?
(b)   For a 3% rate of inflation per year from 1997 on, what would the CPI be in the year 2000?
(c)   Gasoline cost about $1.10 per gallon in mid-1992, when Ross Perot proposed increasing federal tax on gasoline by 10 cents each year for five years in order to pay for investments in U.S.

infrastructure and to reduce dependence on oil imports. Assume that from 1992 to 1997, the price of gasoline (exclusive of Perot's proposed tax) went up according to the CPI and that in 1997 there was in addition the proposed additional $0.50 tax per gallon. What would the price of gasoline have been in 1997?

(d) Compare your answer to part (c) with your answers to Exercise 11, part (d). Discuss possible conclusions that can be made from your calculations.

13. In Germany the fuel efficiency of cars is measured in terms of liters of gasoline used per 100 kilometers traveled. On a recent trip there, driving a subcompact car, we averaged 6.5 liters per 100 km. What is the equivalent in miles per gallon?

Measuring Length, Area, Volume, and Weight

14. A *light-year* is not a measure of time but of distance; it is how far light travels in a year. Light travels at 2.9979×10^8 m/sec in a vacuum.

(a) How long is a light-year in kilometers?
(b) In miles?
(c) In angstroms (1 Å $= 10^{-10}$ m)?

15. Consider a real locomotive that weighs 88 tons and an HO-gauge scale model of it. (See Exercise 3.)

(a) How much would an exact scale model weigh, in tons?
(b) What assumptions are involved in your answer to part (a)?
(c) How much would an exact scale model weigh, in pounds?
(d) In kilograms?
(e) In metric tonnes (1 metric tonne $= 1000$ kg)?

16. An ad for a software package for data analysis on the Apple II included a data set on tropical rain forests and deforestation. The data were given in hectares and were accompanied by the statement "A hectare equals 10,000 mi^2 or 2471 acres." What conversion factors should have appeared instead?

17. Gasoline is sold in the United States by the U.S. gallon and in Canada by the liter (1 U.S. gallon $= 231$ cu in.; 1 liter $= 1000$ cm^3). What is the equivalent cost, in U.S. dollars per U.S. gallon, for gasoline in Canada priced at 65 Canadian cents per liter, when one Canadian dollar exchanges for 78 cents U.S.?

18. In 1991, Edward N. Lorenz, a meteorologist who was an early researcher into chaos and dynamical systems, received the Kyoto Prize in Basic

Sciences, consisting of a gold medal and 45 million Japanese yen. If U.S. $1 = Y125 at the time, what was the value of the cash award in U.S. dollars?

19.   In 200 B.C., Eratosthenes measured the circumference of the earth and expressed the result as 250,000 *stadia* (plural of *stadium*). In *The American Heritage Dictionary,* Second College Edition (Houghton Mifflin, Boston, 1982), we read for the second meaning of "stadium": "An ancient Greek measure of distance . . . equal to about 185 kilometers, or 607 feet." The name of the unit came from the length of a racecourse that was a bit less than an eighth of a mile long. If the numbers in the definition are correct, what are the correct units that should have appeared?

Sorry, No King Kongs

20.   The weight of a 1-ft cube of steel is 500 lb. What is the pressure on the bottom face in

(a)   pounds per square inch?
(b)   atmospheres (1 atm $=$ 14.7 lb/sq in.)?

21.   In an article on adding organic matter to soil, the magazine *Organic Gardening* (March 1983) said, "Since a 6-inch layer of mineral soil in a 100-square-foot plot weighs about 45,000 pounds, adding 230 pounds of compost will give you an instant 5% organic matter."

(a)   What is the density of the mineral soil, according to the quotation?
(b)   How does this density compare with that of steel?
(c)   How do you think the quotation should be revised to be accurate?

22.   A mature gorilla weighs 400 lb and stands 5 ft tall.

(a)   Give an estimate of its weight when it was half as tall.
(b)   What assumptions are involved in your estimate?
(c)   A mature gorilla's two feet together have a combined area of about 1 ft². When the gorilla is standing on its feet, what is the pressure on its feet, in pounds per square inch?

23.   Suppose King Kong is a gorilla scaled up with a linear scaling factor of 10.

(a)   How much does King weigh?
(b)   What is the pressure on King's feet, in pounds per square inch?

24.   You may have wanted to have a waterbed but found that waterbeds were not allowed in your building. Apart from the danger of flood if the bed should puncture or leak, there is the consideration of the weight.

(a)   If a queen-size mattress is 80 in. long by 60 in. wide by 12 in. high, and water weighs 1 kg/liter, how much does the water in the mattress weigh in pounds?

(b)   If the weight of the mattress and frame is carried by four legs, each 2 in. by 2 in., what is the pressure, in pounds per square inch, on each leg?

(c)   How does the pressure on the legs of the waterbed compare with the pressure that a person exerts on their feet—for example, a 130-lb person with a total foot area of about one-quarter of a square foot in contact with the ground?

(d)   If you aren't allowed to have a waterbed, how about a spa (hot tub)? Find the weight of the water in a spa that is in the shape of a cylinder 6 ft in diameter and 3.5 ft deep. (*Hint:* The volume of a cylinder is $\pi r^2 h$, where r is the radius and h is the height.)

25.   What does the largest giant sequoia tree weigh? Model the tree as a (very elongated) cone, supposing that the tree is 360 ft high and has a circumference of 40 ft at the base, and that the density of the wood is 31 lb/cu ft. (The volume of a cone of height h and radius r is $\frac{1}{3}\pi r^2 h$.)

Falls, Dives, Jumps, and Flights

26.   (Adapted from George Knill and George Fawcett, Animal form or keeping your cool, *Mathematics Teacher*, May 1982, 395–397.) The movie *Them* features enormous ants (about 8 m long and about 3 m wide). We can investigate how feasible such a scaled-up insect is by considering its oxygen consumption. A common ant, which is about 1 cm long, needs about 24 milliliters of oxygen per second for each cubic centimeter of its volume. Since an ant does not have lungs, it must absorb the oxygen through its "skin," which it can do at a rate of about 6.2 milliliters per second per square centimeter. We may suppose that the tissues of a scaled-up ant would have the same need for oxygen for each cubic centimeter, and that its skin could absorb oxygen at the same rate, as a normal ant. Compared to a common ant, how many times as large is an enormous ant's

(a)   length?
(b)   surface area?
(c)   volume?
(d)   What proportion of such an ant's oxygen need could its skin supply?

What can you conclude about the existence of such insects?

27.   In the children's story *Peter Pan,* Peter and Wendy can fly. We may suppose that they are 4 ft tall, so they are about 12 times as tall as a sparrow is long. What should their minimum flying speed be?

28.   Icarus of Greek legend escaped from Crete with his father, Daedalus, on wings made by Daedalus and attached with wax. Against his father's advice, Icarus flew too close to the sun; the wax melted, the wings fell off, and he fell into the sea and drowned. What must have been his minimum cruising speed? What assumptions does your answer involve?

Keeping Cool (and Warm)

◆   29.   Smaller birds and mammals generally maintain higher body temperatures than larger ones. Explain why you would expect this to be so. (Adapted from A. Zherdev, Horseflies and Flying Horses, *Quantum,* May/June 1994, 32–37, 59–60.)

◆   30.   Some humans, such as the Bushmen of the Kalahari Desert in Africa, live in desert environments, where it is important to be able to do without water for periods of time. Would you expect such an environment to favor short or tall individuals? (Adapted from A. Zherdev, Horseflies and Flying Horses, *Quantum,* May/June 1994, 32–37, 59–60.)

Allometry

▲   31.   Consider the data below for weight and metabolic rate of various mammals. Graph these data on log-log paper. What can you conclude from your graph? (Note that you may have to "create" your own log scale on the weight axis, since the weights span six orders of magnitude. Commercial log paper is usually limited to 2, 3, or perhaps 5 orders of magnitude ["cycles"].) If you don't have log-log paper available, use a calculator to take the logarithms (LOG_{10}) of all the numbers and graph these values on ordinary graph paper.

	Weight (kg)	Calories per Kilogram
Guinea pig	0.7	223
Rabbit	2	58
Human	70	33
Horse	600	22
Elephant	4000	13
Whale	150,000	1.7

Fit the best line that you can through the log-log data and estimate its slope (if you have a least-squares line-fitting program available, use that). Compare your results with the discussion and data in McMahon and Bonner, *On Size and Life* (pp. 42–47, 67).

▲ 32. Listed below are the winning times in the 1983 World Rowing Championships for rowing shells with one, two, four, and eight oars. The men's times are for 2000 m, the women's for 1000 m. Convert the times to

Event	Number of Oars	Men (2000 m)	Women (1000 m)
Single sculls	1	6:49.75	3:36.51
Pairs without coxswain	2	6:35.85	—
Fours without coxswain	4	6:14.83	3:26.68
Eights (with coxswain)	8	5:34.39	2:56.22

speed, in meters per minute. For men and women separately, plot speed versus number of oars on ordinary graph paper, and then on log-log paper. If you don't have log-log paper available, use a calculator to take the logarithms (LOG_{10}) of all the numbers and graph these values on ordinary graph paper.

Is the relationship proportional? Allometric?

Additional Exercises

For Exercises 33 and 34, refer to the following: An ancient measure of length, the *cubit* was the distance from the elbow to the tip of the middle finger of a person's outstretched arm. So the length of a cubit depended on the person, though there was some attempt at standardization. Most estimates place the length of a cubit between 17 and 22 in.

33. Goliath (of David and Goliath, as related in the Bible [I Samuel 17:4]) was "six cubits and a span." A span was originally the distance from the tip of the thumb to the tip of the little finger when the hand is fully extended, about 9 in. What range of heights would this indicate for Goliath, in feet and inches? In centimeters?

34. According to classical Greek sources, Pythagoras (sixth century B.C.) used geometric scaling to model the height of Hercules, the heroic figure of classical mythology in the epic poems of Homer. Pythagoras compared the

lengths of two racecourses, one (according to tradition) paced off by Hercules and the other by a man of average height. Both were 600 "paces" long, but the one established by Hercules was longer because of Hercules's longer stride (600 "Herculean" paces versus 600 paces by a normal man). A normal man in the time of Pythagoras would have been about 5 ft tall.

(a)   If the distance paced off by Hercules was 30% longer than the other racecourse, how tall was Hercules? What does your calculation assume?

(b)   In fact, the ancient sources do not give the original data but only the two conflicting answers that Hercules was 4 cubits tall and 4 cubits 1 foot tall. What range does this give for the height of Hercules, in feet and inches? In centimeters?

◆   35.   Recent years have seen the beginnings of human-powered controlled flight, in the *Gossamer Condor* and other superlightweight planes. The *Gossamer Condor* is far longer than an ostrich, but it flies at only 12 mph. How can it?

◆   36.   Jonathan Swift's Gulliver also traveled to Lilliput, where the Lilliputians were human-shaped but only about 6 in. tall. In other words, they were geometrically similar in shape to ordinary human beings but only one-twelfth as tall. What would a Lilliputian weigh?

Are Lilliputians ruled out by the size-shape and area-volume considerations in this chapter? If you think they are, what considerations do you find convincing? If not, why not?

37.   [Contributed by Charlotte Chell of Carthage College, Kenosha, Wisconsin.] A 6-ft indoor holiday tree needs four strings of lights to decorate it. How many strings of lights are needed for an outdoor tree that is 30 ft high?

38.   What would you expect an individual *Quetzalcoatlus northropi* to weigh if it had half the wingspan of an adult? If an individual weighed half as much as an adult, what would you expect its wingspan to be?

◆   39.   The TV series *All in the Family* featured Archie Bunker, his wife, Edith, his daughter, Gloria, and her friend (and later husband), Michael (called "Meathead" by Archie). This family provides a way to visualize the changes over the past 20 years in the economic situation of a U.S. blue-collar family. The series began in 1973. Archie is a factory worker, about 50 years old, with at most a high school degree earning about $13,000 then. Gloria, living at home and employed half-time, earns about $2000. Michael earns about $10,000 at his factory job.

We consider a similar family in 1979, the Trapps, with people of the same ages, education, and social background as before. The 50-year-old factory worker Art earns about $12,600. To help support them, his wife, Enid, is employed, earning about $9000. Their daughter, Gina, is married to Martin ("Cheddarhead"), who earns about $15,000 at the cheese factory, while Gina is at home with children.

We move now to 1988 and yet another family with the same age structure, the Sands. Patriarch Arnie earns about $34,000 at his job, while wife, Eve, makes $12,000. Their daughter, Gwen, is married to Matt ("Muttonhead"), who earns $25,000 (when he isn't laid off from the meat-packing plant). Gwen works, too, earning $12,000; but day-care expenses for their son cost $5000 per year.

Use the Consumer Price Index (see Table 17.1) to convert all of these figures to 1997 dollars, and then compare the relative situations of these similar families in 1973, 1979, and 1988.

(Thanks to Paul Solman of PBS's *MacNeil-Lehrer Newshour* for the idea and the data.)

◆ 40. According to various studies, about one-third of adult Americans are overweight. To determine if someone weighs too much, the researchers calculate the body-mass index (BMI): body weight divided by the square of height. When units of meters and kilograms are used, a BMI over 27 qualifies a person as overweight, meaning about 20–24% above the desirable weight as listed in a widely used 1983 table developed by Metropolitan Life Insurance.

(a) Calculate the BMI for a woman 160 cm tall who weighs 65 kg.

(b) Suppose that height is measured in feet and weight in pounds, and BMI in U.S. Customary units is calculated as body weight divided by the square of height. What BMI in those units qualifies a person as overweight?

(c) Since, generally speaking, body weight is average density times body volume, BMI is average density times a quantity that has units of length. Discuss whether BMI makes sense as a measure of being overweight. Would dividing by a different power of height make for a better measure?

WRITING PROJECTS

1 ▶   A human infant at birth usually weighs between 5 and 10 pounds and has a height (length) between 1 and 2 feet, with the shorter babies having the lesser weight. Considering the weight and height of an adult human, give an argument that human growth must not be just proportional growth.

2 ▶   Use algebra to demonstrate that for proportional growth, "area scales as volume to the two-thirds power."

3 ▶   The principle that area scales with the square of length, and volume with the cube, has important consequences for the depiction and interpretation of data in graphic form. Suppose we wish to indicate in an artistic way that the weekly income of a U.S. carpenter is twice that of a carpenter in (mythical) Rotundia. We draw one moneybag for the Rotundian and another one "twice as large" for the American. [Illustration from Darrell Huff, *How to Lie with Statistics,* Norton 1954, p. 69.]

What's the problem? Well, first, people tend to respond to graphics by comparing areas. Since the larger moneybag is twice as high and twice as wide as the smaller one, the image of it on the page has four times the area. Second, we are used to interpreting depth and perspective in drawings in terms of three-dimensional objects. Since the larger bag is also twice as thick as the smaller, it has eight times the volume. The graphic leaves the subconscious impression that the U.S. carpenter earns eight times as much, instead of twice as much.

With these ideas in mind, evaluate the depictions of data below and on page 700. [Illustrations reproduced or adapted from Edward R. Tufte, *The Visual Display of Quantitative Information,* Graphics Press, 1983, pp. 55, 57, and 70.]

(a)

Comparative Annual Cost per Capita for care of Insane in Pittsburgh City Homes and Pennsylvania State Hospitals.

Pittsburgh Civic Commission, *Report on Expenditures of the Department of Charities* (Pittsburgh, 1911), p. 7.

(b)

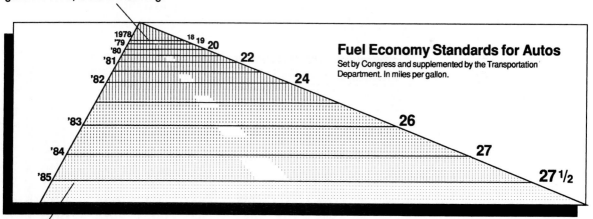

This line, representing 18 miles per gallon in 1978, is 0.6 inches long.

Fuel Economy Standards for Autos
Set by Congress and supplemented by the Transportation Department. In miles per gallon.

This line, representing 27.5 miles per gallon in 1985, is 5.3 inches long.

New York Times, August 9, 1978, p. D2

(c)

1958—Eisenhower

1963—Kennedy

1968—Johnson

1973—Nixon

1978—Carter

1984—Reagan

1990—Bush

1993—Clinton

4   As in Writing Project 3, evaluate the depictions below and on page 701. [Illustrations reproduced or adapted from Edward R. Tufte, *The Visual Display of Quantitative Information,* Graphics Press, 1983, pp. 62 and 69.]

(a)

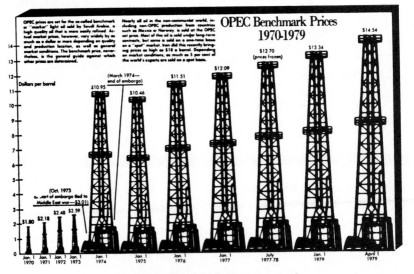

Washington Post, March 28, 1979, p. A-18.

(b)

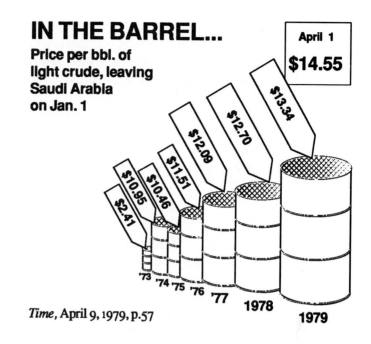

IN THE BARREL...
Price per bbl. of light crude, leaving Saudi Arabia on Jan. 1

April 1
$14.55

$13.34

$12.70

$12.09

$11.51

$10.46

$10.95

$2.41

'73 '74 '75 '76 '77 1978 1979

Time, April 9, 1979, p.57

(c)

THE SHRINKING FAMILY DOCTOR
In California

Percentage of Doctors Devoted Solely to Family Practice

1964	1975	1990
27%	16.0%	12.0%

1: 4,232
6,212

1: 3,167
6,694

1: 2,247 RATIO TO POPULATION
8,023 Doctors

5   On July 11, 1994, the Russian ruble traded at 3736 rubles to the U.S. dollar. One day later, it traded at 3926 rubles to the U.S. dollar. There are two competing practices for stating how much one currency has depreciated (lost value) against another. Option A, used by the International Monetary Fund and the British periodical *The Economist,* takes the difference in the first country (here, 3926 − 3736) and divides it by the new trading value (3926) and multiplies by 100 to get a result in percent (here, 4.84%). Option B, sometimes called the "popular method," does the same but divides by the old trading value instead of the new value (getting 5.09%).

(a)   Calculate the results for both methods for the Jamaican dollar, which traded at J$1.78 to US$1 in January 1983 and at J$5.50 to US$1 for most of 1985. Does it make sense to speak of a currency depreciating more than 100%?

(b)   Calculate the value (in current U.S. dollars) of one Jamaican dollar in January 1983 versus the value of one Jamaican dollar in 1985. By what percentage has this value declined? With which of your answers to part (a) does this number agree?

(c)   Is the percentage obtained from Option A always higher than that from Option B? or always lower? or neither? Which would you expect a person to use who wanted to make a decline seem large?

◆   (d)   Critique the data display below and offer alternatives.

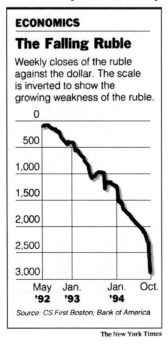

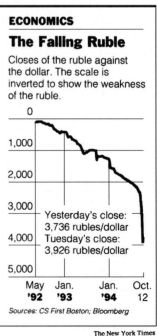

CHAPTER 18

GEOMETRIC GROWTH

M any problems that we face relate to populations and their changes over time. We all have a stake in the problems associated with human population increase, such as hunger and disease. Our food supplies are affected by the growth and behavior of nonhuman populations such as bacteria, locusts, and rats. Even inanimate populations affect us. Growing "populations" of household refuse and nuclear waste pose disposal and storage issues, with accompanying environmental questions, while depletion of natural resources causes other concerns. Similarly, but more favorably, a growing population of dollars in a bank account can provide resources that enrich our lives.

Geometric growth is the key mathematical concept behind the growth of populations of all kinds. Understanding the mathematics of geometric growth is essential for realizing just how severe the economic and social issues caused by growth can become, and for measuring the effectiveness of policies to alter patterns of growth. To analyze the growth of a population, we concentrate on the questions:

- How big is the population?
- How fast is it growing or shrinking?
- How is its structure or makeup changing?

Population growth refers to both increases (positive growth) and decreases (negative growth) in population size. **Population structure** refers to the divisions of a population into subgroups. For example, human populations are frequently categorized by age structure or by economic, social, or educational criteria.

We investigate two models of population growth. In the first, the population grows at a rate proportional to its current size, so that, for example, when it is twice as large, it is growing twice as fast. In the second, the increase is the same in each time interval. We focus on two seemingly different kinds of populations—financial and biological—to illustrate how broadly the models apply.

GEOMETRIC GROWTH AND FINANCIAL MODELS

When you open a savings account, your primary concerns are the safety and the growth of the "population" of your savings. Suppose that you deposit $1000 in an account that, you are told, "pays interest at a rate of 10%, compounded and paid annually." Assuming that you make no other deposits or withdrawals, how much is in the account after 1, 2, or 5 years?

The $1000 is the **principal,** the **initial balance** of the account. At the end of one year, **interest** is added. The amount of interest is 10% of the principal, or

$$10\% \times \$1000 = 10 \times 0.01 \times \$1000 = 0.10 \times \$1000 = \$100$$

in this case. (You can think of the symbol "%" as standing for "$\times 0.01$.") So the balance at the beginning of the second year is $1100. During the second year the interest is also 10%—not of the *initial* balance of $1000 but of the *new* balance of $1100—so at the end of the second year, 10% of $1100, or $110, is added to the account.

Thus, during the second year you earn interest on both the principal of $1000 and on the $100 interest earned during the first year. Interest that is paid on both the principal and on the accumulated interest is known as **compound interest.** You receive more interest during the second year than during the first, that is, the account grows by a greater amount during the second year. At the beginning of the third year the account contains $1210, so at the end of the third year you receive $121 in interest. Again this is larger than the amount you received at the end of the preceding year. Moreover, the increase during the third year,

third-year interest − second-year interest = $121 − $110 = $11

is larger than the increase during the second year,

second-year interest − first-year interest = $110 − $100 = $10

Thus, not only is the account balance increasing each year, but the amount added also increases each year.

Banks and savings institutions often compound interest more often than once a year, for example, quarterly (four times per year). With an interest rate of 10% per year and quarterly compounding, you get one-fourth of the rate, or 2.5%, paid in interest each quarter. The quarter (3 months) is the **compounding period,** or the time elapsing before interest is paid.

Consider again a principal of $1000. At the end of the first quarter, you have the original balance plus $25 interest, so the balance at the beginning of the second quarter is $1025. During the second quarter you receive interest equal to 2.5% of $1025, or $25.63, so the balance at the end of the second quarter is $1050.63. Continuing in this manner, you find that the balance at the end of the first year is $1103.81. (You should "read" all calculations in this chapter by confirming them on your calculator.)

Even though the account was advertised as paying 10% interest (the *nominal rate*), the interest for the year is 10.381% of the principal. The 10.381% is the *effective rate* or *equivalent yield.*

The **nominal rate** is the stated rate of interest on which any compounding is based.

The **effective rate** is the actual percentage increase in the account. It is also called the **equivalent yield,** since an account that earns the same increase without compounding would have this percentage as its nominal rate.

For a compounding period of one year, the nominal rate is called the **annual (percentage) rate (APR),** and the effective rate is called the **annual (equivalent) yield** or **annual percentage yield (APY).**

If interest is compounded monthly (12 times per year) or daily (365 times per year), the resulting balance is even larger. A comparison of yearly, quarterly, monthly, and daily compounding for an interest rate of 10% is shown in Table 18.1. We will shortly summarize results in a general formula.

TABLE 18.1 **Comparing Compound Interest: The Value of $1000, at 10% Annual Interest, for Different Compounding Periods**

Years	Compounded Yearly	Compounded Quarterly	Compounded Monthly	Compounded Daily	Compounded Continuously
1	1100.00	1103.81	1104.71	1105.16	1105.17
5	1610.51	1638.62	1645.31	1648.61	1648.72
10	2593.74	2685.06	2707.04	2717.91	2718.28

From now on, we will express all interest rates as decimals. For example, 10% is 0.10; to convert a percentage to a fraction, divide the percentage by 100, by moving the decimal point two places to the left. An interest rate of 0.10 is the same as 10%, or 10/100; an interest rate r is the same as $100r$%, or (the rate in %)/100.

THE MATHEMATICS OF GEOMETRIC GROWTH

We look for the underlying mathematical pattern of compounding. For quarterly compounding, you have at the end of the first quarter

initial balance + interest = $1000 + $1000(0.025) = $1000(1 + .025)

and at the end of the second quarter

$$
\begin{aligned}
\text{initial balance + interest} &= \$1000(1 + 0.025) \\
&\quad + [\$1000(1 + 0.025)](0.025) \\
&= [\$1000(1 + 0.025)] \times (1 + 0.025) \\
&= \$1000(1 + 0.025)^2
\end{aligned}
$$

The pattern continues in this way, so that you have $1000(1 + 0.025)^4$ at the end of the fourth quarter.

You use the calculator button marked $\boxed{y^x}$ to evaluate expressions like $(1.0125)^2$: Enter 1.012, push $\boxed{y^x}$, enter 2, and push $\boxed{=}$; you get 1.02515625.

More generally, with an initial balance of P and an interest rate r ($= 100r$%) per compounding period, you have at the end of the first compounding period

$$P + Pr = P(1 + r)$$

This amount can be viewed as a new starting balance. Hence, in the next compounding period, the amount $P(1 + r)$ grows to

$$P(1 + r) + P(1 + r)r = P(1 + r)(1 + r)P(1 + r)^2$$

The pattern continues, and we reach the following conclusion:

> **Compound interest formula:** If a principal P is deposited in an account that pays interest at an effective rate r per compounding period, then after n compounding periods the account contains the amount
>
> $A = P(1 + r)^n$

The amount added each compounding period is proportional to the amount present. This type of growth is called *geometric growth.*

> **Geometric growth** (also called **exponential growth**) is growth proportional to the amount present.

EXAMPLE ▶ *Compound Interest*

Suppose that you have a principal of $P = \$1000$ invested at 10% nominal interest per year. Using the compound interest formula $A = P(1 + r)^n$, you can use the compound interest formula to determine the amount in the account after 10 years, which varies according to the compounding period:

- *Annual compounding.* The annual rate of 10% gives $r = 0.10$, and after 10 years the account has

$$\$1000(1 + 0.10)^{10} = \$1000(1.10)^{10} = \$1000(2.59374)$$
$$= \$2593.74$$

- *Quarterly compounding.* Then $r = 0.10/4 = 0.025$, and after 10 years (40 quarters) the account contains

$$\$1000(1 + \tfrac{0.10}{4})^{40} = \$1000(1.025)^{40} = \$1000(2.68506)$$
$$= \$2685.06$$

- *Monthly compounding.* Then $r = 0.10/12 = 0.008333$. The amount in the account after 10 years (120 months) is

$$\$1000(1 + \tfrac{0.10}{12})^{120} = \$1000(1.00833333^{120}) = \$1000(2.70704)$$
$$= \$2707.04$$

These entries are found in the last row of Table 18.1. ◆

Suppose that you want to make a one-time deposit now, of amount P, that will grow to a specific amount A in n years from now by earning interest at an effective rate of $100r\%$ per year. The quantities, A, P, r, and n are related through the compound interest formula, $A = P(1 + r)^n$. The quantity P is called the **present value** of the amount A to be paid n years in the future.

EXAMPLE ▶ *Certificate of Deposit*

Suppose that you will need $15,000 to pay for a year of college for your child 18 years in the future, and you can buy a certificate of deposit whose interest rate of 10% compounded quarterly is guaranteed for that period. How much do you need to deposit?

We have $A = \$15,000$, $r = 0.10/4 = 0.025$, and $n = 72$. The compound interest formula gives

$$\$15,000 = A = P(1 + r)^n = P(1.025)^{72} = 5.91723P$$

so $P = \$15,000/5.91723 = \$2534.94$.   ◆

EXAMPLE ▶ *Money Market Account*

In some cases you know the principal, the current balance, and the interval of time, and you want to learn the interest rate. For example, money market funds typically report earnings to investors each month, based on interest rates that vary from day to day. The investor may want to know an average rate of interest for the month, but often the monthly statement does not report one. We find the equivalent average *daily* effective rate, from which we calculate the annual yield. The compound interest formula gives the end-of-month balance as $A = P(1 + r)^n$, where P is the balance at the beginning of the month, r is the average daily interest rate, and n is the number of days that the statement covers. So we have

$$\frac{A}{P} = (1 + r)^n$$

Changing each quantity to the other side and taking the nth root of each side gives

$$1 + r = \left(\frac{A}{P}\right)^{1/n}, \quad r = \left(\frac{A}{P}\right)^{1/n} - 1$$

Suppose that the monthly statement from the fund reports a beginning balance (P) of $7373.93 and a closing balance (A) of $7416.59 for 28 days ($n$).

We thus have

$$r = \left(\frac{7416.59}{7373.93}\right)^{1/28} - 1 = (1.005785246)^{0.035714286} - 1$$
$$= 0.000206042$$

Thus the average daily effective rate is 0.0206042%. Compounding daily, for a year we would have $(1 + 0.00206042)^{365} = 1.0780972$, for an annual yield of 7.81%. ◆

ARITHMETIC GROWTH

Under another method of paying interest, called **simple interest,** interest is paid only on the original balance, no matter how much interest has accumulated. With simple interest, for a principal of $1000 and a 10% interest rate, you receive $100 interest at the end of the first year; so at the beginning of the second year, the account contains $1100, as before. But at the end of the second year, you again receive only $100; so at the beginning of the third year, the account contains $1200. In fact, at the end of each year you receive just $100 in interest. This method yields less than if interest is compounded.

The amounts in accounts paying interest at 10% per year with compound and simple interest are shown in Table 18.2 and in the graph in Figure 18.1, which dramatically illustrate the growth of money at compound interest compared to simple interest.

TABLE 18.2 The Growth of $1000: Compound Interest vs. Simple Interest		
Years	Amount in Account from Compounded Interest	Amount from Simple Interest
1	1100.00	1100.00
2	1210.00	1200.00
3	1331.00	1300.00
4	1464.10	1400.00
5	1610.51	1500.00
10	2593.74	2000.00
20	6727.50	3000.00
50	117,390.85	6000.00
100	13,780,612.34	11,000.00

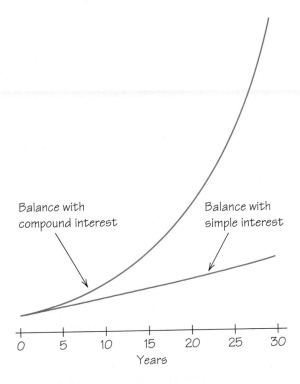

FIGURE 18.1   The growth of $1000: compound interest and simple interest.

Simple interest is seldom used in today's financial institutions. However, we frequently observe the corresponding kind of growth, called *arithmetic growth,* in other contexts.

> **Arithmetic growth** (also called **simple growth**) is growth by a constant amount in each time period.

The population of medical doctors in the United States grows arithmetically, since the fixed number of medical schools each graduate the same total number of doctors each year (and the number of doctors dying is also fairly constant). On the other hand, general human populations tend to grow geometrically because the number of children born—the "interest"—increases as the population—the "balance"—increases.

The distinction between arithmetic growth and geometric growth is fundamental to the major theory of demographer and economist Thomas Malthus (1766–1834). He claimed that human populations grow geometrically but food supplies grow arithmetically, so that populations tend to outstrip their ability to feed themselves (see Spotlight 18.1).

The situation of nuclear waste generated by a nuclear power plant is more complicated. The absolute volume of waste added each year depends on the fixed size and output of the power plant, not on the growing amount of waste in storage. Hence the volume of waste grows arithmetically. What about the total amount of radioactive material in the storage dump? The waste is a mixture of radioactive and nonradioactive substances; over time, the radioactive ingredients decay slowly into nonradioactive ones. While the radioactivity of waste already in storage is decreasing, new amounts of radioactive material are being added each year. The situation requires a hybrid model that incorporates positive arithmetic growth (adding to the dump) accompanied by negative geometric growth (radioactive decay). The situation is like turning on the faucet to the bathtub while leaving the drain hole open; what happens to the height of water in the tub depends on how fast water runs in versus how fast it runs out.

A LIMIT TO COMPOUNDING

The rows in Table 18.1 show a trend: more frequent compounding yields more interest. But, as the frequency of compounding increases, the interest tends to a limiting amount, shown in the far right column.

Spotlight *The Number e*

18.2

The number e is similar to the number π in several respects. Both arise naturally, π in finding the area and circumference of circles, and e in compounding interest continuously (e is also the base for the system of "natural" logarithms). In addition, neither is rational (expressible as the ratio of two integers, such as 7/2) nor even algebraic (the solution of a polynomial equation with integer coefficients, such as $x^2 = 2$); we say that they are *transcendental* numbers. Finally, no pattern has ever been found in the digits of the decimal expansion of either number.

Why is this so? Basically, because the extra interest from more frequent compounding is *interest on interest.* For example, in the first row of Table 18.1, the $3.81 extra interest from compounding quarterly is interest on the $100 yearly interest. The $3.81 is less than 10% of the $100 because the $100 interest is not on deposit for the whole year, since just part of it is credited to the account (and begins earning interest) at the end of each quarter. As compounding is done more and more often, smaller and smaller amounts of interest on interest are added.

How can we determine the limiting amount? Let's suppose that the initial balance is $1, and that we keep track at all stages of even the smallest fractions of a dollar.

We first suppose an interest rate of 100% per year compounded n times per year; later we examine interest rates closer to the ones in stable economies. For an initial balance of $1, the amount at the end of one year is—from the compound interest formula, with $P = \$1$ and $r = 100\%$—

$$A = \$1 \times \left(1 + \frac{100\%}{n}\right) = \$\left(1 + \frac{1.00}{n}\right)^n$$

As n increases, this amount, which is just $(1 + 1/n)^n$, gets closer and closer to a special number called $e \approx 2.71828$ (see Spotlight 18.2). This is illustrated in Table 18.3, where the dots (ellipses) indicate that more decimal places follow.

For a general interest rate r, the amount that $1 grows to when compounded n times during the year is

$$\left(1 + \frac{r}{n}\right)^n$$

As n is made larger and larger, the limiting amount is e^r, and the interest method is called **continuous compounding**. The effective rate is $(e^r - 1)$.

TABLE 18.3 Yield of $1 at 100% Interest, Compounded n Times per Year

n	$\left(1 + \frac{1}{n}\right)^n$
1	2.0000000 . . .
5	2.4883200 . . .
10	2.5937424 . . .
50	2.6915880 . . .
100	2.7048138 . . .
1,000	2.7169239 . . .
10,000	2.7181459 . . .
100,000	2.7182682 . . .
1,000,000	2.7182818 . . .
10,000,000	2.7182818 . . .

(You can calculate powers of e using the $\boxed{e^x}$ button on your calculator; on some calculators, this button is the $\boxed{2\text{nd}}$ function of the button marked $\boxed{\text{LN}}$ or $\boxed{\ln x}$. For example, to calculate $e^{0.10}$, enter 0.10, push $\boxed{2\text{nd}}$, then push $\boxed{\ln x}$; you get 1.105170918.)

EXAMPLE ▶ *Continuous Compounding*

For $1000 at an annual rate of 10%, compounded n times in the course of a single year, the balance at the end of the year is

$$\$1000\left(1 + \frac{0.10}{n}\right)^n$$

This quantity gets closer and closer to $1000e^{0.1} = \$1105.17$ . . . as the number of compoundings n is increased. No matter how frequently interest is compounded—daily, hourly, every second, infinitely often ("continuously")—the original $1000 at the end of one year cannot grow beyond $1105.17. The values after 5 and 10 years are shown in the lower rows of Table 18.1. ◆

According to the **continuous interest formula,** for a principal P, deposited in an account at the nominal rate of $r = 100r\%$ compounded continuously,

- after 1 year, the account contains $A = Pe^r$
- after m years, it contains $A = Pe^{rm}$

It makes virtually no difference whether compounding is done daily or continuously over the course of a year. Most banks apply a daily periodic rate (based on compounding continuously) to the balance in the account each day and post interest daily (rounded to the nearest cent). The daily nominal rate is $r/365$, so each day the balance of the account is multiplied by $e^{r/365}$, the daily effective rate. Except for the rounding in posting interest, the effect is the same as continuous compounding throughout the year, since the compound interest formula gives $A = P(e^{r/365})^{365}$, which is the same as Pe^r from the continuous interest formula.

Also, it makes virtually no difference whether the bank treats a year as

- 365 days with daily nominal interest rate $r/365$ (the *365 over 365 method*)—this is the usual method for daily compounding; or

- 360 days with daily interest rate $r/360$ (the *360 over 360 method*)—this is the usual method for loans with equal monthly installments, since 360 is evenly divisible into 12 equal "months" of 30 days. The daily interest is greater than for the 365 over 365 method, but there are fewer days in the year.

Table 18.4 gives a comparison of different interest methods.

TABLE 18.4 Comparing Methods of Compounding Interest

Method	Compounding Periods per Year	Rate per Period	Formula per One Year	Effective Rate	Effective Rate for $r = 5\%$
360 over 360, daily	360	$r/360$	$P(1 + \frac{r}{360})^{360}$	$(1 + \frac{r}{360})^{360} - 1$	5.12674%
365 over 365, daily	365	$r/365$	$P(1 + \frac{r}{365})^{365}$	$(1 + \frac{r}{365})^{365} - 1$	5.12675%
Continuous			Pe^r	$e^r - 1$	5.12711%

A MODEL FOR ACCUMULATION

The compound interest formula tells the fate over time of a single deposited amount, but another common question that arises in finance is: What size deposit do you need to make *on a regular basis,* in an account with a fixed rate of interest, to have a specified amount at a particular time in the future?

This question is important in planning for a major purchase in the future, accumulating a retirement nest egg, paying off a mortgage, or making install-ment payments on a car. Later we apply the results to calculate the amount of a nonrenewable resource used up over a number of years.

EXAMPLE ▶ *A Savings Plan*

An individual saves $100 per month, deposited directly into her credit union account on payday, the last day of the month. The account earns 5% per year, compounded daily (just simple compounding, without continuous com-pounding during the day). How much will she have at the end of 5 years, as-suming that the credit union continues to pay the same interest rate?

Note that she makes the first deposit at the end of the first month and the last deposit at the end of the sixtieth month. Although the months differ in length and the credit union compounds daily using the 365 over 365 method, we get virtually the same result by assuming instead that each month has 30 days and the 360 over 360 method is used.

The daily interest rate is $5\%/360 = 0.000138889\%$, so the amount earned for a 30-day month is

$$(1.000138889)^{30} - 1 = 1.0041751 - 1 = 0.004175073$$
$$= 0.4175073\%$$

Call this monthly rate r.

It's easier to look at the deposits in reverse time order. The last deposit is deposited on the last day of the 5 years, so it earns no interest and contributes just $100 to the total.

The second last deposit earns interest for 1 month, contributing $100(1 + r)$.

Similarly, the third last deposit is on deposit for 2 months, contributing $100(1 + r)^2$.

Continuing in the same way, we find that the first deposit earns interest for 59 months and contributes $100(1 + r)^{59}$. The total of all of the contributions is

$$\$100 + \$100(1 + r)^1 + \$100(1 + r)^2 + \cdots + \$100(1 + r)^{59}$$
$$= \$100[1 + (1 + r)^1 + (1 + r)^2 + \cdots + (1 + r)^{59}]$$

This expression is known as a **geometric series,** because the successive terms grow geometrically: each succeeding term is a constant common ratio — here, $(1 + r)$ — times the preceding term. There is a formula for the sum of such a series, which we give for a geometric series with ratio x:

$$1 + x + x^2 + x^3 + \cdots + x^{n-1} = \frac{x^n - 1}{x - 1}$$

That this formula works for all x (except $x = 1$) can be confirmed by multiplying both sides by $(x - 1)$ and watching terms on the left cancel (you should do this confirmation for $n = 4$).

In our example, we have $x = 1 + r$, and the formula becomes

$$1 + (1 + r)^1 + (1 + r)^2 + \cdots + (1 + r)^{n-1} = \frac{(1 + r)^n - 1}{r}$$

We have $n - 1 = 59$, or $n = 60$ months, and $r = 0.0041751$, the interest rate per month. The total accumulation is

$$A = \$100 \, \frac{(1 + 0.004175073)^{60} - 1}{0.004175073} = \$6802.36 \quad \blacklozenge$$

In general terms, we have

For a uniform deposit of d per period (deposited at the end of the period) and an interest rate r per period, the amount A accumulated is given by the *savings formula:*

$$A = d \, \frac{(1 + r)^n - 1}{r}$$

The savings formula involves four quantities: A, d, r, and n. If any three are known, the fourth can be found. A common situation is for A, r, and n to be known, with d (the regular payment) to be found, since the practical concern for most people is how much their monthly payment will be.

Sometimes the purpose of a savings plan is to accumulate a fixed sum by a certain date. Such savings plans are called **sinking funds.**

EXAMPLE ▶ *A Sinking Fund*

Suppose that a couple wishes to save for the college education of their child. They begin saving a regular amount d per month after the child is born and want to have \$100,000 available when the child turns 18. How much do they have to save each month, if their account earns 6.5% interest per year, compounded daily?

Again, for simplicity, we can calculate as if each month has 30 days and the savings institution uses the 360 over 360 method. The daily nominal rate is 6.5%/360 = 0.000180556%, so the effective monthly rate (for a 30-day month) is

$$r = (1.000180556)^{30} - 1 = 1.005430885 - 1 = 0.005430885$$
$$= 0.5430885\%$$

We have $A = \$100{,}000$, $r = 0.005430885$, and $n = 12 \times 18 = 216$. Applying the savings formula, we have

$$\$100{,}000 = d\frac{(1.005430885)^{216} - 1}{0.005430885} = d(593.21)$$

so $d = \$100{,}000/593.21 = \$168.57$.   ◆

A common situation that you are likely to encounter is a loan—for a house, a car, or college expenses—to be paid back in equal periodic installments. Your payments are said to **amortize** the loan. Each payment pays the current interest and also repays part of the principal. *As the principal is reduced, less of each payment goes to interest and more toward paying off the principal.*

Let's suppose that Sally buys a house for $100,000 with a loan that she will pay off over 30 years in equal monthly installments. Suppose that the interest rate for her loan is 6.00%. Let's figure out how much her monthly payment needs to be.

Imagine changing the setup slightly so that now Sally is borrowing the entire sum ($100,000) for 30 years, and we think of her monthly payments as the savings fund that she's building up to pay off the loan at the end of the term. The interest rate of 6.00% on the loan is compounded monthly, so the monthly rate is 0.5%. At the end of 30 years, the principal and interest on the loan will (by the compound interest formula) amount to

$$\$100{,}000 \times (1 + 0.005)^{30 \times 12} = \$602{,}257.52$$

On the other hand, saving $d each month for 30 years at 6.00% interest compounded monthly, we know from the savings formula that Sally will accumulate

$$d\frac{(1 + 0.005)^{360} - 1}{0.005}$$

To make *d* just the right amount to pay off the loan exactly, we need to solve the equation

$$d\frac{(1 + 0.005)^{360} - 1}{0.005} = \$100,000 \times (1 + 0.005)^{30\times12} = \$602,257.52$$

for the value of *d*, getting *d* = $599.55 as Sally's monthly payment.

We put this in a more general setting:

Let the principal be *P*, the effective interest rate period *r*, the payment at the end of each period *d*, and let there be *n* periods. Then the **amortization formula** is

$$P(1 + r)^n = d\frac{(1 + r)^n - 1}{r}$$

Examining this equation, you see the compound interest formula on the left and the savings formula on the right. You can think of paying off the loan as making payments to a savings account, earning interest at the same rate as the loan, which will exactly balance principal and interest on the loan at the end of the loan term.

EXAMPLE ▶ *Buying a Car*

You decide to buy a new Wheelmobile car. After a down payment, you need to finance (borrow) $12,000. After comparing interest rates offered by the car dealership, local banks, and your credit union, the best rate that you can find is 7.9% (compounded monthly) over 48 months. What will your monthly payment be?

We have *P* = $12,000, monthly interest rate *r* = .079/12 = .006583333, and *n* = 48. Using the amortization formula, we have

$$\$12,000(1.0065833333)^{48} = d\frac{1.00658333^{48} - 1}{.006583333}$$

$$\$16,442.53 = d(56.2345), \qquad d = \$292.39$$

How much interest do you pay? Altogether, you make payments totaling 48 × $292.39 = $14,034.72, so your interest is $14,034.72 − $12,000 = $2,034.72.

Note that if you had gone for a Plushmobile instead, with $24,000 to be financed, you would have borrowed twice as much, and your monthly payment also would have been twice as much. ◆

EXPONENTIAL DECAY

In times of economic inflation, prices behave like populations undergoing exponential growth. For a period when the rate of inflation is constant, the compound interest formula can be used to project prices.

EXAMPLE ▶ *Inflation*

Suppose that there is 3% annual inflation from mid-1997 through mid-2001. What will be the projected price in mid-2001 of an item that costs $100 in mid-1997?

The compound interest formula applies with $P = \$100$, $r = 3\%$, and $n = 4$. The projected price is $A = P(1 + r)^n = \$100(1 + 0.03)^4 = \$112.55$.

◆

During constant-rate inflation, prices grow exponentially, but the value of the dollar goes down geometrically.

> **Exponential decay** is geometric growth with a negative rate of growth.

Let i represent the rate of inflation; what costs $1 now will cost $\$(1 + i)$ this time next year. For example, if the inflation rate were $i = 25\%$, then what costs $1 now would cost $1.25 this time next year. A dollar next year would buy only 0.8 ($= 1/1.25$) times as much as a dollar buys today. In other words, a dollar next year would be worth only $0.80 in today's dollars—by next year, a dollar would have lost 20% of its purchasing power. We say that the **present value** of a dollar next year would be $0.80. Notice that although the inflation rate is 25%, the loss in purchasing power is 20%. For a general inflation rate i, a dollar a year from now will buy only a fraction of what a dollar today can buy; that fraction, the present value of a dollar a year from now, is

$$\frac{1}{1 + i} = \frac{1 + i - i}{1 + i} = \frac{1 + i}{1 + i} \frac{i}{1 + i} = 1 - \frac{i}{1 + i}$$

In other words, a dollar a year from now is worth $\$(1 - i/(i + 1))$ today, and the loss in purchasing power is the fraction $i/(i + 1)$. (You should calculate what these expressions become for $i = 25\%$.) The quantity $-i(1 + i)$ behaves like a negative interest rate. We can use the compound interest formula to find the value of P dollars n years from now as $A = P(1 + r)^n = P(1 - [i/(i + 1)])^n$.

The actual posted price of an item, at any time, is said to be in **current dollars.** That price can be compared with prices at other times by converting all prices to **constant dollars,** dollars of a particular year.

EXAMPLE ▶ *Deflated Dollars*

Suppose that there is 25% annual inflation from mid-1997 through mid-2001. What will be the value of a dollar in mid-2001 in constant 1997 dollars?

We have $i = 0.25$, so $r = -i/(i + 1) = -0.25/1.25 = -0.20$. This, not the 25%, is the negative interest rate, the rate at which the dollar is losing purchasing power. We have $n = 4$ years, so the value of $1 four years from mid-1997 is, in 1997 dollars,

$$\$1(1 + r)^4 = \$(1 - 0.20)^4 = (0.80)^4 = \$0.41. \quad ◆$$

In the example, we may think of the value of the dollar as "depreciating" 20% per year. Depreciation of the value of equipment is similar.

EXAMPLE ▶ *Depreciation*

If you bought a car at the beginning of 1996 for $10,000 and its value in current dollars depreciates steadily at a rate of 15% per year, what will be its value at the beginning of 2001 in current dollars?

We have $P = \$10,000$, $r = -0.15$, and $n = 5$. The compound interest formula gives $A = P(1 + r)^n = \$10,000(1 - 0.15)^5 = \$4437$ ◆

Exponential decay is characteristic of radioactive materials. A radioactive substance emits particles and decreases in quantity at a predictable continuous rate. The amount of radioactive substance remaining is given by the continuous interest formula. The rate r is usually written instead as $-\lambda$; the positive quantity λ is called the *decay constant*.

The **decay constant** for a substance decaying exponentially is the proportion of the substance that decays per unit time.

The amount of remaining after t time units is

$$A = Pe^{-\lambda t}$$

Alternatively, the rate of decay of the substance can be described in terms of the *half-life*. A substance that is decaying geometrically never completely vanishes, even after millions of years. Since there is no time until it is all gone, we settle for measuring how long it takes until half of it is gone.

The **half-life** of a substance decaying geometrically is the time that it takes for one-half of the substance to decay.

For instance, the radioactive isotope radon-222 is a gas that enters many homes in the United States from the underlying soil and rock. It causes an estimated 10% of lung cancers in the United States. Its half-life is 3.82 days; this means that of 1 g (gram) of radon-222 now, in 3.82 days only 0.5 g remains (the rest decays into other elements). Further, in $2 \times 3.82 = 7.64$ days, only 0.25 g remains; and in $3 \times 3.82 = 11.46$ days, only 0.125 g remains.

In terms of the half-life, represented by $T_{1/2}$, the amount remaining after time t (measured in the same units as the half-life) is

$$A = P\left(\frac{1}{2}\right)^{t/T_{1/2}}$$

This and the earlier expression $A = Pe^{-\lambda t}$ are two exactly equivalent ways of expressing the decay. To convert between decay constant and half-life, set the two expressions equal:

$$P\left(\frac{1}{2}\right)^{t/T_{1/2}} = Pe^{-\lambda t}, \qquad \text{so} \qquad \left(\frac{1}{2}\right)^{t/T_{1/2}} = e^{-\lambda t}$$

At this point we take the "natural logarithm" of both sides. This function is denoted by "ln" here and by either $\boxed{\ln x}$ or $\boxed{\text{LN}}$ on your calculator (not $\boxed{\text{LOG}}$ or $\boxed{\log_{10}}$, which stands for a different kind of logarithm). (Try some values on your calculator and observe that the $\boxed{\ln x}$ button undoes what the $\boxed{e^x}$ button does, and vice versa; in fact, $\ln e^1 = 1$.) We get

$$\frac{t}{T_{1/2}} \ln\left(\frac{1}{2}\right) = -\lambda t \ln e = -\lambda t \times 1 = -\lambda t$$

Observe that $\ln \frac{1}{2} = -\ln 2 \approx -0.693$ (check this on your calculator) and divide through by $(-t)$, getting

$$\frac{1}{T_{1/2}} \ln 2 = \lambda$$

Multiplying through by $T_{1/2}$ gives the relationship

$$\lambda T_{1/2} = \ln 2 \approx 0.693$$

EXAMPLE ▶ *Carbon-14 Dating*

Carbon-14 dating is a method of determining the age of organic materials, including mummies, charcoal from ancient fires, parchment, and cloth. The element carbon, which is present in the food that we eat and in all living things, always has small traces of a radioactive form, called carbon-14. Plants and animals continually absorb carbon-14 during their lives, from the air (for plants) and from food (for animals), so that the concentration in their bodies stays the same while they are alive. Once they die, however, no new carbon-14 gets absorbed, and the carbon-14 already present continues to decay. Because we know the concentration of carbon-14 in living things, and we know how long it takes carbon-14 to decay, we can calculate how long ago a sample of a plant or animal was living.

The half-life of carbon-14 is 5730 years; its decay constant is $\lambda = \ln 2/T_{1/2} \approx 0.693/(5730 \text{ yr}) = 1.209 \times 10^{-4}/\text{yr} = 0.0001209/\text{yr}$. In other words, about 12 in 100,000 carbon-14 atoms decay each year. In each gram of carbon, approximately 814 carbon-14 atoms decay each hour.

An approximate age of a sample can be determined by working backwards by half-lives. Suppose that a sample is decaying at 26 atoms per hour per gram of carbon. Table 18.5 shows that the 814 atoms per hour per gram of carbon would decrease to approximately 26 atoms per hour per gram of carbon in approximately 29,000 years, so that is the approximate age of the sample. (An age of 0 for the sample denotes the time of death of the living body.)

How much of the original carbon-14 would be left after 50,000 years? (This is roughly the practical age limit for carbon-14 dating of the typically small samples available.) We have

$$A = Pe^{-\lambda t} = Pe^{-6.05} = 0.0024$$

so only about 0.24% remains. This remaining amount would be decaying at a rate of $0.024 \times 814 = 19.5$ atoms per hour per gram of carbon.

TABLE 18.5	Estimating the Age of a Sample	
Age in Half-Lives	Age in Years	Decays per Hour per Gram of Carbon
0	0	$(\frac{1}{2})^0 (814) = 814$
1	5,730	$(\frac{1}{2})^1 (814) = 407$
2	11,460	$(\frac{1}{2})^2 (814) = 203.5$
3	17,190	$(\frac{1}{2})^3 (814) = 101.8$
4	22,920	$(\frac{1}{2})^4 (814) = 50.9$
5	28,650	$(\frac{1}{2})^5 (814) = 25.5$

The formula that relates t, the age of the sample in years, and N, the number of carbon-14 atoms disintegrating per gram per hour, is

$$\left(\frac{1}{2}\right)^{t/5730} = \frac{N}{814}$$

Solving for N gives

$$N = 814 \times \left(\frac{1}{2}\right)^{t/5730}$$

Using natural logarithms, we can solve for t as

$$\ln N = \ln 814 + \frac{t}{5730} \ln \frac{1}{2}, \quad t = 55{,}403 - 8267 \ln N$$

Thus, a sample decaying at a rate of 105 atoms per hour per gram of carbon would be $t = 55{,}403 - 8267 \times \ln 105 \approx 17{,}000$ years old.   ◆

GROWTH MODELS FOR BIOLOGICAL POPULATIONS

We now use a geometric growth model to make rough estimates about sizes of human populations, using for the growth rate r the difference between the annual birth rate and the annual death rate, for which the technical term is the **rate of natural increase.** In the terminology that we have used for financial models, this is the effective rate.

Birth and death rates rarely remain constant for very long, so projections must be made with extreme care. In addition, we exclude the effect of net migration. In the short run, predictions based on the model may provide useful information. Let's apply this model to two questions about the population of the United States.

E X A M P L E ▶ *Predicting U.S. Population*

The population of the United States was 266 million at the beginning of 1997. It was increasing then at an average growth rate of 0.7% per year. What is the anticipated population at the beginning of the year 2000? What is it if the rate of natural increase is instead 0.4% per year, or 1.0% per year?

We apply the compound interest formula with initial population size ("principal") 266 million. Using a year as the compounding period and the formula $A = P(1 + r)^n$, where $n = 3$, the projected population size in 2000 for a rate $r = 0.007$ is

$$\text{Population in 2000} = (\text{population in 1997}) \times (1 + \text{growth rate})^3$$
$$= 266,000,000(1 + 0.007)^3$$
$$= 266,000,000(1.02115) \approx 272,000,000$$

Because the original estimates of population and growth rate are approximations, we don't copy down all the digits from the calculator but instead round off; the result of a calculation can't be more precise than the ingredients.

In the same way, with a growth rate of 0.4% per year, we predict a population of 269 million, while a growth rate of 1.0% per year yields a prediction of 274 million. So an uncertainty of three-tenths of one percentage point, or 0.003, in the growth rate has major implications, even over fairly short time horizons. The presence or absence of 5 million people would have a significant impact on our social and economic systems! Indeed, much of the concern over long-range funding of the social security programs results from uncertainties over birth and immigration rates. Figure 18.2 gives a graph of the U.S. population in 2000, structured by age and sex.   ◆

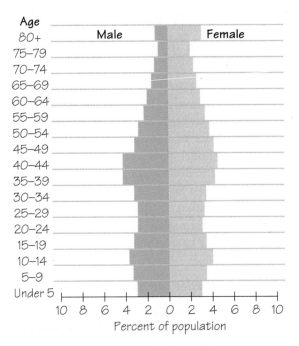

Figure 18.2 Graph of the projected population of the United States in the year 2000 grouped by age and sex and shown as a percentage of the total population. This projection is based on a geometric growth model.

Rates of natural increase in most Third World countries are much higher than in industrialized nations, sometimes 3% per year or more. With its growth rate of 3.1%, Africa's most populous country, Nigeria, whose population was 101 million in mid-1995, will have 162 million by the middle of 2010, an increase of 60%. Projections of this sort are at the root of worldwide concern over the ability to provide sufficient food and other resources for all people.

LIMITATIONS ON BIOLOGICAL GROWTH

According to the geometric growth model, as shown in Figure 18.3, the size of the population keeps increasing over time, to astronomical numbers. Such predictions are unreasonable. No biological population can continue to increase without limit. Its growth is eventually constrained by the availability of resources such as food, shelter, and psychological and social "space." A geometric growth model cannot describe forever the growth of such a population.

The growth rate of a biological population is likely to depend on the size of the population and to decrease as population size increases. (This happened fairly steadily to the U.S. population from 1865 to 1945.) For a sufficiently large population, the "growth" may even be negative.

As the population increases, the resources per individual decrease, and thus the energy available for growth and reproduction decreases. There may in fact be a maximum population size that can be supported by the available resources, the **carrying capacity** of the environment.

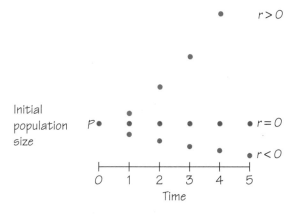

FIGURE 18.3 Projected population over five time units assuming a geometric growth model with growth rate r.

The **logistic model** takes the carrying capacity into account by reducing the natural rate of increase r by a factor of how close the population size P is to the carrying capacity M:

$$\text{growth rate} = r\left(1 - \frac{\text{population size}}{\text{carrying capacity}}\right) = r\left(1 - \frac{P}{M}\right)$$

As the population increases, the growth rate decreases, because the term containing the population P has a negative sign. For a population equal to the carrying capacity ($P = M$), the growth rate is zero.

If at any time the population exceeds the carrying capacity, then the growth rate becomes negative (because $P > M$) and the population decreases. The carrying capacity refers to long-range capacity to support the population, so the population could exceed it for short periods of time. This could happen either because the population grows very rapidly and surges above the carrying capacity, or because of a sudden decrease in the food supply, thus temporarily lowering the carrying capacity, as happens to deer and other animals in winter.

The logistic model provides excellent predictions for the growth of some populations, particularly in laboratory environments. For example, the graph in Figure 18.4 shows the growth of a population of fruit flies in a glass enclosure with a limited food supply. On the same coordinate system we show the predictions of an arithmetic population model (the straight line labeled P_1), a geometric population model (the exponential curve labeled P_2), and a logistic population model (labeled P_3). Predictions based on the logistic model closely describe the actual growth.

FIGURE 18.4 The growth of a population of fruit flies: P_1 is the best approximation by a simple growth model; P_2 is the best approximation by a geometric growth model; and P_3 is the best approximation by a logistic model. The blue points show the actual values.

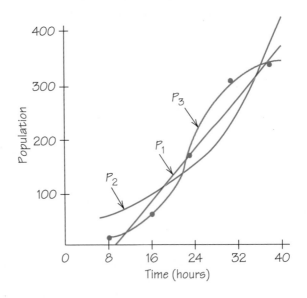

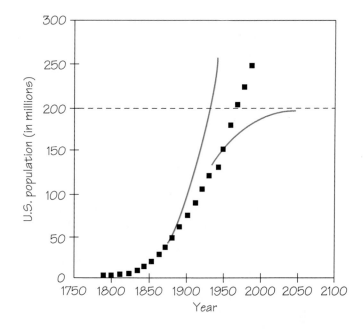

FIGURE 18.5 U.S. population by year. The points show actual growth, the red curve shows exponential (geometric growth) based on the early years of the United States, and the blue curve shows logistic growth based on figures through 1920.

E X A M P L E ▶ *The Logistic Model*

The U.S. population from 1790 to 1950 closely followed a logistic model with $r = 0.031$, $P =$ population in 1790 $= 3,900,000$, and $M = 201$ million. For the early decades after 1790, the population was a small fraction of this "carrying capacity," and it grew at close to the rate r of 3.1% per year (a rate similar to many Third World countries today). By 1920 the U.S. population had reached 106 million, a little more than half of the "carrying capacity," and indeed the growth rate had slowed by about one-half, to 1.5% per year (see Figure 18.5).

The 1997 U.S. population of 266 million far exceeds the hypothesized "carrying capacity" of 201 million. The economic basis of U.S. society changed, from a large proportion of people making their living on family farms to a highly urbanized society. With such a change, the original logistic model was no longer valid. ◆

NONRENEWABLE RESOURCES

People use natural resources, some of which are renewable but others are not. In this section, we model populations of nonrenewable resources; in the next, we treat renewable resources.

A **nonrenewable resource** is one that does not tend to replenish itself; gasoline, coal, and natural gas are important examples. There is no practical way to recover or reconstitute these resources after use. Some substances, such as aluminum or the sand used to make glass, are potentially recyclable; but to the extent that we do not recycle them, they too are nonrenewable.

For a renewable resource, there is only a fixed supply S (in some convenient units) of it that is available to us. Even without human population increases, we are faced with dwindling populations of nonrenewable resources. We are interested in the question: How long will the supply ("population") of the resource last?

As long as the rate of use of the resource remains constant, the answer is easy. If we are using U units per year and continue using U units per year, then the supply will last S/U years. This kind of calculation is the basis for statements such as "at the current rate of consumption, U.S. coal reserves will last 500 years," or that the U.S. strategic reserve of gasoline (stored in salt domes in the South) would last 60 days.

However, the rate of use of resources tends to increase with increasing population and with higher "standard of living." For example, projections for use of electric power are often based on assumptions that the use will increase by some fixed percentage each year. This is the simplest situation (apart from constant usage), and one that we can easily model to give an important perspective.

Suppose that $U_1 = U$ is the rate of use of the resource in the first year (this year), and that usage increases $r = 0.05 = 5\%$ each year. Then the usage in the second year is

$$U_2 = U_1 + 0.05 U_1 = 1.05 U$$

and usage in the third year is

$$U_3 = U_2 + 0.05 U_2 = 1.05 U_2 = 1.05(1.05 U) = (1.05)^2 U$$

Generalizing, we see that usage in year i will be $(1.05)^{i-1} U$. Total usage over the next 5 years, for example, will be

$$U + (1.05)^1 U + (1.05)^2 U + (1.05)^3 U + (1.05)^4 U$$

This situation should remind you of our earlier study of accumulation of regular deposits plus interest. Here the usage U corresponds to a deposit and the increasing rate of use r corresponds to the interest rate. We may think of the situation as making regular withdrawals (with interest) from a fixed supply of the nonrenewable resource. The savings formula gives

$$A = d \frac{(1 + r)^n - 1}{r}$$

In translating to the resource situation, A is the accumulated amount of the resource that has been used up at the end of n years, and U is the initial rate of use. We have

$$A = U\frac{(1 + r)^n - 1}{r}$$

To find out how long the supply S will last, we set the supply S equal to the cumulative use A over n years and then determine what n will be. We have

$$S = U\frac{(1 + r)^n - 1}{r}$$

We perform some algebra to isolate the term involving n, getting

$$(1 + r)^n = 1 + \frac{S}{U} r$$

At this point, to isolate n, we need to take the natural logarithm of both sides. We get

$$\ln [(1 + r)^n] = n \ln (1 + r) = \ln\left(1 + \frac{S}{U} r\right)$$

which gives the final expression

$$n = \frac{\ln [1 + (S/U)r]}{\ln (1 + r)}$$

This expression may look complicated, but it is quite easy to evaluate on a calculator for particular values of S/U and r.

The expression S/U is called the *static reserve*, and n is called the *exponential reserve*.

The **static reserve** is how long the supply S will last at a particular constant rate of use U, namely, S/U units of time.

The **exponential reserve** is how long the supply S will last at an initial rate of use U that is increasing by a proportion r each year, namely

$$\frac{\ln [1 + (S/U)r]}{\ln (1 + r)}$$

units of time.

EXAMPLE ▶ *U.S. Coal Reserves*

We noted earlier that measured reserves of U.S. coal would last about 500 years at the current rate of use, so the static reserve for this resource is 500 years. How long would the supply last if the rate of use increases 5% per year? The corresponding exponential reserve is

$$n = \frac{\ln[1 + (500)(0.05)]}{\ln 1.05} = \frac{\ln 26}{\ln 1.05} = 65 \text{ years}$$

That's quite a difference!   ◆

We must not take such projections as exact predictions. Estimates of supplies of a resource may underestimate how much is available, and previously unknown sources may be discovered or the technology improved to extract previously unavailable supplies. In addition, as supplies dwindle, the economic considerations of supply, demand, and price come into play. We will never completely run out of oil; it will always be available "at a price."

However, we must not take such projections lightly, either, since we are discussing resources that, once used, are gone forever. In any projection, it is very important to examine the assumptions, since small differences in the rate of increase of use can make big differences in the exponential reserve.

RENEWABLE RESOURCES

A **renewable natural resource** is a resource that tends to replenish itself, such as fish, wildlife, and forests. We would like to know how much of a resource we can harvest and still allow for the resource to replenish itself.

We concentrate on the subpopulation of individuals with commercial value. For a forest, this subpopulation might be trees of a commercially useful species and appropriate size. We measure the population size as its **biomass,** the mass of the population expressed in units of equal value. For example, we measure the size of a fish population in terms of pounds rather than numbers of fish, and a forest not by counting the trees but by estimating the number of board feet of usable timber.

Reproduction Curves

In this chapter we have seen several models to analyze and predict population growth. As models, they include many simplifications. Real populations may behave like one of the models we have discussed, or like other known models. But the complicated factors that can affect populations, such as climatic or

The Brazilian rain forest lost hundreds of thousands of acres of forestation when it was discovered that the soil would make good pastureland.

economic change, may mean that the only way to understand the population is to plot a graph of its size over time. Whether the growth of a population can be described by one of the formulas that we have developed or by a table of measurements collected by counting, there are useful techniques to analyze the growth of the population and help us make decisions about managing it. The situation that we will be thinking about most is managing a population of renewable resources, such as trees or fish, to make the most of them.

We use a figure called a **reproduction curve,** which predicts next year's population size (biomass) based on this year's size. Although the precise shape of the curve varies from one population to another, the shape shown in Figure 18.6 is typical. It shows next year's size (on the vertical axis) as a function of

FIGURE 18.6 A typical reproduction curve.

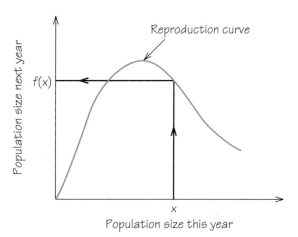

this year's size (on the horizontal axis). For all possible sizes, the reproduction curve shows the change in size from one year to the next, taking into account growth of continuing members, and addition of new members, minus losses due to death and other factors.

Let x on the horizontal axis be a typical size of the population in the current year. The size *next* year is given by the height of the curve above the point marked x. This value is denoted by $f(x)$. (You can think of f as standing for "function of," or even as "forthcoming.")

Figure 18.7 shows the same reproduction curve, plus the broken line $y = x$ (which makes a 45° angle with the horizontal axis). You could trace what happens for various choices for x. For an x for which the curve is above the broken line, next year's size ($f(x)$) is larger than this year's (x). In Figure 18.7, the **natural increase,** or gain in population size, is shown as the length of the green vertical line from the broken line to the curve, which in algebraic terms is $f(x) - x$. For an x for which the curve is below the broken line, next year's size is smaller than this year's and $f(x) - x$ is negative. For the size labeled x_e, for which the curve crosses the broken line, the size is the same next year as this year; this is the *equilibrium population size.*

An **equilibrium population size** is one that does not change from year to year.

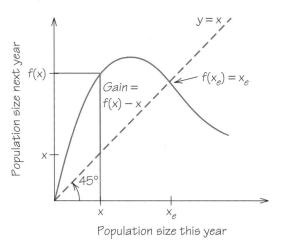

FIGURE 18.7 Depiction of the natural increase (gain) in population from one year to the next. The population size x_e is the equilibrium population size, for which the population one year later is the same, or $f(x_e) = x_e$.

The line $y = x$ provides a convenient way to trace the evolution of the population over several years (see Figure 18.8), by alternating steps vertically to the curve and horizontally to the line $y = x$. Begin with the first year's population on the horizontal axis, go up vertically to the curve; the height is the population in the second year. Proceed horizontally from the curve over to the line $y = x$. Proceeding vertically from there to the curve yields a height that is the population in the third year.

Figure 18.8 shows several traces for the same reproduction curve, each starting from a different initial population on the horizontal axis. The resulting variation is quite surprising—it can even be "chaotic" in a very specific mathematical sense, showing how apparently random behavior can result from strict deterministic rules.

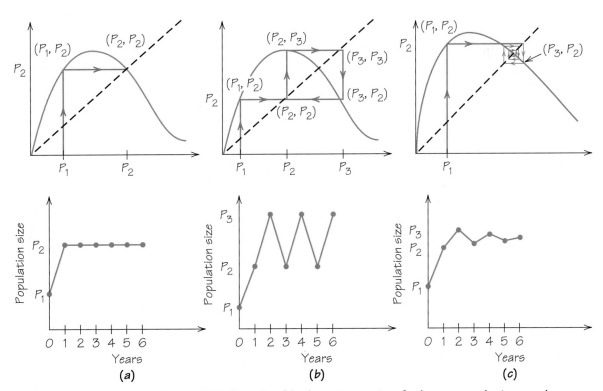

Figure 18.8 Examples of the dynamics, over time, for the same reproduction curve but different starting populations. (a) The population goes in one year to the equilibrium population and stays there year after year. (b) After initial adjustment, the population cycles between values over and under the equilibrium population. (c) The population spirals in toward the equilibrium population.

Sustained-Yield Harvesting

Many biological populations are harvested by predators (including humans). **Yield** is the amount harvested at each harvest. We focus on sustained-yield harvesting, which is important to timber companies and other enterprises that extract a natural resource over a long period.

> A **sustained-yield harvesting policy** is a policy that if continued indefinitely will maintain the same yield.

For a sustainable yield, the same amount is harvested every year and the population remaining after each year's harvest is the same. To achieve this stability, the harvest must exactly equal the natural increase each year, the length of the green vertical line in Figure 18.7.

Each value of x between 0 and x_e determines a different vertical line and corresponding sustained-yield harvest. This harvest can vary from 0 (for $x = 0$ or $x = x_e$) up to some maximum value (for some x between 0 and x_e). A goal for a timber company or a fishery is to harvest the **maximum sustainable yield:** to select an x whose vertical line is as long as possible, marked as x_M in Figure 18.9.

Considerations from Economics

The costs of harvesting should be taken into account in our analysis. We consider two models: one for a cattle ranch and one for either a fishing boat or a tree farm.

FIGURE 18.9 The reproduction curve, with the population size x_M corresponding to the maximum sustainable yield. The maximum sustainable yield is the greatest vertical distance from the 45° line to the reproduction curve.

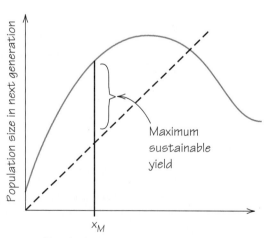

Population size in next generation

Maximum sustainable yield

x_M

Population size in current generation

We assume that the price p received is the same for each harvested unit and does not depend on the size of our harvest. In effect, we assume that our operation is a small part of the total market, not substantially affecting overall supply and hence price.

We want to stay in business, so we do not extinguish the resource for quick profits. For any given population size, we harvest just the natural increase.

EXAMPLE ▶ *Cattle Ranching*

We assume that the cost of raising and bringing a steer to market is the same for every steer and does not depend on how many steers we bring to market. Since the cost does not depend on the population size, the cost curve is a horizontal line (Figure 18.10).

As long as the selling price per unit is higher than the harvest cost per unit, we make a profit. The points of view of economics and biology agree, since the maximum profit occurs for the maximum sustainable yield. ◆

EXAMPLE ▶ *Fishing and Logging*

In this model we assume that the cost of harvesting a unit of the population decreases as the size of the population increases; this is the familiar principle of **economy of sale.** For example, the same fishing effort yields more fish when fish are more abundant. Similarly, a logger's harvest costs per tree are less when the trees are clumped together; this is the logger's motivation to clear-cut large stands.

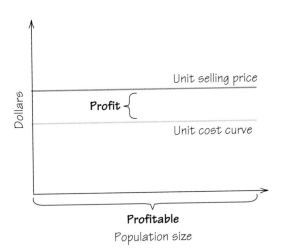

FIGURE 18.10 The unit cost, unit revenue, and unit profit of harvesting one unit, as a function of population size, for the cattle ranch.

Spotlight *Extinction of the Passenger Pigeon*

18.3

Although once numbering in the billions, passenger pigeons are extinct.

Historically, the utilization of a renewable resource has followed a characteristic pattern. First comes a stage of expanding harvests, perhaps based on a new use of the resource or on new harvesting technology. This is followed by concern about overutilization. Conservation measures may then be adopted, and the industry either stabilizes or collapses.

For example, the passenger pigeon *(Ectopistes migratorius)* was once considered the world's most abundant land bird, with billions in huge flocks in eastern North America. But by 1914, the last one died at the Cincinnati Zoo.

This extinction can clearly be traced to expanding harvests brought about by new technologies—the eastern railroad network and the telegraph. These technologies exploited behaviors that were key to the passenger pigeon's earlier success: colonization and nomadism.

Passenger pigeons nested in colonies of millions of pairs (the entire population consisted of perhaps fewer than a dozen flocks). The flocks were sometimes so immense that they were reported to have obscured the sun. (The largest flight recorded was estimated to contain 2.23 billion birds.) In this way, pigeons could "shield" themselves from predators via "predator satiation": no matter where they nested, there were not enough local predators to reduce their numbers significantly.

Passenger pigeons fed on large crops of nuts in deciduous forests. The location of crops large enough to accommodate their numbers varied from year to year, so flocks rarely nested in the same place two years in a row. Hence, it was difficult to predict where flocks would nest.

The arrival of a flock had always meant food for the local people, but the pigeon population began to markedly diminish only after harvesting them for market became a major industry, around 1840.

The railroad and the telegraph increased the efficiency and scope of market harvesting so much that the passenger pigeon was extinguished. The railroad gave professional pigeoners, who numbered about 1000 in their heyday, rapid routes to all major nesting colonies and a fast means of shipping barrels of pigeons to the cities. The telegraph kept the pigeoners informed of the locations of nesting colonies.

The gigantic colonies made pigeons especially vulnerable. People could not understand that so abundant a resource could ever be severely diminished. They did not allow for undisturbed nesting sites so that the pigeons could replenish their numbers. Instead, harvests were complete.

Passenger pigeons were once a renewable resource, but within a period of about 20 years—twice a pigeon's lifetime—they became extinct. Other species, such as the bison, have also been reduced to numbers below economic significance.

Sustained-yield policies involve revenues that will be received, year after year, in the future. The value of these revenues should be discounted to reflect the lost investment income that we could earn if instead we had the revenues today. For funds invested at a return of $100r\%$ per year, compounded annually, the present value P of an amount A to be received in n years in the future is related to A by the compound interest formula $A = P(1 + r)^n$.

The economic goal is to maximize the sum of the present values of all future receipts from harvesting. The optimal harvesting policy thus must depend on the expected rate of return r. We don't delve into the details of the calculations here, but instead just give the results of the analysis.

Again there are several cases to consider:

1. The unit cost of harvesting exceeds the unit price received, for all population sizes. Then it is impossible to make a profit.
2. For some population size x, the unit cost of harvesting equals the unit price received. Then there is a size between x and x_e (the equilibrium population size) for which the present value of the total return is maximized and the population and its yield are sustained.
3. The unit price exceeds the unit cost for *all* population sizes.

   - For r small, the situation is the same as in the second case just cited.
   - For larger r, the economically optimal policy may be to harvest the entire population immediately—extinguish the resource—and invest the proceeds.

Let's put this in the simplest and starkest terms. Suppose that you own a valuable resource, such as a forest, whose cost of harvesting is small relative to the value of the resource. If the rate at which the forest population grows is greater than what you can earn on other investments, it pays to let the forest keep on growing.

On the other hand, if the forest is growing more slowly than the rate of return on other investments, the economically optimal harvesting policy is to cut down all the trees now and invest the money. You could then start raising cattle on the land—and right there you have the scenario that is resulting in deforestation all over the world.

The sobering fact is that *very few economically significant renewable resources can sustain annual growth rates over 10%*. Many, like whales and most forests, have growth rates in the 4% to 5% range. These values—even a growth rate of 10%—are far below the return investors expect on their investment. For example, until recent deregulation, Wisconsin electric utilities were guaranteed 14.25% profit; and venture capital firms expect to exceed 25% profit.

The concept of maximum sustainable yield is an attractive ideal if expectations of investors are low enough. However, there are still difficult problems:

$Spotlight$ *The Tragedy of Easter Island*

18.4

Easter Island

Easter Island is famous for its isolation—1400 miles to the nearest island—and for its hundreds of huge stone statues. For 30,000 years before the arrival of people in about A.D. 400, Easter Island maintained a lush forest, with several species of land birds. By the time of the first visit by Europeans in 1722, the island was barren, denuded of all trees and bushes over 10 feet high, and with no native animals larger than an insect. The 2000 or so islanders had only three or four leaky canoes made of small pieces of wood.

What happened? Careful analysis of pollen in soil samples tells the sad story. The settlers and their descendants cut wood to plant gardens, build canoes, make sledges and rollers to move the huge statues, and burn for cooking and warmth in the winter. In addition to crops they raised and chickens that they had brought to the island and cultivated, they ate palm fruit, fish, shellfish, the meat and eggs of birds, and the meat of porpoises that they hunted from seagoing canoes. The population of the island grew to 7000 (or perhaps even 20,000).

By 1500, the forest was gone. Most tree species, all land birds, half of the seabirds, and all large and medium-sized shellfish had been extinguished. There was no firewood, no wood for sledges and rollers to transport hundreds of statues at various stages of completion, and no wood for seaworthy canoes. Without canoes, fishing declined and porpoises could not be taken. Stripping the trees exposed the soil, which eroded, so crop yields fell off. The people continued raising chickens, but warfare and cannibalism ensued. By 1700, the population had crashed to 10–25% of its former size.

Why didn't the people notice earlier what was happening, imagine the consequences of keeping on as they had been, and act to avert catastrophe? After all, the trees did not disappear overnight.

From one year to the next, changes may not have been very noticeable. The forests may have been regarded as communal property, with no one charged with limiting exploitation or ensuring new growth. There was no quantitative assessment of the resources available and need for conservation versus the long-term needs of the "public works" program of erecting statues. Moreover, the religion of the people, the prestige of the chiefs, and the livelihood of hundreds depended on the statue industry. There was no perceived need to limit the population and no technology for birth control. Once the large trees were gone, there was no means for excess population to emigrate.

Adapted from Jared Diamond, "Easter's end," *Discover*, 16 (8) (August 1995): 63–69.

- One problem is "the tragedy of the commons," discussed by ecologist Garrett Hardin. Several hundred years ago, English shepherds would graze their flocks together on common land. The grass of the commons could support only a fixed number of sheep. Each shepherd could reasonably think that adding just one or two more sheep to his flock would not overtax the commons; yet if each did so, there could be disaster, with all the sheep starving. Many natural-products industries, such as fisheries, are a form of commons; small overexploitation by each harvester can produce disastrous results for all.

- How, in the presence of human needs or greed, can we anticipate and prevent overexploitation and possible extinction of a resource? By and large, it has been politically impossible to force a harvesting industry to reduce current harvests to assure stability in the future.

- In some industries, such as a fishery, growth of the population may be abundant one year but meager another, so that a steady yield cannot be sustained without damaging the resource. A few good years in a row may provoke increased investment in fishing capacity; then attempting to harvest at the same levels in succeeding normal or below-normal years results in overfishing. This exact scenario destroyed the California sardine fishery in the 1930s, the Peruvian anchovy fishery in 1972, and much of the North Atlantic fishery in the 1980s.

REVIEW VOCABULARY

Amortize   To repay in regular installments.

Annual (equivalent) yield, annual percentage yield   Effective rate on an annual basis.

Annual rate   Nominal rate on an annual basis.

Arithmetic growth   Growth by a constant amount in each time period.

Biomass   A measure of a population in common units of equal value.

Carrying capacity   The maximum population size that can be supported by the available resources.

Compound interest   The method of paying interest on both the principal amount and the accumulated interest in an account.

Compounding period   The interval that elapses before interest is calculated on an account.

Constant dollars   Costs are expressed in constant dollars if inflation or deflation has been taken into account by converting all costs to their equivalent in dollars of a particular year.

Continuous compounding   Payment of interest in the amount toward which compound interest tends with more and more frequent compounding.

Current dollars   The actual cost of an item is said to be in current dollars; inflation or deflation has not been taken into account.

Decay constant   For a substance decaying exponentially, the proportion that decays per unit time.

e   The base for continuous compounding, geometric (exponential) growth, and natural logarithms; $e = 2.71828 \ldots$.

Economy of scale   Costs per unit decrease with increasing volume.

Effective rate   The percentage increase in an account after one year.

Equilibrium population size   A population size that does not change from year to year.

Equivalent yield   Effective rate.

Exponential decay   Geometric growth with a negative rate of growth.

Exponential growth   Geometric growth.

Exponential reserve   How long a fixed amount of a resource will last at a constantly increasing rate of use.

Geometric growth   Growth proportional to the (increasing) amount present.

Geometric series   A sum of terms, each of which is the same constant times the previous term, that is, the terms grow geometrically.

Half-life   For a substance decaying geometrically, the time until one half of an initial quantity remains.

Initial balance   Initial deposit in a bank account.

Interest   Money earned on financial investments, such as bank accounts.

Logistic model   A particular population model that begins with near geometric growth but then tapers off toward a limiting population (the carrying capacity).

Maximum sustainable yield   The largest harvest that can be repeated indefinitely.

Natural increase   The growth of a population that is not harvested.

Nominal rate   The stated rate of interest per year, on which any compounding is based.

Nonrenewable resource   A resource that does not tend to replenish itself.

Population growth   Change in population, whether increase (positive growth) or decrease (negative growth).

Population structure   The division of a population into subgroups.

Present value   The value today of money to be received in the future.

Principal   Initial balance.

Rate of natural increase   Birth rate minus death rate, the annual rate of population growth without taking into account net migration.

Renewable natural resource   A resource that tends to replenish itself; examples are fish, forests, wildlife.

Reproduction curve   A curve that shows population size in the next year plotted against population size in the current year.

Simple growth   Arithmetic growth.

Simple interest   The method of paying interest on only the initial balance in an account and not on any accrued interest.

Sinking fund   A savings plan to accumulate a fixed sum by a certain date.

Static reserve   How long a fixed amount of a resource will last at a constant rate of use.

Sustained-yield harvesting policy   A harvesting policy that can be continued indefinitely while maintaining the same yield.

Yield   The amount harvested at each harvest.

SUGGESTED READINGS

ARROW, KENNETH A., ET AL. Economic growth, carrying capacity, and the environment, *Science,* 268 (April 28, 1995): 520–521. Letters: Economic growth and environmental policy (June 16, 1995): 1549–1551.

BARTLETT, ALBERT A. Forgotten fundamentals of the energy crisis, *American Journal of Physics,* 46 (9) (September 1978): 876–888.

CLARK, COLIN. Some socially relevant applications of calculus, *The Two-Year College Mathematics Journal,* 4 (2) (Spring 1973): 1–15. Gives a mathematical approach to animal resource economics.

CLARK, COLIN. The mathematics of overexploitation, *Science,* 181 (August 17, 1973): 630–634.

COHEN, JOEL. *How Many People Can the Earth Support?* Norton, New York, 1995.

COHEN, JOEL. Ten myths of population, *Discover,* 17 (4) (April 1996): 42–47.

DIAMOND, JARED. Easter's end, *Discover,* 16 (8) (August 1995): 63–69.

HASS, PAUL H., ED. Some documents relating to the passenger pigeon, *Wisconsin Magazine of History,* 59 (Summer 1976): 259–281. Reprinted as *Passenger Pigeons,* Wisconsin Stories series, State Historical Society of Wisconsin, Madison.

HARDIN, GARRETT. The tragedy of the commons, *Science,* 162 (1968): 1243–1248.

KASTING, MARTHA. *Concepts of Math for Business: The Mathematics of Finance.* UMAP Modules in Undergraduate Mathematics and Its Applications: Modules 370–372. COMAP, Inc., Arlington, Mass., 1980.

KLEINBAUM, DAVID G., AND ANNA KLEINBAUM. *Adjusted Rates: The Direct Rate.* UMAP Modules in Undergraduate Mathematics and Its Applications: Module 330. COMAP, Inc., Arlington, Mass., 1980. Reprinted in *The UMAP Journal,* 1 (1) (1980): 49–80; and in *UMAP Modules: Tools for Teaching 1980,* Birkhäuser, Boston, 303–334. A beginning exploration into the structure of populations, which explains, for instance, the paradox of how a Third World country can have a lower overall mortality rate than the United States, yet have a higher mortality rate for every age group.

LINDSTROM, PETER A. *Nominal vs. Effective Rates of Interest.* UMAP Modules in Undergraduate Mathematics and Its Applications: Module 474. COMAP, Inc., Arlington, Mass., 1988. Reprinted in *UMAP Modules: Tools for Teaching 1988,* Paul J. Campbell (ed.), COMAP, Inc., Arlington, Mass., 21–53. A learning module about the difference between nominal and effective rates of interest and how to calculate them. Gives examples of banks using particular options for calculating interest.

LUDWIG, DONALD, RAY HILBORN, AND CARL WALTERS. Uncertainty, resource exploitation, and conservation: Lessons from history, *Science,* 260 (April 2, 1993): 17, 36.

MEADOWS, DONELLA H., DENNIS L. MEADOWS, AND JØRGEN RANDERS. *Beyond the Limits: Confronting Global Collapse, Envisioning a Sustainable Future,* Chelsea Green, Post Mills, Vt., 1992.

OLINICK, MICHAEL. Modelling depletion of nonrenewable resources, *Mathematical Computer Modelling,* 15 (6) (1991): 91–95.

OPHULS, WILLIAM, AND A. STEPHEN BOYAN, JR. *Ecology and the Politics of Scarcity Revisited,* W. H. Freeman, New York, 1992.

POPULATION REFERENCE BUREAU. *Annual World Population Data Sheet.* 777 14 St. N.W., Suite 800, Washington, D.C. 20005.

SAFINA, CARL. The world's endangered fisheries. *Scientific American,* 273 (5) (November 1995): 46–53.

SCHWARTZ, RICHARD H. *Mathematics and Global Survival,* 3rd ed., Ginn, Needham Heights, Mass., 1993.

EXERCISES ▲ *Optional.* ■ *Advanced.* ◆ *Discussion.*

The exercises below require a scientific calculator with buttons for powers $\boxed{y^x}$, exponential $\boxed{e^x}$, and natural logarithm $\boxed{\ln x}$.

The Mathematics of Geometric Growth

1. You deposit $1000 at 8% per year. What is the balance at the end of one year, and what is the annual yield, if the interest paid is

    (a) simple interest?
    (b) compounded annually?
    (c) compounded quarterly?
    (d) compounded daily?

2. Repeat Exercise 1, but for $1000 at 3% per year.

3. *Zero-coupon bonds* are securities that pay no current interest but are sold at a substantial discount from redemption value. The difference between purchase price and redemption value provides income to the bondholder at the time of redemption or resale. If the interest rate in the economy is now 7%, what should be the price of a zero-coupon bond that will pay $10,000 eight years from now? (Use daily compounding.)

4. Repeat Exercise 3, but for a current interest rate of 5% and a zero-coupon bond that will pay $10,000 five years from now.

5. *The rule of 72* is a rule of thumb for finding how long it takes money at interest to double: if $100r\%$ is the annual interest rate, then the doubling time is approximately $72/100r$ years.

    (a) Calculate the balance at the end of the predicted doubling time for each $1000, with annual compounding, for the small growth rates of 3%, 4%, and 6%.
    (b) Repeat part (a), for the intermediate interest rates of 8% and 9%.
    (c) Repeat part (a), for the larger interest rates of 12%, 24%, and 36%.
    (d) What do you conclude about the rule of 72?

6. More frequent compounding yields greater interest, but with diminishing returns as the frequency of compounding is increased. For small interest rates, there is little difference in yield for compounding annually, quarterly, monthly, daily, or continuously. Investigating doubling times with continuous

compounding leads to understanding why the rule of 72 of Exercise 5 works. Recall that for continuous compounding at annual rate r, the balance A at the end of m years is Pe^{rm} for an initial principal of P. Let D be the number of years that it takes for the initial principal to double. Then we have $2P = A = Pe^{rD}$, so $e^{rD} = 2$. Taking the natural logarithm of both sides yields $rD = \ln 2$, where ln stands for the natural logarithm, represented on a calculator by a button marked either $\boxed{\ln}$ or $\boxed{\text{LN}}$ (not $\boxed{\log}$ or $\boxed{\log_{10}}$, which stands for a different kind of logarithm). Using the button gives $\ln 2 = 0.693$. So we have $rD = 0.693$, from which we can determine D if we know r.

Calculate the doubling times for continuous compounding at 3%, 6%, and 9%, and compare them with those predicted by the rule of 72. What do you conclude? Why do you think people prefer a rule of 72 over a rule of 69.3?

7.   Suppose that on the report for your money market account this month, the initial balance was $7373.98, the report was for 28 days, and the final balance was $7416.59. Calculate the annual yield.

8.   Repeat Exercise 7, but for the previous month, which had an initial balance of $7331.35, a period of 31 days, and a final balance of $7373.93.

A Limit to Compounding

9.   [Contributed by John Oprea of Cleveland State University.] Use your calculator to evaluate for $n = 1, 10, 100, 1000,$ and $1,000,000$:

(a) $(1 + \frac{1}{n})^n$

(b) $(1 + \frac{2}{n})^n$

(c)   As n gets large, what numbers are the expressions in parts (a) and (b) tending toward?

10.   Use your calculator to evaluate for $n = 1, 10, 100, 1000,$ and $1,000,000$:

(a) $(1 - \frac{1}{n})^n$

(b) $(1 - \frac{2}{n})^n$

(c)   As n gets large, what numbers are the expressions in parts (a) and (b) tending toward?

11.   You have $1000 on deposit at your bank at an annual rate of 4%. How much interest do you receive after one year, if the bank compounds

(a)   continuously?

(b)   daily, using the 360 over 360 method?

(c)   daily, using the 365 over 365 method?

12. Suppose that you have a bank account with a balance of $5432.10 at the beginning of the year and $5632.10 at the end of the year. Your bank advertises "continuous compounding," but in fact compounds continuously over each 24-hour day and posts interest to accounts daily.

    (a) What effective rate did you receive?

    (b) What nominal rate is the calculation based on?

    (c) What difference is there between what the bank is doing and true continuous compounding?

A Model for Accumulation

13. Suppose that you want to save up $2000 for a trip abroad two years from now. How much do you have to put away each month in a savings account that earns 5% interest compounded daily?

14. Repeat Exercise 11, except that you have found a better deal, 7% interest compounded daily.

15. You want to buy a car and you need to borrow $7000 of the cost. You can get a 48-month loan from the car dealer at 8.9%. What is your monthly payment?

16. Suppose that you and two friends decide to live off campus in your senior year. One of them (who has wealthy parents) suggests that instead of renting an apartment, you could buy a house together, live in it for your senior year, then rent it or else sell it. Assuming that you could get a mortgage for $60,000 to buy a house near your college, what would be the monthly mortgage payment for a 30-year loan at 7.5%?

▩ 17. A 1990 advertisement reads, "If you had put $100 per month in this fund starting in 1980, you'd have $37,747 today." Assume that deposits were made on the last day of the month, starting in January 1980, through December 1989, and that interest is paid monthly on the last day of the month (120 months).

    (a) How much money was deposited during this period?

    (b) What annual rate of interest, compounded monthly, would lead to the result described in the advertisement? And what is the annual yield? There is no formula to solve exactly the resulting equation, so you need to use a computer graphing program, a calculator equation solver, or successive guessing to find the monthly interest rate.

▩ 18. A family struggles for the first few years after their child is born but finally is able to start saving toward the child's college education when the child goes to kindergarten at age 5. If the family saves $100 per month in an account paying 5.5% interest compounded continuously, how much will they have for college expenses 13 years hence?

Exponential Decay

19.   Suppose inflation proceeds at a level rate of 4% per year from mid-1997 through mid-2001.

(a)   Find the cost in mid-2001 of a basket of goods that cost $1 in mid-1997.

(b)   What will be the value of a dollar in mid-2001 in constant mid-1997 dollars?

20.   Suppose you bought a car in early 1996 for $10,000. If its value in current dollars depreciates steadily at 12% per year, what will be its value in current dollars in early 2001?

21.   If there is also 3% annual inflation from 1996 through 2001, what will be the value of the car in Exercise 20 in early 2001 in "inflation-adjusted" (mid-1996) dollars?

22.   If a substance contains 20 g of carbon-14 now, in how many years will there be only 5 g of the isotope remaining in the substance?

23.   The isotope plutonium-239 (produced in breeder nuclear reactors and used in atomic bombs) has a half-life of approximately 24,400 years. If a substance contains 10,000 grams of the isotope now, in how many years will there be only 1250 grams of the isotope remaining in the substance?

24.   [Contributed by John Oprea of Cleveland State University.] Radioactive iodine-125 caused great concern when it was released into the atmosphere across Europe in the nuclear reactor disaster at Chernobyl in 1986, since its radiation can cause thyroid cancer. Its half-life is 60 days. How long does it take for a quantity of iodine-125 to decay to 0.1% (one one-thousandth) of the original amount?

25.   [Contributed by John Oprea of Cleveland State University.] The Nuclear Test Ban Treaty of 1963 brought an end to atmospheric testing of nuclear weapons. Testing during the 1950s and 1960s released into the atmosphere the radioactive isotope strontium-90. It settled out of the air onto grass in fields, was eaten by cows, and wound up in children's milk. In the body, strontium-90 is absorbed into the bones, where its radiation can cause cancer; its half-life is 25 years. Of the strontium-90 absorbed into the bones of children in the

1950s, approximately how much will still remain 50 years later, during the first decade of the twenty-first century?

26. The "Ice Man" is the popular name for the body of a man that was found in 1991 preserved in a glacier in the Tyrolean Alps. At first researchers speculated that he had been a medieval messenger who perished in a storm. Carbon-14 dating of the Ice Man surprised everyone when it revealed that he had died about 5000 years ago. About how many atoms of carbon-14 are breaking down today per hour per gram of carbon of his tissue?

Growth Models for Biological Populations

27. The total population of the less developed countries (excluding China) was 3.314 billion in mid-1995, and the growth rate was 2.2% per year (this is an annual yield, so you may think of it as compounded annually). If this growth rate continues until mid-2005, what will be the size of the population then?

28. In its estimates for doubling times for populations in the world, the Population Reference Bureau uses a rule of 70, similar to (but more accurate than) the rule of 72 used in banking. As noted in Exercise 6, a rule of 69.3 would be even more accurate ; but the difference between that and the rule of 70 is only 1%. Apply the rule of 70 to estimate the doubling times for the following populations (figures are for mid-1995):

  (a) Africa, 720 million, 2.8%
  (b) United States, 263 million, 0.7%
  (c) China, 1.218 billion, 1.1%
  (d) The world as a whole, 5.702 billion, 1.5%

Limits on Biological Growth

29. Suppose that a population of size P grows by the amount

$$Pk\left(1 - \frac{P}{M}\right)$$

measured in appropriate biomass units, between observations, where k is the intrinsic growth rate and M is the carrying capacity of the environment. Suppose also that the carrying capacity is 100 units for the first 5 observations and then drops to 70. (There is an environmental catastrophe at that time; for instance, a flood wipes out much of the food supply.)

(a) For an initial population of 20 and $k = 0.9$, find the population sizes for the first 10 observations.

(b) Repeat part (a), but suppose also that the carrying capacity M is increasing steadily. This might be the case if, for example, the food supply is increasing steadily. Suppose that in the nth year the carrying capacity is $100 + 5n$. For an initial population of 10 and $k = 0.7$, find the population sizes for the first 10 observations.

(c) Repeat part (b), but for an initial population of 10 and $k = 2$. Find the population sizes for the first 10 observations.

(d) What do you conclude? Describe what is happening in terms of the setting of the problem.

30. Suppose that a population of size P grows by an amount

$$Pk\left(1 - \frac{P}{100}\right)$$

measured in appropriate biomass units, between observations. We view k as an intrinsic growth rate (the rate of population growth without resource constraints) and 100 biomass units as a carrying capacity of the environment.

(a) For an initial population of 10 units and $k = 0.8$, find the sizes of the population for the next 10 observations.

(b) Repeat part (a) but for an initial population of 110. What differences do you observe between the results for these two initial populations?

(c) Repeat parts (a) and (b), but with $k = 1.8$.

(d) On the basis of your analysis of the situations, what can you say about the dependence of the population growth on the parameter k?

Nonrenewable Resources

31. In 1990 the known global oil reserves totaled 917 billion barrels. Consumption, which had been 53.4 million barrels per day in 1983, rose an average of about 1.7% per year through 1990, when the consumption was about 60 million barrels per day.

(a) What was the static reserve for oil in 1990?

(b) If the rate of increase in consumption stays constant at 1.7%, what was the exponential reserve for oil in 1990?

(c) What considerations may affect these reserves over time?

32. Aluminum is the most abundant structural metal in the earth's crust. The world demand for new supplies of aluminum in 1983 was 16.5 million metric tons, while the known reserves were then 21,000 million metric tons.

    (a) What was the static reserve for aluminum in 1983?

    (b) In 1983, the demand for new aluminum was projected to increase at 4% per year at least through the year 2000. For that rate of increase, what was the exponential reserve for aluminum in 1983?

    (c) What considerations may affect these reserves over time?

Renewable Resources

33. A reproduction curve for a population is shown in the figure below. Estimate the equilibrium population size and the maximum sustainable yield. (The units are in thousands.)

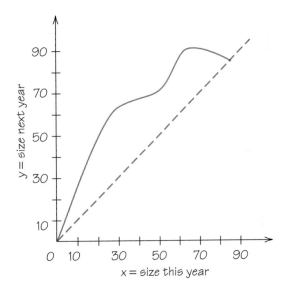

34.   Suppose that a reproduction curve for a certain population is as in the accompanying figure, where the units are in thousands.

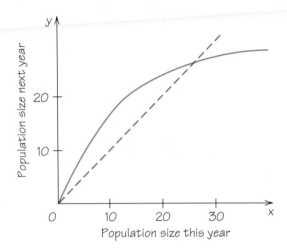

(a)   Estimate the sustainable yield corresponding to a population of size 10,000 remaining after the harvest.

(b)   Estimate the maximum sustainable yield.

Additional Exercises

35.   An issue of *Computer Language,* a magazine for professional programmers, repeated an oft-heard claim that "the amount of information in the world doubles every three days." Presumably the claim refers to the amount of data, which can be quantified in terms of number of bits. (A bit is the smallest unit of storage in a computer.) Show that the claim is absolutely preposterous, by doing a little arithmetic and comparing your result with the estimated number of particles in the universe (10^{70}). In particular:

(a)   Start with one bit of data and double the number of bits every third day. How long does it take to get past 10^{70}? (*Hint:* Don't just keep multiplying by 2 over and over. Convince yourself that since the amount of data increases by a factor of 2 every three days, then it increases by a factor of $2^2 = 4$ every six days, a factor of $4^2 = 16$ every twelve days, a factor of $16^2 = 256$ every twenty-four days, and so forth.)

(b) Part (a) involves a lot of multiplying by 2, even if you do it efficiently. Another approach is to use the fact that $2^{10} = 1024 \approx$ 1000. Thus, the amount of data increases by a factor of more than 1000 every $3 \times 10 = 30$ days, or every month (except February, but the 31-day months make up for it). By when will the total be sure to be past 10^{70}?

36. An old legend tells of a wizard who agreed to save a kingdom provided that the king would agree to a "modest" reward. The wizard asked to be given merely as much grain as would put one kernel on the first square of a chessboard, two kernels on the second, four on the third, eight on the fourth, and so forth, up through the sixty-fourth square. The king agreed, the wizard saved the kingdom, but the king was completely unable to honor the agreement. Why? (*Hints:* Notice that $1 = 2^1 - 1$, and $(1 + 2) = 2^2 - 1$, and $(1 + 2 + 4) = 2^3 - 1$; generalize to arrive at a total for the number of kernels. A kernel of rice is about a quarter of an inch long and about a sixteenth of an inch wide and a sixteenth of an inch high. So about a thousand kernels will fit in a cubic inch (you should verify this calculation). Calculate the total volume of kernels.)

■ 37. Surprise! Just for fun, one of your friends wrote your name on an Illinois State Lottery ticket, and you are the sole winner of $40 million! You discover, however, that you don't get the $40 million all at once; in fact, it is paid in 20 annual installments. All you get right away is the first installment of $2 million (minus 20% withheld against federal income tax due, and whatever you think your friend deserves for the favor). So, what is the prize really worth to you? That depends on the rate of inflation over the years. Assume a constant rate of inflation over the 19 years until your last payment and calculate the present value of your prize winnings by using the formula for present value combined with the formula for the sum of a geometric series. Do the calculation for rates of interest of

(a) 3%
(b) 6%
(c) 9%

Actually, the checks will come from not the state of Illinois, but an insurance company from which Illinois purchases an annuity (a contract to pay a certain amount each year for a specified number of years). The price of the annuity depends on current long-term interest rates.

◆ 38. [Contributed by John T. Montgomery, University of Rhode Island.] "I live in Rhode Island and depend on a well for my water. After living in my house for 20 years, I discovered that the level of radon-222 in my water is about 28,000 picocuries (pc) per liter. (A *picocurie* is a measure of radioactivity: 1 picocurie = 3.7×10^{-2} = 0.037 decay of radioactive atoms per second.) Although they are guessing, some scientists say that the safety level is about 280 pc per liter. The half-life of radon-222 is 3.82 days. The danger of radon is not in drinking it but in breathing it while taking a shower. How long should I let the water stand in my water tanks before taking a shower (with the pump turned off)? If I don't turn the pump off, the water in the tanks is replaced with fresh water as fast as it is being used. How long should I wait if I want to shower with the pump on, and how long can the shower be?"

◆   39.   By the time that there is concern about using up a nonrenewable resource, it may be too late. Suppose that a resource has a static reserve of 10,000 years but consumption is growing at 3.5% per year.

> (a)   How long will the resource last?
> (b)   How long before half the resource is gone?
> (c)   How much longer will the resource last if after half of it is gone, consumption is stabilized at the then-current level?
> (d)   What implications do you see to your answers?

◆   40.   In mid-1995, the population of the world was 5.702 billion and increasing at 1.5% per year.

> (a)   Project the world's population to mid-2000, to mid-2020 (by which time you will likely have finished having children, if you do), and to mid-2040 (by which time you will likely have retired).
> (b)   What are the assumptions involved in your projections?
> (c)   You can make a more refined model, which will give more realistic answers, by dividing the countries of the world into three groups that have differing rates of increase (see the table below).

Group	Population Mid-1995 (billions)	Rate of Growth (%)
More developed countries	1.169	0.2
Less developed countries (excluding China)	3.314	2.2
China	1.219	1.1

To get a projection for the world's population, project each group separately and add the totals. Redo your projections for the years 2000, 2020, and 2040. Do you find the differences from your earlier projections to be significant?

(d) Will the world be able to support the numbers of people that you project? What problems will these greater numbers of people cause? What could be done to avert those problems? Do you think that anything will be done before there is some kind of worldwide crisis?

WRITING PROJECTS

1 ▶ Based on the calculations you did in Exercise 40 and the discussion you had with other members of the class, write a short essay in the form of a guest editorial for a newspaper. Describe your projections and how you arrived at them, how serious a problem you think population growth is, what problems it is likely to cause, what you think needs to be done, and what the implications are for your own life.

2 ▶ Identify a particular regional, national, or world nonrenewable primary resource (e.g., coal) or secondary resource (e.g., electric power). Research how much of it is available now and what the current rate of consumption is. Determine the static reserve. Estimate the growth rate in consumption, taking into account human population increase, and determine the exponential reserve. What social and technological factors contribute to the increasing rate of consumption? Brainstorm how those factors could be changed.

3 ▶ Identify a particular regional, national, or world renewable resource (e.g., timber, clean drinking water). Research how much of it is produced now, how much is harvested now, and what the current rate of consumption is. Estimate the growth rate in consumption, taking into account human population increase. For how long can this resource continue to meet the demand? What social and technological factors contribute to the increasing rate of consumption? Brainstorm how those factors could be changed.

4 ▶ You want to buy a car and you need to borrow $8000 of the cost. You can get a 48-month loan from the car dealer at 7.9%, with a $500 rebate on the price of the car, or else a loan at 4.9%.

(a)   What is your monthly payment with the 7.9% loan? (The $500 rebate comes to you separately from the manufacturer, so you still have to borrow the full $8000.)

(b)   What is your monthly payment with the 4.9% loan?

(c)   Neglecting inflation, which is the better deal for you, in terms of total current dollars?

(d)   Suppose that you anticipate inflation to be 3% per year for the duration of the loan. Which is the better deal for you, in terms of total constant dollars?

5 ▶   You want to buy a house for which you need to borrow $100,000. You have the option of a 15-year mortgage or a 30-year mortgage, both at 7.9%.

(a)   What are your monthly payments in each case?

(b)   How much interest do you pay, and what is the total amount that you pay, in each case?

(c)   Suppose that you anticipate inflation to be 3% per year for the duration of the mortgage. What is the value in today's dollars of the stream of payments that you will make under the 15-year mortgage? Under the 30-year mortgage? (*Hint:* Use the formula for the sum of a geometric series.)

(d)   Use a spreadsheet to prepare an amortization table for each loan, showing for each month how much interest is paid and how much principal is repaid.

CHAPTER 19

NEW GEOMETRIES FOR A NEW UNIVERSE

As an organized body of knowledge, geometry consists of statements (*theorems* and *corollaries*) logically derived from other statements (*postulates,* more commonly called *axioms*) that are assumed to be true. About 500 B.C., Euclid presented five postulates from which he developed a large body of theorems, a system that we call **Euclidean geometry.**

From then until late in the nineteenth century, Euclidean geometry—including its extension to three dimensions—was thought to be the only mathematics of space. Its theorems were thought to be truths about the world in which we live. No one could imagine a different geometry.

New kinds of *non-Euclidean* geometry were conceived during the nineteenth century and came into their own only in the twentieth century, when one of them became the basis for a major revolution in physics and cosmology—the theory of relativity.

A **non-Euclidean geometry** is any collection of postulates, theorems, and corollaries for geometry (concerning points, lines, circles, and angles) that differs from the collection formulated by Euclid.

EUCLIDEAN GEOMETRY

Euclid's five postulates concern points, lines, circles, and angles, and presuppose notions about what it means for a point to be on a line and for two lines to meet in a point. The postulates were intended to be absolute, self-evident truths. However, they are not postulates about points as pencil dots or lines drawn on paper with a ruler, but instead attempt to characterize the ideal concepts behind these physical realizations. Although the points and lines of our (and Euclid's) experience are the inspiration for the postulates, and Euclid gave fuzzy definitions of them (e.g., a point is "that which has no dimension"), the "points" and "lines" of the postulates are basic undefined terms. Their properties are described implicitly by the postulates since only the postulates are used in reasoning about the concepts. Any objects that satisfy the postulates can be taken to be "points" and "lines."

Paraphrased somewhat, Euclid's postulates are as follows:

1. Two points determine a line.
2. A line segment can always be extended.
3. A circle can be drawn with any center and any radius.
4. All right angles are equal.
5. If *l* is any line and *P* any point not on *l*, then there exists exactly one line *m* through *P* that does not meet *l*.

These five statements were supposed to be absolute, self-evident truths. The first four are simple statements, sufficiently unrelated that it can be shown fairly easily that each is *logically independent* of the others. What this means is that we cannot conclude from the other three postulates whether the remaining postulate is false (in which case our system would be self-contradictory) or true (in which case the remaining postulate is superfluous).

> A statement is **logically independent** of a collection of other statements if neither it nor its negation can be deduced from them.

The fifth postulate concerns what are known as *parallel lines:*

> Two lines are **parallel** if they do not meet.

You may remember from high school geometry that in Euclidean geometry in the plane, parallel lines are an equal distance apart at all points. This fact of Euclidean geometry can be proved with the use of additional postulates about distance. However, the definition of "parallel" is that lines do not intersect; parallel lines in non-Euclidean geometries are *not* equidistant everywhere.

The fifth postulate is usually referred to as *Euclid's parallel postulate.*

Euclid's parallel postulate states that for any line *l* and any point *P* not on *l*, there is exactly one line *m* through *P* that is parallel to (that is, does not meet) *l*.

Many early geometers thought that this postulate is not independent of the first four but instead is a logical consequence of them. In fact, Euclid himself may have thought his parallel postulate to be an unnecessary assumption for his geometry, for he derived nearly 30 theorems before using it.

The long history of attempts to derive Euclid's parallel postulate as a consequence of the first four postulates is a fascinating story of failures. Whenever someone proposed a proof, it was found to be tacitly based on some extra assumption in addition to the first four postulates—an assumption *logically equivalent* to the parallel postulate. The reasoning was circular, hence invalid.

Two statements are **logically equivalent** if each can be deduced from the other.

Some hidden assumptions logically equivalent to the parallel postulate are:

* The sum of the angles of a triangle equals 180°.
* There is exactly one circle through any three points that are not on the same line.
* Parallel lines are equidistant.

ELLIPTIC GEOMETRY

G. F. Bernhard Riemann (1826–1866) (see Figure 19.1), a young German mathematician, analyzed postulate 2 ("A line segment can always be extended"). He observed that such a property of lines should be distinguished from "A line is infinite." That is, *unboundedness does not imply infinite extent.*

What Riemann had in mind was the geometry of the surface of the earth, the **spherical geometry** used in navigation. The "points" of this geometry are the points on the surface of the sphere. But what are the "lines"?

FIGURE 19.1   G. F.
Bernhard Riemann

FIGURE 19.1   G. F.
Bernhard Riemann

Think of the surface of the earth. Going "straight" along what you would imagine was a line, you can travel another mile and another mile, and so on, and you would eventually return to your starting point. You have traveled on a finite path, but one that is unbounded—that is, you can keep traveling on and on. When Riemann investigated the consequences of a line coming back on itself, he came to the conclusion that interpreting postulate 2 as referring to *unbounded* lines rather than infinite ones opens the door to a geometry that abandons Euclid's parallel postulate in a striking way. Riemann replaced Euclid's parallel postulate with:

Postulate E: Every two lines intersect.

Postulate E leads to *elliptic geometry*.

An **elliptic geometry** is one in which there are no parallel lines.

From Euclidean geometry in the plane, you are accustomed to the line through two points as determining "the shortest distance between two points." That useful property suggests that a good interpretation of "line" for the surface of the sphere might be the analogous shortest route on the surface of the sphere.

However, there are no "straight" lines that lie on the surface of the sphere. The shortest distance between two points on the surface of the sphere would be along a tunnel between them, but such a route would be "out of bounds": we must stay on the *surface* of the sphere and consider "lines" made up of points on that surface.

> A curve (possibly straight) that gives the shortest path between two points on a surface is called a **geodesic.**

If we intersect the sphere in Figure 19.2 with a plane through *A* and *B*, the cross section is a circle passing through the given points; and the shorter of the two arcs from *A* to *B* of this circle is a candidate for the shortest path from *A* to *B*. Every plane section of the sphere is a circle, each with a different curvature. The larger the circle, the less curving; the less curving, the shorter the path between *A* and *B*. Thus, the largest circle obtained as a cross section of the sphere gives rise to the shortest path. The largest circle is the *great circle,* obtained by the plane determined by *A, B,* and the center of the sphere (see Figure 19.2).

> A **great circle** on a sphere is a circle whose plane includes the center of the sphere.

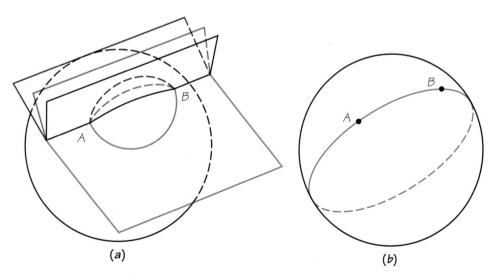

(a)   (b)

FIGURE 19.2   (a) The planes through *A* and *B* intersect the sphere in circles, the largest of which has the smallest curvature and the smallest distance from *A* to *B*. (b) The great circle through *A* and *B*.

Great circles are familiar to pilots who fly long distances and ship captains on long voyages. If an airplane is on the equator and the pilot wishes to fly to another point on the equator, the pilot would simply fly along the equator, which is a great circle. The great circle "lines" of the sphere can indeed be extended indefinitely, since you can keep on going round and round the earth along one. However, if the airplane is at 10° north latitude and the pilot wishes to fly to a destination at the same latitude, then the shortest distance requires going farther north. An airplane flying from New York to Naples, both at the same latitude, would travel quite far north in the Atlantic Ocean, while the shortest flight path from Washington, D.C., to Ho Chi Minh City, Vietnam, is almost directly over the North Pole. In the polar view of Figure 19.3, the shortest path appears as almost a straight line segment, as the great circle between the two cities is almost edge-on to you.

Great circles look curved when we observe the earth from space, but they do not look curved as we move on the surface of the earth. A ship moving directly east along the equator is following a curved path around the earth; but as far as the crew is concerned, the ship is always headed "straight" east.

Riemann's predecessors certainly realized the practicalities of navigation on the sphere, and about great circles intersecting, but they did not think of the geometry of the surface of the sphere as an alternative to Euclidean geometry—despite the fact that we live not on a plane but on a sphere.

In the geometry of the surface of the sphere, if we take the "lines" to be the great circles, then there are no parallel "lines." Every pair of "lines" (great circles) intersect, so postulate E holds.

FIGURE 19.3   The shortest path between two cities is an arc of a great circle.

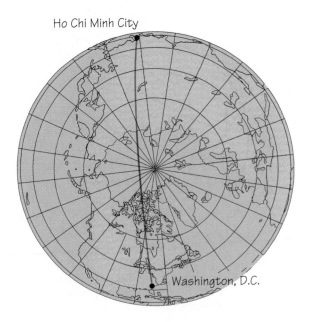

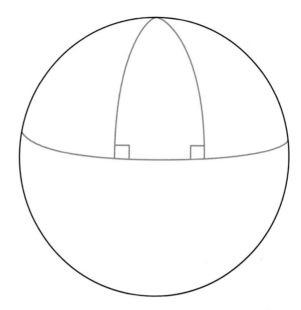

FIGURE 19.4   In elliptic geometry, a triangle can have two or more right angles.

The geometry of the surface of a sphere has other notable differences from Euclidean geometry. Postulate 1 fails for points that are directly opposite each other on the sphere, since not one but an infinite number of elliptic lines (great circles) go through two such points. On the surface of a sphere, triangles can have two or even three right angles: just put two vertices on the equator and one at a pole (see Figure 19.4). In fact, a general theorem of elliptic geometry states that *in an elliptic geometry, the sum of the angles of any triangle is greater than 180°.*

HYPERBOLIC GEOMETRY

The first non-Euclidean geometry was formulated not by Riemann but by Nikolai Ivanovich Lobachevsky (1792–1856) and János Bolyai (1802–1860). (Carl Friedrich Gauss [1777–1855] investigated hyperbolic geometry before Lobachevsky and Bolyai, but did not publish his results.) Unlike many of their predecessors, they did not try to derive Euclid's parallel postulate from the other postulates. Instead, working alone, each decided that Euclid's parallel postulate must be logically independent of the other postulates. They realized that this means that those postulates really have nothing to say about the existence or nonexistence of parallels. Thus, including Euclid's parallel postulate with the other postulates cannot lead to any inconsistencies with them. Similarly, if they were to leave out Euclid's parallel postulate and include instead an alternative postulate about parallels—even one contradictory to Euclid's parallel postulate—they should once again get a consistent system of postulates. They both chose the same alternative postulate and proceeded to invent and explore the resulting "non-Euclidean" geometry, proving theorems from its postulates in the same style that Euclid had proved theorems from his (see Spotlight 19.1).

The alternative postulate that Lobachevsky and Bolyai chose was:

Postulate H: If l is any line and P is any point not on the line, then there exists more than one line through P not meeting l.

The use of postulate H leads to an entirely new system of theorems and corollaries, which we now call *hyperbolic geometry*.

A **hyperbolic geometry** is one in which for any given line and a point not on the line, more than one line passes through the point and is parallel to the given line.

Some of the theorems in hyperbolic geometry are exactly the same as in Euclidean geometry, because theorems derived only from postulates 1 through 4 must be valid in both systems. However, hyperbolic geometry provides some new and very surprising theorems.

Lines in hyperbolic geometry behave differently from what our intuition about Euclidean lines leads us to expect. In representing hyperbolic geometry in the plane of the page, it is useful to represent hyperbolic lines as curved. However, just as great circles appear straight to people living on a sphere, hyperbolic lines would appear perfectly straight, and to be geodesics (giving shortest distances), to people living on a hyperbolic surface.

Figure 19.5 shows a line l (shown curved) and a point P not on l. We drop a perpendicular from P to l, calling A the foot of the perpendicular. Now, consider the line PE, which is perpendicular to PA. The line PE is parallel to l (it does not intersect it). Under Euclid's parallel postulate, it would have to be the only parallel to l through P. However, if we assume H instead, then there is *another* "line" m that passes through P and is parallel to l. It must make a smaller angle with PA, as shown on the left in Figure 19.5 (if it too made a right angle, then it couldn't be a different line from PE, by postulate 4).

A key observation in Figure 19.5 is that with m making an acute angle with PA on the left side of PA, and necessarily an obtuse angle on the right side, there must be yet another line n making the same acute angle with PA on the right side. By symmetry, n too must be parallel to l.

From this argument comes the first astonishing conclusion, that through P there are *infinitely many* parallels to line l. This is clear as soon as you consider all the lines through P and divide them into two classes. One class contains PA and lines making a smaller angle with PA than n and m do. The other class contains PE, n, m, and all lines lying between PE and n and between PE and m (they lie in the shaded region in Figure 19.5). All the lines in the second class are parallel to l.

Using similar reasoning, Bolyai and Lobachevsky discovered many unusual theorems, of which we list three that were important for their shock value to

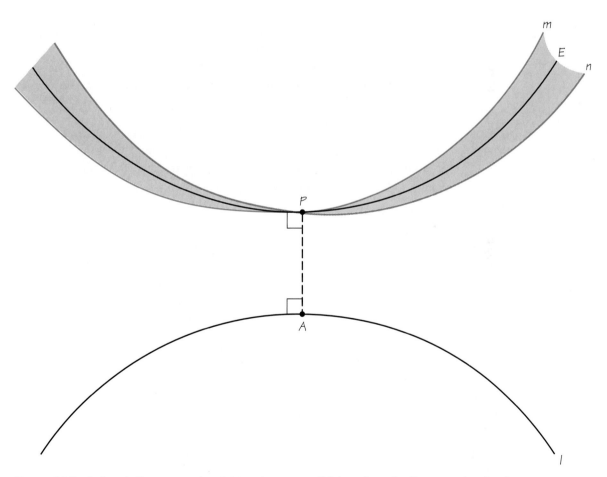

Figure 19.5   In hyperbolic geometry there is more than one parallel through a point P not on a given line l.

the dubious mathematicians who listened to their theories (Figure 19.6 illustrates theorems 1 and 3):

1. The sum of the angles in any triangle is less than 180°.
2. Similar triangles are congruent—that is, triangles having the same shape (angles) must be the same size; the size of a triangle depends on the sum of its angles.
3. Given two parallel lines, there exists a third line perpendicular to one and parallel to the second.

> Hence, in each of the three geometries we have investigated, we have different sums for the angles in a triangle:
>
> - <180° in hyperbolic geometry (angle sums range between 0° and 180°)
> - =180° for all triangles in Euclidean geometry (sometimes called *parabolic geometry*) (see Spotlight 19.2)
> - >180° in elliptic geometry (angle sums range between 180° and 540°)

Thus, having a constant angle sum for all triangles dramatically distinguishes Euclidean geometry from elliptic geometry and hyperbolic geometry. In both elliptic geometry and hyperbolic geometry, any two triangles with the same area have the same angle sum.

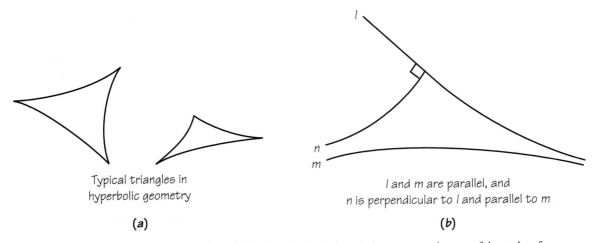

Typical triangles in
hyperbolic geometry

(a)

l and *m* are parallel, and
n is perpendicular to *l* and parallel to *m*

(b)

FIGURE 19.6   (a) Typical triangles in hyperbolic geometry; the sum of the angles of any triangle is less than 180°. (b) Given two parallel lines *l* and *m*, there exists a third line *n* perpendicular to *l* and parallel to *m*.

Spotlight 19.1

The Discovery of Non-Euclidean Geometry

Nikolai Ivanovich Lobachevsky

János Bolyai

Carl Friedrich Gauss

Nikolai Ivanovich Lobachevsky (1792–1856) and János Bolyai (1802–1860) independently discovered non-Euclidean geometry. Lobachevsky was the first to publish an account of it (1829), which he first called "imaginary geometry" and later "pangeometry." His work attracted little attention, largely because it was written in Russian and the Russians who read it were very critical.

Bolyai published his work as a 26-page appendix to a book (the *Tentamen,* 1831) by his mathematician father Wolfgang, who proudly sent the work by his son to Carl Friedrich Gauss (1777–1855), the leading mathematician of his day. Gauss replied to Wolfgang that he himself had earlier discovered non-Euclidean geometry! From correspondence and private papers that became available after his death, we know that Gauss's claim was correct, though Bolyai and Lobachevsky deserve credit for having the courage to publish their discoveries.

MODELS FOR HYPERBOLIC GEOMETRY

For Euclidean geometry, we have as a natural *model* the plane, with "points" and "lines" interpreted as the points and lines that you are accustomed to.

An example that satisfies a collection of axioms is called a **model** of the axioms.

Spotlight 19.2 — Angle Sums in a Triangle

The following simple proof shows that the angle sum in a triangle is equal to a straight angle. Let triangle ABC be an arbitrary triangle. Our goal is to prove that $A + B + C$ = a straight angle ($=180°$).

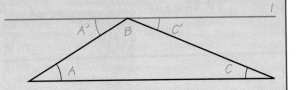

The Proof

At vertex B, draw a line l parallel to side AC. Then, by one of the first properties of parallels, which states that when parallel lines are cut by a transversal, the "alternate interior angles" are equal, angle A = angle A'. Likewise, angle C = angle C'. Now clearly $A' + B + C'$ = a straight angle. Hence, by substituting A for A' and C for C', we have $A + B + C$ = a straight angle, which is what we set out to prove.

We can't use this same proof on the sphere because it has no parallels at all. On the sphere every two straight-line paths, or great circle routes, eventually cross. In fact, on the sphere the sum of the angles of a spherical triangle is always greater than a straight angle.

On the other hand, in the plane of hyperbolic geometry there are "too many" parallels, and when the line l is drawn at B so that angle A' = angle A, then angle C is always less than angle C'. Hence, the angle sum $A + B + C$ in triangle ABC is always less than the straight angle $A' + B + C'$.

Showing that an example satisfies the axioms involves interpreting the terms of the axioms as concrete entities in the particular example.

For elliptic geometry, we have as a natural model the surface of the sphere. The elliptic points of the geometry are points on the sphere and the elliptic lines are its great circles. Since the points and great circles on the sphere behave according to postulates 2–4 and E, the surface of the sphere is a model of the postulates.

To get a good understanding of hyperbolic geometry, you need to look at and ponder various models. We present several intriguing models, each with its own interpretation of "points" and "lines." First we give a simple example of a *finite* hyperbolic geometry and then consider other models.

EXAMPLE ▶ *A Geometry of Political Alliances*

Suppose that a country has five political parties, which we will designate by *A* (Anarchists), *B* (Businessmen), *C* (Conservatives), *D* (Democrats), and *E* (Environmentalists). There are ten possible two-party political alliances: *AB*, *AC*, *AD*, *AE*, *BC*, *BD*, *BE*, *CD*, *CE*, and *DE*.

Consider the individual parties to be "points" and the two-party alliances to be "lines." We say that a "point" (party) is on a "line" (alliance) if the party is part of that alliance; for example, *A* is on *AB*. We say that two "lines" (alliances) intersect if they have a party in common; for example, *AB* and *AC* intersect in *A*. Moreover, two "lines" (alliances) are parallel if they do not have a "point" (party) in common; for example, *AB* and *CD* are parallel.

This is a small geometry indeed! It has 5 points and 10 lines. It satisfies postulate 1 of Euclid ("two points determine a line"), but not postulate 2 ("a line segment can always be extended"), and there is no mention of the circles and right angles of postulates 3 and 4. In terms of parallels, it is not a Euclidean geometry but a hyperbolic one; it satisfies postulate H, in a stronger form: If *l* is any line and *P* is any point not on the line, then there exist *exactly two* lines through *P* not meeting *l*. For example, if the line is *AB* and the point is *C*, then *CD* and *CE* are lines through *C* that are parallel to *AB*.

This geometry can be represented in terms of points and line segments, as in Figure 19.7. The parties are the vertices of a pentagon, and the alliances are line segments joining vertices. In making this interpretation, we disregard what appear as intersections of the diagonals of the pentagon; they are not points of the geometry and would not even appear if we "drew" the figure in higher-dimensional space. ◆

You can get a better understanding of hyperbolic geometry by examining models that conform to the first four postulates of Euclid and differ from ordinary Euclidean geometry only in satisfying postulate H rather than Euclid's parallel postulate (see Spotlight 19.3).

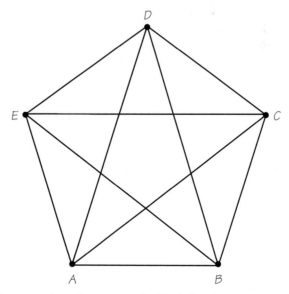

FIGURE 19.7 Representation of a geometry of political alliances.

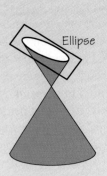

Ellipse

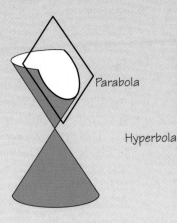

Parabola

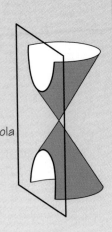

Hyperbola

Spotlight 19.3

What's Hyperbolic About Hyperbolic Geometry?

The word "hyperbola" comes from Greek, meaning "a throwing beyond," or excess (we get the slang word "hyper" from the same root). The word "ellipse" means "a falling short," or defect. The word "parabola" means "falling beside," or being parallel to. These terms are comparing the angle at which the conic section (hyperbola, ellipse, or parabola) cuts the cone, compared to the angle of the cone itself (see the accompanying figure).

How do these terms come to be applied to geometries? Compared to Euclidean geometry, hyperbolic geometry has an "excess" of parallels, since given a line and a point not on the line, more than one line through the point is parallel to the given line. Similarly, in an elliptic geometry, there is a "defect" of parallels, since there are no parallel lines.

In a deeper sense, the distance measure in a hyperbolic geometry has an algebraic form similar to the algebraic form of a hyperbola in analytic geometry coordinates ($y^2 - x^2 = 1$), while the distance measure in an elliptic geometry has a form similar to that of an ellipse ($x^2 + y^2 = 1$).

For the sum of the angles of a triangle, there is a reversal: hyperbolic geometry features triangles whose angle sums are "defective," while those in elliptic geometry are "excessive," compared to Euclidean geometry's 180°.

Euclidean geometry corresponds to a space with no curvature, elliptic geometry describes a space of constant positive curvature, and hyperbolic geometry describes a space of constant negative curvature.

Finally, when it comes to trigonometry, the relevant formulas are different for each geometry. The Pythagorean theorem of Euclidean geometry for right triangles does not hold in the other geometries, which have their own analogues. For spherical geometry, the formulas for trigonometry involve the familiar "circular" trigonometric functions (sine, cosine, tangent); in hyperbolic geometry, the related hyperbolic trigonometric functions apply (sinh, cosh, tanh—you may see keys for these on your calculator).

The models that we describe below have a physical interpretation that explains the fact that distances in these geometries, which we represent in the two dimensions of the Euclidean plane, are measured differently from Euclidean distances. Imagine a universe with a closed boundary that, for whatever reason, you can approach but never get to. The closer you seem to get, the harder it is to get even closer, as if there were a force pushing you away. The boundary might as well be infinitely far away. We can try to draw a picture of such a universe by drawing the boundary as a circle. To make the model work, the boundary must be infinitely far away from any point; so distances in this universe will not correspond to our ordinary notions of distance. What can the lines of this geometry—its geodesics—be? The Poincaré disk model provides an answer.

Software exercises on the models below make clear some of the amazingly different properties of hyperbolic geometry.

EXAMPLE ▶ *The Poincaré Disk Model*

"Points" are again the interior points of a fixed circle. "Lines" are circular arcs that meet the bounding circle at right angles (the endpoints on the bounding circle are not points of the geometry). Figure 19.8 shows how postulate H is fulfilled: for the line *l* and the point *P* not on the line *l*, lines are shown that go through *P* but do not meet *l*.

The Poincaré disk model of hyperbolic geometry was used by the artist M. C. Escher in his fascinating "Circle Limit" prints (see Spotlight 19.4). ◆

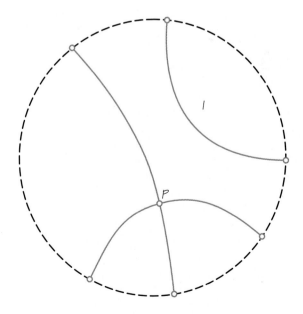

FIGURE 19.8 The Poincaré disk model of hyperbolic geometry.

Spotlight *Angels and Devils*

19.4

The Dutch artist M.C. Escher (1896–1972) was particularly interested in figures that change almost imperceptibly into other figures or into larger or smaller versions of themselves. He knew how to draw, inside a circle or square, figures that gradually get larger as they approach the outside of the enclosure. But it wasn't until he was shown a mathematician's representation of hyperbolic geometry (in which the sum of the angles of a triangle is always less than 180°) that he discovered how to make figures gradually get *smaller* toward the outside of a circle.

Douglas Dunham of the University of Minnesota, Duluth, has devised a computer program based on hyperbolic geometry that can produce an infinite variety of the type of drawings Escher so ingeniously drew. One of these, Dunham's *Circle Limit IV,* is shown here.

The "lines" of this geometry are arcs of circles that are perpendicular to the outside circle. (The "lines" of spherical geometry are great-circle routes on the sphere.) This particular print is based on a regular tiling of the hyperbolic plane. The tiles shown here are regular quadrilaterals— their vertices are the points where the feet of three angels meet the feet of three devils. Six of these tiles meet at each vertex. Thus, the angle of each is 60° instead of 90°, which it would be for the corresponding tiling of the Euclidean plane, where four squares meet at each vertex.

The edges of some of the tiles (of the underlying tiling) have been drawn in so that they can be seen, and two are shaded. Although the tiles appear to get smaller (in the Euclidean sense) toward the edge of the outside circle, the hyperbolic geometry uses a different distance measure in which all of the tiles, including the two that are shaded, are congruent (the same size).

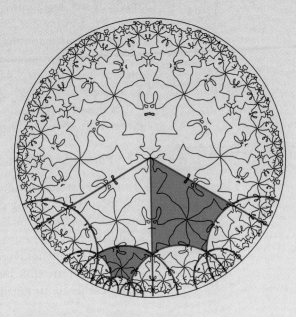

Dunham's *Circle Limit IV* plot.

This computer-generated tiling in the hyperbolic plane creates an image very similar to that of M. C. Escher's *Angels and Devils.* Note how the positions of feet and heads are related by radii and arcs that define the underlying tiling pattern.

(a) M. C. Escher's *Heaven and Hell* (also known as *Angels and Devils*). Photographed from one of Escher's notebooks, this example demonstrates a repeating pattern of the Euclidean plane. Note the uniform size of the figures and the central meeting of the wingtips of four angels and four devils. (b) Escher's pattern of angels and devils carved on an ivory sphere by Masatoshi. This mapping of the pattern onto a sphere shows the different effects of a spherical geometry. Note that in this version, the wingtips of three angels and three devils meet.
(c) M. C. Escher's *Circle Limit IV (Heaven and Hell)*. This example shows the repeating angels and devils pattern mapped onto a hyperbolic geometry. At the center, the feet of three angels and three devils meet. As one moves outward, the figures get smaller. Note that the wingtips of four angels and four devils meet.

A slight variation on the Poincaré disk model produces the Klein disk model of hyperbolic geometry.

EXAMPLE ▶ *The Klein Disk Model*

"Points" are the interior points of a fixed circle and "lines" are chords of the circle with endpoints omitted. This geometry looks more Euclidean to us who view it from outside, because the hyperbolic lines look like straight Euclidean line segments. The more unusual properties of the geometry come out when you draw figures and move them around, as the exercises ask you to do. Figure 19.9 shows an instance of postulate H being fulfilled in a particular instance, for a line *l* and a point *P* not on *l*.   ◆

Another approach to modeling a bounded universe is to let the unapproachable closed boundary be a line that divides the plane into two halves.

EXAMPLE ▶ *The Upper Half-Plane Model*

"Points" are the points in a half-plane excluding the bounding line. There are two kinds of "lines": open semicircles with center on the bounding line and open rays perpendicular to the bounding line. Figure 19.10 shows how postulate H is fulfilled in a particular instance.   ◆

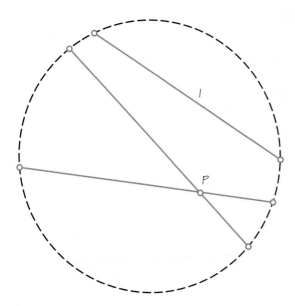

FIGURE 19.9   The Klein disk model of hyperbolic geometry.

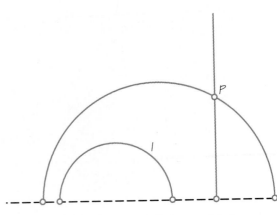

FIGURE 19.10   The upper half-plane model of hyperbolic geometry.

THE THEORY OF RELATIVITY

In 1905, Albert Einstein (Spotlight 19.5) put forth his *special theory of relativity,* a complicated theory that constituted the first step in the greatest revolution in physics since Newton's *Principia Mathematica.*

Einstein proposed a new way of thinking about events in the history of the universe. An event takes place in our three-dimensional space at a specific time in history. Thus, an event is located in *space-time* by four coordinates: three determine its position in space, and the fourth determines its position in time. Of course, these coordinates locate the event relative to a specific coordinate system. Einstein observed that the location of an event in space-time therefore depends on the position of the observer—that is, on the origin and orientation of the coordinate system being used. Different observers may obtain very different views of events, especially if one observer is traveling very fast with respect to the other.

Let's consider these ideas geometrically. The *distance* between two events, usually in relativity theory called an *interval,* is split into two parts: a *space-part* and a *time-part.* The space-part is the part of the interval that comes from the position of the events in three-dimensional space, and the time-part is the length of time that separates the events. This splitting depends on the coordinate system and its orientation, so different results may be obtained by different observers (see Figure 19.11). However, the interval, being a line segment joining the two events in four-dimensional space-time, is absolute—in the sense that it is the same for an observer at rest and for all other observers who are traveling at a constant velocity with respect to the one at rest.

FIGURE 19.11
(a) A coordinate system representing space-time. The *t*-axes show time and the *x*-axes space. The black axes are a system at rest. Note that the blue moving system tilts toward the 45° light ray line. (b) Observers in the system at rest (black axes) will say the events *A* and *B* occur at the same time. Observers in the moving blue system will say that event *B* occurs before event *A*.

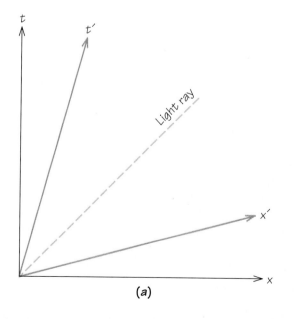

(a)

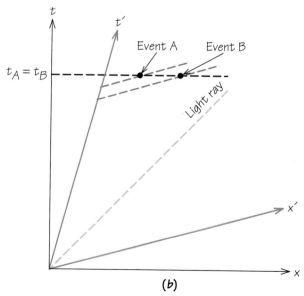

(b)

Spotlight Albert Einstein

19.5

Albert Einstein

Albert Einstein (1879–1955) was born in Ulm to a German Jewish family with liberal ideas. Although he showed early signs of brilliance, he did not do well in school. He especially disliked German teaching methods. In the mid-1890s he went to study in Switzerland, a country much more to his liking, where he went to work as a patent clerk. Einstein burst upon the scientific scene in 1905 with his theory of special relativity. In 1916 he published his theory of general relativity. General relativity was successfully tested in 1919, and his fame grew enormously. Nazism forced Einstein to leave Europe. He settled at the Institute for Advanced Study at Princeton, where he remained until his death at age 76.

EXAMPLE ▶ *"Simultaneous" Events*

Let's imagine that the eruption of Mount St. Helens in Washington in 1980 took place at the very same time that someone on Mount Palomar in California observed an astronomical phenomenon. Perhaps the observation was of a supernova explosion in a galaxy 100 light-years away (a **light-year** is the distance that light travels in a year). The eruption and the explosion appear to us on earth to be simultaneous.

However, knowing that the light from the supernova takes 100 years to reach earth, we realize that the eruption took place 100 years later than the supernova. For those of us on earth (and at rest relative to the earth), the interval between the two events has a space-part of 100 light-years and a time-part of 100 years.

For observers traveling at constant velocity with respect to the earth, say, at 50 light-years away from earth, the space-part and time-part of the interval would be very different. One observer might determine that the two events took place 200 years apart, while another might conclude that the two events happened simultaneously. Their splitting of the interval into space-parts and time-parts would be very different from ours. The geometry of space-time is indeed strange: in its four-dimensional space, the "distance" between two points—the interval between two events—remains invariant (in the sense we have described), but its respective parts vary. ◆

Three years after Einstein published his first paper on the subject, the mathematician Hermann Minkowski (1864–1909) gave Einstein's work a geometric interpretation that accepted Einstein's strange calculation of intervals and greatly simplified the theory. The geometry that was used, justifiably called *Minkowskian geometry,* is certainly non-Euclidean. Further, it makes use of one of Riemann's far-reaching ideas—that the nature of a mathematical space is determined by the way distance is measured; the distance formula therefore determines the nature of the geometry.

In Euclidean plane geometry, distance is determined with the help of the Pythagorean theorem $c^2 = a^2 + b^2$ about the length c of the hypotenuse of a right triangle and the lengths a and b of the legs (see Figure 19.12a), so that the distance $c = \sqrt{a^2 + b^2}$. In three dimensions, the Pythagorean theorem can be applied twice to yield the distance formula $d = \sqrt{a^2 + b^2 + c^2}$ (see Figure 19.12b); and if you imagine that in four dimensions it would be $e = \sqrt{a^2 + b^2 + c^2 + d^2}$, you would be right.

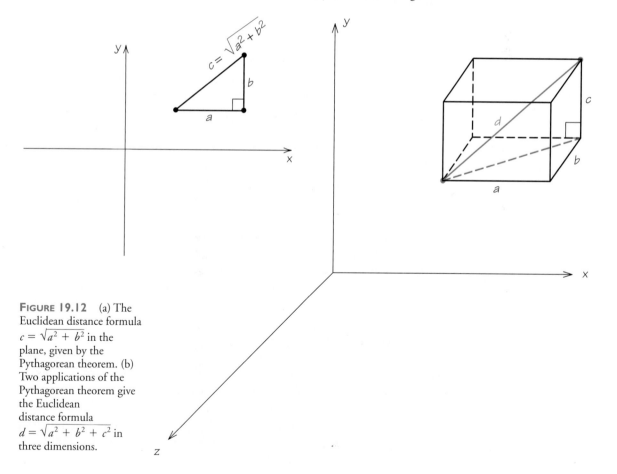

FIGURE 19.12 (a) The Euclidean distance formula $c = \sqrt{a^2 + b^2}$ in the plane, given by the Pythagorean theorem. (b) Two applications of the Pythagorean theorem give the Euclidean distance formula $d = \sqrt{a^2 + b^2 + c^2}$ in three dimensions.

If two events are A apart in the x-direction, B apart in the y-direction, and C apart in the z-direction, and T apart in time, and if the space is Euclidean four-space, then the distance d between the events would be

$$d = \sqrt{c^2 T^2 + A^2 + B^2 + C^2},$$

where c is the speed of light.

In Minkowskian space, the interval I, which measures the distance between the events, is defined by

$$I = \sqrt{c^2 T^2 - A^2 - B^2 - C^2}$$

This formula, with its minus signs, is very different from Euclidean distance and determines a different geometry.

General Relativity

Little more than a decade after introducing his special theory of relativity, Einstein came forth with his *general theory of relativity.* This work, too, astonished the scientific world. Among other revolutionary ideas was his contention that space was "curved." By this he meant that light rays, which are considered to travel on paths of shortest distance, don't actually follow "straight lines" but bend to follow shortest distance paths in the curved space. Light rays even bend to different degrees, depending on where in the universe they are; if they pass through a strong gravitational field, then they bend considerably.

A test of this contention was made in 1919 during a total eclipse of the sun, when the light rays from a distant star passed close to the sun and could be studied. Einstein was right; the rays did bend—and in an amount very close to his predictions. This observation showed that lines in the geometry of general relativity are not of the same character as Euclidean lines.

What sort of geometry was Einstein using? There are several answers to the question. First, the idea of "curved" space smacks of elliptic geometry, in the sense that a line through Einstein's universe comes back on itself. Second, Einstein used a variation on Minkowskian geometry in which the distance formula appropriate to the needs of physics varies from place to place in the universe, depending on the strength of the gravitational field. So, Einstein was using a form of Minkowskian geometry along with some considerably modified ideas of elliptic geometry. An appreciation of these non-Euclidean geometries very likely motivated his remark about the postulates of geometry in a famous 1921 lecture: "[They] are voluntary creations of the human mind. To this interpretation of geometry I attach great importance, for should I not have been acquainted with it, I would never have been able to develop the theory of relativity."

Relativity and Length Contraction

The road that led Einstein to relativity is marked by one crucial experiment, conducted in 1887 by A. Michelson and E. C. Morley in Cleveland at what is now Case Western Reserve University. They were trying to determine whether there was a substance, an "ether," that served as the medium for the transmission of light and electromagnetic radiation. Sound waves do not travel in a vacuum but require a medium (e.g., air or water), relative to which we can measure the speed of sound. Nineteenth-century scientists reasoned that light waves, which do pass through the vacuum of space, must be carried through it in some medium that had not yet been detected.

The essence of their experiment was to send a lightbeam out to a mirror and back and measure the time that elapsed. They did this for two situations. In one, the lightbeam was aligned in the direction of the earth's rotation on its axis, so that on the way out the speed of the earth would add to the speed of light through the ether, and on the way back, it would subtract from the speed of the light. In the other situation, they aligned the lightbeam perpendicular to the earth's rotation on its axis, so that the earth's spinning would not affect the *speed* of the lightbeam in the ether. In this situation, however, the earth's rotation does affect *how far* the lightbeam has to travel.

What Michelson and Morley expected to observe was that the speed of light would be different for the two situations.

EXAMPLE ▶ *Swimming in a Current*

We imagine an analogous situation. You are in a river that is a distance d wide and you decide to conduct a swimming experiment. In place of bouncing lightbeams, you will swim back and forth. Your speed of swimming c corresponds to the speed of light, and the speed v of flow of the river corresponds to the speed of the rotation of the earth.

Corresponding to the first situation, you swim a fixed distance d downstream, then swim back again (see Figure 19.13a); let's say that it takes you time t_1 downstream and time t_2 upstream. On the trip downstream, you are swimming through the water at speed c but you are moving at a speed of $(c + v)$ relative to the bank. Similarly, on the trip upstream, you are really moving only at speed $(c - v)$.

Using the basic formula distance = rate × time, we have $d = (c + v)t_1$ and $d = (c - v)t_2$. Your total time for the round-trip is

$$t_{up+down} = t_1 + t_2 = \frac{d}{c + v} + \frac{d}{c - v}$$

$$= \frac{2dc}{(c + v)(c - v)} = \frac{2d/c}{1 - v^2/c^2}$$

FIGURE 19.13
(a) Swimming with and against the current of a river. (b) Swimming across a river and back to the starting point. (c) To stay even with the starting point, the swimmer needs to head upstream to some degree.

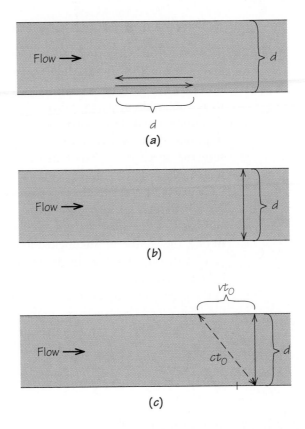

FIGURE 19.13
(a) Swimming with and against the current of a river. (b) Swimming across a river and back to the starting point. (c) To stay even with the starting point, the swimmer needs to head upstream to some degree.

You should check these calculations for a numerical example; for example, the distance d is 1/2 mile, you can swim at a speed c of 2 miles an hour, and the river flows at a speed v of 1 mile per hour. You should find that it takes you one-sixth of an hour downstream and one-half of an hour upstream, for a total time for the round-trip of two-thirds of an hour, or about 0.67 hr.

Corresponding to the second situation, you swim distance d across the river and back to your starting point (see Figure 19.13b). As you swim, however, you have to fight the downstream current, so that you always stay even with your starting point. You not only have to swim the distance across the river, you also need to swim upstream the equivalent distance that the river is carrying you downstream; you need to always head part way upstream, so only part of your swimming effort is directed toward carrying you directly across the river. In effect, you are swimming the length of the hypotenuse of a right triangle, as shown in Figure 19.13c, and being carried downstream by the length of one leg.

Denote the time that it takes you to cross the river by t_0. The relationship between how far you travel and how long it takes is given by the Pythagorean theorem, as mentioned earlier:

$$(ct_0)^2 = vt_0^2 + d^2$$

We solve the equation for the time:

$$c^2 t_0^2 - v^2 t_0^2 = d^2,$$

$$t_0^2 = \frac{d^2}{c^2 - v^2},$$

$$t_0 = \frac{d}{c\sqrt{1 - v^2/c^2}}$$

The time for the return trip is the same, so the total time for the round-trip is

$$t_{\text{back}+\text{forth}} = \frac{2d/c}{\sqrt{1 - v^2/c^2}}$$

Testing this algebra with the same numerical values as before, you should find that the round-trip across the river takes you $1/\sqrt{3} \approx 0.577$ hr.

You observe that the times for the two round-trips are different. The time for the round trip in the direction of motion of the current and back is longer, by a factor of $1/\sqrt{1 - v^2/c^2}$, than the trip across the current and back. ◆

Michelson and Morley had a different experience: the times that they observed in their analogous experiment were *the same*. Because of their experience and reputation as experimenters, they and their colleagues were sure that this astonishing result was not the consequence of measurement or experimental error.

So what was wrong with the theory? One ingenious explanation is that the speed of the rotation of the earth somehow *contracts lengths along the direction of motion*, in exactly the right proportion to cancel the difference in times. In our swimming example, it would be as if distance up and down stream were made shorter by the current. Similarly, a planet moving at nearly the speed of light would flatten in the direction of motion, so as to turn into a pancake. For this explanation to work, the length l' along the direction of motion would have to be related to the length l in the direction perpendicular to the motion by

$$l' = l\sqrt{1 - v^2/c^2} = l/\gamma$$

where

$$\gamma = \frac{1}{\sqrt{1 - v^2/c^2}}$$

is known as the **Lorentz-Fitzgerald factor.**

Because the speed of light is so great—186,000 miles per second, or 3.0×10^{10} cm/sec—the value of the Lorentz-Fitzgerald factor is very nearly 1 until v reaches about 10% of the speed of light.

Why couldn't Michelson and Morley detect this shrinkage in the direction of motion? Because their measuring rods, when aligned along the direction of motion, shrank too. So the contraction theory could never be verified by direct measurements.

You would think that photography would help—for example, that you could see a ball moving at nearly the speed of light as the pancake shape that it must be. However, most surprisingly, even a camera can't see the contraction! An optical distortion compensates for the shrinkage, as we now explain.

You and the camera see by means of "particles" (photons) of light reflected from an object. Light from faraway objects can take a long time to reach us: light from the sun takes about eight minutes, and the light that reaches us now from distant stars was emitted billions of years ago. Similarly, in the case of a moving object, we see at the same time images of close-up parts and (because of time delay from the fixed speed of light) earlier images of faraway parts of an object. Hence, the object appears (to us or on the film) stretched in the direction of motion. This stretching compensates for the contraction. Figure 19.14 shows a computer reconstruction of how an object moving at close to the speed of light would appear. Notice that it does not appear to shrink in the direction of motion (even though it actually does).

The Lorentz-Fitzgerald contraction theory was based on a complicated theory of matter interacting with the ether. Scientists eventually were forced to conclude that there is no medium in which light waves move, no "ether" relative to which we can measure the speed of light. The appealing analogy between a light wave and a swimmer swimming through water is misleading.

Twenty-four years after the Michelson-Morley experiment, Einstein hypothesized that the speed of light is not affected by the motion of the source nor of the observer. The rotation of the earth cannot add to or subtract from the speed of light in the Michelson-Morley experiment. Einstein's theory predicts the same time, $2d/c$, for the round-trip in either alignment of their equipment.

Einstein's theory of relativity also predicts a contraction of length in the direction of relative motion, by exactly the Lorentz-Fitzgerald factor. The reason for this relativistic contraction, however, has nothing to do with ether or with Lorentz's theory to explain the Michelson-Morley results on the basis of it.

RenderMan Image by Pixar (c) 1990

FIGURE 19.14   (a) A teapot at rest. (b) A view from the same angle of the teapot passing by at 99% of the speed of light.

Einstein's theory eliminates the need to suppose that there is an ether. The underlying reason for the relativistic length contraction is relativity itself: the relative motion of object and observer. As viewed by you, the teapot is moving at close to the speed of light, and its length must contract in the direction that it appears to you to be traveling (even though you can't see the contraction, as we've explained). As viewed by an ant on the teapot, *you* appear to be moving at close to the speed of light and are thin as a pancake in the direction that you appear to be moving (even though the ant can't see you that way).

Another consequence of relativity is that *time, too, contracts with motion.* Consider two observers moving at a constant velocity relative to each other. Each will observe that the other's clock runs slow compared to their own, by a factor of γ. This strange result is known as the *clock paradox.*

WHICH GEOMETRY IS TRUE?

As far as measurement and travel on the surface of the earth go, we know that we live on a world with an elliptic geometry. When it comes to travel at velocities near the speed of light, the geometry that applies to space-time is Minkowskian geometry, a kind of non-Euclidean geometry. But what about the universe of space beyond the earth's surface, without regard to time? Do we really live in a spatial universe that is Euclidean?

Spotlight
19.6

The Implications of Non-Euclidean Geometry and Relativity

The creation of non-Euclidean geometry affected scientific thought in two ways. First of all, the major facts of mathematics, that is, the axioms and theorems about triangles, squares, circles, and other common figures are used repeatedly in scientific work. Since these facts could no longer be regarded as truths, all conclusions of science that depended upon strictly mathematical theorems also ceased to be truths.

Second, the debacle in mathematics led scientists to question whether they could ever hope to find a true scientific theory.

Even on the level of engineering, a serious question emerged. Since bridges, buildings, dams, and other works were based on Euclidean geometry, was there not some danger that these structures would collapse? But this thought did not alarm the scientists and engineers of the nineteenth century, who did not believe that the geometry of physical space could be other than Euclidean. However, the advent of the theory of relativity drove home the point that Euclidean geometry is not necessarily the best geometry for applications. For engineering involving motion with high velocities, such as modern accelerators of electrons or neutrons, the theory of relativity is used.

Past ages have sought absolute standards in law, ethics, government, economics, and other fields. They believed that by reasoning one could determine the perfect state, the perfect economic system, the ideals of human behavior, and the like. This belief in absolutes was based on the conviction that there were truths in the respective spheres. But in depriving mathematics of its claim to truth, the non-Euclidean geometries shattered the hope of ever attaining any truths.

The view that mathematics is a body of truths was accepted at face value by every thinking being for 2000 years. This view, of course, proved to be wrong. We see, therefore, on the one hand, how powerless the mind is to recognize the assumptions it makes. Apparently we should constantly re-examine our firmest convictions, for these are most likely to be suspect. They mark our limitations rather than our positive accomplishments. On the other hand, non-Euclidean geometry also shows the heights to which the human mind can rise. In pursuing the concept of a new geometry, it defied intuition, common sense, experience, and the most firmly entrenched philosophical doctrines just to see what reasoning would produce.

Source: Adapted from Morris Kline, *Mathematics for the Nonmathematician*, Dover, New York, 1985, pp. 474–476.

One way to tell would be to measure the angles in a large triangle, to see how their sum compares with 180°. Because all measurements have some imprecision, we cannot prove that the angles of a measured triangle add up to exactly 180°. Sufficiently precise measurements of very large triangles, however, could conceivably prove that space is not Euclidean.

Gauss was employed for a time by the government of Hanover in a geodetic survey, in the course of which he measured the angles in a triangle formed by three mountain peaks roughly 50 miles apart. The deviation from

180° was less than the error estimate for the measurement, so the sum could be equal to 180°, or greater, or less—the sum was consistent with all three hypotheses. In fact, if there is any difference from 180° for this triangle of mountain peaks, it was far too small for Gauss to detect—as he probably realized—and far too small even for us to detect today.

Lobachevsky considered even larger triangles and looked into the parallax of stars (the apparent relative motion, as the earth orbits the sun, of nearer stars compared to more distant ones). But neither he nor others since have found a triangle whose angle sum is definitely different from 180°, despite the fact that in hyperbolic geometry, the larger the area of the triangle, the larger the defect must be.

If space does have a hyperbolic geometry, then there is a lower bound for the parallax of stars (though that would not mean that there is a limit to how far away stars can be). Although there is certainly a smallest observed parallax among the thousand or so stars whose parallax we know, there may be stars yet unmeasured whose parallax is even smaller.

Space could have an elliptic geometry, with triangles having angular excesses rather than defects. The universe could then be the three-dimensional analogue of the two-dimensional surface of a sphere. Just as the surface of a sphere has a finite area, the universe could have a finite volume, despite having no boundaries—just as the surface of a sphere has no boundaries. Such a space would have positive curvature.

Regardless of the true situation for actual three-dimensional space, we appear to perceive space visually as hyperbolic. Common visual illusions, classic experiments in perception, and the empirical truth of *Brentano's hypothesis* (that humans tend to overestimate small angles and underestimate large ones) all lead to the conclusion that "perceived space" is hyperbolic.

The question of which geometry is true would not have occurred to anyone before the nineteenth century. The discovery of non-Euclidean geometry and the theory of relativity have had profound intellectual implications in all fields (see Spotlight 19.6).

REVIEW VOCABULARY

Elliptic geometry   A geometry in which there are no parallel lines.
Euclidean geometry   The "ordinary" system of geometry based on the five postulates Euclid used, including Euclid's parallel postulate.
Euclid's parallel postulate   If *l* is any line and *P* any point not on *l*, then there exists exactly one line through *P* that does not meet *l*—in other words, given a line and a point not on the line, there is exactly one other line that passes through the point and is also parallel to the given line.

Geodesic   A curve (possibly straight) that gives the shortest path between two points on a surface.

Great circle   The set of points that is the intersection of a sphere and a plane containing its center.

Hyperbolic geometry   A geometry in which for any given line and a point not on the line, more than one line both passes through the point and is parallel to the given line.

Light-year   The distance that light travels in a year.

Logically equivalent   Two statements are logically equivalent if each can be deduced from the other.

Logically independent   A statement is logically independent of a collection of other statements if neither it nor its negation can be deduced from them.

Lorentz-Fitzgerald factor   The factor by which length and time are contracted by motion.

Model   An example that satisfies a collection of axioms is a model of the axioms.

Non-Euclidean geometry   Any collection of postulates, theorems, and corollaries for geometry (concerning points, lines, circles, and angles) that differs from the collection formulated by Euclid.

Parallel lines   Two lines are parallel if they do not meet.

Parallel postulate   A basic assumption of geometry that states whether through a point P, not on a given line l, there exists none, one, or more than one line parallel to given line l.

Postulate E   Every two lines intersect.

Postulate H   If l is any line and P is any point not on the line, then there exists more than one line through P not meeting l.

Spherical geometry   The geometry of a sphere or the earth's surface.

SUGGESTED READINGS

ABBOTT, EDWIN A.   *Flatland: A Romance of Many Dimensions,* Princeton University Press, Princeton, N.J., 1989.

BOLOTOVSKY, B. M.   What's that you see? On the perceived shape of rapidly moving objects, *Quantum,* 3 (4) (March/April 1993): 5–8.

BOLTYANSKY, VLADIMIR.   Turning the incredible into the obvious: How many geometries do you know? *Quantum,* 3 (1) (September/October 1992): 19–23.

BURGER, DIONYS.   *Sphereland: A Fantasy About Curved Spaces and an Expanding Universe,* Crowell, New York, 1965. Reprinted with a new introduction, Harper & Row, New York, 1983.

CARVER, MAXWELL. Brain bogglers: Chicken a la king, *Discover* (March 1988): 96, 92.

CROWE, DONALD W. Some exotic geometries, in Anatole Beck, Michael Bleicher, and Donald W. Crowe, eds., *Excursions into Mathematics,* Worth, New York, 1969, pp. 211–314. A comparison of spherical, Euclidean, and hyperbolic geometries, with sections on finite geometries.

DAVIS, A. S. The relevance of mathematics for the remnant, University of Oklahoma, Department of Mathematics Preprints No. 48, 1968.

DUBROVSKY, VLADIMIR. Inversion: A most useful kind of transformation, *Quantum,* 3 (1) (September/October 1992): 40–46; and (November/December 1992): 80–81.

GINDIKIN, SIMON. The wonderland of Poincaria, *Quantum,* 3 (2) (November/December 1992): 21–26, 58–59.

GREENBERG, M. J. *Euclidean and Non-Euclidean Geometries,* Freeman, New York, 1980.

HSIUNG, PING-KANG, ROBERT H. THIBADEAU, AND ROBERT H. P. DUNN. Ray-tracing relativity, *Pixel,* 1 (1) (January/February 1990): 10–18.

JACOBS, HAROLD R. *Geometry,* 2nd ed., Freeman, New York, 1987. Chapters 6 and 16 treat parallel lines and non-Euclidean geometries.

MAURER, STEPHEN B. The king chicken theorems, *Mathematics Magazine,* 53 (1980): 67–80.

PARKER, GEORGE D. *Poincaré: An Excursion into Hyperbolic Geometry,* IBM-PC program for doing hyperbolic geometry in the Poincaré upper-half-plane model. Available from the author at 1702 West Taylor, Carbondale, IL 62901.

PETERSON, IVARS. Space-time odyssey: Visualizing the effects of traveling near the speed of light, *Science News,* 137 (April 14, 1990), 232–233, 237.

PETIT, JEAN-PIERRE. *Everything Is Relative: The Adventures of Archibald Higgins,* William Kaufmann, Inc. Relativity as told in a cartoon tale, "for adults of any age."

SMART, JAMES R. *Modern Geometries,* 4th ed., Brooks/Cole, Pacific Grove, Calif., 1994.

SVED, MARTA. *Journey into Geometries,* Mathematical Association of America, Washington, D.C., 1991. Introduces hyperbolic geometry in a delightfully informal style.

TRUDEAU, RICHARD J. *The Non-Euclidean Revolution,* Birkhäuser, Boston, 1987.

WEEKS, JEFFREY R.   *The Shape of Space: How to Visualize Surfaces and Three-Dimensional Manifolds,* Dekker, New York, 1985. Contains chapters on the hyperbolic plane (including how to make hyperbolic paper), spherical geometry, hyperbolic space, and three-dimensional spherical geometry.

ZAGE, WAYNE M. The geometry of binocular visual space. *Mathematics Magazine,* 53 (1980): 289–294.

EXERCISES ▲ *Optional.* ■ *Advanced.* ◆ *Discussion.*

Euclidean Geometry

1.   In triangle *ABC* in the Euclidean plane, the measure of ∠*A* is twice the measure of ∠*B*, and the measure of ∠*C* is three times the measure of ∠*B*. Determine all three measures.

Hyperbolic Geometry

For Exercises 2 and 3, refer to the following: You are working in hyperbolic geometry, and the quadrilateral *ABCD* in the following figure is drawn to suggest the situation of a quadrilateral in hyperbolic geometry with right angles at *A* and *B* and acute angles at *C* and *D*.

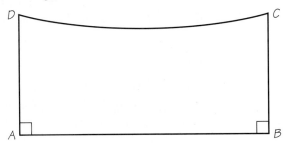

2.   Prove that either the sum of the angles of triangle *ABD* is less than 180°, or the sum of the angles of triangle *BCD* is less than 180°, or both. (*Hint:* What can you say about the sum of the angles of the quadrilateral?) (In fact, both triangles have angle sum less than 180°, but that is harder to prove.)

3.   Prove that if *AD* = *BC,* then angles *C* and *D* are of equal measure. (*Hint:* Draw the diagonals. Then use the Euclidean theorems on the congruence of triangles, which do not depend on Euclid's parallel postulate.)

Elliptic Geometry

4.   By producing a specific example, show that there is a triangle in elliptic geometry in which all three angles are right angles, so that the sum of the angles of the triangle is 270°. (*Hint:* Spherical geometry is an elliptic geometry.)

Models for Hyperbolic Geometry

Exercises 5–10 require a Macintosh computer and the freeware program SnapPeaFunAndGames.sea.hqx

- from the World Wide Web source http://archives.math.utk.edu/software/mac/geometry/.directory.html, or
- by anonymous ftp transfer from softlib.rice.edu in the directory pub/NonEuclid, or
- by using a Gopher client to archives.math.utk.edu, proceeding to Software (Packages, Abstracts and Reviews), to Macintosh Software arranged by subjects, to Geometry.

From any of these sources, you must use BinHex 4.0, StuffIt Expander, or similar helper program to convert the downloaded file into application programs. The program that you are interested in is hyperbolic MacDraw. After launching the program, you will be presented with a window entitled "Poincaré Disk Model" containing a circle. Within this circle you can use the mouse to draw lines of hyperbolic geometry, including lines that extend to the boundary of the circle. Other tools in the Tool menu allow you to draw closed figures, such as triangles, quadrilaterals (four-sided polygons), and regular polygons (all sides the same length) with any number of sides. Changing tools to the hand tool, you can move figures around inside hyperbolic space. Moreover, the Model menu gives you the option to also view your figures in the Klein disk model and in the upper half-plane model.

5. When you first launch the program, you are in the Poincaré disk model and the tool for drawing a line is active. Start with the pen cursor on the boundary of the circle, hold the mouse button down, and move the cursor to elsewhere on the boundary. Notice that wherever you end the line, both ends of the line make a 90° angle with the boundary.

6. The program does not provide for erasing figures. Under File select New to get a new hyperbolic plane. Start with the pen cursor inside the circle, hold the mouse button down, and move to another point inside the circle, passing through the center of the circle. You have drawn a line segment, which should look fairly straight to you. Change to the hand tool and move the segment around. Notice that as you move the segment it appears to curve. Also, as you move the segment closer to the boundary, it appears to get shorter. That is your perspective from viewing the hyperbolic plane from the outside; from inside, the segment moves around, but always lies along some line of the geometry (all lines make 90° angles with the boundary) and remains the same length.

7. Again get a new hyperbolic plane. Use the tool in the lower left-hand corner of the Tool menu to draw a triangle, which the program will automatically shade. Use the hand tool to move the triangle around. As you move it, its

sides appear to change shape and it appears to change area. (Careful! If you move it too close to the boundary, you can lose it!) Again, that is a perspective from outside the geometry; from inside, the sides always lie along lines and the area remains the same.

8.   Get a new hyperbolic plane and select the polygon tool on the right-hand side of the Tool menu. Click in the exact center of the circle. The program will prompt you for the number of sides for the polygon (enter 8) and for the angle between two adjacent sides (enter 4 to get $2\pi/4$; 2π radians corresponds to 360°, so the sides of this polygon will be at 90° to each other). What happens as you move this octagon around?

9.   As in part (d), but let the polygon have four equal sides. Try various values for the divisor for 2π. Which values give you something that when positioned in the middle of the hyperbolic plane looks most like a square? What are the angles between its adjacent sides? What is the smallest divisor for 2π that will produce a hyperbolic polygon with four equal sides, and what is the angle between adjacent sides?

10.   Take each of the hyperbolic planes in which you have drawn figures in Exercises 5–9 and convert first to the Klein disk model and then to the upper half-plane model. What can the sides of a hyperbolic polygon with four equal sides look like in the latter model?

Exercises 11–16 require a Macintosh computer. The Macintosh program NonEuclid is available without fee for educational use from the sources below. (Similar and additional capabilities are provided by the MSDOS program Poincaré for doing hyperbolic geometry in the upper half-plane model instead of the Poincaré disk model. That program is available for a fee from the author, George D. Parker, 1702 West Taylor, Carbondale, IL 62901.) NonEuclid can be obtained

- from either of the World Wide Web sources http://archives.math.utk.edu/software/mac/geometry/.directory.html, or http://riceinfo.rice.edu:80/projects/NonEuclid/NonEuclid.html, or
- by anonymous ftp transfer from softlib.rice.edu in the directory pub/NonEuclid, or
- by using a Gopher client to archives.math.utk.edu, proceeding to Software (Packages, Abstracts and Reviews), to Macintosh Software arranged by subjects, to Geometry.

From any of these sources, you must use BinHex 4.0, StuffIt Expander, or similar helper program to convert the downloaded file into the application program and associated files. Unlike hyperbolic MacDraw, this program does not allow you to move figures once you have constructed them; but it does do measurements for you of lengths, angles, and areas of figures.

11. Under Help, read Introduction and My First Triangle. Follow the instructions of My First Triangle to draw a triangle, measure its angles, and find its angle sum.

12. Most of the remaining options under Help ask you to use the program to perform constructions to test whether theorems of Euclidean geometry are true or not in hyperbolic geometry. Read the text under the corresponding title and do constructions to try to answer the questions there about whether theorems of Euclidean geometry carry over. Begin with What To Do, which tells how to go about your experiments, and then investigate

    (a)   parallel lines
    (b)   circles
    (c)   angles

13. Repeat Exercise 12, but for

    (a)   parallelograms
    (b)   rhombuses

14. Repeat Exercise 12, but for

    (a)   rectangles
    (b)   squares

15. Repeat Exercise 12, but for

    (a)   triangles
    (b)   isosceles triangles
    (c)   equilateral triangles
    (d)   right triangles

16. Use Open under File to open and explore files in the folder NonEuclid Examples, which discuss area, equilateral triangles, radii of circles, tilings by triangles, and whether a coordinate system is possible in hyperbolic geometry. The tiling of Web of Congruence is the pattern for M. C. Escher's print *Circle Limit IV (Heaven and Hell)* in Spotlight 19.4 (page 772).

For Exercises 17–24, refer to the following: A collection of trees is arranged in rows so that the following axioms are satisfied.

    A   There is at least one tree.
    B   Each row contains exactly two trees.
    C   Each tree belongs to at least one row.
    D   Any two trees have exactly one row in common.
    E   For any row, there is exactly one other row with no trees in common with the first row.

17. Show the following:

(a) There is at least one row.
(b) There are at least two rows.
(c) There are at least four trees.

18. Show the following:

(a) Every tree belongs to at least two rows.
(b) There are at least six rows.

19. Show the following:

(a) There are exactly four trees.
(b) There are exactly six rows.
(c) Each tree belongs to exactly three rows.

20. In this problem, you create models of the axiom system by interpreting the terms used in the axioms:

(a) Interpreting "tree" as a point and "row" as a line, draw a model of the axiom system. Note that the lines that you draw in the plane have many other planar points that belong to them but that we don't consider, as they are not part of this particular axiom system. (This gives a *four-point geometry*.)
(b) Interpreting "tree" as a line and "row" as a point, draw a model of the axiom system. (This gives a *six-point geometry*.)

21. In this problem, you create further models of the axiom system:

(a) Interpreting "tree" as student and "row" as committee, construct a model of the axiom system. Name or label the students and list all of the members of each committee.
(b) Interpreting "tree" as committee and "row" as student, construct a model of the axiom system. Name or label the students and list all of the members of each committee.

22. Interpreting "tree" as a line and "row" as a point:

(a) How many triangles are there? (A triangle consists of three points and three lines joining them in pairs.)
(b) Are there two lines that are parallel? (Two lines are parallel if they do not have a point in common.)
(c) Does Euclid's parallel postulate hold?

23. Interpreting "tree" as a point and "row" as a line:

(a) How many triangles are there? (A triangle consists of three points and three lines joining them in pairs.)

    (b)   Are there two lines that are parallel? (Two lines are parallel if they do not have a point in common.)

    (c)   Does Euclid's parallel postulate hold?

24.   [Adapted from Crowe (1969), p. 289.] Suppose that "point" means the location of a store where groceries are sold—say, Albany, Birmingham, Chicago, or Denver. Suppose that "line" means a particular line of groceries—say, apples, bananas, cheese, doughnuts, eels, or figs. Finally, suppose that "a tree belongs to a row" or "a row contains a tree" means that a particular line of goods is sold at a particular location. Give an explicit distribution of the six commodities in the four cities so that the axioms apply.

For Exercises 25–28, refer to the following: Suppose the following set of axioms concerns two classes of objects K and L, whose nature is left undetermined:

    A   Any two members of K are contained in exactly one member of L.

    B   No member of K is contained in more than two members of L.

    C   There is no member of L that contains all of the members of K.

    D   Any two members of L contain exactly one member of K in common.

    E   No member of L contains more than two members of K.

We are interested in possible models of these axioms. Check the axioms to verify that if neither K nor L has any members at all, the axioms still hold. We say that they are vacuously true for this "empty" model.

25.   For each of the following situations, is there a model? If so, verify that the axioms all hold; if not, show what axioms would have to be violated.

    (a) K has no members and L has at least one.

    (b) K and L each have one member.

26.   Is there a model in which:

    (a) K has one member and L has two members?

    (b) K has one member and L has three or more members?

27.   Add the additional axiom

    F K has at least two members.

    (a)   Show that K has at least three members.

    (b)   Show that K cannot have more than three members, so it has exactly three members. (This gives a *three-point geometry.*)

    (c)   How many members can L have?

    (d)   Does this axiom system satisfy Euclid's parallel postulate?

28.   As in Exercise 27,

(a)   Using points and lines, give a model of the set of axioms.
(b)   If axiom C is now omitted, are other models possible?

For Exercises 29–32, refer to the following: Consider the following system of axioms concerning two classes of objects M and N.

A N has at least one member.
B   Every member of N contains exactly three members of M.
C   Any two members of M are contained in exactly one member of N.
D   There is no member of N that contains all of the members of M.
E   Any two members of N have at least one member of M in common.

29.   We explore how many members M and N may have.

(a)   Show that each two members of N have exactly one member in common.
(b)   Show that each of M and N has at least seven members. To help your intuition, you may want to try to construct a model of the axioms out of points and lines. Remember, though, that you must reason from the axioms, not from any picture that you draw.

30.   We continue with determining the number of members of M and N.

(a)   Suppose that M has an eighth member. Show that this supposition leads to a contradiction.
(b)   Show that M and N must have exactly seven members each.

31.   What are some models for this set of axioms?

(a)   Interpret the members of M as students and the members of N as committees. Name or label the students and list all of the members of each committee.
(b)   Interpret the members of M as points and the members of N as lines, and draw an appropriate model. (This is *Fano's seven-point geometry.*)

32.   For the model in terms of points and lines in Exercise 30(b):

(a)   Are there two lines that are parallel? (Two lines are parallel if they do not have a point in common.)

(b)   Does Euclid's parallel postulate hold?

For Exercises 33–37, refer to the following: Consider the set {0, 1, 2} under a "clock" arithmetic, with 3 taking the role of the 12 in the usual clock arithmetic, so that $1 + 2 = 0$ and $2 + 2 = 1$. We can even introduce multiplication, with the usual results except that $2 \times 2 = 1$ (we would expect 4, which converts to 1 in our clock arithmetic). Using the set {0, 1, 2} as the possible x- and y-coordinates, we can form the nine points in the figure below. These will be the points of a "miniature" analytic geometry. The lines will consist of points that satisfy linear equations, that is, equations of the form $ax + by = c$, with a, b, and c from the set {0, 1, 2}.

33.   List all of the points on the line

   (a) $x = 1$
   (b) $y = 2$
   (c) $x + y = 1$
   (d) $x + 2y = 1$

34.   Find the intersection of the lines $x + y = 1$ and $2x + y = 2$

   (a)   by using algebra. (*Hint:* Subtract one equation from the other.)
   (b)   by listing all of the points on each line and comparing the lists.

35.   We explore how many lines and points are in this geometry:

   (a)   How many different lines are there? (*Hint:* A line is either a vertical line, with equation $x = c$, or else it can be written in the form $y = mx + b$, with each of c, m, and b being either 0, 1, or 2.)
   (b)   How many points lie on each line?
   (c)   How many lines pass through each point?

36.   This miniature analytic geometry is known as the *affine two-dimensional geometry over 3 elements,* an example of an *affine plane.* It satisfies the following three axioms:

   A   For any pair of distinct points, there is exactly one line containing both of them.

   E   Given a line and a point not on the line, there is exactly one line through the point that does not contain any points of the given line. (This is the parallel postulate, so this is a Euclidean geometry.)

   B   There are at least four points, no three of which lie on the same straight line.

(a)   Prove that axiom A is satisfied. (*Hint:* Use the two-point formula for the equation of a straight line,

$$\frac{y - y_1}{x - x_1} = \frac{y_2 - y_1}{x_2 - x_1}$$

to exhibit one such line. Suppose that there are two distinct lines through the same two points. Take one of the points and count all of the points on the lines that pass through that point. Show that a contradiction results.)

(b)   Prove that axiom E is satisfied. (*Hint:* Join the given point to every point on the given line. How many lines is that? Are there any lines left over, available to be "parallels"?)

(c)   Prove that axiom B is satisfied.

37.   Even in so miniature a geometry, we can do more than just play with points and lines. We can define circles, ellipses, hyperbolas, parabolas, and even tangents to circles! Here we'll just whet your appetite by introducing circles. A circle centered at (a, b) will consist of all of the points that satisfy an equation of the form $(x - a)^2 + (y - b)^2 = r$, where each of a, b, and r is one of 0, 1, or 2.

(a)   Find the points on the circle $x^2 + y^2 = 0$.
(b)   Find the points on the circle $x^2 + y^2 = 1$.
(c)   Find the points on the circle $x^2 + y^2 = 2$.

38.   [Adapted from Boltyansky (1992).] Take as the "points" of a geometry all points of the plane except a single point O. Take as "lines" all circles and straight lines that pass through the deleted point O.

(a)   Given two distinct "points" A and B, show how to construct a "line" that passes through both of them. Is this the only line with this property?

(b)   Given a "line" l and a "point" P not on l, show that there is a unique "line" through P that does not intersect l. In other words, the parallel postulate holds.

This geometry can be further furnished with angle measure, distance measure, "circles," triangles, and so forth, so that it satisfies all of the postulates of Euclidean geometry and hence is a model of those postulates.

Relativity and Length Contraction

39. For each of the following objects, calculate at what fraction of the speed of light it is traveling, the corresponding value of γ, and how much it is shortened by Lorentz-Fitzgerald contraction.

   (a) Automobile at 65 mph
   (b) Rifle bullet at 1 km/sec
   (e) Comet Hyakutake as it passed the earth at the end of March 1996, 200,000 mph

40. Repeat Exercise 39, but for

   (a) an object moving at 99% of the speed of light
   (b) subatomic particles in an accelerator at 99.94% of the speed of light

WRITING PROJECTS

For Projects 1 to 4, refer to the following (adapted from Davis [1968]): Folklore has it that everybody is (or should be) an enemy of their friends' enemies and a friend of their friends' friends, as well as a friend of their enemies' enemies and an enemy of their enemies' friends. If this is the case, what patterns of friendship are possible in a stable society? We investigate this question by converting the folklore into an axiom system and then seeing what conclusions can be drawn. We use capital letters to denote people, together with the symbols =, $\heartsuit$, and ♯. Our interpretations of the symbols are

- $X = Y$ means X and Y are the same person.
- $X \heartsuit Y$ means that X is an immediate friend of Y.
- $X ♯ Y$ means that X is an immediate enemy of Y.

We also make some definitions. We say that X is *immediately involved with* Y if either $X \heartsuit Y$ or $X ♯ Y$. Also, we define a *chain of involvements* from X to Y as a sequence of people $Z_0, Z_1, \ldots, Z_n$ (for some n) such that $Z_0 = X$, $Z_n = Y$, and each Z_{i-1} is immediately involved with Z_i, for $i = 1, 2, \ldots, n$. The Z_i don't have to be distinct. In fact, we count $X \heartsuit X$ and $X ♯ X$ as chains of length one.

Our axioms are:

A   For all X and Y, if $X \heartsuit Y$, then $Y \heartsuit X$.
B   For all X and Y, if $X \sharp Y$, then $Y \sharp X$.
C   Every pair of people is connected by a chain of involvements.

Axiom C asserts that no person is ever completely isolated from another. We call a chain of involvements *positive* if the number of immediate enmities in it is even, and *negative* if it is odd. We also say that X is a *(distant) friend* of Y if there is a positive chain of involvements from X to Y, and is a *(distant) enemy* if there is a negative chain. Finally, we say that X is *ambivalent* toward Y if X is both a friend and an enemy of Y.

1 ▶   Suppose that $X \heartsuit Z \sharp W \heartsuit Y$. Is X a friend of Y?

2 ▶   Show that if X is a friend of, an enemy of, or ambivalent toward Y, then Y is likewise related to X.

3 ▶   A *society* is a set of people, each pair of which is connected by a chain of involvements consisting of members of that set.

(a)   Show that the set of all people is a society.
(b)   Can one person alone be a society?

4 ▶   A society is *stable* if there exist no ambivalences in it: No pair of people are friends and enemies both.

(a)   Show that in an unstable society, everyone is everyone's friend and enemy both. In other words, everybody is ambivalent toward everybody, including themselves.
(b)   Show that a society is stable if and only if it divides into two sets of people so that everyone is a friend of just the people in their own set and an enemy of exactly those in the other. (One of the sets may be empty, in which case everyone is a friend of everyone and an enemy of none.)

5 ▶   Examine the axioms, terminology, and theorems of this axiomatic system. Discuss how well you think this theory fits human attitudes and behavior.

CHAPTER 20

SYMMETRY AND PATTERNS

"The senses delight in things duly proportional." So said the famous philosopher-theologian Thomas Aquinas more than 700 years ago, in noting human aesthetic appreciation. In this chapter we examine some of the elements of that aesthetic appreciation, particularly what we call *symmetry*.

Symmetry, like beauty, is very difficult to define. Dictionary definitions talk about "correspondence of form on opposite sides of a dividing line or plane or about a center or an axis," "correspondence, equivalence, or identity among constituents of an entity," and "beauty as a result of balance or harmonious arrangement" (*American Heritage Dictionary,* 3rd ed.).

In the narrowest sense, symmetry refers to "mirror-image" correspondence between parts of an object. Crystals, in both their appearance and their atomic structure, provide examples of symmetry in this sense. Taken in a wider sense, though, symmetry includes notions of *balance, similarity,* and *repetition.*

It is our sense of symmetry that leads us to appreciate patterns. As we noted in the introduction to this part of the book, mathematics is the study of patterns, and we will see here that mathematics gives important insights into symmetry.

Patterns abound in nature. The successive sections of the beautiful chambered nautilus grow according to a very strict and specific spiral pattern, a broader kind of symmetry. This spiral has the property that it has the same shape at any size: a photographic enlargement superimposed on it would fit exactly.

Botanists have long appreciated other spirals. In plant growth from a central stem, the shoots, leaves, and seeds often occur in a spiral pattern known as **phyllotaxis.** For instance, the scales of a pineapple or a pinecone are arranged in spirals (Figure 20.1), as are the seeds of a sunflower (Figure 20.2a) and the petals on a daisy. Like the chambers of the nautilus in Figure 20.2b, the spirals on these plants are geometrically similar to one another, and they are arranged in a regular way, with balance and "proportion." These plants have a kind of symmetry we would naturally call **rotational.**

FIGURE 20.1   Spirals of scales on a pinecone: 8 right, 13 left.

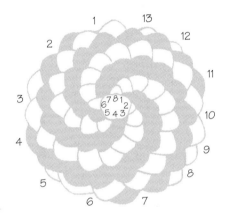

FIBONACCI NUMBERS

Associated with the geometric symmetry of phyllotaxis, there is also a kind of *numeric symmetry,* with a "proportion" in the sense of a ratio of numbers. Strangely, the number of spirals in plants with phyllotaxis is not just any whole number but always comes from a particular sequence of numbers, called the *Fibonacci numbers.*

Fibonacci numbers occur in the sequence

1, 1, 2, 3, 5, 8, 13, 21, 34, 55, 89, 144, 233, 377, . . .

This sequence begins with the numbers 1 and 1 again, and each next number is obtained by adding the two preceding numbers together.

FIGURE 20.2 (a) This sunflower has 55 spirals in one direction and 89 spirals in the other direction. (b) A chambered nautilus shell.

(a)   *(b)*

Sometimes a sequence of numbers is specified by stating the value of the first term or first several terms and then giving an equation to calculate succeeding terms from preceding ones. This is called a *recursive rule,* and the sequence is said to be defined by **recursion.** Let's denote the nth Fibonacci number by F_n; then the Fibonacci sequence can be defined by

$$F_1 = 1, F_2 = 1, \quad \text{and} \quad F_{n+1} = F_n + F_{n-1} \quad \text{for } n \geq 2$$

The recursive rule just expresses in algebraic form that the next Fibonacci number is the sum of the previous two.

Look at the sunflower in Figure 20.2a. You see a set of spirals running in the counterclockwise direction and another set in the clockwise direction. It is (just barely) possible to count the number of spirals in both directions; in the sunflower there are 55 in one and 89 in the other direction—two consecutive Fibonacci numbers. In the case of the pineapple, there are three sets of spirals, one each along the three directions through each hexagonally shaped scale. For the common grocery pineapple (*Ananas comosus*), there are always 8 spirals to the right, 13 to the left, and 21 vertically—again, consecutive Fibonacci numbers.

Why are the numbers of spirals in plants the same numbers that appear next to each other in a purely mathematical sequence? The question has been the subject of extensive research, and there is no easy answer; there are several intricate theories about the dynamics involved in the plant's growth.

Leonardo Pisano Bigollo ("Fibonacci")

Leonardo Pisano ("Fibonacci")
A portrait of unlikely authenticity.

Born in Pisa in 1170, Leonardo Pisano Bigollo has been known as "Fibonacci" for the past century and a half. This nickname, which refers to his descent from an ancestor named Bonaccio, is modern, and there is no evidence that he was known by it in his own time.

Leonardo was the greatest mathematician of the Middle Ages. His stated purpose in his book *Liber abbaci* (1202) was to introduce calculation with Hindu-Arabic numerals into Italy, to replace the Roman numerals then in use. Other books of his treated topics in geometry, algebra, and number theory.

We know little of Leonardo's life apart from a short autobiographical sketch in the *Liber abbaci*:

I joined my father after his assignment by his homeland Pisa as an officer in the customhouse located at Bugia [Algeria] for the Pisan merchants who were often there.

He had me marvelously instructed in the Arabic-Hindu numerals and calculation. I enjoyed so much the instruction that I later continued to study mathematics while on business trips to Egypt, Syria, Greece, Sicily, and Provence and there enjoyed discussions and disputations with the scholars of those places. [L. E. Sigler, *Leonardo Pisano Fibonacci, The Book of Squares: An Annotated Translation into Modern English*, Academic Press, New York, 1987.]

The *Liber abbaci* contains a famous problem about rabbits, whose solution is now called the Fibonacci sequence. Leonardo did not write further about it.

THE GOLDEN RATIO

During the last several centuries, an attractive myth arose that the ancient Greeks considered a specific numerical proportion essential to beauty and symmetry. Known variously in modern times as the *golden ratio,* **golden mean,** or even **divine proportion,** this proportion was investigated by Euclid in Book II of his *Elements*. Recent research reveals little evidence connecting this proportion to Greek aesthetics, but we pursue the golden ratio briefly because of its intimate connection to the Fibonacci sequence and because it does have appeal as a standard for beautiful proportion.

The value of the **golden ratio,** which is usually denoted by the Greek letter phi (ϕ), is

$$\phi = \frac{1 + \sqrt{5}}{2} = 1.618034 \ldots$$

The basic aesthetic claim is that a **golden rectangle**—one whose height and width are in the ratio of 1 to ϕ—is the most pleasing of all rectangles. The Greeks treated lengths geometrically, so for them it was important to construct lengths using straightedge and compass; in Spotlight 20.2 we show how to construct a golden rectangle that is 1 unit by ϕ units.

What would make anyone think that this is such an attractive ratio? And where did it come from? The answer lies not in Fibonacci numbers but in the Greeks' pursuit of balance in their study of geometry.

Given two line segments of different lengths, one way to find another length that "strikes a balance" between the two is to average them. For lengths l (the larger) and w (the smaller), their *arithmetic mean* (average) is $m = (l + w)/2$, and it satisfies

$$l - m = m - w$$

The length m strikes a balance between l and w, in terms of a common difference from the two original lengths. More generally, the arithmetic mean of n numbers or lengths is their sum divided by n.

The Greeks, however, preferred a balance in terms of ratios rather than differences. They sought a length s, the *geometric mean,* that gives a common ratio

$$l \div s = s \div w \quad \text{or} \quad \frac{l}{s} = \frac{s}{w}$$

Spotlight *How the Greeks Constructed a Golden Rectangle*

20.2

They started from a one-by-one square (shown in black in the figure), which they made by constructing perpendiculars at the two ends of a horizontal segment of unit length. To extend the square to a golden rectangle, bisect the original segment to get a new point that divides it into two pieces of length one-half each. Using this new point and a compass opening equal to the distance from it to a far corner of the square, you can cut off an interval of length ϕ.

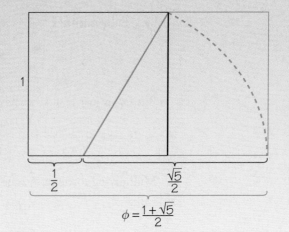

$$\phi = \frac{1 + \sqrt{5}}{2}$$

Hence $lw = s^2$, which expresses the geometric fact that s is the side of a square, the area of which is the same as the area of a rectangle that is l by w (the Greeks thought in terms of geometrical objects). In geometry, the geometric mean s is called the *mean proportional* between l and w (see Figure 20.3).

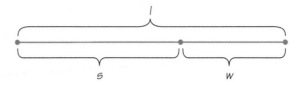

FIGURE 20.3   The line segment of length l is divided so that the length of s is the geometric mean between l and $w = l - s$; the dividing point divides the length l in the golden ratio.

The quantity $s = \sqrt{lw}$ is the **geometric mean** of l and w. More generally, the geometric mean of n numbers is the nth root of the product of all n factors: the geometric mean of $x_1, \ldots, x_n$ is $\sqrt[n]{x_1 \cdots x_1}$. For example, the geometric mean of 1, 2, 3, and 4 is $\sqrt[4]{1 \times 2 \times 3 \times 4} = \sqrt[4]{24} = 24^{1/4} \approx 2.213$.

The Greeks were interested in the problem of cutting a single line segment of length l into lengths s and w, where $l = w + s$, so that s would be the mean proportional between w and l. Surprisingly, the ratio ϕ arises, as we show. Denote the common ratio

$$\frac{l}{s} = \frac{s}{w}$$

by x. Substituting $l = s + w$, we get

$$x = \frac{l}{s} = \frac{s + w}{s} = \frac{s}{s} + \frac{w}{s} = 1 + \frac{w}{s}$$

But w/s is just $1/x$, so we have

$$x = 1 + \frac{1}{x}$$

Multiplying through by x gives

$$x^2 = x + 1 \qquad \text{or} \qquad x^2 - x - 1 = 0$$

This is a quadratic equation of the form

$$ax^2 + bx + c = 0$$

with $a = 1$, $b = -1$, and $c = -1$. We apply the famous quadratic formula

$$x = \frac{-b \pm \sqrt{b^2 - 4ac}}{2a}$$

to get the two solutions

$$x = \frac{1 + \sqrt{5}}{2} = 1.618034 \ldots \quad \text{and} \quad \frac{1 - \sqrt{5}}{2} = -0.618034 \ldots$$

We discard the negative solution since it does not correspond to a length. The first solution is the golden ratio ϕ. It occurs often in other contexts in geometry, for example, ϕ is the ratio of a diagonal to a side of a regular pentagon (see Figure 20.4).

Thanks to recent work of Roger Herz-Fischler (Wilfrid Laurier University) and George Markowsky (University of Maine), we now know that the term "golden ratio" was not used in antiquity and that there is no evidence that the Great Pyramid was designed to conform to ϕ, that the Greeks used ϕ in the proportions of the Parthenon, or that Leonardo da Vinci used ϕ. Moreover, experiments show that people's preferences for dimensions of rectangles cover a wide range, with golden rectangles not holding any special place.

FIGURE 20.4 In a pentagon with equal sides, ϕ is the ratio of a diagonal to a side. The five-pointed star formed by the diagonals was the symbol of the followers of the ancient Greek mathematician Pythagoras.

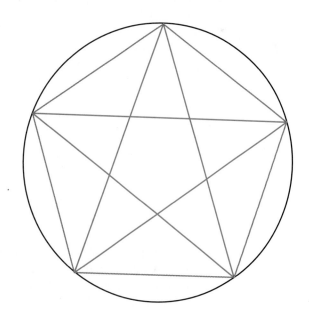

FIGURE 20.5 A logarithmic spiral determines a sequence of golden rectangles and corresponding squares.

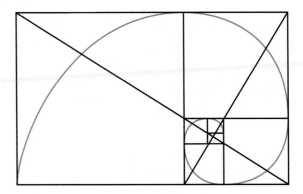

The impressionists Gustave Caillebotte (1848–1894) and Georges Seurat (1859–1891) may have used the golden ratio to design some of their paintings, but we do not have any historical evidence that they claimed or intended to do so.

It is true that human bodies exhibit ratios close to the golden ratio, as you can see by comparing your overall height to the height of your navel. The twentieth-century Swiss-born architect Le Corbusier (Charles-Edouard Jeanneret [1887–1965]) used the golden ratio (including a navel-height feature) as the basis for his "Modulor" scale of proportions.

There are intriguing connections between the spiral of the nautilus and the spirals of the sunflower and between the golden ratio and the Fibonacci sequence. The nautilus shape follows what is known as an *equiangular* or *logarithmic* spiral, which in its turning determines a sequence of golden rectangles (see Figure 20.5). The spirals of the sunflower are in fact approximations to a logarithmic spiral. The mathematical reason for this connection is that the ratios of consecutive Fibonacci numbers

$$\frac{1}{1} \quad \frac{2}{1} \quad \frac{3}{2} \quad \frac{5}{3} \quad \frac{8}{5} \quad \frac{13}{8} \quad \frac{21}{13} \cdots$$

$$1.0 \quad 2.0 \quad 1.5 \quad 1.666\ldots \quad 1.6 \quad 1.625 \quad 1.615\ldots$$

provide alternately under- and overapproximations to $\phi = 1.618034\ldots$

BALANCE IN SYMMETRY

The spiral distribution of the seeds in a sunflower head and the spiraling of leaves around a plant stem are instances of *similarity* and *repetition*, two key aspects of symmetry; they also illustrate *balance*, which refers to regularity in *how* the repetitions are arranged.

In considering patterns with repetition, we distinguish the individual element or figure of the design (sometimes called the *motif*) from the *pattern* of the design—*how the copies of the motif are arranged.*

The problem that we will work on for the rest of this chapter is to classify the fundamentally different ways that a flat design can be symmetric. The ideas that we discuss were used by chemists and crystallographers to discover how many different crystalline forms are possible. Although there is a limitless number of different chemicals, and of motifs that people can make, what is quite surprising is that there is only a limited number of ways that atoms of a chemical or motifs of a design can be arranged in a symmetrical way.

How can we possibly enumerate the ways that designs can be put together without counting all the actual designs themselves? The key mathematical idea is to look not at the motifs that make up the patterns, but what you can *do* to the pattern without changing its appearance. This is what we pursue in the next section.

RIGID MOTIONS

Mathematicians describe a variety of kinds of symmetry by using the geometric notion of a *rigid motion,* also known as an **isometry** (which means "same size"). A rigid motion is a specific kind of variation on the original pattern: we pick it up and move it, perhaps rotate it, possibly flip it over—but we *don't change its size or shape.* (To connect this concept with the language of Chapter 17, the original figure and its image are not just geometrically similar but congruent—the same size.)

Figure 20.6 shows the results of various motions applied to the rectangle in Figure 20.6a. In Figure 20.6b, each side is shrunk by 50%: not a rigid motion, because the size of the rectangle changes. For Figure 20.6c, we imagine that the rectangle has rigid sides but hinges at the corner; like an unbraced bookshelf, it sags: again, this is not a rigid motion because the shape of the rectangle changes. In Figure 20.6d we rotate the rectangle 90° (a quarter turn) clockwise around the center of the rectangle: this is a rigid motion. Similarly, in Figure 20.6e, rotating by 180° (a half turn) is a rigid motion.

In Figure 20.6f we reflected the rectangle along a vertical mirror down the middle: could you tell? The right and left halves have exchanged places.

Figure 20.6g shows the result of reflecting across a diagonal of the rectangle. All reflections and all rotations are rigid motions. So are all **translations,** which move every point in the plane a certain distance in the same direction.

The only remaining kind of rigid motion in the plane is a hybrid of reflection and translation. Known as a **glide reflection,** it is the kind of pattern your footprints make as you walk along: each successive element of the design (foot-

FIGURE 20.6   Results of various motions applied to a rectangle: (a) the original rectangle; (b) 50% reduction (not a rigid motion); (c) sagging (not a rigid motion); (d) quarter turn; (e) half turn; (f) reflection along the vertical line down the middle; (g) reflection along a diagonal line.

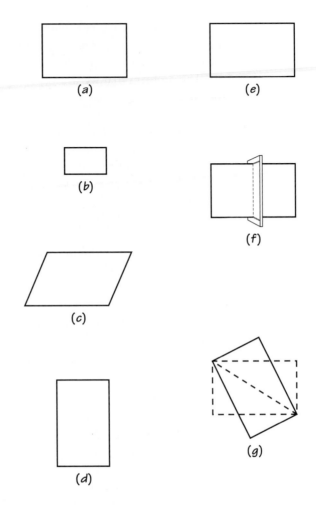

print) is a reflection of the previous one (see Figure 20.7). The motion combines in an integral way a translation ("glide") with a reflection across a line that is parallel to the direction of the translation.

A **rigid motion** is one that preserves the size and shape of figures; in particular, any pair of points is the same distance apart after the motion as before. Any rigid motion of the plane must be one of:

- Reflection (across a line)
- Rotation (around a point)
- Translation (in a particular direction)
- Glide reflection (across a line)

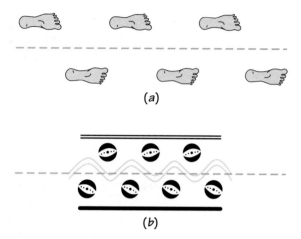

FIGURE 20.7 Glide reflection of (a) footprints; (b) design elements on a pot from San Ildefonso Pueblo.

Performing one rigid motion after another results in a rigid motion that (surprisingly) must be one of the four types that we have just explored.

PRESERVING THE PATTERN

In terms of symmetry, we are especially interested in rigid motions like those of Figures 20.6e and 20.6f that **preserve the pattern**—that is, ones for which the pattern looks exactly the same, *with all the parts appearing in the same places,* after the motion is applied.

You might enjoy thinking of applying these motions as a game, "The Pattern Game": you turn your back, I apply a transformation, then you turn back and see if you can tell if anything is changed.

The 90° rotation of Figure 20.6a into Figure 20.6d does not preserve the pattern. The moved rectangle doesn't fit exactly over the original rectangle.

On the other hand, the 180° rotation in Figure 20.6e does preserve the pattern. It's true that the top of the original rectangle is now on the bottom of the transformed version, but you can't tell that that has happened, because you can't distinguish the two. If you had turned your back while the motion was applied, you wouldn't be able to tell that anything had been done. A rotation by any multiple of 180° would also preserve the pattern.

Similarly, the reflection across the vertical line in Figure 20.6f preserves the pattern, while the one in Figure 20.6g, where the reflection line is along a diagonal, does not. (For an illustration of rotation and left-right reflection in calligraphy, see Spotlight 20.3.)

The pattern of footsteps in Figure 20.7a is not preserved under just reflection along the direction of walking—there is not a left footprint directly across from a right footprint. The pattern is preserved under a glide reflection along the direction of walking, as well as by a translation of two steps, or one of four steps, and so on—but not by a translation of one step.

ANALYZING PATTERNS

Given a pattern, we analyze it by determining which rigid motions preserve the pattern. These are often referred to as the **symmetries of the pattern.** We then can classify the pattern by which rigid motions preserve it.

We may think of a pattern as a recipe for repeating a figure (motif) indefinitely. Of course, any pattern we see in nature or art has only finitely many

Spotlight 20.3

Symmetry in Modern Design: Scott Kim, an Artist in Symmetric Forms

Scott Kim—whom author Isaac Asimov called "the Escher-of-the-Alphabet"—created in the 1980s a new art form with words. His calligraphy is playful, surprising, elegant, and fun. He calls the results *inversions*: words that can be read right side up, upside down, and every which way. His inversions use symmetries, distorting letters a little here or there, reflecting them as in mirrors, or rotating them around central points. The two illustrations shown here, taken from his book *Inversions* (W. H. Freeman, 1989), illustrate rotation (*MAN*) and left-right reflection (*WOMAN*). He comments:

Deceptively simple constructions involving the letter M. Adding a single crossbar to a symmetric zigzag is enough to distinguish three asymmetrically placed letters: *M, A,* and *N*. A lowercase *a* is used in *WOMAN*. Notice that the two words, although closely related, have different symmetries: if you look at this design in a mirror, *WOMAN* will look the same but *MAN* will not. If you turn this design 180°, *MAN* will look the same but *WOMAN* will not.

FIGURE 20.8 (*left*) Flower with petals with reflection symmetry. (*right*) Pinwheel.

copies of the figure; but if the recipe for repetition is clear, we may imagine that we are looking at just a part of a pattern that extends indefinitely.

Patterns in the plane can be divided into those that have indefinitely many repetitions in

- no direction — the **rosette patterns**
- exactly one direction (and its reverse) — the **strip patterns**
- more than one direction — the **wallpaper patterns**

A rosette pattern describes the possible symmetries for a flower. There is just one flower in the pattern; the repetition aspect of symmetry consists of the repetition of the petals around the stem. Translations and glide reflections do not come into play. The pattern is preserved under a rotation by certain angles, corresponding to the number of petals. There may or may not be reflections that preserve it, depending on whether the petal itself has reflection symmetry. Most flowers do (Figure 20.8, left), but some do not. An everyday example of the rosette pattern — a human-made one — that does not have reflection symmetry is a pinwheel (Figure 20.8, right). If there is no reflection symmetry, the motif of the pattern (the element that is repeated) is an entire petal; if there is reflection symmetry, the motif is just half a petal, because the entire pattern can be generated by rotation and reflection of a half petal. The fact that these are the only possibilities is sometimes called *Leonardo's theorem,* after Leonardo da Vinci, who, in the course of planning the design of churches, needed to decide if chapels and niches could be added without destroying the symmetry of the central design.

Leonardo realized that there were two different classes of rosettes, the ones without reflection symmetry *(cyclic rosettes)* and the ones with reflection symmetry *(dihedral rosettes)* (see Figure 20.8). The respective notations for the patterns are *cn* and *dn*, where *n* is the number of times that the rosette coincides with its original position in one complete turn around the center. It coincides with itself for every rotation of 360°/*n*. A cyclic pattern has no lines of reflection symmetry, while the dihedral pattern *dn* has *n* different lines of reflection symmetry. The flower in Figure 20.8 has a dihedral pattern, because each petal has reflection symmetry. The pinwheel in Figure 20.8 has pattern *c8*.

STRIP PATTERNS

We illustrate the different kinds of strip patterns, and their "ingredient" symmetries, with patterns in the art of the Bakuba people of Zaire, who are noted for their fascination with pattern and symmetry (see Spotlight 20.4).

All of the strip patterns offer repetition and **translation symmetry** along the direction of the strip. For simplicity, we will always position the pattern so that its repetition runs horizontally.

It may be that the pattern has no other rigid motions that preserve it apart from translation, as in Figure 20.9a.

The simplest other rigid motion to check for preservation of the pattern is reflection across a line. For a strip pattern, the center line of the strip may be a reflection line; if so, as in Figure 20.9b, we say that the pattern has symmetry across a horizontal line. There may instead be reflection across a *vertical* axis, such as the vertical lines through or between the V's in Figure 20.9c.

What kind of rotational symmetry can a strip pattern have? The only possibility for a strip pattern is a rotation by 180° (a half-turn), since any other angle won't even bring the strip back into itself. (We don't count rotations of 360° or integer multiples [full turns], since any pattern is preserved under these.) Figure 20.9d shows a strip pattern that is unchanged by a 180° rotation about any point at the center of the small crosshatched regions.

What about glide reflections? A row of alternating p's and b's has glide reflection:

Glide	p	p	p	p	p	p	p	p	p
Reflection	p	p	p	p	p	p	p	p	p
	b	b	b	b	b	b	b	b	b
Glide reflection	p	b	p	b	p	b	p	b	p

FIGURE 20.9 Bakuba patterns. (a) Carved stool; (b) pile cloth; (c) pile cloth; (d) embroidered cloth; (e) embroidered cloth; (f) carved back of wooden mask; (g) carved box.

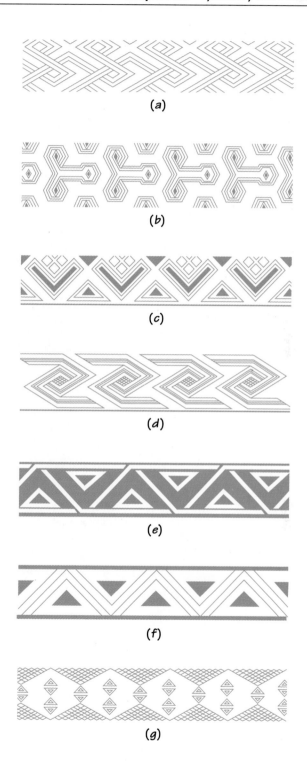

(a)

(b)

(c)

(d)

(e)

(f)

(g)

Spotlight 20.4

Patterns Created by the Bakuba People

Among the Bakuba people of Zaire (shaded area of map), it is considered an achievement to invent a new pattern, and every Bakuba king had to create a new pattern at the outset of his reign. The pattern was displayed on the king's drum throughout his reign and, in the case of some kings, on his dynastic statue.

When missionaries first showed a motorcycle to a Bakuba king in the 1920s, he showed little interest in it. But the king was so enthralled by the novel pattern the tire tracks made in the sand that he had it copied and gave it his name.

Source: Adapted from Jan Vansina, *The Children of Woot,* University of Wisconsin Press, Madison, 1978, p. 221.

Two women with raffia cloths from the Bakuba village of Mbelo, July 1985; Mpidi Muya with embroidered raffia cloth (left) and Muema Kenye with plush and embroidered raffia cloth (right).

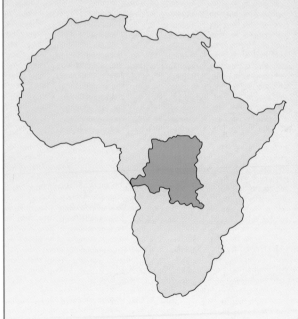

The pattern made by tire tracks fascinated the Bakuba people.

For glide reflection, a **p** is translated as far as the next **b** and is then reflected upside down. Figure 20.9e shows a Bakuba pattern whose only symmetry (except for translation) is glide reflection.

Having examined symmetries on strip patterns, we can ask: What *combinations* of the four are possible? It turns out that apart from the five kinds of patterns we have already seen, there are only two other possibilities: we can have vertical line reflection, half-turns, and glide reflection, either with (Figure 20.9f), or without horizontal line reflection (Figure 20.9g).

Mathematical analysis reveals:

There are only seven ways to repeat a pattern along a strip.

That this number is so small is quite surprising, since there are myriad different design elements (motifs). The key idea is that two designs may look entirely different yet share the same pattern of reproducing their design elements.

SYMMETRY GROUPS

We mentioned earlier that the key mathematical idea about detecting and analyzing symmetry is to look not at the motifs of a pattern but at its symmetries, the transformations that preserve the pattern.

The symmetries of a pattern have some notable properties:

- If we combine two symmetries by applying first one and then the other, we get another symmetry.
- There is an identity, or "null," symmetry that doesn't move anything, but leaves every point of the pattern exactly where it is.
- Each symmetry has an inverse or "opposite" that undoes it and also preserves the pattern. A rotation is undone by an equal rotation in the opposite direction, a reflection is its own inverse, and a translation or glide reflection is undone by another of the same distance in the opposite direction.
- In applying a number of symmetries one after the other, we may combine consecutive ones without affecting the result.

These properties are common to many kinds of mathematical objects; they characterize what mathematicians call a *group*. Various collections of numbers and numerical operations that are already familiar to you are groups.

EXAMPLE ▶ *A Group of Numbers*

The positive real numbers form a group under multiplication:

- Multiplying two positive real numbers yields another positive real number.
- The positive real number 1 is an identity element.
- Any positive real number x has an inverse $1/x$ in the collection.
- In multiplying several numbers together, it doesn't matter if we first multiply together some adjacent pairs of numbers, that is, it doesn't matter how we group or parenthesize the multiplication. For instance, $2 \times 3 \times 4 \times 5$ is equal to $2 \times (3 \times 4) \times 5 = 2 \times 12 \times 5$ and also to $(2 \times 3) \times 4 \times 5 = 6 \times 4 \times 5$. ◆

A **group** is a collection of elements $\{A, B, \ldots\}$ and an operation $\circ$ between pairs of them such that the following properties hold:

closure: The result of one element operating on another is itself an element of the collection ($A \circ B$ is in the collection).

identity element: There is a special element I, called the identity element, such that the result of an operation involving the identity and any element is that same element ($I \circ A = A$ and $A \circ I = A$).

inverses: For any element, A, there is another element, called its inverse and denoted A^{-1}, such that the result of an operation involving an element and its inverse is the identity element ($A \circ A^{-1} = I$ and $A^{-1} \circ A = I$).

associativity: The result of several consecutive operations is the same regardless of grouping or parenthesizing, provided the consecutive order of operations is maintained: $A \circ B \circ C = A \circ (B \circ C) = (A \circ B) \circ C$.

The group of symmetries that preserve a pattern is called the **symmetry group of the pattern.**

EXAMPLE ▶ *The Symmetry Group of a Rectangle*

Consider the rectangle of Figure 20.10. Its symmetries, the rigid motions that bring it back to coincide with itself (even as they interchange the labeled corners), are

- the identity symmetry I, which leaves every point where it is
- a 180° (half-turn) rotation R around its center
- a reflection V in the vertical line through its center
- a reflection H in the horizontal line through its center

FIGURE 20.10 A rectangle, with reflection symmetries and 180° rotation symmetry marked.

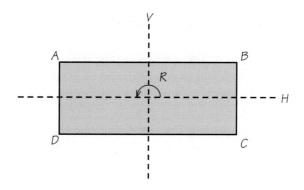

You should convince yourself that these four elements form a group. Combining any pair by applying first one and then the other is equivalent to one of the others; for example, applying first H and then V is the same as applying R, that is, $H \circ V = R$ (check this by following where the corner A goes to under the operations). The element I is an identity element. Each element is its own inverse. Try some examples to verify that associativity holds. For instance, $R \circ H \circ V = (R \circ H) \circ V = R \circ (H \circ V)$; in other words, applying R followed by H followed by V, we get the same result if we combine the first two and then apply the third, or if we combine the second two and apply the first followed by that combination. ◆

EXAMPLE ▶ *Symmetry Groups of Strip Patterns*

Each of the strip patterns of Figure 20.9 is distinguished by a different group of symmetries. The pattern of Figure 20.9a is preserved only by translations. If we let T denote the smallest translation to the right that preserves the pattern, then the pattern is also preserved by $T \circ T$ (which we write as T^2), by $T \circ T \circ T = T^3$, and so forth. Although the pattern looks the same after each of these translations by different distances, we can tell these translations apart if we number each copy of the motif and observe which other motif it is carried into under the symmetry. For instance, T^2 takes each motif into the motif two to the right. The symmetry T has an inverse T^{-1} among the symmetries of the pattern: the smallest translation to the *left* that preserves the pattern; and $T^{-1} \circ T^{-1}$ (which we write as T^{-2}), $T^{-1} \circ T^{-1} \circ T^{-1} = T^{-3}$, and so forth also are symmetries. The entire collection of symmetries of the pattern is

$$\{ \ldots, T^{-3}, T^{-2}, T^{-1}, I, T, T^2, T^3, \ldots \}$$

From this listing, you see that it is natural to think of the identity I as being T^0. All of the strip patterns are preserved by translations, so the symmetry group of each includes the *subgroup* of all translations in this list. We say that the group is generated by T, and we write the group as $<T>$, where between the angle brackets we list symmetries (generators) that, in combination, produce all of the group elements.

The symmetry group of Figure 20.9e includes in addition a glide reflection G and all combinations of the glide reflection with the translations. Doing two glide reflections is equivalent to doing a translation, which we express as $G^2 = T$; the glide is only "half as far" as the shortest translation that preserves the pattern. Check that $G \circ T = T \circ G$. The symmetry group of the pattern is

$$\{. \ . \ . \ , G^{-3}, G^{-2} = T^{-1}, G^{-1}, I, G, G^2 = T, G^3, . \ . \ .\} = <G>$$

The pattern of Figure 20.9c is preserved by vertical reflections at regular intervals. If we let V denote a reflections at a particular location, the other reflections can be obtained as combinations of V and T. The symmetry group of the pattern is

$$\{. \ . \ . \ , T^{-3}, T^{-2}, T^{-1} I, T, T^2, T^3, . \ . \ . \ ;$$
$$. \ . \ . \ , V \circ T^{-3}, V \circ T^{-2}, V \circ T^{-1}, V, V \circ T, V \circ T^2, V \circ T^3, . \ . \ .\}$$

This group is notable because not all of its elements satisfy the *commutative property* that $A \circ B = B \circ A$, which you are used to for numerical operations ($a + b = b + a$, $a \times b = b \times a$). In fact, we do not have $T \circ V = V \circ T$, but instead $T \circ V = V \circ T^{-1}$, a fact that you should verify by labeling one of the V shapes in the pattern and observing where it is carried by each of these three combinations of symmetries. We can express the group as $<T, V \mid TV = VT^{-1}>$, where we list the generators and indicate relations among them. ◆

We have made a transition from thinking about patterns in geometrical terms to reasoning about them in algebraic notation—in effect, applying one branch of mathematics to another. This kind of cross-fertilization is characteristic of contemporary mathematics.

The concept of a group is a fundamental one in the mathematical field of abstract algebra. The generality ("abstractness") is exactly why groups and other algebraic structures arise in so many applications, in areas ranging from crystallography, quantum physics, and cryptography, to error-correcting codes (see Chapters 9 and 10), anthropology (describing kinship systems; see Ascher [1991])—and analyzing symmetries of patterns.

NOTATION FOR PATTERNS

It's useful to have a standard notation for patterns, for purposes of communication. Crystallographers' notation is the one most commonly used. For the strip patterns, it consists of four symbols (an example is *pma2*):

1. The first symbol is always a *p*, which indicates that the pattern repeats (is "periodic") in the horizontal direction.
2. The second symbol is *m* if there is a vertical line of reflection, or a *1* otherwise.
3. The third symbol is

   - *m* (for "mirror") if there is a horizontal line of reflection (in which case there is also glide reflection),
   - *a* (for "alternating") if there is a glide reflection but no horizontal reflection, or
   - *1* if there is no horizontal reflection or glide reflection.

4. The fourth symbol is *2* if there is half-turn rotational symmetry, or *1* otherwise.

A *1* always means that the pattern does not have the symmetry corresponding to that position.

In the notation

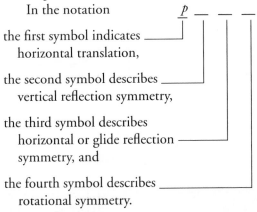

the first symbol indicates horizontal translation,

the second symbol describes vertical reflection symmetry,

the third symbol describes horizontal or glide reflection symmetry, and

the fourth symbol describes rotational symmetry.

Figure 20.11 gives a flowchart for identifying the seven ways patterns repeat, together with the notations for them.

EXAMPLES ▶ *Bakuba Patterns*

We use the flowchart of Figure 20.11 to analyze some of the Bakuba patterns of Figure 20.9.

Figure 20.9a does not have a vertical reflection, so we branch right, and the pattern notation begins to take shape as *p1__*. The figure does not have a horizontal reflection, nor a glide reflection, so we branch right again, filling in the third position in the notation, to get *p11_*. A half-turn preserves part but not all of the pattern, so we conclude that we have a *p111* pattern.

Figure 20.9b does not have vertical reflection, so we branch right, to *p1__*. The figure does have horizontal reflection, so we branch left and left, concluding that the pattern is *p1m1*.

We follow the flowchart in Figure 20.11 and get the following:

- Figure 20.12a: Is there a vertical reflection? *No.* Is there a horizontal reflection or glide reflection? *No.* Is there a half-turn? *No.* Hence the pattern is *p111*.
- Figure 20.12b (narrow interpretation of the diagonal lines): Is there a vertical reflection? *No.* Is there a horizontal reflection or glide reflection? *No.* Is there a half-turn? *Yes* (e.g., around the center of each cross). The pattern is *p112*.
- Figure 20.12b (liberal interpretation—diagonal lines as shading, their direction doesn't have to be preserved): Is there a vertical reflection? *Yes* (e.g., on a vertical line through the center of a cross). Is there a horizontal reflection? *Yes* (e.g., through the center of a cross). The pattern is *pmm2*.
- Figure 20.12c: Is there a vertical reflection? *No.* Is there a horizontal reflection or glide reflection? *No.* Is there a half-turn? *Yes* (e.g., around the center of each jagged white line). The pattern is *p112*. (This pitcher has the interesting feature that the patterns on the neck and the body are mirror images of each other.)

Women made the pots at Starkweather; they strongly preferred the symmetry of half-turns; very few of the pots have any reflection symmetry, either reflection or glide. The avoidance of reflection symmetry was a consistent feature of pottery of the indigenous peoples of the Western Hemisphere. Spotlight 20.5 discusses the significance of pattern classification for anthropologists.

FURTHER POSSIBILITIES

So far we have classified the patterns with no translation repetition (the rosette patterns) and those with repetition in one direction (the strip patterns). What about those that have repetition in more than one direction—the wallpaper patterns? It turns out that there are exactly 17 of those. Illustrations, notation, and a flowchart are given in Spotlight 20.6.

The method we have developed here for classifying patterns, by the combinations of symmetry elements present, was originally developed by crystallographers in the nineteenth century. They wanted to classify and recognize the three-dimensional patterns associated with crystal structure. They proved—after several years of different crystallographers coming up with different totals!—that there are exactly 230 crystal patterns.

We emphasize again that our analysis of patterns does not refer to the design in the pattern, but to how its repetition is structured across the plane.

Spotlight

Symmetry in Ancient Design

20.5

Commenting on the significance of design as an indicator of a culture's history and how design can signal change in a culture, archeologist Dorothy Washburn states:

Human beings do things in a very consistent fashion. Over the years, within a given cultural group, they repeat behavior patterns, which we can observe in their material culture. What we did not see before the mathematics of symmetry was that the *structure* of a culture's decorative designs is consistent over time and through space within a given cultural group.

Although some people produce random patterns, by far the largest number of decorative designs in a cultural group are based on symmetry. And any time you repeat a motif in a systematic fashion, you're using one of the four rigid motions.

Early descriptions of design were largely idiosyncratic and dealt just with the individual types of material—types of textiles or types of pottery—and the features that typified these types. But now we can study and compare how motifs in design are *arranged,* in material from all contemporary or past cultures throughout the world, and see how

these designs are put together and how they change through time and space.

One of the most interesting things we've found is that a given cultural group, a tribe or a band unit, will choose just a few symmetries to structure its designs. I've tested this observation through studies of California Indian baskets, with the work of Bakuba cloth weavers, and I've even found a consistency in the archeological record among the Anasazi, one of the prehistoric traditions of the American Southwest.

Let's take an example from material found on Crete. The site of Knossos had 3000 years of uninterrupted prehistory. For 1500 of those years, only two of the seven one-dimensional symmetries were used. Then suddenly five more symmetries came into use. The design motifs were the same, but they were rearranged into different patterns.

That rearrangement suggested to us that something really interesting was happening, which turned out to be the beginning of trade in the Aegean. We could see that simply by noting the introduction of new symmetry patterns from cultures outside that island. The increase in the number and variety of symmetries indicated that trade was coming in; the change in design structure was an incredibly sensitive marker of change.

There is an infinite variety of possible designs that artists can devise. You should imagine that the artist has created one copy of the design and is then contemplating how to place equal-sized copies of it in other parts of the (infinite) plane, in a way that is symmetric. It is those strategies for placement of which there are very few.

A slightly more involved analysis allows mathematicians to refine the classification of patterns to take into account colors that are repeated in a symmetric way.

Spotlight

The 17 Wallpaper Patterns

20.6

There are exactly 17 wallpaper patterns. We give an example of each, together with a flowchart for identifying the patterns.

The International Crystallographic Union has established a standard notation for the wallpaper patterns. The full notation consists of four symbols:

1.   The first symbol is c (for "centered") if all rotation centers lie on reflection lines, or p (for "primitive") otherwise.

2.   The second symbol indicates rotational symmetry. It is either $1, 2, 3, 4,$ or 6, corresponding to rotational symmetry of, respectively, 360°, 180°, 120°, 90°, or 60°. The symbol is the largest applica-

ble number. For example, if symmetries of 360°, 120°, and 60° are present, the symbol is 6.

3.   The third symbol is either $m, g,$ or 1, corresponding to the presence of "mirror," glide, or no reflection symmetry.

4.   The fourth symbol ($m, g,$ or 1) is for describing symmetry relative to an axis at an angle to the symmetry axis of the third symbol.

(*Note:* The patterns $p31m$ and $p3m1$ are an exception to this scheme.)

Below each pattern illustration we give both the standard abbreviation (on top) and the full notation (below).

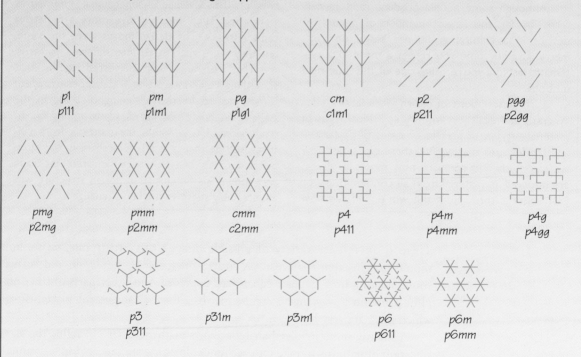

The 17 wallpaper patterns, with abbreviations and the full notation used by crystallographers.

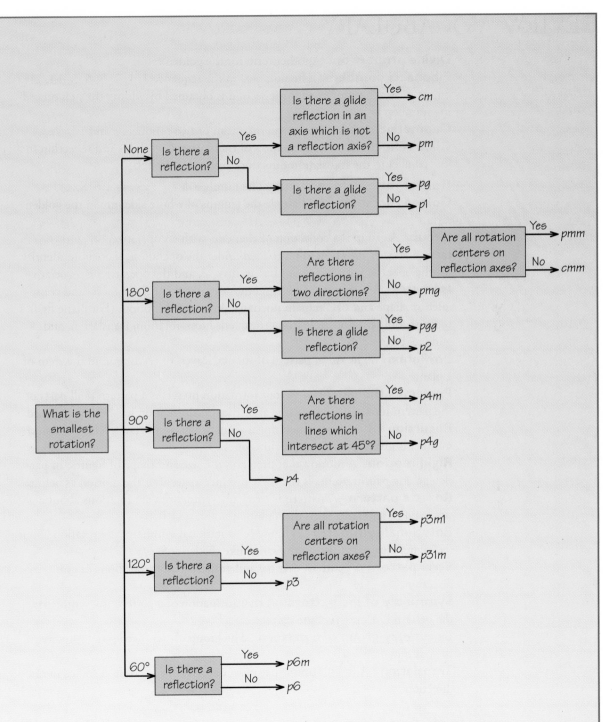

Flowchart for the 17 wallpaper patterns, with abbreviated International Crystallographic Union notation.

REVIEW VOCABULARY

Divine proportion   Another term for the golden ratio.

Fibonacci numbers   The numbers in the sequence 1, 1, 2, 3, 5, 8, 13, 21, 34, . . . (each number after the second is obtained by adding the two preceding numbers).

Geometric mean   The geometric mean of two numbers a and b is $\sqrt{ab}$.

Glide reflection   A combination of translation (= glide) and reflection in a line parallel to the translation direction. Example: **pbpbpb**.

Golden ratio, golden mean   The number $\phi = \frac{1 + \sqrt{5}}{2} = 1.618. . . .$

Golden rectangle   A rectangle the lengths of whose sides are in the golden ratio.

Group   A group is a collection of elements with an operation on pairs of them such that the collection is closed under the operation, there is an identity for the operation, each element has an inverse, and the operation is associative.

Isometry   Another word for rigid motion. Angles and distances, and consequently shape and size, remain unchanged by a rigid motion. (For plane figures there are only four possible isometries: reflection, rotation, translation, and glide reflection.)

Phyllotaxis   The spiral pattern of shoots, leaves, or seeds around the stem of a plant.

Preserves the pattern   A transformation preserves a pattern if all parts of the pattern look exactly the same after the transformation has been performed.

Recursion   A method of defining a sequence of numbers, in which the next number is given in terms of previous ones.

Rigid motion   A motion that preserves the size and shape of figures; in particular, any pair of points is the same distance apart after the motion as before.

Rosette pattern   A pattern whose only symmetries are rotations about a single point and reflections through that point.

Rotational symmetry   A figure has rotational symmetry if a rotation about its "center" leaves it looking the same, like the letter S.

Strip pattern   A pattern that has indefinitely many repetitions in one direction.

Symmetry of the pattern   A transformation of a pattern is a symmetry of the pattern if it preserves the pattern.

Symmetry group of a pattern   The group of symmetries that preserve the pattern.

Translation   A rigid motion that moves everything a certain distance in one direction.

Translation symmetry   An infinite figure has translation symmetry if it can be translated (slid, without turning) along itself without appearing to have changed. Example: AAA

Wallpaper pattern   A pattern in the plane that has indefinitely many repetitions in more than one direction.

SUGGESTED READINGS

ASCHER, MARCIA. *Ethnomathematics: A Multicultural View of Mathematical Ideas,* Brooks/Cole, 1991. Chapter 3, The logic of kin relations (pp. 66–83) shows that kinship systems have the structure of dihedral groups.

BOLES, MARTHA, AND ROCHELLE NEWMAN. *The Golden Relationship: Art, Math & Nature, Book 1: Universal Patterns; Book 2: The Surface Plane,* Pythagorean Press, Bradford, Mass., 1992.

CRISLER, NANCY. *Symmetry & Patterns,* COMAP, Inc., Lexington, Mass., 1995.

CROWE, DONALD W. *Symmetry, Rigid Motions and Patterns,* HiMAP Module 4, COMAP, Inc., Lexington, Mass., 1987. Reprinted in smaller format in *The UMAP Journal,* 8 (1987): 207–236. Instructional module on rigid motions of the plane, strip patterns, and wallpaper patterns, with worksheets.

GALLIAN, JOSEPH A. Symmetry in logos and hubcaps. *American Mathematical Monthly,* 97 (3) (March 1990): 235–238.

GALLIAN, JOSEPH A. Finite plane symmetry groups, *Journal of Chemical Education,* 67 (7) (July 1990): 549–550. Hubcap examples.

HARGITTAI, ISTVÁN, AND MAGDOLNA HARGITTAI. *Symmetry: A Unifying Concept,* Shelter Publications, Bolinas, Calif., 1994.

HERZ-FISCHLER, ROGER. *A Mathematical History of Division in Extreme and Mean Ratio,* Wilfrid Laurier University Press, Waterloo, Ont., Canada, 1987.

HOGGATT, VERNER E., JR. *Fibonacci and Lucas Numbers,* Houghton Mifflin, New York, 1969.

HUNTLEY, H. E. *The Divine Proportion,* Dover, New York, 1970.

MARKOWSKY, GEORGE. Misconceptions about the golden ratio, *College Mathematics Journal,* 23 (1) (January 1992): 2–19.

MARTIN, GEORGE E.   *Transformation Geometry: An Introduction to Symmetry,* Springer-Verlag, New York, 1982.

O'DAFFER, PHARES G., AND STANLEY R. CLEMENS.   *Geometry: An Investigative Approach,* Addison-Wesley, Reading, Mass., 1976, chapters 1–5. A gentle introduction to the geometry of symmetry, with lots of examples and illustrations. Chapter 4 gives an elementary proof that there are only four kinds of rigid motions in the plane.

RUNION, GARTH E.   *The Golden Section and Related Curiosa,* Scott, Foresman, Glenview, Ill., 1972.

SIBLEY, THOMAS Q.   *Geometric Patterns: A Study in Symmetry,* Saint John's University, Collegeville, Minn., 1989.

STEWART, IAN.   Mathematical recreations: Daisy, Daisy, give me your answer, do, *Scientific American,* 272 (1) (January 1995): 96–99. Explains the occurrence of Fibonacci numbers in plants based on the dynamics of plant growth and the efficient packing of seeds into spiral faces.

WASHBURN, DOROTHY K., AND DONALD W. CROWE.   *Symmetries of Culture: Theory and Practice of Plane Pattern Analysis,* University of Washington Press, Seattle, 1988. An introduction to the mathematics of symmetry, splendidly illustrated with photographs of patterns from cultures all over the world. Includes a complete analysis of patterns with two colors. Appendixes contain proofs of the facts that there are only four rigid motions in the plane and that there are exactly seven strip patterns.

EXERCISES

▲ *Optional.*   ■ *Advanced.*   ◆ *Discussion.*

Fibonacci Numbers

1.   Examine the "scales" on the surface of a pineapple, which are arranged in spirals (parastichies) around the fruit. Note that there are spirals in three distinct directions. For each direction, how many spirals are there?

2.   Repeat Exercise 1, but for a pinecone from your area.

3.   Repeat Exercise 1, but for a sunflower.

4.   Here are two primitive models of natural increase of biological populations, similar to those Fibonacci hypothesized around the year 1200. A pair of newborn male and female rabbits is placed in an enclosure to breed.

(a) Suppose that the rabbits start to bear young one month after their own birth. This may be unrealistic for rabbits, but we could substitute another species for which it is realistic; Fibonacci used rabbits. At the end of each month, they have another male–female pair, which in turn matures and starts to bear young one month later. Assuming that none of the rabbits dies, how many pairs of rabbits will there be at the end of six months from the start (just before any births for that month)? (*Hint:* Draw a month-by-month chart of the situation at the end of the month, just before any births.)

(b) Repeat part (a), but assume instead that the rabbits start to bear young exactly two months after their own birth.

The Golden Ratio

5. Put the golden ratio $\phi = (1 + \sqrt{5})/2$ into the memory of your calculator.

(a) Look at the value of ϕ. Now square it (either use the x^2 button or multiply it by itself). What do you observe?

(b) Back to ϕ. Now take its reciprocal (either use the $1/x$ button or divide it into 1). What do you observe?

(c) What formula explains what you saw in part (a)?

(d) What formula explains what you saw in part (b)?

6. The golden ratio satisfies the equation $x^2 = x + 1$.

(a) Show that $(1 - \phi)$ also satisfies the equation.

(b) Use part (a) to show that $(1 - \phi) = (1 - \sqrt{5})/2$ is the other solution to $x^2 - x - 1 = 0$.

7. The geometric mean has both arithmetic and geometric interpretations.

(a) Find the geometric mean of 3 and 27.

(b) Find the length of a side of a square that has the same area as a rectangle that is 4 by 64.

8. Here's further practice on arithmetic and geometric interpretations of the geometric mean:

(a) Find the geometric mean of 4 and 9.

(b) You are to make a golden rectangle with 6 inches of string. How wide should it be, and how high?

9.   Another sequence closely related to the Fibonacci sequence is the Lucas sequence, which is formed using the same recursive rule but different starting numbers. The nth Lucas number L_n is given by

$$L_1 = 1, L_2 = 3, \text{ and } L_{n+1} = L_n + L_{n-1} \text{ for } n \geq 2$$

(a)   Calculate L_3 through L_{10}.
(b)   Calculate the ratio of successive terms of the Lucas sequence:

$$\frac{L_2}{L_1}, \frac{L_3}{L_2} \ldots, \frac{L_{10}}{L_9}$$

What do you notice?

10.   For a sequence specified by a recursive rule, finding an explicit expression for the nth term is not easy, nor is the form necessarily simple. An exact expression for the nth term of the Fibonacci sequence is given by the Binet formula:

$$F_n = \frac{1}{\sqrt{5}} \left(\frac{1 + \sqrt{5}}{2} \right)^n - \frac{1}{\sqrt{5}} \left(\frac{1 - \sqrt{5}}{2} \right)^n$$

(a)   Verify the formula for $n = 1$ and $n = 2$ (by multiplying out, not by using a calculator).
(b)   Use the Binet formula and your calculator to find F_5.
(c)   In fact, the second term on the right of the equation gets closer and closer to 0 as n gets large. Since we know that the Fibonacci numbers are integers, we can just round off the result of calculating the first term. Find F_{13} by calculating the first term with your calculator and rounding.

Preserving the Pattern

11.   Determine whether each of the following statements is always true or sometimes false. (Drawing some sketches may be helpful.)

(a)   A line reflection preserves collinearity of points. That is, if the points A, B, and C are in a straight line (collinear), then their images reflected in some other line also lie in a straight line.
(b)   A line reflection preserves betweenness. That is, if the collinear points A, B, and C (with B between A and C) are reflected about a line, then the image of B is between the images of A and C.

(c) The image of a line segment under a line reflection is a line segment of the same length.

(d) The image of an angle under a line reflection is an angle of the same measure.

(e) The image of a pair of parallel lines under a line reflection is a pair of parallel lines.

12. Determine whether each of the following statements is always true or sometimes false. (Drawing some sketches may be helpful.)

(a) The image of a pair of perpendicular lines under a line reflection is a pair of perpendicular lines.

(b) The image of a square under a line reflection is a square.

(c) Label the vertices of a square *A*, *B*, *C*, and *D* in a clockwise direction. Then their images *A′*, *B′*, *C′*, and *D′* under a line reflection also follow a clockwise direction.

(d) The perimeter of a geometric figure is equal to the perimeter of its image under a line reflection.

(e) The image of a vertical line under a line reflection is always a vertical line.

13. Which of the 26 capital letters of the alphabet have

(a) a horizontal line of reflection symmetry?

(b) a vertical line of reflection symmetry?

(c) rotational symmetry? (Assume that each letter is drawn in the most symmetric way. For example, the upper and lower loops of "B" should be the same size.)

14. Repeat Exercise 13, but for the lowercase letters.

15. In *The Complete Walker III* (3rd ed., Knopf, 1984, p. 505), Colin Fletcher's answer to "What games should I take on a backpacking trip?" is the game he calls "Colinvert": "You strive to find words with meaningful mirror (or half-turn) images." Some of the words he found are

MOM   WOW   pod   MUd   bUM

(a) Which of his words reflect into themselves?

(b) Which of his words rotate into themselves?

(c) Find some more words or phrases of these various types—the longer, the better.

16. Repeat Exercise 15, but for words written vertically instead of horizontally.

Analyzing Patterns

17.   Give the notation (e.g., *d4* or *c5*) for the symmetry patterns of the rosettes in hubcaps (a) through (c), disregarding the logos in the centers. (Can you identify the make of car and year for each hubcap?)

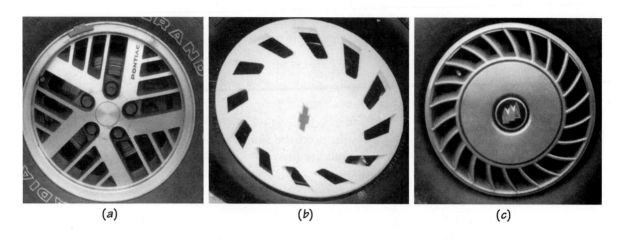

(a)   (b)   (c)

18.   Repeat Exercise 17, for hubcaps (d) through (f).

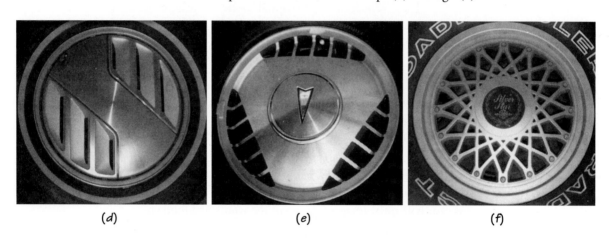

(d)   (e)   (f)

19.   Repeat Exercise 17, for corporate logos (a) through (c). (Can you identify the corporations?)

(a)   (b)   (c)

20. Repeat Exercise 17, for corporate logos (d) through (f).

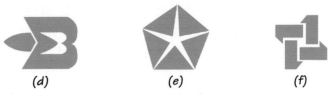

(d) *(e)* *(f)*

21. For each of the shapes in parts (a) through (e) of the accompanying figure, determine all lines of symmetry.

(a) *(b)* *(c)* *(d)* *(e)*

22. Repeat Exercise 21, but for the shapes in parts (f) through (j).

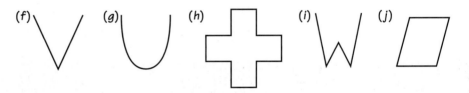

(f) *(g)* *(h)* *(i)* *(j)*

Strip Patterns

23. For each of the following strip patterns, identify the rigid motions that preserve the pattern:

    (a) AAAAAAAAAA;   (c) 0000000000;

    (b) BBBBBBBBBB;   (d) FFFFFFFFFF.

24. Repeat Exercise 23, but for

    (a) NNNNNNNNNN;   (c) dbpqdbpqdbpq.

    (b) bdbdbdbdbd;

Symmetry Groups

25. What is the group of symmetries of

    (a) an equilateral triangle (all three sides equal)?

    (b) an isosceles triangle (two equal sides) that is not equilateral?

    (c) a scalene triangle (no pair of sides equal)?

26. What is the group of symmetries of a square?

27.   Explain, by referring to the properties of a group, whether the collection of all real numbers is a group under the operation of (a) addition; (b) multiplication.

28.   (a)   Give a numerical example to show that the operation of subtraction on the integers is not associative.
        (b)   Repeat part (a), but for division on the positive real numbers.

29.   What are the elements of the group of symmetries of (a) Figure 20.9b? (b) Figure 20.9f?

30.   What are the elements of the group of symmetries of (a) Figure 20.9d? (b) Figure 20.9g?

31.   What are the elements of the group of symmetries of the dihedral pattern *d8* (see the flower in Figure 20.8)?

32.   What is the group of symmetries of the cyclic pattern *c8*?

Notation for Patterns

33.   Use the flowchart in Figure 20.11 to identify (by International Crystallographic Union notation) the types of the strip patterns from Hungarian needlework, shown in the accompanying illustration.

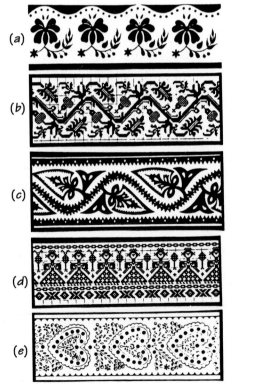

(a)

(b)

(c)

(d)

(e)

(f)

(g)

Hungarian needlework designs. (a) Edge decoration of table cover from Kalocsa, southern Hungary. (b) Pillow end decoration from Tolna County, southwest Hungary. (c) Decoration patched onto a long embroidered felt coat of Hungarian shepherds in Bihar County, eastern Hungary. (d) Embroidered edge decoration of bed sheet from the eighteenth century. (Note the deviations from symmetry in the lower stripes of the pattern.) (e) Shirt from Karád, southwest Hungary. (f) Pillow decoration pattern from Torockó (Rimetea), Transylvania, Romania. (g) Grape leaf pattern from the territory east of the river Tisza.

34. [Contributed by Margaret A. Owens, California State University, Chico.] In each of the four accompanying examples, two adjacent triangles of an infinite strip are shown.

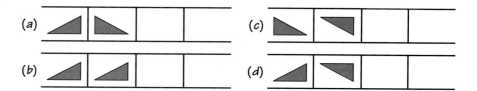

For each example

    (a)   Determine a motion (translation, reflection, rotation, or glide reflection) that takes the first (= left) triangle to the second (= right) one.

    (b)   Draw the next four triangles of the infinite strip that would result if the second triangle is moved to the next space by another motion of the same kind, and so on.

    (c)   Identify (by notation) the resulting strip as one of the seven possible strip patterns.

Imperfect Patterns

35. Repeat Exercise 33, for the accompanying eight strip patterns, all of which appear on the brass straps for a single lamp from nineteenth-century Benin in West Africa. [From H. Ling Roth, *In Great Benin.*] Note that the patterns are roughly carved, so you will need to discern the intent of the artist.

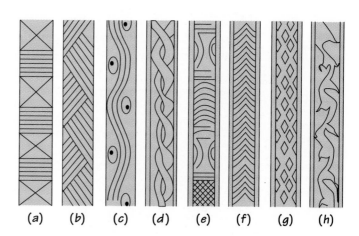

36.   Repeat Exercise 33, for the accompanying patterns from San Ildefonso Pueblo, New Mexico.

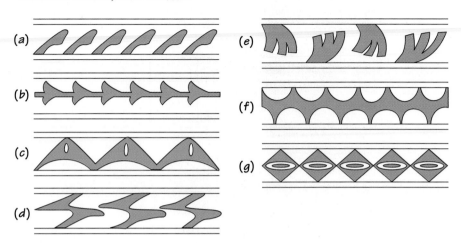

Further Possibilities

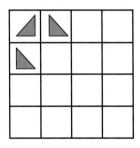

▲ 37.   The triangles in the grid at the top of the figure at the left show beginning steps in forming instances of several of the wallpaper patterns, by putting together a vertical motion and a horizontal motion.

(a)   Identify the horizontal motion.
(b)   Identify the vertical motion.
(c)   Fill in the remaining empty squares.
(d)   Identify the wallpaper pattern.

▲ 38.   For each of the Yoruba cloths shown in the illustrations below and on page 837, use the flowchart in Spotlight 20.6 to identify (by notation) the type of wallpaper pattern.

Patterns on Yoruba (West Africa) *adire* cloth, made by starching a pattern onto white cloth, then dyeing the cloth blue before rinsing out the starch, so that the starched portion remains as a white design against a blue background.

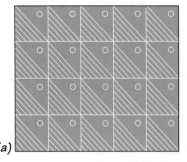

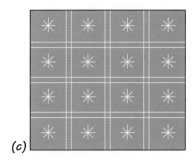

(c)

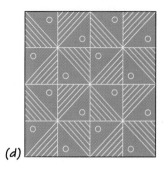

(d)

Additional Exercises

39.   For positive integers a and n, the expression a and n means remainder when a is divided by n. Thus, 23 mod 4 = 3 since $15 = 5 \cdot 4 + 3$, and we say that "23 is equivalent to 3 modulo 4" (see Chapter 10 for further details about this *modular arithmetic*). Every positive integer is equivalent to either 0, 1, 2, or 3 modulo 4. Consider the collection of elements {0, 1, 2, 3} and the operation $\oplus$ on them defined by $a \oplus b = (a + b)$ mod 4. Show that under this operation, the collection forms a group.

40.   The following table shows comparative data about the frequency of occurrence of strip designs of various types on pottery (Mesa Verde, United States) and smoking pipes (Begho, Ghana, Africa) from two continents.

Frequency of Strip Designs on Mesa Verde Pottery and Begho Smoking Pipes

	Mesa Verde		Begho	
Strip Type	Number of Examples	Percentage of Total	Number of Examples	Percentage of Total
p111	7	4	4	2
p1m1	5	3	9	4
pm11	12	7	22	10
p112	93	53	19	8
p1a1	11	6	2	1
pma2	27	16	9	4
pmm2	19	11	165	72
Totals	174		230	

(a)   Which types of motions appear to be preferred for designs from each of the two localities?

(b)   What other conclusions do you draw from the data of this table?

(c)   On the evidence of the table alone, in which locality is each of the strip patterns in the figure on the next page most likely to have been found?

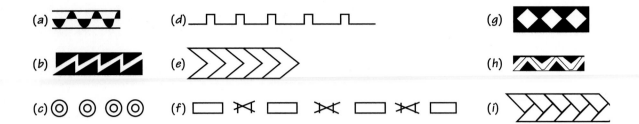

(a)   (d)   (g)

(b)   (e)   (h)

(c)   (f)   (i)

WRITING PROJECTS

1 ▶   The Fibonacci Association is devoted to fostering interest in Fibonacci and related numbers. In November 1988, the society's journal, *The Fibonacci Quarterly,* published "Suppose More Rabbits Are Born" (pp. 306–311), by Shari Lynn Levine (a high school student when she wrote it). The article begins: "How would Fibonacci's age-old sequence be redefined if, instead of bearing one pair of baby rabbits per month, the mature rabbits bear two pairs of baby rabbits per month?" The article goes on to discuss properties of the resulting "Beta-nacci" sequence and the sequences that result from even greater rabbit fertility. Here we ask you to rediscover some of Shari's results about the Beta-nacci sequence:

   (a)   How many rabbits will there be each month for the first 12 months?

   (b)   What is the recursive rule for the nth Beta-nacci number B_n?

   (c)   For the terms of the sequence in part (a), calculate the ratios B_{n+1}/B_n of successive terms. (Motivating hint: It's not the golden ratio this time.)

   (d)   Suppose that the ratio of successive terms approaches a number x. We show how to find x exactly. For very large n, we have $B_{n+1} \approx xB_n \approx x^2 B_{n-1}$. Substituting these values into the recursive rule for the sequence and dividing by B_{n-1} gives us the equation $x^2 = x + 2$. Solve this equation for x (you can use the quadratic formula). Make a table of values of $3B_n$ versus 2^n. From the evidence, can you suggest a formula for B_n?

2 ▶   Generalize Writing Project 1, parts (a) through (d)

   (a)   to the case of each pair of rabbits having three pairs of rabbits (the "Gamma-nacci" sequence).

   (b)   to the case of each pair of rabbits having q pairs of rabbits.

3 ▶   Which wallpaper patterns can be formed by the technique of Exercise 37?

CHAPTER 21

TILINGS

When our ancestors used stones to cover the floors and walls of their houses, they selected shapes and colors to form pleasing designs. We can see the artistic impulse at work in mosaics, from Roman dwellings to Muslim religious buildings (see Figure 21.1). The same intricacy and complexity arise in other decorative arts—on carpets, fabrics, baskets, and even linoleum.

Such patterns have one feature in common: they use repeated shapes to cover a flat surface, without gaps or overlaps. If we think of the shapes as tiles, we can call the pattern a *tiling*, or *tessellation*. Even when efficiency is more

FIGURE 21.1 Mosaic tile dome built by Abbas I, Safavid dynasty (1611–1638), Iran.

important than aesthetics, designers value clever tiling patterns. In manufacturing, for example, stamping the components from a sheet of metal is most economical if the shapes of the components fit together without gaps—in other words, if the shapes form a tiling.

> A **tiling** is a covering of the entire plane with nonoverlapping figures.

The major mathematical question about tilings is: Given one or more shapes (in specific sizes) of tiles, can they tile the plane? And, if so, how?

The surprising answer to the first question is that it is undecidable. That is, given any arbitrary set of tile shapes, there is no way to determine for certain if they can tile the plane or not. For some particular sets of tiles, we can exhibit tilings, and for others, we can prove that there can't be any tiling. In this chapter we will see examples of both situations. But mathematicians have proved that there is no algorithm (mechanical step-by-step process) that can tell for any set of tile shapes which of the two situations happens. (See Spotlight 11.1, pages 428–429, for other examples of "unattainable ideals.")

Given this sobering (and puzzling) limitation, we begin our investigation of tilings by considering the simplest kinds of tiles and tilings.

REGULAR POLYGONS

The simplest tilings use only one size and shape of tile, and they are known as *monohedral tilings*.

> A **monohedral tiling** is a tiling that uses only one size and shape of tile.

In particular, we are interested especially in tiles that are **regular polygons,** figures all of whose sides are the same length and all of whose angles are equal. A square is a regular polygon with four equal sides and four equal interior angles; a triangle with all sides equal (an **equilateral triangle**) is also a regular polygon. A polygon with five sides is a pentagon, one with six sides is a hexagon, and one with n sides is an **n-gon.** Regular polygons are especially interesting because of their high degree of symmetry; each has the reflection and rotation symmetries of a dihedral rosette pattern (see Chapter 20).

By a convention dating back to the ancient Babylonians, angles are measured in degrees. An **exterior angle** of a polygon is one formed by one side and the extension of an adjacent side (Figure 21.2). Proceeding around the polygon

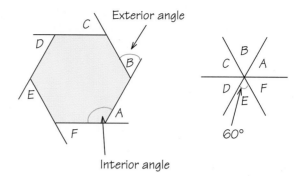

Exterior angle

Interior angle

FIGURE 21.2 The exterior angles of a regular hexagon, like those of any regular polygon, add up to 360°. Each interior angle measures 60°.

in the same direction, we see that each **interior angle** (the angle inside a polygon formed by two adjacent sides) is paired with an exterior angle. If we bring all the exterior angles together at a single point, they will add up to 360° (see Figure 21.2). If the polygon has n sides, then each exterior angle must measure $360/n$ degrees. For example, a square with $n = 4$ sides has 4 exterior angles, each measuring 90°; a pentagon with $n = 5$ sides has 5 exterior angles, each measuring 72°; whereas a regular hexagon with $n = 6$ sides has 6 exterior angles, each measuring 60°. Notice that each exterior angle plus its corresponding interior angle make up a straight line, or 180°. For a regular polygon with more than six sides, the interior angle is between 120° and 180°. This last consideration will prove crucial shortly.

REGULAR TILINGS

A monohedral tiling whose tile is a regular polygon is called a **regular tiling.**

A square tile is the simplest case. Apart from varying the size of the square, which would change the scale but not the pattern of the tiling, we can get different tilings by offsetting one row of squares some distance from the next.

However, there is only one tiling that is edge-to-edge:

In an **edge-to-edge tiling,** the edge of a tile coincides entirely with the edge of a bordering tile (see Figure 21.3 for a tiling that is not edge-to-edge and another that is).

Figure 21.3 (a) A tiling that is not edge-to-edge; the horizontal edges of two adjoining squares do not exactly coincide. (b) A tiling by right triangles that is edge-to-edge.

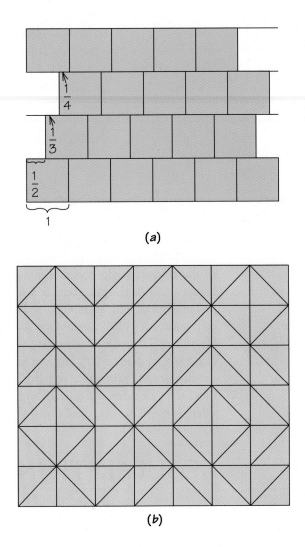

(a)

(b)

For simplicity, from now on we consider only edge-to-edge tilings. In them (even in ones with tiles of different shapes and sizes), edges of different tiles meet at points that are surrounded by tiles and their edges.

Any tiling by squares can be refined to one by triangles by drawing a diagonal of each square; but these triangles are not regular (equilateral). Equilateral triangles can be arranged in rows by alternately inverting triangles; as with squares, there is only one pattern of equilateral triangles that forms an edge-to-edge tiling.

What about tiles with more than four sides? An edge-to-edge tiling with regular hexagons is easy to construct (see the upper right pattern in Figure 21.5, on page 846).

However, if we look for a tiling with regular pentagons, we won't find one. How do we know whether we're just not being clever enough or there really isn't one to be found? This is the kind of question that mathematics is uniquely equipped to answer. In the other sciences, phenomena may exist even though we have not observed them; such was the case for bacteria before the invention of the microscope. In the case of an edge-to-edge tiling with regular pentagons, we can conclude with certainty that there is no edge-to-edge tiling with regular pentagons.

The proof is very easy. As we calculated earlier, the interior angles of a pentagon are each 108°. At a point where several pentagons meet, how many can meet there? The total of all of the angles around a point must be 360°. As you can see in Figure 21.4, four pentagons at a point would be too many (their angles would add to $4 \times 105° = 420°$, so they'd have to overlap), and three would be too few (their angles would add to $3 \times 108° = 324°$, so some of the area wouldn't be covered). Since 108 does not evenly divide 360, *regular pentagons can't tile the plane.*

With this argument, we can do something that is a favorite with mathematicians: we can generalize it. Its main idea is a criterion for when a regular pentagon can tile the plane: when the size of its interior angles divides 360 evenly. We can apply this criterion to determine exactly which other regular polygons can tile the plane.

FIGURE 21.4 Polygons that come together at a vertex in a tiling must have interior angles that add up to 360°—no less, no more.

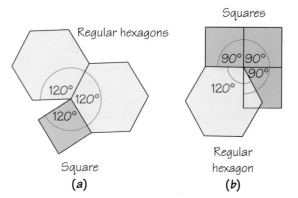

EXAMPLE ▶ *Identifying the Edge-to-Edge Regular Tilings*

A regular hexagon has interior angles of 120°; 120 divides 360 evenly, and 3 regular hexagons fit together exactly around a point. A regular 7-gon—or any regular polygon with more than six sides—has interior angles that are larger than 120° but smaller than 180°. Now 360 divided by 120 gives 3, and 360 divided by 180 gives 2—and there aren't any other possibilities in between.

Spotlight

Regular Polyhedra and Buckyballs

21.1

The three-dimensional analogue of a regular polygon is a regular polyhedron, a convex solid whose faces are regular polygons all alike (same number of sides, same size), with each vertex of the polyhedron surrounded by the same number of polygons. Although there are infinitely many regular polygons, there are only five regular polyhedra, a fact proved by Theaetetus (414–368 B.C.); they were called the *Platonic solids* by the ancient Greeks.

If the restriction that the same number of polygons meet at each vertex is relaxed, five additional convex polyhedra are obtained, all of whose faces are equilateral triangles. If we allow more than one kind of regular, thirteen further convex polyhedra are obtained, known as the *semiregular polyhedra* or *Archimedean solids* (although there is no documented evidence that Archimedes studied them—but Kepler did catalogue them all). The truncated icosahedron, whose faces are pentagons and hexagons, is known throughout the world (once inflated) as a regulation soccer ball. Drawings of it appear in the work of Leonardo da Vinci.

The truncated icosahedron is also the structure of C_{60}, the new form of carbon known as buckminsterfullerene, and more familiarly, "buckyball." Sixty carbon atoms lie at the 60 vertices of this molecule, which was discovered in 1985. It is named after R. Buckminster Fuller (1895–1983), inventor and promoter of the geodesic dome. The molecule resembles a dome.

The buckyball is part of a family of carbon molecules, the *fullerenes,* in which each carbon atom is joined to three others. Thirty years before the discovery of fullerenes, mathematicians had shown a convex polyhedron in which every vertex that has three edges must have 12 pentagon faces, but may have any number of hexagon faces, from 0 on up, except for 1.

That there must be 12 pentagons follows from a famous equation due to Leonhard Euler (1707–1783). For any convex polyhedron, it must be true that $v - e + f = 2$, where v is the number of vertices, e is the number of edges, and f is the number of faces of the polyhedron.

A truncated icosahedron, which represents the structure of carbon atoms in a buckyball. The earliest drawing of this polyhedron appeared in a work by Piero della Francesca. This particular figure appeared as one of a series of illustrations by Leonardo da Vinci for the book *The Divine Proportion* by Luca Pacioli, published in 1509.

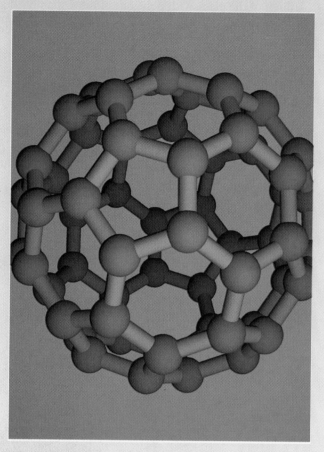

A computer-generated model of a (C_{60}) buckminsterfullerene.

The five regular polyhedra.

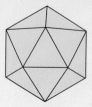

Tetrahedron   Cube   Octahedron   Dodecahedron   Icosahedron

Angles between 180° and 120° divided into 360° will give a result between 2 and 3, and consequently not an integer. So there are no edge-to-edge regular tilings of the plane with polygons of more than 6 sides. ◆

The only edge-to-edge regular tilings are the ones with equilateral triangles, with squares, and with regular hexagons.

The follow-up question, of course, is which *combinations* of regular polygons of different numbers of sides can tile the plane edge-to-edge? The arrangement of polygons around a vertex in an edge-to-edge tiling is the **vertex figure** for that vertex.

A systematic tiling that uses a mix of regular polygons with different numbers of sides but in which all vertex figures are alike—the same polygons in the same order—is called a **semiregular tiling** (see Figure 21.5).

FIGURE 21.5 The three regular tilings and the eight semiregular tilings, plus one tiling that does not belong to either group. Can you identify it?

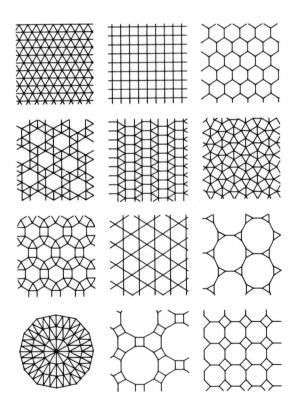

As before, the technique of adding up angles at a vertex (to be 360°) can eliminate some impossible combinations, such as "square, hexagon, hexagon" (Figure 21.4). Once we have found an arrangement that is numerically possible, we must confirm the actual existence of each tiling by constructing it (i.e., show that it is geometrically possible). For example, even though a possible arrangement of regular polygons around a point is "triangle, square, square, hexagon," it is not possible to construct a tiling with that vertex figure at every vertex.

The result of such an investigation is that in a semiregular tiling no polygon can have more than 12 sides. In fact, polygons with 5, 7, 9, 10, or 11 sides do not occur either. Figure 21.5 exhibits all of the semiregular tilings.

If we abandon the restriction about the vertex figures being the same at every vertex, then there are *infinitely many* systematic edge-to-edge tilings with regular polygons, even if we continue to insist that all polygons with the same number of sides have the same size.

TILINGS WITH IRREGULAR POLYGONS

What about edge-to-edge tilings with irregular polygons, which may have some sides longer than others, or some interior angles larger than others? We will look just at monohedral tilings (in which all tiles have the same size and shape) and investigate in turn what triangles, **quadrilaterals** (four-sided polygons), hexagons, and so forth, can tile the plane.

The most general shape of triangle has all sides of different lengths and all interior angles of different sizes. Such a triangle is called a **scalene triangle,** from the Greek word for "uneven." We can always take two copies of a scalene triangle and fit them together to form a **parallelogram,** a quadrilateral whose opposite sides are parallel (Figure 21.6a). It's easy to see that we can then use such parallelograms to tile the plane, by making strips and then fitting layers of strips together edge-to-edge (Figure 21.6b). So:

Any triangle can tile the plane.

What about quadrilaterals? We have seen that squares tile the plane, and rectangles certainly will, too; and we have just noted that any parallelogram will tile. What about a quadrilateral (four-sided polygon) with its opposite sides not parallel, as in Figure 21.7a? The same technique as for triangles will

FIGURE 21.6 (a) A scalene triangle. (b) Every scalene triangle tiles the plane.

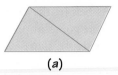

(a)

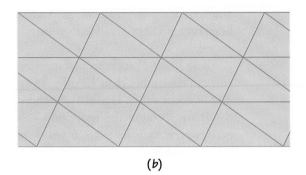

(b)

FIGURE 21.7 (a) A general quadrilateral. (b) Any quadrilateral tiles the plane.

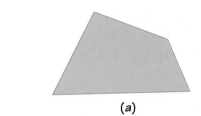

(a)

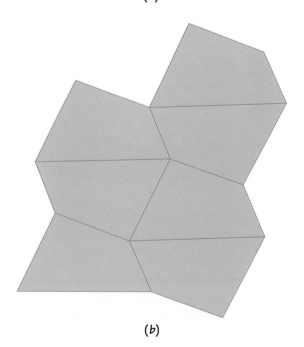

(b)

work. We fit together two copies of the quadrilateral, forming a hexagon whose opposite sides are parallel. Such hexagons fit next to each other to form a tiling, as in Figure 21.7b.

The quadrilaterals shown in Figure 21.7 are all **convex.** If you take any two points on the tile (including the boundary), the line segment joining them lies entirely within the tile (again, including the boundary). The quadrilateral of Figure 21.8a is not convex, but the same approach works for using it to form a tiling (Figure 21.8b). So:

Any quadrilateral, even one that is not convex, can tile the plane.

We could hope that such success would extend to irregular polygons with any number of sides, but it doesn't. The situation for convex hexagons was determined by K. Reinhardt in his 1918 doctoral thesis. He showed that for a convex hexagon to tile, it must belong to one of three classes, and that every hexagon in those classes will tile. Examples of the three classes are shown in Figure 21.9, together with their characterizations. Notice that tilings with a hexagon of Type 2 (Figure 21.9b) use both ordinary and mirror-image versions of the hexagon.

Exactly three classes of convex hexagons can tile the plane.

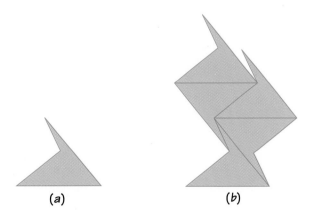

(a)   (b)

FIGURE 21.8   (a) A general nonconvex quadrilateral. (b) Any quadrilateral, convex or not, tiles the plane.

FIGURE 21.9 The three
types of convex hexagon
tile.

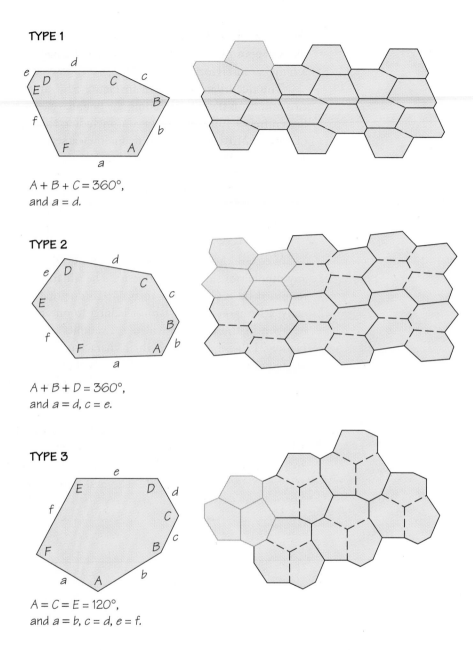

TYPE 1

$A + B + C = 360°$,
and $a = d$.

TYPE 2

$A + B + D = 360°$,
and $a = d, c = e$.

TYPE 3

$A = C = E = 120°$,
and $a = b, c = d, e = f$.

Reinhardt also explored convex pentagons and found five classes that tile. For example, any pentagon with two parallel sides will tile. Reinhardt did not complete the solution, as he did for hexagons, by proving conclusively that no other pentagons could tile; he claimed that it would be very tedious to finish the analysis. Still, he felt that he had found them all. In 1968, after 35 years of working on the problem on and off, R. B. Kershner, a physicist at Johns

Hopkins University, discovered three more classes of pentagons that will tile. Kershner was sure that he had found all pentagons that tile, but again did not offer a complete proof, which "would require a rather large book."

When an account of the "complete" classification into eight types appeared in *Scientific American* (July 1975), the article provoked an amateur mathematician to discover a ninth type! A second amateur, Marjorie Rice, a housewife with no formal education in mathematics beyond high school "general mathematics" 36 years earlier, devised her own mathematical notation and found four more types over the next two years (see Spotlight 21.2). A fourteenth type was found by a mathematics graduate student in 1985. Since then, no new types have been discovered, yet no one knows if the classification is complete.

With the situation so intricate for convex pentagons, you might think that it must be still worse for polygons with seven or even more sides. In fact, however, the situation is remarkably simple, as Reinhardt proved in 1927:

A convex polygon with seven or more sides cannot tile.

M. C. ESCHER AND TILINGS

The Dutch artist M. C. Escher (1898–1972) was inspired by the great variety of decoration in tilings in the Alhambra, a fourteenth-century palace built during the last years of Islamic dominance in Spain. He devoted much of his career of making prints to creating tilings with tiles in the shapes of living beings (a practice forbidden to Muslims). Those prints of interlocking animals and people have inspired awe and wonder among people all over the world. Figures 21.10–21.13 (see pages 854–857) illustrate a few of his drawings and finished works. Like Marjorie Rice, he too developed his own mathematical notation for the different kinds of patterns for the tilings.

TILING BY TRANSLATIONS

You may wonder just how much liberty can be taken in shaping a tile, and how you might be able to design an Escher-like tiling yourself.

The simplest case is when the tile is just *translated* in two directions, that copies are laid edge-to-edge in rows, as in Figure 21.10. Each tile must fit exactly into the ones next to it, including its neighbors above and below. We say that each tile is a **translation** of each other one, since we can move one to coincide with another without doing any rotation or reflection.

Spotlight

In Praise of Amateurs

21.2

R.B. Kershner's claim to have found all convex pentagons that tile was reported by Martin Gardner in his column in *Scientific American,* which was read by many amateur puzzle enthusiasts, including Richard James III and Marjorie Rice. James found a tiling that Kershner had missed, a discovery that Gardner reported in a later column.

Rice, a San Diego housewife and mother of five, read about James's new tile, "I thought I would like to understand these fascinating patterns better and see if I could find still another type. It was like a delightful new puzzle to me." Her search became a full-scale assault on the problem, lasting two years.

Rice had no formal education in mathematics beyond a high school general mathematics course. She not only worked out her own method of attack but also invented her own notation as well, both of which were far from the conventional ways that mathematicians use.

"I began drawing little diagrams on my kitchen counter when no one was there, covering them up quickly if someone came by, for I didn't wish to have to explain what I was doing to anyone. Soon I realized that many interesting patterns were possible but did not pursue them further, for I was searching for a new type and a few weeks later, I found it." Over the next two years, she found three additional new tilings.

What makes a person pursue a problem so steadfastly as Marjorie Rice? She was not trained to do this, nor paid to do it, but obviously gained

Marjorie Rice

personal satisfaction in her patient and persistent search.

She was born in 1923 in St. Petersburg, Florida, a first child. At age 5, she began school in a one-room country school with eight grades and two dozen pupils.

"When I was in the 6th or 7th grade, our teacher pointed out to us one day the Golden Section in the proportions of a picture frame. This immediately caught my imagination and though it was just a passing incident, I never forgot it. I've . . . been especially interested in architecture and the ideas of architects and planners such as Buckminster Fuller. I've come across the Golden Section again in my reading and considered its use in painting and design."

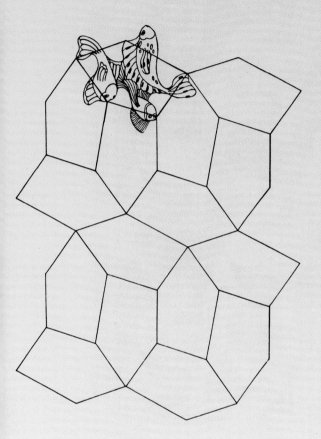

Underlying grid for Marjorie Rice's *Fish,* based on one of her unusual tilings by pentagons.

After high school, Marjorie Rice worked until her marriage in 1945. She was drawn back into mathematics by her children, finding solutions to their homework problems "by unorthodox means, since I did not know the correct procedures." She became especially interested in textile design and the works of M. C. Escher. As she pursued the pentagonal tilings, she produced some beautiful geometric designs and imaginative Escher-like patterns (see Figure 21.19 and the figure here.)

"I enjoy puzzles of all kinds, crosswords, jigsaw puzzles, mathematical puzzles and games, and have purchased books of mathematical puzzles over the years. Those of a geometric nature are a special delight."

The intense spirit of inquiry and the keen perception of all they encounter are the forte of all such amateurs. No formal education provides these gifts. Lack of a mathematical degree separates these "amateurs" from the "professionals," yet their curiosity and ingenious methods make them true mathematicians.

Source: Adapted from Doris Schattschneider, "In Praise of Amateurs," in *The Mathematical Gardner,* edited by David A. Klarner, pp. 140–166, plus Plates I–III, Wadsworth, Belmont, Calif., 1981.

FIGURE 21.10   Escher No. 128 [*Bird*], from Escher's 1941–1942 notebook.

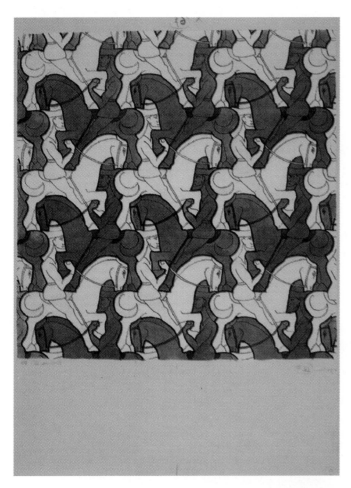

FIGURE 21.11A Escher
No. 67 [*Horseman*], from
Escher's 1941–1942
notebook.

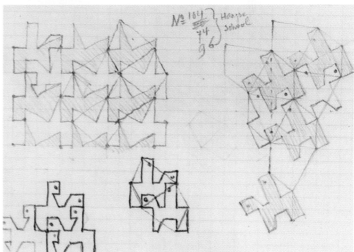

FIGURE 21.11B Sketch
by Escher showing the
design of the tile for the
Horseman print.

FIGURE 21.12   Escher No. 6 [*Camel*], from Escher's 1941–1942 notebook.

FIGURE 21.13 (a) Escher No. 88 [*Sea Horse*], and (b) the skeleton for No. 88 from Escher's 1941–1942 notebook.

When is it possible for a tile to cover the plane in this manner? The boundary of the tile must be divisible into matching pairs of opposing parts that will fit together. Figures 21.10 and 21.11 illustrate two basic ways that this can happen. In the first, two opposite pairs of sides match; in the second, three opposite pairs of sides match.

A tile can tile the plane by translations if either

- there are four consecutive points *A*, *B*, *C*, and *D* on the boundary such that

  the boundary part from *A* to *B* is congruent by translation to the boundary part from *D* to *C*, and
  the boundary part from *B* to *C* is congruent by translation to the boundary part from *A* to *D* (see Figure 21.14a)

- or there are six consecutive points *A*, *B*, *C*, *D*, *E*, and *F* on the boundary such that the boundary parts *AB*, *BC*, and *CD* are congruent by translation, respectively, to the boundary parts *ED*, *FE*, and *AF* (see Figure 21.15b).

The tiles for each of Figures 21.10 and 21.11 are shown in outline form in Figure 21.14, together with points marked to show how the tiles fulfill the criterion.

To create tilings, you can proceed exactly as Escher did. His notebooks show that he designed his patterns in just the way that we now describe.

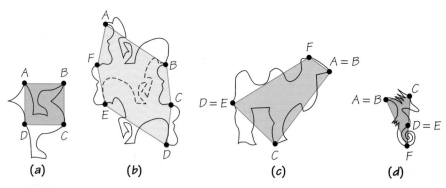

FIGURE 21.14   Individual tiles traced from the Escher prints of Figures 21.10–21.13, with points marked to show how they fulfill the criteria for tiling by translations or by translations and half-turns.

FIGURE 21.15 How to make an Escher-like tiling by translations, from a parallelogram base.

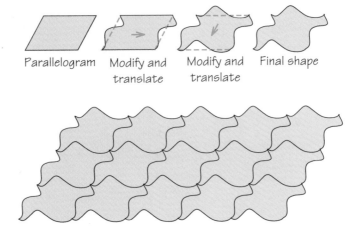

Parallelogram   Modify and translate   Modify and translate   Final shape

EXAMPLE ▶ *Tiling the Plane Using a Parallelogram*

For the first case of the theorem, start from a parallelogram, make a change to the boundary on one side, then copy that change to the opposite side. Similarly, change one of the other two sides and copy that change on the side opposite it (Figure 21.15). Revise as necessary, always making the same change to opposite sides. You might find it useful (as Escher did) to make your designs on graph paper, or you can work by cutting and taping together pieces of heavy paper. ◆

EXAMPLE ▶ *Tiling the Plane Using a Hexagon*

For the second case, start from a **par-hexagon,** one whose opposite sides are equal and parallel; this is one of the kinds of hexagons that tile the plane. Again, make a change on one boundary and copy the change to the opposite side, and do this for all three pairs of opposite sides (Figure 21.16). ◆

Of course, there is a real art to being able to make the resulting tile resemble an animal or human figure!

TILING BY TRANSLATIONS AND HALF-TURNS

If the tiling is to allow half-turns, so that some of the figures are "upside down," the part of the boundary of a right-side-up figure has to match the corresponding part of itself in an upside-down position. For that to happen, that part of the boundary must be **centrosymmetric,** that is, symmetric about (un-

FIGURE 21.16   How to
make an Escher-like tiling
by translations, from a par-
hexagon base.

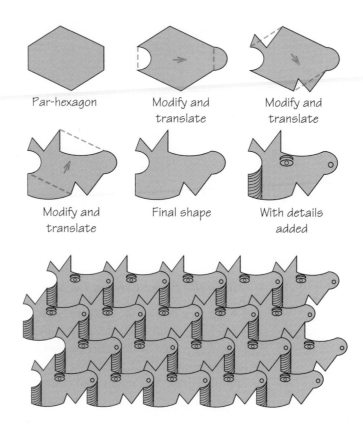

FIGURE 21.16   How to make an Escher-like tiling by translations, from a par-hexagon base.

Par-hexagon

Modify and translate

Modify and translate

Modify and translate

Final shape

With details added

altered by) a 180-degree rotation around its midpoint. The key to some of Escher's more sophisticated monohedral designs, and the fundamental principle behind some further easy recipes for making Escher-like tilings, is the **Conway criterion,** formulated by John H. Conway of Princeton University:

A tile can tile the plane by translations and half-turns if there are six consecutive points on the boundary (some of which may coincide, but at least three of which are distinct)—call them *A, B, C, D, E,* and *F*—such that

- the boundary part from *A* to *B* is congruent by translation to the boundary part from *E* to *D,* and
- each of the boundary parts *BC, CD, EF,* and *FA* is centrosymmetric.

The first condition means that we can match up the two boundary parts exactly, curve for curve, angle for angle. The second condition means that each of the remaining boundary parts is brought back into itself by a half-turn around its center. Either condition is automatically fulfilled if the boundary part in question is a straight-line segment.

The tiles for each of Figures 21.12 and 21.13 are shown in outline form in Figure 21.14, together with points marked to show how the tiles fulfill the Conway criterion.

Once again, you can make Escher-like tilings by starting from simple geometric shapes that tile. This time, the starting geometric tile can be any triangle or any quadrilateral.

EXAMPLE ▶ *Tiling the Plane Using a Triangle*

For a triangle, modify half of one side, then rotate that side around its center point to extend the modification to the rest of the side, thereby making the new side centrosymmetric. Then you may do the same to the second and third sides (Figure 21.17). ◆

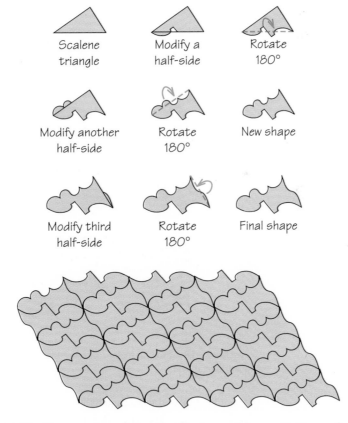

FIGURE 21.17 How to make an Escher-like tiling by translations and half-turns, from a scalene triangle base.

EXAMPLE ▶ *Tiling the Plane Using a Quadrilateral*

For the quadrilateral, do the same, modifying each of the four sides, or as many as you wish (Figure 21.18). ◆

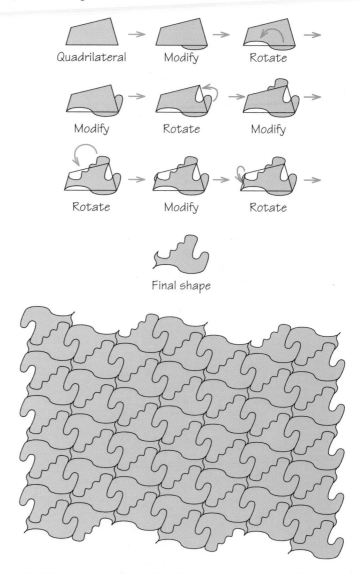

Figure 21.18 How to make an Escher-like tiling by translations and half-turns, from a quadrilateral base.

The same approach will work with some of the sides of some pentagons and hexagons that tile. Because not all sides can be modified, there is less freedom for designing tiles, so it is more difficult to make the resulting tiles resemble intended figures. Figure 21.19 shows the beautiful results achieved by Marjorie Rice, using one of the unusual tilings by pentagons that she discovered.

FIGURE 21.19 *Fish,* by Marjorie Rice, based on one of her unusual tilings by pentagons.

The sketches in Escher's notebook in Figures 21.10–21.13 indicate how he designed the prints whose tiles you see in Figure 21.14. For Figure 21.14a, he modified the two pairs of sides of a square. For Figure 21.14b, he modified the pairs of sides of a par-hexagon that became a tile made up of a pair of dark and light knights. This figure also has a reflection symmetry, taking a leftward-facing light knight to a rightward-facing dark knight. However, we have not discussed criteria for when you can start with a tile (e.g., a single knight) and produce a tiling with this symmetry. In Figure 21.14c, the blue overlay shows how the tile could be made by modifying half of every side of a general quadrilateral, though Figure 21.12 shows that Escher actually designed the tiling from a parallelogram base. Regarding Figure 21.14d, Figure 21.13 shows that Escher used a triangle base. He did not use the procedure that we noted earlier, in which half of every side is modified. Instead, he treated the triangle as a quadrilateral, in which two adjacent sides (*CD* and *DF*) happen to continue on in a straight line.

FURTHER CONSIDERATIONS

Some of the most impressive of Escher's prints use two or more interlocking tile shapes. Such is the case with *Heaven and Hell* (Angels and Devils) in Spotlight 19.4 (pages 772–773). All of his prints of tilings of the plane have underlying symmetries that are the wallpaper patterns whose classification is discussed in Spotlight 20.6 (pages 824–825). Often those symmetries are enhanced further by the use of color.

Several of Escher's more remarkable prints are tilings not of the Euclidean plane but of the Poincaré disk model of hyperbolic geometry, discussed in Chapter 19. Two examples, one by Escher and one by Douglas Dunham, are shown in Spotlight 19.4.

All the patterns that we have exhibited and discussed so far have been **periodic tilings.** If we transfer a periodic tiling to a transparency, it is possible to slide the transparency a certain distance horizontally, without rotating it, until the transparency exactly matches the tiling everywhere. We can also achieve the same result by moving the transparency some second direction (possibly vertically) a certain (possibly different) distance.

In a periodic tiling you can identify a **fundamental region**—a tile, or a block of tiles—with which you can cover the plane by translations at regular intervals. For example, in Figure 21.10, a single bird forms a fundamental region. In Figure 21.12, two adjacent camels, one right-side up and one upside down, form a fundamental region. In Escher's *Heaven and Hell* of Spotlight 19.4, a pair of foot-to-foot angels and the pair of foot-to-foot devils between them form a fundamental region. In the terminology of Chapter 20, the periodic tilings are ones that are preserved under translations in more than two directions. (The wallpaper patterns of Spotlight 20.6 are sometimes called *periodic plane patterns*. In this chapter we are concerned with the design elements more than with the patterns, which were the main topic of Chapter 20.)

NONPERIODIC TILINGS

> A **nonperiodic tiling** is a tiling in which there is no regular repetition of the pattern by translation.

The lower left pattern in Figure 21.5 (page 846), with its expanding rings of triangles, does not have any regular repetition by translation.

In Figure 21.3a (page 842), the second row from the bottom is offset one-half of a unit to the right from the bottom row, the third row from the bottom is offset one-third of a unit further, and so forth. Since the sum $\frac{1}{2} + \frac{1}{3} + \frac{1}{4} + \cdots + \frac{1}{n}$ never adds up to exactly a whole number, there is no direction (horizontal, vertical, or diagonal) in which we can move the entire tiling and have it coincide exactly with itself.

EXAMPLE ▶ *A Random Tiling*

Consider the usual edge-to-edge square tiling. For each square, flip a coin; depending on the result, divide the square into two right triangles by adding either a rising or a falling diagonal (see Figure 21.3b, page 842). Because what happens in each individual square is unconnected to what happens in the rest of the tiling, the tiling by right triangles that is produced by this procedure has no chance of being periodic. ◆

THE PENROSE TILES

For all known cases, if a single tile can be used to make a nonperiodic tiling of the plane, then it can also be used to make a periodic tiling. It is still an open question whether this property is true for every possible shape. In 1993,

Conway discovered an example in three dimensions of a single convex polyhedron that tiles space nonperiodically but cannot be used to make a periodic tilogy.

For a long time they also tended to believe the more general assertion that if you can construct a nonperiodic tiling with a set of one *or more* tiles, you can construct a periodic tiling from the same tiles. But in 1964 a set of tiles was found that permits only nonperiodic tiling. It contains 20,000 different shapes! Over the next several years, smaller sets were discovered with the same property, with as few as 100 shapes. But it was still amazing when in 1975 Roger Penrose, a mathematical physicist at Oxford, announced a set that would tile only nonperiodically—consisting of just two tiles! (See Figure 21.20 and Spotlight 21.3.)

Penrose called his tiles "darts" and "kites" and both of these **Penrose tiles** can be obtained from a single rhombus. (A **rhombus** is a quadrilateral with four equal sides and equal opposite interior angles.) The particular rhombus from which the Penrose tiles are constructed has interior angles of 72° and 108°. If we cut the longer diagonal in two pieces so that the longer piece is the golden ratio ($(1 + \sqrt{5})/2 \approx 1.618$) times as long as the shorter (see Chapter 20), and connect the dividing point to the remaining corners, we split the rhombus into a dart and a kite (Figure 21.20).

Since the two Penrose pieces come from a rhombus, and a rhombus can be replicated to tile the plane periodically, the rules for fitting the Penrose pieces together do not allow the periodic rhombus arrangement. We may label the front and back vertices of the dart with H (for head) and its two wing tips with T (for tail), and do the reverse for the kite. Then the rule is that only vertices with the same letter may meet: heads must go to heads, and tails to tails.

FIGURE 21.20
Construction of Penrose's "dart" (beige area) and "kite" (blue area). The length $1/\phi \approx 0.618$ is the golden ratio.

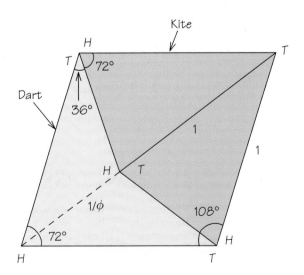

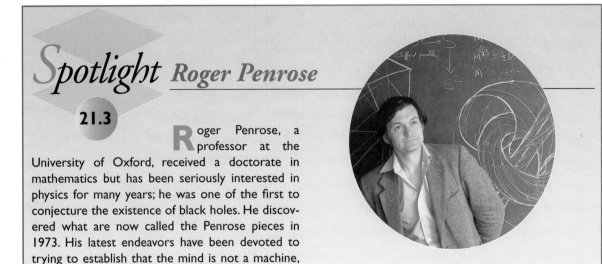

A prettier method of enforcing the rules, proposed by Conway, is to draw circular arcs of different colors on the pieces and require that adjacent edges must join arcs of the same color. The result is the pretty patterns of Figure 21.21. In fact, Conway thinks of the darts as children, each with two hands. The rule for fitting the pieces together is that children are forced to hold hands. Penrose patterns become dancing circles of children.

FIGURE 21.21 A Penrose tiling with specially marked tiles, forming what is known as the cartwheel tiling.

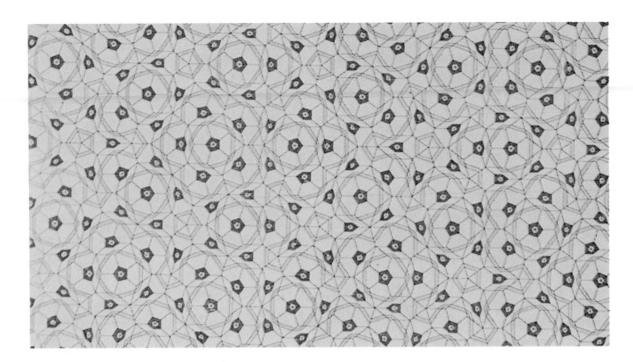

FIGURE 21.22 A Penrose nonperiodic tiling made with two rhombus shapes.

Figure 21.22 shows a tiling by a different pair of pieces, both rhombuses, that tile the plane only nonperiodically. Figure 21.23 shows a modification of the Penrose pieces into two bird shapes. Figure 21.24 shows a coloring of one particular tiling with the Penrose pieces so that no two adjacent pieces have the same color.

Although tilings with Penrose's pieces cannot be periodic, the tilings possess unexpected symmetry. As you recall, we have explored our intuitions of symmetry in terms of *balance, similarity,* and *repetition.* Patterns made with the Penrose pieces certainly involve repetition, but it is the balance in the arrangement that we seek. What balance can there be in a nonperiodic pattern? It turns out that some Penrose patterns have a single line of reflection. But most surprising of all, every Penrose pattern has arbitrarily large regions with fivefold rotational symmetry!

EXAMPLE ▶ *Fivefold Symmetry*

Consider, for example, any one of the 10 colored pieces of Figure 21.25. If we rotate the pattern around its center through one-fifth of a turn, the region surrounded by the colored pieces looks exactly the same as before (but parts of the pattern farther away may not exactly match). Note that rotation through one-tenth of a turn would not preserve all of that region. In Conway's metaphor, whenever a chain of children closes, the region inside has fivefold symmetry. ◆

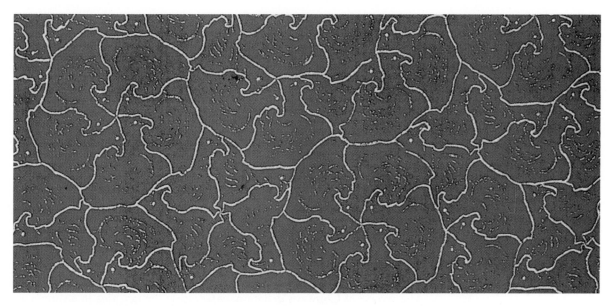

FIGURE 21.23 A modification of a Penrose tiling by refashioning the kites and darts into bird shapes.

FIGURE 21.24 A Penrose tiling by kites and darts, colored with five colors. A Penrose tiling can always be colored using four colors, in such a way that two tiles that share an edge have different colors. Whether a Penrose tiling can be colored in such a way using only three colors is an unsolved problem; we know, though, that if one Penrose tiling can be colored using three colors, all Penrose tilings can.

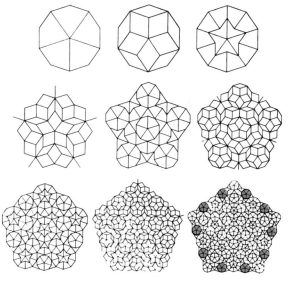

FIGURE 21.25 Successive deflation (that is, the systematic cutting up of large tiles into smaller ones) of patches of tiles of a Penrose nonperiodic tiling.

There are, in fact, two—and only two—Penrose tilings for which the entire pattern has the fivefold rotation symmetry of a rosette. Figure 21.25 shows how to construct both of these, beginning with five kites meeting at a vertex. We cut the darts and kites up into smaller darts and kites and then enlarge the new ones to the same size as the old, that way covering more area each time (the figure doesn't show the enlargement). Conway calls this operation *deflation.* As we proceed with successive steps, we get partial tilings alternately by two different patterns, and each has a fivefold rotational center: one has five kites at the center, the other has five darts.

Where does this rotational symmetry come from? The original rhombus that we split up has the angles shown in Figure 21.20. Except in the recess of the dart and matching part of the kite, all the internal angles of the kite and dart are either 72° or 36°. Now, 72 goes into 360 five times, and 36 goes 10 times. If we recall that it is the interior angles that matter in arranging polygons around a point, we see that fivefold or tenfold symmetry could conceivably result from using such tiles.

The reverse of Conway's deflation, *inflation,* is the key idea in a simple argument to show that a Penrose pattern must be nonperiodic. For the inflation process, cut each dart down its middle and put glue on the short edges of the resulting triangles (but not on the cut itself). The result is a pattern of larger kites and darts!

We show that a Penrose pattern is nonperiodic by proceeding by contradiction. Suppose (contrary to what we want to establish) that some Penrose pattern is periodic, that is, it has translation symmetry. Let d be the distance along the translation direction to the first repetition. Performing inflation does the same thing to each repetition, so the inflated pattern must still have translation symmetry and a distance d along the translation direction to the first repetition. Keep on performing inflation, time after time, until the darts and kites are so large that they are more than d across. The pattern, as we have just argued, must still have translation symmetry at a distance d, but it can't, because there's no repetition inside a single tile! We reach a contradiction. So what's wrong? Our initial supposition, that the pattern was periodic in the first place, must have been erroneous. We conclude that all Penrose tilings are nonperiodic.

Despite their being nonperiodic, all Penrose patterns are somewhat alike, in the following sense:

Any finite region in one Penrose pattern is contained somewhere inside every other Penrose pattern; in fact, it occurs infinitely many times in every Penrose pattern.

Penrose tilings have another feature that allows us to characterize them as **quasiperiodic,** or somewhere between periodic and random. (Noting the precise definition of this term would take us too far afield.) Robert Ammann introduced onto the two rhombic Penrose pieces used in Figure 21.22 lines that are now known as *Ammann bars.* In any Penrose tiling, these bars line up into five sets of parallel lines, each set rotated 72° from the next, forming a pentagonal grid (Figure 21.26). The distance between two adjacent parallel bars is one of only two values, either *A* or *B.* Do you want to guess what the ratio of the longer *A* is to the shorter *B?* You don't think it could possibly be anything but the golden ratio, do you? And so it is.

FIGURE 21.26 Penrose tilings with Ammann bars. Specially placed lines on the tiles produce five sets of parallel bars in different directions.

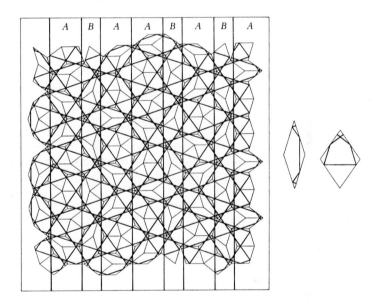

EXAMPLE ▶ *Musical Sequences*

What about the order in which the *A*'s and *B*'s occur, as we move from left to right in Figure 21.26? Is there any pattern to that? From the limited part of the pattern we can observe, we see the sequence as

> *A B A A B A B A A B A B A*

You might think from the figure that the pattern continues repeating the group

> *A B A A B*

indefinitely; after all, there are five symbols in this group. But such is not the case. Known as a musical sequence, the sequence of intervals between

Ammann bars is nonperiodic—it cannot be produced by repeating any finite group of symbols. We can think of it as a one-dimensional analogue of a Penrose tiling.

There is some regularity in musical sequences. Two *B*'s can never be next to each other, nor can we have three *A*'s in a row. Just as any finite part of any Penrose tiling occurs infinitely often in any other Penrose tiling, any finite part of any musical sequence appears infinitely often in any other one. The order of the symbols is neither periodic nor random, but between the two—quasi-periodic.

These sequences are called *musical* sequences because musicians represent the large-scale structure of songs in terms of the letters *A* and *B*. For example, a common pattern for popular songs is *AABA,* indicating that the first, second, and fourth verses have the same melody, but the third verse has a different melody.   ◆

The ratio of darts to kites in an infinite Penrose tiling, or of *A*'s to *B*'s in a musical sequence, is exactly the golden ratio, approximately 1.618. So if you are going to play with sets of Penrose pieces and see what kinds of patterns you can create, you will need about 1.6 times as many darts as kites.

As pointed out by geometers Marjorie Senechal (Smith College) and Jean Taylor (Rutgers University), Penrose tilings have three important properties:

- They are constructed according to rules that force nonperiodicity.
- They can be obtained from a substitution process (inflation and deflation) that features self-similarity.
- They are quasiperiodic.

Research of the late 1980s indicates that these properties are somewhat independent, meaning that one or two may be true of a tiling without all three being true.

QUASICRYSTALS AND BARLOW'S LAW

Although Penrose's discovery was a big hit among geometers and in recreational mathematics circles in the mid-1970s, few people thought that his work might have practical significance. In the early 1980s some mathematicians even generalized Penrose tilings to three dimensions, using solid polyhedra to fill space nonperiodically. Like the two-dimensional Penrose patterns, these have orderly fivefold symmetry but are nonperiodic.

Yet in 1982 scientists at the U.S. National Bureau of Standards discovered unexpected fivefold symmetry while looking for new ultrastrong alloys of aluminum (mixtures of aluminum with other metals).

Manganese doesn't ordinarily alloy with aluminum, but the experimenters were able to produce small crystals of alloy by cooling mixtures of the two metals at a rate of millions of degrees per second. Following routine procedures, chemist Daniel Shechtman began a series of tests to determine the atomic structure of the special crystals. But there was nothing routine about what he found: the atomic structures of the manganese-aluminum crystals were so startling that it took Shechtman three years to convince his colleagues they were real.

Why did he encounter such resistance? His patterns—and the crystals that produced them—defied one of the fundamental laws of crystallography. Like our discovery that the plane cannot be tiled by regular pentagons, **Barlow's law,** also called the **crystallographic restriction,** says that a crystal can have only rotational symmetries that are twofold, threefold, fourfold, or sixfold. Since crystals are periodic, if there were a center of fivefold symmetry, there would have to be many such centers. Barlow proved this impossible.

Peter Barlow was a nineteenth-century British mathematician whose name survives today in the name of a book of mathematical tables. His argument was a very simple proof by contradiction, similar to Conway's proof in which we saw earlier that Penrose patterns are not periodic. Suppose (contrary to what we intend to show) that there is more than one fivefold rotation center. Let A and B be two of these that are closest together (see Figure 21.27). Rotate the pattern of Figure 21.27 by one-fifth of a turn clockwise around B, which carries A to some point A'. Since the pattern has fivefold symmetry around B, the

FIGURE 21.27 Barlow's proof that no pattern can have two centers of fivefold symmetry.

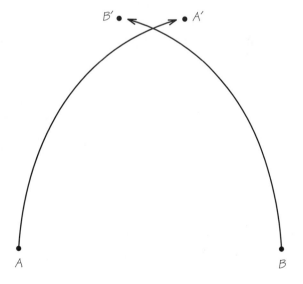

Spotlight Quasicrystals

21.4

In 1984, working at the University of Pennsylvania, Paul Steinhardt and Don Levine calculated the diffraction patterns that three-dimensional Penrose patterns would produce if the building blocks were real atoms instead of geometric tiles.

When a beam of electrons or X rays passes through a solid material, they are diffracted, or scattered, by the atoms inside. The diffracted beams can be photographed head-on, and the images they form reflect the atomic architecture of the solid.

The diffraction patterns of crystals contain sharp, isolated dots, due to the periodicity of the underlying structure. In a few directions, depending on the arrangement of the atoms, the diffracted beams reinforce one another, producing bright spots on the film. A crystal is a little like an orchard planted in a rigid geometric grid. Most lines of sight are blocked by trees, but you can see right through to the other side in a few directions.

For another class of diffraction patterns, for glassy materials, the dots are either spread out into fuzzy rings or altogether absent. Glasses, in contrast to crystals, are made up of atoms or molecules stuck together randomly; they're more like random forests than well-planned orchards. Because they offer no preferred directions for diffraction, the patterns they produce contain no sharp dots.

The computed diffraction pattern for Levine and Steinhardt's imaginary solid contained a surprise: unmistakable sharp points. Since the atomic arrangement of their solid was nonperiodic, it should have produced the fuzzy diffraction pattern characteristic of glassy substances. Since the dots in the pattern were arranged with fivefold symmetry, the solid wasn't a crystal either. Steinhardt decided to call it a quasicrystal.

In the fall of 1984 a colleague of Steinhardt's showed him a diffraction image made from a real substance, Shechtman's alloy of aluminum and manganese. The picture looked amazingly similar to Steinhardt and Levine's computer simulation.

In short order sevenfold, ninefold, and other symmetries proved to be possible. But no one could think of a mechanism by which millions upon millions of real atoms could arrange themselves spontaneously in those intricate patterns.

Anyone who tries to assemble Penrose pieces into tilings quickly realizes that it's not easy. You have to think ahead and keep the whole pattern in mind when adding a tile; otherwise, there is trouble. Local rules, or instructions for fitting a tile into a particular niche, don't seem sufficient to build the entire pattern without global rules that force you to plan ahead and check the configuration of tiles at distant points.

In 1988 playfulness paid off once more. George Onoda, an IBM ceramics expert, toying with about 200 Penrose tiles, learned how to assemble flawless tilings of any size using only local rules.

For a complete theory of quasicrystals, the local rules will have to be generalized to three dimensions, and they must be shown to correspond to actual atomic forces. In the meantime, experimentalists continue to report bigger, more perfect quasicrystals.

Source: Adapted from Hans C. von Baeyer, "Impossible Crystals," *Discover*, 11 (2) (February 1990): 69–78, 84.

(a) A nonperiodic tiling by rhombuses. (b) A three-dimensional crystallike structure based on this tiling. (c) The crystal pattern observed by chemist Daniel Shechtman in a special manganese-aluminum alloy. Note the similarity to the pattern in (a). (d) A scanning electron microscope image of the quasicrystal alloy $Al_{5???}Li_3Cu$. The fivefold symmetry can be seen in the five rhombic faces that meet at a single point in the center of the photograph, forming a starlike shape. (e) This image of the quasicrystal material $AL_{65}Co_{20}Cu_{15}$ was obtained with a scanning tunneling microscope; the resulting image has been overlaid with a nonperiodic tiling to display the local fivefold symmetry.

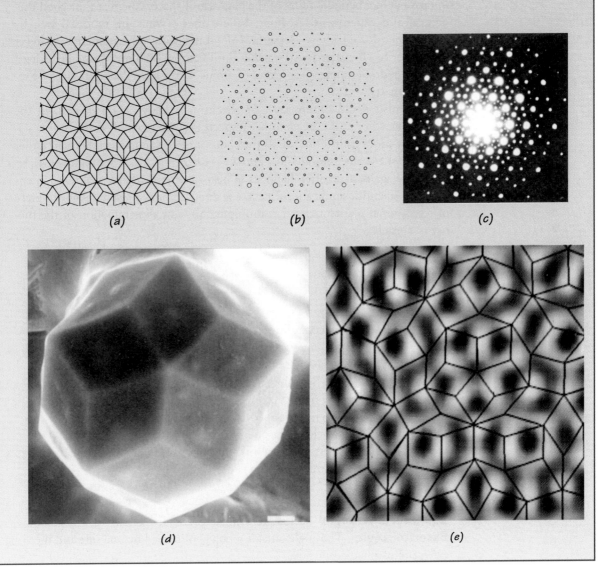

(a)   *(b)*   *(c)*

(d)   *(e)*

point A', which is the image of the fivefold center A, must itself be a fivefold center. Now use A as a center and rotate the pattern by one-fifth of a turn counterclockwise, which carries B to some point B'; as we just argued in the case of A', B' must also be a fivefold center. But A' and B' are closer together than A and B, which is a contradiction. Hence our original supposition must be false, and a pattern can have at most one fivefold rotation center (as the patterns in Figure 21.25 in fact do) and so cannot be periodic.

Barlow's law, as a mathematical theorem, shows that fivefold symmetry is impossible in a periodic tiling of the plane or of space. Chemists, for good theoretical and experimental reasons, believe that crystals are modeled well by three-dimensional tilings. An array of atoms with no symmetry whatever would not be considered a crystal. Yet until Penrose's discovery, no one realized that nonperiodic tilings—or arrays of atoms—can have the regularity of fivefold symmetry.

Chemists could simply say that Shechtman's alloys aren't crystals. In the classic sense they aren't, but in other respects they do resemble crystals. It is scientifically more fruitful to extend the concept of crystal to include them rather than rule them out; they are now known as *quasicrystals* (see Spotlight 21.4).

Once again, as so often happens in history, pure mathematical research anticipated scientific applications. Penrose's discovery, once just a delightful piece of recreational mathematics, has prompted a major reexamination of the theory of crystals.

REVIEW VOCABULARY

Barlow's law, or the **crystallographic restriction**   A law of crystallography that states that a crystal may have only rotational symmetries that are twofold, threefold, fourfold, or sixfold.

Centrosymmetric   Symmetric by 180 degrees rotation around its center.

Convex   A geometric figure is convex if for any two points on the figure (including its boundary), all the points on the line segment joining them also belong to the figure (including its boundary).

Conway criterion   A criterion for determining whether a shape can tile by means of translations and half-turns.

Edge-to-edge tiling   A tiling in which adjacent tiles meet only along full edges of each tile.

Equilateral triangle   A triangle with all three sides equal.

Exterior angle   The angle outside a polygon formed by one side and the extension of an adjacent side.

Fundamental region   A tile or group of adjacent tiles that can tile by translation.

Interior angle The angle inside a polygon formed by two adjacent sides.

Monohedral tiling A tiling with only one size and shape of tile (the tile is allowed to occur also in "turned-over," or mirror-image, form).

***n*-gon** A polygon with *n* sides.

Nonperiodic tiling A tiling in which there is no repetition of the pattern by translation.

Parallelogram A convex quadrilateral whose opposite sides are equal and parallel.

Par-hexagon A hexagon whose opposite sides are equal and parallel.

Periodic tiling A tiling that repeats at fixed intervals in two different directions, possibly horizontal and vertical.

Quadrilateral A polygon with four sides.

Regular polygon A polygon all of whose sides and angles are equal.

Regular tiling A tiling by regular polygons, all of which have the same number of sides and are the same size; also, at each vertex, the same kinds of polygons must meet in the same order.

Rhombus A parallelogram all of whose sides are equal.

Scalene triangle A triangle no two sides of which are equal.

Semiregular tiling A tiling by regular polygons; all polygons with the same number of sides must be the same size.

Tiling A covering of the plane without gaps or overlaps.

Translation A rigid motion that moves everything a certain distance in one direction.

Vertex figure The pattern of polygons surrounding a vertex in a tiling.

SUGGESTED READINGS

CHOW, WILLIAM W. Automatic generation of interlocking shapes, *Computer Graphics and Image Processing,* 9 (1979): 333–353. Shows how to design a computer program to draw interlocking patterns.

CHOW, WILLIAM W. Interlocking shapes in art and engineering, *Computer Aided Design,* 12 (1980): 29–34. Discusses applications to sheet material manufacturing (e.g., fabrication of gloves, can openers, forks, key blanks, and bunk bed brackets).

CHUNG, FAN, AND SHLOMO STERNBERG. Mathematics and the buckyball, *American Scientist,* 81 (1993): 56–71.

FLAHERTY, TERRY. *Escher-Sketch,* Intellimation Library for the Macintosh, Box 219, Santa Barbara, CA 93116. Apple Macintosh program that allows you to design Escher-like patterns, using any of the 17 wallpaper patterns discussed in this chapter.

FOSTER, LORRAINE. *The Alhambra Past and Present: A Geometer's Odyssey.* Illustrates strip and wallpaper patterns from the Alhambra in Spain.

GARDNER, MARTIN. Mathematical games: Extraordinary nonperiodic tiling that enriches the theory of tiles, *Scientific American* (January 1977): 110–121, 132, and front cover. Reprinted with additional material in Martin Gardner, *Penrose Tiles to Trapdoor Ciphers,* Freeman, New York, 1989, pp. 1–29.

GARDNER, MARTIN. Mathematical games: On tessellating the plane with convex polygon tiles, *Scientific American* (July 1975): 112–117, 132. Reprinted with additional material in Martin Gardner, *Time Travel and Other Mathematical Bewilderments,* Freeman, New York, 1988, pp. 163–176.

GRÜNBAUM, BRANKO, AND G. C. SHEPHARD. *Tilings and Patterns,* Freeman, New York, 1987. Abbreviated edition: *Tilings and Patterns: An Introduction.*

KORYEPIN, V. Penrose patterns and quasi-crystals: What does tiling have to do with a high-tech alloy? *Quantum,* 4 (4) (January/February 1994): 13–19, 4 (5) (March/April 1994): 59–62.

MINNESOTA EDUCATIONAL COMPUTING CONSORTIUM (MECC). *Tesselmania!* Computer program for Macintosh (System 6.0.7 or later), with 100-page teacher's guide with lesson plans.

RANUCCI, ERNEST, AND JOSEPH TEETERS. *Creating Escher-Type Patterns,* Creative Publications, Oak Lawn, Ill.

SCHATTSCHNEIDER, DORIS. Will it tile? Try the Conway criterion! *Mathematics Magazine,* 53 (1980): 224–233.

SCHATTSCHNEIDER, DORIS. In praise of amateurs, in David A. Klarner, ed., *The Mathematical Gardner,* Wadsworth, Belmont, Calif., 1981, pp. 140–166, plus Plates I–III.

SCHATTSCHNEIDER, DORIS. *Visions of Symmetry: Notebooks, Periodic Drawings, and Related Work of M. C. Escher,* Freeman, New York, 1990.

SCHATTSCHNEIDER, DORIS. Penrose puzzles, *SIAM News,* 28 (6) (July 1995): 8, 14. Review of various commercial puzzles based on variations on the Penrose pieces, distributed by Kadon Enterprises (1227 Lorene Drive, Suite 16, Pasadena, MD 21122; 410-437-2163) and World of Escher (14542 Brook Hollow Boulevard, no. 250, San Antonio, TX 78232-3810; 800-237-2232).

SEYMOUR, DALE, AND JILL BRITTON. *Introduction to Tessellations,* Dale Seymour Publications, Palo Alto, Calif., 1989. An excellent introduction to tessellations, including how to make Escher-like tessellations.

TEETERS, JOSEPH L.   How to draw tessellations of the Escher type, *Mathematics Teacher,* 67 (1974): 307–310.

Tessellation Winners: Original Student Art, 2 vols. *Book One: The First Contest, 1989–90,* and *Book Two: The Second Contest, 1991–92,* Dale Seymour Publications, Palo Alto, Calif., 1991, 1993.

EXERCISES   ▲ *Optional.*   ■ *Advanced.*   ◆ *Discussion.*

Hint: For the exercises about determining whether a shape will tile the plane, you should make a number of copies of the shape and experiment with placing them. One easy way to make copies is to trace the shape onto a piece of paper, staple half a dozen other blank sheets behind that sheet, and use scissors to cut through all the sheets along the edges of the traced shape on the top sheet.

Regular Polygons

    1.   Determine the measure of an exterior angle and of an interior angle of a regular octagon (eight sides).

    2.   Determine the measure of an exterior angle and of an interior angle of a regular decagon (ten sides).

    3.   Discover a formula for the measure of an interior angle of a regular *n*-gon.

    4.   Using the formula from Exercise 3 and either your calculator or a short computer program, make a chart of the interior angle measures of regular polygons with 3, 4, . . . , 12 sides.

Regular Tilings

    5.   Give a numerical reason why a semiregular tiling could not include both polygons with 12 sides and polygons with 8 sides (with or without any polygons with other numbers of sides).

    6.   The lower left corner of Figure 21.5 shows a tiling by isosceles triangles.

        (a)   Use the center vertex to determine the measures of the angles of the isosceles triangle tile.

        (b)   Every vertex except the center vertex has the same vertex figure, in terms of the measures of the angles surrounding the vertex. What is that vertex figure?

Tilings with Irregular Polygons

7.   For each of the tiles below, show how it can be used to tile the plane.

Exercise 7 (Adapted from *Tilings and Patterns,* by Branko Grünbaum and G. C. Shephard, Freeman, New York, 1987, p. 25.)

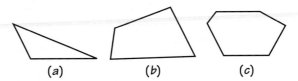

(a)   (b)   (c)

8.   You know that a regular pentagon cannot tile the plane. Suppose you cut one in half. Can this new shape tile the plane? (See page 805 for a regular pentagon that you can trace.)

Tiling by Translations

Refer to tiles (a) through (g) in doing Exercises 9 and 10.

Exercise 9 (From *Tiling the Plane,* by Frederick Barber et al., COMAP, Lexington, Mass., 1989, pp. 1, 8, 9.)

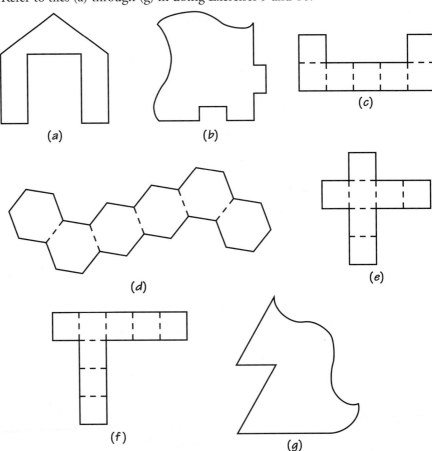

(a)   (b)   (c)

(d)   (e)

(f)   (g)

9.   For each of the tiles (a) through (c), determine if it can be used to tile the plane by translations.

10.   Repeat Exercise 9, but for the tiles (d) through (g).

11.   Start from a parallelogram of your choice and modify it to tile the plane by translations. (You will probably find it useful to do your work on graph paper.) Can you draw a design on the tile so as to make an Escher-like pattern?

12.   Start from a par-hexagon of your choice and modify it to tile the plane by translations. (You will probably find it useful to do your work on graph paper. If you choose a regular hexagon, there is special graph paper, ruled into regular hexagons, that would be particularly useful.) Can you draw a design on the tile so as to make an Escher-like pattern?

Tiling by Translations and Half-Turns

Refer to tiles (a) through (g) in doing Exercises 13 and 14.

13.   For each of the tiles (a) through (c), determine if it can be used to tile the plane by translations and half-turns.

14.   Repeat Exercise 13, but for the tiles (d) through (g).

15.   Show how an arbitrary pentagon with two parallel sides can tile the plane.

16.   Shown below is a pentagonal tile of type 13, discovered by Marjorie Rice. Show how it can tile the plane. (*Hint:* Carefully trace and cut out a dozen or so copies and try fitting them together.) The parts of this pentagon satisfy the following relations: $A = C = D = 120°$, $B = E = 90°$, $2A + D = 360°$, $2C + D = 360°$, $a = e$, and $a + e = d$.

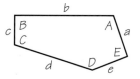

Exercise 16 (Adapted from "In Praise of Amateurs," by Doris Schattschneider, in *The Mathematical Gardner,* edited by David A. Klarner, Wadsworth, Belmont, Calif., 1981, p. 162.)

17. Start from a triangle of your choice and modify it to tile the plane by translations and half-turns. (You will probably find it useful to do your work on graph paper.) Can you draw a design on the tile so as to make an Escher-like pattern?

18. Start from a quadrilateral of your choice and modify it to tile the plane by translations and half-turns. (You will probably find it useful to do your work on graph paper.) Can you draw a design on the tile so as to make an Escher-like pattern?

Additional Exercises

19. Use the chart of interior angle measures from Exercise 4 to determine all of the possible vertex figures of regular polygons (with at most 12 sides) surrounding a point.

20. Which of the vertex figures of Exercise 19 do not occur in a semiregular tiling?

21. In addition to the vertex figures of Exercise 19, exactly five others are possible, each involving one polygon with more than 12 sides. None of these vertex figures leads to a semiregular tiling. The five many-sided polygons involved in these five vertex figures have 15, 18, 20, 24, and 42 sides. Determine the other polygons in these vertex figures.

▲ 22. In the text we discuss criteria and methods for generating Escher-like patterns that involve just translations or translations and half-turns. A slight variation on one of those methods allows construction of tilings that feature a tile and its mirror image.

Begin with a parallelogram made from two congruent isosceles triangles. Each of these triangles has two sides equal; be sure that the two triangles are arranged so that they have one of the equal sides in common, forming a diagonal of the parallelogram.

Start by modifying the two opposite sides of the parallelogram that are the third, unequal sides of their triangles. Make any modification to half of one side, then mirror-reflect that side across its center to extend the modification to the rest of the side. Take that entire modified side and reflect it across the original side. Let the mirror image be the pattern for modifying the opposite side.

Now modify in any way one of the other two remaining sides. Make the same change to the opposite side, without any rotation or reflection. The key step is to mirror-reflect one of these sides across its center and make this mirror-reflection change to the diagonal.

The result is a modified parallelogram that will tile by translation (though in a different way from the special cases that we used to introduce the Conway criterion) and that can be split into two pieces that are mirror images of each other. Escher used a similar technique, but starting from a par-hexagon made from two quadrilaterals, in his Horseman print, as shown in his sketch in Figure 21.11a.

Use this technique to produce a tiling of your own design. Can you draw a design on the tile so as to make an Escher-like pattern?

▲ 23.   Show that the modified parallelogram in Exercise 22 fulfills the Conway criterion, by identifying the six points of the criterion.

▲ 24.   The rabbit problem in Chapter 20 (Exercise 4, pages 828–829) can lead us directly into nonperiodic patterns and musical sequences. Let A denote an adult pair of rabbits and B denote a baby pair. We will record the population at the end of each month, just before any births, in a particular systematic way—as a string of A's and B's. At the end of their second month of life, a rabbit pair will be considered to be adult. At the end of the first month, the sequence is just A; and the same is true at the end of the second month. When an adult pair A has a baby pair B, we write the new B immediately to the right of the A. So at the end of the third month, the sequence is AB; at the end of the fourth, it is ABA, since the first baby pair is now adult; at the end of the fifth month we have $ABAAB$.

Mathematicians and computer scientists call this manner of generating a sequence a *replacement system*. At each stage we replace each A by AB and each B by A.

  (a)   What is the sequence at the end of the sixth month?
  (b)   Why can't we ever have two B's next to each other?
  (c)   Why can't we ever have three A's in a row?
  (d)   Show that from the fourth month on, the sequence for the current month consists of the sequence for last month followed by the sequence for two months ago.

WRITING PROJECT

1 ▶   Just as for Penrose patterns, we will define inflation and deflation for any sequence of A's and B's. Both of these operations will preserve musicality: If we inflate or deflate a musical sequence, we get another musical sequence.

(a)   Inflation can be used to generate musical sequences. Start at the first stage with just B. Inflation consists of replacing each A and AB and each B by A. Show that at the nth stage there are F_n (the nth Fibonacci number) symbols in the sequence.

(b)   Deflation can be used to check whether a finite block of A's and B's can belong to a musical sequence or not. Each deflation stage has two parts: first replace each A by $\frac{A}{2} B \frac{A}{2}$ and each B by $\frac{A}{2} \frac{A}{2}$, then combine pairs of adjacent $\frac{A}{2}$'s into a single A so that no fractional A's remain. Another way to get the same result is to proceed from left to right, replacing B by A, replacing AA by B, and deleting single A's. The deflated block will be shorter. If at any stage we have a block with two or more B's in a row, or three or more A's in a row, then the original block could not be part of a musical sequence; otherwise, the original block will eventually deflate to a single symbol, at which point we conclude that the original block is a part of a musical sequence (in fact, infinitely often, a part of every musical sequence). Check the two sequences $ABAABABAAB$ and $ABAABABABA$.

ANSWERS TO ODD-NUMBERED EXERCISES

CHAPTER 1

1. A:1; B:3; C:3; D:3; and E:0. The graph shows that geographically E is isolated, perhaps on an island or a different continent than the other cities.

3.

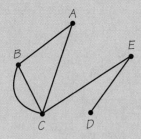

The valences are A:2; B:3; C:4; D:1; and E:2. The real-world consequences are that if you are at one city, you will be able to travel to another particular city because the graph is connected.

5. Remove the edges dotted in the figure below and the remaining graph will be disconnected.

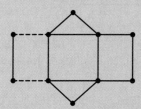

7. Discussion. Answers will vary.

9. The supervisor is not satisfied because all of the edges are not traveled upon by the postal worker. The worker is unhappy because the end of the worker's route wasn't the same point as where the worker began. The original job description is unrealistic because there is no Euler circuit in the graph.

11.

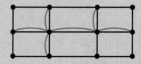

13.

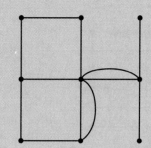

15. Edge 3 or 6 could be chosen, but not edge 2.

17.

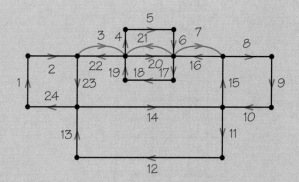

19. Remove the vertical edge in the middle

A-1

21.

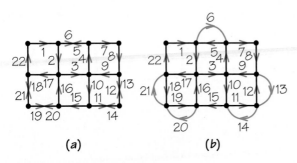

(a)   (b)

23.

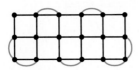

25.

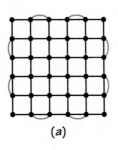

(a)

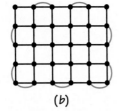

(b)

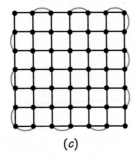

(c)

27.

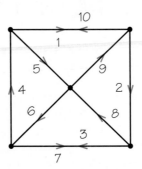

29. The graph is similar to the one in Exercise 28, except 3 of the bridges are represented by 2 edges, as shown below. This graph needs eulerization, but we don't show it. A circuit with one repeated edge is shown on this graph.

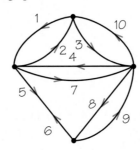

31.

33. Both graphs (b) and (c) have Euler circuits.

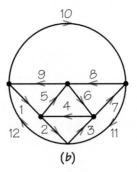

(b)

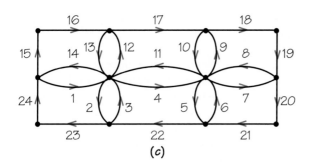

(c)

The valences of all of the vertices in (a) are odd, which makes it impossible to have an Euler circuit there.

35. There are many such graphs. Here is one:

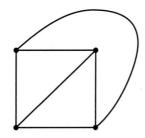

37. Graph (b).

39. Yes, because each street is represented by two edges, every vertex has even valence.

41.

43. When you attach a new edge to an existing graph, it gets attached at two ends. At each of its ends, it makes the valence of the existing vertex go up by one. Thus the increase in the sum of the valences is two. Therefore, if the graph had an even sum of the valences before, it still does, and if its valence sum was odd before, it still is.

45.

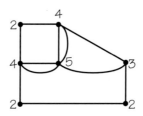

The graph is connected.

47. Discussion. Answers will vary.

CHAPTER 2

1. (a) $X_1, X_6, X_5, X_2, X_3, X_4, X_1$.
   (b) $X_1, X_6, X_7, X_8, X_9, X_{10}, X_{11}, X_{12}, X_5, X_4, X_3, X_2, X_1$.
   (c) $X_1, X_4, X_5, X_8, X_9, X_6, X_7, X_2, X_3, X_1$.
   (d) $X_1, X_2, X_5, X_8, X_3, X_4, X_7, X_6, X_1$.
   (e) $X_1, X_{10}, X_7, X_6, X_9, X_5, X_4, X_3, X_2, X_8, X_1$.

3. Other Hamiltonian circuits include $ABIGDCEFHA$ and $ABDCEFGIHA$.

5. $ACEDCA$ revisits the vertex C. This graph has no Hamiltonian circuit.

7. (a) Yes.
   (b) No Hamiltonian circuit.
   (c) No Hamiltonian circuit.

9. The n-cube has 2^n vertices, and the number of edges of the n-cube is equal to twice the number of edges of an $(n-1)$-cube plus 2^{n-1}. A formula for this number is $n2^{n-1}$.

11. Examples include inspection of traffic control devices at corners and placing new hour stickers on mailboxes located at street corners.

13. (a) $9 \cdot 8 \cdot 7 \cdot 6 \cdot 5 = 15{,}120$.
    (b) $(26)(26)(26) = 17{,}576$.

15. (a) $26(26)(26)(10)(10)(10) - (26)(26)(26) = (26)^3(10^3 - 1)$.
    (b) Answers will vary.

17. These graphs have 6, 10, and 15 edges, respectively. The n vertex complete graph has $[n(n-1)]/2$ edges. The number of TSP tours is 3, 12, and 60, respectively.

19. (a)

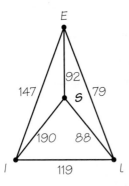

(b) (1) *UISEU*; mileage $= 119 + 190 + 92 + 79 = 480$.
   (2) *USIEU*; mileage $= 88 + 190 + 147 + 79 = 504$.
   (3) *UIESU*; mileage $= 119 + 147 + 92 + 88 = 446$.
(c) *UIESE* (Tour 3).
(d) No.
(e) Starting from *U*, one gets (see answer (b)) Tour 1. From *S* one gets Tour 2; from *E* one gets Tour 2, and from *I* one gets Tour 1.
(f) Tour 2. No.

21. *MACBM* takes 345 minutes to traverse.

23. A traveling salesman problem.

25. The complete graph shown has a different nearest-neighbor tour that starts at *A* (*AEDBCA*), a sorted-edges tour (*AEDCBA*), and a cheaper tour (*ADBECA*).

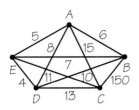

27. Optimal tour is the same, but its cost is now $3460 + 15(30) = 3910$.

29. (a) a. Not a tree since there is a circuit. Also, the wiggled edges do not include all vertices of the graph.
      b. The circuit does not include all the vertices of the graph.
   (b) a. The tree does not include all vertices of the graph.
      b. Not a circuit.
   (c) a. Not a tree.
      b. Not a circuit.
   (d) a. Not a tree.
      b. Not a circuit.

31. (a) 1, 2, 3, 4, 5, 8.
   (b) 1, 1, 1, 2, 2, 3, 3, 4, 5, 6, 6.
   (c) 1, 1, 1, 2, 2, 2, 2, 2, 3, 3, 3, 3, 4, 4, 4, 5, 5, 6, 7.
   (d) 1, 2, 2, 3, 3, 3, 4, 5, 5, 5, 6, 6.

33. Yes. At each stage of applying the algorithm for this graph, there is no choice about which edge to choose.

35. Yes. Change all the weights to negative numbers and apply Kruskal's algorithm. The resulting tree works, and the maximum cost is the negative of the answer you get. If the numbers on the edges represent subsidies for using the edges, one might be interested in finding a maximum-cost spanning tree.

37. A negative weight on an edge is conceivable, perhaps a subsidization payment. Kruskal's algorithm would still apply.

39. (a) True.   (b) False.   (c) True.   (d) False.
   (e) Not necessarily.

41. Two different trees with the same cost are shown:

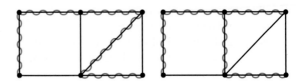

43.

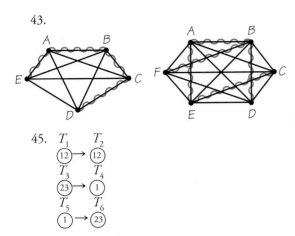

45.

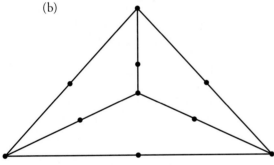

47. Answers will vary.

49. T_3, T_5, T_7 is currently critical. If T_5 or T_7 is reduced by just more than one, T_3, T_4 becomes critical.

51. (a) a. Add edge AB.
       b. Add edge $X_1 X_3$.
   (b)

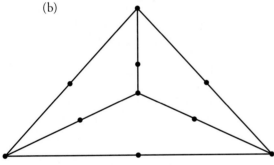

   (c) Yes: $ADGFBEC$. Figure (a).
       Yes: X_1, X_2, X_9, X_8, X_7, X_6, X_5, X_4, X_3.
       Figure (b).
   (d) Yes. A worker who must do an inspection of sites on her way to work.

53. With no other restrictions, 10^7. With no other restrictions, 9×10^2.

55. Optimal tour is the same, but its cost has now doubled to 9040.

57. (a) Stamp sizes: 5, 9, 10; denomination: 12.
   (b) Stamp sizes: 1, 8, 9, 15; denomination: 17. The greedy algorithm yields $15 + 1 + 1$, but uses three stamps. Using $8 + 9$ uses two stamps.

(c) Stamp sizes: 1, 2, 4, 8. For any denomination, the greedy choice is optimal.

CHAPTER 3

1. Jocelyn must pack, get to the airport, make various connections, perhaps kennel her dog, etc. Processors include plane, bus, taxi (to get to the airport), and Jocelyn herself. Unless she can get a friend to help her pack or take her dog to the kennel, none of the tasks can be done simultaneously.

3. (a) Operating room schedules, doctor schedules, emergency room staffing schedules, etc.
   (b) Schedules for the trains or buses and their crews, etc.
   (c) Scheduling runway use, reservation agents, food service for planes, etc.
   (d) Schedules for each mechanic, radiator repair, etc.
   (e) Schedules for drilling, welding, etc.
   (f) Schedules for washing clothes, cleaning rooms, dusting, etc.
   (g) Schedules for bus run, recess duty, etc.
   (h) Day, night, afternoon shift schedules, etc.
   (i) Firefighter shift schedules, schedules for checking if equipment on trucks is in repair, etc.

5. (a) Processor 1: T_1, T_6, idle 13 to 15; T_5, T_7, T_{11}, idle 34 to 38; T_{10}. Processor 2: T_2, T_9, idle 21 to 27; T_8, idle 38 to 45. Processor 3: T_3, T_4, idle 15 to 45.
   (b) Processor 1: T_1, T_3, T_4, T_6, T_7, T_9, T_{11}, idle 42 to 45. Processor 2: T_2, T_5, T_8, T_{10}.

7. (a) The critical path, which has length 17, is $T_1 T_2 T_3$.
   (b) T_1, T_4, T_5, T_2, T_6, T_7, T_3 is the list to be used. The one processor would have the tasks scheduled on it: T_1, T_4, T_5, T_2, T_6, T_7, T_3.
   (c) T_6, T_1, T_7, T_2, T_4, T_3, T_5 would be the list. The resulting schedule on one processor would be: T_1, T_7, T_4, T_2, T_5, T_6, T_3.

(d) No idle time. Their completion times are the same.

(e) Earlier completion of tasks giving rise to cash payments.

(f) The required schedule is Processor 1: T_1, T_2, T_7; Processor 2: T_4, T_5, T_6, T_3, idle 19 to 21.

(g) The completion time does not halve. As the number of processors goes up, the completion time may decrease, but at some point the length of the critical path will govern the completion time rather than the number of processors.

(h) (i) Completion time goes down by 7.

     (ii) Completion time is 19 for two processors using the decreasing time list.

9. Such criteria include decreasing length of the times of the tasks, order of size of financial gains when each task is finished, and increasing length of the times of the tasks.

11. (a) Task times: $T_1 = 3$, $T_2 = 3$, $T_3 = 2$, $T_4 = 3$, $T_5 = 3$, $T_6 = 4$, $T_7 = 5$, $T_8 = 3$, $T_9 = 2$, $T_{10} = 1$, $T_{11} = 1$, and $T_{12} = 3$. This schedule would be produced from the list: T_1, T_3, T_2, T_5, T_4, T_6, T_7, T_8, T_{11}, T_{12}, T_9, T_{10}.

(b) Task times: $T_1 = 3$, $T_2 = 3$, $T_3 = 3$, $T_4 = 2$, $T_5 = 2$, $T_6 = 4$, $T_7 = 3$, $T_8 = 5$, $T_9 = 8$, $T_{10} = 4$, $T_{11} = 7$, $T_{12} = 9$, $T_{13} = 3$. This schedule would be produced from the list: T_1, T_5, T_7, T_4, T_3, T_6, T_{11}, T_8, T_{12}, T_9, T_2, T_{10}, T_{13}.

13. (a) One reasonable possibility is (time in min):

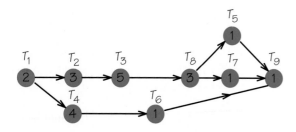

The earliest completion time is 15.

(b) The decreasing-time list is T_3, T_4, T_2, T_8, T_1, T_5, T_6, T_7, T_9. The schedule is Processor 1: T_1, T_4, T_6, idle 7 to 10, T_8, T_5, T_9; Processor 2: idle 0 to 2, T_2, T_3, idle 10 to 13, T_7, idle 14 to 15.

15. The task times total to 34. Since (34/3) rounded up is 12, the earliest completion time is 12.

17. (a) One such rule, admittedly artificial, could be: if by keeping a machine voluntarily idle, there is a longer task that becomes ready one time unit later, keep the machine idle.

(b) If a longer task becomes ready at a certain time than the remaining time on a task currently scheduled, schedule the longer task and reschedule the interrupted task later. An assumption must be made whether or not an interrupted task must be resumed on the machine it was originally scheduled on or can be rescheduled on any machine that becomes free.

19. (a) The tasks are scheduled on the machines as follows: Processor 1: 12, 13, 45, 34, 63, 43, 16, idle 226 to 298; Processor 2: 23, 24, 23, 53, 25, 74, 76; Processor 3: 32, 23, 14, 21, 18, 47, 23, 43, 16, idle 237 to 298.

(b) The tasks are scheduled on the machines as follows: Processor 1: 12, 24, 14, 34, 25, 23, 16, 16, 76, idle 183 to 240; Processor 2: 23, 23, 21, 63, 43, idle 173 to 240; Processor 3: 32, 23, 53, 74, idle 182 to 240; Processor 4: 13, 45, 18, 47, 43, idle 166 to 240.

(c) The decreasing-time list is 76, 74, 63, 53, 47, 45, 43, 43, 34, 32, 25, 24, 23, 23, 23, 23, 21, 18, 16, 16, 14, 13, 12.

     The tasks are scheduled on three machines as follows: Processor 1: 76, 45, 43, 24, 23, 18, 16, 13; Processor 2: 74, 47, 34, 32, 23, 21, 14, 12, idle 257 to 258; Processor 3: 63, 53, 43, 25, 23, 23, 16, idle 246 to 258.

The tasks are scheduled on four machines as follows: Processor 1: 76, 43, 24, 23, 16, idle 182 to 194; Processor 2: 74, 43, 25, 23, 16, 13; Processor 3: 63, 45, 32, 23, 18, 12, idle 193 to 194; Processor 4: 53, 47, 34, 23, 21, 14, idle 192 to 194.

(d) The new decreasing-time list is 84, 82, 71, 61, 55, 45, 43, 43, 34, 32, 25, 24, 23, 23, 23, 23, 21, 18, 16, 16, 14, 13, 12.

The tasks are scheduled as follows: Processor 1: 84, 45, 43, 25, 23, 23, 16, 12; Processor 2: 82, 55, 34, 32, 23, 18, 14, 13; Processor 3: 71, 61, 43, 24, 23, 21, 16, idle 259 to 271.

21. Examples include jobs in a videotape copying shop, data entry tasks in a computer system, scheduling nonemergency operations in an operating room. These situations may have tasks with different priorities, but there is no physical reason for the tasks not to be independent, as would be the case with putting on a roof before a house had walls erected.

23. (a) Each task heads a path of length equal to the time to do that task.

(b) (i) The worst finish time is
$$(2 - \tfrac{1}{3})(450) = 750$$

(ii) The worst finish time, if the decreasing-time list is used, is
$$[\tfrac{4}{3} - 1/(3)(3)](450) = 550.$$

25. The times to photocopy the manuscripts, in decreasing order, are 120, 96, 96, 88, 80, 76, 64, 64, 60, 60, 56, 48, 40, 32. Packing these in bins of size 120 yields Bin 1: 120; Bin 2: 96; Bin 3: 96; Bin 4: 88, 32; Bin 5: 80, 40; Bin 6: 76; Bin 7: 64, 56; Bin 8: 64, 48; Bin 9: 60, 60. Nine photocopy machines are needed to finish within 2 minutes using FFD. The number of bins would not change, but the placement of the items in the bins would differ for worst-fit decreasing.

27. (a) Using the next-fit algorithm, the bins are filled as follows: Bin 1: 12, 15; Bin 2: 16, 12; Bin 3: 9, 11, 15; Bin 4: 17, 12; Bin 5: 14, 17; Bin 6: 18; Bin 7: 19; Bin 8: 21; Bin 9: 31, Bin 10: 7, 21; Bin 11: 9, 23; Bin 12: 24; Bin 13: 15, 16; Bin 14: 12, 9, 8; Bin 15: 27; Bin 16: 22; Bin 17: 18.

(b) The decreasing list is 31, 27, 24, 23, 22, 21, 21, 19, 18, 18, 17, 17, 16, 16, 15, 15, 15, 14, 12, 12, 12, 12, 11, 9, 9, 9, 8, 7.

The next-fit decreasing schedule is Bin 1: 31; Bin 2: 27; Bin 3: 24; Bin 4: 23; Bin 5: 22; Bin 6: 21; Bin 7: 21; Bin 8: 19; Bin 9: 18, 18; Bin 10: 17, 17; Bin 11: 16, 16; Bin 12: 15, 15; Bin 13: 15, 14; Bin 14: 12, 12, 12; Bin 15: 12, 11, 9; Bin 16: 9, 9, 8, 7.

(c) The worst-fit schedule using the original list is Bin 1: 12, 15, 9; Bin 2: 16, 12; Bin 3: 11, 15; Bin 4: 17, 12; Bin 5: 14, 17; Bin 6: 18, 7; Bin 7: 19, 9; Bin 8: 21, 15; Bin 9: 31; Bin 10: 21, 9; Bin 11: 23, 8; Bin 12: 24; Bin 13: 16, 12; Bin 14: 27; Bin 15: 22; Bin 16: 18.

(d) The worst-fit decreasing schedule would be Bin 1: 31; Bin 2: 27, 9; Bin 3: 24, 12; Bin 4: 23, 12; Bin 5: 22, 12; Bin 6: 21, 15; Bin 7: 21, 15; Bin 8: 19, 17; Bin 9: 18, 18; Bin 10: 17, 16; Bin 11: 16, 12, 8; Bin 12: 15, 14, 7; Bin 13: 11, 9, 9.

29. Discussion. Answers will vary.

31. The total performance time exceeds what will fit on 4 disks. Using FFD, one can fit the music on 5 disks.

33. The proposed heuristic may fill many bins to capacity, but the computation to find weights summing to exactly W may be very time-consuming.

35. (a) Packing boxes of the same height into crates; packing want ads into a newspaper page.

(b) We assume, without loss of generality, $p \geq q$. One heuristic, similar to first-fit, orders the rectangles $p \times q$ as in a dictionary (i.e., $p \times q$ listed prior to $r \times s$ if $p > r$ or $p = r$ and $q \geq s$). It then puts the rectangles in place in layers in a first-fit manner; that is, do not put a rectangle into a second layer until all positions on the first layer are filled. However, extra room in the first layer is "wasted."

(b)

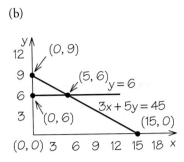

Finding point of intersection:

$y = 6$, so $3x + 5(6) = 45$, and $x = 5$

(c)

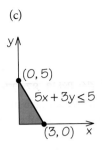

5. (a)

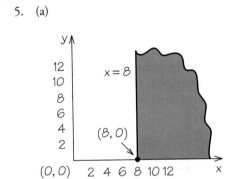

(b)

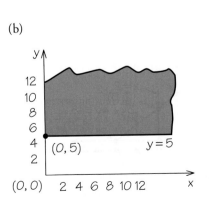

(d)

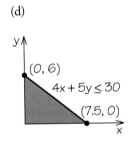

7. $4x + 2y \leq 28$.

9. $6x + 4y \leq 240$.

11.

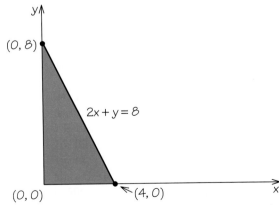

13.

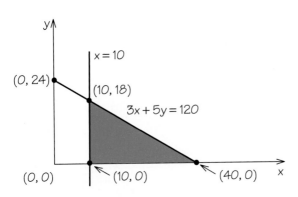

Finding point of intersection:
$x = 10$, so $3(10) + 5y = 120$, and $y = 18$

(b) $(10, 6)$ is a point of the feasible region of Exercise 13, but not of Exercises 11 and 15.

19. $(0, 0)$: 0 skateboards and 0 dolls.
   Profit $= \$2.30(0) + \$3.70(0) = \$0$.
   $(0, 30)$: 0 skateboards and 30 dolls.
   Profit $= \$2.30(0) + \$3.70(30) = \$111$.
   $(12, 0)$: 12 skateboards and 0 dolls.
   Profit $= \$2.30(12) + \$3.70(0) = \$27.60$.
Optimal production policy: make 0 skateboards and 30 dolls for a profit of \$111.

15.

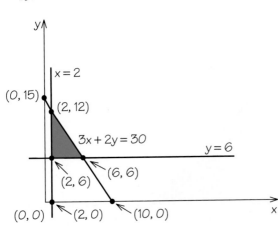

Finding points of intersection:
$x = 2$, so $3(2) + 2y = 30$, and $y = 12$
$y = 6$, so $3x + 2(6) = 30$, and $x = 6$

21. (a)

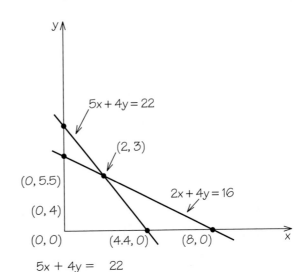

$$5x + 4y = 22$$
$$-1(2x + 4y = 16)$$
$$5x + 4y = 22$$
$$-2x - 4y = -16$$
$$3x \quad\quad = 6$$
$$x \quad\quad = 2$$

$$5(2) + 4y = 22$$
$$so \quad y \quad = 3$$

17. (a) $(2, 4)$ is a point of the feasible region of Exercise 11, but not of Exercises 13 and 15.

(b)

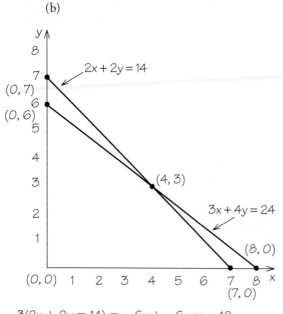

$$3(2x + 2y = 14) = \quad 6x + \quad 6y = \quad 42$$
$$-2(3x + 4y = 24) = -6x + -8y = -48$$
$$\overline{\qquad\qquad\qquad\qquad -2y = \quad -6}$$
$$\qquad\qquad\qquad\qquad\quad y = \quad 3$$

$2x + 2(3) = 14$, so $x = 4$

23.

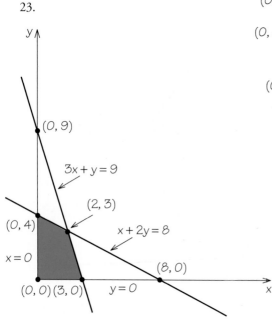

25.

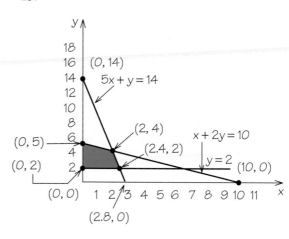

27.

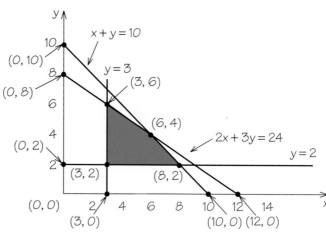

29. (a) (4, 2) is a point of the feasible region of Exercise 27, but not of Exercises 23 and 25.

(b) (1, 3) is a point of the feasible region of Exercises 23 and 25, but not of Exercise 27.

31. (a)

	Cloth (600 yds)	Mins	Profit
Shirts, x items	3	100	$5
Vests, y items	2	30	$2

(b) Constraint inequalities   Profit formula
Cloth: $3x + 2y \leq 600$ $\$5x + \$2y$
Mins: $x \geq 100, y \geq 30$

(c)

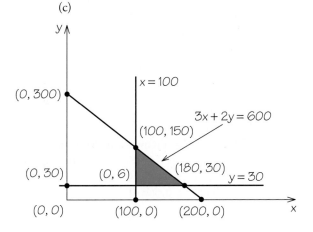

(d) Profit
At (100, 30), profit = $560.
At (100, 150), profit = $800.
At (180, 30), profit = $960.*
Make 180 shirts and 30 vests.

With zero minimums, the feasible region looks like this:

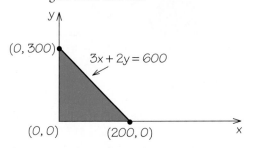

At (0, 0), profit is $0.
At (0, 300), profit is $600.
At (200, 0), profit is $1000.*
Make 200 shirts and no vests.

33. (a)

	Space (90 sq ft)	Profit
Refrigerator, x	10	$30
Stove, y	15	$40

(b) Constraint inequalities   Profit formula
Space: $10x + 15y \leq 90$ $\$30x + \$40y$
Mins: $x \geq 0, y \geq 0$

(c)

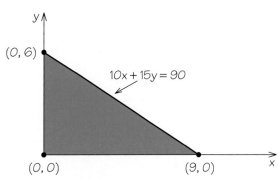

(d) Profit
At (0, 0), profit is $0.
At (0, 6), profit is $240.
At (9, 0), profit is $270.*
Display 9 refrigerators and no stoves.

With nonzero minimums of $x \geq 3$ and $y \geq 2$, the feasible region looks like:

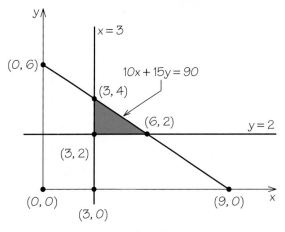

At (3, 2), profit is $170.
At (3, 4), profit is $250.
At (6, 2), profit is $260.*
Display 6 refrigerators and 2 stoves.

35. (a)

	Bushels (600)	Mins	Profit
Oysters, x bushels	1	100	$8
Clams, y bushels	1	200	$10

(b) **Constraint inequalities**   **Profit formula**

Bushels: $x + y \leq 600$ $\$8x + \$10y$

Mins: $x \geq 100, y \geq 200$

(c)

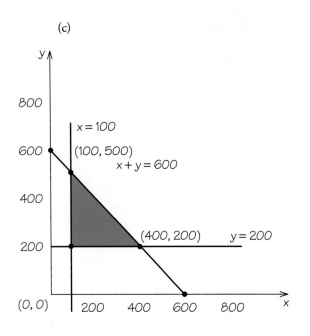

(d) **Profit**

At (100, 200), profit = $2800.

At (100, 500), profit = $5800.*

At (400, 200), profit = $5200.

Process 100 bushels of oysters and 500 bushels of clams.

   With zero minimums, $x \geq 0$, $y \geq 0$, feasible region is:

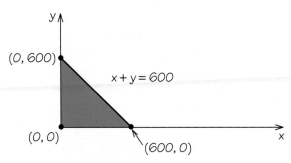

At (0, 0), profit = $0.

At (0, 600), profit = $6000.*

At (600, 0), profit = $4800.

Process no oysters and 600 bushels of clams.

37. (a)

	Oven (12 hr)	Prep (16 hr)	Profit
Bread, x	1.5	1	$0.50
Cakes, y	1	2	$2.50

(b) **Constraint inequalities**   **Profit formula**

Oven: $1.5x + 1y \leq 12$ $\$0.50x + \$2.50y$

Prep: $1x + 2y \leq 16$

Mins: $x \geq 0, y \geq 0$

(c)

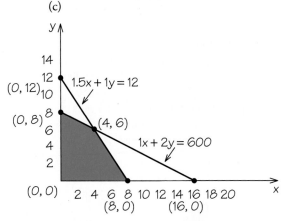

(d) **Profit**

At (0, 0), the profit is $0.00.

At (0, 8), the profit is $20.00.*

At (4, 6), the profit is $17.00.

At (8, 0), the profit is $4.00.

Make no breads and 8 cakes.

   With nonzero minimums of $x \geq 2$ and $y \geq 3$, we get this feasible region:

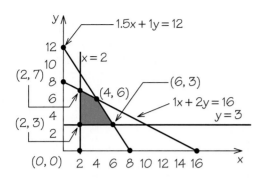

At (2, 3), profit is $8.50.
At (2, 7), profit is $18.50.*
At (4, 6), profit is $17.00.
At (6, 3), profit is $10.50.
Make 2 breads and 7 cakes.

39. (a)

	Space (100 acres)	Money ($2600)	Profit
Potatoes, x acres	1	$20	$25
Broccoli, y acres	1	$40	$60

(b) **Constraint inequalities**   **Profit formula**
Space: $1x + 1y \leq 100$ $\$25x + \$60y$
\$: $20x + 40y \leq 2600$
Mins: $x \geq 0, y \geq 0$

(c)

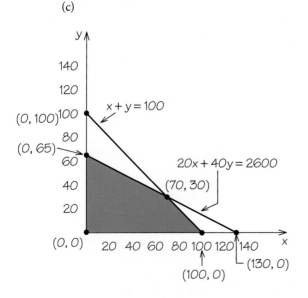

(d) **Profit**
At (0, 0), profit = $0.
At (0, 65), profit = $3900.*
At (70, 30), profit = $3550.
At (100, 0), profit = $2500.
Plant no acres of potatoes and 65 acres of broccoli.

With nonzero minimums of $x \geq 20$ and $y \geq 20$, the feasible region looks like this:

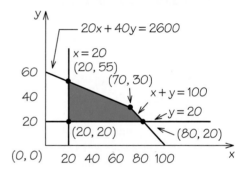

At (20, 20), profit is $1700.
At (20, 55), profit is $3800.*
At (70, 30), profit is $3550.
At (80, 20), profit is $3200.
Plant 20 acres of potatoes and 55 acres of broccoli.

41. (a)

	Bird Count (100)	Cost $ (2400)	Profit
Pheasants, x	1	$20	$14
Partridges, y	1	$30	$16

(b) **Constraint inequalities**   **Profit formula**
Count: $1x + 1y \leq 100$ $14x + 16y$
\$: $20x + 30y \leq 2400$
Mins: $x \geq 0, y \geq 0$

(c) **Feasible Region**

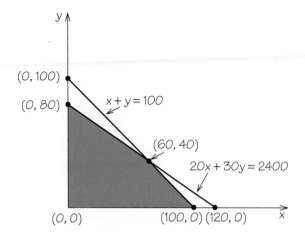

(d) **Profit**
At (0, 0), profit is $0.
At (0, 80), profit is $1280.
At (60, 40), profit is $1480.*
At (100, 0), profit is $1400.
Raise 60 pheasants and 40 partridges.
  With nonzero minimum of $x \geq 20$ and $y \geq 10$, there is no change because the optimal production policy obeys these minimums.

43. (a)

	Shaper (50)	Smoother (40)	Painter (60)	Profit
Toy A, x	1	2	1	$4
Toy B, y	2	1	3	$5
Toy C, z	3	2	1	$9

(b) **Constraint inequalities**
Shaper: $1x + 2y + 3z \leq 50$
Smoother: $2x + 1y + 2z \leq 40$
Painter: $1x + 3y + 1z \leq 60$
Mins: $x \geq 0, y \geq 0, z \geq 0$

Profit formula
$4x + 5y + 9z$
(c) Optimal product policy
Make 5 toy A, no toy B, 15 toy C for a profit of $155.

45. (a)

	Chocolate (1000 lb)	Nuts (200 lb)	Fruit (100 lb)	Profit
Special, x boxes	3	1	1	$10
Regular, y boxes	4	0.5	0	$6
Purist, z boxes	5	0	0	$4

(b) **Constraint inequalities**   **Profit formula**
$3x + 4y + 5z \leq 1000$ $10x + 6y +$
$1x + 0.5y + 0z \leq 200$ $4z$
$1x + 0y + 0z \leq 100$
(c) Make 100 boxes of Special, 175 boxes of Regular, and 0 boxes of Purist.

47. A feasible region has infinitely many points; the corner point principle tells us we only need to evaluate the profit formula at a few of those points (the corner points), not all of them.

49. (a)

	Time (240 mins)	Profit
Tee-shirts, x pieces	4	$5
Jeans, y pieces	6	$4

(b) **Constraint inequalities**   **Profit formula**
Time: $4x + 6y \leq 240$ $5x + 4y$
Mins: $x \geq 0, y \geq 0$

(c)

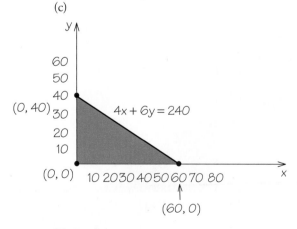

(d) **Profit**
At (0, 0), profit = $0.

At $(0, 40)$, profit $= \$160$.
At $(60, 0)$, profit $= \$300.^*$
Help 60 tee-shirt customers and no jeans customers.

With nonzero minimums $x \geq 12$, $y \geq 10$, feasible region looks like:

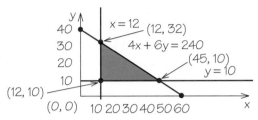

At $(12, 10)$, profit is $\$100$.
At $(12, 32)$, profit is $\$188$.
At $(45, 10)$, profit is $\$265.^*$
Help 45 tee-shirt and 10 jeans customers.

51. (a)

	Machine (12 hr)	Point (16 hr)	Profit
Bikes, x	2	4	$12
Wagons, y	3	2	$10

(b) **Constraint inequalities**   **Profit formula**
Mach: $2x + 3y \leq 12$ $\$12x + \$10y$
Paint: $4x + 2y \leq 16$
Mins: $x \geq 0, y \geq 0$

(c)

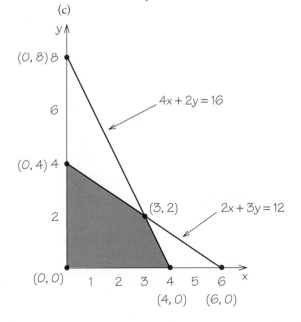

(d) **Profit**
At $(0, 0)$, profit is $\$0$.
At $(0, 4)$, profit is $\$40$.
At $(3, 2)$, profit is $\$56.^*$
At $(4, 0)$, profit is $\$48$.
Make 3 bikes and 2 wagons.

With nonzero minimums of $x \geq 2$ and $y \geq 2$ there will be no change because the optimal production policy obeys these minimums.

CHAPTER 5

1. *Population:* Employed adult women. *Sample:* The 48 women who returned the questionnaire.

3. *Population:* All chips of this type made by the supplier, including future production. *Sample:* The 400,000 chips inspected.

5. Probably higher, because those who believe they have benefited are more likely to respond.

7. Voluntary response; women with strong negative opinions are more likely to write. Probably lower.

9. If labels 00 to 29 are assigned to the 30 students in alphabetical order, the sample consists of 28 = Trevino, 14 = Hughes, and 01 = Anderson.

11. (a) The faculty. (b) Label the faculty (say, in alphabetical order) 000 to 379. Read three-digit groups from Table 5.1. (c) 290, 304, 307, 312, and 276.

13. (a) If labels 00 to 24 are assigned to the members in alphabetical order, the tickets go to 21 = Wang, 15 = Myrdal, 01 = Binet, 03 = Blum, and 17 = Spencer. Note that 21 is repeated and that it is necessary to continue to the next line of the table. (b) The answers obtained will depend on the parts of the table used. In the long run, we expect an average of two of the five tickets to go to women. (c) A rough empirical answer is based on how many of your 20 samples included cases in which no women received tickets. (In fact, the probability that no tickets go to women is about .056. It is somewhat unlikely that no women will receive tickets.)

15. 46% of the sample believes in life on other planets. (A recent opinion poll found this result.) We can be confident that between 43% and 49% of all adults believe in extraterrestrial life.

17. The effect (if any) of the tea is confounded with the effect of visits and conversation with college students. The visits alone might make the residents more cheerful.

19. The unemployment rate is most strongly influenced by general economic conditions; it might have been still higher without the training program. The effect (if any) of the program on unemployment is confounded with the stronger effect of economic conditions.

21. Observational study. The subjects had already been processed by the housing authority. They were not assigned to housing as part of the study. (Note that the effect of public housing on family stability is confounded with the effects of all factors that influenced the housing authority to accept some and reject others.)

23. Subjects who do not receive the drug should receive a placebo to avoid confounding the placebo effect with the effect of the drug. In the absence of a reason to do otherwise, it is best to assign equal numbers of subjects to each treatment. Here is the design:

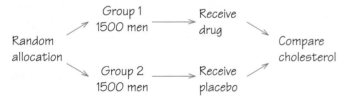

25. Because the diagnosis of mild heart attacks is somewhat subjective, it is best if the diagnosing physicians are blind in both experiments. The study of Exercise 23 should be double-blind; in Exercise 24c the subjects know whether or not they are in an exercise program, so they cannot be blind.

27. (a) Because students choose which version to take, there may be systematic differences between the two groups of students. Any such differences are confounded with the method of instruction. (b) The outline is similar to Figure 5.4, with 15 students in each group and the two teaching methods as the treatments. If we label the subjects 00 to 29, Group 1 contains 10, 07, 24, 25, 04, 14, 27, and so on until 15 are chosen. The remaining 15 subjects make up Group 2.

29. Because of the time required for the randomization, this problem is best done by assigning a different starting point to each of several students and combining their results. In the randomization of the previous problem (with labels in alphabetical order), two of the eight asterisks are assigned to Group 1.

31. To do this, choose a simple random sample of 5 to form Group 1. Then choose a simple random sample of 5 of the remaining 15 for Group 2 and another simple random sample of 5 of the remaining 10 for Group 3. The 5 who still remain are Group 4. It is best to relabel the 10 who remain after the second stage as 0 to 9 to speed the third sample. If the initial labels are 00 to 19, Group 1 contains 03, 04, 12, 13, and 15. Group 2 contains 17, 07, 10, 05, and 00. The members of Group 3 depend on relabeling.

33. The average earnings of men exceeded those of women by so much that it is very unlikely that the chance selection of a sample would produce so large a difference if there were not a difference in the entire student population. But the black–white difference was small enough that it might be due to the accident of which students were chosen for the sample.

35. (a) The sample can contain only people with phones. This will underrepresent poor people. (b) Only registered voters are queried; they are on the average older and richer than adults who are not registered.

37. If labels 00 to 31 are used for the 32 candidates in alphabetical order, the sample contains 07 = Drasin, 29 = Vlasov, 10 = Garcia, and 05 = Chan.

39. Take as subjects a large group of young people who have never smoked. Divide them at random into (say) three groups. Group 1 will begin to smoke at age

16 and smoke two packs a day for life. Group 2 will smoke until age 40 and then stop. Group 3 will never smoke. Compulsion will be necessary. The design is similar to Figure 5.6.

41. (a) Follow the model of Figure 5.4, with 20 women in each group. (b) If we label the subjects 00 to 39 in alphabetical order, the 20 who read the version that mentions child care are: Gerson, Cortez, Rosen, Gutierrez, Morse, Iselin, Rivera, Durr, Danielson, Howard, Adamson, Cansico, Lippman, Ullmann, Wong, Curzakis, Chen, Martinez, Abrams, Hwang.

43. Many designs are possible. One variable is with/without ZIP code. Another might be typed versus handwritten address. It is best to mail all letters on the same day of the week and to the same city to remove variation due to these factors.

45. (a) The factors, or experimental variables, are type of corn (normal or floury-2) and protein level (12%, 16%, or 20%). (b) The experimental units are 60 chicks; these are divided at random into 6 groups of 10 chicks each; each group is fed one of the 6 diets; weight gains after 21 days are measured and compared. The outline is similar to Figure 5.6, but with 6 groups rather than 3.

47. The average ethnocentrism score was higher among church attenders by an amount so large that it would only rarely occur just by chance variation.

CHAPTER 6

1. There are no outliers or other unusual features.

3. The distribution is roughly symmetric, with center near noon (12 hours from midnight). There are no outliers or gaps.

5. (a) Your histogram will depend a bit on your choice of cells. Here is one choice:

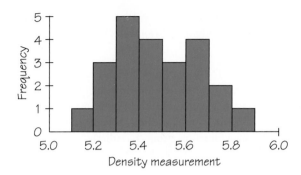

(b) The distribution is roughly symmetric, with no outliers or other unusual features. (With only 23 observations, we cannot insist on a close approach to exact symmetry.)

7. (a) $\bar{x} = 2.45$, $M = 2$. (b) $Q_1 = 1$, $Q_3 = 4$.

9. (a) The 1968 and 1992 results may be low outliers. In both years, third-party candidates drew over substantial shares of the vote.

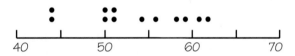

(b) $M = 52.3\%$. (c) $Q_3 = 58.1\%$, so that the 1964 (Johnson defeats Goldwater), 1972 (Nixon defeats McGovern), and 1984 (Reagan defeats Mondale) elections were landslides.

11. (a) $\bar{x} = 2.7$, $\Sigma(x - \bar{x})^2 = 28.1$, $s = 1.767$.

13. (a) The five-number summary is 210, 220, 223, 230.5, 283. The middle 50% of the distribution falls in the range between the quartiles, 220 to 230.5 millimeters. (b) Because of the high outlier and right skewness, the mean will be larger than the median. In fact, $\bar{x} = 8145/36 = 226.25$. (c) $s = 12.52$. Because of the right skewness and outliers, s should not be used to describe this distribution.

15. $\bar{x} = 126.12/23 = 5.483$ and $s = 0.190$.

17. The distribution of salaries is strongly skewed to the right, so the median is smaller than the mean.

19. The five-number summary for the home run data is 22, 36.5, 40.5, 44.5, 51. The middle half falls

between the quartiles 36.5 and 44.5, or because home run counts must be whole numbers, between 37 and 44 inclusive.

21. (a) The explanatory variable x is power boat registrations. (c) There is an increasing and roughly linear overall pattern, but the points are not tightly clustered about a straight line. There are no clear outliers (1983 is closest to being an outlier).

23. (a) Speed is the explanatory variable, plotted horizontally. (b) The relationship is curved. At first, increasing speed uses less fuel as the car shifts to higher gears; eventually air resistance and friction mean that going faster uses more fuel. (c) There is no constant direction. (d) The points follow a clear pattern with little wobble, so the relationship is quite strong.

25. (a) The plot against year has a clear increasing linear pattern until the latest three years. Both 1992 and 1993 stand out as low outliers, and 1994 is slightly low. (b) The mean observed count is 42; predictions will vary somewhat.

27. For $x = 20$, $y = 5.27$ hundred cubic feet per day; for $x = 40$, $y = 9.31$ hundred cubic feet per day.

29. (a) There is a general increasing straight-line pattern. Spaghetti and snack cake both fall above the overall pattern; that is, the estimated calories for these two foods are unusually high relative to their true calories. (b) If the outliers are ignored, the line will pass close to the other eight points, including the points for macaroni and a candy bar. Use the "up and over" graphical method for prediction as in Figure 6.8b; the value of y depends on the line drawn. (c) For $x = 200$, the least-squares line predicts $y = 318.6$ calories.

31. Using all data points, the line is $y = -35.1786 + 0.112x$. The prediction for $x = 716$ is $y = 45.0$ manatees killed per year.

33. The least squares regression line is $y = 260.56 - 22.969x$.

35. The five-number summary for males is 46.9, 47.4, 51.9, 62.0, 62.9; for females, it is 33.1, 38.2, 42.0, 49.55, 54.6. The generally higher lean body mass of males is apparent in the boxplots. The extremes in

the male distribution do not extend far beyond the quartiles, but this is not surprising in a boxplot of only $n = 7$ observations.

37. (a) Here is a histogram, using classes of width 2%.

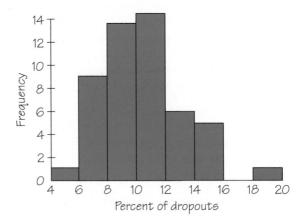

(b) The distribution is very roughly symmetric, or at least not clearly skewed. There is an outlier: the District of Columbia, which is a city rather than a state. Dropout rates are higher in urban schools.

39. Choose the south to consist of AL, AR, DC, FL, GA, KY, LA, MS, NC, SC, TN, and VA. (Other choices are possible.) Then the five-number summary for these states is 10.4, 11.8, 12.8, 13.85, 19.1, and the five-number summary for the remaining 39 states is 4.3, 7.9, 9.6, 11, 14.9. As the boxplot shows, dropouts are much more common in the south.

41. The histogram shows that the distribution is strongly skewed to the right. So we prefer the five-number summary. The five-number summary is 43, 82.5, 102.5, 151.5, 598. The right-skewness of the distribution is reflected in the fact that Q_3 is much farther from the median than is Q_1, and the largest observation is much farther from the median than is the smallest.

43. (a) The smallest possible s is 0; any four identical numbers have $s = 0$. (b) 0, 0, 10, 10 (largest possible spread). (c) In part (a), yes. In part (b), no.

45. The high scores of the top students will pull the mean above the median. So colleges that recruit

some top students prefer the mean. Colleges with few such students will look better if the median (the typical student's score) is used.

47. A single high outlier is enough. For the data 1, 1, 2, 3, 3, 4, 28, the third quartile is 4 and the mean is 42/7 = 6.

49. Make a histogram or dotplot of the SAT mathematics scores. The distribution has two peaks, showing two clusters of states. The cluster with lower scores contains states where most students take the SAT.

51. (a) Higher. The equation of the regression line shows that math scores are roughly 27 points higher. (b) For $x = 422$, the prediction is $y = 461.7$. (c) Hawaii's math score is higher (or its verbal score lower) than the general relationship suggests. Over 60% of Hawaii's population consists of Asians and Pacific Islanders, so it is possible that less use of English may lower verbal scores.

CHAPTER 7

1. Results will vary.

3. Results will vary.

5. (a) $S = \{0, 1, 2, 3, 4, 5, 6, 7, 8, 9, 10\}$.
   (b) $S = \{0, 10, 20, 30, 40, 50, 60, 70, 80, 90, 100\}$.
   (c) $S = \{Yes, No\}$.

7. With obvious abbreviations, $S = \{TMS, TMN, TFS, TFN, CMS, CMN, CFS, CFN\}$.

9. Model A has entries with sum 1.1, so is not legitimate. Model B has entries with sum 0.9, so is not legitimate. Model C assigns probabilities that are all between 0 and 1 and have sum 1, so is legitimate.

11. P(Chavez is promoted) = 0.55.

13. The model is

Sum	2	3	4	5	6	7	8	9	10	11	12
Probability	$\frac{1}{36}$	$\frac{2}{36}$	$\frac{3}{36}$	$\frac{4}{36}$	$\frac{5}{36}$	$\frac{6}{36}$	$\frac{5}{36}$	$\frac{4}{36}$	$\frac{3}{36}$	$\frac{2}{36}$	$\frac{1}{36}$

(a) $\frac{8}{36}$. (b) $\frac{21}{36}$.

15. Repeats allowed: $(20 \times 20 \times 20)/(26 \times 26 \times 26) = 0.455$. No repeats allowed: $(20 \times 19 \times 18)/(26 \times 25 \times 24) = 0.438$.

17. No x: $(35 \times 35 \times 35)/(36 \times 36 \times 36) = 0.919$. No digits: $(26 \times 26 \times 26)/(36 \times 36 \times 36) = 0.377$.

19. $\frac{21}{6} = 3.5$.

21. $\frac{12}{8} = 1.5$.

23. (a) 0.81. (b) 2.68.

25. Group exercise; results will vary.

27. Draw a normal curve, then mark the axis so that the center (mean) is at 69 and the change-of-curvature points are at 66.5 and 71.5. The curve should reach the horizontal axis at about 61.5 and 76.5.

29. The first quartile is 67.325 inches and the third quartile is 70.675 inches.

31. (a) 50%. (b) 2.5% (half of 5%).

33. $10/\sqrt{3} = 5.77$ milligrams.

35. We want $10/\sqrt{n} = 5$, so $n = 4$ times. The average of several measurements is less variable than a single measurement.

37. Results will vary.

39. $S = \{0, 1, 2, 3, \ldots\}$ (all nonnegative whole numbers) is the simplest choice because we do not have to decide a largest possible amount.

41. In a shuffled deck, all 13 possible outcomes are equally likely, so each has probability 1/13. Any assignment that obeys Laws 1 and 2 is legitimate.

43. There are $10^4 = 10,000$ possible four-digit sequences and 10 ways of getting all four identical, so the probability is 0.001.

45. The possibilities are *ags, asg, gas, gsa, sag, sga*, of which "gas" and "sag" are English words. The probability is 2/6 = 0.33.

47. (a) The probabilities are all between 0 and 1 and have sum 1. (b) $A = \{9, 10, 11, 12\}$, $P(A) = 0.931$. (c) 11.251.

49. Results will vary.

51. (a) 2.5% (half of 5%). (b) 0.15% (half of 0.3%).

53. 628 or above.

CHAPTER 8

1. 64.5 is a statistic, 63 a parameter.

3. Both are parameters.

5. Mean = 35%, standard deviation $\sigma_{\hat{p}}$ = 3.37%.

7. 1.18%, 1.26%, 1.29%, 1.26%, 1.18%.

9. If we repeated Gallup's sampling process, we might get a different answer. But 95% of all samples will give an answer that is within ± 3% of the percent of all adults who jog. Gallup announces the margin of error to allow for this variation and indicate how accurate his result will usually be.

11. 59.3% ± 8.02%. Because the interval falls entirely above 50%, we are 95% confident that more than half of all visitors are in favor.

13. Only (c). The margin of error includes only the random-sampling error described by the sampling distribution of the statistic.

15. The margin of error would be greater than ± 3 points. Smaller sample sizes result in wider intervals for the same level of confidence. (Because the poll did not use a simple random sample, we cannot calculate the margin of error.)

17. Mean = 12 cm, standard deviation $\sigma_{\bar{x}}$ = 0.002 cm.

19. (a) 0.001 ± 0.0002 inch. (b) 0.001 ± 0.0001 inch.

21. (a) Normal with mean 0 and standard deviation 0.0173 meter. (b) Between ± 0.0346 meter.

23. 53.123 ± 0.035.

25. 36.9 ± 1.74 points.

27. 0.7505 ± 0.00032 inch. The interval does not contain 0.75, so we are confident that μ is not 0.75.

29. Center line = 50, control limits = 50 ± 0.035.

31. The control charts in Exercises 31 to 33 have center line (drawn solid) at 101.5 and control limits (drawn dashed) at 101.2 and 101.8. Here there are no points outside the limits and no run of 8 or more on the same side of the center line.

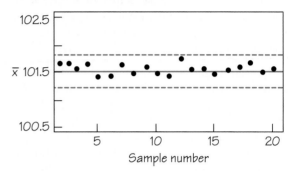

33. Samples 15, 18, and 19 fall above the upper control limit. The last six points lie above the center line. There is a clear upward drift in the plot.

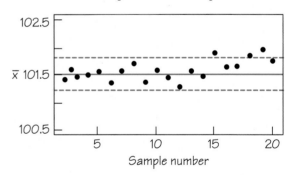

35. The percents of students who smoke for each parent condition are: Both parents smoke, 22.5%; one parent smokes, 18.6%; neither parent smokes, 13.9%. There is a clear association between parent smoking and student smoking, with students with smoking parents being more likely to smoke themselves.

37. (a) The two-way table is

	Bill	Will
Hits	120	130
Outs	380	370
At bats	500	500

Bill gets a hit .240 of his times at bat, while Will's average is .260. So Will has the higher batting average. (b) Bill hits better against right-handers (.400 vs. .300) and also against left-handers (.200 vs. .100). (c) Both players hit much better against right-handers; Will bats against right-handers much more often than does Bill. So even though Bill does better against both types of pitchers than Will, Will has a higher average overall. We should choose Bill for our team.

39. (a) 47% ± 3%. (b) Another sample would probably give a different percent. We must take into account this sampling variability when announcing a conclusion about the population. (c) This particular result (47% ± 3%) was obtained by a method that in 95% of all samples gives an interval covering the true population percent.

41. (a) 20% ± 5.66%. (b) 80% ± 8%.

43. (a) The distribution is roughly symmetric; the overall shape depends on the choice of classes, but more advanced methods show that the distribution is approximately normal. (b) 35.1 ± 3.32. (c) No. One class in one school is probably not a random sample of children in the district.

45. 59.3% ± 4.01%. The 68% confidence interval is half as wide as the 95% confidence interval (one standard deviation rather than two) because a smaller margin of error is sufficient if we allow lower confidence in the result.

47. (a) $\hat{p} \pm 1.28\sqrt{\hat{p}(100 - \hat{p})/n}$. (b) 59.3% ± 5.13%.

CHAPTER 9

1. (a) 51593-2067; 2. (b) 50347-0055; 1. (c) 44138-9901; 1.

3. (a) 20782-9960. (b) 55435-9982. (c) 52735-2101.

5. If a double error in a block results in a new block that does not contain exactly two long bars, we know this block has been misread. If a double error in a block of five results in a new block with exactly two long bars, the new block gives a different digit than the original one. If no other digit is in error, the check digit

catches the error, since the sum of the 10 digits will not end in 0. So, in every case an error has been detected. Errors of the first type can be corrected just as in the case of a single error. When a double error results in a legitimate code number, there is no way to determine which digit is incorrect.

7. 3.

9. 3.

11. 5.

13. 2.

15. X.

17. 9.

19. 2.

21. F.

23. No. The computer only needs to know which digit is the check digit.

25. (a) 7. (b) 4. (c) 7. (d) 2.

27. 0-669-09325-4.

29. The check digit would be the same.

31. S000, S200, L550, L300, E663, O451.

33. 42758.

35. March 29, female; September 17, male.

37. If you replace each short bar in Figure 9.8 with an *a* and each long bar in Figure 9.8 with a *b*, the resulting strings are listed in alphabetical order.

39. Substitution of *b* for *a* where $|b - a| = 5$ in positions 1, 5, 7, 9, and 11 is undetected; all errors in position 3 are undetected; substitution of *b* for *a* where $b - a$ is even in position 8 is undetected; the transposition $ab \rightarrow ba$ increases the check digit by $(b - a)$ if $b \le a$ and by $b - a + 10$ if $b > a$.

41. Z05428925245913640.

43. Since many people don't like to make their age public, this method is used to make it less likely that people would notice that the license number encodes year of birth.

45. The combination 72 contributes $7 \cdot 1 + 2 \cdot 3$ = 13 or $7 \cdot 3 + 2 \cdot 1 = 23$ (depending on the location of the combination) toward the total and 3 toward the check digit (since only the last digit of the total matters). The combination 27 also contributes 3 toward the check digit. So, the error is not detected. When the combination 26 contributes 0 to the check digit, the combination 62 contributes 2 to the check digit. Also, when the combination 26 contributes 2 to the check digit, the combination 62 contributes 0 to the check digit. In both cases the check digit for the number resulting from the transposition error will not match the check digit for the correct number.

47. The Canadian scheme detects any transposition error involving adjacent characters. Also, there are 21,866,000 possible Canadian codes, but only 100,000 U.S. 5-digit ZIP codes. Hence the Canadian scheme can target a location more precisely.

CHAPTER 10

3. (a) 6. (b) 3.

5. 1001101.

7. 000000, 100011, 010101, 001110, 110110, 101101, 011011, 111000.

9. 0000000, 1000001, 0100111, 0010101, 0001110, 1100110, 1010100, 1001111, 0110010, 0101001, 0011011, 1110011, 1101000, 1011010, 0111100, 1111101. No, since 1000001 has weight 2.

11. 000000, 100101, 010110, 001011, 110011, 101110, 011101, 111000. 001001 is decoded as 001011; 011000 is decoded as 111000; 000110 is decoded as 010110.

13. a and b.

15. 23, 49, 16.

17. 13.

19. The integer x is encrypted as $(x + 3) \mod 26$.

21. 111101000111001010; AABAACAEADB.

23. t, n, and r; e.

25. In the Morse code a space is needed to determine where each code word ends. In a fixed-length code of length k, a word ends after each k digit.

27. 00000000, 00010111, 00101110, 01001011, 10001101, 11000110, 10100011, 10011010, 01100101, 01011100, 00111001, 11101000, 11010001, 10110100, 01110010, 11111111. The code will detect any three errors or correct any single error.

29. $2^5 = 32$.

31. 0000, 1012, 2021, 0111, 0222, 1120, 2210, 2102, 1201.

33. $3^4 = 81; 3^6 = 729$.

CHAPTER 11

1. A dictatorship satisfies condition (2): If a new election is held and every voter (in particular, the dictator) reverses his or her ballot, then certainly the outcome of the election is reversed. A dictatorship also satisfies condition (3): If a single voter changes his or her ballot from being a vote for the loser of the previous election to being a vote for the winner of the previous election, then this single voter could not have been the dictator (since the dictator's ballot was not a vote for the loser of the previous election). Thus, the outcome of the new election is the same as the outcome of the previous election. A dictatorship, however, fails to satisfy condition (1): If the dictator exchanges marked ballots with any voter whose marked ballot differs from that of the dictator, then the outcome of the election is certainly reversed.

3. Minority rule satisfies condition (1): An exchange of marked ballots between two voters leaves the number of votes for each candidate unchanged, so whichever candidate won on the basis of having fewer votes before the exchange still has fewer votes after the exchange. Minority rule also satisfies condition (2): Suppose candidate X receives n votes and candidate Y receives m votes, and candidate X wins because $n < m$. Now suppose that a new election is held, and every voter reverses his or her vote. Then candidate X has m votes and candidate Y has n votes, and so candidate Y is the new winner. Minority rule, however, fails to satisfy condition (3): Suppose, for example, that there are 3 voters, and that candidate X wins with 1 out of the 3 votes. Now suppose that one of the 2 voters who voted for candidate Y reverses his or her vote. Then candidate X would have 2 votes, and candidate Y would have 1 vote, thus resulting in a win for candidate Y.

5. (a) Plurality: A (with 4 first-place votes).
   (b) Borda: B (with 15 points).
   (c) Hare: C (first D is eliminated, then B).
   (d) Sequential pairs with A, B, C, D agenda: D.

7. (a) Plurality: A.
   (b) Borda: B.
   (c) Hare: E.
   (d) Sequential pairs with B, D, C, A, E agenda: E.

9. (a) Plurality: A.
   (b) Borda: D.
   (c) Hare: B.
   (d) Sequential pairs with B, D, C, A agenda: B.

11. (a) Plurality satisfies Pareto: If everyone prefers B to D, then D has no first-place votes at all. Thus, D cannot be among the winners in plurality voting.
    (b) Plurality satisfies monotonicity: If an alternative wins on the basis of having the most first-place votes, then moving that alternative up one spot on some list (and making no other changes) neither decreases the number of first-place votes for the winning alternative nor increases the number of first-place votes for any other alternative. Hence, the original winner remains a winner in plurality voting.

13. (a) A Condorcet winner always wins the kind of one-on-one contest that is used to produce the winner in sequential pairwise voting.
    (b) Moving an alternative up one spot on some list only improves that alternative's chances in one-on-one contests.

15. (a) If the first election is contested using plurality voting, then A wins with two first-place votes. If the second election is contested the same way, then A and B tie with two first-place votes. Thus, B has gone from nonwinner status to winner status even though no voter reversed the order in which he or she had B and the winning alternative from the previous election (i.e., A) ranked.
    (b) If the Hare system is used instead of plurality, we still have A winning the first election, and A and B tying for the win in the second election. Thus, the argument here is the same as in part (a).

17. Alternative A wins if the agenda is D, C, B, A. Alternative B wins if the agenda is A, C, D, B. Alternative C wins if the agenda is D, B, A, C. Alternative D wins if the agenda is B, A, C, D.

19.

A	J	I	H	G	F	E	D	C	B
B	A	J	I	H	G	F	E	D	C
C	B	A	J	I	H	G	F	E	D
D	C	B	A	J	I	H	G	F	E
E	D	C	B	A	J	I	H	G	F
F	E	D	C	B	A	J	I	H	G
G	F	E	D	C	B	A	J	I	H
H	G	F	E	D	C	B	A	J	I
I	H	G	F	E	D	C	B	A	J
J	I	H	G	F	E	D	C	B	A

21. Consider the following sequence of preference lists:

Number of Voters

Rank	1	1	1
First	A	B	A
Second	B	C	C
Third	C	A	B

Alternative A is a Condorcet winner, and thus it must be the unique winner of the election contested under our hypothetical voting rule. Therefore, A is a winner and B is a nonwinner (for *this* sequence of preference lists).

Because our hypothetical voting rule satisfies independence of irrelevant alternatives, we know that alternative B will remain a nonwinner as long as no voter reverses his or her ordering of B and A. But to arrive at the preference lists from the voting paradox, we can move C (the alternative that is irrelevant to A and B) up one slot in the third voter's list. Thus, because of IIA, we know that alternative B is a nonwinner when our voting rule is confronted by the preference lists from the voting paradox of Condorcet.

The argument showing that C is a nonwinner is similar.

23. (a) $3! + 3(2!) + 3(1!) + 0! = 16$.
    (b) $3!$ (no ties) $+ 3(2)$ (two tied) $+ 1$ (three tied) $= 13$.

25. Answers will vary.

27. (a) Plurality: C.
    (b) Borda: E.
    (c) Sequential pairs with A, B, C, D, E agenda: E.
    (d) Hare: D.

29. (a) A.
    (b) B.
    (c) There would be no difference in the ranking of the nominees.

CHAPTER 12

1. In (a), a coalition is winning if and only if it has at least five members, and it is blocking if and only if it has at least five members. Therefore if the voters in a blocking coalition all vote Y, they will form a winning coalition. In (b), winning coalitions need seven Y votes, but blocking coalitions only need six N votes. A blocking coalition with six members could not be transformed into a winning coalition. In (c), a winning coalition requires all nine members of the jury to vote Y. Every juror has veto power, so all coalitions except the empty coalition are blocking coalitions; and all coalitions except the full jury have insufficient power to pass a motion. We conclude that the answer to this question is that it depends on the voting system.

3. (a) (i), $\{A\}$, $\{A, B\}$; (ii), $\{A\}$, $\{A, B\}$; (iii), none; (iv), B.

(b) (i), $\{A, B\}$, $\{A, C\}$, $\{A, B, C\}$; (ii), $\{A, B\}$, $\{A, C\}$, $\{A, B, C\}$; (iii), $\{A\}$; (iv), none.
(c) (i), $\{A, B\}$, $\{A, C\}$, $\{A, B, C\}$; (ii), $\{A, B\}$, $\{A, C\}$, $\{A, B, C\}$; (iii), $\{A\}$; (iv), none.
(d) (i), $\{A, B\}$, $\{A, C\}$, $\{A, B, C\}$; (ii), $\{A, B\}$, $\{A, C\}$, $\{A, B, C\}$; (iii), $\{A\}$; (iv), none.
(e) (i), $\{A, B\}$, $\{A, C\}$, $\{A, B, C\}$, $\{A, B, D\}$, $\{A, C, D\}$, $\{A, B, C, D\}$; (ii), $\{A, B\}$, $\{A, C\}$, $\{A, B, C\}$, $\{A, B, D\}$, $\{A, C, D\}$, $\{A, B, C, D\}$; (iii), $\{A\}$, $\{A, D\}$; (iv), D.
(f) (i), $\{A, B\}$, $\{A, C\}$, $\{A, B, C\}$, $\{A, B, D\}$, $\{A, C, D\}$, $\{A, B, C, D\}$; (ii), $\{A, B\}$, $\{A, C\}$, $\{A, B, C\}$, $\{A, B, D\}$, $\{A, C, D\}$, $\{A, B, C, D\}$; (iii), $\{A\}$, $\{A, D\}$; (iv), D.
(g) (i), $\{A, B\}$, $\{A, C\}$, $\{A, B, C\}$, $\{A, B, D\}$, $\{A, C, D\}$, $\{A, B, C, D\}$; (ii), $\{A, B\}$, $\{A, C\}$, $\{A, B, C\}$, $\{A, B, D\}$, $\{A, C, D\}$, $\{A, B, C, D\}$; (iii), $\{A\}$, $\{A, D\}$; (iv), D.
(h) (i) All coalitions including A and at least two others; (ii) all coalitions including A and at least two others; (iii) $\{A\}$, and all coalitions of A and one other voter; (iv), none.

5. $\{A, B\}$, $\{A, C\}$, $\{A, D\}$; and $\{B, C, D\}$.

7. (a) $\{A, B\}$, $\{A, C\}$, $\{A, D\}$, $\{A, B, C\}$, $\{A, B, D\}$, and $\{A, C, D\}$.
   (b) $\{A, B\}$, $\{B, C, D\}$.

9. (a) $(4, 0)$.
   (b) $(4, 4, 4)$.
   (c) $(4, 4, 4)$.
   (d) $(4, 2, 2)$.
   (e) $(8, 8, 8, 0)$.
   (f) $(8, 8, 8, 0)$.
   (g) $(8, 8, 8, 0)$.
   (h) $(12, 12, 12, 12, 12)$.

11. Every voting combination in which k voters vote Y is also a combination with $n - k$ voters voting N. Thus if everyone changes his or her vote, we will have a voting combination with $n - k$ voters voting Y.

13. (a) 15.
    (b) 4950.
    (c) 4950.
    (d) 252.

15. (a) (24, 8, 8, 0, 0) (The Glen Cove and Long Beach supervisors are dummy voters.)

17. $[q: w(C), w(M_1), \ldots, w(M_6)] = [4:1, 1, 1, 1, 1, 1, 1]$.

19. The four ordinary members can pass a motion that the chair opposes, since she does not have a veto. Thus the minimal winning coalitions are $\{C, M_1\}$, $\{C, M_2\}$, $\{C, M_3\}$, $\{C, M_4\}$, and $\{M_1, M_2, M_3, M_4\}$, all of which have a total weight of 4.

21. (a) $[q: w(D), w(F_1), w(F_2), w(F_3)] = [4:2, 1, 1, 1]$.
    (b) $[q: w(P), w(D), w(F_1), w(F_2), w(F_3)] = [6:2, 2, 1, 1, 1]$.
    (c) This system is not equivalent to a weighted system.

23. All four voter systems can be presented as weighted voting systems.

Minimal winning coalitions	Weights
$\{A,B,C,D\}$	[4:1,1,1,1]
$\{A,B\}, \{A,C,D\}$	[5:3,2,1,1]
$\{A,B,C\}, \{A,B,D\}$	[5:2,2,1,1]
$\{A,B\}, \{A,C\}, \{A,D\}$	[4:3,1,1,1]
$\{A,B\}, \{A,C\}, \{B,C,D\}$	[5:3,2,2,1]
$\{A,B\}, \{A,C,D\}, \{B,C,D\}$	[4:2,2,1,1]
$\{A,B,C\}, \{A,B,D\}, \{A,C,D\}$	[4:2,1,1,1]
$\{A,B\}, \{A,C\}, \{A,D\}, \{B,C,D\}$	[4:3,2,1,1]
$\{A,B,C\}, \{A,B,D\}, \{A,C,D\}, \{B,C,D\}$	[3:1,1,1,1]

25. $(\frac{1}{2}, \frac{1}{6}, \frac{1}{6}, \frac{1}{6})$.

27. (a) The dean has a Shapley–Shubik index of $\frac{1}{2}$, and each faculty member has an index of $\frac{1}{6}$.
    (b) The dean and provost have Shapley–Shubik indices of $\frac{7}{20}$, and each faculty member has an index of $\frac{1}{10}$.
    (c) Each of the administrators has a Shapley–Shubik index of $\frac{13}{105}$, while each faculty member has an index of $\frac{11}{70}$.

29. Among the 100 small shareholders, there will be an average of 50 yes votes, with a standard deviation of 5. The big shareholder will be a critical voter if the number of small shareholders voting the yes is between 41 and 60; that is, within two standard deviations from the mean. By the 68–95–99.7 percent rule, the big shareholder will be a critical voter in 95% of the voting combinations. A small shareholder can be a critical voter only when joined by 60 other small shareholders (and not the big shareholder), or when joined by 40 small shareholders and the big shareholder. The probability that the small shareholder vote will break 60–40 and the big shareholder will be on the right side is, by the same reasoning, less than 2.5%. Thus, the big shareholder is more than $95/2.5 = 38$ times as powerful as the small shareholder.

31. $(\frac{7}{12}, \frac{1}{4}, \frac{1}{12}, \frac{1}{12})$.

33. (e).

35. In a set of n voters, there are no voting combinations with more than n Y votes.

37. If there are $2m$ voters, there are $2C_m^{2m-1}$ swings out of 2^{2m} voting combinations. The following table lists the probabilities.

Number of Voters	2	4	6	8	10	12	14
Probability	$\frac{1}{2}$	$\frac{3}{8}$	$\frac{5}{16}$	$\frac{35}{128}$	$\frac{63}{256}$	$\frac{231}{1024}$	$\frac{429}{2048}$

39. $[q: w(M), w(C), w(P), w(K), w(H), w(Q), w(X), w(S)] = [9:4, 4, 4, 1, 1, 1, 1, 1]$.

41. (a) 16.
    (b) 12.
    (c) 4.
    (d) If the senators vote independently, each will be a critical voter in 12 of the 32 voting combinations. If we assume the combinations are equally likely, each senator has probability $\frac{3}{8}$ of being a critical voter. The pact makes $\{A, B\}$ a critical voter in $\frac{3}{4}$ of the voting combinations in which A and B vote together. C has probability $\frac{1}{4}$ of being a critical voter.

43. (a) (6, 2, 2).
    (b) There are four voting combinations in which C and D take opposite sides. C is a critical voter in two of the combinations,

and D is a critical voter in two. $\{A, B\}$ is a critical voter in all of the combinations, so the Banzhaf index is (4, 2, 2). Every voter has increased probability of being a critical voter, but the power of C and D is increased relative to that of $\{A, B\}$ by the quarrel!

CHAPTER 13

1. Calvin gets the cannon, 43% of the unopened chest, the doubloon, the sword, the cannon ball, the wooden leg, the flag, and the crow's nest (for a total of 61.5 of his points). Hobbes gets the rest.

3. Answers will vary.

5. Mary receives the car and gives John $15,081.25.

7. Mary receives the car, and John receives the house and pays Mary $13,668.75.

9. E receives the Duesenberg and Cord and pays $8500, F receives the Bentley and Aston-Marton and pays $7500, and G receives the Ferrari plus $16,000.

11. The chooser.

13. (a)

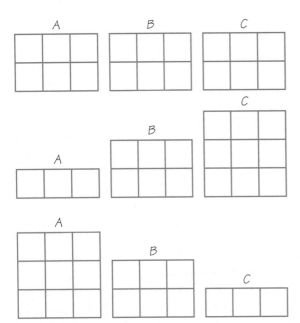

(b) Player 2 finds B acceptable (6 square units) and C acceptable (9 square units). Player 3 finds A acceptable (9 square units) and B acceptable (6 square units).

(c) Player 3 chooses A (9 square units). Player 2 chooses C (9 square units). Player 1 chooses B (6 square units). Yes. Player 2 chooses C (9 square units). Player 3 chooses A (9 square units). Player 1 chooses B (6 square units).

15. (a)

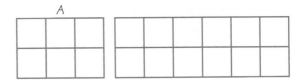

(b) Player 2 will view A as being only 3 square units, and thus he will pass.

(c) Player 3 will view A as being 9 square units, and will thus diminish it to yield A' as follows:

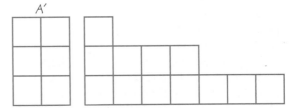

(d) Player 3 gets the piece cut off the cake (A'), because he was the last player to diminish it. He thinks it is 6 square units. (The player receiving the first piece always thinks it is $1/n$ of the cake, assuming that all n players interpret "acceptable" in this way, and all follow the prescribed strategies.)

(e) If Player 1 cuts the rest he will make each piece 7 square units. Player 2 will choose the rightmost piece, which he thinks is 10 square units.

(f) If player 2 cuts the rest, he will make each piece 8 square units. Player 2 will choose the leftmost piece, which he thinks is $8\frac{2}{3}$ square units.

(g) Player 1 will cut off 6 squares again. Player 2, thinking it is 5 square units, will pass. Player 1 receives the piece, and Player 2 gets what is left (which he thinks is 11 square units).

(h) Player 2 will cut off 6 squares. Player 1, thinking it is 7 square units, will trim it and take it. Player 2 will then receive what is left, which he thinks is 11 square units.

17. (a)

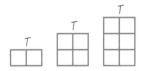

(b)

$^2/_3$ square wide

(c) Player 3 chooses any one of the three; he thinks they are all 2 square units. Player 1 chooses either of the remaining two; he thinks both are $\frac{2}{3}$ square unit. Player 2 receives the remaining piece; he thinks it is $\frac{4}{3}$ square units.

19. Player 2 trims four pieces to create a five-way tie for largest. Player 3 trims two pieces to create a three-way tie for largest. Finally, player 4 trims one piece to create a two-way tie for largest.

21. Player 1 will think the leftover from the eleventh stage is at most $(\frac{4}{5})^{11}$ of the whole cake, which is approximately 0.086. Hence, this is less than one-tenth of the cake.

23. (a) Ted thinks he is getting at least one-third of the piece that Bob initially received, and at least one-third of the piece that Carol initially received. Thus, Ted thinks he is getting at least one-third of part of the cake (Bob's piece) plus one-third of the rest of the cake (Carol's piece).

(b) Bob gets to keep exactly two-thirds (in his own view) of the piece that he initially received and thought was at least of size one-half. Two-thirds times one-half equals one-third.

(c) If, for example, Ted thinks the "half" Carol initially gets is worthless, then Ted may wind up thinking that he (Ted) has only slightly more than one-third of the cake, while Bob has (in Ted's view) almost two-thirds of the cake. In such a case, Ted will envy Bob.

25. (a) If a player follows the suggested strategy, then clearly he or she will receive a piece of size exactly one-fourth *if* he or she does, in fact, call cut at some point. How could a player (Bob, for example) fail to call cut when using this strategy? Only if each of the other three players "preempted" Bob by calling cut before he did each time the knife was set in motion. But this means that each of the other three left with a piece that Bob considered to be of size less than one-fourth. Hence, when the other three players have left with their shares, there is, in Bob's view, over one-fourth of the cake left for him.

(b) If you call cut first—and thus exit the game with a piece of size exactly one-fourth in your estimation—you will envy the next player to receive a piece *if* no one calls cut until the next piece is larger than one-fourth in your estimation.

(c) If there are four players and the first player has exited with his or her piece, then you could wait to call cut until the knife reaches the point where one half of the original cake is left. Alternatively, you could wait until the knife passed over one-third of what was left.

CHAPTER 14

1.

Party	Vote	Quota	Lower Quota	Appor- tionment
Social Democrats	323,829	23.0369	23	23
Democratic Socialists	880,702	62.6524	62	63
Christian Democrats	5,572,614	396.4309	396	396
Greens	1,222,498	86.9674	86	87
Communists	111,224	7.9124	7	8
Totals	8,110,867	577.0000	574	577

3. The apportionments are given in the following table:

State	Original Census	Corrected Population
A	42	43
B	27	26
C	30	29
D	1	2
Totals	100	100

States B and C had population increases and decreased apportionments; although the population of state D decreased slightly, its apportionment increased. This is the population paradox.

5. If the fractional part of q is less than 0.5, then $q + 0.5$ is still less that $\lceil q \rceil$. Therefore $\lfloor q + 0.5 \rfloor = \lfloor q \rfloor = \langle q \rangle$ in this case. If the fractional part of q is greater than or equal to 0.5, then $q + 0.5 \geq \lceil q \rceil$ so $\lfloor q + 0.5 \rfloor = \lceil q \rceil = \langle q \rangle$.

7.

Class	Enrolled	Apportionments Hamilton	Jefferson	Webster
Geometry	76	3	3	3
Algebra	19	1	0	1
Calculus	20	1	1	1
Total	115	5	4	5
Divisor	—		19	30.4

Divisors are not unique; answers in the following ranges are valid:

$$\text{Jefferson} \quad 19$$
$$\text{Webster} \quad 31.8 - 30.5$$

With the Jefferson method, geometry and algebra are tied for the fifth section.

9. The quotas are 15.5, 10.5, and 10 seats, which round to 16, 11, and 10 seats, respectively, to give tentative apportionments whose sum is 37. The divisors necessary to reduce each state's apportionment by one are found by dividing the respective populations by 15.5, 10.5, and 9.5. The divisors are therefore 10,000, 10,000, and 10,527. We are to take the smallest of these; there is an obvious tie.

11.

State	Population (%)	Hamilton	Jefferson	Webster
National	92.15	92	95	90
Splinter #1	1.59	2	1	2
Splinter #2	1.58	2	1	2
Splinter #3	1.57	2	1	2
Splinter #4	1.56	1	1	2
Splinter #5	1.55	1	1	2
Total 100	100	100	100	100
Divisor	—		0.97	1.02

Divisors are not unique; answers in the following ranges are valid:

$$\text{Jefferson} \quad 0.960 - 0.970$$
$$\text{Webster} \quad 1.019 - 1.029$$

Each of the divisor methods violates the quota condition.

13. 40%.

15. 6.25%.

17. (a) Ohio, 573,017; Kansas, 621,400.
(b) 8.44%
(c) 21.67%

19. The average district populations for states A and B are p_A/a_A and p_B/a_B, respectively. Assuming that

the average district population for A is greater than that of B, the absolute difference is

$$\frac{p_A}{a_A} - \frac{p_B}{a_B} = \frac{p_A a_B - p_B a_A}{a_A a_B}.$$

To obtain the relative difference, divide the absolute difference by the smaller average district population, p_B / a_B. This yields

$$\frac{p_A a_B - p_B a_A}{a_A a_B} \div \frac{p_B}{a_B} = \frac{p_A a_B - p_B a_A}{a_A p_B}. \tag{1}$$

Upon multiplying this expression by 100%, we arrive at the desired formula.

In calculating the relative differences of representative share, note that the state with the larger average district population will have the smaller representative share. Thus the absolute difference in representative share for these states is

$$\frac{a_B}{p_B} - \frac{a_A}{p_A} = \frac{p_A a_B - p_B a_A}{p_A p_B}.$$

The relative difference is again obtained by dividing the absolute difference by the smaller representative share, a_A / p_A. The result is

$$\frac{p_A a_B - p_B a_A}{p_A p_B} \div \frac{a_A}{p_A} = \frac{p_A a_B - p_B a_A}{a_A p_B}.$$

Since this expression is identical to the expression (1) for relative difference in average district size, the two relative measures of inequity are the same.

21. With 10 seats for Massachusetts, and 6 for Oklahoma, the inequity in representative share is 0.2488 microseats, in favor of Oklahoma. If Massachusetts had 11 seats, and Oklahoma 5, the inequity would be less, at 0.2350 microseats, in favor of Massachusetts. Therefore, the Webster apportionment would give Massachusetts the seat.

23. Geometry and Algebra, 2 sections each; Calculus, 1 section. Divisors between 17.6 and 29.6 can be used.

25. 147,232.

27. (a) The sum of the quotas is h, and, unless each quota is a whole number, the total will increase when rounding occurs. (b) Small states would benefit more than the large states, since rounding up gives them a larger percentage increase in representation than large states get.

29.

State	Population	Hamilton	Jefferson	Webster	Hill
A	27,774	10	11	10	10
B	25,178	9	9	9	9
C	19,947	7	7	8	7
D	14,614	5	5	5	6
E	9,225	4	3	3	3
F	3,292	1	1	1	1
Total	100,030	36	36	36	36
Divisor		—	2,524	2,659	2,668

Divisors are not unique; answers in the following ranges are valid:

Jefferson	2517.9–2524.9
Webster	2651.1–2659.5
Hill	2665.5–2668.1

31. (a) One quota will be rounded up, and the other down to obtain the Webster apportionment. The quota that is rounded up will have a fractional part greater than 0.5, and greater than the fractional part of the quota that is rounded down. The Hamilton method will give the party whose quota has the larger fractional part an additional seat. Thus the apportionments will be identical.

(b) These paradoxes never occur with the Webster method, which gives the same apportionment in this case.

(c) The Hamilton method, which always satisfies the quota condition, gives the same apportionment.

(d) No. Suppose one party gets only 0.6% the the vote, and the other party gets 99.4%. With a divisor of 0.994, Jefferson would apportion all 100 seats to the second party, while Hamilton would apportion the seats 1 to 99. On the other hand, Hill–Huntington will give at least one seat to any party that receives at least one vote. Thus their apportionment would differ from Hamilton's if the vote were to split 0.4% to 99.6%.

33. (a) Lowndes favors small states, because in computing the relative difference, the fractional part of the quota will be divided by the lower quota. If a large state had a quota of 20.9, the Lowndes relative difference works out to be 0.045. A state with a quota of 1.05 would have priority for the next seat.

(b) Yes.

(c) Yes. Since the method satisfies the quota condition, the population paradox is inevitable.

(d) Let r_i denote the relative difference between the quota and lower quota for state i.

State	p_i	q_i	$\lfloor q_i \rfloor$	r_i (%)	Rank	a_i
VA	630,560	18.310	18	1.7	14	18
MA	475,327	13.803	13	6.2	8	14
PA	432,879	12.570	12	4.8	9	12
NC	353,523	10.266	10	2.7	13	10
NY	331,589	9.629	9	7.0	7	10
MD	278,514	8.088	8	1.1	15	8
CT	236,841	6.877	6	14.6	6	7
SC	206,236	5.989	5	19.8	5	6
NJ	179,570	5.214	5	4.3	10	5
NH	141,822	4.118	4	3.0	11	4
VT	85,533	2.484	2	24.2	4	3
GA	70,835	2.057	2	2.9	12	2
KY	68,705	1.995	1	99.5	1	2
RI	68,446	1.988	1	98.8	2	2
DE	55,540	1.613	1	61.3	3	2
Totals	3,615,920	105	97	—	—	105

35. Let $f_i = q_i - \lfloor q_i \rfloor$ denote the fractional part of the quota for state i. Since the Hamilton method assigns to each state either its lower or its upper quota, each absolute deviation is equal to either f_i (if state i received its lower quota) or $1 - f_i$ (if it received its upper quota). For convenience, let's assume that the states are

ordered so that the fractions are decreasing, with f_1 the largest and f_n the smallest. If the lower quotas add up to $h - k$, where h is the house size, then states 1 through k will receive their upper quotas. The maximum absolute deviation will be the larger of $1 - f_k$ and f_{k+1}. To achieve any other apportionment that satisfies the quota condition, we would have to start by transferring a seat from a state j, where $j \leq k$ to a state l, where $l > k$. The new absolute deviations would be f_j (since now state j has its lower quota) and $1 - f_l$ (because state l has its upper quota). Because of the way the fractions have been ordered, we have $f_j \geq f_{k+1}$ and $1 - f_l \geq 1 - f_k$. Therefore, the absolute deviation for one of states j and l will be equal to or exceed the maximum absolute deviation of the Hamilton apportionment. This proves that no apportionment that meets the quota condition has a lower maximum absolute deviation. Apportionments that do not satisfy the quota condition always have maximum absolute deviation in excess of 1, while Hamilton apportionments always have maximum absolute deviation less than 1.

CHAPTER 15

1. (a) and (b) Saddlepoint at row 1 (maximin strategy), column 2 (minimax strategy), giving value 5. (c) Row 2 and column 1.

3. (a) No saddlepoint. (b) Rows 1 and 2 are both maximin strategies; column 1 is the minimax strategy. (c) None.

5. (a) and (b) Saddlepoint at row 3 (maximin strategy), column 2 (minimax strategy), giving value -20. (c) Columns 1 and 2.

7. Batter's optimal mixed strategy is (1/4, 3/4), and pitcher's is (1/4, 3/4), giving value .275.

9. Saddlepoint is "not cheat" and "audit," giving value $-\$100$.

11. (a)

	Officer Does Not Patrol	Officer Patrols
You park in street	0	-40
You park in lot	-32	-16

(b) You: $(\frac{2}{7}, \frac{5}{7})$; officer: $(\frac{3}{7}, \frac{4}{7})$; value: $-\$22.86$.

(c) It is unlikely that the officer's payoffs are the opposite of yours—that she always benefits when you do not.

(d) Use some random device, such as a die with seven sides.

13. (a) Move first to the center box; if your opponent moves next to a corner box or to a side box, move to a corner box in the same row or column. There are now six more boxes to fill, and you have up to three more moves (if you or your opponent does not win before this point), but the rest of your strategy becomes quite complicated, involving choices like "move to block the completion of a row/column/diagonal by your opponent."

(b) Showing that your strategy is optimal involves showing that it guarantees at least a tie, no matter what choices your opponent makes.

15. Player I should play H, winning 1 on average.

17. (a) Player II should avoid "call" because "fold" dominates it.

(b) Player I: $(\frac{1}{3}, \frac{2}{3}, 0)$; player II: $(\frac{2}{3}, 0, \frac{1}{3})$; value: $-\frac{1}{12}$.

(c) Player II. Since the value is negative, player II's average earnings are positive and player I's are negative.

(d) Yes. Player I bets first while holding L with probability $\frac{2}{3}$. Player II raises while holding L with probability $\frac{1}{3}$, so sometimes player II raises while holding L.

19. The Nash equilibria are $(4, 3)$ and $(3, 4)$. [It would be better if the players could flip a coin to decide between $(4, 3)$ and $(3, 4)$.]

21. These choices give x as an outcome. X certainly would not want to depart from a strategy that yields a best outcome; furthermore, neither Y's departure to another outcome in the first column, nor Z's departure to another outcome in the second row, can improve on x for these players. It seems strange, however, that Z would choose x over z, since z is sincere and dominates

x. Thus, there seem few if any circumstances in which this Nash equilibrium would be chosen.

23. X no longer has a dominant strategy, so one must consider two possibilities: (1) that X votes for x, as assumed in the 3×3 outcome matrix in Figure 15.4; and (2) that X votes for y (X would never vote for his worst choice, z), as assumed in the 3×3 outcome matrix in Figure 15.5. In the case of (1), Y's strategy y dominates x and z, and Z's strategy x is dominated by y and z, giving a reduced 1×2 matrix, in which Z would choose y, yielding y as the outcome if X chose x. In the case of (2), Y's strategy of y and Z's strategy of z are dominant, yielding y as the outcome if X chose y. In both cases, note that Y would choose y; knowing this, Z would also choose y if Z wanted to prevent the possibility that x would be chosen (x is Z's worst outcome). X's tacit deception of announcing for y, and then voting for it, would induce Z to vote for z, but y would still be the outcome. If X actually voted for x, Y's deception would be revealed, but then there would be a three-way tie, leaving unclear what would be the outcome, and therefore whether revealed deception was worthwhile.

Additional Exercises

25. (a) No saddlepoint. (b) Row 2 is the maximin strategy; column 1 is the minimax strategy. (c) None.

27. (a) Four saddlepoints at the four 5s in the payoff matrix. (b) Rows 1 and 3 are the maximin strategies; columns 2 and 4 are the minimax strategies; the saddlepoints at the four intersections of these rows and columns give the same value of 5. (c) Rows 2 and 4.

29. (a) Whatever box the first player chose, choose a box as close as possible to that box. If there are several equally close boxes (e.g., that are all adjacent to the box the first player chose), choose one of these closest boxes at random. (b) No.

31. (a) Leave umbrella at home if there is a 50% chance of rain; carry umbrella if there is a 75% chance of rain.

(b) Carry umbrella in case it rains.

(c) Saddlepoint at "carry umbrella" and "rain," giving value -2.

33. Player II's first strategy is dominant; (3, 4) is a Nash equilibrium.

35. Consider the 7-person voting game in which 3 voters have preference *xyz* (one of whom is chair), 2 voters have preference *zxy*, and 2 voters have preference *zyx*. Then for the 3 *xyz* voters, voting for both *x* and *y* dominates voting for only x; and for the 2 *zyx* voters, voting for only *z* dominates voting for both *z* and *y*. With the dominated strategies of *x* and *zy* eliminated, in the second-reduction matrix *z* dominates *zy* for the 2 *zyx* voters, yielding the sophisticated outcome *z*, which is the chair's worst outcome.

CHAPTER 16

1. The ordering of payoffs in a zero-sum game is the same as that of a total-conflict game.

3. Backward induction from each state in the first game leads to the NME of (2, 3) and in the second game to the NME of (3, 2), wherever play starts. In the third game, backward induction from (2, 3) and (1, 4) leads to the NME of (2, 3), whereas backward induction from (3, 2) and (4, 1) leads to the NME of (3, 2). The NMEs in the first two games coincide with Nash equilibria, but the NMEs in the third game are not Nash equilibria.

5. The other games in Exercise 4 are the following:

Symmetric: $\begin{vmatrix} (3, 3) & (4, 2) \\ (2, 4) & (1, 1) \end{vmatrix}$ Asymmetric: $\begin{vmatrix} (3, 3) & (4, 2) \\ (1, 1) & (2, 4) \end{vmatrix}$

The unique NME in the symmetric game is (3, 3), whereas both (3, 3) and (4, 2) are NMEs in the asymmetric game. (Only if play starts at (4, 2) is this the NME in the latter game.) In both the symmetric and asymmetric games, (3, 3) is the unique Nash equilibrium.

7. This game assumes that the victim's primary goal is to avoid injury, whereas her secondary goal is to keep her money and valuables. Her tertiary goal is to facilitate the capture of the mugger, the likelihood of which will be increased if the victim resists and, by doing so, attracts public attention. As for the mugger, this game assumes that his primary goal is to obtain money or other valuables from the victim, whereas his secondary goal is to avoid attention from passersby. His tertiary goal is not to use force, so that if caught he will not face charges of assault as well as robbery. Other goals are possible and could be the basis of an alternative game.

9. Assess the argument in the "Note." (Statistics show that most victims do submit, and that the mugger does not have to use force.)

11. NMEs and moving-power outcomes are based on different rules, so it is not surprising that they give different outcomes. In fact, *C* can induce her preferred NME of (2, 4) if play starts at (3, 3), (4, 1), or—if *C* has order power—(1, 1). Curiously, if play starts at (2, 4), the game will not stay there but instead move to (3, 3).

13. Yes. Note in each case that the 4-player is being magnanimous, in a manner of speaking, by sacrificing 4 for 3. But this move prevents a worse outcome (i.e., 2) from occurring, making his or her magnanimity rational.

15. Nobody will shoot.

17. The possibility of retaliation deters earlier shooting.

Additional Exercises

19. The four "almost" total-conflict games are as follows:

$\begin{vmatrix} (3, 3) & (4, 1) \\ (1, 4) & (2, 2) \end{vmatrix}$ $\begin{vmatrix} (3, 3) & (4, 1) \\ (2, 2) & (1, 4) \end{vmatrix}$ $\begin{vmatrix} (2, 2) & (4, 1) \\ (3, 3) & (1, 4) \end{vmatrix}$ $\begin{vmatrix} (2, 2) & (4, 1) \\ (1, 4) & (3, 3) \end{vmatrix}$

Dominant strategies: row 1 and column 1 in the first game; row 1 in the second; none in the third; row 1 and column 1 in the fourth. *Nash equilibria* (in pure strategies): (3, 3) in the first; (3, 3) in the second; none in the third; (2, 2) in the fourth.

21. One might compare the characters, motivations, and circumstances of Delilah and Lady Macbeth.

23. Variation cycles in a counterclockwise direction, which would require Samson at some point to switch from telling his secret to retracting it, which is not plausible in this story.

25. Because the resulting game is neither Prisoners' Dilemma nor Chicken but a completely different game, it is not surprising that it does not share the property of having more than one NME.

27. If a player threatens to switch from C to $\overline{C}$ if his or her opponent initially chooses $\overline{C}$ in either game, the threatened player would have good reason to choose C initially. Such threats in the form of "mutual assured destruction" (MAD) seemed to have been instrumental in deterring the superpowers (the United States and the Soviet Union) from launching nuclear attacks against each other in the so-called Cold War that began in the late 1940s and lasted until the late 1980s.

CHAPTER 17

1. (a) 1. (b) 3; 9 times as large. (c) 4; 24 in.^{2}. (d) Almost, but not exactly. (e) The 4-by-6 prints are almost twice as expensive per square inch of paper. (f) 79 cents; $1.46.

3. (a) $\frac{1}{87} \approx 0.0115$. (b) The volume of the real boxcar is $87^3 = 658,503$ times as large as the volume of the model. (c) $\frac{1}{48}$.

5. (a) Always. (b) Sometimes. (c) Always. (d) Sometimes. (e) Always. (f) Sometimes. (g) Sometimes (when the rectangle is also a square). (h) Always. (i) Never. (j) Sometimes.

7. (a) The new altar would have a volume 8 times as large—not "8 times greater than" or "8 times larger than," and definitely not twice as large—as the old altar. (b) $\sqrt[3]{2} \approx 1.26$.

9. The writer of the ad meant that the volume was 2.5 times as much before packaging. Since 2.5 bags have been compressed to one bag, the new volume is $1/2.5 = 0.4$ "times as much as" before. We could also correctly say that the peat moss has been compressed "to 40% of its original volume" or "by 60%," or that the compressed volume is "60% less than" the original volume.

11. (a) $25.37. (b) $458.57. (c) $50.05. (d) 1970: $1.03; 1974: $2.28.

13. 36 mpg.

15. (a) 0.00013 ton. (b) We assume that all parts of the scale model are made of the same materials as the real locomotive. (c) 0.27 lb. (d) 0.12 kg. (e) 0.00012 metric ton.

17. U.S. $1.92/gal.

19. 185 m, or 607 ft.

21. (a) 900 lb/ft^3. (b) Almost twice as dense. (c) Since 230 lb of compost is supposed to add about 5%, the original should be about 230 lb divided by 0.05, or 4600 lb. The revised quotation should say that the mineral soil weighs about 4500 lb.

23. (a) 400,000 lb. (b) 28 lb/in.^{2}.

25. 470,000 lbs, or almost 240 tons.

27. $\sqrt{12} \times 20$ mph = 69 mph.

29. A small warm-blooded animal has a large surface-area-to-volume ratio. Pound for pound, it loses heat more rapidly than a larger animal, hence must produce more heat per pound, resulting in a higher body temperature.

31. On log-log plot with calories/kilogram on the y-axis and weight on the x-axis, the middle four points lie fairly close to a straight line of slope -0.192 (found by fitting the least squares line). The guinea pig uses more calories per kilogram than that line would predict, and the whale uses fewer. The straight line on the log-log plot translates into the power relationship calories/kg = $70(\text{weight})^{-0.192}$.

33. 9 ft 3 in. to 11 ft 9 in. (in modern times there have been men over 9 ft tall); 282 cm to 358 cm.

35. It has disproportionately large wings compared to geometric scaling up of a bird, hence lower wing loading and lower minimum flying speed. Also, in part it glides rather than flies.

37. It is the outside of the tree branches that the lights are strung around, so that in effect you are covering the outside "area" of the tree (thought of as a cone)

with strings of lights. Hence, the number of strings needed grows in proportion to the square of the height: a 30-ft tree will need $5^2 = 25$ times as many strings as a 6-ft tree. However, you could also argue that a 30-ft tree is meant to be viewed from farther away, so that the strings of lights would produce the same effect as on the shorter tree if they were strung farther apart, so you wouldn't need quite so many.

39. 1973: $41,200, $6300, $31,700. 1979: $24,400, $17,400, $29,000. 1988: $40,400, $14,300, $29,700, $14,300, $5900.

WP1. A human grows from a height (length) of between 1 and 2 ft to a height usually between 5 and 6 ft, hence by a scaling factor of between 2.5 and 6. Under proportional scaling, its weight would have to go up by the cube of the scaling factor, hence by a factor of between $2.5^3 = 15.6$ and $6^3 = 216$; so it would have a weight between $15.6 \times 10 = 156$ lb and $216 \times 5 = 1,080$ lb. But the vast majority of human adults weight between 100 and 200 lbs.

WP3. (a) Both the width and the height of the buildings are proportional to the cost, so that Warren, with a cost about 1.5 times as much as South Mountain, has a building whose area on the page is about $(1.5)^2 = 2.25$ times as large, and whose implied volume is $(1.5)^3 = 3.4$ times as large. (b) Simply monstrous! The line for 27.5 mpg, which is about $1\frac{1}{2}$ times as much as 18 mpg, is about 9 times as long as the line for 18 mpg. (c) The picture shows the dollar bill shrinking in both length and width, even though the value shrinks only once. To use area to reflect the purchasing power of the dollar, the 1978 dollar should have about twice the area shown (and the other depictions also adjusted accordingly).

WP5. (a) 68%; 209%. No, it doesn't make sense. (b) $0.56; $0.18; 68%, agreeing with the Option A calculation. (c) For currencies that depreciate, Option B always gives a higher percentage. (d) Answers will vary.

CHAPTER 18

1. (a) $1080.00; 8.000%. (b) $1080.00; 8.000%. (c) $1082.43; 8.243%. (d) $1083.28; 8.328%.

3. $5712.39.

5. (a) $2032.79; $2025.82; $2012.20. (b) $1999.00; $1992.56. (c) $1973.82; $1906.62; $1849.60. (d) For small and intermediate interest rates, the rule of 72 gives good approximations to the doubling time.

7. 7.81%

9. (a) 2, 2.59, 2.705, 2.7169, 2.718280469. (b) 3, 6.19, 7.245, 7.3743, 7.389041321. (c) $e = 2.718281828 \ldots$; $e^2 = 7.389056098 \ldots$. Your calculator may give slightly different answers, because of its limited precision.

11. In all cases, $40.81, not taking into account any rounding to the nearest cent of the daily posted interest.

13. Using either "360 over 360" and 30-day months, or "365 over 365" and $30\frac{5}{12}$-day months: $79.40.

15. $173.87 (rounded up to the nearest cent, so as to pay the complete amount).

17. (a) $12,000. (b) The resulting equation is $100[(1 + r)^{120} - 1]/r = 37,747$. Replacing r by $1 + x$ and rearranging gives $x^{120} - 377.47x + 376.47 = 0$. The solution is $x = 1.016714122$, for an annual nominal interest rate of $12(0.016714122) = 20.06\%$. The effective annual yield is $(1.016714122)^{12} - 1 = 22.01\%$.

19. (a) $(1.04)^4 = $1.17. (b) $1/1.17 = $0.85.

21. $10,000(1 - 0.12)^5(1 - 0.03)^5 \approx $4500.

23. Three half-lives, or $3 \times 24,400$ yr = 73,200 yr.

25. Two half-lives will have elapsed, so one-fourth of the strontium-90 will remain.

27. 4.12 billion.

29. (a) 34.4, 54.7, 77.0, 92.9, 98.8, 62.2, 68.4, 69.8, 70.0, 70.0. (b) 16.3, 26.2, 40.2, 58.9, 80.7, 102.1, 119.5, 131.8, 140.2, 146.6. (c) 28.1, 69.9, 124.8, 114.9, 133.5, 126.3, 142.6, 137.3, 151.8, 148.1. (d) For $k = 0.7$ the population is increasing; it

approaches an upward-sloping trend line from below. For $k = 2$ the population oscillates about an upward-sloping trend line. In neither case is there an upper bound to the size of the population.

31. (a) 42 yr. (b) 32 yr. (c) Would tend to increase the indexes: more oil may be discovered; oil-exporting companies may reduce production, thereby raising prices and reducing consumption; societies may move more to other sources of energy. Would tend to decrease the indexes: a lower price for oil may increase the growth rate of consumption.

33. Equilibrium population size 25, maximum sustainable yield 7 for an initial population of 10.

35. (a) 698 days. (b) After 24 months.

37. (a) $\$2,000,000 \times (1 + x + \cdots + x^{19})$, with $x = 1/1.03$, giving \$30.6 million. (b) \$24.3 million. (c) \$19.9 million.

39. (a) 170 yr. (b) 150 yr. (c) 28 yr. (d) By the time half the resource is gone, freezing the consumption level will not extend the life of the resource by very much.

WP1. Answers will vary.

WP3. Answers will vary.

WP5. (a) \$726.81 (30-year); \$949.89 (15-year).
(b) \$261,651.60, of which \$161,651.60 is interest (30-year); \$170,980.20, of which \$70,980.20 is interest (15-year).
(c) \$90,461.19 (30-year); \$93,985.33 (15-year).
(d) Answers will vary.

CHAPTER 19

1. $\angle A = 60°$; $\angle B = 30°$; $\angle C = 90°$.

3. $\triangle BAD$ is congruent to $\triangle ABC$ by SAS. Hence $BD = CA$. Then $\triangle ACD$ is congruent to $\triangle BDC$ by SSS. Hence $\angle C = \angle D$.

5.–15. Answers will vary.

17. (a) By **A**, there is at least one tree, and by **C**, that tree must belong to at least one row; so there is at least one row. (b) By part (a), there is at least one row;

applying **E** to this row tells us that there is another row. (c) By part (b), there are at least two rows with no trees in common. By **B**, each row contains exactly two trees. So there are at least four trees.

19. (a) By Exercise 17c, there are at least four trees; call them T_1, T_2, T_3, and T_4. Suppose that there is a fifth tree, T_5. By **D**, T_5 and T_4 have a row in common. But neither the row containing T_1 and T_2 nor the row containing T_1 and T_3 has any tree in common with the row containing T_4 and T_5, which fact contradicts **E**. So there can be no fifth tree, and there must be exactly four trees. (b) By **B**, each row contains exactly two trees. We can make exactly six pairs of trees from the four trees, so there are exactly six rows. (c) Consider one of the four trees. By **D**, it has rows in common with each of the other three points; by **B**, these must be three different rows. If there were a fourth row through our given tree, by **B** it would have to contain one of the three other trees. But then, by **D**, it would be the same as one of the three rows we have already described. So each tree belongs to exactly three rows.

21. (a) Let the students be A, B, C, and D. Then the committees are AB, AC, AD, BC, BD, and CD. (b) Let the students be a, b, c, d, e, and f. One set of possible committees is abc, ade, bdf, and cef.

23. (a) 4. (b) Yes. (c) Yes; with this interpretation, axiom **E** is the parallel postulate.

25. (a) No; any member of L would (vacuously) contain all of the members of K, contradicting **C**. (b) Yes, provided the member of L does not contain the member of K, so as not to contradict **C**.

27. (a) By **F**, there are at least two members of K. By **A**, these two members of K are contained in a member of L. To avoid contradicting **C**, there must be yet another member in K. (b) By part (a), K has at least three members; suppose there are four or more. Take two pairs of these members, with no member common to both pairs. By **A**, each pair of members of K is contained in a member of L. The two members of L corresponding to our two pairs have no member of K in common, contradicting **D**. (A variety of other proofs

are possible, appealing to contradictions of other axioms.) (c) 3. (d) No.

29. (a) By **E**, each two members of N have at least one member of M in common. If two members of N had two or more members of M in common, **C** would be violated. (b) By **A** and **B**, there are at least three members of M; call them a, b, and c. By **D**, there must be an additional member of M, call it d. By **C**, there must be three more distinct members of N containing d and a (by **B**, together with e, a new member of M), d and b (and f), and d and c (and g). By **C** and **E**, there must be three more members of N containing a, f, and g; c, f, and e; and e, b, and g.

31. (a) Let the students be a, b, c, d, e, f, and g. One set of possible committees is abc, ade, afg, bdf, beg, cdg, and cef; other answers are possible. (b) See figure below.

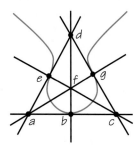

33. (a) $(1, 0)$, $(1, 1)$, $(1, 2)$. (b) $(0, 2)$, $(1, 2)$, $(2, 2)$. (c) $(1, 0)$, $(0, 1)$, $(2, 2)$. (d) $(1, 0)$, $(2, 1)$, $(0, 2)$.

35. (a) 12. (b) 3. (c) 4.

37. (a) $(0, 0)$. (b) $(1, 0)$, $(0, 1)$, $(2, 0)$, and $(0, 2)$. (c) $(1, 1)$, $(1, 2)$, $(2, 1)$, and $(2, 2)$.

39. The limited precision of a standard scientific calculator cannot produce the values for the slower objects in the following table:

Object	v (km/sec)	v/c	Gamma (γ)	Length (m)	Amount Shortened (m)
Car	0.029	0.0000001	1.000000000	6	2.8×10^{14}
Bullet	1	0.000003	1.000000000	0.03	1.7×10^{-13}
Hya-kutake	89.4	0.0003	1.000000050	1×10^9	4.4×10^{-1}

WP1. By **A** and **B**, we can reverse any chain from X to Y to get one from Y to X, and the new chain will be negative or positive as the original is.

WP3. (a) Suppose a and b are ambivalent toward each other. In any chain of involvements involving a, we can replace one occurrence of a by $a \heartsuit b \sharp a$, thereby producing a chain that is positive if the original was negative, and vice versa. (b) Suppose we have a stable society. Let a be a person in the society, let $\mathcal{F}$ be the set of friends of a and $\mathcal{E}$ the set of enemies of a. Because the society is stable, $\mathcal{F}$ and $\mathcal{E}$ do not overlap. By definition of a society, everyone is related to a in some way, so everyone is in either $\mathcal{F}$ or $\mathcal{E}$. We need to show that the people in each of these sets are all friends of each other and enemies of the people in the other set. If b and c are both in $\mathcal{F}$, then there is a positive chain from b to a and a positive chain from a to c. Joining the two chains gives a positive chain from b to c. The same idea can be used to show that any two people in $\mathcal{E}$ are friends and that any person in $\mathcal{E}$ is an enemy of any person in $\mathcal{F}$. To show the other half, that if the society divides as described, then it must be stable, we proceed by contradiction. Suppose that the society is not stable; then there are two people, a and b, who are ambivalent toward each other. Because they are friends, the two must both be in the same set; but because they are enemies, they must be in different sets, which is a contradiction, since the sets do not overlap. So the society must be stable.

WP5. Answers will vary.

CHAPTER 20

1. 5, 8, and 13.

3. Answers will vary.

5. (a), (b) The digits after the decimal point do not change. (c) $\phi^2 = \phi + 1$. (d) $1/\phi = \phi - 1$.

7. (a) 9. (b) 16.

9. (a) 4, 7, 11, 18, 29, 47, 76, 123. (b) 3, 1.333, 1.75, 1.571, 1.636, 1.611, 1.621, 1.617, 1.618. The ratios approach ϕ.

11. All are true.

13. (a) B, C, D, E, H, I, K, O, X. (b) A, H, I, M, O, T, U, V, W, X, Y. (c) H, I, N, O, S, X, Z.

15. (a) MOM, WOW (both either horizontally or vertically); MUd and bUM reflect into each other. (b) pod rotates into itself; MOM and WOW rotate into each other. (c) Here are some possibilities: NOW NO; SWIMS; ON MON; CHECK BOOK BOX; OX HIDE.

17. (a) $c5$. (b) $c12$. (c) $c22$.

19. (a) $c6$. (b) $d2$. (c) $c16$.

21. (a) Vertical. (b) Vertical and every multiple of 45°. (c) Vertical and every multiple of 72°. (d) Vertical and horizontal. (e) None.

23. For all parts, translations. (a) Reflection in vertical lines through the centers of the **A**s or between them. (b) Reflection in the horizontal midline. (c) Reflection in the horizontal midline, reflections in vertical lines through the centers of the **O**s or between them; 180° rotation around the centers of the **O**s or the midpoints between them; glide reflections. (d) None, other than translations.

25. (a) $d3$. (b) $d1$. (c) $c1$.

27. (a) Yes. (b) Yes.

29. (a) $\langle T,\ H | H^2 = I,\ T \circ H = H \circ T \rangle =$ $\{ \ldots,\ T^{-1},\ I,\ T^1,\ \ldots;\ \ldots,\ H \circ T^{-1},\ H,\ H \circ T,\ \ldots \}$. (b) $\langle G,\ R | R^2 = I,\ R \circ G = G^{-1} \circ R \rangle$ $= \{ \ldots,\ G^{-2} = T^{-1},\ G^{-1},\ I,\ G^1,\ G^2 = T,\ \ldots;$ $\ldots,\ R \circ G^{-1},\ R,\ R \circ G,\ \ldots \}$.

31. $\langle R,\ H | R^4 = I,\ H^2 = I,\ R \circ H = H \circ R^{-1} \rangle$ $= \{ I,\ R,\ R^2,\ R^3,\ H,\ H \circ R,\ H \circ R^2,\ H \circ R^3 \}$, where R is a rotation by 90° and H is a reflection across a line of symmetry.

33. $p111, p1a1, p112, pm11, p1m1, pma2, pmm2$.

35. $pmm2, p1a1, pma2, p112, pmm2$ (perhaps), $p1m1, pma2, p111$.

37. (a) Reflection in a vertical line. (b) Glide reflection. (c) See figure in next column above.

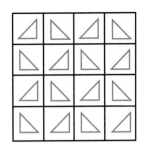

(d) cm.

39. Answers will vary.

41. F_{61} and F_{62} both end in 1.

WP1. (a) 1, 1, 3, 5, 11, 21, 43, 85, 171, 341, 683, 1365. (b) $B_n = B_{n-1} + 2B_{n-2}$. (c) 1, 3, 1.667, 2.2, 1.909, 2.048, 1.977, 2.012, 1.994, 2.003, 1.999. (d) $x = 2,\ -1$; we discard the -1 root. (e) $B_n = [2^n - (-1)^n]/3$.

WP3. Answers will vary.

CHAPTER 21

1. Exterior: 135°. Interior: 45°.

3. $180° - \dfrac{360°}{n}$.

5. A regular polygon with 12 sides has interior angles of 150°, and a regular polygon with 8 sides has interior angles of 135°. No integer combination of these numbers can add up to 360°.

7. See figures below and on the following page.

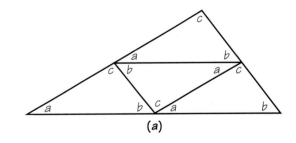

(a)

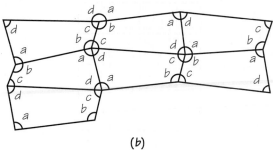

(b)

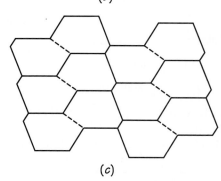

(c)

9. (a) No. (b) No. (c) No.

11. Answers will vary.

13. (a) Yes. (b) No. (c) No.

15. See figure below.

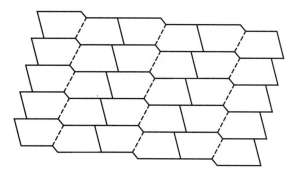

17. Answers will vary.

19. The usual notation for a vertex figure is to denote a regular n-gon by n, separate the sizes of polygons by periods, and list the polygons in clockwise order starting from the smallest number of sides, so that, for example, 3.3.3.3.3.3 denotes six equilateral triangles meeting at a vertex. The possible vertex figures are 3.3.3.3.3.3, 3.3.3.3.6, 3.3.3.4.4, 3.3.4.3.4, 3.3.4.12, 3.4.3.12, 3.3.6.6, 3.6.3.6, 3.4.4.6, 3.4.6.4, 3.12.12, 4.4.4.4, 4.6.12, 4.8.8, 5.5.10, and 6.6.6.

21. 3.7.42, 3.9.18, 3.8.24, 3.10.15, and 4.5.20.

23. Answers will vary.

WP1. Let S_n, A_n, and B_n be the total number of symbols, the number of As, and the number of Bs at the nth stage. We note that the only Bs at the nth stage must have come from As in the previous stage, so $B_n = A_{n-1}$. Similarly, the As at the nth stage come from both As and Bs in the previous stage, so $A_n = A_{n-1} + B_{n-1}$. Using both of these facts together, we have $A_n = A_{n-1} + A_{n-2}$. We note that $A_1 = 0$, $A_2 = 1$, $A_3 = 1$, $A_4 = 2, \ldots$. The A_n sequence obeys the same recurrence rule as the Fibonacci sequence and starts with the same values one step later; in fact, it is always just one step behind the Fibonacci sequence: $A_n = F_{n-1}$. Consequently, $B_n = A_{n-1} = F_{n-2}$, and $S_n = A_n + B_n = F_{n-1} + F_{n-2} = F_n$.

CREDITS

PART I (*from left to right*): *Background grid:* City lights at dusk, Masa Uemura/Tony Stone Images. Inspection of pistons on conveyor belt, Martin Rogers/Tony Stone Worldwide. Space shuttle and mission control, NASA. Businessmen discussing plans, Charles Thatcher/Tony Stone Images. Frame of house, Zigy Kaluzny/Tony Stone Images. House framework, B/W Archive Photos/American Stock. *Spotlight:* Robert Freitag.

Chapter 1. *page 5:* Masa Uemura/Tony Stone Images. *Spotlight 1.1:* Portrait by Emanuel Handmann, *Bildnis des Mathematikers,* 1753, Oeffentliche Kunstsammlung Basel, Kunstmuseum. *Figure 1.20a:* Superstock, Inc. *Figure 1.20b:* Courtesy of Sidney B. Bowne & Son, Consulting Engineers.

Chapter 2. *page 35:* Charles Thatcher/Tony Stone Images. *page 37:* Archive Photos/Lambert. *page 51:* AT&T Archives. *page 55:* Roger Tully/Tony Stone Images.

Chapter 3. *page 83:* NASA. *page 84:* Roger Tully/Tony Stone Images. *page 86:* Roger Tully/Tony Stone Images. *page 101:* Archive Photos.

Chapter 4. *page 123:* B/W Archive Photos/American Stock. *page 124:* General Motors. *Figures 4.22, 4.23:* Courtesy of AT&T Bell Laboratories.

PART II (*from left to right*): Scoreboard, Focus on Sports. Looking at map, Busco/The Image Bank. Roulette wheel, Archive Photos/Lambert. Baseball scoreboard, Paul Bereseill © 1993 Newsday. Thinking Machines' Connection machine CM-200, © Thinking Machines Corporation 1987, photo by Steve Grohe.

Chapter 5. *page 179:* Focus on Sports. *page 182:* Andy Sacks/Tony Stone Images.

Chapter 6. *page 221:* Focus on Sports. *page 223:* Charles Thatcher/Tony Stone Images. *Spotlight 6.1:* Kip Brundage/Woodfin Camp & Associates, all rights reserved. *page 247:* CD-ROM from *Earth,* February 1996, p. 64, Rhonda Sherwood, *Earth* magazine.

Chapter 7. *page 267:* Archive Photos/Lambert. *Figure 7.1:* Superstock, Inc. *page 268:* Ron Avery/Superstock, Inc. *Figure 7.13:* Ken Whitmore/Tony Stone Images.

Chapter 8. *page 309:* Busco/The Image Bank. *Spotlight 8.2:* AT&T.

Part III (*from left to right*): *Background grid:* Code line reader, Alain Altair/The Image Bank. 4 × 5 chip, Intel Corporation. Music score, Johann Sebastian Bach, *Prelude in G Minor* (early version), BWV 535a Autograph, c. 1705–7, Staatsbibliothek zu Berlin-Preussischer Kulturbesitz, Muzikabteilung mit Mendelssohn-Archiv. Navajo code talkers, UPI/The Bettmann Archives.

Chapter 9. *page 359:* Stewart Cohen/Tony Stone Images. *page 364:* Scott Camazine.

Chapter 10. *page 379:* UPI/The Bettmann Archives. *Spotlight 10.1:* Photo by Rex Ridenouse, NASA/JPL. *page 394:* UPI/The Bettmann Archives. *Spotlight 10.5:* Photo by Matthew Mulbry.

Part IV (*from left to right*): *Background grid:* Chess player, Raphael D'Lugoff. Republican convention, B. Daemmrich/Stock Boston. Voting machine, The Bettmann Archives. UN recognition of Beijing as sole representative of China, UPI/Corbis-Bettmann. Electronic voting, Rick Maiman/Sygma. Barbara Jordan, Fred Ward/Black Star.

Chapter 11. *page 411:* B. Daemmrich/Stock Boston. *page 412:* Martin Simon/Archive Photos. *page 414:* Tom Horan/Sygma.

Chapter 12. *page 443:* The Bettmann Archives. *page 444:* Larry Freed/The Image Bank. *page 447:* White-Packard/The Image Bank.

Chapter 13. *page 487:* UPI/Corbis-Bettmann. *Spotlight 13.1:* Superstock, Inc. *page 491:* Courtesy of Alan Taylor. *page 509:* Courtesy of German Information Center.

Chapter 14. *page 527:* B. Daemmrich/Stock Boston. *Figure 14.1: Signing the Constitution of the United States,* Thomas Pritchard Rossiter, 1867, Fraunces Tavern Museum, New York. *page 531:* National Portrait Gallery. *page 536:* National Portrait Gallery. *page 541:* National Portrait Gallery. *Spotlight 14.2:* Willcox, Department of Manuscripts and University Archives, Cornell University Libraries; Huntington, Courtesy of Harvard University Archives.

Chapter 15. *page 561:* Raphael D'Lugoff. *Spotlight 15.1:* Photos courtesy of the Institute for Advanced Study, Princeton University Archives. *page 571:* Focus on Sports. *Spotlight 15.2:* Harsanyi, Reuters/Pressens Bild/Archive Photos; Nash, Reuters/The Bettmann Archives; Selten, The Bettmann Archives. *page 597:* The Bettmann Archives.

INDEX